# ATOMIC PHYSICS 6

**1969 — Atomic Physics 1**

Proceedings of the First International Conference on
Atomic Physics, June 3—7, 1968, New York City
V. W. Hughes, Conference Chairman
B. Bederson, V. W. Cohen, and F. M. J. Pichanik, Editors

**1971 — Atomic Physics 2**

Proceedings of the Second International Conference on
Atomic Physics, July 21—24, 1970, Oxford, England
G. K. Woodgate, Conference Chairman
P. G. H. Sandars, Editor

**1973 — Atomic Physics 3**

Proceedings of the Third International Conference on
Atomic Physics, August 7—11, 1972, Boulder, Colorado
S. J. Smith and G. K. Walters, Conference Chairmen and Editors

**1975 — Atomic Physics 4**

Proceedings of the Fourth International Conference on
Atomic Physics, July 22—26, 1974, Heidelberg, Germany
G. zu Putlitz, Conference Chairman
E. W. Weber and A. Winnacker, Editors

**1977 — Atomic Physics 5**

Proceedings of the Fifth International Conference on
Atomic Physics, July 26—30, 1976, Berkeley, California
Richard Marrus, Conference Chairman
Michael Prior and Howard Shugart, Editors

**1979 — Atomic Physics 6**

Proceedings of the Sixth International Conference on
Atomic Physics, August 17—22, 1978, Riga, USSR
A. M. Prokhorov, Conference Chairman
R. Damburg, Editor

# 6th INTERNATIONAL CONFERENCE ON ATOMIC PHYSICS PROCEEDINGS

Conference Chairman
A. M. PROKHOROV
P. N. Lebedev Physics Institute,
USSR Academy of Sciences, Moscow

Editor
ROBERT DAMBURG
Physics Institute, Latvian SSR
Academy of Sciences, Riga
with technical assistance by

OLGA KUKAINE
Physics Institute, Latvian SSR
Academy of Sciences, Riga

ZINĀTNE
RIGA

PLENUM PRESS
NEW YORK AND LONDON

**Proceedings of the Sixth International Conference on Atomic Physics, August 17—22, 1978, Riga, USSR**

ISBN 0-306-40217-3



19, Turgeņeva Street, Riga 226018, Latvian SSR, USSR
Printed in the USSR

S $\frac{20408—023}{M\ 811(11)}$

S. E. FRISH (1899—1977)

## PREFACE

The present volume contains the invited papers of the VI International Conference on Atomic Physics held in August 17-22, 1978 in Riga. We are grateful to all authors who have presented for publication these valuable papers. We would like also to thank the emoloyees of "Zinatne" Publishing House for the diligent and careful way in which they prepared the book, as well as for their effective assistance in publishing the volume containing abstracts of about 300 contributed papers.

To retain the spirit and traditions of previous International Conferences on Atomic Physics, the Organizing Committee aimed at including into the programme original reports related to the problems of atomic physics not considered before.

A considerable part of the volume make the contributions of Soviet scientists working in different fields of atomic physics which offers a comprehensive review of current development of atomic physics in the USSR.

The volume opens with the introductory word by Academician A.M.Prokhorov given in honour and memory of S.E.Frish, the Corresponding Member of the USSR Academy of Sciences to whom the VI ICAP was dedicated.

The volume is meant for all those working in various fields of atomic physics. It could be also recommended to students of senior courses, especially to those planning to take up scientific activities.

The texts were mainly reproduced directly from the original contributions received from the authors, which caused certain difficulties in editorial work. The same reason would serve as an excuse for the lack of uniformity in text appearance, notations and formula expressions.

Robert Damburg

Physics Institute,
Latvian SSR Academy of Sciences
Riga, Salaspils, USSR

Contents

To the Memory of S.E.Frish .............................. 13
A.M.Prokhorov

Excitation Energy Transfer Processes in Metal Vapours .............................. 15
E.Kraulinya

Experimental Determination of Electronic Transition Probabilities and the Lifetimes of the Excited Atomic and Ionic States .............................. 33
N.P.Penkin

The Development of Ideas of V.Fock in Atomic Pnysics .............................. 65
Yu.N.Demkov

The Multiconfiguration Hartree-Fock Method for Atomic Levels and Transition Probabilities .............................. 77
Ch.F.Fisher

Peculiarities of the Theoretical Investigation of the Spectra of Many-Electron Atoms and Multiple-Charged Ions .............................. 92
Z.B.Rudzikas

Relativistic Perturbation Theory for Atoms and Ions .............................. 111
M.A.Braun and L.N.Labsovsky

Ionization and Excitation of Atomic Shells by Electron Impact in Semiclassical Approximation .............................. 133
M.A.Braun and V.I.Ochkur

Spectroscopy of Multicharged Ions in Laboratory Plasma .............................. 159
L.A.Vainstein, V.A.Boiko, E.Ya.Kononov

Mesic Atomic and Mesic Molecular Processes in the Hydrogen Isotope Mixtures .............................. 182
L.I.Ponomarev

Determination of the Lamb Shift (H, n=2) in the "Atomic Interferometer" Method .......................... 207
Yu.L.Sokolov

Spectroscopy of Negative Ions ............................ 223
J.S.Risley

Threshold Behaviour of Elementary Atomic Processes ................................................ 249
M.Gailitis

Quasi-Resonance Processes in Atomic Collisions ................................................ 267
E.E.Nikitin and B.M.Smirnov

Correlated Transitions in Atom with Two Inner-Shell Vacancies .................................... 289
V.V.Afrosimov and A.P.Shergin

Optical Pumping Investigation of Atomic Interactions ................................................ 308
R.A.Zhitnikov

The Role of Atomic and Molecular Processes in the Computing Modeling of Gas Lasers .......................... 335
K.Smith and S.A.Roberts

Molecular Formation and Destruction in the Interstellar Medium ........................................ 368
H. van Regemorter, A.Giusti Suzor, E.Roueff

Laser-Photochemical Dissociation of Small Molecules: Calculation of Isotope Effects with Respect to Dissociation Probabilities ................................................ 383
H.Johansen, K.Johst

Decay of Atomic Polarization Moments ...................... 410
M.I.Dyakonov, V.I.Perel

Alignment of Excited Atoms in a Gas Discharge .............. 423
M.P.Chaika

Applications of Anticrossing Spectroscopy .................. 435
H.J.Beyer and H.Kleinpoppen

Resonant Interaction of Atom with Intense Radiation Fields ........ 462
M.L.Ter-Mikaelian, M.A.Sarkissian

Investigation of Collisions by Nonlinear Spectroscopy Methods ........ 493
S.G.Rautian

Quantum Beats ........ 521
E.B.Alexandrov

Selective Multistep Laser Action upon Atoms and Molecules ........ 535
N.V.Karlov

Laser Detection of Single Atoms ........ 565
V.I.Balikin, G.I.Bekov, V.S.Letokhov, V.I.Mishin

Coherent Phenomena in Superhigh Resolution Spectroscopy ........ 585
V.P.Chebotayev

Laser Spectroscopy in Molecular Beams ........ 612
W.Demtröder

Spectroscopic and Ionization Properties of Atomic Rydberg States ........ 626
S.Feneuille and P.Jacquinot

Observation of Parity-Nonconservation in Atomic Transition ........ 648
L.M.Barkov, M.S.Zolotorev

Parity Non-Conservation in Atomic Bismuth ........ 653
P.E.G.Baird

Index ........ 663

# TO THE MEMORY OF S.E.FRISH

A.M.Prokhorov

The Lebedev Physical Institute, USSR Academy of Sciences, Moscow

We decided to dedicate the present Conference to the memory of the late Corresponding Member of the USSR Academy of Sciences Prof. S.E.Frish who died in November, 1977.

Prof. S.E.Frish was a prominent specialist in optics and spectroscopy, one of the leaders of the world-wide known Leningrad school of optics. As a vice-chairman of the International Organizing Committee he had done much for the preparation of the present Conference.

The beginning of Prof. Frish's scientific career coincided with the time when atomic spectroscopy was in fact the main source of data needed to create the theory of atom. The progress in the development of nuclear physics stimulated Prof.Frish's pioneer research on the interaction of the nucleus with the electron shell of atom resulting in the hyperfine structure of the spectrum. In the middle of the thirties Prof.Frish was among the initiators of the spectroscopic studies of gas-discharge plasma. One of the accomplishments of Prof.Frish met with general recognition was a wide application of atomic spectroscopy in gas-discharge physics. It is this branch of research that constituted the principal field of scientific work of the Chair of Optics in the Leningrad University headed by Prof.Frish. A series of his studies related to the interferometric analysis of line contours of atoms and ions and the determination of their concentration at various energetic states. In this respect the methods of reabsorption and emission were of wide use. Prof. S.E.Frish also had been conducting a thorough investigation of the high frequency discharge. It resulted in the development of a widely used method for the spectral analysis of gases. In a series of his works on optical functions of excitation Prof.Frish was the first to state the idea of the existence of the fine structure of these functions. He also explained the significance of cascade transitions and found excitation cross-sections of energetic levels.

Prof.Frish is the author of a well known "General Physics" edition in three volumes which was repeatedly printed in the Soviet Union and translated into several languages. His monographs were and actually are used by all spectroscopists of our country.

Until his very last days Prof.S.E.Frish headed the editorial board of the Journal "Optika i Spektroskopija". He also participated actively in the Bureau of the Department of General Physics and Astronomy of the USSR Academy of Sciences.

We have lost an outstanding scientist, a very good and kind person. The memory of Prof.Frish will endure in our hearts forever.

# EXCITATION ENERGY TRANSFER PROCESSES IN METAL VAPOURS.

E. Kraulinya

P.Stuchka Latvian State University, Riga, USSR

One of the problems which attracted Prof.Frish's attention was investigation of second kind collisions between atoms according to the following scheme:

$$A^{*} + B \rightarrow A + B^{*} \mp \Delta E \qquad (I)$$

where excited atoms A transfer their energy to atoms B. As a result atoms B emit spectral lines, but atoms A are de-excited into the ground-state. $\Delta E$ is the energy difference between excited states of atoms A and B which is assumed to be negative if the excited state of the atom B lies above the level of $A^*$, and $\Delta E$ is positive if the excited level of B lies below the level of $A^*$ (Fig.I). Process (I) is also called sensitized fluorescence.

The first experiments concerning sensitized fluorescence were carried out in the I920's. They only corroborated the existence of the phenomenon. There were, however, no quantitative investigations until the late I940's. S.Frish was the first at that time to propose more detailed studies of sensitized fluorescence in metal vapours. Some new papers on the subject appeared in the early I950's, but systematic research was initiated only in the I960's when investigations of inverse population of atomic levels became important. In those years interpretation of some phenomena in low-temperature plasma and in astrophysics required better understanding of interaction between atoms. At present there are many experimental and theoretical papers on sensitized fluorescence in alkali vapours. The cross-sections of energy transfer between fine-structure components $P_{I/2}$ and $P_{3/2}$ have been obtained for collisions between alkali atoms and between alkali and noble-gas atoms. The greatest part of this work was performed by L. Krause in Canada /I,2, et al./, but theoretical work was carried out by E.E.Nikitin and his collaborators in Moscow /3-6 , et al./. Some interesting results concerning intramultiplet mixing in highly excited caesium atoms in collision with noble gas atoms were obtained in the early 70's by Berlande and collaborators in France /7-9, et al./. Now it is clear that the cross-sections of intramultiplet mixing of alkali fine-structure components as well as other inelastic collision processes can be explained by an analysis of the potential curves of the so-called quasi-molecules formed by colliding atoms.

The mixtures of heavier atoms investigated at the Latvian State University have been studied only experimentally because, as far as we know, no potential curves for such atomic pairs are available at present. For example, mercury-sodium /IO/, mercury-cadmium /II,I2/, mercury-zinc /II/, mercury-indium/I3/, /I4/, mercury-thallium /I5,I6/ and cadmium-caesium /I7-I9/mixtures have been investigated. In all these mixtures mercury atoms were optically excited to the resonance level $6^3P_I$, but cadmium atoms to the resonance level $5^3P_I$. Sodium, cadmium, zinc, indium, thallium and caesium spectral lines were observed in sensitized radiation. All experiments were carried out under conditions when no secondary processes are present, the concentration of atoms being $IO^{II} \div IO^{I4}$ $cm^{-3}$, and kinetic energy of atoms about 0.0I - 0.I3 eV. The results of our experiments have shown that in spite of various behaviour of different atomic mixtures in excitation energy transfer processes some general conclusions can be made.

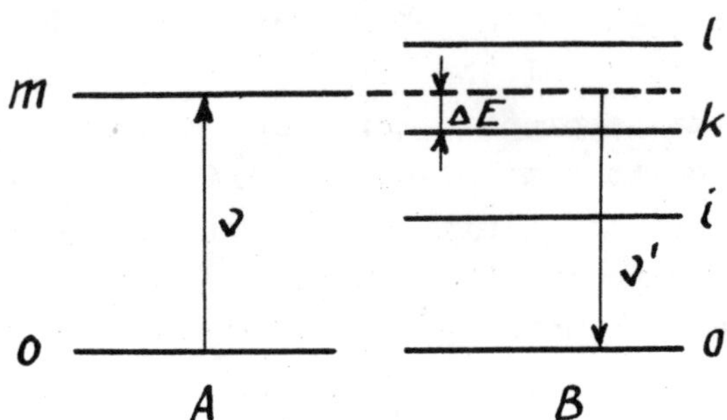


FIG. I.

Before passing over to the main results of our experiments I should like to make a remark. Further, I shall mention the energy difference $\Delta$ E as a characteristic of the observed effects. However, at present we know very well that this $\Delta$ E is not the only quantity determining the efficiency of excitation energy transfer. This efficiency depends on the behaviour of the potential curves of the quasi-molecule. But since these curves are not known in our case, $\Delta$ E will be used to describe our results as it was used historically in earlier works.

I. The effect of resonance can be observed in excitation energy transfer processes. It can be demonstrated very well by the investigation of $Hg^{*}$-Na and $Cd^{*}$-Cs mixtures because sodium and caesium have many energy levels lying close to the exciting levels Hg $6^3P_I$ and Cd $5^3P_I$, Cd $5^3P_0$ (Figs. 2, 3, 4, 5).

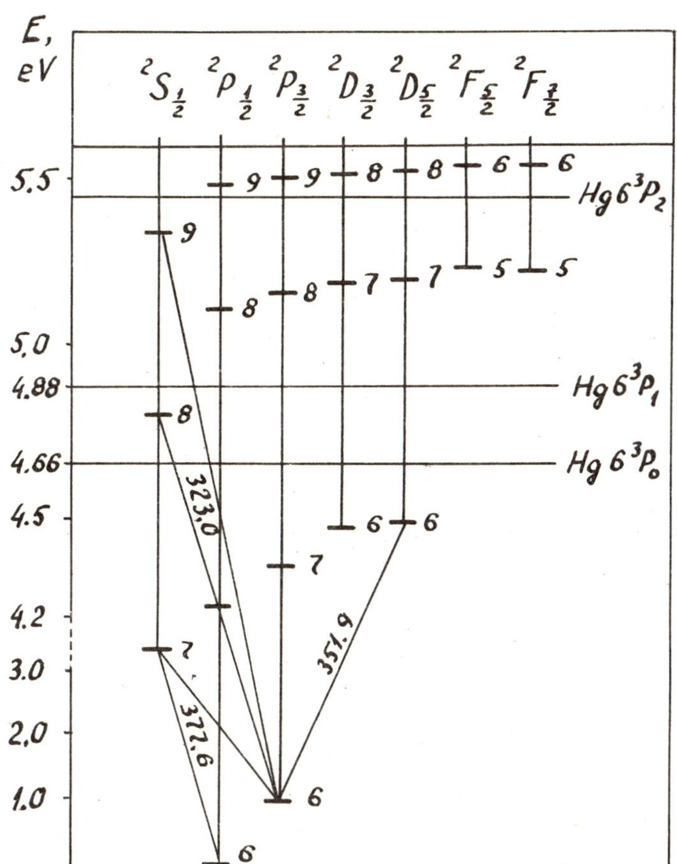


FIG. 2. Energy levels for atomic pairs Hg-Tl.

As can be seen from Figs.6, 7 , levels corresponding to smaller $\Delta$ E are excited more efficiently.

That $\Delta$ E is not the quantity determining the efficiency of energy transfer can be seen from the example of sodium 8D level.

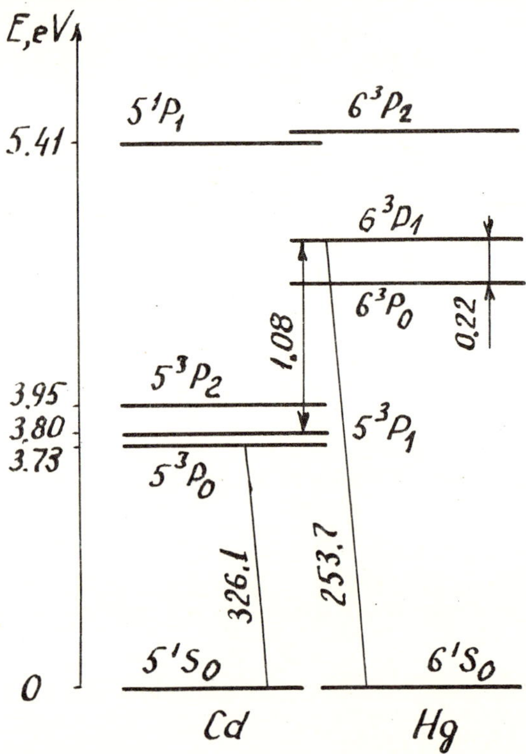


FIG. 3. Energy levels for atomic pairs Hg-Cd.

Its energy difference is larger than that of 7D, but the corresponding energy transfer cross-section exceeds that of 7D. In the Hg*-In mixture the indium $^4P_{5/2}$ level lies very close to the mercury $5^3P_0$ level, but it is not excited by collision /I3/.

2. It is an interesting fact that the energy levels lying above the exciting level are excited more efficiently than those lying below this level at the same distance. For example, sodium 8D level which lies above the exciting Hg $6^3P_I$ level has an excitation cross-section higher than that of the levels having energy difference of the opposite sign. In the Cd*-Cs mixture Cs I5 $^2D_{3/2,5/2}$ and Cs I2 $^2D_{3/2,5/2}$ are excited with higher cross-sections than those caesium levels which are below the exciting level of cadmium (Fig. 7). Such asymmetry has been observed in energy transfer to other sodium and caesium spectral series, as well as in some other mixtures.

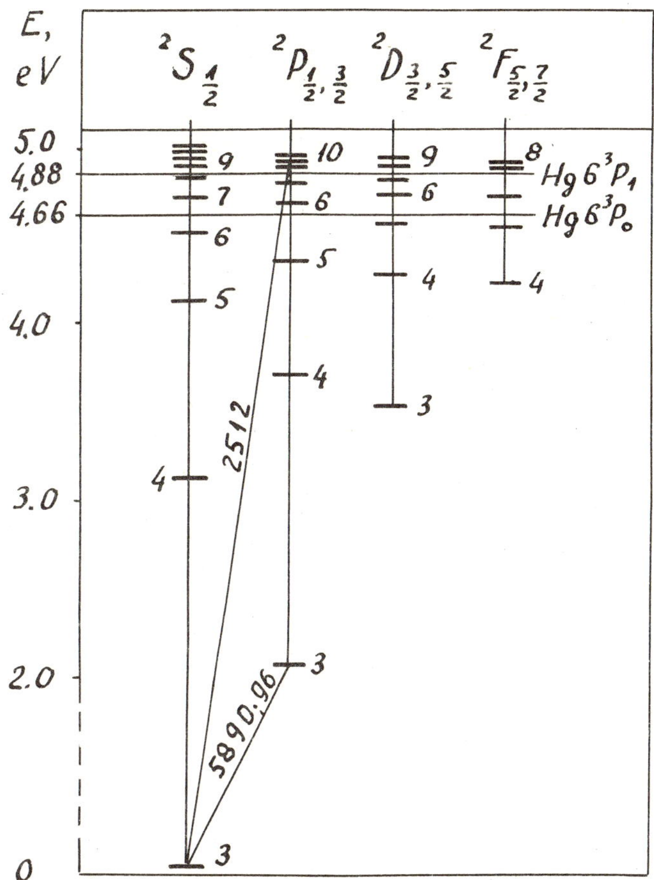


FIG. 4. Energy levels for atomic pairs Hg-Na.

3. In the sensitized fluorescence radiation some spectral lines are observed which originate from levels corresponding to large $\Delta$ E. Thus, in Hg*-Cd and Hg*-Zn mixtures levels were excited with energy differences from the exciting level as high as 0.85 eV and I.08 eV. Their cross-sections have considerable order of magnitude - about $10^{-17}$ - $10^{-18}$ cm$^2$. In the Hg*-Tl mixture spectral lines originating from $8^2P_{1/2,3/2}$ levels were observed with $\Delta$ E being 0.29 eV and corresponding excitation transfer cross-section reaching the value of $6.10^{-16}$ cm$^2$. It is difficult at present to explain the excitation of such distant levels. An attempt has been made to explain the phenomenon by means of an ionic complex $Hg^-Cd^+$ or $Hg^-Zn^+$ /20/, but further work is needed to obtain more reliable estimates.

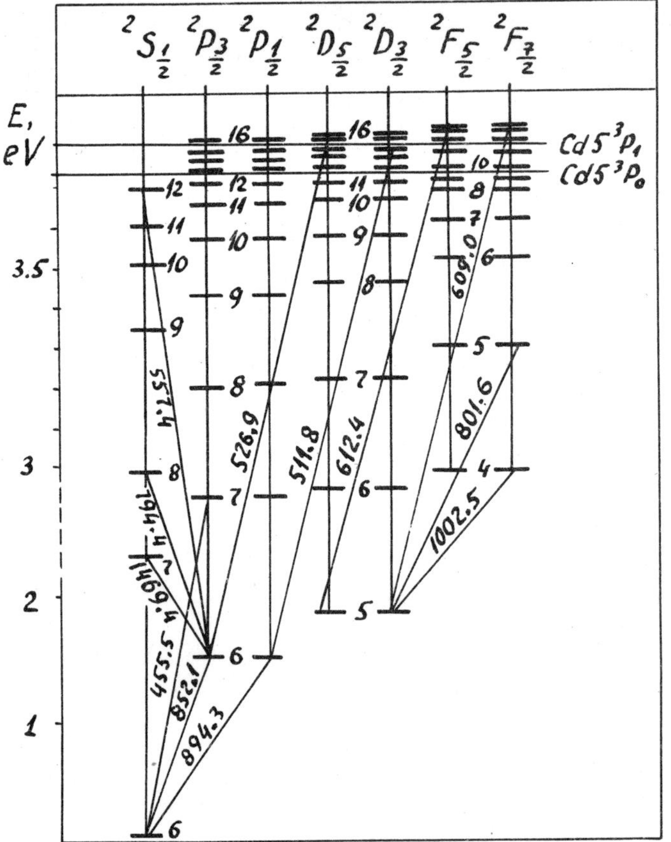


FIG.5. Energy levels for atomic pair Cd-Cs.

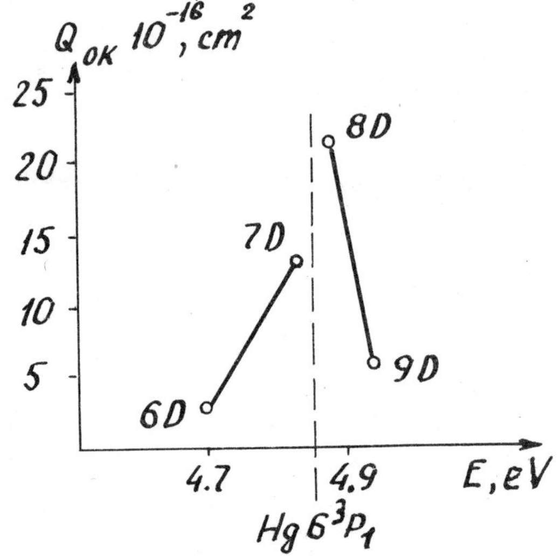


FIG.6. Dependence of absolute values of cross-sections $Q_{ok}$ on excitation energy of the diffuse series of sodium /IO/.

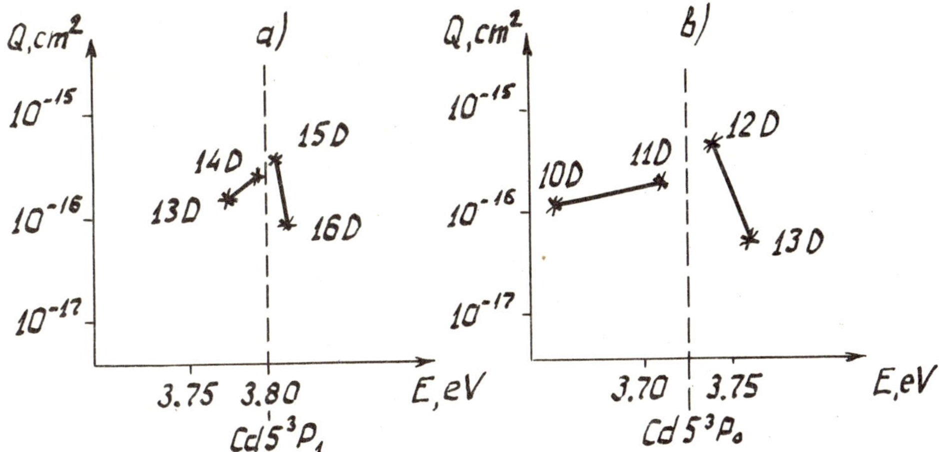


FIG. 7. Dependence of absolute values of cross-sections $Q_{ok}$ on excitation energy of the diffuse series:a) for Cd $5^3P_I$ level , b) for Cd $5^3P_0$ level /I9/.

4. Different spectral series are excited with differing efficiency. In the case of $Hg^*$-Na and $Cd^*$-Cs mixtures excitation of P-series is weak. Excitation cross-sections for S, D, F series are also differing as well. For example, the 8P level is excited I0 times more efficiently than the 7D level although their energies are very close.

5. The shapes of spectral lines in sensitized fluorescence have also been studied.

Excitation energy transfer taking place due to second kind collisions according to (I), the positive or negative energy difference $\triangle$ E changes the kinetic energy of colliding atoms, and the shape of spectral lines varies owing to the Doppler effect. Spectral lines of component B can be broadened or narrowed. This was actually observed in $Hg\text{-}Tl^{205}$ mixture by means of a scanning Fabry-Perot interferometer and a photon counter. The results are given in the Table I /I6/.

The results obtained indicate that positive or negative energy difference is distributed between colliding partners. It corroborates the fact that thallium $8^2S_{I/2}$, $6\ ^2D5/2$ ,

TABLE I. Measured and calculated Doppler broadening of thallium lines.

| λ,nm | Transition | ΔE,eV | $N_0$(Tl), $cm^{-3}$ | Doppler broadening, $cm^{-1}$: Measured at sensit. fluoresc. | Calc. for fluoresc. cell temper. | Measured at reson. fluoresc. |
|---|---|---|---|---|---|---|
| 323.0 | $8^2S_{1/2}-6^2P_{3/2}$ | +0.083 | $1.2x10^{13}$ | 0.056 | 0.048 | 0.048 |
| 351.9 | $6^2D_{5/2}-6^2P_{3/2}$ | +0.398 | $1.0x10^{11}$ | 0.086 | 0.045 | - |
| 352.9 | $6^2D_{3/2}-6^2P_{3/2}$ | +0.408 | $2.8x10^{11}$ | 0.046 | 0.044 | 0.044 |
| 655.0 | $8^2P_{3/2}-7^2S_{1/2}$ | -0.289 | $5.0x10^{13}$ | 0.020 | 0.024 | - |
| 671.4 | $8^2P_{1/2}-7^2S_{1/2}$ | -0.242 | $5.0x10^{13}$ | 0.020 | 0.023 | - |
| 535.0 | $7^2S_{1/2}-6^2P_{3/2}$ | +1.603 | $2.0x10^{11}$ | 0.029 | 0.029 | 0.029 |

$8^2P_{3/2}$ and $8^2P_{1/2}$ levels are excited in second kind collisions with Hg $6^3P_1$ atoms. The $6^2D_{3/2}$ level is populated by collisions with the optically excited mercury, but an additional Doppler broadening is not present because the $6^2D_{3/2}$ state is destroyed through two channels, one of them being the transition to the ground-state. The relaxation of the distribution of excited atoms was discussed in detail at the VI International Conference on Atomic Physics /21/.

The coincidence of Doppler broadening of the 535.0 nm line in resonance and sensitized fluorescence are due to the fact that the thallium $7^2S_{1/2}$ level is populated by cascade transitions from higher thallium levels as it was also shown in the paper /15/.

6. Metastable atoms take part in excitation energy transfer in mixtures of metal vapours.

Let us consider, for example, a Cd*-Cs mixture. The Cd atom has a metastable $5^3P_0$ level close to the excited $5^3P_1$ level (their energy difference is about 0.07 eV). That is why the metastable Cd level is populated from the resonance level and can take part in energy transfer. The absolute values for energy transfer cross-sections from cadmium $5^3P_0$ and $5^3P_1$ levels to Cs atoms are given in Table 2.

TABLE 2. Cross-sections of Energy Transfer from Cadmium $5^3P_o$ and $5^3P_I$ to Caesium $^2D$, $^2F$ States.

| Caesium states | Q ($5^3P_o$), $cm^2$ | Q ($5^3P_I$), $cm^2$ |
|---|---|---|
| I6 $^2D_{3/2}$ | $I.0xI0^{-I7}$ | $8.0xI0^{-I7}$ |
| I5 $^2D_{3/2}$ | $I.0xI0^{-I7}$ | $4.5xI0^{-I6}$ |
| I4 $^2D_{3/2}$ | $I.6xI0^{-I7}$ | $2.5xI0^{-I6}$ |
| I3 $^2D_{3/2}$ | $5.0xI0^{-I7}$ | $I.7xI0^{-I6}$ |
| I2 $^2D_{3/2}$ | $4.2xI0^{-I6}$ | $8.0xI0^{-I7}$ |
| II $^2D_{3/2}$ | $I.8xI0^{-I6}$ | $2.6xI0^{-I7}$ |
| I0 $^2D_{3/2}$ | $I.IxI0^{-I6}$ | $4.2xI0^{-I7}$ |
| I5 $^2D_{5/2}$ | $< I.0xI0^{-I7}$ | $7.0xI0^{-I6}$ |
| I4 $^2D_{5/2}$ | $I.2xI0^{-I6}$ | $4.2xI0^{-I6}$ |
| I3 $^2D_{5/2}$ | $I.3xI0^{-I6}$ | $2.8xI0^{-I6}$ |
| I2 $^2D_{5/2}$ | $2.2xI0^{-I6}$ | $I.7xI0^{-I6}$ |
| II $^2D_{5/2}$ | $I.4xI0^{-I6}$ | $I.4xI0^{-I6}$ |
| I0 $^2D_{5/2}$ | $7.0xI0^{-I7}$ | $I.6xI0^{-I6}$ |
| I2 $^2F_{5/2}$ | $< I0^{-I9}$ | $I.6xI0^{-I6}$ |
| II $^2F_{5/2}$ | $9.0xI0^{-I8}$ | $I.6xI0^{-I6}$ |
| I0 $^2F_{5/2}$ | $6.0xI0^{-I7}$ | $I.4xI0^{-I6}$ |
| 9 $^2F_{5/2}$ | $8.0xI0^{-I7}$ | $I.3xI0^{-I6}$ |
| I2 $^2F_{7/2}$ | $< I0^{-I9}$ | $3.7xI0^{-I6}$ |
| II $^2F_{7/2}$ | $7.0xI0^{-I8}$ | $3.0xI0^{-I6}$ |
| I0 $^2F_{7/2}$ | $I.5xI0^{-I6}$ | $2.4xI0^{-I6}$ |
| 9 $^2F_{7/2}$ | $I.6xI0^{-I6}$ | $I.8xI0^{-I6}$ |

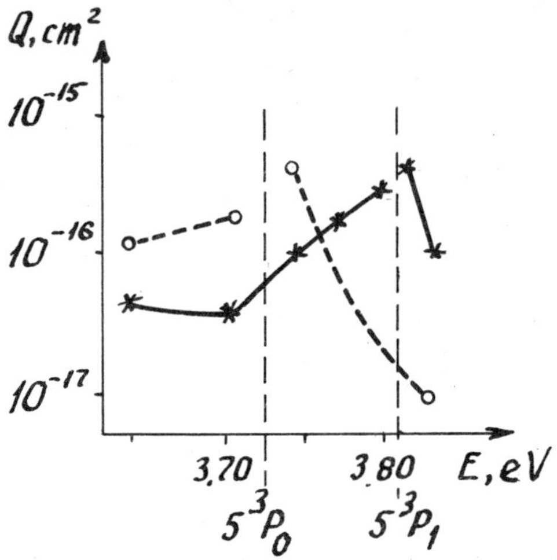


FIG. 8. Energy transfer cross-section from $5^3P_o$ to $5^3P_I$ cadmium as a function of the excitation energy ∘ ∘ ∘ – $Q(5^3P_o)$, ∗ ∗ ∗ – $Q(5^3P_I)$.

The plot (Fig.8) shows better than the table that cross-sections of energy transfer from metastable Cd $5^3P_o$ atoms in the $Cd^*$-Cs mixture are comparable to cross-sections of energy transfer from resonance $5^3P_I$ levels, and the role of metastable states should always be taken into account.

In the mixtures where the exciting level is the mercury $6^3P_I$ resonance level the metastable mercury $6^3P_o$ level is weakly populated because it lies at a relatively large distance ($\Delta E = 0.22$ eV), and the energy transfer cross-section between these levels is IOO times less that in the case of cadmium. The metastable mercury $6^3P_o$ level plays a considerable role only in the case when nitrogen has been added. It is known that nitrogen transfers mercury atoms from the $6^3P_I$ state into the $6^3P_o$ one.

Energy transfer cross-sections from metastable and resonance levels in the $Hg^*$-Tl-$N_2$ case are given in Table 3.

The second group of experiments carried out at the Latvian State University is connected with transfer of electronic energy from excited alkali molecules to alkali atoms. Some theoretical research has also been made to achieve better understanding of the experimental results.

Alkali molecules are excited by laser radiation to resonan-

TABLE 3. Cross-sections of energy transfer from Hg $6^3P_I$ and Hg $6^3P_o$ to thallium atoms.

| States of Tl atoms | Tl - Hg | | | | | Tl - Hg - $N_2$ | | | |
|---|---|---|---|---|---|---|---|---|---|
| | ΔE (Hg $6^3P_I$ as zero), eV | Contribution of cascades, % | Q(Hg $6^3P_I$–Tl), $10^{-16}$ cm² | | | ΔE (Hg $6^3P_o$ as zero), eV | Q(Hg $6^3P_o$ – Tl), $10^{-16}$ cm² | | |
| | | | Q | $\alpha$ | Q/$\alpha$ | | Q | $\alpha$ | Q/$\alpha$ |
| $7^2S_{1/2}$ | + 1.603 | 100 | ≤ 0.1 | 1 | ≤0.1 | + 1.384 | | | |
| $7^2P_{1/2}$ | + 0.651 | 35 | 2.2* | 1 | 2.2 | + 0.432 | 25 | 1 | 25 |
| $7^2P_{3/2}$ | + 0.527 | 19 | 6.6 | 1 | 6.6 | + 0.308 | | | |
| $6^2D_{3/2}$ | + 0.408 | 8 | 2.8 | 1 | 2.8 | + 0.189 | 3.9 | 1 | 3.9 |
| $6^2D_{5/2}$ | + 0.398 | 1 | 9.2 | 1 | 9.2 | + 0.179 | 7.8 | 1 | 7.8 |
| $8^2S_{1/2}$ | + 0.083 | 21 | 5.8 | 1 | 5.8 | - 0.136 | 0.5 | 0.59 | 0.85 |
| $8^2P_{1/2}$ | - 0.242 | ≤ 1 | 1.7 | 0.287 | 6.0 | - 0.462 | ≤0.1 | 0.051 | - |
| $8^2P_{3/2}$ | - 0.289 | ≤ 1 | 1.3 | 0.203 | 6.4 | - 0.508 | ≤0.1 | 0.034 | - |
| $7^2D_{3/2}$ | - 0.322 | 3 | 0.17 | 0.156 | 1.1 | - 0.541 | 0.035 | 0.026 | 1.3 |
| $7^2D_{5/2}$ | - 0.327 | 5 | 0.09 | 0.151 | 0.6 | - 0.546 | 0.058 | 0.025 | 2.3 |

* Our estimation.

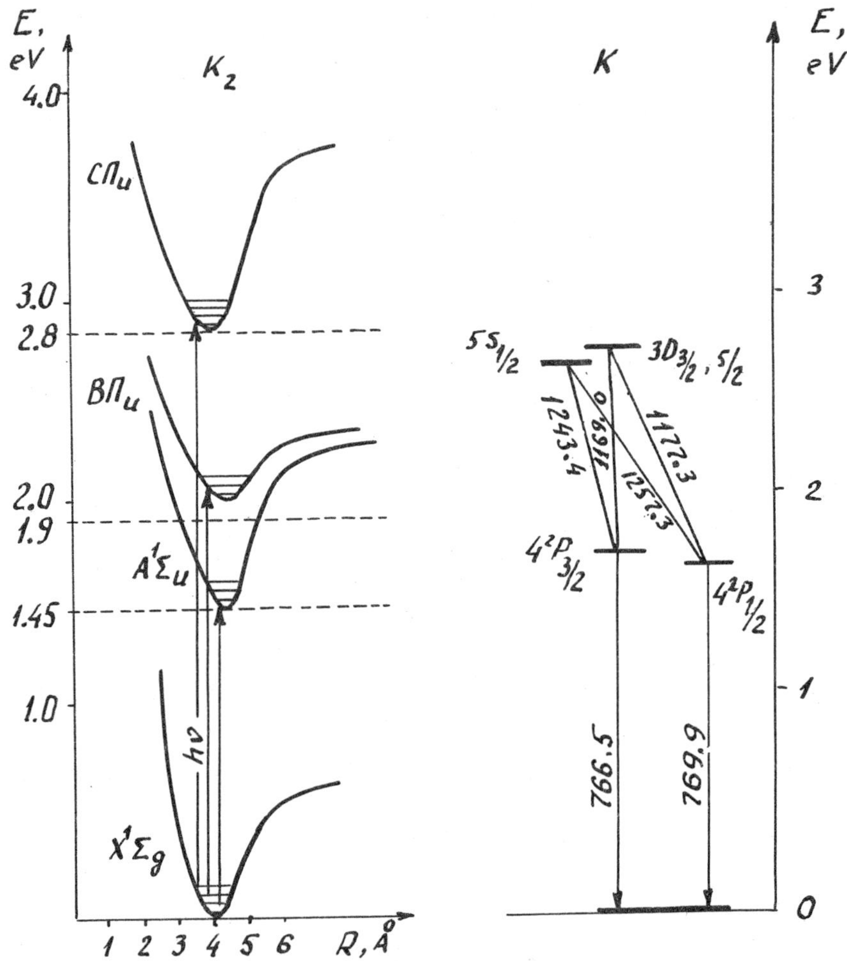


FIG. 9. A simplified diagram of atomic and molecular potassium energy levels.

ce states $Na_2(B^I\Pi_u)$, $Na_2(A^I\Sigma)$, $K_2(B^I\Pi_u)$, $NaK(D^I\Pi)$ or to higher states $K_2(C^I\Pi_u)$, $Cs_2(C^I\Pi_u)$, $Rb_2(^I\Pi_u)$.In the radiation spectrum molecular fluorescence is observed as well as several atomic lines originating from resonance levels and higher atomic levels (Fig.9).

The problem is what elementary processes are responsible for energy transfer from excited molecules to atoms.

The experimental and theoretical investigation which has been performed allows us to make some conclusions about the mechanism of energy transfer.

It turned out that several processes are involved in excitation of atoms. The role of each one depends on the state of the excited molecule, on the relative position of molecular and atomic energy terms as well as on the relative concentration of atoms and molecules. It is essential that different processes can be involved simultaneously. Their relative role depends on excitation frequency and concentration.

The problems were discussed in detail in three poster reports at the VI International Conference on Atomic Physics /22-24/.

Here I shall give only the main conclusions.

In the case when stable resonance states of molecules $Na_2(B^I\Pi_u)$, $Na_2(A^I\Sigma)$, $K_2(B^I\Pi_u)$, $NaK(D^I\Pi_u)$ are excited the excitation of atomic levels $n^2P_j$ is due to collisional energy transfer according to the following reactions

$$Na_2(B^I\Pi_u\ v'J')+Na(3S) \rightarrow Na_2(X^I\Sigma_g^+)+Na(3P_j)+\Delta E\ (\sim 0.50\ eV), \quad (2)$$
$$k = (I - 2)xIO^{-9}\ cm^3s^{-I} \qquad Q = (I - 2)xIO^{-I4}\ cm^2$$

$$Na_2(A\Sigma_u^+\ v'J')+Na(3S) \rightarrow Na_2(X^I\Sigma_g^+)+Na(3P_j)-\Delta E\ (\sim 0.05\ eV), \quad (3)$$
$$k = 3xIO^{-II}\ cm^3s^{-I} \qquad Q = 3xIO^{-I6}\ cm^2$$

$$K_2(B^I\Pi_u\ v'J')+K(4S) \rightarrow K_2(X^I\Sigma_g^+)+K(3P_j)+\Delta E\ (\sim 0.50\ eV), \quad (4)$$
$$k = 7xIO^{-IO}\ cm^3s^{-I} \qquad Q = IxIO^{-I4}\ cm^2$$

$$Na_2(B^I\Pi_u\ v'J')+K(4S) \rightarrow Na_2(X^I\Sigma_g^+)+K(4P_j)+\Delta E\ (\approx I\ eV), \quad (5)$$
$$k = (3 - 5)xIO^{-9}\ cm^3s^{-I} \qquad Q = (4 - 6)xIO^{-I4}\ cm^2$$

$$Na_2(A^I\Sigma_u^+\ v'J')+K(4S) \rightarrow Na_2(X^I\Sigma_g^+)+K(4P_j)+\Delta E\ (\approx I\ eV), \quad (6)$$
$$k = 5xIO^{-9}\ cm^3s^{-I} \qquad Q = 6xIO^{-I4}\ cm^2$$

$$Na_2(B^I\Pi_u\ v'J')+K(4S) \rightarrow Na_2(X^I\Sigma_g^+)+K(3D_j)+\Delta E\ (\leq 0.I\ eV), \quad (7)$$
$$k = (I - 5)xIO^{-IO}\ cm^3s^{-I} \qquad Q = (I - 6)xIO^{-I5}\ cm^2$$

$$NaK(D^I\Pi_u\ v'J')+K(4S) \rightarrow NaK(X^I\Sigma_g^+)+K(3D_j,5S)+\Delta E\ (\leq 0.I\ eV). \quad (8)$$
$$k = (2 - 39)xIO^{-II}\ cm^3s^{-I} \qquad Q = (2 - 45)xIO^{-I6}\ cm^2$$

For all exothermic processes 2, 4, 5, 6, 7, 8 large values of rate constants of energy transfer have been obtained in spite of considerable energy differences between atomic and molecular states.

For processes 5 and 7 rate constants have been determined for higher K(5S) and K(3D) levels as well as for the K (4P) resonance level taking into account infra-red cascade transitions. The most important is that excitation of resonance K(4P) levels with an excitation energy (about I eV) considerably smaller than that of the electronic-vibrational excitation of the sodium molecule occurs with a higher cross-section($5 x 10^{-14}$ $cm^2$), as compared to excitation of higher K(3D) and K (5S) levels which are in good energetic resonance.

Excitation mechanisms for resonance and higher level excitation of the atom are different.

One of atomic level excitation processes in reactions 5,6 and 7 with different partners ($Na_2$+K) is excitation by second kind collisions instead of another possible process - a bimolecular exchange reaction. Another excitation process of atomic levels can be photodisintegration of the excited molecules, photodissociation in the continuum of $B^1\Pi_u$ term or spontaneous pre-dissociation from higher stable $C^1\Pi_u$ states, perturbed by molecular resonance states.

a) Photodissociation -

$$Na_2(X^1\Sigma_g^+ \; v''^*) + h\nu_{las.} \rightarrow Na(3S) + Na(3P_{3/2}), \qquad (9)$$

$$K_2(X^1\Sigma_g^+ \; v''^*) + h\nu_{las.} \rightarrow K(4S) + K(4P_{3/2}). \qquad (10)$$

b) Predissociation -

$$K_2(C^1\Pi_u) \rightarrow K(4S) + K(4P_{3/2}) + \Delta E \quad (\sim 0.65 \; eV), \qquad (11)$$

$$Cs_2(C^1\Pi_u) \rightarrow Cs(6S) + Cs(6P_{3/2}) + \Delta E \quad (\sim 0.1 \; eV), \qquad (12)$$

$$Rb_2(C^1\Pi_u) \rightarrow Rb(5S) + Rb(5P_{3/2}) + \Delta E \quad (\sim 0.5 \; eV). \qquad (13)$$

Excited vibrational levels of the ground states are marked by asterisk.

It ought to be stressed that photodisintegration leads to excitation of atomic n $P_{3/2}$ states.

The photodissociation cross-section is strongly dependent on excitation frequency in good agreement with the resonance condition for vertical transitions. An excellent agreement between experimental results and theoretical calculations /25/, /26/ has been obtained. The measurements of relative intensities of fine-structure components for Na, K, Rb, Cs atoms have provided useful information. In the case of photodissociation of $Na_2(B^1\Pi_u)$, $K_2(B^1\Pi_u)$ molecules or in the case of predissociation of $K_2(C^1\Pi_u)$, $Cs(C^1\Pi_u)$ and $Rb(C^1\Pi_u)$ the population of the n $^2P_{3/2}$ component prevailed over that of $n^2P_{1/2}$(Table 4)

TABLE 4. Alkali atom D-line intensity ratio $I(n^2P_{3/2} \rightarrow n^2S_{I/2})$ / $I(n^2P_{I/2} \rightarrow n^2S_{I/2})$ for laser photodisintegration of alkali dimers.

| Elements | $\lambda$, nm | $I(n^2P_{3/2} \rightarrow n^2S_{I/2})/I(n^2P_{I/2} \rightarrow n^2S_{I/2})$ | |
|---|---|---|---|
| | | Experimental | Calculated |
| Na | 476.5 | 8.0 ∓ 0.5 | 9 - II |
| | 488.0 | 6.0 ∓ 0.5 | |
| K | 44I.6 | >I4 | 30 |
| Rb | 476.5 | >60 | exp (I5) |
| | 488.0 | >50 | |
| Cs | 632.8 | > 22 | exp (65) |

It may be explained as follows. After photodisintegration atoms of the disintegrating molecule are separated at a certain velocity through the corresponding molecular term which correlates with a certain atomic term, $n^2P_{3/2}$ or $n^2P_{I/2}$ for example.

The relative position of quasi-molecular terms of separated atoms at large internuclear distances determines, which of the following two processes takes place.

The first process may be an adiabatic evolution of the system, i.e. the atomic terms is excited which correlates directly with the molecular term from which the molecule disintegrates. The second one includes non-adiabatic transitions to molecular terms correlating with other resonance level components. Theoretical calculations based on the above-mentioned model have been performed to determine probabilities of non-adiabatic transitions. Experimental results are in good agreement with calculated ones. This agreement corroborates the feasibility of our approach.

Conclusions can be made about the molecular photodisintegration process, and molecular terms can be determined where from photodissociation takes place, as well as perturbing terms can be established in the case of predissociation.

The third mechanism of atomic level excitation is as follows. When potassium vapours were excited by non-resonant 476.5 nm, 488.0 nm, 5I4.5 nm Ar laser lines not corresponding to the absorption region of $K_2(X^I\Sigma_g^+)$ stable molecules and not capable of initiating photodissociation with subsequent populating of higher potassium levels, a considerable excitation of K $(3\ D_j)$ and K(5S) levels was observed.

Infra-red lines (II689 nm and II77I nm, i.e. $3\ D_j - 4\ P_j$. and I2523 nm and I2434 nm, i.e. $5S - 4P_j$) were detected in the emission spectrum, no molecular transitions being present. The intensity measurements of those infra-red lines with varying potassium atom concentration and intensity of laser radiation showed the excitation process of atomic levels to be due to absorption of laser radiation by two interacting potassium atoms with formation of a quasi-molecule, as follows

$$K(4S_{I/2})+K(4S_{I/2})+h\nu_{las} \rightarrow K_2(X^3\Sigma)+h\nu_{las} \rightarrow K(5S_{I/2},3D_j)+K(4S). \qquad (I4)$$

The corresponding rate constants were determined experimentally (Table 5).

TABLE 5.

| $\lambda$, nm | $\alpha$, $cm^5$ | $\Delta$ E, eV |
|---|---|---|
| 5I4.5 | $(I \div 2) \times IO^{-40}$ | $\sim -0.2$ |
| 476.5 | $2 \times IO^{-39}$ | $\sim -0.0I$ |

Large rate constants $\alpha \sim IO^{-39}$ $cm^5$ were obtained for the 476.5 nm line with wave-length close to the $K(5S,3D_j)$ level excitation energy. For other lines $\alpha \sim IO^{-4I}$ $cm^5$. Comparision with theory can give at present a qualitative agreement with experimental results and make quantitative estimates making use of available dipole moment values and the behaviour of quasi-molecular terms.

Apart from experimental studies of atom-atom collisions,some theoretical work is in progress at the Latvian State University Laboratory of Spectroscopy. An attempt has been made to improve the approximate calculations of the diatomic potential curve in the region of intermediate interatomic distances where asymptotic methods are unreliable.

The work on sensitized fluorescence discussed in this report are limited mainly to those performed in the Latvian State University Laboratory of Spectroscopy.

## REFERENCES


I. L.Krause,Appl.Opt.5,I375 (I966).

2. L.Krause, Physics of Electronic and Atomic Collisions (North-Holland,Amsterdam,I972),p.65.

3. E.E.Nikitin,Opt.Spektrosk. 22, 689 (I967).

4. E.I.Dashevskaya and E.E.Nikitin, Opt. Spektrosk. 22, 866 (I967).

5. E.I.Dashevskaya,A.I.Voronin and E.E.Nikitin,Can.J.Phys.47, I237 (I969).

6. E.I.Dashevskaya,E.E.Nikitin and A.I.Reznikov,J.Chem.Phys. 53, II75 (I970).

7. M.Pimbert,J.L.Rocchiccioli and J.Cuvellier,C.R.Acad. Sci. B 270, 684 (I970).

8. J.Cuvellier,P.R.Fournier,F.Gounand and J.Berlande, C. R. Acad. Sci. B 276, 855 (I973).

9. J.Berlande,J.Cuvellier,P.R.Fournier,F.Gounand and J.Pascale, Phys.Rev. A II, 846 (I975).

IO. E.K.Kraulinya and M.L.Janson, in Sensitized Fluorescence of Metal Vapour Mixtures (P.Stuchka Latvian State University, Riga, USSR, I97I), p. 52.

II. E.K.Kraulinya and M.G.Arman,Opt.Spectrosc.26,5II (I969).

I2. J.A.Spigulis,D.A.Ozolinsh and M.L.Janson, in Sensitized Fluorescence of Metal Vapour Mixtures (P.Stuchka Latvian State University,Riga,USSR,I975), p.35.

I3. E.K.Kraulinya and M.L.Janson,Opt.Spektrosk. 29,445 (I970).

I4. E.K.Kraulinya and M.L.Janson,Opt.Spektrosk. 29,827 (I970).

I5. E.K.Kraulinya and A.E.Lezdin,Opt.Spektrosk. 42,783 (I977).

I6. E.K.Kraulinya,S.Ya.Liepa and A.J.Skudra, Opt. Spektrosk. 40, 767 (I976).

I7. E.K.Kraulinya,A.P.Bryukhovetskiy and L.I.Kartasheva, Opt. Spektrosk. 4I, 903 (I976).

I8. J.A.Spigulis and L.I.Kartasheva,in Sensitized Fluorescence of Metal Vapour Mixtures (P.Stuchka Latvian State University,Riga,USSR,I977), p.24.

I9. S.B.Zagrebin and L.I.Kartasheva,in Sensitized Fluorescence of Metal Vapour Mixtures (P. Stuchka Latvian State University,Riga,USSR,I977), p.33.

20. E.N.Morozov,L.P.Presnyakov and A.D.Ulantsev,Kratkiye soobshcheniya po fizike,FIAN AN SSSR,3,I2 (I972).

2I. A.E.Bulyshev,A.E.Suvorov,E.K.Kraulinya,S.Ya.Liepa and A. Y.Skudra, in Abstracts of the Sixth International Conference on Atomic Physics,Riga, August I7-22,I978,p.270.

22. J.P.Klavins,M.L.Janson and G.V.Shlyapnikov, in Abstracts of the Sixth International Conference on Atomic Physics, Riga, August I7-22, I978, p.285.

23. I.I.Ostroukhova,G.V.Shlyapnikov,S.M.Papernov and M.L.Janson, in Abstracts of the Sixth International Conference on Atomic Physics,Riga, August I7-22, I978, p.478.

24. S.M.Papernov and M.L.Janson, in Abstracts of the Sixth International Conference on Atomic Physics,Riga,August I7 - - 22, I978, p.480.

25. E.Gordeev, V.B.Grushevsky,E.E.Nikitin and A.I.Shushin, in Sensitized Fluorescence of Metal Vapour Mixtures (P.Stuchka Latvian State University, Riga,USSR,I977),p.87.

26. V.B.Grushevsky,S.M.Papernov and M.L.Janson,Opt.Spektrosk. 44, 809 (I978).

# EXPERIMENTAL DETERMINATION OF ELECTRONIC TRANSITION PROBABILITIES AND THE LIFETIMES OF THE EXCITED ATOMIC AND IONIC STATES

N.P. Penkin

State Leningrad University, Leningrad B-164 USSR

In recent years in the connection with a rapid development of physics of plasma and its technical applications, lasers, investigations of the upper atmosphere and space the interest has greatly increased as to the determination of the electronic transition probabilities (f-values) and the lifetimes of excited atomic states.

The experimental data are also necessary for the improvement of the quantum-mechanical methods for the calculation of atomic structures, since a theory needs the comparison of its results with the experimental data.

It is known that there is a simple relationship between the probability of spontaneous radiation $A_{ki}$, Einstein's coefficients $B_{ki}$ and $B_{ik}$ and $f_{ik}$-values:

$$A_{ki} = \frac{8\pi h\nu^3}{c^3}\frac{g_i}{g_k} B_{ik} = \frac{8\pi h\nu^3}{c^3} B_{ki} = \frac{g_i}{g_k}\frac{8\pi^2 e^2}{mc\lambda_{ik}^2} f_{ik} \qquad (1)$$

Therefore, having determined one of these values one can calculate all the rest.

It is also known that the radiation lifetime of the excited level k $\tau_{k\tau}$ is related to the spontaneous transition probabilities starting from this level as follows:

$$\tau_{k\tau} = \frac{1}{\sum_i A_{ki}} \qquad (2)$$

## The Experimental Methods of Determination of Transition Probabilities (f-values)

At present the electronic transition probabilities (f-values) in atoms and ions are measured, mainly using the following optical methods: dispersion, emission and absorption.

Each method has its own advantages and disadvantages as well as its peculiar field of application. Each method gives the opportunity of determining the values for the $N_i f_{ik}$ or

$N_k A_{ki}$ products. The products mentioned can be determined with a high degree of accuracy by any of these methods provided that in measuring the conditions for their applicability are fulfilled. Unfortunately, while using the emission and absorption methods one is not always successful in ensuring the fulfilment of the required conditions, which leads to erroneous values for quantities measured.

Dispersion Method. There are various ways of determining Nfl values by measuring dispersion in vapours and gases. Puccianti /1/ suggested to obtain Nfl-values by measuring the interference band path near the absorption line. This method requires numerous measurements for it has a low degree of accuracy. Roshdestvensky /2/ developed a method called "hook method". This method, created more than 60 years ago, has been widely used in the laboratories of different countries (USSR, USA, Great Britain, France, Japan) both for the determination of f-values and the investigations of gas-discharge plasma and the processes occuring in shock tubes. This method allows with a large degree of accuracy (to one or several per cent) to get the Nfl values by reducing the measurement of the distances between the hook peaks ( $\Delta_{ik}$ ) situated on either side of the spectral line $\lambda_{ik}$ $(N_i f_{ik} \ell \sim \Delta^2_{ik})$

Fig.1 represents the photographs of hooks and dispersion near the absorption line of violet triplet of MnI $(a^6S_{5/2} - z^6P_{3/2,5/2,7/2})$ and shows that even in the case of closely situated spectral lines $\Delta_{ik}$-values may be measured with a high degree of accuracy. Dispersion interdependence near closely situated absorption lines is provided by simple calculations/2/. The advantage of the hook method is that the results derived with this method are independent of the profile and hyperfine structure of spectral lines. This method gives one the possibility to readily measure the f-values for the lines widely separated in the spectrum. Therefore, this method is most suitable for the determination of transition probabilities in the long series of spectral lines. The main drawback of the method is its small sensitivity and resolution as compared to the absorption and, particularly, emission methods.

In recent years there have been proposed and developed photoelectric methods of measuring Nfl values /3/ based on the

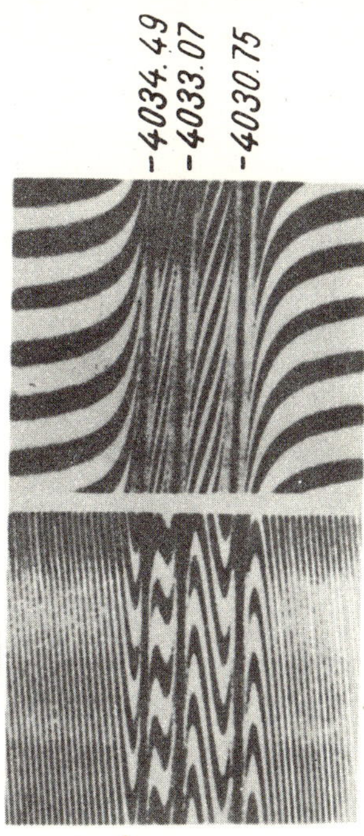


Fig.I. The photographs of hooks and dispersion near the absorption line of violet triplet of MnI ($a^6S_{5/2} - z^6P_{3/2, 5/2, 7/2}$).

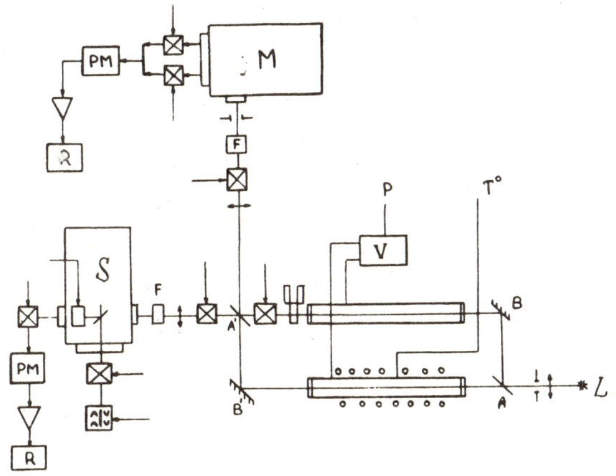


Fig.2. The scheme of an experimental arrangement for determining by simultaneous measuring hooks and total absorption.

hook method of Roshdestvensky and the application of different types of interferometers. It has greatly widened the possibilities of the application of this method.

The Nfl values can also be measured by the shift of the interference lines due to vapour or gas introduced into one of the interferometer beams. The papers of Frish et al. /4/ should be mentioned here.

The investigation of Grance et al. /5/ is also of interest. In this a new technique of measuring relative oscillator strengths along with the application of dye laser has been developed.

In the hook method, as well as in the absorption and emission methods passing from the measured quantities $N_i f_{ik}$ to the f-values themselves, it is necessary to know either the population of the energy levels or the law of distribution of the atoms over the energy levels (in measurements of relative f-values). This is the process that generates the chief source of error, because it is always difficult to determine accurately concentrations of absorbing or emitting atoms. It puts before the experimentators the task to to develop such methods for measurements of transition probabilities when it is not necessary to produce a vapour or a gas column with a known concentration of atoms. There are two ways to approach the problem:

1) the simultaneous application of different optical methods for one and the same vapour column (combined methods), and 2) the determination of the lifetime of excited states.

## Method for Determination Absolute f-values Based on the Simultaneous Measurement of Dispersion and Total Absorption

A method of determining the damping constant $\gamma$ ($\gamma = \gamma_i + \gamma_k + \gamma_{col}$; $\gamma_i = \frac{1}{\tau_i}$; $\gamma_k = \frac{1}{\tau_k}$) based on the simultaneous measurement of total absorption ($A_\lambda$) and dispersion has been devised in our laboratory /6/. Using this method we can determine not only the absolute f-values for the resonance lines, but also the effective cross-sections of the collisions, broadening the atomic resonance lines and the interaction of colliding particles / 7/.

It is known that for a large optical thickness, the following equation will hold:

$$A_\lambda \sim \sqrt{N_i f_{ik} l \gamma} \qquad (3)$$

On the other hand, from the theory of hook methods it follows that

$$N_i f_{ik} l \sim \Delta_{ik}^2 \qquad (4)$$

Hence,

$$\gamma = C'\left(\frac{A_\lambda}{\Delta_{ik}}\right)^2 \qquad (5)$$

where C' is a known constant.

By studying the relationship between $\gamma$ and Nfl and the pressure of a foreign gas one can obtain $\gamma_k = A_{ki}$ and the damping constants due to the collisions with the atoms of the proper ($\gamma_{col_1} = 8\sqrt{\frac{RT}{\mu}}\sigma_1 N$ ) and foreign ($\gamma_{col_2} = 2N_A\sqrt{\frac{8}{\pi RT}\left(\frac{1}{\mu}+\frac{1}{\mu_g}\right)}\sigma_2 p$) gases.

In this method of measurement it is not necessary to have the vapour column with a known concentration of absorbing atoms.

Fig.2 represents the scheme of an experimental arrangement for determining $\gamma$ by simultaneous measuring hooks and total absorption. It consists of an interferometric device allowing to register the Nfl-values by photographic or photoelectric methods and a monochromator with a photoelectric attachment for measuring the equivalent width $A_\lambda$.
Measurement errors of $\gamma_k = \frac{1}{\tau_k} = A_{ki}$ due to this method do not generally exceed 10 per cent, and in many cases are in the range of 5%.

The results of our combined method measurements of lifetime of resonance levels for a series of atoms are presented in Table 1. Included into the table are also lifetimes of resonance levels obtained by beam-foil, level-crossing, phase--shift and delayed-coincidence methods.

It is seen from the table that in most cases data obtained by different methods are in agreement. When comparing the data it should be born in mind that the combined method is free of cascading and radiation imprisonment.

Table 1

| Atom | Level | $\tau$, ns | | | | |
|---|---|---|---|---|---|---|
| | | HA | BFS /8,9/ | Hanle /10-16/ | PS /17-19/ | DC /20/ |
| Mg | $3^1P_1^o$ | 2.35±0.04 | 2.2±0.2 | 2.03±0.06 | 1.90±0.3 | |
| Ca | $4^1P_1^o$ | 5.4 ±0.3 | 6.2±0.5 | 4.62±0.15 | 4.6 ±0.6 | |
| Sr | $5^1P_1^o$ | 6.4 ±0.3 | - | 5.29±0.10 | 4.56±0.21 | 5.4±0.5 |
| Ba | $6^1P_1^o$ | 8.8 ±0.5 | - | 8.37±0.14 | 8.36±0.25 | 8.4±0.6 |
| Eu | $6y\,^8P_{9/2}$ | 5.5 ±0.2 | - | 5.5 ±0.3 | - | - |
| | $6y\,^8P_{7/2}$ | 5.7 ±0.3 | - | 5.8 ±0.3 | - | - |
| | $6y\,^8P_{5/2}$ | 5.4 ±0.3 | - | 6.7 ±0.7 | - | - |
| Ga | $5\ ^2S_{1/2}$ | 6.9 ±0.3 | 6.9±0.5 | 6.8 ±0.3 | 7.7 ±0.4 | 6.8±0.5 |
| | $4\ ^2D_{3/2}$ | 5.9 ±0.5 | 6.4±0.5 | - | 7.7 ±0.3 | 6.6±0.5 |
| | $6\ ^2S_{1/2}$ | 6.3 ±0.3 | 7.5±0.7 | 7.0 ±0.3 | 7.5 ±0.3 | 7.2±0.5 |
| In | $5\ ^2D_{3/2}$ | 7.4 ±0.5 | 6.3±0.3 | - | 7.9 ±0.5 | 6.9±0.5 |
| | $5\ ^2D_{5/2}$ | 7.4 ±0.5 | 7.6±0.5 | 7.1 ±0.6 | 7.9 ±0.5 | 6.9±0.5 |
| Tl | $7\ ^2S_{1/2}$ | 7.4 ±0.4 | 7.7±0.5 | 7.4 ±0.2 | 7.6 ±0.2 | 7.4±0.5 |
| | $6\ ^2D_{3/2}$ | 5.9 ±0.4 | 6.8±0.5 | 6.2 ±1.0 | 6.9 ±0.4 | 6.9±0.5 |

Table 2

| | I | II | III | IV | V | VI | VII | VIII | | |
|---|---|---|---|---|---|---|---|---|---|---|
| 2 | Li | | | | | | | Ne | | |
| 3 | Na | Mg | Al | Si | | | | Ar | | |
| 4 | K | Ca | Sc | Ti | V | Cr | Mn | Fe | Co | Ni |
| | Cu | Zn | Ga | Ge | | | | | | |
| 5 | Rb | Sr | | | | | | | | |
| | Ag | Cd | In | Sn | | | | | | |
| 6 | Cs | Ba | | | | | | | | |
| | Au | Hg | Tl | Pb | | | | | | |

| | | Nd | | Sm | Eu | Gd | | Dy | Ho | Er | Tm | Yb | |
|---|---|---|---|---|---|---|---|---|---|---|---|---|---|

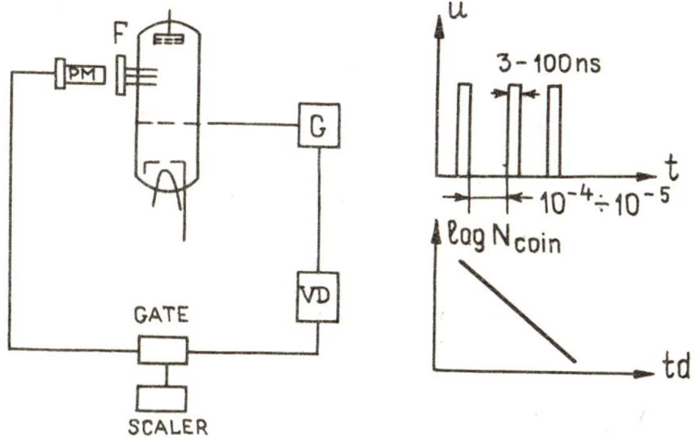


Fig.3. The principle scheme of the one-channel delayed coincidence method.

Table 2 presents the atoms (43 total) for whose spectral lines f-values are measured by the hook or combined methods. These measurements are mainly carried out by Soviet scientists. By the same methods we measured the oscillator strengths of the spectral lines in atoms belonging to the elements of the 1, II, III and IV groups of the Mendeleev table and atoms with the 3d (from Sc to Ni) and 4f(Nd, Sm, Eu, Gd, Dy, Ho, Er, Tm, Yb) unfilled shells.

Emission Method is the most sensitive one. It is based on the measurement of spectral line intensities, which is rather difficult since the source of linear radiation are more or less non-homogenous, and the spectral lines are re-absorbed and self-reversed. Although various methods for counting space non-homogenety of light source and making corrections of re-absorption are developed, it is, in fact, very difficult to foresee their effect on the measurement results.

Therefore, the measured intensity may differ greatly from its true value. It is to be noted that by measuring the re--absorption value and the dip width or the distance between the maxima of self-reversed lines one can get the Nf products. These methods are widely used in the study of gas-discharge plasma /22/.

But even if the spectral line intensity is measured accurately, and, therefore, correct values of $N_kA_{ki}$ are obtained, an experimentator faces a difficult tas to determine the absolute or relative populations of radiating levels. Many examples can be given when, due to wrong determination of $N_k$-values, transition probabilities obtained varied greatly from their true values.

Allen and Asaad /23/ suggested and developed a method for measurement of transition probabilities. It was based on determination of spectral relative intensities in the plasma of electric arc, its electrodes being made of an alloy with a known concentration of its components. Oscillator strengths of the spectral lines of one of the components must be measured independently or calculated. The accuracy of the data obtained by this method is estimated by the authors as 60-100%.

A method similar to that of Allen and Asaad was used by Gorliss and Bozman for measuring gA(gf)-values of 25,000 spectral lines in atoms and ions of 70 elements. The results of their measurements are given in a well-known book of these authors /24/. They think that the absolute gf-values are measured by them with the accuracy of ratio 2. However, as the comparison of Corliss and Bozman data with the most trustworthy ones shows the deviation ratio often exceeds 2. The analysis of the data obtained by emission method indicates that one can get reliable and accurate results by measuring relative transition probabilities of the weak lines or those lines which are ending on the excited levels, particularly, if the excitement conditions in the light source are under control and the latter is homogeneous, and the intensities of the spectral lines are determined by photoelectric method.

Absorption Method for measurements of f-values is used in different forms. Lately, its variant based on the measurement of the total absorption has been chiefly used:

$$A_\nu = \int \frac{J_{\nu 0} - J_\nu}{J_{\nu 0}} d\nu = \int (1 - e^{-\kappa_\nu \ell}) d\nu \qquad (6)$$

In case of a low optical thickness

$$A_\lambda = \frac{\pi e^2 \lambda^2}{m c^2} N_i f_{ik} \ell \qquad (7)$$

Using this variant American physicists obtained f-values for a greater number of lines. In their measurements the column of absorbing atoms was created in quartz cells, King's vacuum furnace or an atom beam was used.

To measure the $N_i f_{ik}$-values, particularly in gas-discharge plasma, the following variants of the absorption method are used: a) method of two tubes; b) method of various lengths of positive column; c) method of applying one or two mirrors. All of them are considered by Frish /22/ in details.

In order to apply the absorption method correctly one must know the contour shape of the spectral line. It requires additional measurements. Besides, if a spectral line has hyperfine structure, the latter should be taken into account. To understand the effect of hyperfine structure on the results of measurements one must know the position of hyperfine structure components and their relative intensity. Unfortunely, an experimentator very seldom possesses such information. If hyperfine structure is not made allowance for, it leads to wrong values of Nf, and, hence, to those of f. Lvov /25/ developed his method of measuring the absolute and relative oscillator strengths of the spectral lines on the base of linear and total absorption for the light sources used in atom--absorption analysis. In this latter case it is not necessary to know the concentration of absorbing atoms. Lvov measured the relative f-values for a great number of spectral lines and the absolute f-values for 30 lines in 23 atoms. For most spectral lines absolute f-values are in good agreement (within 20-30%) with the most reliable data obtained by other methods.

Methods for measurement of lifetimes in the excited atoms and ions. At present lifetimes of atoms and ions are measured by the delayed-coincidence, beam-foil (BFS), level-crossing and phase-shift methods. The reviews of Corney /26/, Imhof and Read /27/ discuss all these methods, their merits and drawbacks. Consequently, we shall restrict our attention to general observations, and consider the results of the measurements made mainly at the Institute of Physics, the University of Leningrad.

Delayed-coincidence method brought from nuclear physics to optical spectroscopy was first used by Heron /28/ in 1954 for the determination of the lifetime of He atom. In the Soviet Union the work of this kind started in 1958 by Osherovitch/29/ at the Institute of Physics, University of Leningrad and is successfully under way now. Delayed-coincidence technique may be used in its two variants: a) in pulsed excitation of atoms or ions; b) for the observation of the photon-photon coincidence in cascade in the stationary light source. Most numerous data were obtained by the pulsed variant of the method. At first one-channel method was used. The principal scheme of the experimental device is given in Fig.3.

With the delay given, the number of coincidences ($N_{coin}$) of the impulses from photomultiplier and pulse generator which is registered by a counter is proportional to the number of photons emitted at the moment of the transition of atoms from k-state to i-state. Apparently, the diagram of the log dependence between the delayed-coincidence values $(N)_{coin}$ and the delay time ( $\tau_{d}$ ) will be a straight line the slope of which gives the mean lifetime of the k-level.

In his experimental device Bennett /30/ turned one channel method into multichannel one. This not only sped up the measuring procedure but made it more accurate. The measurement results may be distorted due to cascade transitions, atom-atom collisions and radiation imprisonment. Cascading being the principal source of errors, leads to the complex decay curve. Special methods have been developed allowing to display an exponent corresponding to the decay of the given level and thus to determine the lifetime of k-level. In order to get rid of cascade transitions, lasers have come to be used recently for the selective excitation of the level studied, as well as the electron-photon and photon-photon coincidence methods. The photon-photon coincidence technique in cascade helps to measure the lifetime of the intermediate state. Although such measurements face certain technical difficulties, nevertheless it is now possible to measure the lifetime of the $7^3S_1$-level and the $6^3P_1$-state of mercury.

The effects of collisions and radiation imprisonment on the measurements are either taken into account or minimized by decreasing the pressure in an excitation cell or by replacing a vapour cell for an atom beam. On the other hand, studying the lifetime dependence upon the concentration of collided particles one can obtain effective cross-sections of collisions.

Delayed-coincidence technique has been used to measure the lifetimes of a great number of levels in atoms, ions and molecules. Osherovitch et al./31/ using this method were able to measure the lifetimes for many states in atoms and the first ions of elements of the I, II and IV groups of the Mendeleev table as well as rare gases. The authors of the works stated that the lifetime ($\tau_n$) dependence of the sequence of any atom upon the effective quantum number $n^*$ is thus expressed:

$$\tau_n = c\, n^{*\alpha} \qquad (8)$$

The degree exponent $\alpha$ is different for various atoms and series. For instance, the $\ln\tau_n$-$\ln n^*$ dependence for the $n^3D_3$ level of SrI is shown in Fig.4. It is seen from the figure that the experimental points have formed a straight line. The establishment of $\tau_n$-$n^*$ dependence is of great interest since

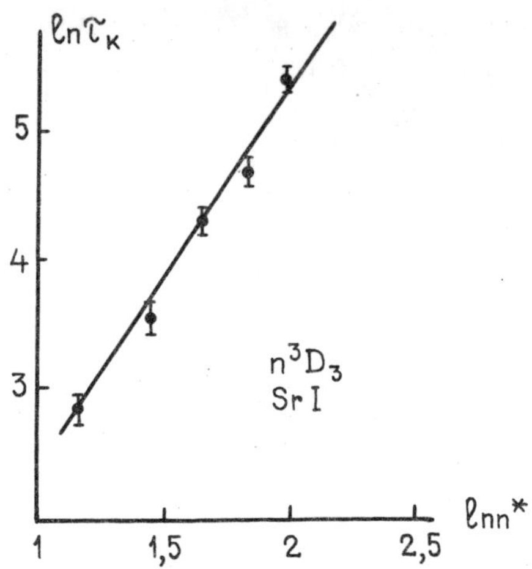


Fig.4. Dependence of $\ln \tau_k$ on ln n* for the $n^3D_3$ level of S

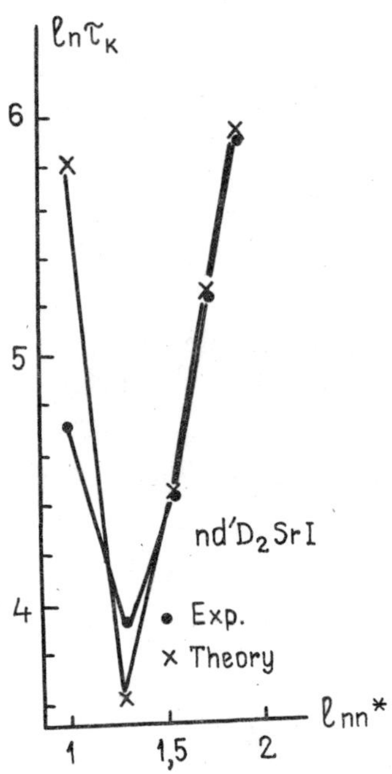


Fig. 5. Dependence of $\ln \tau_k$ on ln n* for the nd'$D_2$ level of SrI.

it allows to extrapolate the measurement results to the levels where for some reasons it is difficult to get the lifetime values. Afanas'eva and Gruzdev /32/ have shown by means of theoretical calculations that the degree dependence of $\tau_n$ upon main quantum number is valid for neon atoms. In case there is an interaction of configurations, $\tau_n$ changes non-monotonically together with the n -change. Fig.5 represents $\ln \tau_n - \ln n^*$ dependence for the $nd^1D_2$-level of SrI. A good agreement between experimental data and the results obtained by MCQDT method may be observed, except for the lower levels.

In recent years the beam-foil method (BFS) has got a wide distribution for measuring the lifetimes in atoms and chiefly in multicharged ions. Four international conferences were devoted to the studies carried out by this method. Such great interest to these investigations may be due to the fact that only by this method it is possible to measure the lifetimes of excitated states of multicharged ions. The information about them is necessary for astrophysical and hot plasma researches.

The principle of the method is that a beam of ions accelerated to several tens of thousand or million volts passes through a graphite or a metal foil. The beam will consist of atoms and ions a part of which is in various excited states. At their glow spectral lines appear; their intensity changes with the distance (l) from the foil according to the law: $J = J_0 e^{-\frac{l}{v\tau_k}}$. Changing the distance l by the foil replacement, measuring J and v one can get the value of the lifetime $\tau_k$. Here it is not possible to consider numerous results obtained by the beam-foil method. We shall note only the most important ones: 1) the lifetimes of atoms and ions of more than 40 elements (from H to U) have been measured. Hence, the f--values of spectral lines are determined. In many cases they are compared to the theoretical ones. 2) The f-values have been investigated and compared to the computed ones for many isoelectronic sequences, which was impossible before. 3) By means of BFS new levels corresponding to double-excited states in light atoms are established and their lifetimes are measured.

The data obtained by BFS method gave an impetus to a great

many new investigations in the theory of atom structures. The method requires to take into account cascade transitions, and also the accurate determination of the beam speed and its contents after it had passed the foil target. The correct determination of cascading and of beam speed is rather difficult, which reduces the accuracy of lifetime measurements. However, the comparison of BFS data to those obtained by other methods and to quantum-mechanical methods shows that beam-foil method gives reliable and exact results.

Proceeding to the other methods of lifetime measurements let us confine ourselves to some notes.

The double-resonance method

By this method the lifetimes of atoms of alkali metals, zink, cadmium, mercury, strontium, barium, thallium, lead and in ions of zink and cadmium have been measured. A relatively small number of objects with the lifetimes measured is accounted for some difficulties which take place in double-resonance technique, namely, 1) the width of a signal depends on the number of particles in resonance cell due to imprisonment of radiation 2) the increase in power of high-frequency field evoking induced transitions leads to the broadening of the curve from which $\tau$ can be determined, its maximum is replaced, and with a further field strength increase the curve may have two maxima. It is very difficult to exclude the effect mentioned above. The double-resonance technique is chiefly used for determining constant A (magnetic dipole interaction between the nucleus and electronic shell) and B (electric quadrupole interaction).

The phase-shift method is known to measure the phase-shift between an excited light source and fluorescent radiation. Modern methods for measuring the phase-shift have a high degree of accuracy. Radiation of lasers, when they appeared came to be used as a source of optical pumping, while a discharge tube might serve as a resonance cell. It is clear that the application of lasers and gas-discharge plasma extends the possibilities of the method, because the lifetime of greater number of levels may be measured. The use of a modulated electronic beam for exciting the spectral lines expands the possibilities of the phase-shift method as well. A difficult problem, however, arises as to checking the effects of cascade transitions on measurement results.

The phase-shift method, in its modern variants, has a high degree of accuracy (1-5%). Unfortunately, its potentialities are not yet fully used.

Level-crossing method is widely used for lifetime determination. Its basic principles were developed by Hanle (crossing in zero-fields) and Kollergov and Franken (non-zero-fields). At the Institute of Physics, of the Leningrad University the lifetimes in some atoms were measured by Kaliteevsky, Chayka et al. With the invention of lasers the possibilities of the method have also increased.

The contour of Hanle's signal depends upon imprisonment of radiation, Faraday effect and collisions with atoms. From the width of Hanle's signal (in the absence of inprisonment of radiation) one is able to get effective sections for depolarizing collisions. This method also allows to determine the A and B constants.

In conclusion we shall present a diagram (Fig.6) showing the number of works carried out using different methods.

## Discussion on Experimental Results of Transition Probabilities (f-values)

The number of investigations where transition probabilities and f-values were measured is so great that to discuss them in one report is not possible. Therefore, we shall deal with those works on atoms and ions which have been brought about at our laboratory by Blagoyev, Komarovsky, Ostrovsky, Parchevsky, Red'ko, Slavenas, Shabanova and many others. Included into the discussion will be the results obtained in other laboratories as well.

1. The absolute f-values of the resonance lines in atoms of elements of the I, II, III and IV groups measured at our laboratory are given in the 4-th column of Table 3. The table also includes the absolute f-values obtained both by averaging the most valid , in our opinion, experimental ones and those obtained by different quantum-mechanical methods. Special reference should be made of 1) a good agreement (in most cases) between the measurement results and a satisfactory

Table 3

| Atom | Transition | $\lambda$, nm | f | | |
|---|---|---|---|---|---|
| | | | HA | other meas. | theor. |
| Li | $2^2S_{1/2} - 2^2P_{1/2,3/2}$ | 670.8 | - | 0.71 | 0.74 |
| Na | $3^2S_{1/2} - 3^2P_{3/2}$ | 589.0 | 0.687 | 0.652 | 0.96 |
| | $- 3^2P_{1/2}$ | 589.6 | 0.343 | | |
| K | $4^2S_{1/2} - 4^2P_{3/2}$ | 766.5 | 0.688 | 0.682 | 0.66 |
| | $- 4^2P_{1/2}$ | 769.9 | 0.342 | 0.339 | 0.33 |
| Rb | $5^2S_{1/2} - 5^2P_{3/2}$ | 780.0 | | 0.668 | 0.69 |
| | $- 5^2P_{1/2}$ | 794.8 | | 0.329 | 0.34 |
| Cs | $6^2S_{1/2} - 6^2P_{3/2}$ | 852.1 | | 0.721 | 0.70 |
| | $- 6^2P_{1/2}$ | 894.3 | | 0.346 | 0.34 |
| Cu | $4^2S_{1/2} - 4^2P_{3/2}$ | 324.7 | 0.62 | 0.43 | 0.36 |
| | $- 4^2P_{1/2}$ | 327.4 | 0.32 | 0.22 | 0.18 |
| Ag | $5^2S_{1/2} - 5^2P_{3/2}$ | 328.1 | 0.51 | 0.50 | 0.45 |
| | $- 5^2P_{1/2}$ | 338.3 | 0.25 | 0.23 | 0.22 |
| Au | $6^2S_{1/2} - 6^2P_{3/2}$ | 267.6 | 0.41 | 0.39 | 0.47 |
| | $- 6^2P_{1/2}$ | 242.8 | 0.19 | | 0.22 |
| Be | $2^1S_0 - 2^1P_1$ | 234.8 | | 1.38 | 1.26 |
| Mg | $3^1S_0 - 3^1P_1$ | 285.2 | 1.56 | 1.79 | 1.57 |
| Ca | $4^1S_0 - 4^1P_1$ | 422.7 | 1.49 | 1.59 | 1.73 |
| Sr | $5^1S_0 - 5^1P_1$ | 460.7 | 1.52 | 1.85 | 1.62 |
| Ba | $6^1S_0 - 6^1P_1$ | 553.6 | 1.50 | 1.58 | 1.64 |
| Zn | $4^1S_0 - 4^1P_1$ | 213.9 | | 1.34 | 1.52 |
| Cd | $5^1S_0 - 5^1P_1$ | 228.8 | | 1.25 | 1.60 |
| Hg | $6^1S_0 - 6^1P_1$ | 185.0 | | 1.19 | 1.54 |
| Al | $3^2P_{3/2} - 4^2S_{1/2}$ | 396.1 | 0.135 | 0.12 | 0.124 |
| | $3^2P_{1/2} - 4^2S_{1/2}$ | 394.4 | 0.135 | 0.13 | 0.123 |
| Ga | $4^2P_{3/2} - 5^2S_{1/2}$ | 417.2 | 0.130 | 0.13 | 0.137 |
| | $4^2P_{1/2} - 5^2S_{1/2}$ | 403.3 | 0.130 | 0.12 | 0.131 |
| In | $5^2P_{3/2} - 6^2S_{1/2}$ | 451.1 | 0.154 | 0.15 | 0.153 |
| | $5^2P_{1/2} - 6^2S_{1/2}$ | 410.2 | 0.142 | 0.12 | 0.137 |
| Tl | $6^2P_{3/2} - 7^2S_{1/2}$ | 535.0 | 0.143 | 0.14 | 0.176 |
| | $6^2P_{1/2} - 7^2S_{3/2}$ | 377.6 | 0.129 | 0.13 | 0.131 |

Table 3 (continuation)

| Atom | Transition | λ, nm | f: HA | f: other meas. | f: theor. |
|---|---|---|---|---|---|
| Al | $3\,^2P_{3/2} - 3\,^2D_{5/2}$ | 309.3 | 0.209 | 0.18 | 0.69 |
| | $- 3\,^2D_{3/2}$ | 309.3 | | | |
| | $3\,^2P_{1/2} - 3\ D_{3/2}$ | 308.2 | 0.201 | - | 0.62 |
| Ga | $4\,^2P_{3/2} - 4\ D_{5/2}$ | 294.4 | 0.27 | 0.34 | 0.43 |
| | $- 4\ D_{3/2}$ | 294.4 | 0.032 | - | 0.045 |
| | $4\,^2P_{1/2} - 4\ D_{3/2}$ | 287.4 | 0.30 | 0.25 | 0.43 |
| In | $5\,^2P_{3/2} - 5\,^2D_{5/2}$ | 325.6 | 0.36 | 0.31 | 0.51 |
| | $- 5\,^2D_{3/2}$ | 325.9 | 0.042 | - | 0.056 |
| | $5\,^2P_{1/2} - 5\,^2D_{3/2}$ | 303.9 | 0.35 | 0.37 | 0.49 |
| Tl | $6\,^2P_{3/2} - 6\ D_{5/2}$ | 351.9 | 0.33 | 0.37 | 0.41 |
| | $- 6\ D_{3/2}$ | 352.9 | 0.038 | - | 0.056 |
| | $6\,^2P_{1/2} - 6\ D_{3/2}$ | 276.8 | 0.30 | 0.29 | 0.47 |
| C | $2p^{2\,3}P - 2p3s\,^3P_1^o$ | 165.7 | | 0.14 | 0.11 |
| Si | $3p^{2\,3}P - 3p4s\,^3P_1^o$ | 251.4 | 0.157 | 0.157 | 0.14 |
| Ge | $4p^{2\,3}P - 4p5s\,^3P_1^o$ | 265.2 | 0.163 | | 0.16 |
| Sn | $5p^{2\,3}P - 5p6s\,^3P_1^o$ | 286.3 | 0.230 | 0.186 | 0.17 |
| Pb | $6p^{2\,3}P - 6p7s\,^3P_1^o$ | 283.3 | 0.212 | 0.197 | 0.21 |

agreement with the data of calculations and 2) the f-values of the resonance lines of atoms having similar electronic shells coincide each other in the range of 10-20%. The latter allows, if necessary, to extrapolate very effectively the data of measurements or calculations to those atoms for which f-values are not yet determined.

The f-values of the spectral lines corresponding to P-D transitions in atoms of the III group are nearly two times larger than the f-values of the lines in P-S-transitions.

2. <u>On the transition probabilities in the principal series of atoms of the I, II and III groups</u>. The results of measurements of the transition probabilities along the principal series of the spectral lines in atoms of elements of the I, II and III groups of the Mendeleev table are given in Figs.7, 8 and 9. For atoms of the II, and III groups the measurements were carried out by Shabanova and myself /33/. The dashed curves refer to the case when $A_{ki}$ changes as $1/n^3$. As is shown in the figures, in all atoms of alkali metals, except for litium, a monotonic decrease of transition probabilities is observed as the principal quantum number increases. In the principal series ($^1S_0-^1P_1$) CaI, SrI and BaI the transition probabilities non-monotonically change with the increase of the principal quantum number of the upper level. The same may be seen in the diffuse series and AlI ($^2P^\circ_{1/2,3/2}-^2D_{3/2,5/2}$) and sharp series TlI ($^2P^\circ_{1/2,3/2}-^2S_{1/2}$).

The comparison of the change type of the transition probabilities along the principal series with the curves of quantum defects shows that a non-monotonic change of the transition probabilities corresponds to the non-monotonic change of the quantum defect. It is known that the presence of perturbations (configuration interaction) leads to a non-monotonic or strong change in the value of quantum defect in a series sequence of terms. Consequently, non-monotonic change of the transition probabilities is also a result of the configuration interaction. The investigations of such kind of non-monotonic run in the transition probabilities and the lifetime is of certain experimental and theoretical interest.

Table 4

| Transition | λ, Å | f |
|---|---|---|
| $3d^{10}\,4s\;^2S_{1/2} - 3d^9\,4s4p\;^4P^o_{3/2}$ | 2492 | 0.0091 |
| $^4P^o_{1/2}$ | 2442 | 0.0038 |
| $^4D^o_{3/2}$ | 2244 | 0.0066 |
| $^4D^o_{1/2}$ | 2226 | 0.059 |
| $^2P^o_{1/2}$ | 2182 | 0.125 |
| $^2P^o_{3/2}$ | 2179 | 0.198 |
| $^2D^o_{3/2}$ | 2165 | 0.125 |

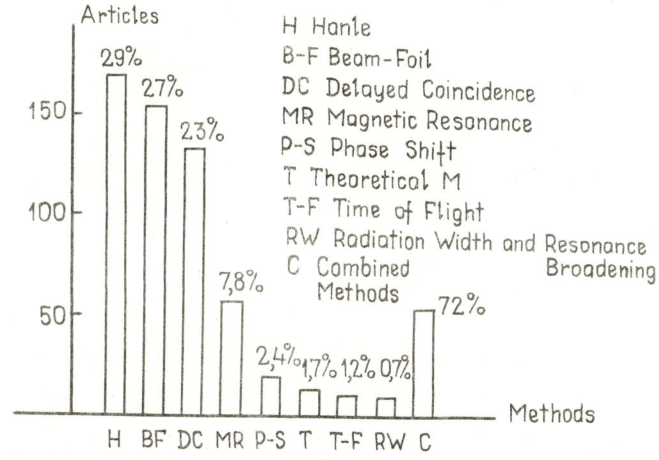


Fig. 6. Distribution of number of works of atom lifetimes measurement (up to October 1977).

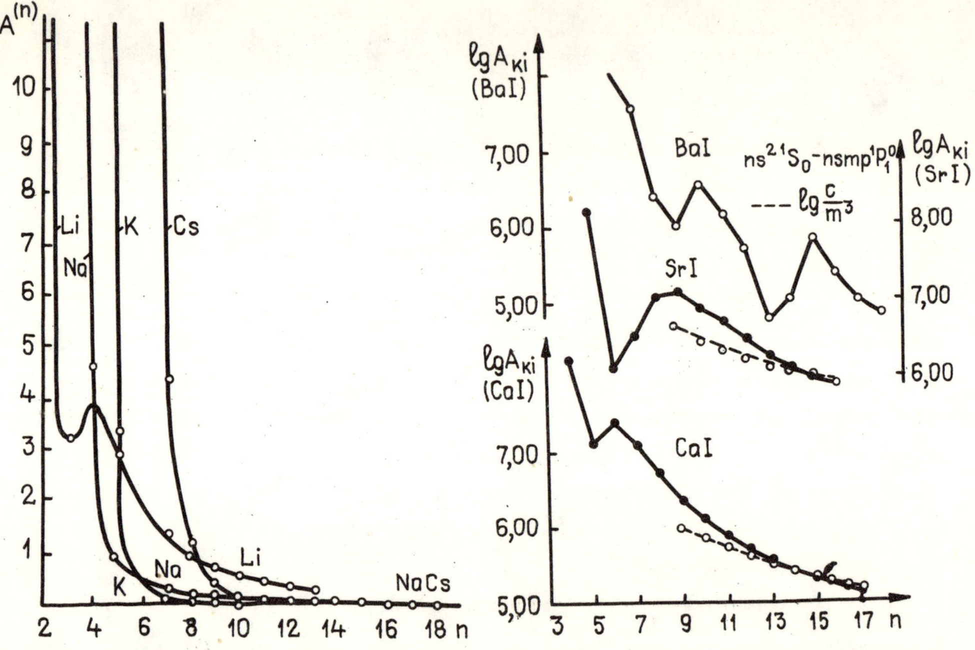


Fig.7. Dependence of $A_{ki}$ on n in the main series of alkaline metals.

Fig.8. Dependence of $lgA_{ki}$ on n in the main series of elements of II group.

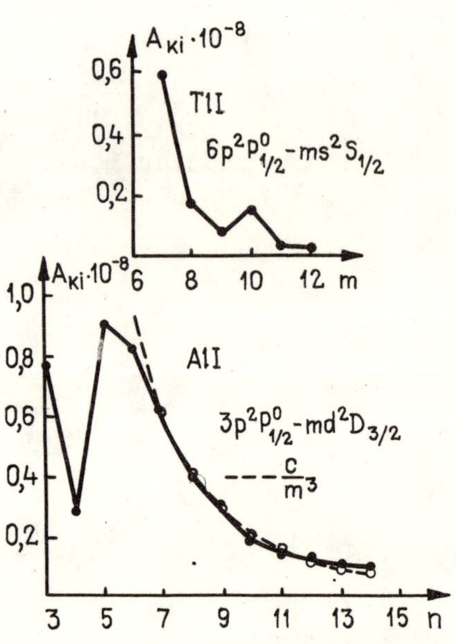


Fig.9. Dependence of $A_{ki}$ on n of $3p^2P^o_{1/2}$ - $nd^2D_{3/2}$ series AlI; $6p^2P^o_{1/2}$ - $ns^2S_{1/2}$ series TlI.

### 3. Oscillator strengths of the spectral lines arising in two-electron transitions and Beutler lines.

The oscillator strengths of these lines are insufficiently investigated both experimentally and theoretically. At the same time, available data obtained in particular in our laboratory /21/ indicate that oscillator strengths may have a considerable value ($10^{-1}$)

The cross-sections of the excitation of the Beutler levels and the formation of ions by tearing out electrons from their filled shells have a big value. As an example, Table 4 gives the f-values of the spectral lines in Cu atoms emerging at the $3s^{10}4S-3d^{9}4S4P$-transitions. The table shows that the f-values of the strongest lines in such transition combination are the values of the same order as those of the resonance douplet.

The measurements of the effective cross section of generation of the excited states of ions of zink, cadmium and mercury as a result of collisions of normal atoms with electrons are given in Table 5. It follows that the effective cross section of Beutler states have the same or sometimes greater values compared with the generation of the excited states of ions corresponding to the tearing out of a valent electron.

The data given above confirm the necessity to investigate two-electron transitions and the processes connected with the electron excitation in the filled shells both in the atom--photon and atom-electron interactions. The investigations of these processes may have not only purely scientific but also some practical interest.

### 4. The oscillator strengths of the spectral lines and the lifetimes of the excited states in atom and ions of the rare--earth elements with the unfilled 4 f shell.

In recent years an interest have greatly increased to calculating and measuring the oscillator strengths and the lifetimes of the excited states in the atoms and ions of the rare--earth elements (REE) It is mainly due to 1) the invention of laser on the vapours of some rare-earth elements; 2) the need to apply the spectral analysis of rare-earth elements since the chemical analysis of REE suffers from the similarity

Table 5

$$A + e \rightarrow A^{++} + 2e$$

$Q$ ( $10^{-18}$ $cm^2$)

| Level | | Zn II | Cd II | Hg II |
|---|---|---|---|---|
| The level of the usual system of terms | $^2P_{3/2}$ | 6.8 | 11.0 | - |
| | $^2P_{1/2}$ | 4.7 | 10.0 | 7.9 |
| | $^2S_{1/2}$ | 1.1 | 2.0 | 1.4 |
| | $^2D_{5/2}$ | 1.2 | 1.8 | 1.2 |
| | $^2D_{3/2}$ | - | 1.8 | - |
| The Beutler levels | $^2D^*_{3/2}$ | 7.0 | 17.0 | - |
| | $^2D^*_{5/2}$ | 7.5 | 30.0 | - |
| | $1^*_{5/2}$ | - | - | 0.9 |
| | $2^*_{7/2}$ | - | - | 1.7 |
| | $3^*_{5/2}$ | - | - | 1.2 |

of their properties; and 3) the necessity to determine the REE contents in the atmosphere of the Sun and stars in the spectra of which the lines of atoms and the ions of these elements are recorded.

Komarovsky and myself /35/ have compiled and systematized the data on the classification of spectral lines and energy levels in atoms and the first ions of REE with the unfilled 4f shell. The review results are given in Tables 6 and 7. Table 6 shows that for all atoms of REE the ionization potentials and the ground atomic states are determined, with a high degree of accuracy the wavelengths of more than 156,000 spectral lines measured of which only some 20,000 lines have been classified. We were in a similar situation when dealing with the classification of levels and spectra of the first ions of REE (see Table 7). For all first ions the ionization potentials and their ground states have been determined, the wavelengths of more than 71.000 spectral lines measured, 21.500 being classified. From the said above it follows that a great work is to be done on the spectral line classification in

Table 6

| Designation Element | I.P. (eV) | Ground level | Range class. wavelength(Å) | Range class. wavelength(Å) | Number meas. wavelength | Number class wavelength |
|---|---|---|---|---|---|---|
| Ce I Cerium | 5.47 | $4f5d6s^2\ {}^1G^o_4$ | 24200-3200 | 24200-3200 | 25000 | 1100 |
| Pr I Praseodymium | 5.42 | $4f^3 6s^2\ {}^4I^o_{9/2}$ | 30000-1740 | 8500-3500 | 37000 | 130 |
| Nd I Neodymium | 5.49 | $4f^4 6s^2\ {}^5I_4$ | 40500-2450 | 40500-3350 | 27000 | 1200 |
| Pm I Promethium | 5.55 | $4f^5 6s^2\ {}^6H^o_{5/2}$ | 10000-3100 | 10000-3100 | 10000 | 900 |
| Sm I Sa marium | 5.63 | $4f^6 6s^2\ {}^7F_0$ | 41000-2600 | 41000-3200 | 3500 | 1300 |
| Eu I Europium | 5.67 | $4f^7 6s^2\ {}^8S^o_{7/2}$ | 10200-1350 | 10200-1350 | 2300 | 1200 |
| Gd I Gadolinium | 6.14 | $4f^7 5d6s^2\ {}^9D^o_2$ | 25000-2100 | 25000-2700 | 4000 | 2000 |
| Tb I Terbium | 5.85 | $4f^9 6s^2\ {}^6H^o_{15/2}$ | 11630-2335 | 11630-2335 | 2300 | 5394 |
| Dy I Dysprosium | 5.93 | $4f^{10} 6s^2\ {}^5I_8$ | 11400-2300 | 11400-2300 | 1000 | 2000 |
| Ho I Holmium | 6.02 | $4f^{11} 6s^2\ {}^4I^o_{15/2}$ | 41000-2300 | 9000-2500 | 1800 | 80 |
| Er I Erbium | 6.10 | $4f^{12} 6s^2\ {}^3H_6$ | 12500-2300 | 12500-2700 | 5000 | 2000 |
| Tm I Thulium | 6.18 | $4f^{13} 6s^2\ {}^2F^o_{7/2}$ | 25000-1350 | 25000-1350 | 5500 | 3000 |
| Yb I Ytterbium | 5.43 | $4f^{14} 6s^2\ {}^1S_0$ | 12000-1060 | 8000-1060 | 2200 | 300 |

Table 7

| Designation | Element | I.P. (eV) | Ground level | Range measur. wavelength(Å) | Range class. wavelength(Å) | Number meas. wavelength | Number cl. wavelength |
|---|---|---|---|---|---|---|---|
| Ce II | Cerium | 10.85 | $4f5d^2\ ^4H^o_{7/2}$ | 24200–2450 | 24200–2450 | 11000 | 7500 |
| Pr II | Praseodymium | 10.55 | $4f^3 6s\ ^5I^o_4$ | 12000–2000 | 8800–3600 | 1500 | 350 |
| Nd II | Neodymium | 10.73 | $4f^4 6s\ ^6I_{7/2}$ | 40500–2450 | 40500–2450 | 1500 | 1350 |
| Pm II | Promethium | 10.90 | $4f^5 6s\ ^7H^o_2$ | 10000–3100 | 10000–3100 | 17500 | 1500 |
| Sm II | Samarium | 11.07 | $4f^6 6s\ ^8F_{1/2}$ | 41060–2600 | 41060–2600 | 1600 | 1550 |
| Eu II | Europium | 11.24 | $4f^7 6s\ ^9S^o_4$ | 7500–2500 | 7500–2500 | 2000 | 500 |
| Gd II | Gadolinium | 12.09 | $4f^7 5d6s\ ^{10}D^o_{5/2}$ | 25000–2100 | 25000–2100 | 3000 | 1500 |
| Tb II | Terbium | 11.52 | $4f^9 6s\ ^7H_8$ | 8800–2500 | 8800–2500 | 5500 | 200 |
| Dy II | Dysprosium | 11.67 | $4f^{10} 6s\ ^6I_{17/2}$ | 11400–2300 | 11400–2500 | 9000 | 1000 |
| Ho II | Holmium | 11.80 | $4f^{11} 6s\ ^5I^o_8$ | 12000–2500 | 4200–3300 | 800 | 10 |
| Er II | Erbium | 11.93 | $4f^{12} 6s\ ^4H_{13/2}$ | 12000–2300 | 12000–2300 | 12000 | 1200 |
| Tm II | Thulium | 12.05 | $4f^{13} 6s\ ^3F^o_4$ | 25000–2200 | 25000–2200 | 800 | 700 |
| Yb II | Ytterbium | 12.18 | $4f^{14} 6s\ ^2S_{1/2}$ | 25000–1370 | 25000–1370 | 5432 | 4154 |

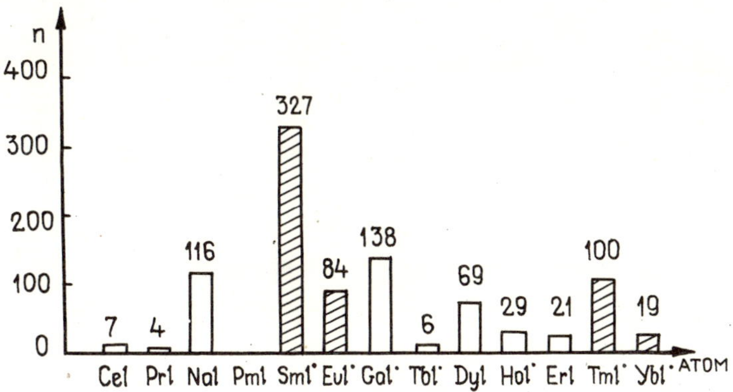


Fig.10. Diagram of the oscillator strengths of atomic spectral lines for REE with the unfilled 4 f shell.

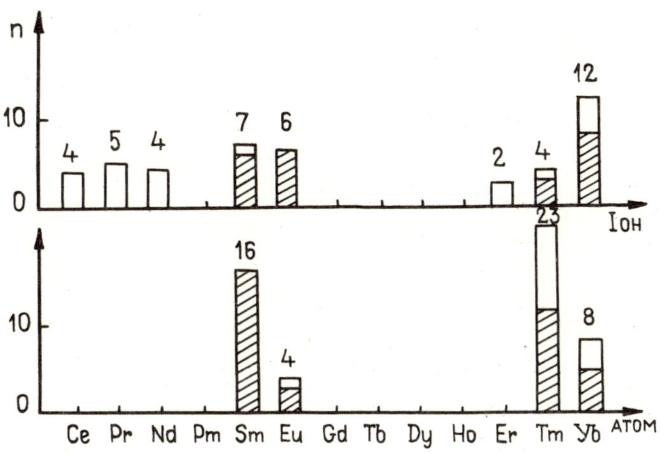


Fig.11. Diagram of lifetimes of excited states of the atoms and first ions of REE with the unfilled 4 f shell.

atoms and ions of REE.

The diagram in Fig.10 illustrates the state of our knowledge as to f-values of REE atomic spectral lines. It is seen that the oscillator strengths are measured for a relatively few number of lines. For prometium atom they are not measured at all. The absolute f-values are known only for small number of spectral lines in atoms of samarium, europium, tulium and ytterbium (the corresponding columns are shaded). Let us note by the way that out of the data given in Fig.10 the f-values of 889 absorption lines situated in the range from 2.400 to 7.800 Å are determined in our laboratory by the hook method /35, 36/. For the elements marked with asterisk the data on the transition probabilities (the f-values) are in the well-known tables of Corliss and Bozman. But as we explained earlier they are not included into the diagram.

Fig. 11 represents graphically lifetimes $\tau$ of excited states of REE atoms and ions.
The shaded parts of columns in the diagram corresponds to the number of those levels for which the values of $\tau$ were measured in our laboratory /37/ by the multichannel delayed-coincidence method. The results are given in the tables 8 and 9. The lifetimes of a number of levels in atoms and ions of REE were measured by other authors by means of level-crossing and BFS techniques. Note that the data obtained by different methods show a good agreement. As Fig.11 shows, generally speaking, the lifetimes are determined for only 40 levels in ions and for 51 levels in REE atoms. For many atoms and ions such measurements were not performed.

The analysis of the f-values and the lifetimes in atoms and ions of REE allows us to conclude:

1) The absolute f-values of the resonance lines of Yb I and Yb II are close in their value to those in atoms and ions of alkali earth metals having electronic shells similar to YbI and YbII.

2) In spectra of a number of REE atoms (e.g. in Yb I, GdI) strong intercombination lines may be observed and relative line intensities in multiplets do not follow the intensity rules based on the L-S coupling. All this affirms a strong spin-orbital interaction and consequently the deviation from

Table 8

| Atom | λ, nm | Level, K | τ, ns Our data | Measur. other authors | | | |
|---|---|---|---|---|---|---|---|
| Sm I | 565.9 | 18475 | 69±4 | | | | |
| | 527.1 | 19777 | 41±2 | | | | |
| | 520.0 | 20713 | 36±2 | 34±3 /38/ | | | |
| | 497.5 | 20091 | 49±3 | 44±4 | | | |
| | 472.8 | 22632 | 9.1±0.5 | | | | |
| | 471.7 | 21194 | 8.8±0.5 | 7.9±0.7 | | | |
| | 464.5 | 21813 | 10±0.6 | | | | |
| | 459.6 | 22041 | 12±1 | | | | |
| | 448.0 | 22314 | 12±1 | | | | |
| | 436.2 | 22914 | 8.3±0.5 | 7.1±0.7 | | | |
| | 433.0 | 22381 | 13±1 | | | | |
| | 397.4 | 27425 | 12±1 | | | | |
| | 392.5 | 26281 | 13±1 | | | | |
| Eu I | 564.5 | 17707 | | 1300±300 /39/ | | | |
| | 466.1 | 21445 | 5.6±0.2 | 6.7±0.7 | | | |
| | 462.7 | 21605 | 5.4±0.2 | 5.8±0.3 | | | |
| | 459.4 | 21761 | 5.4±0.2 | 5.5±0.3 | | | |
| Tm I | 589 | 16957 | | 820±80 /38/ | 16500 /40/ | 2300 /41/ | 1400±400 /41/ |
| | 576.4 | 17343 | | | 710 | 2000 | 2100±600 |
| | 567.5 | 17614 | | 580±50 | 450 | 790 | 760±70 |
| | 563.1 | 17753 | | 650±50 | 490 | 760 | 760±70 |
| | 530.7 | 18837 | | 410±40 | 130 | 410 | 450±40 |
| | 473.3 | 21121 | | 470±50 | 870 | 600 | 500±50 |
| | 459.9 | 21738 | | | 440 | 1500 | 1700±500 |
| | 438.6 | 22791 | | 250±20 | 100 | 180 | 250±20 |
| | 435.9 | 22930 | | 74±7 | 55 | 97 | 74±7 |
| | 420.3 | 23782 | 43.2±0.9 | 41±4 | 38 | 66 | 40±4 |
| | 418.7 | 23873 | 14.9±0.8 | 17±2 | 13 | 38 | 17±1.7 |
| | 410.5 | 24349 | 16.3±0.9 | 16±2 | 19 | 26 | 16±1.6 |
| | 409.4 | 24418 | 11.0±0.8 | 9.6±1 | 10 | 15 | 9.6±1 |
| | 394.9 | 34085 | 8.4±0.9 | | | | |
| | 391.6 | 34297 | 6.4±1.4 | | | | |
| | 389.6 | 25656 | | | 270 | | |
| | 388.7 | 25717 | 28.4±5.1 | | 230 | | |
| | 388.3 | 25745 | 11.5±0.8 | | 12 | | |
| | 382.6 | 26127 | | | 210 | | |
| | 379.8 | 35090 | 11.7±0.8 | | | | |
| | 375.1 | 26646 | 55.2±2.8 | | | | |
| | 374.4 | 25701 | 10.9±0.4 | | 9.5 | | |
| | 371.7 | 26889 | 8.1±0.2 | 18±2 | 11 | 17 | 17.7±1.8 |
| Yb I | 572.0 | 45155 | 14±1 | /42/ | /43/ | /44/ | /45/ |
| | 555.6 | 17992 | 850±80 | 820±20 | | 760±80 | 827±40 |
| | 398.7 | 25068 | 6.8±0.8 | 5.1±0.1 | 5.2±0.6 | | 5.5±0.3 |
| | 377.0 | 43805 | 15±1 | | | | |
| | 346.4 | 28857 | 15±2 | 14.4±0.5 | 17±2 | 17±2 | |
| | 267.1 | 37415 | 82±6 | 77.4±6 | | | |
| | 246.4 | 40564 | 9.8±0.6 | 9.3±0.6 | | | |
| | 227.1 | 44018 | 47±4 | 39.1±3.5 | | | |

Table 9

| Ion | λ, nm | Level, K | τ, ns<br>Our data | Measur. other authors | |
|---|---|---|---|---|---|
| Ce II | 422.2 | 24668 | | 10.0±1.5 /43/ | |
| | 456.2 | 25766 | | 5.8±0.8 | |
| | 418.6 | 30847 | | 5.9±0.8 | |
| | 416.5 | 31340 | | 6.9±1.0 | |
| Pr II | 399.4 | 25468 | | /46/ | 9.5±1.5 /43/ |
| | 414.3 | 27128 | | 8.4±1.2 | |
| | 410.0 | 28816 | | 8.0±1.2 | |
| | 417.9 | 25569 | | 7.8±1.0 | |
| | 406.2 | 23010 | | 6.0±0.8 | |
| Nd II | 430.3 | 23230 | | 13 ± 3 /43/ | |
| | 394.1 | 25877 | | 10 ± 3 | |
| | 406.1 | 28419 | | 12 ± 3 | |
| | 401.2 | 30002 | | 13 ± 3 | |
| Sm II | 446.7 | 27696 | | 16 ± 4 /43/ | |
| | 367.1 | 28072 | 14± 1.5 | | |
| | 366.1 | 27631 | 11± 1.5 | | |
| | 363.4 | 28997 | 12± 2 | | |
| | 360.9 | 29935 | 8± 1.5 | | |
| | 359.2 | 30880 | 14± 2 | 10 ± 3 | |
| | 356.8 | 31926 | 15±2 | 11 ± 3 | |
| Eu II | 420.5 | 23774 | 11.4± 1.5 | | |
| | 413.0 | 24208 | 10.3± 0.5 | | |
| | 397.2 | 26838 | 10.9±0.8 | | |
| | 393.0 | 27104 | 7.1± 0.4 | | |
| | 390.7 | 27256 | 9.9± 0.8 | | |
| | 382.0 | 26172 | 7.8± 0.8 | | |
| Er II | 389.6 | 26099 | | 14.5 ± 3 /43/ | |
| | 338.5 | 29973 | | 9 ± 1 | |
| Tm II | 384.8 | 25980 | 17 ±2 | | |
| | 370.1 | 27009 | | 34 /48/ | |
| | 370.0 | 27254 | | | |
| | 346.2 | 28875 | 20 ± 3 | 26 | 21 ± 2 /46/ |
| | 336.2 | 29967 | 12 ± 1 | 11 | 9.8± 1.0 |
| | 313.1 | 31927 | 12 | | 9.0± 1.0 |
| Yb II | 418.0 | 54304 | 35 ± 3 | /43/ | |
| | 369.4 | 27062 | 7.1±0.4 | 6.9±0.6 | |
| | 352.0 | 58961 | 15 ± 1 | 7.4±0.4 | 5.8±0.6 /42/ |
| | 310.7 | 62559 | 4.9±0.2 | | |
| | 303.1 | 32981 | 16 ± 1 | | |
| | 297.0 | 33653 | 42 ± 3 | | |
| | 289.1 | 34575 | 31 ± 3 | 18±3 | |

the L-S coupling.
3) The lifetimes of the levels in the studied terms Sm II ($4f^5 5d6S$), Eu II ($4f^7 6p$) and TmII ($4f^{13} 6p$) are close to each other and are 10-20 ns. Most SmI levels lying from 20.500 to 23.500 $cm^{-1}$ independent of a term and electronic configuration possess the lifetime less than 100 ns. These levels may be used as upper laser ones, and then lower levels belong to the configuration $4f^6 5d$ 6S. The wavelengths of the expected laser transitions are 1, 1.5 and 1.7 mkm long. In the field of energy below 20.000 $cm^{-1}$ the configuration levels $4f^6 6S6p$SmI are predominantly longlived. They have $\tau$ about 1 mks.

The lifetimes of TmI levels belonging to the same term are similar to each other.

4) It is of interest to note that the experimental values of the transition probabilities for DyI, ErI and TmI are, in most cases, in satisfactory correlation with the data obtained by quantum-mechanical methods /40,41/. It points out that the latter give good results also for such complex atoms as the rare-earth atoms.

In conclusion we should like to emphasize that recently, thanks to the efforts of experimentators and theoreticians, much progress has been achieved in the investigations of transition probabilities and lifetimes of the excited states in atoms and ions. However, available data are too scarce to meet the needs of science and practice. Therefore, intensive work in this field of atomic physics must be continued. One should bear in mind that full information about transition probabilities and lifetimes may be obtained by rational combination of various experimental and theoretical methods.

## R e f e r e n c e s


1. L.Puccianti, Nuov.Cim, 2, 257 (1901)
2. D.S.Roshdestvensky Raboty po anomalnoj dispersii v parah metallov. AN SSSR (1951).
3. M.Crance. Rev.Phys.Appl., 8, 325 (1973).
   V.A.Moskalev, J.N.Nagibina, N.A.Polushkina, Opt.Spektrosk. (USSR) 36, 979 (1974).
   Tokubo, D.Jnjne, M.Shimaza, Jap.J.Appl.Phys. 14, 1633(1975).
   G.V.Zhuvikin, L.N.Shabanova, Prikladnaja spektroskopija. M. p.p.42 (1977). Trudy XVIII Vsesojusnogo S'esda po spektroskopii (USSR).
4. S.E.Frish, Opt.Spectrosk. (USSR) 27, 542 (1969).
   L.P.Rasumovskaya, N.S.Ryasanov, S.E.Frish, Opt.Spektrosk(USSR) 41, 353 (1976).
5. M.Crance, P.Juncar, J.Pinard, J.Phys.B:Atom and Mol.Phys. 8, 2461 (1975).
6. Yu.J.Ostrovsky, N.P.Penkin, L.W.Shabanova, DAN SSSR, 120 66 (1953).
7. N.P.Penkin, L.N.Shabanova. Opt.Spektrosk. (USSR), 23, 22 (1967).
8. A.Omount, C.R.Acad.Sci., 26, 213 (1966).
9. T.Andersen et al., J.A.S.R.T., 10, 1143 (1970).
10. T.Andersen, G.Sørensen, Phys.Rev.A5, 2497 (1972).
11. W.Smith, A.Gallager, Phys.Rev., 145, 26 (1966).
12. L.O.Dickie, J.M.Kelly, Can.J.Phys., 48, 879 (1970).
13. L.O.Dickie et al., Can.J.Phys., 51, 1088 (1973).
14. E.Handrich et al. Opt.Pump and Atom Line Shape Warsaw 417(1969).
15. A.Gallager, M.Norton, Phys.Rev., 3, 741 (1971).
16. P.Zimmermann, Z.Phys., 233, 21 (1970).
17. A.Gallager, A.Lurio, Phys.Rev., 136, 87 (1964).
18. W.H.Smith, H.S.Liszt, J.Opt.Soc.Am., 61, 938 (1971).
19. E.Hulpke, E.Paul, W.Paul, Z.Phys., 177, 257 (1964).
20. P.T.Hanningham, J.A.Link, J.Opt.Soc.Am., 57, 1000 (1967).
21. N.P.Penkin, Spektroskopiya gasorasriadnoj plasmy.(USSR) Nauka (1970).
22. S.E.Frish, Spektroskopija gasorasriadnoj plasmy. (USSR ) Nauka (1970).
    N.G.Preobrazhensky, Spektroskopija opticheski plotnoj plasmy (USSR) Nauka (1971).

23. C.W.Allen, M.N.Assad, Mon Not R Astron.Soc., 117, 36 (1957); 117, 622 (1957).

24. C.H.Corliss, W.R.Bozmann, Experimental Transition Probabilities for Spectral Line of Seventy Elements, (1962)

25. B.V.Lvov J.Q.S.R.T., 12, 651 (1972).

26. A.Corney, Adv.Electron. Electron Phys., 29, 115 (1969).

27. R.Imhof, F.H.Read, Rep.Prog.Phys., 40, 1 (1977).

28. S.Heron, Mc Whirter, E.H.Roderick, Nature, 174, 564 (1954).

29. A.L.Osherovitch, I,G.Savich, Opt.Spektrosk.(USSR) 4,715,1958.

30. W.R.Bennett, P.I.Kindlmann, G.N.Mercer, Appl.Opt. Supplement 2, 34 (1965).
W.R.Bennet, P.I.Kindlmann, Phys.Rev., 145, 38 (1966).

31. A.L.Osherovitch, Ya.F.Verolainen, A.Ya.Nikolaich, V.J.Privalov, S.A.Pulkin, V.V.Tesikov. Prikladnaja spektroskopija, 33 (1977). Trudy XVIII Vsesojusnogo S'esda po spektroskopii (USSR).

32. N.V.Afanasjeva, P.F.Gruzdev, Opt.Spektrosk. (USSR) 38, 1013 (1975).

33. N.P.Penkin, L.W.Shabanova, Opt.Spektrosk.(USSR) 12, 3 (1962); 15, 12 (1963); 14, 167 (1963); 15, 228 (1963); 18, 896 (1965); 18, 941 (1965); 23, 22 (1967).

34. A.A.Mityureva, N.P.Penkin, Opt.Spektrosk. (USSR), 33, 1023 (1972).

35. N.P.Penkin, V.A.Komarovsky, J.Q.S.R.T. 16, 217 (1976).

36. K.B.Blagoev, V.A.Komarovsky, N.P.Penkin, Opt.Spektrosk. (USSR) 40, 622 (1976).

37. K.B.Blagoev, V.A.Komarovsky,Opt.Spektrosk.(USSR) 42,407 (1977); 42, 594 (1977).
V.A.Komarovsky, N.P.Penkin, L.N.Shabanova, Opt.Spektrosk. (USSR) 25, 155 (1969).
M.L.Burshtein, Ya.F.Verolainen, V.A.Komarovsky, A.L.Osherovitch, N.P.Penkin, Opt.spektrosk. (USSR) 37, 617 (1974)
K.B.Blagoev, V.A.Komarovsky, N.P.Penkin, Opt.Spektrosk. (USSR) 42, 424 (1977).
K.B.Blagoev, M.L.Burshtein, Ya.F.Verolainen, V.A.Komarovsky, A.L.Osherovitch, N.P.Penkin, Opt.Spektrosk. (USSR) 44, 32, 1978.
K.B.Blagoev, V.A.Komarovsky, N.P.Penkin, Opt.Spektrosk. (USSR) 44, 224 (1978).

38. E.Handrich, A.Stendel, R.Wallenstein, H.Walther, J.de Phys., 30, N 1, Suppl., 18 (1969).

39. N.Lange, J.Luther, A.Stendel, H.Walther, Phys.Lett., 20, 166 (1966)

40. P.Camus, J.Phys., 31,985 (1970).

41. R.Wallenstein, Z.Phys., 251, 57 (1972)

42. F.H.K.Rambow, L.D.Schearer, Phys.Rev.A, 14 738 (1976)

43. T.Andersen, O.Poulsen, P.S.Ramamijam, Petrakiev Petkov, Solar Physics, 44, 257 (1975)

44. Budik B., J.Snir, Phys.Rev. 1, 545 (1970).

45. M.Baumann, G.Wandel, Phys.Lett., 22, 283 (1966)

46. T.Andersen, G.Sorensen, Solar Physics, 38, 343 (1974)

47. B.Engman, J.O.Stoner, Jr., and J.Martinson, N.E.Cerne, Phys.Scr., 13, 363 (1976).

48. L.J.Curtis, J.Martinson, R.Buchta. Nucl.Instr. and Meth., 110, 391 (1973).

# THE DEVELOPMENT OF IDEAS OF V.FOCK IN ATOMIC PHYSICS

Yu.N.Demkov

Leningrad State University

In December 1978 we mark the 80-th anniversary of V.Fock's birthday (1898-1974).

Vladimir Alexandrovich Fock was one of the leading Soviet theoretical physicists. His works in various fields of theoretical physics such as diffraction theory and gravitation are well known everywhere.

However, the main subject of his activity consists in the atomic physics and the quantum theory. Just here the achievements of V.Fock are most fundamental and best known as they have influenced the atomic physics as a whole.

The purpose of this talk is to trace this influence, especially because many of Fock's results are so general and so inherently included into the general scheme of quantum theory that the authorship of V.Fock is sometimes almost forgotten. Some of the works of Fock appeared ahead of the time. Their importance begins to be recognized and real use starts only now, decades afterwards.

Inherent to V.Fock was his own irreproducible style connected considerably with his briilliant mathematical education - with the traditions of the Petersburg mathematical school - with the names of Ostrogradsky, Tchebyshev, Ljapounov, Steklov, Tamarkin, Smirnov. His striking mathematicsl abilities are reflected in many anecdotes about his apparently effortless solution of some problems considered as unsoluble by such prominent mathematicians as V.I.Smirnov. There is a rumour that Max Born called V.Fock "Die grosse mathematische Kanone". Consequently, V.Fock's works bear the reflection of these abilities; therefore, some people sometimes stress especially this side of his papers, considering him first of all as a mathematician and only secondly as a physicist. But his physical intuition was also remarkably deep and allowed him to obtain outstanding results in various fields of physics. Thus, certainly, we must consider V.Fock first as a physicist and only then as a mathematician, although we can find brilliant mathematical results in his papers as well.

We shall consider here only the subject of atomic physics; here Fock's contribution was especially important, partly because the beginning of his scientific activity coinsided with the creation of quantum mechanics - with this romantic and remarcable period when all our ideas about the micro-world have been changed fundamentally within a few years.

V.Fock met this period armed with heavy mathematical amunition. To characterize his physical intuition and its connection with mathematics it is worth-while to mention (as he told us at the time of the 50-th anniversary of quantum mechanics) that beeing a student of Petrograd University in the very beginning of the twenties and attending the lectures of Professor Tamarkin on integral equations he considered the idea that the spectra of atoms and those of integral equations are somehow connected.

Being sent to Germany (Gottingen) in 1928 V.Fock happened to be in the very center of events, and his extraordinary individuality allowed him to participate actively in the creation of quantum physics.

His first achievement obtained even before the German period was the establishment of the relativistic second order wave equation for a spin-less charged particle in the electromagnetic field [1]. Here, chronologically, Fock was the second, six weeks later than Klein [2]. So it is natural to call this equation the Klein-Fock one (what is accepted now by many theorists) and not the Klein-Gordon equation as some people do. Actually, Gordon is at least the third [3], he sent the paper to the Zeitschrift fur Physik after the publication of Klein's paper, and he did not consider the electromagnetic field.

The second and, perhaps, the best known result of V.Fock is the self-consistent field method including exchange - the Hartree-Fock method [4], its rigorous derivation from the variational pronciple, and the first calcu-lations performed together with his disciple Mary I.Petrashen [5] demonstrat-ing the efficiency of the method. So, the original semi-intuitive idea of Hartree was developed, put on the rigorous mathematical basis, generalized

VLADIMIR ALEXANDROVICH FOCK (1898-1974)

Main fields of interest: Quantum Theory, Atomic Physics, Quantum Field Theory, Propagation and Diffraction of Waves, Gravitation, Papers in Photometry, Electrotechnics, Acoustics, Elasticity, Ballistics, Mathematics.

Academician (1940). The member of foreign Academies: Norwegian (1958), Danish (1965), GDR "Leopoldina" (1967). Orders of Lenin in 1945, 1953, 1958, 1968. Hero of Socialist Labor, 1968.

The Lenin Prize for the works in quantum field theory (1960). The Mendeleev Award for the quantum theory of complex atoms. The Leningrad University Prize for the works in the theory of gravitation (1956).

About 300 scientific papers. Out of them more than 70 in atomic physics.

and completed.

During the 45 years, passed since the formulation of the method, it was generalized in different directions and applied practically to all complicated many-particle microsystems. One of the first generalizations of the method, where the two-electron correlations were taken into account, has been done by Fock, Veselov and Petrashen [6]. Further works in this direction (accounting many-electron correlations) were done by O.Sinanoglu [7] and others much later. In the works of A.P.Yucys [8] and his co-workers the multoconfiguration and relativistic generalizations have been considered. In the works of Brueckner [9] and Bogoljubov [10] the further generalizations of the method have been formulated, allowing to apply it to such new systems and phenomena as the nuclei and the superconductivity.

The applications of the method in the molecular theory (the molecular orbitals method, the Roothan method [11]), to the electron-atom and atom-atom collisions, in solid state theory and in nuclear physics allow us to characterize the Hartree-Fock method as a universal approximate method for the description of complex quantum systems.

One of the remarcable properties of the method is the apparent absence of a small parameter, allowing one to present the approximate method as the expansion of the exact results in a power series of this parameter. So, the applicability of the method to various systems is not so easy to prove, and, actually, very often only the comparison between the theory and the experiment verifies the quality of the method. So, for instance, the applicability of the method to the nuclei has been originally negated by Bohr, due to the absence of the strong center of force and the short-range interaction in contrast to the case of the electronic cloud of the atom. Only in the forties by Maria Goeppert-Mayer and others this applicability was verified, explained from the standpoint of the Pauli principle and led to the establishment of the nuclear shell structure.

An important step in the understanding of the nature of the method was the investigation of its connection with the perturbation theory by using the graph's representation. Then the method could be considered as a partial summation of an infinite subset of the whole set of graphs [12].

The third important result of V.Fock and M.Born has been given in their paper, about the adiabatic approximation in the case of degenerate energy levels [13]. They have shown that the adiabatic approximation is valid also for this more complicated case. Recently these results were applied to the atomic collision processes and particularly to the process of the mixing of the degenerate atomic states with nonzero angular momentum in slow collisions between the atoms and ions [14].

It is important to emphasize here the further development of the adiabatic approximation by Landau who has introduced the idea of analyticity, the integration over the complex time plane, and has proved the exponential smallness of nonadiabatic transition probabilities [15]. This result is one of the most important ones for the theory of slow atomic collisions and for some other aspects of the atomic and molecular theory.

The works of V.Fock on the second quantization [16] and the method of functionals [17] brought complete clarity into the theory of systems, allowing nonconservation of the number of particles. Prior to this work there was a vague idea that the second quantization was something qualitatively new and different from the ordinary quantum theory. Fock made it clear that it was not the case. In every textbook considering the second quantization in all details the material is presented according to the classical paper of Fock. The term "Fock's space" used to indicate the hierarchy of the configuration spaces for the description of such systems is now gererally accepted. Using the second quantization V.Fock found, for instance, one of the simplest ways to derive the general formulas of the Hartree-Fock method [16].

The further development of the second quantization method led Fock to the formulation of the method of functionals [17] used in quantum electrodynamics and in the quantum field theory, i.e. beyond the essential atomic physics.

The last of the most important results obtained by V.Fock in atomic physics was the discovery of the internal symmetry of the hydrogen atom [18].

It is curious to note that Fock himself was at that time (1935) a little bit sceptical as to the quantum mechanical applications of the group theory in contrast to the general enthusiasm which arose due to the remarcable works of Wigner, Weyl, von-Neumann and others. However, it was just Fock who introduced the concept of the internal symmetry and found the first example: the hydrogen atom - the system investigated from every side by many most prominent theorists. After this discovery the internal symmetry has been found for several other simple quantum systems, such as the isotropic oscillator and others.

The hydrogen atom symmetry can be used to simplify several calculations of the properties of the hydrogen atom, especially when one has to perform the summation over all states with the same principal quantum number n . Recently the connection between Fock's symmetry of the hydrogen atom and the well known problem of the geometrical optics - the so called "Maxwell fish-eye problem" (formulated by Maxwell in 1854) has been established [19]. Actually, the transition from one problem to the other is equivalent to the interchange of the coordinates and momenta, i.e. to the Fourier transformation. Finally, quite an unexpected connection between the Maxwell fish-eye problem and that of the consecutive filling of electronic states in the Mendeleev periodic system has been found [19].

The idea about the hydrogen atom symmetry group has been also recently used in the theory of elementary particles.

V.Fock used this group to calculate the closed shell density matrix of many-electron atoms [18]. Here the concept of the approximate internal symmetry has been introduced - the concept which is now widely used in several cases. The further development of this approach has been considered in [20].

One of the simplest problems where Fock's symmetry enables the exact solution is the first order perturbation splitting of the hydrogen energy levels in arbitrary oriented, crossed, uniform electric and magnetic fields [21].

In the theory of collisions Fock's results allow one to consider the scattering of a particle by the Coulomb field using the Lorentz group symmetry of the hydrogen for the positive energy case.

The hydrogen symmetry can be also used to calculate the mixing of the energy levels in collisions betwen the excited hydrogen atoms and ions. For this problem it was possible to introduce a new concept of the dinamical, angular momentum dependent states of the quasimolecule and the corresponding dinamical potential curves [22]

Among the less important papers of V.Fock one can mention the derivation of the virial theorem for the bound states in quantum mechanics by using the scaling transformation and the variational principle [23] Later, using the same approach the formulation of the virial theorem for the collisional processes has been found [24]. There are also interesting papers of Fock considering the Dirac equation, the semiclassical approximation, the Thomas-Fermi method, etc.

Let us mention finally his paper on the helium atom [25], where for the first time a rigorous series expansion of the wave function has been obtained in the vicinity of the origin of the configuration space, where the electrons are close to the nucleus. The series contains in this case the powers of logarithms of the distance from the origin. The correct account of this expansion is important in numerical variational high accuracy calculations of the energy and wave functions of many-electron atoms. This paper becomes more and more important at present when we can investigate both experimentally and theoretically the states of many-electron systems where two or more electrons are excited, and the states are unstable relative to the Auger-ionization.

Considering the work of V.Fock in atomic physics as a whole, we can find there some basic ideas, which have been systematically used, and are characteristic of his theoretical style.

First of all, it is the investigation and use of exact or approximate symmetry properties of physical systems. This idea is evident in the hydrogen atom case, but also the Hartree-Fock method or, better to say, its good applicability to the real systems reflects their objective feature. Actually, the description of the whole system by the one-electron quantum numbers is a good description in most cases. This means that the eletronic cloud is not an amorphous and structureless one, but has quite a distinct structure, i.e. the internal symmetry which is reflected in different experimental properties and can be used in calculations.

The second characteristic feature of Fock's papers (also connected with the symmetry properties of the system) is the tendency to use, where possible, the generating functions. Such a method is usually the shortest and mathematically the most elegant one elucidating the internal properties of the system.

The third property of several Fock's papers is the wide use of the heuristic possibilities of the variational principles. The importance of the variational principles is often underestimated. The papers of Fock show the significance of variational approach for the development of new approximate methods, for the derivation of various equalities, for the investigation of the most general properties of the physical systems.

Many Fock's papers in other fields of theoretical physics influenced indirectly some results in atomic physics. For instance, the theory of the electron detachment in slow collisions between atoms and ions can be formulated as the detachment of the surface wave from the surface with variable properties [26], and this problem can be solved by means of the methods developed by Fock in diffraction theory.

V.Fock paid a great attention to the basic problems of atomic physics, particularly to the problem of a correct materialistic interpretation of quantum mechanics criticizing both the naive mechanistic views and the positivistic ones. His discussions with Bohr on this subject in 1957 were fruitful, and Bohr apparently accepted some of Fock's arguments, for instance, he stopped to use the term "the principally uncontrollable interaction" criticized by Fock. I think that his interpretation of quantum mechanics formulated in the last edition of his textbook [27] is at present the most exact one corresponding to the generally accepted point of view, but specified in several aspects.

In conclusion I want to mention one, may be less known, paper of Fock [28] which is of phylosophical nature and which stems essentially from his theoretical works. He writes there about the role of the approximate methods in the formation of our conceptions on the physical objects. Its main idea is that as far as we consider the object in the most general way, using the most general theory, we have not enough "food" for our imagination and the object "slips away" from us. Only by and by, developing the approximate methods for the description, we can introduce the new concepts and properties describing the system. So, from the general Schrodinger equation for the many-electron atom it is difficult to extract any simple and clear results, but when passing to the Hartree-Fock approximation such properties as one-electron states and quantum numbers, the shell structure, etc. become quite clear and describe the real atom much better. So the approximate method acquires a new meaning: it is not only a tool for the calculations, but also a generator of new conceptions. All the works of V.Fock illustrate in the best way this thesis.

Almost all scientific activity (more than 60 years) of V.Fock was connected with the Leningrad University. There he founded a scientific school embracing now four Chairs. Hundreds of his students and the students of his students work now in all parts of the Soviet Union. His remarcable personal traits, high sense of responsibility, friendliness, constant readiness to help and to discuss the new results - all that led to the formation of our large and friendly theoretical family - the Fock theoretical school. It is our duty to keep the high standards of the theoretical works established by Fock and we shall always consider all our achievements to be the continuation and an inalienable part of works of V.Fock.

## References

1. V.A.Fock, Zs.f.Phys. 38, 242, 1926; received 11-th June 1926.
2. O.Klein, Zs.f.Phys. 37, 895, 1926; received 28-th April 1926.
3. W.Gordon, Zs.f.Phys. 40, 117, 1926; received 29-th September 1926.
4. V.A.Fock, Zs.f.Phys. 75, 622, 1932.
5. V.A.Fock, M.I.Petrashen, Zh.Eksp.Teor.Fiz. 4, 295, 1934.
6. V.A.Fock, M.G.Veselov, M.I.Petrashen, Zh.Eksp.Teor.Fys. 10, 727, 1940. See also the review M.G.Veselov, L.N.Labzovsky, Problems of Theoretical Physics, v.I, Leningrad University, 1973.
7. O.Sinanoglu, Many-electron Theory of Atoms, Molecules and their Interactions. Moscow. "Mir", 1964.
8. A.P.Yucys, Adv. in Chem.Phys. 14, 191, 1969.
9. K.A.Brueckner, Phys.Rev., 100, 36, 1955.
10. N.N.Bogoljubov. Lectures on Quantum Statistics (in Ukranian) Sov. Shkola, 1949.
11. C.C.J.Roothaan, J.Chem.Phys. 19, 1445, 1951.
12. H.P.Kelly, Adv. in Chem.Phys. 14, 129, 1969.
13. M.Born, V.Fock, Zs.f.Phys. 51, 165, 1928.
14. Yu.N.Demkov, V.N.Rebane, Th.K.Rebane, Problems of Theoretical Physics, Leningrad University, v.I, p.263, 1973; Yu.N.Demkov, V.N.Ostrovsky, ibid.,p.279.
15. L.D.Landau, E.M.Lifshiz, Quantum Mechanics, Fizmatgiz, 1963.
16. V.Fock, Zs.f.Phys. 75, 622, 1932.
17. V.Fock, Phys.Zs.d.Sow.Union. 6, 368, 1934.
18. V.Fock, Izv.AN SSSR, ser.fiz. 2, 169, 1935; Zs.f.Phys. 98, 145, 1935.
19. Yu.N.Demkov, V.N.Ostrovsky, Zh.Eksp.Teor.Fyz. 60, 2011, 1971; 62, 125, 1972.
20. P.P.Pavinsky, A.I.Sherstjuk, Problems in Theoretical Physics, Leningrad University, p.66, 1973.
21. Yu.N.Demkov, B.S.Monozon, V.N.Ostrovsky, Zh.Eksp.Teor.Fyz. 57, 1431, 1969.

22. Yu.N.Demkov, V.N.Ostrovsky, E.A.Solovjev, Zh.Eksp.Teor.Fyz. 66, 125, 1974.

23. V.Fock, Zs.f.Phys. 63, 850, 1930.

24. Yu.N.Demkov, Dokl.AN SSSR. 89, 249, 1953.

25. V.Fock, Izv. AN SSSR, ser.fiz. 18, 161, 1954.

26. Yu.N.Demkov, A.Z.Devdariany, Teor.Mat.Fiz. 21, 74, 1975.

27. V.Fock, Foundations of Quantum Mechanics, Nauka, 1976.

28. V.Fock, Usp.Fiz.Nauk. 16, 1070, 1936;
Phylosophical Problems in Physics, Leningrad University.,1974, p.3.

# THE MULTICONFIGURATION HARTREE-FOCK METHOD FOR ATOMIC ENERGY LEVELS AND TRANSITION PROBABILITIES*

C. F. Fischer

Department of Computer Science, The Pennsylvania State University

University Park, PA 16802, U.S.A.

## 1. Introduction

The effect of correlation in the motion of electrons in a many electron system is an important factor in the theoretical determination of atomic properties. When Hartree[1] derived his equations, he assumed the electrons moved in the field of the nucleus screened by the spherically averaged distribution of all the other electrons. Thus the effect of correlation in the motion of electrons was neglected. In fact, in his model there was a finite probability that two electrons might occupy the same region of space. Later Fock[2] modified these equations. Starting with an antisymmetric total wavefunction and applying a variational procedure, he obtained what Hartree called the "equations with exchange" now referred to as the Hartree-Fock (HF) equations. Electrons with the same spin co-ordinates are repelled in this model but other correlation effects remain. For reasons such as these, Löwdin[3] defined the error <u>in the HF approximation</u> as the correlation error.

Let $\psi(\gamma LS)$ be a total wavefunction for a state labelled $\gamma LS$, and let $E(\gamma LS)$ be the total energy of that state. Furthermore, let $\psi(\gamma LS)$ and $E(\gamma LS)$ satisfy Schrödinger's equation

$$H\psi = E\psi \tag{1}$$

where $H$ is the non-relativistic Hamiltonian. Thus $\psi$ and $E$ are exact solutions of the non-relativistic problem and a <u>state</u> of the system is an eigensolution of Eq. (1). Let $\psi^{HF}$ and $E^{HF}$ be HF approximations. Then the equations

$$\begin{aligned} &\psi = \psi^{HF} + \psi^{corr}, \ \langle \psi^{HF} | \psi^{corr} \rangle = 0 \\ &E = E^{HF} + E^{corr} \end{aligned} \tag{2}$$

define the correlation function, $\psi^{corr}$, and correlation energy, $E^{corr}$, respectively. Thus relativistic effects must be neglected. Of course, exact non-relativistic results are not known for many electron systems and usually one compares Hartree-Fock results with observed values to estimate the correlation effects. The observed values include relativistic effects but, by restricting the discussion to outer processes of neutral atoms or ions of low degree of ionization, the relativistic effects should be small.

Correlation effects can be computed in a variety of ways and one of the simplest, conceptually, is a configuration interaction (CI) calculation. Spectroscopists have been able to explain many phenomena in terms of configurations. In fact the spectroscopic label, $\gamma LS$, for an observed state usually designates the configuration, the coupling of the various angular momenta, as well as the total angular and spin momenta of the dominant component. Though the configuration is not a quantum number for the state, it often describes the system remarkably well and it is reasonable, therefore, to express not only the Hartree-Fock approximation but also the correlation effects in terms of configurations.

Let $(n\ell)$ represent a configuration for an N-electron system, and $|(\ell)\nu LS\rangle$ the coupling of the angular and spin momentum functions according to the coupling scheme, $\nu LS$. Then the <u>configuration state function</u> for the <u>configuration state</u>, $\gamma LS \equiv (n\ell)\nu LS$, is defined as the antisymmetric function

$$\Phi(\gamma LS) = \mathcal{A}\{\prod_{i=1}^{N} R(n_i\ell_i;\, r_i)\,|(\ell)\gamma LS\rangle\}$$

Here $R(n\ell;\, r) = (1/r)\, P(n\ell;\, r)$ and $P(n\ell;\, r)$ is called a <u>radial function</u>. Thus $\psi^{HF}$ is a configuration state function but constructed from special radial functions, namely functions which are solutions of the Hartree-Fock equations. By augmenting the Hartree-Fock radial functions to form a complete, orthonormal basis of radial functions for each $\ell$ and forming all possible configuration states, we obtain a basis of configuration state functions for the N-electron problem. Then

$$\psi(\gamma LS) = \sum_{\gamma'} c_{\gamma'}\, \Phi(\gamma' LS)$$

where the sum is over all configuration states, $\gamma' LS$.

Though simple conceptually, the CI approach soon becomes unwieldy. For example, in a study of correlation in $2p^5\ ^2P$ of F I, Sasaki and Yoshimine[4] used a total of 2,649 configuration state functions. Sinanoğlu[5] showed that for closed shell systems the correlation problem for many-electron systems can be reduced to first-order to a series of much simpler two-electron problems. He and his co-workers[6] extended the pair correlation theory to more general systems. Other methods, too, have been developed, each of which introduces correlation effects in a prescribed manner so as to achieve greater computational efficiency--the superposition of configuration (SOC) method used successfully by Weiss[7], the many-body perturbation theory (MBPT) applied by Kelly[8] and his co-workers to a wide variety of atomic properties, the random phase approximation with exchange (RPAE) used extensively by Amusia[9] and his co-workers for photoionization cross-section calculations, the multi-configuration Hartree-Fock method[10], and others.

Unlike many of the other methods, the MCHF method determines the total wavefunction for each state independently. This has the advantage that the radial orbital basis can be tailored to the state under consideration and frequently a relatively small number of configurations will suffice. Rearrangement effects, which play a crucial role in properties such as the transition probability of two-electron one-photon $K_{\alpha\alpha}$ transitions $(2s\ 2p \rightarrow 1s^2)$[11], can readily be accounted for. In the calculation of photoionization of Xe, Cs and Ba, Amusia, Ivanov and Chernysheva[12] obtained better agreement with experiment when RPAE calculations were modified to use "relaxed" excited states. However, the lack of a common orbital basis introduces some complications. In practice, in the MCHF approach, core orbitals are usually constrained so that outer orbitals are orthogonal to the same core. In some cases the only complication then is the addition of overlap integrals. As an alternative, to overcome the lack of a common orbital basis for transition studies, Goscinski et al[13] proposed a transition state whose energy is the average of the initial and final state. The optimized orbitals for this "state" may be viewed as a compromise between initial and final state orbitals. The method has been extended to include correlation and is a variation of the MCHF method[14].

An MCHF approximation to a state, $\gamma LS$, that includes the configuration states $\{\gamma_1, \gamma_2, \ldots, \gamma_m\}LS$ is a total wavefunction of the form

$$\psi^{MCHF}(\gamma LS) = \sum_{i=1}^{m} c_i \Phi(\gamma_i LS)$$

where the radial functions as well as the mixing coefficients are such that the energy,

$$E^{MCHF} = \langle \psi^{MCHF}(\gamma LS) | H | \psi^{MCHF}(\gamma LS) \rangle / \langle \psi^{MCHF}(\gamma LS) | \psi^{MCHF}(\gamma LS) \rangle$$

is stationary with respect to variations in these quantities, subject to certain orthogonality constraints. The stationary condition leads to a system of coupled integro-differential equations for the radial functions and a secular problem for the mixing coefficients. In this paper MCHF results will be presented for several cases where correlation plays an important role.

## 2. A First Order Theory

Fortunately, many atomic structure problems do not require a full correlation study to yield moderately accurate results. A first order theory often can predict those correlation effects that are important for a particular problem.

Let us partition the total wavefunction

$$\psi = \psi_0 + \psi_1 + \ldots, \quad \langle\psi|\psi\rangle = 1$$

so that $\psi_0$ is an approximate wavefunction of reasonable accuracy, $\psi_1$ the first order correction, and so on. Often $\psi_0$ is proportional to $\psi^{HF}$, but when important configuration interaction effects are present an accurate first-order theory requires that these be included in $\psi_0$. For example, in Mg I, we should define

$$\psi_0(3s\ 3d\ ^1D) = c_0\Phi(3s\ 3d\ ^1D) + c_1\Phi(3p^2\ ^1D)$$

since the interaction here is strong with $c_0 = 0.88$, $c_1 = -0.47$. The first order correction can then be defined as a sum over all configuration states $\gamma'$, which interact with $\psi_0$. Thus

$$\psi_1 = \sum_{\gamma'} c_{\gamma'}\Phi(\gamma'LS)$$

where

$$\langle\psi_0|H|\Phi(\gamma'LS)\rangle \neq 0 \quad \text{and} \quad \langle\psi_0|\Phi(\gamma'LS)\rangle = 0.$$

The mixing coefficients are given to first-order by the expression

$$c_{\gamma'} \approx \frac{\langle\psi_0|H|\Phi(\gamma'LS)\rangle}{E_0(\gamma LS) - E(\gamma'LS)}$$

Similarly

$$E_1(\gamma LS) \approx \sum_{\gamma'} \frac{\langle\psi_0|H|\Phi(\gamma'LS)\rangle^2}{E_0(\gamma LS) - E(\gamma'LS)}.$$

Thus it would appear that all configurations contributing to $\psi_1$ are important in an energy calculation, but frequently energy differences are required, and an analysis of the difference may show that many contributions cancel.

Often atomic properties are defined in terms of a one-electron operator, say Op, connecting an initial state, $\gamma_iL_iS_i$ to a final state, $\gamma_fL_fS_f$. Let $\mathcal{O} = \langle\psi(\gamma_fL_fS_f)|Op|\psi(\gamma_iL_iS_i)\rangle$. Then $\mathcal{O}$ also has an expansion, namely

$$\mathcal{O} = \mathcal{O}_0 + \mathcal{O}_1 + \ldots$$

where

$$\mathcal{O}_0 = \langle\psi_0(\gamma_fL_fS_f)|Op|\psi_0(\gamma_iL_iS_i)\rangle$$

and

$$\mathcal{O}_1 = \langle\psi_0(\gamma_fL_fS_f)|Op|\psi_1(\gamma_iL_iS_i)\rangle + \langle\psi_1(\gamma_fL_fS_f)|Op|\psi_0(\gamma_iL_iS_i)\rangle.$$

Selection rules for the operator now apply. Consequently many configurations which contribute to $\psi_1$ make no contribution to $\mathcal{O}_1$ and if higher order effects are not important, they may be omitted. A similar theory has been proposed by Nicolaides and Beck[15] for oscillator strengths.

The configurations $\gamma'$ contributing to $\psi_1$ can readily be enumerated using pair correlation theory. By decoupling pairs of electrons from the occupied configurations in $\psi_0$, coupling these two electrons to form a two electron state, say $\alpha\beta L'S'$, and making all allowed replacements $\alpha\beta L'S' \rightarrow \alpha'\beta'L'S'$, excluding those which lead to configurations already occupied in $\psi_0$, configurations are generated which enter into the expansion of the $\alpha\beta L'S'$ pair correlation function. This process is repeated for all possible pairs. For example $1s^2\ 2s\ {}^2S$ has <u>three</u> pair correlation functions -- $1s^2\ {}^1S$, $1s\ 2s\ {}^1S$, and $1s\ 2s\ {}^3S$.

The multi-configuration Hartree-Fock (MCHF) method is particularly well suited to pair correlation calculations. As in the Hartree-Fock approximation, an MCHF total wavefunction has the property that interactions with certain configurations differing by one electron from those included in the MCHF approximation are zero. This is an extension of Brillouin's theorem. Alternatively one may view this result as implying that by determining the radial basis variationally these configurations have been included to first order with respect to a certain variation. In fact it can be shown[10] that the doubly infinite sums entering into the definition of the pair correlation can be <u>reduced</u> to a single sum. For example, there exists a transformation of the radial basis, such that

$$\sum_{n,m} c_{nm}\Phi(ns\ mp\ {}^1P) = \sum_i d_i\Phi(s_i\ p_i\ {}^1P)$$

where on the right hand side now each configuration differs by two electrons. Often this series converges sufficiently rapidly that only one configuration need be included. The radial functions representing the correlation effects in the reduced form generally differ markedly from the "spectroscopic" radial functions and for this reason they have been labelled simply $s_i$ and $p_i$ respectively. The MCHF method determines these radial functions directly.

In the case of one-electron operators the important configurations are usually of the form $\alpha(L'S')m\ell\ LS, m = n, n+1, \ldots$ where $\alpha(L'S')$ represents a constant core. Infinite sums over these configurations can be reduced to a single configuration, for example,

$$\sum c_m\Phi(\alpha(L'S')m\ell\ LS) \equiv d\ \Phi(\alpha(L'S')\underline{\ell}\ LS)$$

where the radial function is given by the transformation

$$P(\underline{\ell};\ r) = \{\sum c_m\ P(m\ell;\ r)\}/(\sum c_m^2)^{1/2}.$$

The MCHF method determines $P(\underline{\ell}, r)$, in effect replacing the computation of $c_m$ and the evaluation of the sum by the solution of an integro-differential equation, and the solution of a small, possibly $2 \times 2$ eigenvalue problem.

### 3. The 3p 3d $^3F$ State of Aℓ II

The multi-configuration Hartree-Fock method is particularly convenient for cases which can be treated as two electron systems. The magnesium sequence is such an example. An interesting case in Aℓ II is the 3p 3d $^3F$ state whose Hartree-Fock energy lies between 3s 6f $^3F$ and 3s 7f $^3F$. In fact 3p 3d interacts with the whole 3s nf series.

In theory the MCHF method can represent the interaction with the series by a single configuration state function so that

$$\psi^{MCHF} = c_1 \, \Phi(3p\ 3d) + c_2 \, \Phi(3s\ f_1).$$

In practice the variational procedure for this state is unstable. Part of the 3s nf series raises the energy for the state, part of it lowers the energy and so the energy of 3s $f_1$ fluctuates about the Hartree-Fock energy of 3p 3d and SCF iterations do not converge.

The MCHF method converges best when an energy minimum is sought. By the Hylleraas-Undheim-MacDonald[16] theorem, the fourth eigenvalue of an interaction matrix will be an upper bound to the fourth exact eigenvalue. With the approximation

$$\begin{aligned}\psi^{MCHF} &= c_1 \, \Phi(3p\ 3d) + c_2 \, \Phi(3s\ 4f) \\ &\quad + c_3 \, \Phi(3s\ 5f) + c_4 \, \Phi(3s\ 6f) \\ &\quad + c_5 \, \Phi(3s\ f_1)\end{aligned}$$

$$\langle 4f|f_1\rangle = \langle 5f|f_1\rangle = \langle 6f|f_1\rangle = 0$$

the problem becomes a minimization problem and 3s, 4f, 5f, 6f orbitals can be kept fixed. The variational procedure for 3p, 3d, and $f_1$ converges rapidly.

The resulting interaction matrix, which also includes some other configurations and was derived for the <u>fourth</u> eigenvalue, was diagonalized. In Table 1 some results are presented for the third and fourth eigenstate. Notice now that 3s 6f is the dominant component of <u>two</u> eigenvectors, that 3p 3d is not dominant in either. Weiss'[17] model CI calculation for the interaction of 3p 3d with 3s nf , n = 4, 5, ..., 10 , did not exhibit this phenomenon though he observed a similar situation for the $^2D$ state of Aℓ I when more accurate SOC calculations were performed. There the 3s $3p^2$ $^2D$ component was distributed over the $3s^2$ nd $^2D$ series with the dominant components being 3d, 4d, 4d, 5d, ... In no state did 3s $3p^2$ appear as a dominant component. These examples illustrate some of the difficulties which may be encountered when states are labelled according to their dominant component.

Table 1. Wavefunctions and ionization energies (IE) for $^3F$ states of Al II[a)]

| Configuration | Mixing Coefficient | |
|---|---|---|
| | 3rd eigenstate | 4th eigenstate |
| 3s 4f | -0.120 | -0.088 |
| 3s 5f | -0.306 | -0.163 |
| 3s 6f | 0.683 | -0.720 |
| 3p 3d | 0.588 | 0.554 |
| 3s $f_1$ | 0.282 | 0.372 |
| 3d $f_2$ | -0.021 | -0.024 |
| 3p 5g | 0.003 | 0.030 |
| 4p 4d | -0.019 | -0.020 |
| IE | 0.0621 | 0.0507 |
| $IE_{obs}$ [b)] | 0.0607 | 0.0490 |
| $IE_{Weiss}$ [c)] | 0.0606 | 0.0476 |

a) As a consequence of near degeneracy, the relativistic corrections assume greater importance, but have not been included.

b) Ref. [17]

c) Ref. [18]

## 4. Correlation in the $3d^n$ Shell

A study of the oscillator strength for the $4s\ ^2S - 4p\ ^2P$ transition of the Cu I sequence[19] suggested that correlation in the $3d^{10}$ shell might be important, at least towards the neutral end of the sequence. Indeed, for Cu II the Hartree-Fock ionization energy is $IE^{HF} = E^{HF}(3d^9) - E^{HF}(3d^{10}) = 0.6437$ a.u. , whereas the observed value is 0.7457 a.u., a rather large discrepancy of about 14%.

Since the two states both have the same core, the important corrections to the ionization energy come from correlation with and within the $3d^n$ group. An estimate of the correlation effect can be obtained from a study of correlation in $3d^{10}$. This is a complete shell and only two electron replacements need be considered. In Table 2, the contribution to the energy from various configurations is presented. In each case, all possible couplings were included. Notice that the most important contributions come from $4d^2$ and $4f^2$ replacements. Since the mean radius of the 3d orbital in $3d^9$ was considerably smaller than in $3d^{10}$ , part of the calculation was repeated with occupied orbitals from a Hartree-Fock calculation for $3d^9$. The main effect was to decrease the $4d^2$ contribution and increase the $4f^2$ contribution somewhat.

Table 2. Contributions to the energy from correlation within $3d^{10}$ and between $3p^6$ and $3d^{10}$ subshells in Cu II

| Configuration | 3d from Cu II ($\bar{r}$ = 0.978) | 3d from Cu III ($\bar{r}$ = 0.900) |
|---|---|---|
| $3p^6\ 3d^8\ 4s^2$ | -0.0015 a.u. | -0.0016 a.u. |
| $4p^2$ | -0.0120 | -0.0117 |
| $4d^2$ | -0.2032 | -0.1809 |
| $4f^2$ | -0.1284 | -0.1416 |
| 4s 4d | -0.0035 | |
| 4p 4f | -0.0156 | |
| $5s^2$ | -0.0003 | |
| $3p^5\ 3d^9$ 4s 4f | -0.0064 | |
| 4s 4p | -0.0067 | |

<u>Total Correlation Energy</u>

i) Within: $3d^{10}$ -0.3646

ii) Between: $3p^6$ and $3d^{10}$ -0.0131

In pair correlation theory as analyzed by Layzer et al[20], the correlation energy is approximately proportional to the number of pairs, and subsequent calculations have shown that the correlation per pair depends largely on the proximity of the two electrons. In the present case, the correlation energy per (3d, 3d) pair from the Cu II results is about -0.0081 a.u. whereas the correlation per (3p, 3d) pair is about -0.0002 a.u. Thus contributions from the other subshells can be expected to be small. If we now assume that correlation in $3d^9$ is the sum of the number of (3d, 3d) or (3p, 3d) pairs times the per pair correlation energy, using the results in the second column of Table 2 when available, we obtain a revised ionization energy of 0.725, a value in error by 2.8%. A more accurate study requires a detailed analysis of correlation in $3d^9$, a greater task since many more coupling schemes are now allowed.

This study has shown the importance of correlation within the $3d^n$ shell when configurations with a different number of 3d electrons are present. In Fig. 1 some energy levels of Mn I are depicted. On the left, the Hartree-Fock energies relative to the $3d^6$ ($^5D$) ionization limit are compared with observed. The Hartree-Fock values for $3d^6$ ($^5D$) 4s $^6D$ and $3d^6$ ($^5D$) 4p $^6P$ are upper bounds and could be decreased by the inclusion

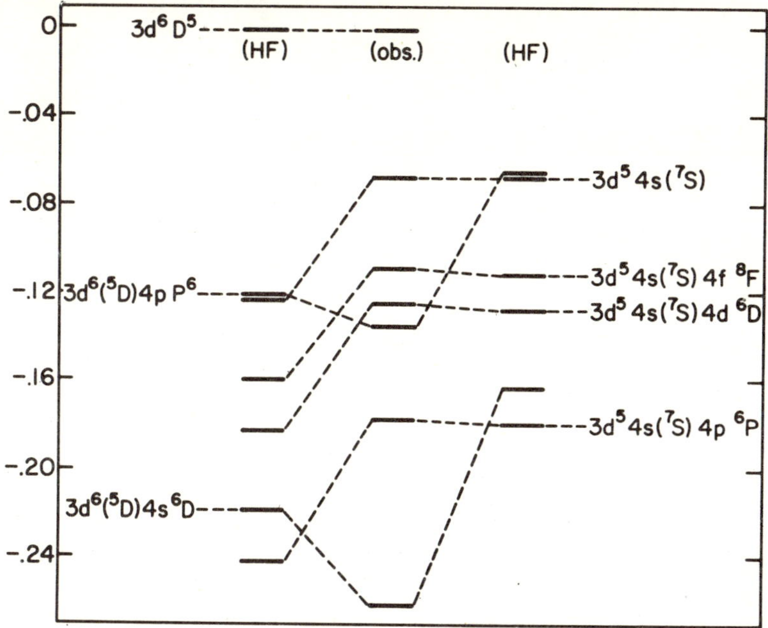


Fig. 1. Term structure in Mn I.

of correlation between the outer electron and $3d^6$ ($^5D$). The $3d^5$ 4s ($^7S$) $n\ell$ levels are much too low; in fact $3d^5$ 4s ($^7S$) lies <u>below</u> $3d^6$ ($^5D$) 4p $^6P$. On the right, the Hartree-Fock values are shifted so that the $3d^5$ 4s ($^7S$) ionization limit agrees with the observed. The energy levels of $3d^5$ 4s ($^7S$) $n\ell$ now agree well with the observed. Thus the discrepancies between the energies of $3d^6$ ($^5D$) $n\ell$ and $3d^5$ 4s ($^7S$) $n\ell$ are due largely to greater correlation energy in $3d^6$ ($^5D$) than in $3d^5$ 4s ($^7S$).

It is well known that in the iron series $3d^n$ 4p often interacts strongly with $3d^{n-1}$ 4s 4p. In Mn I the Hartree-Fock $3d^6$ 4p $^6P$ and $3d^5$ 4s 4p $^6P$ energy levels are too far apart and configuration interaction between the two will increase the separation. However, in Cr I the Hartree-Fock energy of $3d^5$ ($^6S$) 4p $^7P$ lies above the $3d^4$ ($^5D$) 4s 4p $^7P$ energy, contrary to the observed order as seen in Fig. 2. Again, configuration interaction will not change the order and will merely increase the discrepancy, but the greater correlation energy in $3d^5$ can account for this discrepancy. In Table 2 it was evident that most of the correlation arose from $3d^2 \rightarrow 4d^2$ replacements, sometimes referred to as <u>radial</u> correlation. This type of correlation can be included by the extended Hartree-Fock (EHF) method first proposed by Eckart[21] for the Helium ground state and generalized by Jucys[22] and his collaborators[23] to many electron systems. A

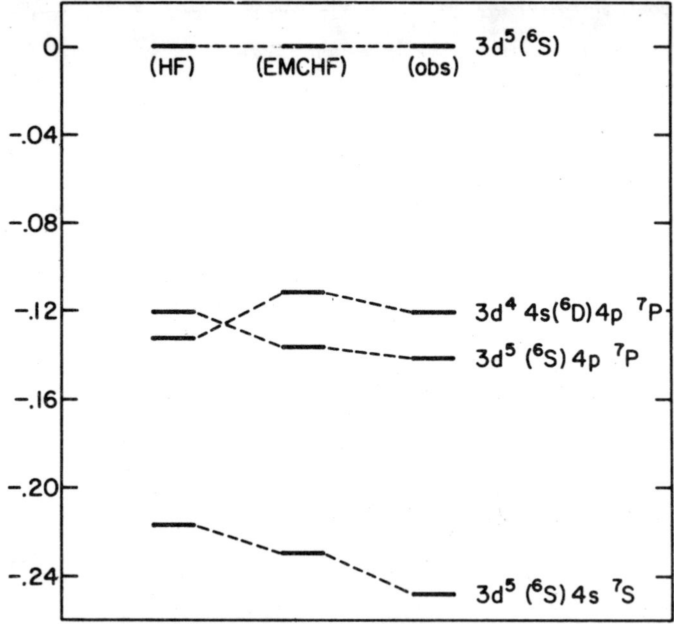


Fig. 2. Term structure in Cr I.

partially extended model in which the first four 3d electrons are considered to be equivalent and the fifth non-equivalent already reverses the order. When configuration interaction between the $^7P$ configuration states is included in the EHF calculation (denoted by EMCHF) the ionization energies are in fair agreement with observation, as seen in Fig. 2. The f-values for the $^7S$ - $^7P$ transitions also are in much better agreement with experiment. An earlier calculation[24], that included all configurations contributing to the transition matrix element to first-order but ignored correlation in the 3d subshell, did not predict the dominant components of the two $^7P$ states correctly. In Table 3, the results of the two calculations are compared with observation, each time comparing data for transitions to states with the same dominant component. Notice that the earlier calculation (denoted as MCHF) associated the larger f-value with the wrong transition. Because the energies of $3d^5$ 4p and $3d^5$ 4s 4p were in the wrong order, the configuration interaction calculation produced mixing coefficients for which the relative phases had the incorrect sign. Thus an f-value calculation which should have been enhanced through configuration interaction was damped, and vice versa. Proper relative phases are preserved if eigenstates are ordered according to their eigen energy, regardless of the composition of the state. Under such a scheme the earlier

MCHF results would be reversed and the f-values would appear to be in better agreement with observation. Some of these matters are discussed in a recent review article on oscillator strength calculations by Hibbert[26].

Table 3. A comparison of f-values for some $^7S$ - $^7P$ transitions in Cr I. Both length ($\ell$) and velocity (v) forms of the f-value are given.

| λ(Å) | | | f-values | | |
|---|---|---|---|---|---|
| EMCHF1) | MCHF2) | obs3) | EMCHF1) | MCHF2) | obs4) |
| a) $3d^5\ 4s\ ^7S$ - $3d^5\ 4p\ ^7P$ | | | | | |
| 4882. | 4010. | 4254. | 0.343($\ell$) | 1.12($\ell$) | 0.242 |
| | | | 0.283(v) | 1.08(v) | |
| b) $3d^5\ 4s\ ^7S$ - $3d^4\ 4s\ 4p\ ^7P$ | | | | | |
| 3825. | 4758. | 3579. | 1.036($\ell$) | 0.089($\ell$) | 0.83 |
| | | | 0.666(v) | 0.000(v) | |

1) partial EHF including configuration interaction between $^7P$ configuration states

2) Ref. [24] ignoring correlation within $3d^n$

3) Ref. [18]

4) Ref. [25]

The experimental f-value trend for $3d^n$ 4s - $3d^n$ 4p transitions of the iron series shows anomalous behavior particularly for chromium and manganese. Stewart and Rotenberg first noted this behavior and conjectured that the mixing of $3d^n$ 4p with $3d^{n-1}$ 4s 4p reallocated the line strength, but a simple MCHF calculation[27] did not support the conjecture. It appears that correlation in the $3d^n$ shell is important and a more extensive correlation study is required.

## 5. f-values in the presence of cross-overs

Currently many f-value trends are studied along an isoelectronic sequence and plotted as a function of $1/Z$, where $Z$ is the nuclear charge. Usually the theoretical f-values vary smoothly though the experimental values often exhibit considerable scatter. As Weiss[28] has shown, theoretical f-value trends may have irregularities when energy levels are crossing, but the practice of computing f-values for integral Z-values only, often obscures discontinuities which may be present.

Consider the $^2S$ states of the boron sequence. In B I, the lowest state is $2s^2\ 3s\ ^2S$ and the next $2s\ 2p^2\ ^2S$, but in C II the two states already have reversed order. Thus the energy levels of these two interacting

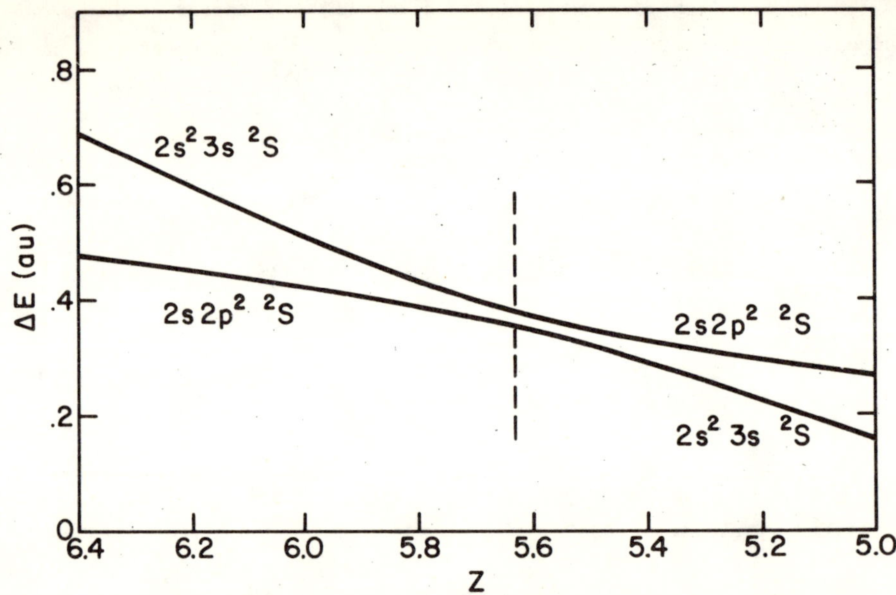


Fig. 3. Transition energies as a function of Z from model CI calculations for crossing configurations $2s^2\ 3s\ ^2S$ and $2s\ 2p^2\ ^2S$. (The broken vertical line indicates the cross-over point.)

configurations cross between Z = 5 and Z = 6. Theoretical calculations, of course, can be performed for non-integral Z values and in Fig. 3 the transition energies obtained from a model configuration interaction calculation are plotted for transitions from the $2s^2\ 2p\ ^2P$ ground state. The eigen energy of the upper/lower eigen state varies continuously and the two curves do not actually cross. When the diagonal energies of the interaction matrix cross, the dominant component of the eigenfunction changes though the relative phases remain the same along the curve.

In Fig. 4, f-values are plotted for transitions from the ground state to these two excited states. Notice that a discontinuity is present when f-value trends are plotted for a given dominant component. But in this case, since the cross-over occurs near the middle of the interval, f-values for integral Z-values can be joined by a relatively smooth curve. The observed term energies of the $^2S$ states actually cross closer to Z = 6.0 than in our model CI calculation. For this reason, Weiss' more extensive calculation for this f-value trend shows considerable irregularity, though our model study would indicate that the minimum in the f-value curve should coincide with the cross-over point for the energies.

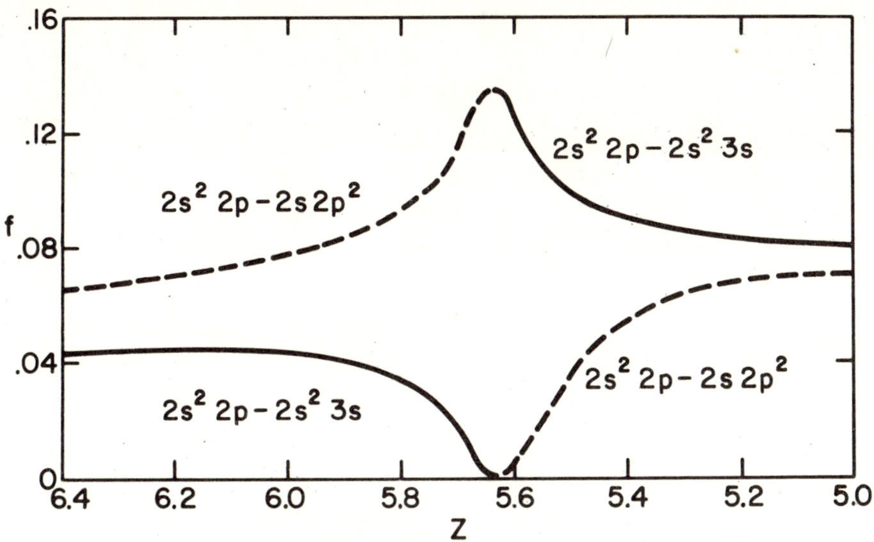


Fig. 4. The $2s^2\ 2p\ ^2P\ -\ ^2S$ f-values as a function of Z from model CI calculations.

In Fig. 5 f-values are plotted for transitions from the $2s^2\ 3p\ ^2P$ state which has an allowed transition only to the $2s^2\ 3s$ component of the

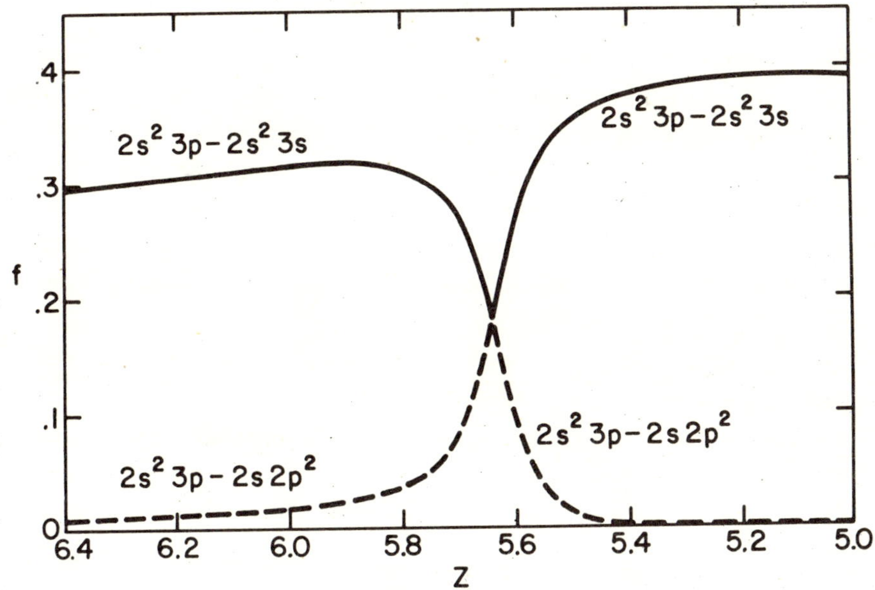


Fig. 5. The $2s^2\ 3p\ ^2P\ -\ ^2S$ f-values as a function of Z from model CI calculations.

$^2S$ states. At the cross-over point, the amount of $2s^2\ 3s$ is exactly 50%. As a result the f-value for the $2s^2\ 3p\ ^2P - 2s^2\ 3s\ ^2S$ transition is reduced by a factor of one half, and a dip appears in what otherwise is a fairly smooth trend.

From these figures it is clear that the smoothness of the f-value trend for integral Z values will depend largely on where the cross-over occurs and how rapidly the energy levels separate. The f-value trends for $4s\ ^2S - 4p\ ^2P$ transitions in the Aℓ sequence[29] exhibit several irregularities such as those in Fig. 5 even when only integral values are considered.

Atomic properties such as f-values can often be improved significantly by an MCHF calculation with a relatively small number of configurations, the latter being determined by a first-order theory for the transition operator. However, when two interacting configurations are nearly degenerate, sufficient energy correlation effects must be included so that the relative phases of components are predicted correctly.

## 6. References

*This work was supported in part by the U. S. Department of Energy under contract EG-77-02-4264.


1. D. R. Hartree, Proc. Camb. Phil. Soc. 24, 89, 111 (1927).
2. V. Fock, Z. Phys. 61, 126 (1930); 62, 795 (1930).
3. P. O. Löwdin, Adv. Chem. Phys. 2, 207 (1959).
4. F. Sasaki and M. Yoshimine Phys. Rev. A9, 17, 26 (1974).
5. O. Sinanoğlu, Proc. Roy. Soc. (London) A260, 379 (1961).
6. O. Sinanoğlu, Adv. Atom. Molec. Phys. 14, 237 (1969).
7. A. W. Weiss, Adv. Atom. Molec. Phys. 9, 1 (1973).
8. H. P. Kelly, Adv. Chem. Phys. 14, 129 (1969).
9. M. Ya. Amusia, Proceedings of the IV International Conference on Vacuum Ultraviolet Radiation Physics (eds. E. E. Koch, R. Haensel, and C. Kunz, Pergamon Vieweg 205, (1974).
10. C. Froese Fischer, The Hartree-Fock Method for Atoms, (Wiley, New York, 1977).
11. M. Gavrila and J. E. Hansen, J. Phys. B: Atom. Molec. Phys. 11, 1353 (1978).
12. M. Ya. Amusia, V. K. Ivanov, L. V. Chernysheva, Phys. Lett. 59A, 191 (1976).
13. O. Goscinski, B. T. Pickup, and G. Purvis, Chem. Phys. Lett. 22, 167 (1973).
14. M. Godefroid, J. J. Berger, and G. Verhaegen, J. Phys. B: Atom. Molec. Phys. 9, 2181 (1976).
15. C. Nicolaides and D. R. Beck, Chem. Phys. Lett. 36, 79 (1975).
16. E. A. Hylleraas and B. Undheim, Z. Phys. 65, 759 (1930); also J. K. L. MacDonald, Phys. Rev. 43, 830 (1933).

17. A. W. Weiss, Phys. Rev. A 9, 1524 (1974).

18. C. E. Moore, Atomic Energy Levels, NBS Circular 467, U. S. Government Printing Office, Washington, D. C., (1949).

19. C. Froese Fischer, J. Phys. B: Atom. Molec. Phys. 10, 1241 (1977).

20. D. Layzer, Z. Horak, M. N. Lewis, and D. P. Thompson, Ann. Phys. 29, 101 (1964).

21. C. Eckart, Phys. Rev. 36, 828 (1930).

22. A. P. Jucys, Int. J. Quantum Chem. 1, 311 (1967).

23. A. P. Jucys, E. P. Našlênas and P. S. Žvirblis, Int. J. Quantum Chem. 6, 465 (1972).

24. C. Froese Fischer, J. E. Hansen, and M. Barwell, J. Phys. B: Atom. Molec. Phys. 9, 1841 (1976).

25. T. M. Bieniewski, Astrophys. J. 208, 228 (1976).

26. A. Hibbert, Phys. Scripta 16, 7 (1977).

27. C. Froese Fischer, J. Quant. Spectrosc. Radiat. Transfer 13, 201 (1973).

28. A. W. Weiss, Beam-Foil Spectroscopy 1 (ed. I. A. Sellin and D. J. Pegg, Plenum, New York, (1976) p. 51.

29. C. Froese Fischer, Can. J. Phys. (1978) to appear.

# PECULIARITIES OF THE THEORETICAL INVESTIGATION OF THE SPECTRA OF MANY-ELECTRON ATOMS AND MULTIPLE-CHARGED IONS

Z.B.Rudzikas

Institute of Physics of the Academy of Sciences of the Lithuanian SSR, K.Poželos 54, Vilnius 232600, USSR

## 1. Introduction

At present the number of papers in which many-electron atoms and ions are studied both theoretically and experimentally is increasing rapidly. The greatest attention is paid to the multiple-charged ions, observed both in astrophysical and laboratory high-temperature plasma. They are particularly important in the investigation of the solar spectra in the short wavelength region as well as in the study of the thermonuclear fusion.

The contemporary theory of many-electron atoms is based mainly on the ideas of Hartree and Fock. Their self-consistent field method, together with the methods of the irreducible tensorial operators, as well as the genealogical coefficients, is a powerful tool for the theoretical investigation of many-particle systems. A substantional contribution to the development and application of this method was made by Academician of the Academy of Sciences of the Lithuanian SSR A.P.Jucys. He was always proud of the fact that he was the only researcher who had worked both under Hartree and Fock.

In this paper we shall try to discuss briefly some new results obtained in this branch of physics during last years. In the main, we shall be referring to the works of a group of physicists at the Institute of Physics of the Academy of Sciences of the Lithuanian SSR guided by the author of this contribution.

## 2. Some aspects of the theory of many-electron atom

Nowadays the elements practically from all the regions of the periodical table are investigated. Special difficulties arise when studying the iron group and rare earth elements with unfilled shells of d- and f-electrons. These elements have complicated energy spectra consisting of a lot of tightly lying levels. Using powerful computers one is in position to calculate quite realistic models of atoms and ions as well as to obtain the results in many cases fairly well coinciding qualitatively and even quantitatively with the corresponding experimental data.

The development of methods for theoretical investigation of atoms and ions and the preparation of corresponding general computer programs which could cover the widest possible regions of the periodical table allow us to carry out the extensive investigations of the spectra of atoms and ions, including the case of the isoelectronic sequences up to very high ionization degrees.

It is especially worth-while to point out the necessity to combine both theoretical and experimental investigations when studying the energy spectra of complicated systems. For identification and classification of the experimentally measured energy spectra and for choosing optimal coupling schemes as well as quantum numbers, the theoretical modelling of the corresponding quantum-mechanical systems is absolutely essential.

Usually the single-configuration Hartree-Fock method serves as the starting point[1]. When refining the theory of many-electron atoms and ions one has first to take into account correlation and relativistic effects. The former effects can be taken into account by the configuration superposition method or by multiconfiguration approximation while solving Hartree-Fock-Jucys equations. The role of the relativistic effects grows rapidly with the increase of the ionization degree. Later on these effects predominate, therefore they have to be taken into account first. For neutral and slightly ionized atoms they can be regarded as corrections. For extremely highly ionized atoms a full relativistic treatment ought to be used. In the case of many-electron atoms various coupling schemes, including an intermediate one, should be used.

The so-called Hartree-Fock-Pauli (HFP) approximation is usually used to take account of the relativistic effects as corrections[2]. Its Hamiltonian is presented below[3]. It consists of well-known one-particle operators as well as two-particle operators, including the corresponding relativistic corrections of the order $\alpha^2$ ($\alpha$ is the fine structure constant):

$$\mathcal{H}=\mathcal{H}_0+\mathcal{H}_1+\mathcal{H}_2+\mathcal{H}_3+\mathcal{H}_4+\mathcal{H}_5 .$$

The zero-order Hamiltonian $H_0$ consists of operators of the kinetic energy of the electrons with respect to the nucleus and the electrostatic interaction of the electrons with the nucleus and with one another:

$$\mathcal{H}_0=\sum_i \frac{\vec{p}_i^{\,2}}{2m}-\sum_i \frac{Ze^2}{r_i}+\sum_{i>j} \frac{e^2}{r_{ij}} .$$

The relativistic corrections of the order $\alpha^2$ are considered as perturbation and are taken into account as a first order correction using the radial wave functions of the zero-order Hamiltonian. These terms include the correction due to the dependence of the electron mass on the velocity

$$\mathcal{H}_1=-\frac{1}{8m^3c^2}\sum_i \vec{p}_i^{\,4},$$

the orbit-orbit interaction

$$\mathcal{H}_2=-\frac{e^2}{2m^2c^2}\sum_{i>j}\frac{1}{r_{ij}}\left\{(\vec{p}_i\cdot\vec{p}_j)+\frac{(\vec{r}_{ij}(\vec{r}_{ij}\cdot\vec{p}_i)\cdot\vec{p}_j)}{r_{ij}^2}\right\},$$

the contact interactions

$$\mathcal{H}_3=\mathcal{H}_3'+\mathcal{H}_3''=\frac{Z\pi e\hbar^2}{2m^2c^2}\sum_i \delta(\vec{r}_i)-\frac{\pi e\hbar^2}{m^2c^2}\sum_{i>j}\delta(\vec{r}_{ij}),$$

the spin-orbit interaction

$$\mathcal{H}_4=\frac{e^2\hbar}{2m^2c^2}\left(\left\{\sum_i \frac{Z}{r_i^3}[\vec{r}_i\times\vec{p}_i]-\sum_{i>j}\frac{1}{r_{ij}^3}[\vec{r}_{ij}\times\vec{p}_i]+\right.\right.$$

$$\left.\left.+\sum_{i>j}\frac{2}{r_{ij}^3}[\vec{r}_{ij}\times\vec{p}_j]\right\}\cdot\vec{s}_i\right),$$

the spin-contact and spin-spin interactions

$$\mathcal{H}_5=\mathcal{H}_5'+\mathcal{H}_5''=-\frac{e^2\hbar^2}{m^2c^2}\frac{8\pi}{3}\sum_{i>j}(\vec{s}_i\cdot\vec{s}_j)\delta(\vec{r}_{ij})+$$
$$+\frac{e^2\hbar^2}{m^2c^2}\sum_{i>j}\frac{1}{r_{ij}^3}\left[(\vec{s}_i\cdot\vec{s}_j)-\frac{3(\vec{s}_i\cdot\vec{r}_{ij})(\vec{s}_j\cdot\vec{r}_{ij})}{r_{ij}^2}\right].$$

A general program for computer BESM-6 to calculate the energy spectra of atoms and ions in the above-mentioned approximation by means of an intermediate coupling scheme starting with the LS one has been worked out at the Institute of Physics of the Academy of Sciences of the Lithuanian SSR. Another program uses the data obtained to calculate oscillator strengths and transition probabilities in the same approximation starting with two forms ("length" and "velocity") of the electric dipole radiation operator. The program is able to display the results in numerical or graphic form. This form of presentation of the data obtained is very convenient when interpreting the corresponding experimental results.

An analogous program is being prepared for the case of relativistic Dirac-Hartree-Fock (DHF) approximation, the Hamiltonian of which may be written as follows (in atomic units):

$$\mathcal{H}=\sum_i(H_i^1+H_i^2+H_i^3)+\sum_{i>j}(H_{ij}^e+H_{ij}^m+H_{ij}^r),$$

$$H_i^1=c(\vec{\alpha}_i\cdot\vec{p}_i),\quad H_i^2=\beta_i c^2,\quad H_i^3=V(r_i),$$

$$H_{ij}^e=\frac{1}{r_{ij}},\qquad H_{ij}^m=-\frac{(\vec{\alpha}_i\cdot\vec{\alpha}_j)}{r_{ij}},$$

$$H_{ij}^r=-\frac{1}{2}(\vec{\alpha}_i\cdot\vec{\nabla}_i)(\vec{\alpha}_j\cdot\vec{\nabla}_j)r_{ij}.$$

One-particle operators describe kinetic energy, one-electronic part of the spin-orbit interaction, the mass effect, as well as the potential energy. Two-particle operators represent the energies of the electrostatic, magnetic and retardation interactions. The last two terms

$$\mathcal{H}^{Br} = \mathcal{H}^{m} + \mathcal{H}^{r}$$

may be presented also in the following form:

$$\mathcal{H}^{Br} = \mathcal{H}^{5} + \mathcal{H}^{6},$$

where

$$\mathcal{H}_{ij}^{5} = -\frac{1}{2r_{ij}} (\vec{\alpha}_i \cdot \vec{\alpha}_j),$$

$$\mathcal{H}_{ij}^{6} = -\frac{1}{2r_{ij}^{3}} (\vec{\alpha}_i \cdot \vec{r}_{ij})(\vec{\alpha}_j \cdot \vec{r}_{ij}).$$

$\vec{\alpha}$ and $\beta$ are defined by the formulas:

$$\vec{\alpha} = \begin{pmatrix} 0 & \vec{\sigma} \\ \vec{\sigma} & 0 \end{pmatrix}, \qquad \beta = \begin{pmatrix} I & 0 \\ 0 & -I \end{pmatrix},$$

where $\vec{\alpha}$ and $\beta$ are Dirac matrices, $\vec{\sigma}$ and $I$ are Pauli and unit matrices of the second order, correspondingly.

One-electronic relativistic wave function used is of the form:

$$|nljm\rangle = \begin{pmatrix} |ljm\rangle f(nlj) \\ -i|l'jm\rangle g(nl'j) \end{pmatrix},$$

where $l' = 2j - l$, f and g are the radial parts of the wave function. For its spin-angular part the following new expression is found[4]:

$$|ljm\rangle = \sqrt{\frac{2j+1}{8\pi}} \begin{pmatrix} D^{(j)}_{m\,1/2} \\ (-1)^{l-j+1/2} D^{(j)}_{m,-1/2} \end{pmatrix}.$$

Here $D^{(j)}_{m\,1/2}$ denotes the generalized spherical function. A very interesting feature of this wave function is that its dependence on the orbital quantum number is only in the form of the phase multiplier defining the parity of a configuration to the non-conservation of which much attention is paid now. In similar fashion, a more simple expression is obtained for the case of the non-relativistic wave function in jj coupling scheme as well. The use of these new types of the wave functions simplifies a great deal the procedure of finding the matrix elements of the physical operators and leads to more convenient expressions, in which only radial integrals and phase multipliers depend on orbital quantum numbers.

Another very effective way to simplificate the many-electron atom theory is the application of the Wigner-Eckart theorem in orbital, spin and quasi-spin spaces as well as the introduction of the fully reduced matrix elements and subgenealogical coefficients[5]. So, for the tensor $T^{\{K_1 K_2 K_3\}}_{M_1 M_2 M_3}$ one obtains

$$\left(l^N \alpha Q L S M_Q M_L M_S \middle| T^{\{K_1 K_2 K_3\}}_{M_1 M_2 M_3} \middle| l^{N'} \alpha' Q' L' S' M_{Q'} M_{L'} M_{S'}\right) =$$

$$= \frac{(-1)^{2K_1+2K_2+2K_3}}{\sqrt{(2Q+1)(2L+1)(2S+1)}} \begin{bmatrix} Q' & K_1 & Q \\ M_{Q'} & M_1 & M_Q \end{bmatrix} \begin{bmatrix} L' & K_2 & L \\ M_{L'} & M_2 & M_L \end{bmatrix} \begin{bmatrix} S' & K_3 & S \\ M_{S'} & M_3 & M_S \end{bmatrix} \times$$

$$\times \left(l \alpha Q L S ||| T^{\{K_1 K_2 K_3\}} ||| l \alpha' Q' L' S'\right),$$

where

$$Q_i = \frac{1}{2}(2l+1-v_i), \qquad M_{Q_i} = -\frac{1}{2}(2l+1-N_i).$$

For the one-particle genealogical coefficient one has

$$\left(l \alpha Q L S ||| l \alpha' Q' L' S'\right) = \left(l \alpha Q L S ||| a^{(q l s)} ||| l \alpha' Q' L' S'\right) =$$

$$= (-1)^{(N+1)[\varphi(N)+1]} \sqrt{N(2Q+1)(2L+1)(2S+1)} \begin{bmatrix} Q' & q & Q \\ M_{Q'} & 1/2 & M_Q \end{bmatrix}^{-1} \times$$

$$\times \left(l^N \alpha Q L S || l^{N-1} \alpha' Q' L' S', l\right),$$

where

$$\varphi(N) = \begin{cases} 1, & \text{if } N \geq 2l+1, \\ 0, & \text{if } N < 2l+1. \end{cases}$$

Here the tensor $a^{(q l s)}$ is written in the standard phase system, defined by the equation[6]

$$T^{(k)+}_{q} = (-1)^{k-q} T^{(k)}_{-q}.$$

The use of the reduced in quasi-spin space genealogical coefficients and matrix elements of the tensorial operators is of great importance because the tables of their numerical values are very small and the properties of the spin and quasi-spin symmetry may be easily taken into account. This method can be used to take into account correlation effects, too. It is valid for any coupling scheme. It is especially efficient in the case of complex electronic configurations, including the matrix elements non-diagonal with respect to configurations. Using the vectorial coupling of the momenta in quasi-spin space one has a possibility to introduce the notion of a total quasi-spin quantum number and treat simultaneously a set of electronic configurations $(\ell_1+\ell_2)^N$, defining the corresponding wave function as follows[7]:

$$|(\ell_1+\ell_2)^N \alpha_1 Q_1 L_1 S_1 \alpha_2 Q_2 L_2 S_2 Q L S M_L M_S) = \sum_{M_{Q_1} M_{Q_2}} \begin{bmatrix} Q_1 & Q_2 & Q \\ M_{Q_1} & M_{Q_2} & M_Q \end{bmatrix} \times$$

$$\times | \ell_1^{N_1} \ell_2^{N_2} \alpha_1 Q_1 L_1 S_1 \alpha_2 Q_2 L_2 S_2 L S M_L M_S).$$

The investigation of the properties of genealogical coefficients as well as of the reduced matrix elements of the irreducible tensorial operators with respect to the transposition of spin and quasi-spin quantum numbers leads to a set of new relations between the afore-mentioned quantities. So, for the genealogical coefficients with one detached electron the following formula is valid[8]:

$$(\ell^N \alpha v L S || \ell^{N-1} \alpha' v' L' S', \ell) = (-1)^{\varphi} \sqrt{\frac{N_S}{N}} \begin{bmatrix} Q' & 1/2 & Q \\ M_{Q'} & 1/2 & M_Q \end{bmatrix} \times$$

$$\times \begin{bmatrix} S' & 1/2 & S \\ M_{S'} & 1/2 & M_S \end{bmatrix}^{-1} (\ell^{N_S} \alpha v_S L Q || \ell^{N_S-1} \alpha' v'_S L' Q', \ell).$$

Here
$$S = \frac{1}{2}(2\ell+1-v_S), \quad M_S = -\frac{1}{2}(2\ell+1-N_S).$$

In some cases it is possible to obtain analytical expressions for the quantities considered, e.g.,

$$(\ell^N v L S || \ell^{N-1}\, {}^{2\ell}_{1}\ell, \ell) = (-1)^{\varphi(N+1)+\ell+1} \times$$

$$\times\sqrt{\frac{2(4\ell+2-N)!!\,(N-v)!!\,(4\ell-v)!!}{N(4\ell)!!\,(N-2)!!\,(4\ell+2-N-v)!!\,(2-v)!!}}\,,$$

$$(\ell^N vLS\|U^{k_e}\|\ell^N v'L'S)=(-1)^L\left[1-\delta(v,v')+(-1)^{\varphi(N)}\delta(v,v')\right]\times$$

$$\times 2\sqrt{(2L+1)(2L'+1)(2S+1)}\begin{Bmatrix}\ell & L & \ell\\ L' & \ell & k_e\end{Bmatrix}\begin{bmatrix}Q' & 1 & Q\\ M_Q & 0 & M_Q\end{bmatrix}\begin{bmatrix}Q' & 1 & Q\\ \frac{1-2\ell}{2} & 0 & \frac{1-2\ell}{2}\end{bmatrix}^{-1}.$$

In the last formula $k_e$ is even.

The use of a conception of irreducible tensorial sets is a powerful tool for the simplification of the formulas of the operators, already expressed in terms of irreducible tensors, as well as of the expressions for the complex matrix elements[9]. So, taking into account the tensorial properties of the wave function $\Psi^{(j)}_m$, one can obtain the following relation as an example ($J^{(1)}$ — the angular momentum operator):

$$\left[J^{(1)}\times\Psi^{(j)}\right]^{(k)}_m=-i\sqrt{j(j+1)}\,\Psi^{(j)}_m\,\delta(k,j).$$

Making use of this idea, it is possible to express a reduced matrix element of the irreducible tensorial operator in terms of a double tensorial product of the corresponding operators:

$$(j\|T^{(k)}\|j')=(-1)^{j'-k+j}\,4\pi\left[\Psi^{(j)}\times\left[T^{(k)}\times\Psi^{(j')}\right]^{(j)}\right]^{0}.$$

Analogous formulas are found in cases of more complicated tensorial products as well.

Using the method of the second quantization the formulas expressing the genealogical coefficients with two detached electrons in terms of the reduced matrix elements of the irreducible tensorial operators $U^k$ and $V^{k1}$ have been obtained[10]

$$2\sqrt{2(2k+1)Q(N,v_1)}\,(\ell^N\alpha vLS\|V^{k1}\|\ell^N\alpha_1v_1L_1S_1)=\sqrt{N(N-1)(2L+1)(2S+1)}\times$$

$$\times(-1)^{\varphi(N)}\left[Q(N,v_1)-Q(N+2,v)\right](\ell^N\alpha vLS\|\ell^{N-2}\alpha_1v_1L_1S_1,\ell^2\,{}^3k);$$

where $$Q(N,v)=\frac{1}{4}(N-v)(4l+4-N-v).$$

A similar expression for $U^k$ can be obtained from the above formula by multiplying its left side by 1/2.

In many cases, when investigating the energy spectra of many-electron atoms and ions, one has to use different coupling schemes, including an intermediate one. The following formula which connects the weights of the wave function in two different coupling schemes may be used for the classification of the energy levels with the help of various sets of quantum numbers:

$$c_{jk}=\sum_r a_{jr}(\varphi_k|\psi_r),$$

$$a_{ij}=\sum_k c_{ik}(\psi_j|\varphi_k).$$

Explicit formulas of the transformation matrices for the configurations $l^N$ and $l_1^{N_1}l_2^{N_2}$ are presented in the paper [11].

The use of the afore-mentioned result in the theory of angular momentum as well as of the irreducible tensorial operators simplifies considerably the theory of complex atomic spectra and makes it possible to investigate more efficiently the properties of the most complicated atoms and ions. These results may be easily rewritten for the case of jj coupling scheme.

During the last years interesting results were obtained both in the non-relativistic theory and relativistic theory of electronic transitions. Now we are in a position to calculate electronic transitions in relativistic approximation for the general case of complex atoms and ions.

A probability of an electric radiation of multipolarity k for the case of electronic transitions between the levels $\beta_1 J_1$ and $\beta_2 J_2$ can be written as follows in relativistic approximation[12]:

$$W^k_{1\to 2}=\frac{2e^2\omega(2k+1)}{k(2J_1+1)}\left|\langle\beta_2 J_2\|{}^1Q^{(k)}\|\beta_1 J_1\rangle\right|^2=$$

$$=\frac{2e^2\omega^3(k+1)}{k(2k+1)(2J_1+1)}\left|\langle\beta_2 J_2\|{}^2Q^{(k)}\|\beta_1 J_1\rangle\right|^2,$$

where the operators ${}^1Q^{(k)}$ and ${}^2Q^{(k)}$, obtained under the gauge conditions of the electromagnetic field potential in the form $C=-\sqrt{(k+1)/k}$ and $C=0$, respectively, are given by the following expressions:

$$ {}^1Q^{(k)}_q=\sqrt{k+1}\,C^{(k)}_q g_k(\omega r)+i\sqrt{2k+3}\left[C^{(k+1)}\times\alpha^{(1)}\right]^{(k)}_q g_{k+1}(\omega r), $$

$$ {}^2Q^{(k)}_q=C^{(k)}_q r\left[g_{k-1}(\omega r)-\frac{k}{k+1}g_{k+1}(\omega r)\right]+\frac{i}{k+1}\left\{\sqrt{k(2k-1)}\left[C^{(k-1)}\times\alpha^{(1)}\right]^{(k)}_q+\right. $$
$$ \left.+\sqrt{(k+1)(2k+3)}\left[C^{(k+1)}\times\alpha^{(1)}\right]^{(k)}_q\right\} r\, g_k(\omega r). $$

In these equations

$$ g_k(\omega r)=\sqrt{\frac{\pi}{2\omega r}}\,J_{k+1/2}(\omega r), $$

where $J_{k+1/2}$ denotes the Bessel function. The numerical values of the angular part of these matrix elements for some cases of electronic transitions are presented in the paper[13]. For the case of non-relativistic wave functions, general expressions of the electric multipole transition probability may be written in the following way[14]:

$$ W^k_{a\to b}=\frac{2(k+1)(2k+1)}{k[(2k+1)!!]^2}\frac{\omega^{2k+1}}{c^{2k+1}}\times $$
$$ \times\left|\left(b\left|\left\{Q^{(k)}_{-q}+C\sqrt{\frac{k}{k+1}}\left[\frac{1}{\omega}Q'^{(k)}_{-q}-Q^{(k)}_{-q}\right]\right\}\right|a\right)\right|^2, $$

$$ W^k_{a\to b}=\frac{2(k+1)(2k+1)}{k[(2k+1)!!]^2}\frac{\omega^{2k-1}}{c^{2k+1}}\times $$
$$ \times\left|\left(b\left|\left\{Q'^{(k)}_{-q}+C\sqrt{\frac{k}{k+1}}\left[Q'^{(k)}_{-q}-\omega\, Q^{(k)}_{-q}\right]\right\}\right|a\right)\right|^2, $$

where
$$ Q^{(k)}_{-q}=-r^k C^{(k)}_{-q}, $$
$$ Q'^{(k)}_{-q}=-r^{k-1}\left\{k\,C^{(k)}_{-q}\frac{\partial}{\partial r}+\frac{i}{r}\sqrt{k(k+1)}\left[C^{(k)}\times L^{(1)}\right]^{(k)}_{-q}\right\}. $$

Here $\omega = E_a - E_b$. These formulas are obtained from the corresponding relativistic expressions without specifying the gauge condition value. $Q'^{(k)}_{-q}$ is the generalization of the "velocity" form of an electric dipole radiation operator for the case of any multipolarity. Taking into account the commutation relation

$$[H, Q^{(k)}_{-q}] = -Q'^{(k)}_{-q}$$

one can see that in case of the exact wave functions the transition probabilities will not depend on the gauge condition. However, otherwise, choosing an appropriate gauge condition one can obtain any value of the transition probability.

These formulas show that there is no simple relation between the expressions for the corresponding relativistic and non-relativistic operators. One and the same relativistic expression depending on the gauge condition chosen leads both to the "length" and "velocity" forms of the corresponding non-relativistic operator.

The next three formulas describe relativistic corrections to the operator of the electric dipole transition in the "velocity" form:

$$W^{1}_{a\to b} = \frac{4}{3}\frac{\omega}{c^3}\left|\left(b\left\|\left\{Q'^{(1)}_{-q} + \frac{i}{c^2\sqrt{2}}\omega^2\left[Q^{(1)}\times S^{(1)}\right]^{(1)}_{-q}\right\}\right\|a\right)\right|^2,$$

$$W^{1}_{a\to b} = \frac{4}{3}\frac{\omega}{c^3}\left|\left(b\left\|\left\{Q'^{(1)}_{-q} + \frac{i}{c^2\sqrt{2}}\omega\left[Q'^{(1)}\times S^{(1)}\right]^{(1)}_{-q}\right\}\right\|a\right)\right|^2,$$

$$W^{1}_{a\to b} = \frac{4}{3}\frac{\omega}{c^3}\left|\left(b\left\|\left\{Q'^{(1)}_{-q} - \frac{i}{c^2\sqrt{2}}\left[Q''^{(1)}\times S^{(1)}\right]^{(1)}_{-q}\right\}\right\|a\right)\right|^2.$$

As you can see, these corrections may be written in three forms as well, - "length", "velocity" and "acceleration". Below a similar formula is presented for the "length" form of the corresponding operator[15]

$$W^{1}_{a\to b} = \frac{4}{3}\frac{\omega^3}{c^3}\left|\left(b\left\|\left\{Q^{(1)}_{-q} - \frac{i}{c^2\sqrt{2}}\left(\omega\left[Q^{(1)}\times S^{(1)}\right]^{(1)}_{-q} - \left[Q'^{(1)}\times S^{(1)}\right]^{(1)}_{-q}\right)\right\}\right\|a\right)\right|^2 =$$

$$= \frac{4}{3}\frac{\omega^3}{c^3}\left|\left(b\left\|\left\{\vec{Q} - \frac{i}{2c^2}\left[\left\{[H,\vec{r}] + \vec{\nabla}\right\}\times\vec{S}\right]\right\}\right\|a\right)\right|^2.$$

In this case, when using the exact wave functions, relativistic corrections are equal to zero. Therefore, the equivalency of the "length" and "velocity" forms of the electric dipole transition operators is valid only when we take into account relativistic corrections to the "velocity" form.

## 3. Numerical results

Further on we shall briefly discuss the application of the methods described to the theoretical investigation of the energy spectra and electron transitions in multiple-charged ions as well as the regularities in the values considered along isoelectronic sequences. At the beginning we shall deal with the investigations of the structure and properties of the multiple-charged ions and their dependence on the nuclear charge.

Figure 1 shows the dependence of the mean values of the distance of the $1s^2$, $2s^2$, $2p^4$ and 3d electrons from the nucleus (in atomic units) on the nuclear charge Z for the case of the fluorine isoelectronic sequence. It can be seen that with the increase of the ionization degree the size of the ion decreases very rapidly, the electronic shells approaching the nucleus. It is very interesting to point out that with the increase of the ionization the distance of the 2p-shell from the nucleus becomes smaller than 2s. A separate diagram presents in more details the values considered for the large Z values. The solid lines correspond to the Hartree-Fock-Pauli (HFP) approximation[2], whereas the dashed ones refer to the Dirac-Hartree-Fock (DHF) approximation[16]. For the large Z values, the splitting of electronic shells into subshells occurs (e.g., $2p_+$ and $2p_-$).

Figure 2 illustrates the dependence of the total electronic density of the ions $Ne^{+4}$, $Fe^{+20}$ and $W^{+68}$ (configuration $1s^22s^22p^2$) on the logarithmic distance from the nucleus. Again, with the increase of the ionization degree we can see its sharp shift towards the nucleus and its clearly pronounced concentration in the narrow region of r values, i.e. with the increase of Z the electronic shells of ions become thinner and thinner.

Figure 3 illustrates the change in the distribution of the total electronic density with the excitation of one electron (configurations $1s^22s^22p^2$, $1s^22s^22p3s$ and $1s^22s2p^3$).

Energy spectrum is a fundamental characteristic of an atom or ion.

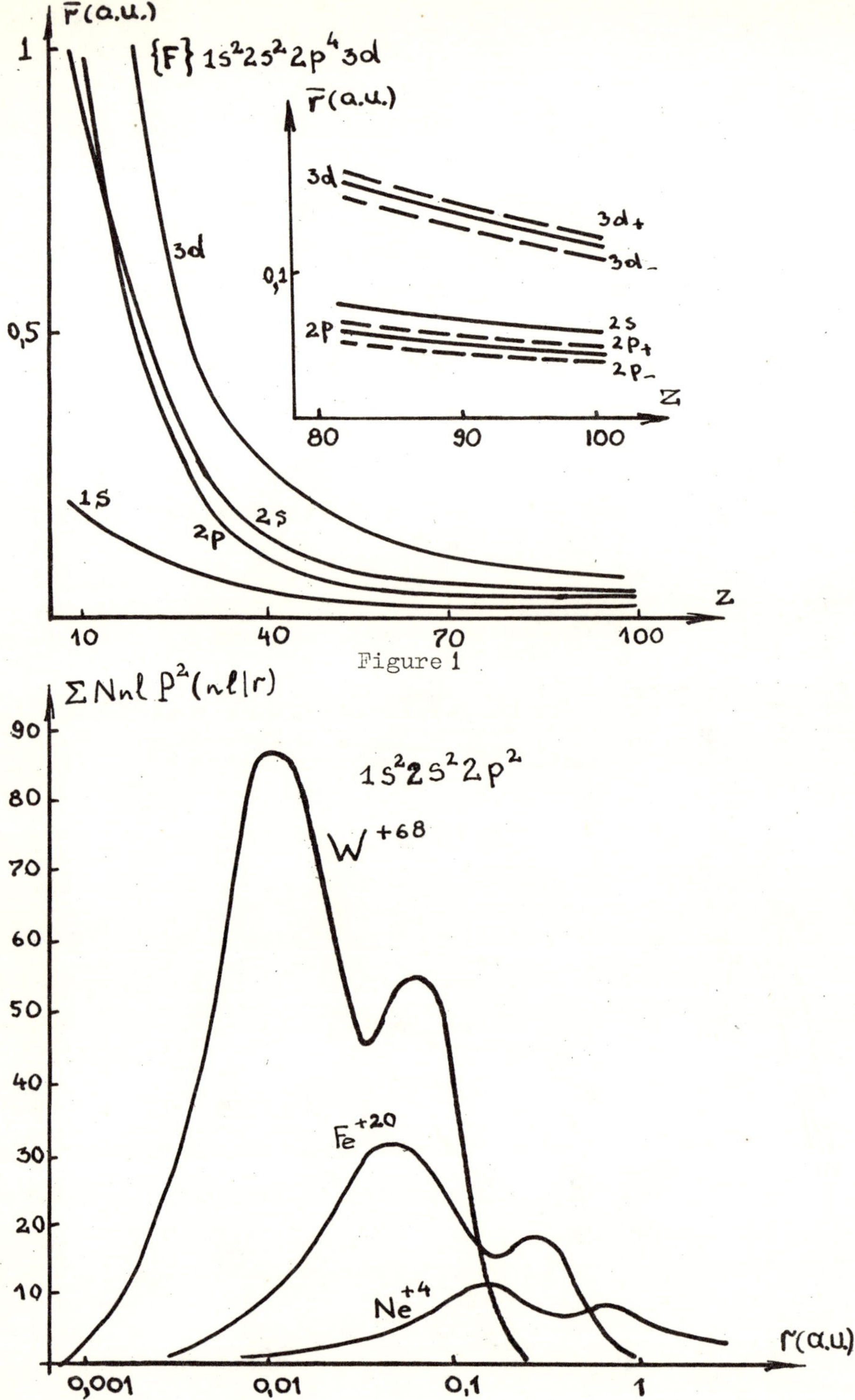


Figure 1

Figure 2

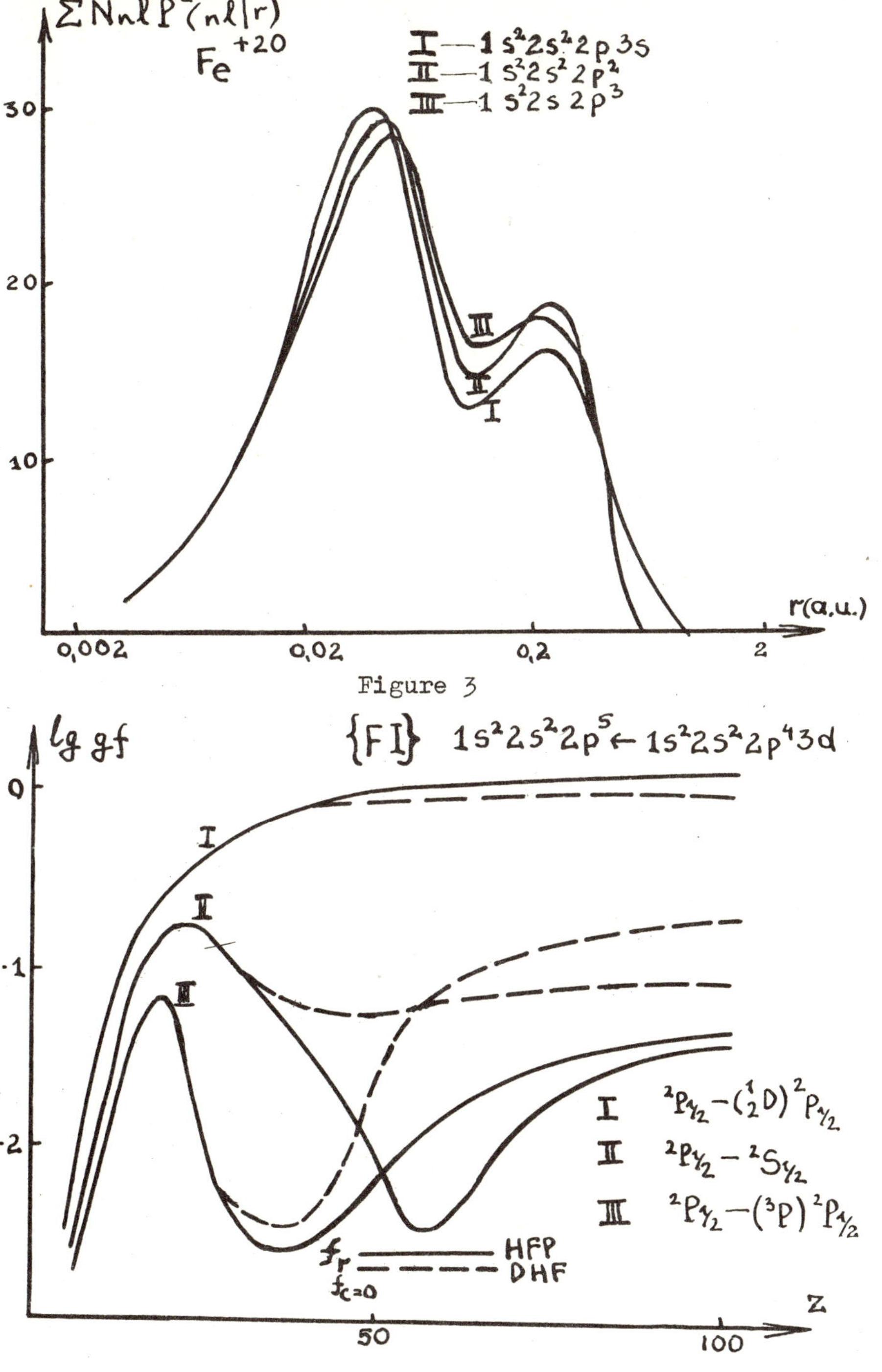


Figure 3

Figure 4

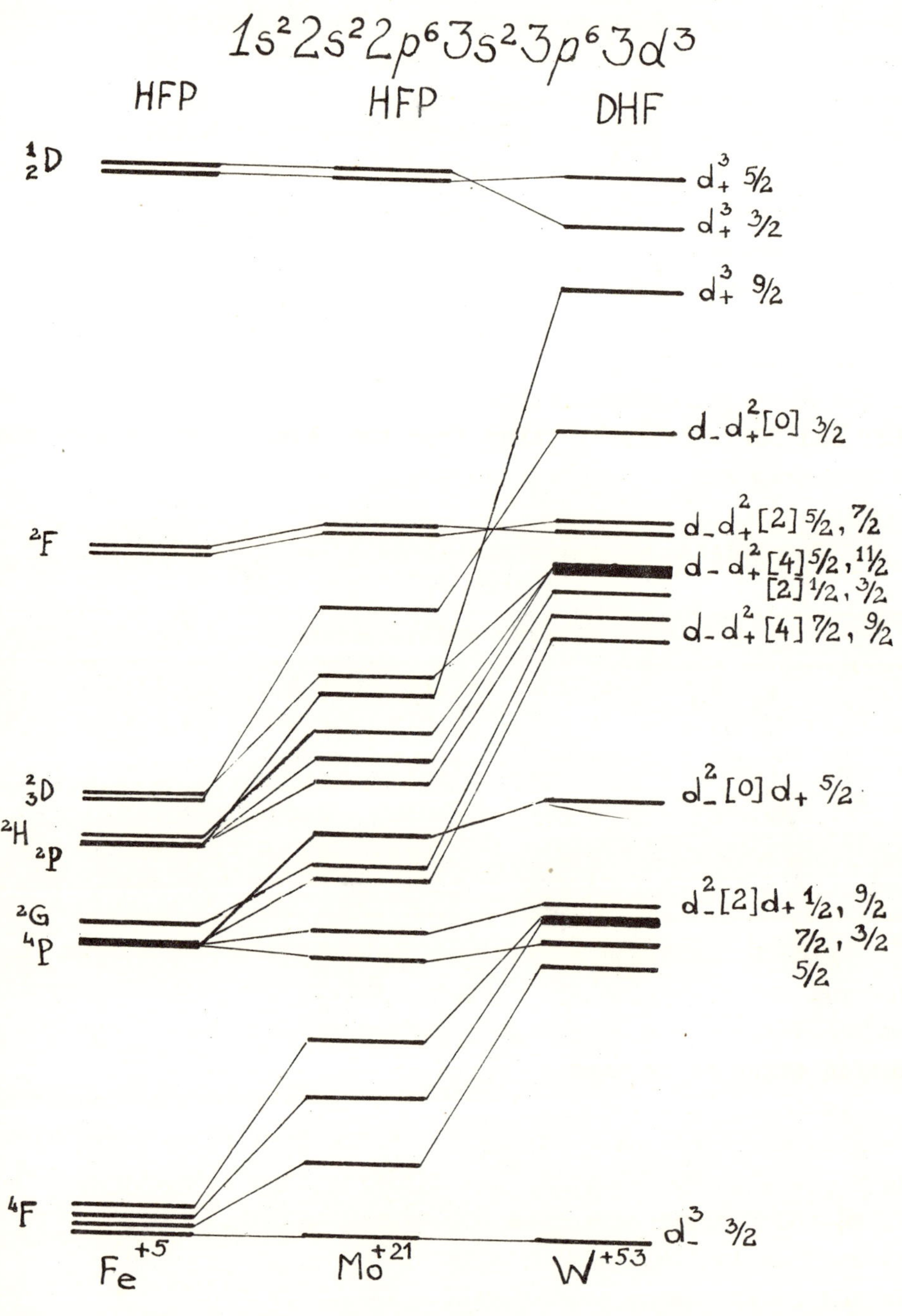
$1s^2 2s^2 2p^6 3s^2 3p^6 3d^3$
HFP
HFP
DHF
$^1_2D$
$^2F$
$^2_3D$
$^2H$
$^2P$
$^2G$
$^4P$
$^4F$
$d_+^3$ 5/2
$d_+^3$ 3/2
$d_+^3$ 9/2
$d_-d_+^2[0]$ 3/2
$d_-d_+^2[2]$ 5/2, 7/2
$d_-d_+^2[4]$ 5/2, 11/2
[2] 1/2, 3/2
$d_-d_+^2[4]$ 7/2, 9/2
$d_-^2[0]d_+$ 5/2
$d_-^2[2]d_+$ 1/2, 9/2
7/2, 3/2
5/2
$d_-^3$ 3/2
$Fe^{+5}$
$Mo^{+21}$
$W^{+53}$

Figure 5

Figure 5 demonstrates the energy spectra of the ions $Fe^{+5}$, $Mo^{+21}$ and $W^{+53}$ in the configuration $1s^22s^22p^63s^23p^63d^3$ (in relative units, taking the same value for the whole width of the energy spectra). For the first two ions the Hartree-Fock-Pauli approximation is used (the coresponding formulas may be found in the monograph[17]), whereas for the third one the Dirac-Hartree-Fock approximation is applied. These spectra illustrate very clearly the changes in the coupling schemes. Pure LS coupling and the distinct multiplets in the case of the ion $Fe^{+5}$ are replaced for $W^{+53}$ by jj coupling. The levels form subshells in the case of the $W^{+53}$ very distinctly. The levels of a certain multiplet may find themselves in different subshells.

The last part of this paper deals with the investigations of the electronic transitions in the approximations considered (HFP and DHF). Table 1, taken from the paper[16], gives some idea of the accuracy of the methods used.

Table 1

Wavelengths of the transitions $1s^22s^22p^53d-1s^22s^22p^6$ (in Å) for $Fe^{+16}$ and $Mo^{+32}$

| | $2p^53d-2p^6$ | PT | HFP | DHF | Exp. |
|---|---|---|---|---|---|
| $Fe^{+16}$ | $^3P_1-^1S_0$ | 15,484 | 15,457 | 15,460 | 15,453 |
| | $^3D_1-$ | 15,274 | 15,263 | 15,268 | 15,261 |
| | $^1P_1-$ | 15,023 | 14,996 | 15,004 | 15,012 |
| $Mo^{+32}$ | $^3P_1-^1S_0$ | 4,869 | 4,860 | 4,854 | 4,847 |
| | $^3D_1-$ | 4,812 | 4,802 | 4,803 | 4,804 |
| | $^1P_1-$ | 4,643 | 4,637 | 4,630 | 4,630 |

Each of the three approximations (perturbation theory[18], HFP and DHF) leads to the results, fairly close to the experimental data[19]. For very highly ionized atoms, the DHF approximation ought to be used.

Table 2 lists the results of the calculations of the wavelengths and oscillator strengths of the transition $1s^22s^22p^4 3d\,^2S_{1/2}-1s^22s^22p^5\,^2P_{1/2}$ in the fluorine isoelectronic sequence (Z=26-100), obtained in the DHF approximation ($\lambda_{DHF}$, $gf_C$), as well as in the HFP approximation ($\lambda_{HFP}$, $gf_L$ and $gf_V$). In the relativistic approximation two gauge conditions C=0 and $C=-\sqrt{2}$ are used, whereas in HFP approximation the "length"

( $gf_L$ ) and "velocity" ( $gf_V$ ) forms of the transition operator are considered. The relativistic corrections to $gf_V$ are taken into account again in the "length" and "velocity" forms.

Table 2

Oscillator strengths of the transition $1s^22s^22p^43d\ ^2S_{1/2}$–$1s^22s^22p^5\ ^2P_{1/2}$

| Z | $\lambda_{DHF}$(Å) | $gf_{C=0}$ | $gf_{C=-\sqrt{2}}$ | $\lambda_{HFP}$(Å) | $gf_L$ | $gf_V$ | $gf_{VV}$ | $gf_{VL}$ |
|---|---|---|---|---|---|---|---|---|
| 26 | 14,43 | 0,175 | 0,178 | 14,47 | 0,173 | 0,169 | 0,168 | 0,169 |
| 34 | 7,616 | 0.103 | 0,105 | 7,614 | 0,101 | 0,100 | 0,099 | 0,101 |
| 42 | 4,691 | 0,046 | 0,047 | 4,691 | 0,046 | 0,045 | 0,046 | 0,045 |
| 54 | 2,655 | 0,057 | 0,061 | 2,680 | 0,003 | 0,003 | 0,003 | 0,004 |
| 74 | 1,347 | 0,057 | 0,064 | 1,351 | 0,019 | 0,019 | 0,021 | 0,017 |
| 83 | 1,054 | 0,056 | 0,065 | 1,056 | 0,037 | 0,036 | 0,039 | 0,033 |
| 92 | 0,846 | 0,054 | 0,066 | 0,847 | 0,053 | 0,052 | 0,056 | 0,047 |
| 100 | 0,709 | 0,053 | 0,066 | 0,709 | 0,066 | 0,064 | 0,070 | 0,057 |

Both approximations lead to similar results. It should be noted that with the increase of Z the radiation wavelength decreases very rapidly. The numerical values of the relativistic corrections are rather small. It seems that they have to be taken into account only when using the more exact wave functions.

Figure 4 illustrates the dependence of the logarithm of the oscillator strength of some transitions $1s^22s^22p^43d$–$1s^22s^22p^5$ in the fluorine isoelectronic sequence on the nuclear charge Z. The solid and dashed lines represent the HFP and DHF approximations, respectively. The results obtained using different gauge conditions practically coincide. The discrepancies between the results of the above-mentioned methods are considerable only for large ionization degrees.

In conclusion, it is necessary to emphasize that during the last years interesting new results have been obtained in the theory of many-electron atoms and ions. The use of these results has enriched our knowledge about the structure and properties of the afore-mentioned systems. This concerns especially the multiple-charged ions. Their theoretical investigation has enabled us to find a number of peculiarities in their structure which differ considerably from the properties of the corresponding neutral atoms and ions of the moderate ionization degrees. In many cases, the accuracy of the values of the wavelengths and oscillator strengths is sufficient to use these results for identification and classification of the experimentally measured energy spectra of the highly ionized atoms[20].

## R e f e r e n c e s


1 A.P.Jucys, Atomic Physics, Proceedings of the Third ICAP, Plenum Press, New York, 1973, p. 185.

2 P.O.Bogdanovich, S.D.Šadžiuvienė, J.J.Boruta and Z.B.Rudzikas, Liet.fiz.rinkinys, 16, 505 (1976) (English Translation - Soviet Physics Collection).

3 H.A.Bethe, E.E.Salpeter, Quantum Mechanics of One- and Two-Electron Atoms, Springer-Verlag, Berlin-Göttingen-Heidelberg, 1957.

4 Z.B.Rudzikas, J.M.Kaniauskas, Intern.J.Quantum Chem., X, 837 (1976).

5 V.V.Špakauskas, J.M.Kaniauskas, Z.B.Rudzikas, Liet.fiz. rinkinys, 18, 293 (1978).

6 A.P.Jucys, A.A.Bandzaitis, The Theory of Angular Momentum in Quantum Mechanics, Mintis Publishing House, Vilnius, 1977 (in Russian).

7 J.M.Kaniauskas, V.V.Špakauskas, Z.B.Rudzikas, Liet.fiz. rinkinys, 18 (1978) (to be published).

8 V.V.Špakauskas, I.S.Kičkin, Z.B.Rudzikas, Liet.fiz.rinkinys, 16, 201 (1976).

9 J.M.Kaniauskas, Z.B.Rudzikas, Liet.fiz.rinkinys, 13, 657 (1973).

10 I.S.Kičkin, Z.B.Rudzikas, Liet.fiz.rinkinys, 11, 743(1971).

11 I.S.Kičkin, A.A.Slepcov, V.I.Sivcev, Z.B.Rudzikas, Liet. fiz.rinkinys, 16, 217 (1976).

12 J.M.Kaniauskas, I.S.Kičkin, Z.B.Rudzikas, Liet.fiz.rinkinys, 14, 463 (1974).

13 Z.B.Rudzikas, A.A.Slepcov, I.S.Kičkin, Atomic Data and Nuclear Data Tables, 18, 223 (1976).

14 J.M.Kaniauskas, G.V.Merkelis, Z.B.Rudzikas, Liet.fiz.rinkinys, 19,(1979) (to be published).

15 Z.B.Rudzikas, J.M.Kaniauskas, Liet.fiz.rinkinys, 13, 849 (1973).

16 I.S.Kičkin, V.I.Sivcev, P.O.Bogdanovich, Z.B.Rudzikas, Liet.fiz.rinkinys, 18, 165 (1978).

17 A.P.Jucys, A.J.Savukynas, Mathematical Foundations of the Atomic Theory, Mintis Publishing House, Vilnius, 1973 (in Russian).

18 U.I.Safronova, Z.B.Rudzikas, J.Phys. B: Atom.Molec.Phys., 10, 7 (1977).

19 V.A.Boiko, S.A.Pikuz, A.Ya.Faenov, Preprint FIAN Nr 20, Moscow, 1976.

20 V.A.Boiko, S.A.Pikuz, A.S.Safronova, A.Ya.Faenov, P.O.Bogdanovich, G.V.Merkelis, Z.B.Rudzikas, S.D.Šadžiuvienė, Preprint FIAN Nr 175, Moscow, 1977.

# RELATIVISTIC PERTURBATION THEORY FOR ATOMS AND IONS

M.A.Braun and L.N.Labsovsky

Leningrad State University

This report consists of two parts. In the first part (M.A.B.) a general formalism is developed for perturbation treatment of relativistic atomic systems. The second part (L.N.L.) is devoted to applications of the relativistic perturbation theory of Gell-Mann and Low for the study of multiply charged ions.

## I. Relativistic Perturbation Theory for Degenerate Levels

1. The latest experimental results have stimulated interest in energetic spectra and transition probabilities for highly ionized atoms with few electrons. In their theoretical analysis it is possible to use perturbation theory in the electron-electron interaction which turns out to be an expansion in powers of 1/Z, where Z is the charge of the nucleus. This is also true for inner electrons of neutral atoms with large Z. In such systems the electronic motion is relativistic. This makes it necessary to take fully into account radiation effects using the apparatus of quantum electrodynamics with the corresponding renormalization of mass and charge technique. To do this one needs a perturbation theory for the discrete spectrum which would operate with renormalized quantities. For the particular case of a non-degenerate ground state such a perturbation theory can be obtained as a direct generalization to arbitrary external fields of the standard quantum electrodynamics in the vacuum /1,2/. Vacuum-vacuum Feynman diagrams which correspond to the level shift in this case should undergo an additional renormalization, which together with the usual one removes all ultraviolet divergencies and introduce the physical mass and charge of the electron. A straightforward generalization of this method to degenerate levels, however, leads to multiparticle operator contractions in Wick's theorem or to multiparticle interactions /3/, which

make both the theoretical analysis and practical applications hardly feasible. A series of papers by different authors devoted lately to perturbation theory for degenerate levels have not at all dealt with relativistic problems but solely discussed the problem of the so-called cluster expansion /4/. They all have used the adiabatic transition operator $S(0,\pm\infty)$, which makes the analysis very close to that of the ordinary nonrelativistic quantum mechanics and can hardly be extended to relativistic systems.

In this part of the report a perturbation theory for relativistic many-electron systems is developed which can be applied to degenerate levels. It operates with renormalized quantities and is therefore free from ultraviolet divergences. It has been proposed in /5,6/ and is based on the splitting of the Bethe-Salpeter kernel for the multielectron Green function into two parts, one being a factorized kernel and the other small in some sense. Once the latter is dropped, the Bethe-Salpeter equation can be solved exactly. Further improvement of the solution can be made by perturbation theory. The splitting of the kernel can be done in different ways. The case of two electrons (or holes) outside closed shells appears especially simple and will be studied in most details.

As a result we shall demonstrate that the level shift can be obtained by diagonalizing a finite dimensional matrix $W_{\alpha\beta}(E)$ which is defined in the subspace of degenerate levels and depends on the exact energy E (in close analogy with the quantum-mechanical Wigner-Brillouin perturbation theory). The matrix $W_{\alpha\beta}$ is made first of the matrix elements of the S-matrix for transitions $\alpha\to\beta$ without intermediate states of the same energy. Such approximations seem to be reasonable owing to the analogy with the nonrelativistic perturbation theory and with the Gell-Mann-Low theory for nondegenerate levels /7/. However $W_{\alpha\beta}$ contains an additional contribution from transitions with intermediate states of the same energy, which does not exist in the nonrelativistic case and proceeds from dependence of the interaction on the exchange energy (retardation). Such retardation terms appear in the expression for transition probabilities as well. We explicitly show the retardation corrections for the case of two electrons in the open shell.

Note that for multielectron atoms with small Z the level shifts up to the orders of $\alpha^5$ with radiation corrections were studied in [8] where explicit expressions were given for the effective interaction of electrons, and from where we borrowed the technique of the expansion around the mass shell.

2. Let us study the case of two electrons outside closed shells. We assume that the closed shells are described by choosing properly the contour of integration in the complex energy plane of the one-electron Green function. In the Hartree-Fock approximation each electron can be characterized by the standard set of quantum numbers $a = \{n, j, m, l = j \pm 1/2\}$. The two electrons will be described by a pair of such sets $\alpha = \{a_1 a_2\}$. The degenerate levels of the two electrons form an open shell $A = \{\alpha\}$ with the same non-perturbed energy $E_\alpha^{(0)} = E_{a_1}^{(0)} + E_{a_2}^{(0)}$. In the following we consider only open shells which do not contain electronic levels with all quantum numbers identical but the principal quantum number n. In this case the one-electron renormalized Green-function G is a diagonal matrix between the states $a$ and $b$ which belong to A:

$$G_{ab} = G_a \delta_{ab} , \qquad a, b \in A, \tag{1}$$

because radiative one-particle transitions between states with different j, m or l are impossible. In such shells the radiation corrections due to atomic field do not result in the mixing of the one particle levels. This simplifies the consideration. Shells with electronic states that differ in n only can be considered by analogy with a preliminary diagonalization of the one-particle Green-function. Isolating the pole term in G we write

$$G_a = Z_a / (E_a - E) + \tilde{G}_a \equiv G_a^{(p)} + \tilde{G}_a \tag{2}$$

where $E_a$ is an electronic level in the atomic field with all radiative corrections. The residue $Z_a$ is determined as usual through the derivative of the renormalized self-mass of the electron: $Z_a^{-1} = 1 + \Sigma_a'(E_a)$.

Consider now the two-electron Green function G. It depends on the total energy E and initial and final relative energies of the electrons $\varepsilon_1$ and $\varepsilon_2$ respectively. We limit ourselves with matrix elements $G_{\alpha\beta}$ between the states $\alpha$ and $\beta$ belonging to A. They form a finite dimensional matrix (denoted by G for brevity) satisfying an equation which is a projection onto A of the standard Bethe-Salpeter equation;

$$G = G_0 + G_0 M G \tag{3}$$

Here $G_0$ is the contribution from the disconnected diagrams:

$$G_{0\alpha\beta} = G_{0\alpha}\,\delta_{\alpha\beta}\,\delta(\varepsilon_1-\varepsilon_2)\,;\quad G_{0\alpha} = G_{a_1}G_{a_2}\;,\quad \alpha = \{a_1 a_2\}\,. \tag{4}$$

The kernel $M$ likewise depends on E, $\varepsilon_1$ and $\varepsilon_2$ . The integration over relative energies is implied in (3). Contributions to $M_{\alpha\beta}$ come from all Feynman diagrams for the transition $\alpha \to \beta$ with no intermediate two-electron state from the same open shell A. $M_{\alpha\beta}$ includes standard irreducible diagrams for the Bethe-Salpeter kernel shown in Fig.1 in the first two orders, as well as reducible ones, shown in Fig.2, with intermediate states that do not belong to A. Eq.(3) is in fact a

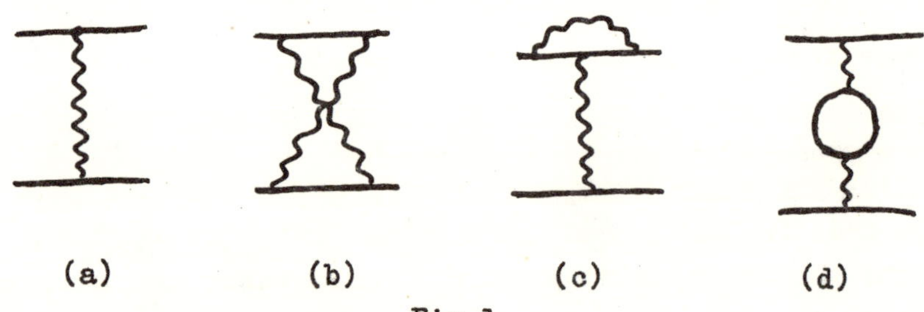

(a) (b) (c) (d)

Fig.1

system of one-dimensional integral equations in the relative energy $\varepsilon$ . In a similar nonrelativistic problem without retardation the dependence on $\varepsilon$ is trivial, and one is left with a system of algebraic equations, which immediately leads to the well-known secular equation. The retardation evidently greatly complicates the problem and would lead to insurmountable difficulties, were it not possible to use perturbation theory.

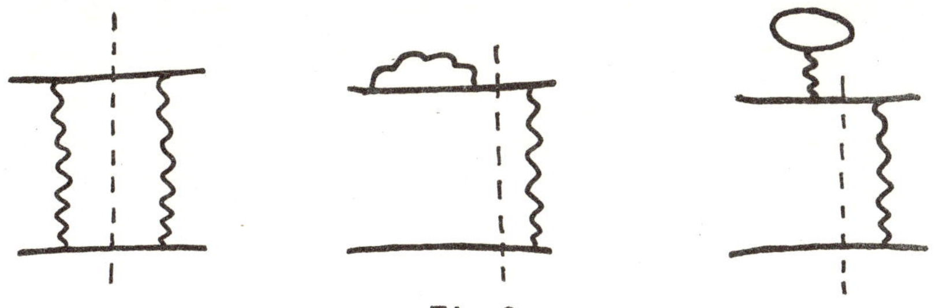

Fig.2

Following[5] we isolate from $M$ its value on the mass shell, i.e. at the point corresponding to the individual energies of the electrons without interaction (but with the radiative corrections in the atomic field taken into account);

$$M = M_0 + M_1\,, \quad M_{0\alpha\beta}(\varepsilon_1,\varepsilon_2) = M_{\alpha\beta}(\varepsilon_\alpha,\varepsilon_\beta)\,, \tag{5}$$

where for $\alpha=\{a_1 a_2\}$ $\varepsilon_\alpha = \frac{1}{2}(E_{a_1} - E_{a_2})$. We separate further the most singular contribution to $G_0$ coming from the product of the pole terms in one-particle Green-functions:

$$G_0 = G_0^{(P)} + \widetilde{G}_0\,; \quad G_{0\alpha}^{(P)} = G_{a_1}^{(P)} G_{a_2}^{(P)}\,, \quad \alpha = \{a_1 a_2\}. \tag{6}$$

Eq.(3) can be rewritten as

$$G = G_0^{(P)} + G_0^{(P)} M_0 G + \widetilde{G}_0 + \widetilde{G}_0^{(P)} \widetilde{M} G \tag{7}$$

with

$$\widetilde{M} = M_1 + G_0^{(P)-1} \widetilde{G}_0 M. \tag{8}$$

The first two terms on the rhs of Eq.(7) have the same structure as in the nonrelativistic theory since the retardation is eliminated by fixing the relative energies according to (5). The third term has nothing to do with the pole terms in $G$ and the last, most complicated term can be shown to admit the perturbation treatment[5], since the vanishing of denominators in it is compensated by the smallness of numerators coming from the subtraction from $M$ of its value on the mass shell or its multiplication by a small quantity $G_0^{(P)-1}\widetilde{G}_0$. This allows one to construct an algorithm to solve Eq.(7) by perturbations. As a first step Eq.(7) without the last two terms is used, which is then solved exactly.

To formalize this procedure, note that in the second term on the rhs of Eq.(7) the Green function G enters integrated over the left relative energy. Denote the result of such an integration multiplied by $(2\pi i)^{-1}$ by $G_L$. Expressing G via $G_L$ and integrating this equation over the left relative energy we obtain a system of algebraic equations for $G_L$ whose solution allows to find the Green function G explicitly:

$$G = G_1\left(1+G_0^{(P)-1}\tilde{G}_0\right) + G_{1R} M_0 \left(1-G_{1LR} M_0\right)^{-1} G_{1L}\left(1+G_0^{(P)-1}\tilde{G}_0\right) \quad (9)$$

where we have used the definition

$$G_1 = G_0^{(P)}\left(1-\tilde{M}\, G_0^{(P)}\right)^{-1} \quad (10)$$

and the symbol $R$ means integration over the right relative energy with a factor $(2\pi i)^{-1}$. The poles of $G$ come from the second term in Eq.(9) and are determined as roots of the algebraic equation

$$\det \mathcal{D}(E) = 0\,, \qquad \mathcal{D} = G_{1LR}^{-1} - M_0 \quad (11)$$

Eq.(11) plays the role of the secular equation in the relativistic theory. Note that $G_{0LR}^{(P)} = Z\left(E-E^{(0)}\right)^{-1}$ with Z and $E^{(0)}$ being diagonal matrices with elements $Z_{a_1} Z_{a_2}$ and $E_{a_1} + E_{a_2}$ respectively. When retardation and radiative corrections are absent ( $Z_a = 1$, $\tilde{M} = 0$ ) Eq. (11) goes over into the usual nonrelativistic secular equation. The matrix D in Eq(11) may be replaced by other matrices obtained from D by multiplying it by nonsingular matrices from both sides. As was mentioned in /5/, the most suitable choice is to multiply D from both sides by $\left(G_{1LR}\, G_{0LR}^{(P)-1} Z^{-1}\right)^{-1/2}$. The resulting matrix $\Delta$ is given in the first two orders by

$$\Delta = E - E^{(0)} - W \quad (12)$$
$$W = Z^{-1/2} M_0 Z^{-1/2} + \tfrac{1}{2} G_{0LR}^{-1} G_{0L} \tilde{M} G_{0R} M_0 + \tfrac{1}{2} M_0 G_{0L} \tilde{M} G_{0R} G_{0LR}^{-1}\,.$$

3. The matrix $W$ has a meaning of the self-energy of the shell A. Contributions to $W$ are of two types. First, they come from $M_0$ which, as we saw, is a sum of all Feynman diagrams for the corresponding transitions without intermediate states of the same energy. The relative energies have to be fixed on the mass shell according to (5). Apart from this contribution $W$ involves other terms coming precisely from intermediate states of the same energy and not represented by Feynman diagrams themselves. This contribution proceeds from $\tilde{M}$ and the

difference Z - 1. In both cases it originates from the dependence of the corresponding Feynman diagrams on the exchange energy, i.e. from the retardation in the broad sense. Correspondingly we call these terms the retardation contribution $W^{(R)}$.

The part of $W^{(R)}$ that comes from $\tilde{M}$ has been investigated in our paper /5/. In the lowest order it may be represented in the following form

$$W_1^{(R)} = \frac{1}{4}\{B,\Delta\}\frac{1}{\Delta}V + \frac{1}{4}V\frac{1}{\Delta}\{B,\Delta\} \tag{13}$$

where $V$ is $M_0$ in the lowest order, i.e. the sum of the Coulomb and Breit potentials, and $B$ is an additional electron-electron potential on the shell A due to retardation. It can be written as

$$B_{\alpha\beta} = -\frac{2\alpha}{\pi}\left\{\left(\vec{\alpha}_1\vec{\alpha}_2 - (\vec{\alpha}_1\vec{r}_{12})(\vec{\alpha}_2\vec{r}_{12})r_{12}^{-2}\right)\mathcal{F}_{\alpha\beta}(r_{12})\right\}_{\alpha\beta} \tag{14}$$

Here $\vec{\alpha}_i$ are Dirac's matrices, $\vec{r}_{12}$ is the distance between the electrons. The function $\mathcal{F}_{\alpha\beta}$ is defined differently for transitions between the states with equal or different unperturbed relative energies. In the first case

$$\mathcal{F}_{\alpha\beta}(r) = C + \ln r\sqrt{\varepsilon_{\alpha\beta}^2 - \delta_{\alpha\beta}^2} + \frac{\pi}{4}r\delta_{\alpha\beta} - 1 + \tag{15}$$
$$\varepsilon_\alpha^{(0)} = \varepsilon_\beta^{(0)} \qquad + \frac{\varepsilon_{\alpha\beta}}{\delta_{\alpha\beta}}\ln\frac{\varepsilon_{\alpha\beta}+\delta_{\alpha\beta}}{\varepsilon_{\alpha\beta}-\delta_{\alpha\beta}} - \frac{\pi i}{2\delta_{\alpha\beta}}\left(|\varepsilon_{\alpha\beta}+\delta_{\alpha\beta}| - |\varepsilon_{\alpha\beta}|\right)$$

with the definitions $\varepsilon_{\alpha\beta} = \varepsilon_\alpha - \varepsilon_\beta$, $\delta_{\alpha\beta} = \frac{1}{2}\Delta_\alpha + \frac{1}{2}\Delta_\beta$.

Branches of the logarithm in (15) are to be taken according to $\varepsilon_{\alpha\beta}$ possessing an infinitely small negative imaginary part. In the second case

$$\mathcal{F}_{\alpha\beta}(r) = \cos x\cdot ci\,x + \sin x\cdot si\,x - \frac{\pi}{2}ie^{i|x|}\operatorname{sign}x \tag{16}$$
$$\varepsilon_\alpha^{(0)} \neq \varepsilon_\beta^{(0)}$$

with $\quad x = r(\varepsilon_\alpha - \varepsilon_\beta)$.

The second part of the retardation contribution $W_2^{(R)}$ comes from the difference Z - 1 and may be represented through the derivative of the electron self-energy. This term acts as an additional renormalization of the vertex part for the emission of a photon from the shell A. If we denote this vertex part as $\Gamma_{ab}^{\mu}(k)$ with k the momentum of the photon, a,b$\in$A, then in the second order we have the following Ward identity

$$\Gamma_{ab}^{(0)}(0) = -\Sigma'_{ab}(E_a) = -\Sigma'_a(E_a)\,\delta_{ab} \tag{17}$$

Among Feynman diagrams for $M_0$ in the second order we find the contribution from the vertex part (Fig.1c). According to (17) the retardation contribution $W_2^{(R)}$ will renormalize $\Gamma^\mu$ replacing it with $\tilde{\Gamma}^\mu$ :

$$\tilde{\Gamma}^{(2)\mu}(k) = \Gamma^{(2)\mu}(k) - \frac{1}{2}\left\{\Gamma^{(2)0}(0), \Gamma^{(0)\mu}(k)\right\}. \tag{18}$$

Here $\Gamma$ is taken as a matrix between electronic states. The numbers in parentheses above show orders of the perturbation expansion. The vertex part (18) has the property $\tilde{\Gamma}^{(0)}_{0}(0)=0$. Thus, the main part of the diagram Fig.1c coming from the Coulomb interaction at long distances has been removed by the additional renormalization (18).

A few words on transition probabilities. Total probabilities. Total probabilities can be calculated in terms of imaginary parts of the corresponding energy levels by diagonalizing the matrix $W$. To study probabilities for the transition into a given state one has to find eigenvalues of the matrix $W$ and using those - the mixing coefficients according to (9). Explicit formulae for transition probabilities are given in /5/. As to energy levels, apart from terms to be expected and given by standard Feynman diagrams, additional terms appear which originate from retardation and can be expressed either via the additional potential $B$ introduced above or via the additional renormalization of the vertex part.

4. The described technique cannot be generalized in a straightforward manner to the case with three or more electrons in an open shell. The point is that for three or more electrons the Bethe-Salpeter kernel $M$ involves disconnected diagrams and is not a smooth function of the relative energies. In this case one may choose a different way of splitting the kernel /6/. Take the many-electron Green function $G_0$ without interaction between electrons. Split it according to the equation

$$G_0 = G_{0R}\,(G_{0LR})^{-1}\,G_{0L} + \tilde{G}_0 \tag{19}$$

where all notations are as before, the integrations being performed over all right or all left relative energies. The function $\tilde{G}_0$ has the property $\tilde{G}_{0L}=\tilde{G}_{0R}=0$. Therefore in the integrals involving $\widetilde{G}_0$ the contribution comes only from the dependence of the interaction on relative energies, i.e. from retardation. As a result, such integrals have no poles in the vicinity of $E = E^{(0)}$. It means that the part of the exact Green function G associated with $\widetilde{G}_0$ can be treated by perturbations. Substituting (19) into (3) we rewrite (3) in the form

$$G = G_1 + G_{1R}(G_{0LR})^{-1} G_{0L} M G = G_1' + G M G_{0R} (G_{0LR})^{-1} G_{1L}' \tag{20}$$

Now we use the notations

$$G_1 = (1 - \widetilde{G}_0 M)^{-1} G_0 \; ; \; G_1' = G_0 (1 - M \widetilde{G}_0)^{-1}. \tag{21}$$

Using (20) one can express G through $G_{LR}$ (see /6/) and show that the pole structure of G is completely determined by that of $G_{LR}$. Multiplying the first Eq.(20) by $G_0 M$ from the left and integrating the result over relative energies from both sides we get an equation for $G_{0L} M G_R$ , the solution of which gives $G_{LR}$:

$$G_{LR} = [G_{0LR}^{-1} - G_{0LR}^{-1} G_{0L} M G_{1R} G_{0LR}^{-1}]^{-1} \equiv D^{-1} \tag{22}$$

Energy levels are determined by Eq.(11) where D is now defined as the expression in square brackets in (22). It can be shown that for two electrons using Eq.(22) is equivalent to the formalism developed in the previous paragraphs. From (22) and the connection between G and $G_{LR}$ one can find the mixing coefficient and thus construct the formalism for transition amplitudes between arbitrary states /6/.

## II. The Application of the Gell-Mann and Low Theory

1. In the second part of the work the quantumelectrodynamical theory of multicharged ions, based on the Gell-Mann-Low formalism, is developed and the applications of this theory to the calculation of various properties of the two-electron ions are discussed.

We consider the many-electron atom or ion as a set of the electrons, interacting with one another and moving in the Coulomb field of the nucleus.

The Hamiltonian of the atom in the representation of the second quantization is:

$$H = H_0 + H_{int} \tag{23}$$

$$H_0 = \int \psi^+(\vec{x})\, h(\vec{x})\, \psi(\vec{x})\, d\vec{x} \tag{24}$$

$$h(\vec{x}) = \vec{\alpha}\vec{p} + \beta - e\mathcal{U}(\vec{x}) \tag{25}$$

$$H_{int} = -\int j_\mu(\vec{x})\, A_\mu(\vec{x})\, d\vec{x} \tag{26}$$

where $\psi(\vec{x}), \psi^+(\vec{x})$ are the electron-positron, field operators, $\vec{p} \equiv -i\nabla$, $\vec{\alpha}, \beta$ are Dirac's matrices, $\mathcal{U}(\vec{x})$ is the potential of the nucleus, $e$ is the charge of the electron, $H_{int}$ is the operator of the interaction of electrons with the electromagnetic field, $j_\mu(x)$ is the 4-vector of the electron current, $A_\mu(x)$ is the vector potential of the electromagnetic field. We use the relativistic units $\hbar$ =c=m=1 where m is the mass of the electron. In these units $e^2 = \alpha \simeq 1/137$, where $\alpha$ is the fine structure constant. We shall consider below the operator $H_{int}$ as the perturbation. Two different parameters exist in our problem: $\alpha Z$ and $Z^{-1}$, where Z is the charge of the nucleus. We shall use only the expansions in powers of $Z^{-1}$, but not $\alpha Z$. Our theory will be applicable, consequently, mostly to the multicharged ions with $Z \gg 1$.

2. We consider first the ground, nondegenerate state of an atom, the wave function of which in zero order approximation may be represented in the form

$$|\Phi^0\rangle = a^+_{n_1}, \ldots, a^+_{n_N} |\Phi_{vac}\rangle \tag{27}$$

where $\Phi_{vac}$ is the vacuum state and $a^+_{n_i}$ are the creation operators for the electrons in one-electron states $\varphi_{n_i}$. These states are the eigenstates of the operator $h(\vec{x})$:

$$h(\vec{x})\, \varphi_n(\vec{x}) = \epsilon_n \varphi_n(\vec{x}) \tag{28}$$

where $\epsilon_n$ are the corresponding eigenvalues. The set of $n_1, \ldots, n_N$ (N is the number of electrons in atom or ion) determines the ground electron configuration of an atom. Zero order energy of an atom is

$$E^{(0)} = \sum_{n=1}^{N} \epsilon_n \tag{29}$$

The energy shift due to the interaction between the electrons is given by the Gell-Mann and Low (adiabatic) formula[7]

$$\Delta E = \lim_{\lambda \to 0} \langle \Phi_0 | S_\lambda | \Phi_0 \rangle^{-1} \frac{1}{2} i \lambda e \frac{\partial}{\partial e} \langle \Phi_0 | S_\lambda | \Phi_0 \rangle \tag{30}$$

$$S_\lambda = 1 + \sum_{n=1}^{\infty} S_\lambda^{(n)} \tag{31}$$

$$S_\lambda^{(n)} = (-i)^n e^n \int_{-\infty}^{\infty} H_{int}(t_1) e^{-\lambda |t_1|} dt_1, \dots, \int_{-\infty}^{t_{n-1}} H_{int}(t_n) e^{-\lambda |t_n|} dt_n \tag{32}$$

$$H_{int} = e^{iH_0 t} H_{int} e^{-iH_0 t} \tag{33}$$

where $\exp(-\lambda|t|)$ is the adiabatic factor, which switches on the interaction.

Below we shall restrict ourselves to the calculation of the corrections of the order up to $e^4$. Therefore we expand the formula (30) and write down first four terms:

$$\begin{aligned} \Delta E = \lim_{\lambda \to 0} \frac{1}{2} i \lambda \Big\{ & \langle \Phi_0 | S_\lambda^{(1)} | \Phi_0 \rangle + \Big[ 2 \langle \Phi_0 | S_\lambda^{(2)} | \Phi_0 \rangle - \\ & - \langle \Phi_0 | S_\lambda^{(1)} | \Phi_0 \rangle^2 \Big] + \Big[ 3 \langle \Phi_0 | S_\lambda^{(3)} | \Phi_0 \rangle - 3 \langle \Phi_0 | S_\lambda^{(2)} | \Phi_0 \rangle \cdot \\ & \cdot \langle \Phi_0 | S_\lambda^{(1)} | \Phi_0 \rangle + \langle \Phi_0 | S_\lambda^{(1)} | \Phi_0 \rangle^3 \Big] + \\ & + \Big[ 4 \langle \Phi_0 | S_\lambda^{(4)} | \Phi_0 \rangle - 4 \langle \Phi_0 | S_\lambda^{(3)} | \Phi_0 \rangle \langle \Phi_0 | S_\lambda^{(1)} | \Phi_0 \rangle + \\ & + 4 \langle \Phi_0 | S_\lambda^{(2)} | \Phi_0 \rangle \langle \Phi_0 | S_\lambda^{(1)} | \Phi_0 \rangle^2 - \\ & - 2 \langle \Phi_0 | S_\lambda^{(2)} | \Phi_0 \rangle^2 - \langle \Phi_0 | S_\lambda^{(1)} | \Phi_0 \rangle^4 \Big] \Big\} \end{aligned} \tag{34}$$

For the calculation of the energy corrections we need only the even terms in (34), because these corrections contain always even powers of e. So up to the fourth order of the perturbation theory, we have to calculate two matrix elements:

$$\langle \Phi_0 | S_\lambda^{(2)} | \Phi_0 \rangle \, , \, \langle \Phi_0 | S_\lambda^{(4)} | \Phi_0 \rangle$$

3. Now we may use the Wick theorem and Feynman's techniques in the Furry representation for the calculation of these matrix elements[9]. It is convenient also to use the Coulomb gauge. The relative importance of the different energy corrections for the two-electron ions in ground state is illustrated in the Figure 3. The charge of the nucleus Z is plotted on the horizontal axis. and the quantity $\lg |\Delta E_i / \Delta E_0|$ is plotted on the vertical axis, where $\Delta E_i$ are the various corrections

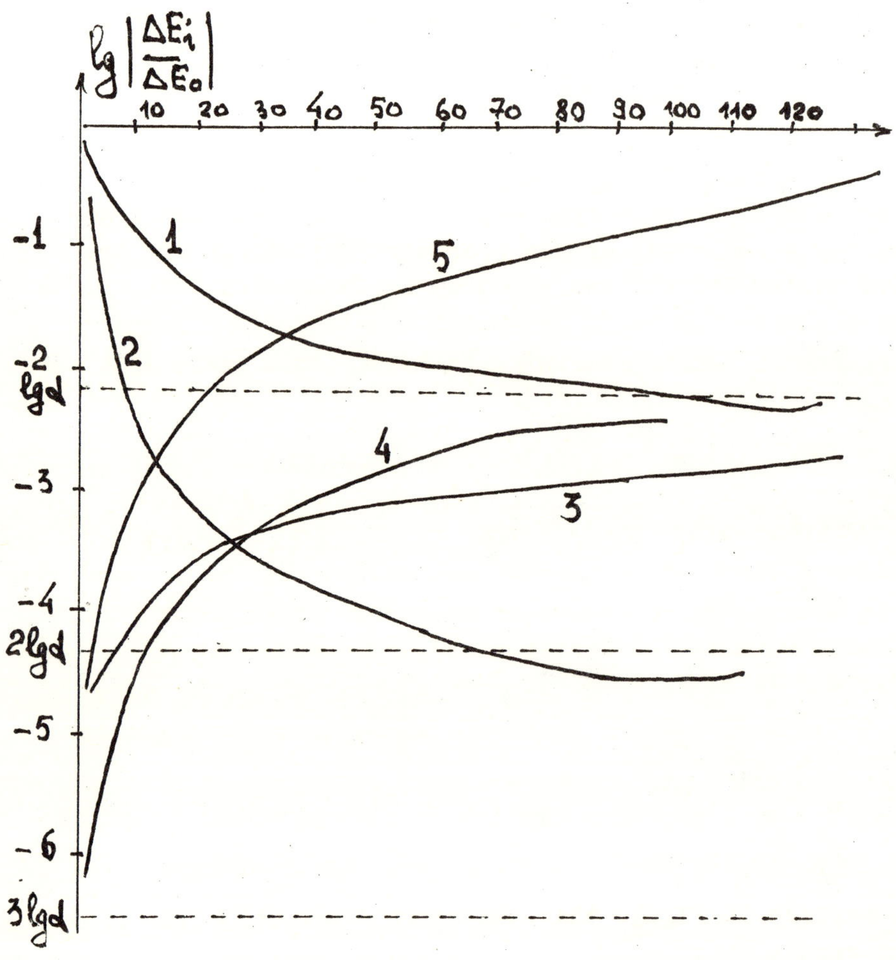
lg |ΔEi / ΔE0|
10 20 30 40 50 60 70 80 90 100 110 120
-1
-2
lgd
-3
-4
2lgd
-5
-6
3lgd
1
2
3
4
5

Fig. 3.

to the energy and $\Delta E_0$ is the binding energy of the two noninteracting electrons in atom. The curves 1, 2 correspond to the corrections, accounting for the first and second order Coulomb interactions between the electrons (i.e. the second and fourth order corrections in powers of e). The expressions for these corrections were derived in /9/:

$$\Delta E_1 = \alpha \left(\frac{1}{r_{12}}\right)_{AB;AB} \tag{35}$$

$$\Delta E_2 = \frac{1}{2}\alpha^2 \sum_{\substack{n_1 n_2 \\ (n_1 n_2 \neq AB)}} \frac{(\Lambda_1^{(+)}\Lambda_2^{(+)} - \Lambda_1^{(-)}\Lambda_2^{(-)})}{\epsilon_A + \epsilon_B - \epsilon_{n_1} - \epsilon_{n_2}} \left|\left(\frac{1}{r_{12}}\right)_{n_1 n_2; AB}\right|^2 \tag{36}$$

where

$$\left(\frac{1}{r_{12}}\right)_{AB;AB} = \left(\frac{1}{r_{12}}\right)_{ABAB} - \left(\frac{1}{r_{12}}\right)_{ABBA}, \tag{37}$$

$\Lambda_{1,2}^{(\pm)}$ are the projection operators on the positive and negative energy eigenstates of the operator $h(\vec{r})$ and A,B are the occupied states for the two-electron configuration. The curve 3 corresponds to the Breit interaction between the electrons (exchange with one transverse photon). This correction is described by the expression /9/

$$\Delta E_3 = -\alpha\left\{\left(\frac{\vec{\alpha}_1\vec{\alpha}_2}{r_{12}} + \frac{1}{2}(\nabla_1\vec{\alpha}_1)(\nabla_2\vec{\alpha}_2)r_{12}\right)_{ABAB} - \left(\frac{\vec{\alpha}_1\vec{\alpha}_2}{r_{12}}\cos(\epsilon_A-\epsilon_B)r_{12} - (\nabla_1\vec{\alpha}_1)(\nabla_2\vec{\alpha}_2)\frac{1}{r_{12}}\frac{\cos(\epsilon_A-\epsilon_B)r_{12}-1}{(\epsilon_A-\epsilon_B)^2}\right)_{ABBA}\right. \tag{38}$$

The curve 4 corresponds to the Lamb shift. All the corrections, but $\Delta E_2$ are of the second order in powers of e, and $\Delta E_2$ is of the fourth order. The curves 1, 2, 3 are taken from /10/, and the curve 4 is taken from the work of Mohr /11/. The curve 5, which corresponds to one-electron relativistic corrections, is given for comparison:

$$\Delta E_5 = \Delta E^0 - \Delta E^0_{nr} = 2\left[\sqrt{1-(\alpha Z)^2} - 1 + \frac{1}{2}(\alpha Z)^2\right], \tag{39}$$

where $\Delta E^0_{nr}$ is the nonrelativistic zero order energy.

We may see in Fig.3, that the corrections $\Delta E_2$ and $\Delta E_5$ become equal, when $Z = Z_1 \simeq 10$. It means, naturally that the relativistic corrections become superior to the electron correlation, when the charge Z increases. When $Z = Z_2 \simeq 30$, the correction $\Delta E_5$ becomes equal to $\Delta E_1$, i.e. relativistic corrections become equal to the Coulomb interaction between the electrons. These two important critical values of Z may be obtained, of course, by simple estimates: $(\alpha Z_1)^2 \sim \frac{1}{Z_1^2}$ and $(\alpha Z_2)^2 \sim \frac{1}{Z_2}$.

4. We consider now the calculation of the energy corrections for the excited degenerate states. The generalization of the Gell-Mann and Low formula for the degenerate states was performed by Dmitriev[12]. The generalized formula has rather complicated form and we shall confine ourselves in this work to the simple cases, when there is no need in such a generalization.

The one-electron states are described in the relativistic theory by the four quantum numbers njlm, where n is principal quantum number, jm determines the electron moment and its projection, and l determines the parity of the state. Formula (27) must be changed now to:

$$\Phi^{\circ}_{JMj_1l_1,\ldots,j_Nl_N} > = \sum_{m_1,\ldots,m_N} C^{JM}_{j_1l_1,\ldots,j_Nl_N}(m_1,\ldots,m_N)\, a^{+}_{n_1j_1l_1m_1},\ldots,a^{+}_{n_Nj_Nl_Nm_N}|\Phi_{vac}> \tag{40}$$

where $JM$ are the total moment and its projection and $C^{JM}_{j_1l_1,\ldots,j_Nl_N}$ are the vector coupling coefficients. In the case, when the symmetry conditions fully determine the coefficients in (40), we may use again expression (30) for the calculation of the energy corrections. In the lowest order of perturbation theory this expression is valid for the arbitrary configurations.

If we want to develop a theory which is valid also for small values of Z, we must use the intermediate coupling scheme and construct the new wave functions:

$$\Phi^{\circ}_{JM} > = \sum_{j_1l_1,\ldots j_Nl_N} a^{JM}_{j_1l_1,\ldots,j_Nl_N}\, \Phi^{\circ}_{JMj_1l_1,\ldots,j_Nl_N} \tag{41}$$

where $a^{JM}_{j_1l_1,\ldots,j_Nl_N}$ are the mixing coefficients.

The evaluation of these coefficients in the first order of perturbation theory can be performed simply by diagonalizing S-matrix.

This method was applied to the calculation of the energy levels for the configurations 1s2s + 1s2p, 1s2s + 1s3p + + 1s3d, $2s^2 + 2p^2 + 2p2s$ /13-15/. The results are shown in Figures 4, 5. The Coulomb and Breit interactions are taken into account in these calculations in the lowest order of perturbation theory. The curves in Fig.4 are improved also by nonrelativistic second order Coulomb corrections, so that the relative error for the intermediate values of Z ($10 \leqslant Z \leqslant 50$) is smaller than $10^{-3}$. The behaviour of the curves in Fig.4,5 demonstrates clearly, that LS-coupling breaks when Z is about 30, quite in agreement with previous discussion. For $Z \sim 50$ we have nearly pure jj-coupling.

5. The calculation of the intrinsic width of the energy levels of an atom can be performed also with the aid of the S-matrix theory. The only difference between these and our previous calculations is that now we are interested in the imaginary part of the total energy shift, described by formulae (30) or (34). Consider first the radiative one-photon width for the one-electron ion. This width is equal to the imaginary part of the self-energy of the electron in an atom. Unlike the real part, the imaginary part of the self-energy is finite, the evaluation gives /16/

$$\Gamma_R = -\frac{e^2}{4}\sum_n \left(1-\frac{\beta_{nA}}{|\beta_{nA}|}\right)\left(1+\frac{\epsilon_n}{|\epsilon_n|}\right)\left(\frac{1-\vec{\alpha}_1\vec{\alpha}_2}{r_{12}}\sin(|\beta_{nA}|r_{12})\right)_{AnnA} \tag{42}$$

where $\beta_{nA} \equiv \epsilon_n - \epsilon_A$. It can be seen from (42) that the negative energy states give no contribution to the $\Gamma_R$ neither do the positive energy states, lying higher than the state A. In the non-relativistic limit the expression (42) gives the usual formula for the radiative width.

For the two-electron ions we must take into account also the contribution of the Breit exchange correction:

$$\Gamma_{Br} = e^2\left(\frac{1-\vec{\alpha}_1\vec{\alpha}_2}{r_{12}}\sin(|\beta_{nA}|r_{12})\right)_{ABBA} \tag{43}$$

The total radiative width for the two-electron ion is equal now to

$$\Gamma_R = \Gamma_R(A) + \Gamma_R(B) + \Gamma_{Br} \tag{44}$$

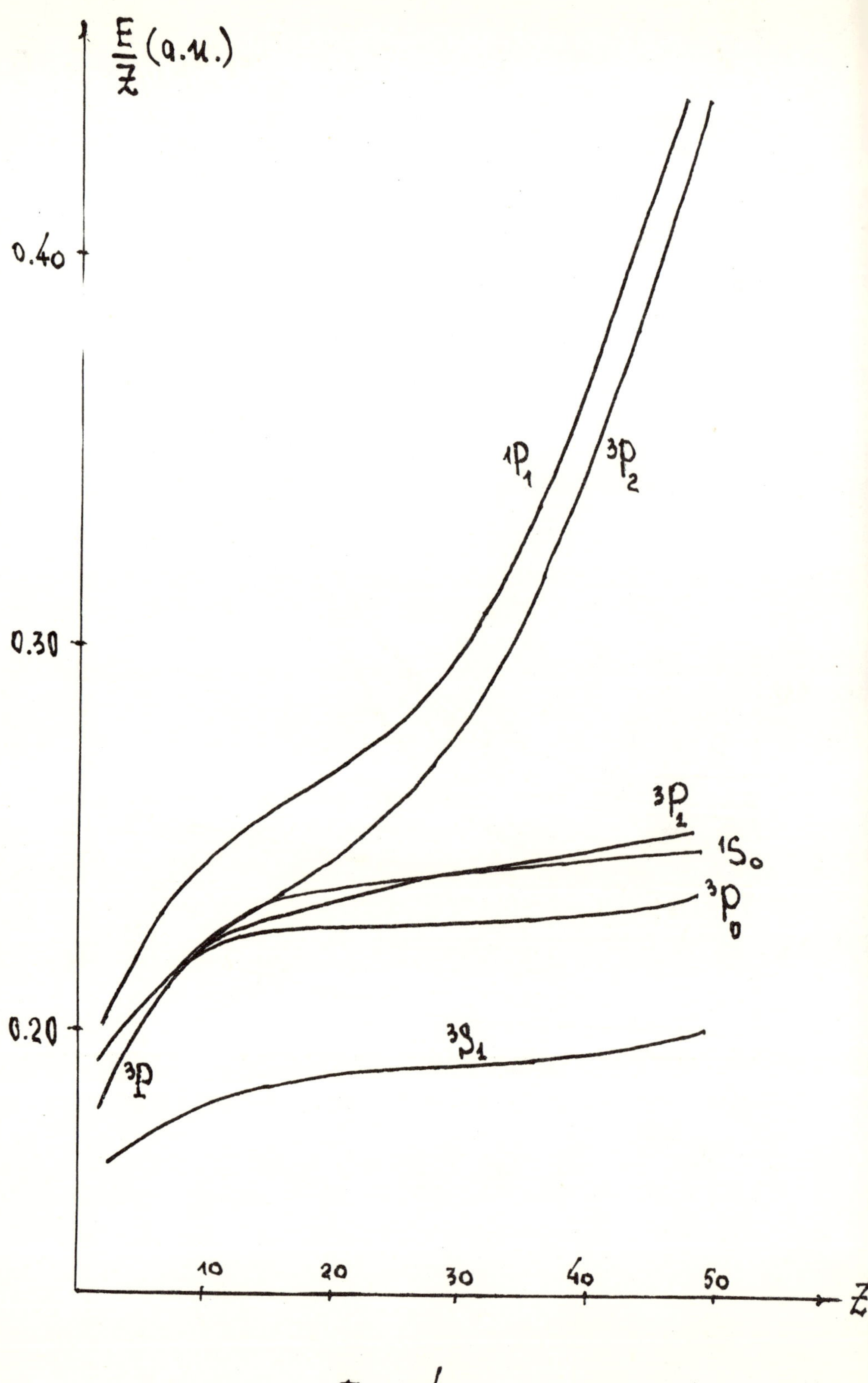


Fig. 4.

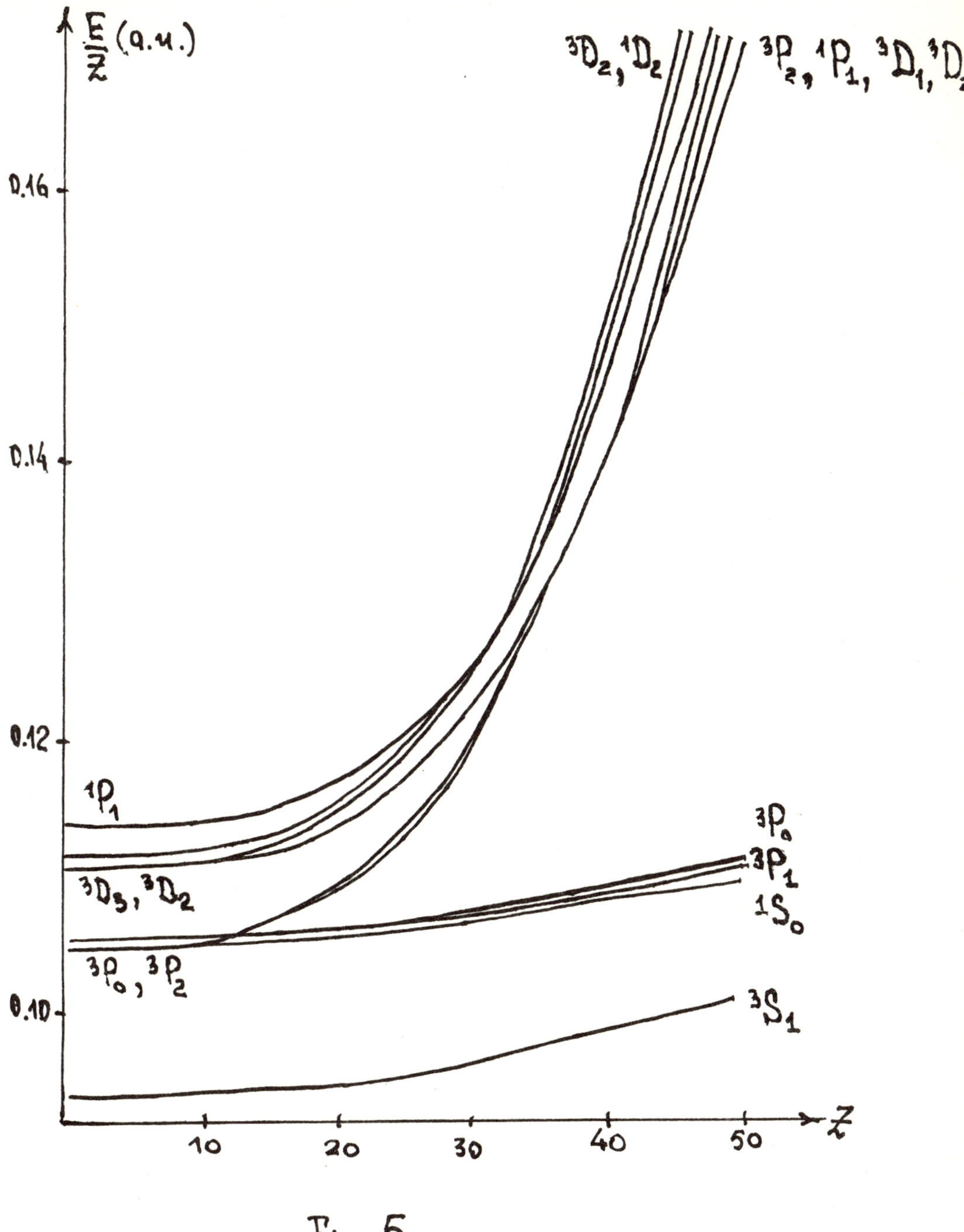
E/Z (a.u.)
0.16
0.14
0.12
0.10
10
20
30
40
50
Z
3D2, 1D2
3P2, 1P1, 3D1, 3D2
1P1
3D3, 3D2
3P0, 3P2
3P0
3P1
1S0
3S1

Fig. 5

If, for example, $\epsilon_A < \epsilon_B$, then the term $\Gamma_{Br}$ cancel the term n=A in $\Gamma_R$ (8) and so prevents the violation of the Pauli principle: there should be no transitions to the occupied states.

For the calculation of the Auger width of the autoionizing state we must turn to the expression of the second order Coulomb energy shift. The imaginary part of this shift gives us the Auger width [16]:

$$\Gamma_A = \pi e^4 \sum_{n_1 n_2} \left| \left( \frac{1}{r_{12}} \right)_{n_1 n_2; AB} \right|^2 \delta(\epsilon_{n_1} + \epsilon_{n_2} - \epsilon_A - \epsilon_B) \qquad (45)$$

Comparing the magnitudes of $\Gamma_A$ and $\Gamma_R$ we obtain for the small values of Z $\frac{\Gamma_R}{\Gamma_A} \sim \alpha^3 Z^4$ and for the high values $\frac{\Gamma_R}{\Gamma_A} \sim \frac{1}{\alpha}$.

The expression (42) for the width $\Gamma_R$ may be represented also as the sum of the probabilities of all the one-photon transitions to the states, lying lower than the state A. This allows us to calculate independently the particular transition probability just in the same way as in the nonrelativistic theory.

The calculations were performed for various transitions between the ground state and excited states mentioned above. Some results of these calculations, taken from [14,17] are shown in Fig.6,7. In Fig. 7 we show the total widths for excited states of the configuration 1s2s+1s2p[18]. The behaviour of different widths with the increase of Z is quite different. The width $\Gamma(^3P_0)$ increases only by a factor of $10^2$ when Z changes from 2 to 80, but the width $\Gamma(^1P_1)$ increases by a factor of $10^8$, and the width $\Gamma(^3S_1)$ increases by a factor of $10^{20}$.

7. In the last section of this report we discuss the application of the theory developed above to the problem of the parity nonconservation in the spectra of multicharged ions. This problem in connection with the multicharged ions was discussed in works of Gorshkov and one of the authors (L.N.L)[19,20] and also in succeeding works [14,15].

The parity nonconservation connected with weak interactions, leads to the appearance of the circular polarization of the light, emitted by the ion, or, when the ion is polariz-

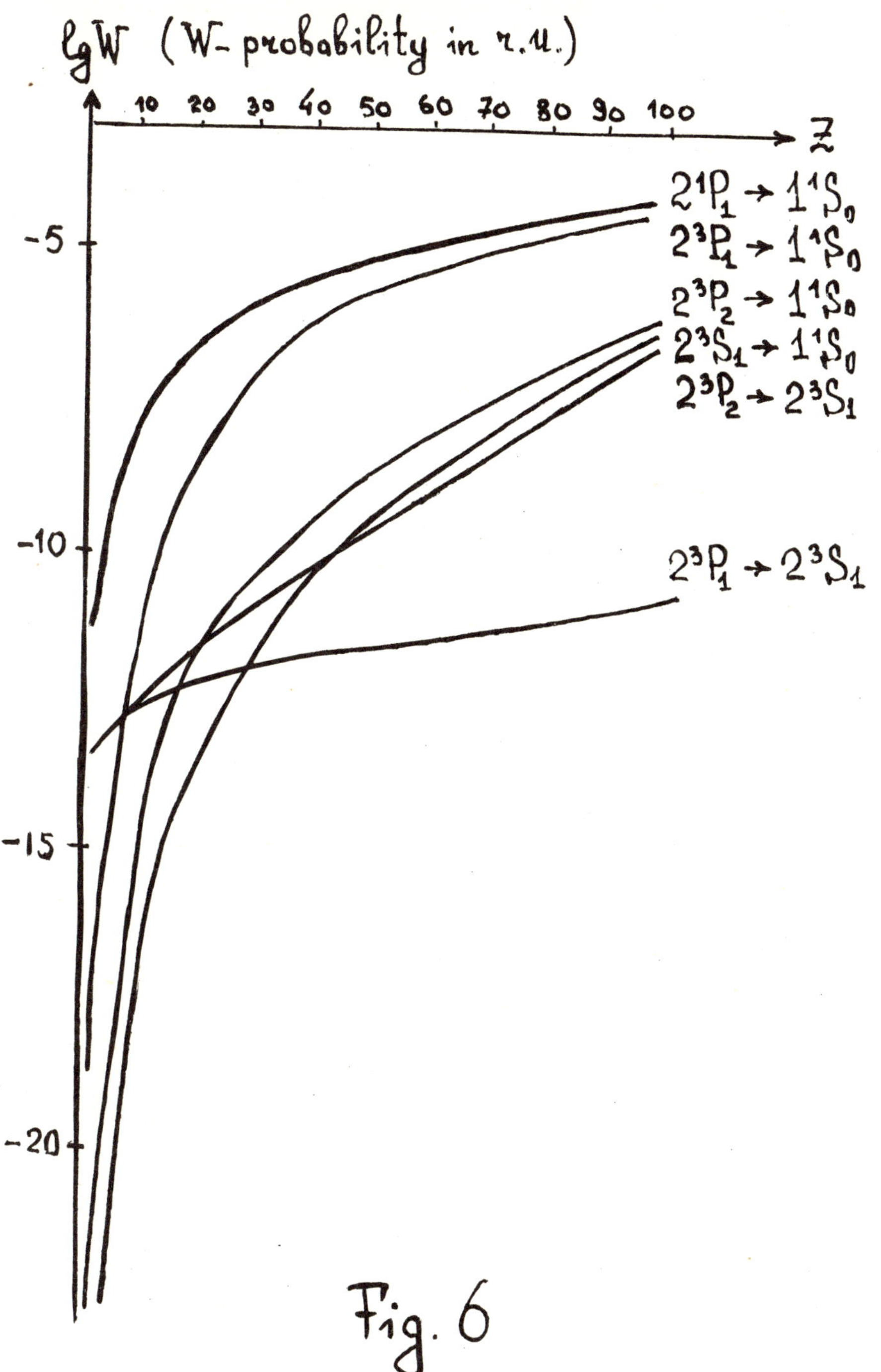
lgW (W- probability in r.u.)
10 20 30 40 50 60 70 80 90 100
Z
-5
-10
-15
-20
$2^1P_1 \to 1^1S_0$
$2^3P_1 \to 1^1S_0$
$2^3P_2 \to 1^1S_0$
$2^3S_1 \to 1^1S_0$
$2^3P_2 \to 2^3S_1$
$2^3P_1 \to 2^3S_1$

Fig. 6

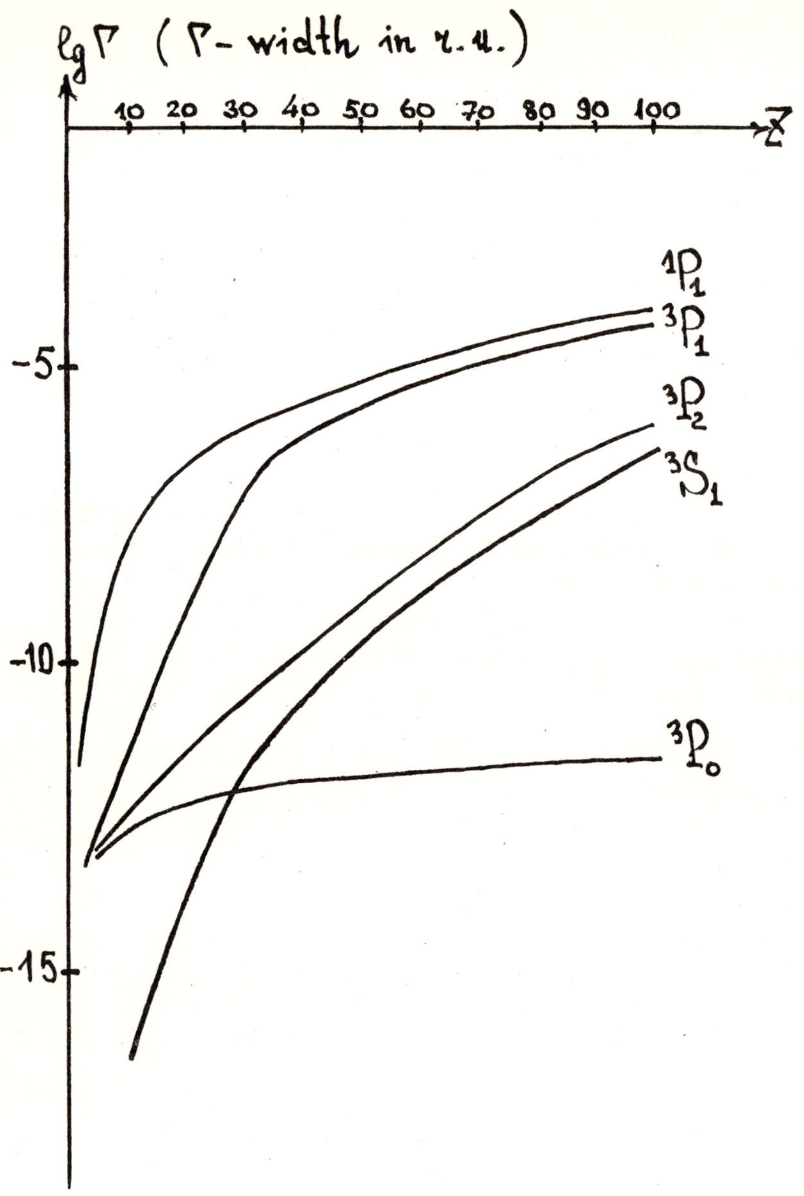
lg Γ ( Γ- width in r.u.)
10 20 30 40 50 60 70 80 90 100
Z
-5
-10
-15
$^1P_1$
$^3P_1$
$^3P_2$
$^3S_1$
$^3P_0$

Fig. 7

ed, to the asymmetry of the emitted radiation relatively to the direction of polarization. The degree of the parity violation may be determined as

$$P = \frac{2\langle V \rangle}{\Delta E}\sqrt{\frac{W_1}{W_0}} \qquad (46)$$

where $\langle V \rangle$ is the matrix element of the effective weak interaction operator, which mixes the states with the opposite parity, $\Delta E$ is the energy interval between the mixed states, $W_0$ is the probability of the main transition, and $W_1$ is the probability of transition allowed by the weak interaction. It is evident from (46) that it is important to find the situations when $\Delta E$ is small, i.e. when two energy levels with opposite parities are close.

Such situations arise every time, when the curves with opposite parities intersect as in Figures 4,5. The most convenient intersections are the intersections $2^1S_0 \times 2^3P_1$ by Z=6 and Z=29 in Figure 4. In these two cases the degree of the parity nonconservation is rather high: $P \simeq 10^{-3}-10^{-4}$.

The corresponding ions CV and CuXXVIII are obtainable with the use of the beam-foil technique /21,22/

## R e f e r e n c e s


1. M.A.Braun, T.N.Sibirkina, JETP, bb, 2065, 1974.
2. M.A.Braun, A.N.Vassiliev, A.V.Kitanin, Yu.S.Pis'mak,TMP, 34, 163, 1978
3. M.A.Braun, V.I.Fomichev, A.V.Shirokov, Vestnik of Leningrad Univ., No.22, 7, 1976.
4. V.V.Tolmachev "Theory of Fermi-Gas",Moscow Univ., 1973
5. M.A.Braun, TMP, 34, 59, 1978.
6. M.A.Braun, A.V.Shirokov, Izvestia of Acad. Sci. of USSR, 41, 2585, 1977.
7. M.Gell-Mann, F.Low, Phys.Rev.,84, 350, 1951
8. M.A.Braun, L.N.Labsovsky, JETP, 53, 1776, 1967
9. L.N.Labzowsky Zh.Exp. i Teor.Fiz 59, 168 (1970)
10. G.L.Klimchitskaya, L.N.Labzowsky. Zh.Exp. i Teor.Fiz 60, 2019 (1971)
11. P.J.Mohr. Ann of Phys. 88, 26 (1974)
12. Dmitriev Y. Int.J.Quant.Chem. 9, 1033 (1975)
13. G.L.Klimchitskaya, L.N.Labzowsky. Optika i Spektr. 34, 633 (1973)
14. V.G.Gorshkov, Klimchitskaya G.L., L.N.Labzowsky, M.Melibaev. Zh.Esp. i Teor. Fiz 72, 1268 (1977)
15. V.G.Gorshkov, G.L.Klimchitskaya, Labzowsky L.N., M.Melibaev. Izv. Akad. Nauk SSSR ser fiz 41, 2502 (1977)
16. Y.Y.Dmitriev, L.N.Labzowsky "The Theory of Atoms and Atomic Spectra" The collection of papers, Riga 1972,p.89--93.
17. G.L.Klimchitskaya, Safronova U.I., L.N.Labzowsky. Optika i Spectr. 38, 838 (1975).
18. U.I.Safronova, G.L.Klimchitskaya, L.N.Labzowsky. J.Phys. B 7, 2471 (1974)
19. V.G.Gorshkov, L.N.Labzowsky. Zh.Esp i Teor.Fiz Pis'ma 19, 768 (1974).
20. V.G.Gorshkov, L.N.Labzowsky. Zh.Eksp i Teor. Fiz. 69, 1141 (1975)
21. Marrus R. Nuch.Instr. and Meth. 110, 333 (1973)
22. Bromaneler J. Nucl.Instr. and Meth 110, 11 (1973)

# IONIZATION AND EXCITATION OF ATOMIC SHELLS BY ELECTRON IMPACT IN SEMICLASSICAL APPROXIMATION

M.A. Braun, V.I. Ochkur

Leningrad State University

## § 1. Introduction

In this paper we present some results concerning the calculation of the cross-sections of inelastic electron - atom collisions in the semiclassical approximation. We start with the formulas of the first order perturbation theory of quantum mechanics, but our aim will be not to calculate the corresponding matrix elements as exactly as possible, but rather to trace the transition from quantum to purely classical description of the collision process. So we use semiclassical wave functions for atomic electrons, stationary phase method for evaluation of integrals and so on.

This approach is interesting from several points of view. First we will get formulas which permit calculations of the cross-sections for transitions between highly excited levels. Such calculations cannot be performed by the straightforward use of the conventional Born-approximation formulas because a great number of terms arise, and strong cancellation between them takes place [1].

Further, our approach permits us to get a better insight into the nature of purely classical calculations of the ionization cross-sections and first of all of those made in the so-called binary encounter approximation which has become popular since the early 60th [2-4]. At the same time we obtain simple formulas for the excitation cross-sections, for which classical binary results were up to now scarce and rather artificial because the classical binary approximation includes no satisfactory concept of initial and final states of the atomic electron.

The semiclassical approach developed below includes not only the interaction between the colliding electron and the atomic one but also the interaction of the colliding electron and the atom as a whole. So it permits to discuss such effects as focusing of the incident beam by atomic field and the acceleration effects, which are sometimes discussed in attempts to improve the binary approximation. And what is probably more important, the allowance for this interaction permits to use our formulas not only to calculate the probability of ionization and excitation for neutral atoms, but for ions as well.

Another good quality of the semiclassical approach is that it gives very simple formulas for the cross-sections, much simpler than those of the Born-approximation, not to speak of the Born-Coulomb approximation. This not only makes practical calculations easier, but also permits to see some features of the Born-type calculations, which are known only from numerical tables and in some cases have not been pointed out as yet.

## § 2. Ionization cross-section

For the initial approximation in the quantum mechanical description of the ionization of an atom by electron impact it is reasonable to choose the approximation of independent particles moving in the self-consistent field. The ionization occurs because of the interaction $V$ of the colliding electron with the atomic ones, which is regarded as perturbation. Let the momentum of the incoming electron be $\vec{p}_1$. After the collision we will define its state by energy $\varepsilon_1'$, orbital momentum $l_1'$ and its projection on $z$ -axis $m_1'$. The corresponding quantum numbers for the atomic electron before and after collision we denote as $\varepsilon_2, l_2, m_2$ and $\varepsilon_2', l_2', m_2'$. Then assuming that atomic shells are completely filled up, we can find for the differential cross-section of ionization from the shell with energy $\varepsilon_2$ and angular momentum $l_2$:

$$d\sigma = 2\cdot 2\pi \sum_{m_2} |\langle \varepsilon_1' l_1' m_1' \varepsilon_2' l_2' m_2' | V | \vec{p}_1 \varepsilon_2 l_2 m_2 \rangle|^2 \times \delta(\varepsilon_1 + \varepsilon_2 - \varepsilon_1' - \varepsilon_2')\, d\varepsilon_1' d\varepsilon_2' \qquad (1)$$

Here it is assumed, that the wave function of the colliding electron is normalized to the unity flux and $\hbar$ is put equal to 1. Factor 2 accounts for the spin degeneration. For ionization the requirement $\varepsilon_2' > 0$ must be fulfilled. The exchange will be ignored now, but can be considered separately. Formula (1) can be rewritten in more simmetric form using the following expansion:

$$|\vec{p}_1\rangle = \frac{\pi\sqrt{2}}{p_1}\sum_{\ell=0}^{\infty} i^{\ell_1} e^{i\delta_{\ell_1}} \sqrt{2\ell_1+1}\,|\varepsilon_1 \ell_1 0\rangle \qquad (2)$$

where $\varepsilon_1 = p_1^2/2m$ and phases $\delta_{\ell_1}$ correspond to the scattering of the colliding electron on the selfconsistent field of the atom. Substitution of (2) into (1) gives:

$$d\sigma = \frac{4\pi^3}{m\varepsilon_1}\sum_{\ell_1 m_2}(2\ell_1+1)\,|\langle\varepsilon_1'\ell_1'm_1'\,\varepsilon_2'\ell_2'm_2'|V|\varepsilon_1\ell_1 0\,\varepsilon_2\ell_2 m_2\rangle|^2 \qquad (3)$$
$$\times\,\delta(\varepsilon_1+\varepsilon_2-\varepsilon_1'-\varepsilon_2')\,d\varepsilon_1'\,d\varepsilon_2'$$

Let us represent potential $V$ in the form

$$V(|\vec{r}_1-\vec{r}_2|) = \sum_{\ell=0}^{\infty}(2\ell+1)\,V_\ell(r_1,r_2)\,P_\ell(\cos\theta_{12}) \qquad (4)$$

where $\theta_{12}$ is the angle between $\vec{r}_1$ and $\vec{r}_2$. For the Coulomb interaction

$$V_\ell(r_1,r_2) = \frac{e^2}{2\ell+1}\cdot\frac{r_<^{2\ell}}{r_>^{2\ell+1}} \qquad (5)$$

where $r_<$ ($r_>$) is the smallest (the largest) of the two quantities $r_1$ and $r_2$. Substituting (4) into (3), performing the angular integrations and the summations over magnetic quantum numbers we arrive at the following expression

$$d\sigma = \frac{4\pi^3}{m\varepsilon_1}\sum_{\ell,\ell_1}(2\ell_1+1)(2\ell_2+1)(2\ell_1'+1)(2\ell_2'+1)(2\ell+1)^{-3}\times$$
$$|C^{\ell}_{\ell_1\ell_1'}|^2\cdot|C^{\ell}_{\ell_2\ell_2'}|^2\,|\langle\varepsilon_1'\ell_1'\varepsilon_2'\ell_2'|V_\ell|\varepsilon_1\ell_1\varepsilon_2\ell_2\rangle|^2\times \qquad (6)$$
$$\times\,\delta(\varepsilon_1+\varepsilon_2-\varepsilon_1'-\varepsilon_2')\,d\varepsilon_1'\,d\varepsilon_2'$$

Here $\langle\varepsilon_1'\,\ell_1'\,\varepsilon_2'\,\ell_2'\,|\,V_\ell\,|\,\varepsilon_1\,\ell_1\,\varepsilon_2\,\ell_2\rangle$ is the radial part of the matrix element of $V_\ell$ defined by (5); $C^{\ell}_{\ell_1\ell_1'}$ stands for the Clebsh-Gordan coefficient with all magnetic quantum numbers equal to zero.

## § 3. Radial matrix element

As it is well known, in the region of classically allowed motion the semiclassical wave function $\rho_{\varepsilon l} = r \cdot R_{\varepsilon l}(r)$ can be represented in the form

$$\rho_{\varepsilon l}(r) = \frac{C}{\sqrt{p(r)}} \cdot \sin(\sigma(r) + \pi/4) \tag{7}$$

with

$$\sigma(r) = \int_{r_1}^{r_2} p(r)\, dr \tag{8}$$

$$p(r) = [2m(\varepsilon - U(r) - (l+\tfrac{1}{2})^2/2mr^2)]^{1/2}. \tag{9}$$

The turning points are defined by $p(r)=0$, $U(r)$ being the potential in which the particle moves.

The normalization factor $C$ for a discrete level is

$$C^2 = \frac{2m}{\tau}; \qquad \tau = m\int_{r_1}^{r_2} \frac{dr}{p(r)} \tag{10}$$

and for continuous spectrum

$$C^2 = 2m/\pi \tag{11}$$

Substituting these functions in our matrix element $\langle \varepsilon_1' l_1' \varepsilon_2' l_2' | v_l | \varepsilon_1 l_1 \varepsilon_2 l_2 \rangle$ and performing the integration over both variables $r_1$ and $r_2$, according to the standard technique of the stationary phase method, we obtain the following result:

$$|\langle \varepsilon_1' l_1' \varepsilon_2' l_2' | v_l | \varepsilon_1 l_1 \varepsilon_2 l_2 \rangle|^2 = \frac{m^2}{4\pi \tau_2} \cdot \frac{r_1 r_2}{\varepsilon^2 p_1 p_2}\, v_l^2(r_1, r_2). \tag{12}$$

Here $\varepsilon = \varepsilon_1 - \varepsilon_2$ is the loss of energy by the incoming electron. Now $r_1$ and $r_2$ are the stationary points defined by

$$r_i^2 = [(l_i'+\tfrac{1}{2})^2 - (l_i+\tfrac{1}{2})^2]/2m(\varepsilon_i' - \varepsilon_i). \tag{13}$$

It is clear that the stationary point exists only if the condition

$$(l_i' - l_i)(\varepsilon_i' - \varepsilon_i) > 0 \tag{14}$$

is met. So we can see that in the semiclassical approximation the increase of energy is always accompanied by the increase of angular momentum .

Finally $p_i \ (i=1,2)$ are the momenta in the stationary points:

$$p_i^2 = 2m\left(\varepsilon_i - U(r_i) - (\ell_i+\tfrac{1}{2})^2/2mr_i^2\right) = 2m\left(\varepsilon_i' - U(r_i) - (\ell_i'+\tfrac{1}{2})^2/2mr_i^2\right) \tag{15}$$

§ 4. <u>Semiclassical ionization cross-section</u>

Substituting (12) and (5) into (6) we get:

$$d\sigma = \frac{\pi^2 m e^4}{\varepsilon_1\, \varepsilon^2} \sum_{\ell \ell_1} \frac{(2\ell_1+1)(2\ell_2+1)(2\ell_1'+1)(2\ell_2'+1)}{\tau_2\,(2\ell+1)^3} \cdot \left|C^{\ell}_{\ell_1 \ell_1'}\right|^2 \times \tag{16}$$

$$\left|C^{\ell}_{\ell_2 \ell_2'}\right|^2 \frac{r_1 \cdot r_2}{p_1(r)\, p_2(r)} \; \frac{r_<^{2\ell}}{r_>^{2\ell+2}} \cdot \delta(\varepsilon_1 + \varepsilon_2 - \varepsilon_1' - \varepsilon_2')\, d\varepsilon_1' d\varepsilon_2'$$

To obtain the cross-section, which can be compared with experiment, we have to perform additional summations over angular momenta $\ell_1'$ and $\ell_2'$ . Such a cross-section will depend on the energy loss $\varepsilon$ only.

In the semiclassical approximation we approximate all summations by integrations and get:

$$\frac{d\sigma}{d\varepsilon} = \frac{\pi m e^4}{16\, \varepsilon_1\, \varepsilon^2} \int ds_1\, ds_1'\, ds_2'\, ds \left|C^{\ell}_{\ell_1 \ell_1'}\right|^2 \cdot \left|C^{\ell}_{\ell_2 \ell_2'}\right|^2 \times \tag{17}$$

$$r_1 \cdot r_2 / s^2 \cdot p_1 \cdot p_2 \times \; r_<^{2\ell} / r_>^{2\ell+2}$$

where $s_1 = (\ell_1 + 1/2)^2$ and so on.

Now we have to introduce explicit expressions for the Clebsh-Gordan coefficients $C^{\ell}_{\ell_1 \ell_1'}$ and $C^{\ell}_{\ell_2 \ell_2'}$. The asymptotic formula for $\left|C^{\ell}_{\ell_1 \ell_1'}\right|^2$ can be found without difficulty:

$$\left|C^{\ell}_{\ell_1 \ell_1'}\right|^2 = \frac{2\ell}{\pi \cdot \Delta(\ell \ell_1 \ell_1')}\, ; \quad \ell, \ell_1, \ell_1' \gg 1 \tag{18}$$

$$\Delta(\ell \ell_1 \ell_1') = 2\ell^2\ell_1^2 + 2\ell^2\ell_1'^2 + 2\ell_1^2\ell_1'^2 - \ell^4 - \ell_1^4 - \ell_1'^4, \tag{19}$$

and we obtain:

$$\frac{d\sigma_{\varepsilon \ell_2}}{d\varepsilon} = \frac{m e^4 (2\ell_2+1)}{\varepsilon_1\, \varepsilon^2\, \tau_2} \int \frac{r_1 \cdot r_2}{p_1 \cdot p_2 \cdot s} \cdot \frac{r_<^{2\ell}}{r_>^{2\ell+2}} \, \frac{ds_1\, ds_1'\, ds_2'\, ds}{\Delta(\ell, \ell_1, \ell_1')\,\Delta(\ell, \ell_2, \ell_2')} \tag{20}$$

Let us now introduce new variables $r_1^2=(s_1-s_2)/2m\varepsilon$ and $\sigma_1 = s_1+s_1'$. We see that $ds_1\,ds_1' = m\varepsilon\, d\sigma_1\, dr_1^2$ and after the integration over $\sigma_1$ which is elementary, we have

$$\frac{d\sigma}{d\varepsilon}=\frac{\pi m^3 e^4(2\ell_2+1)}{\varepsilon_1\cdot\tau_2}\int\frac{dr_1^2\,dr_2^2\,ds\,r_1 r_2\, r_<^{2\ell}}{s^{3/2}p_2\,\Delta(\ell\ell_2\ell_2')\,r_>^{2\ell+2}} \tag{2I}$$

$$\theta\left(4mr_1^2(\varepsilon_1/2+\varepsilon_1'/2-U(r_1))-s/2-2m^2\varepsilon^2 r_1^4/s\right),$$

where $\theta(x)$ is the step-function.

In the semiclassical approach it is natural to think that the main contribution to the cross-section comes from large $\ell$ corresponding to macroscopic changes of angular momentum. It is also clear that the main contribution to (2I) comes from the region $r_1\sim r_2$ where the function $(r_</r_>)^{2\ell}$ approaches 1. Let us consider this point in detail. The integral over $r_1$ and $r_2$ has the form

$$\int_0^\infty dr_1^2\int_0^{r_1^2} dr_2^2\,(r_2^2/r_1^2)^{\ell+1/2}\cdot f(r_1,r_2)+(1\rightleftarrows 2). \tag{22}$$

If we introduce new variables $x$ and $z$ according to definitions $r_1^2=x;\ r_2^2=z\cdot x$, then the first term in (22) can be rewritten as

$$\int_0^\infty x\,dx\int_0^1 dz\cdot z^{\ell+\frac{1}{2}}\cdot f(x,zx). \tag{23}$$

Integration by parts over $z$ shows that

$$\int_0^1 dz\cdot z^{\ell+\frac{1}{2}}\cdot f(x,zx)=\frac{1}{\ell+3/2}\cdot f(x,x)+O\left(\frac{1}{\ell}\right), \tag{24}$$

and all the expression (22) is equal to

$$\frac{2}{\ell}\int_0^\infty f(r,r)\,r^2\,dr^2. \tag{25}$$

Farther one can prove that the step-function $\theta(x)$ in (2I) actually does not work, because in the region where $p_2^2(r)>0$ its argument is always positive. So we secure

$$\frac{d\sigma}{d\varepsilon}=\frac{\pi m^3 e^4(2\ell_2+1)}{\varepsilon_1\quad\tau_2}\int\frac{r^3\,dr^2\,ds}{s^2 p_2\,\Delta(\ell\ell_2\ell_2')}, \tag{26}$$

where in our variables

$$\Delta^2 = -s^2 + 4s(s_2 + m\varepsilon r^2) - 4m^2\varepsilon^2 r^4. \tag{27}$$

The integration over $s$ gives

$$\int \frac{ds}{s^2 \Delta(\ell \ell_2 \ell_2')} = \frac{\pi}{2} \cdot \frac{s_2 + m\varepsilon r^2}{2m^3\varepsilon^3 r^6} , \tag{28}$$

and we obtain finally

$$\frac{d\sigma_{\varepsilon_2 \ell_2}(\varepsilon)}{d\varepsilon} = \frac{\pi^2 m e^4 (2\ell_2+1)}{2\,\varepsilon_1\, \varepsilon^2 \tau_2} \int \frac{dr}{P_2(r)} \left(1 + \frac{s_2}{m\varepsilon r^2}\right). \tag{29}$$

Having in mind the definition of $\tau_2$ and the technique of the evaluation of integrals with semiclassical wave functions we can rewrite (29) also in the form

$$\frac{d\sigma_{\varepsilon_2 \ell_2}(\varepsilon)}{d\varepsilon} = \frac{\pi^2 e^4 (2\ell_2+1)}{2\,\varepsilon_1\, \varepsilon^2} \left(1 + \frac{s_2}{m\,\varepsilon} \left\langle \frac{1}{r^2} \right\rangle\right), \tag{30}$$

where $\langle 1/r^2 \rangle$ is, sure, the semiclassical mean value. Let us remind that this result relates to the case of a completely filled shell, containing $2(2\ell_2+1)$ electrons.

It is pertinent to compare this expression with the differential cross-section for the ionization of the shell with $N = 2\cdot(2\ell_2+1)$ electrons resulting from the classical binary theory [2]:

$$\frac{d\sigma}{d\varepsilon} = \frac{\pi e^4 N}{\varepsilon_1\, \varepsilon^2} \left[1 + \frac{4}{3} \frac{\varepsilon_2}{|\varepsilon|}\right]. \tag{31}$$

We see that both formulas differ only by the factor and instead of 4/3 x kinetic energy, the doubled centrifugal part of the kinetic energy appears. If we consider a specific case of atomic hydrogen we can see that the similarity goes farther. In the classical formula (31) $\varepsilon_2$ now becomes equal to the ionization potential I, and in the quantum formula (30) we know $\langle 1/r^2 \rangle$ for any state:

$$\left\langle \frac{1}{r^2} \right\rangle = \frac{m e^2 Z}{n_2^3 (\ell + \frac{1}{2})} . \tag{32}$$

Using the result we can find that

$$\frac{d\sigma_{n_2}(\varepsilon)}{d\varepsilon} = \frac{1}{2(2l_2+1)} \sum_{l_2} \frac{d\sigma_{n_2 l_2}}{d\varepsilon} = \frac{\pi}{4} \cdot \frac{\pi e^4}{d\varepsilon^2}\left[1+\frac{4}{3}\frac{I}{\varepsilon}\right] \quad (33)$$

and the only difference with (31) remains in the factor $\pi/4$.

In the general case the difference is more serious. It is well known that in most applications of the classical formulas the ionization potential **I** was substituted for $\varepsilon_2$ and for many electron atoms this value is often considerably smaller than the mean value $\langle \varepsilon_2 \rangle$ and than our result (30). More details on the influence of this parameter on the cross-section can be found in [4]. Here we only note that using ( 31 ) the best fit to experimental data can be usually obtained when $\varepsilon_2 = I$. The increase of this parameter leads to a more or less marked disagreement with experiment.

§ 5. Semiclassical ionization cross-section for small $l_2$

The derivation of formula (30) was performed under the assumption that all quantum numbers and their changes are large. This was necessary to obtain the quantum cross-section, which would be as close to the result of purely classical considerations as possible. But available experimental data usually relate to s-, p-, or d-electrons and the condition $l_2 \gg 1$ is never fulfilled. So having in mind to compare the results of the semiclassical calculations with experiment, we first have to rederive (30) using a new approximation for $C^{l}_{l_2 l_2'}$ suitable for the case $l_2 = 0,1,2$, and assuming now that only $l$ is large. This can be done and gives

$$\frac{d\sigma_{\varepsilon_2,0}}{d\varepsilon} = \frac{\pi^2 e^4}{2\varepsilon_1 \varepsilon^2}\left(1+\frac{7}{8m\varepsilon}\left\langle\frac{1}{r^2}\right\rangle\right) \quad (34)$$

$$\frac{d\sigma_{\varepsilon_2,1}}{d\varepsilon} = \frac{\pi^2 e^4}{2\varepsilon_1 \varepsilon^2}\left(1+\frac{23}{8m\varepsilon}\left\langle\frac{1}{r^2}\right\rangle\right)$$

$$\frac{d\sigma_{\varepsilon_2,2}}{d\varepsilon} = \frac{\pi^2 e^4}{2\varepsilon_1 \varepsilon^2}\left(1+\frac{55}{8m\varepsilon}\left\langle\frac{1}{r^2}\right\rangle\right).$$

If we compare these expressions with (30) we find them identical with the only exception that now coefficients 7, 23 and 55 appear instead of 2, 18 and 50, that would stand in (30). So the asymptotic formula makes the contribution of the "centrifugal" term somewhat lower for small $l_2$. Besides we see that already for $l_2 = 2$ its accuracy becomes quite satisfactory.

It should be stressed once again that the derivation of (34) implies that $\ell_2/\ell \ll 1$. If this ratio is not small, but $\ell_2 \gg 1$, we are again in the range of validity of (30).

Let us apply (34) to the calculation of ionization cross-sections for various shells of various atoms. Some examples are given in Table 1. For $\sigma^{theor}_{max}$ Hartree Fock values of $\langle 1/r^2 \rangle$ have been used.

Table 1

| | Maximum of the ionization cross-section in $Å^2$ | | | | | |
|---|---|---|---|---|---|---|
| | Na2s | Na2p | Ne2s | Ne2p | Ag 5s | Ag 4d |
| $\sigma^{theor}_{max}$ | 6.7 | 4.1 | 0.23 | 11.1 | 6.0 | 313 |
| $\sigma^{exp}_{max}$ | 7 | | 0.8 | | 4 | |

In the last line we show $\sigma^{exp}_{max}$ - the maximum value of the experimental cross-section for the atom as a whole.
The comparison with experiment clearly shows that the calculated cross-sections for shells with $\ell_2 \neq 0$ are much too high and for such shells as 3d in Zn or 4d in Ag the discrepancy becomes drastic.

The reason for this discrepancy seems to lie in the incorrect account of the contribution from small angular momentum transfers and we have to reconsider the calculations once again.

For this purpose let us return back to expression (16), which was the basis of all what followed. The summation over $\ell$ in this formula begins with $\ell = 0$. But in our approximation this first term should be omitted because, according to (13), the stationary point for $\ell = 0$ is $r_0 = 0$ and the term itself is zero. But the condition $\ell \geq 0$ was not accounted for in the subsequent treatment, and the limits for the integration over $s = (\ell + \frac{1}{2})^2$ were imposed only by the requirement that the Clebsh-Gordan coefficients $C^{\ell}_{\ell_2 \ell_2'}$ be not equal to zero; in other words, by the inequalities $|\ell_2 - \ell_2'| \leq \ell \leq |\ell_2 + \ell_2'|$, or in our "continuous" variables

$$\sqrt{s_2 + 2m\varepsilon r^2} - \sqrt{s_2} \leq \sqrt{s} \leq \sqrt{s_2 + 2m\varepsilon r^2} + \sqrt{s_2} \qquad (35)$$

If $\varepsilon r^2$ is small enough, then the lower limit in (35) can take on values less than $1$, and such values should be excluded. This elimination of the monopole term considerably damps just the contribution from small $r$, and, hence, will mostly affect the "centrifugal" term in (34) proportional to $\langle 1/r \rangle$ making the theoretical values closer to the experimental ones.

The elimination of the monopole term can be easily achieved by the introduction of the step-function $\theta(s-1)$ into the integrand of (28). So we have

$$\int \frac{ds\,\theta(s-1)}{s^2 \Delta(\ell, \ell_2 \ell_2')} = \frac{\pi (s_2 + m\varepsilon r^2)}{4 m^3 \varepsilon^3 r^6} \Psi(r, s_2, \varepsilon), \qquad (36)$$

and for the differential cross-section

$$\frac{d\sigma_{\varepsilon_2 \ell_2}}{d\varepsilon} = \frac{\pi^2 m e^4 (2\ell_2+1)}{2\varepsilon_1 \varepsilon^2 \tau_2} \int \frac{dr}{p_2(r)} \left(1 + \frac{s_2}{m\varepsilon r^2}\right) \Psi_1(r, s_2, \varepsilon). \qquad (37)$$

The total cross-section is obtained by integration over $\varepsilon$ from the ionization potential up to $\varepsilon_1$ :

$$\sigma_{\varepsilon_2 \ell_2} = \frac{\pi^2 m e^4 (2\ell_2+1)}{2\varepsilon_1 \tau_2} \int_I^{\varepsilon_1} \frac{d\varepsilon}{\varepsilon^2} \int \frac{dr}{p_2(r)} \left(1 + \frac{s_2}{m\varepsilon r^2}\right) \Psi_1(r, s_2, \varepsilon). \qquad (38)$$

Introducing the semiclassical mean value according to

$$\langle A(r) \rangle = \frac{m}{\tau_2} \int \frac{dr}{p(r)} A(r), \qquad (39)$$

we can rewrite (38) in the form

$$\sigma_{\varepsilon_2 \ell_2} = \frac{\pi}{2\varepsilon_1} (2\ell_2+1) \langle F_1(r) \rangle (\pi a_0^2), \qquad (40)$$

where

$$F_1(r) = \int_I^{\varepsilon_1} \frac{d\varepsilon}{\varepsilon^2} \left(1 + \frac{s_2}{m\varepsilon r^2}\right) \Psi_1(r, s_2, \varepsilon). \qquad (41)$$

Calculation of $\Psi_1$ can be done without difficulty performing the straightforward integration in (36).

Having in mind further improvement of the theory we calculate a more general integral

$$\int \frac{ds\,\theta(s-n)}{s^2 \Delta(\ell\, \ell_2\, \ell_2')} = \frac{\pi (s_2 + m\varepsilon r^2)}{4 m^3 \varepsilon^3 r^6} \cdot \Psi_n \qquad (42)$$

Using (13) we find that if

$$\sqrt{s_2 + 2m\varepsilon r^2} - \sqrt{s_2} > n, \tag{43}$$

then $\Psi_n = 1$. So it is nontrivial only in the region

$$\sqrt{s_2 + 2m\varepsilon r^2} - \sqrt{s_2} < n. \tag{44}$$

It is convenient to introduce new variables

$$\xi = \frac{m\varepsilon r^2}{n\sqrt{s_2}} \quad \text{and} \quad \lambda = \frac{\sqrt{s_2}}{n}. \tag{45}$$

In these variables condition (44) takes the form:

$$\xi < 1 + \frac{1}{2\lambda} \tag{46}$$

The exact calculation gives for $\Psi_n$ the following expression:

$$\Psi_n \equiv \Psi(\xi,\lambda) = 1 - \frac{2}{\pi}\arccos y - \frac{2\xi}{\pi}\sqrt{1 - \xi^2 + \frac{\xi}{n} - \frac{1}{4\lambda^2}}, \tag{47}$$

and

$$y^2 = \frac{2\xi^2 - 1 - \xi/\lambda + \sqrt{1 + 2\xi/\lambda}}{2\sqrt{1 + 2\xi/\lambda}}. \tag{48}$$

Introducing $\xi$ as the variable of integration in (4I) we have:

$$F_n(r) = \frac{mr^2}{n\sqrt{s_2}} \int_\alpha^\beta \frac{d\xi}{\xi^2}\left(1 + \frac{\lambda}{\xi}\right)\Psi(\xi,\lambda), \tag{49}$$

and

$$\alpha = \frac{mr^2 I}{n\sqrt{s_2}}; \quad \beta = \frac{mr^2\varepsilon_1}{n\sqrt{s_2}}. \tag{50}$$

From these formulas one can find that after the elimination of the monopole term function $F_n(r)$ not only stops to grow when $r \to 0$ but, on the contrary, approaches zero as $r^2$ Numerical calculations using formulas (40, 47, 50) are more tedious than via (30) because now the integration over $r$ should be done separately for every value of $\varepsilon_1$, while in (30) we had to calculate only the mean value of $1/r^2$.

## § 6. Contribution from the dipole term

Further improvement of the theory can be achieved by a more accurate consideration of a few first terms in the sum over $\ell$ and first of all of the dipole term. For this term all our assumptions based on the inequality $\ell \gg 1$ obviously make little sense. In particular, the supposition that the function $r_<^{2\ell}/r_>^{2\ell}$ has a sharp maximum for $r_1 \approx r_2$ becomes invalid. So we will isolate the dipole term with $\ell = 1$ from the sum (16) and will calculate it separately, while other terms with $\ell \geq 2$ will be treated as before, assuming that $\ell \gg 1$ for all of them.

Thus we write

$$\frac{d\sigma}{d\mathcal{E}} = \frac{d\sigma^{(1)}}{d\mathcal{E}} + \frac{d\sigma^{(2)}}{d\mathcal{E}} \tag{51}$$

where $d\sigma^{(1)}/d\mathcal{E}$ is the dipole term and $d\sigma^{(2)}/d\mathcal{E}$ is the rest of the sum with $\ell \geq 2$. This last part can be obtained by the same method that was used above to calculate the sum of terms with $\ell \geq 1$. It is obvious that we will get the differential and integral cross-section just of the same form as in (37) and (40) with the only difference that factor $\Psi_2$, defined according to (42), should be used.
In particular for the integral cross-section we have

$$\sigma^{(2)} = \frac{\pi}{2\mathcal{E}_1}(2\ell_2+1)\langle F_2(r)\rangle(\pi a_0^2) \tag{52}$$

where $F_2(r)$ is given by (49).
The dipole term $d\sigma^{(1)}/d\mathcal{E}$ has the form:

$$\frac{d\sigma^{(1)}}{d\mathcal{E}} = \frac{\pi^2 m e^4}{3\mathcal{E}_1\mathcal{E}^2\tau}(\ell_2+1)\sum_{\ell_1}\frac{\ell_1 r_1 r_2}{P_1(r)P_2(r)}\frac{r_<^2}{r_>^4} \tag{53}$$

Here we have used the Clebsh-Gordan coefficients for $\ell = 1$. Radii $r_1$ and $r_2$ are defined by

$$r_1^2 = \ell_1/m\mathcal{E}\,, \qquad r_2^2 = (\ell_1+1)/m\mathcal{E} \tag{54}$$

and $\ell_1$ in (53) runs over all values beginning with $\ell_1 = 1$. Converting the summation over $\ell_1$ into the integration over $r_1$ we find:

$$\frac{d\sigma^{(1)}}{d\varepsilon} = \frac{\pi^2 m^3 e^4}{3\,\varepsilon_1 \tau_2}(\ell_2+1)\frac{r_2}{p_2(r_2)}\int_{1/m\varepsilon_1'}^{\infty}\frac{r_1^3 dr_1}{p_1(r_1)}\frac{r_<^2}{r_>^4} \qquad (55)$$

Here the momenta $p_1(r)$ and $p_2(r)$ are defined by

$$p_1^2(r_1) = 2m\left(\varepsilon_1 - U(r_1) - \tfrac{1}{2}m\varepsilon^2 r_1^2 - \tfrac{1}{2}\varepsilon\right) \qquad (56)$$

$$p_2^2(r_2) = 2m\left(\varepsilon_2 - U(r_2) - (\ell_2+\tfrac{1}{2})^2/2mr_2^2\right) \qquad (57)$$

The integral cross-section can be obtained again by the integration of (55) over $\varepsilon$ from ionization potential $I$ up to $\varepsilon_1$. Choosing $r_2$ as the variable of integration we get finally

$$\sigma^{(1)}_{\varepsilon_2 \ell_2} = \frac{4\pi^2 m^2 e^4}{3\,\varepsilon_1 \tau_2}(\ell_2+1)\int_{\sqrt{(\ell_2+1)/m\varepsilon_1}}^{\sqrt{(\ell_2+1)/mI}}\frac{dr_2}{r_2^2 p_2(r_2)}\int_{r_2/\sqrt{\ell_2+1}}^{\infty}\frac{r_1^4 dr_1}{p_1(r_1)}\frac{r_<^2}{r_>^4} \qquad (58)$$

## § 7. High energies

Above we have considered the role of different momentum transfers and presented three versions of the theory. Now we will inspect more closely the stationary phase approximation that we used throughout for the evaluation of the radial matrix element. Making use of this approximation from the formal point of view means that we retain the main part of the integral, proportional to $h^{1/2}$, and neglect the background, proportional to $\hbar$. This procedure is correct, unless the difference $\delta\sigma$ between the phases (8) of initial and final wavefunctions is small, and the background becomes as large as the main term itself. For the atomic electron this means that either $\delta\ell = \ell_2 - \ell_2'$ or $\delta n = n_2 - n_2'$ should be large in the transition. And for the incoming electron for fixed $\delta\ell$ and $\varepsilon$ we get some limitation on the collision energy $\varepsilon_1$.

If we neglect the field of the atom (what is reasonable for $\varepsilon_1 \gg I$ ) then the phase is just

$$\sigma(r) = \sqrt{p_1 \cdot r^2 - s_1^2} - \sqrt{s_1}\arccos\frac{\sqrt{s_1}}{p \cdot r} \qquad (59)$$

The variation $\delta\sigma$ in the limit of $p_1 \to \infty$ is then

$$\delta\sigma(r) = \sqrt{\frac{mr^2}{\varepsilon_1}}\,\delta\varepsilon_1 - \frac{\pi}{2}\cdot\delta\sqrt{s_1} \qquad (60)$$

So if $\delta\sqrt{s_1}$ is not large and if $\delta\sigma(r)$ should be large, then it is obvious that $\varepsilon/\varepsilon_1$ should not be too small and hence the energy of collision should not be too high. In other words we see that for transitions with small change of angular momentum of the atomic electron the stationary phase method becomes invalid for high collision energies.

But for high energies the wave function of the colliding electron can be described with good accuracy by the plane wave and we have for the cross section the well-known Born approximation.

$$\frac{d\sigma}{d\varepsilon} = \frac{8\pi m e^4}{\varepsilon_1} \sum_{m_2 l_2' m_2'} \int |\langle \varepsilon_2' l_2' m_2' | e^{i\vec{q}\vec{r}} | \varepsilon_2 l_2 m_2 \rangle|^2 \frac{dq}{q^3} \tag{61}$$

Performing the angular integration and summation over $m_2$ and $m_2'$ we get

$$\frac{d\sigma}{d\varepsilon} = \frac{4\pi^2 m e^4}{\varepsilon_1} \sum_{l_2' l} (2l_2+1)(2l_2'+1) |C^{l}_{l_2 l_2'}|^2 \times \int \frac{dq}{q^5} |\langle \varepsilon_2' l_2' | \sqrt{\tfrac{q}{r}}\, J_{l+\frac{1}{2}}(qr) | \varepsilon_2 l_2 \rangle|^2 \tag{62}$$

The matrix element in (62) can be calculated again using the semiclassical wave functions for the atomic electron and the stationary phase method, considering the Bessel function as a smooth function. This is justified for small $q$, contributing the main part to the integral cross-section. Applying our old technique we get:

$$\frac{d\sigma}{d\varepsilon} = \frac{2\pi^2 m^2 e^4}{\varepsilon_1 \varepsilon \tau_2} (2l_2+1) \sum_{l l_2'} (2l_2'+1) |C^{l}_{l_2 l_2'}|^2 \int J^2_{l+1/2}(qr) \frac{dq}{p_2 \cdot q^4} \tag{63}$$

where $p_2$ is given by (57) and $r_2$, as above, by (13). The sum over $l$ begins with $l=1$.

As before, we divide the sum in two parts - the dipole term and the rest and consider first the latter. The integral in (63) converges rapidly when $q$ increases. So we can replace the upper limit $p+p'$ by infinity. The lower limit $q_0 = 2m\varepsilon/(p+p') \to 0$ when $\varepsilon_1 \to \infty$ and, as the sum runs only over $l \geq 0$, the integral converges in the limit of small $q$ as well. So we can approximate it by

$$\int_0^\infty J^2_{\sqrt{s}}(q r_2)\,\frac{dq}{q^4} = \frac{2 r_2^3}{3\pi}\,\frac{1}{(s-9/4)(s-1/4)}\,, \tag{64}$$

and $s \geqslant 4$. Supposing that $s \gg 1$ we have for the sum of all terms but the dipole:

$$\frac{d\sigma^{(2)}}{d\varepsilon} = \frac{8 m^3 e^4}{3\varepsilon_1 \tau_2}(2\ell_2+1)\int \frac{ds_2^2\, ds\, \Theta(s-4)\, r_2^3}{s^2 \Delta(\ell \ell_2 \ell_2')\, P_2(r_2)}\,. \tag{65}$$

This formula is seen to be our old formula (26) for the nondipole contribution, where factor $\pi$ is now replaced by 8/3 and, consequently, if we disregard this minor difference, we obtain the same result (52) for the integral cross-section.

Now let us turn our attention to the dipole term:

$$\frac{d\sigma^{(1)}}{d\varepsilon} = \frac{6\pi^2 m^2 e^4}{\varepsilon_1\, \varepsilon_2\, \tau_2}\,\frac{\ell_2+1}{P_2(r_2)}\int_{q_0}^\infty J_{3/2}(q r_2)\,\frac{dq}{q^4} \tag{66}$$

For high energies $q \to 0$, and we can use the asymptotic representation:

$$\int_{q_0}^\infty J_{3/2}(q r_2) = \frac{2 r^3}{9\pi}\left(\ln\frac{1}{q_0 r} + \frac{7}{4} - C - \ln 2\right) + O(q_0)\,,$$

where $C = 0.5772$ is the Euler constant.

So for this term we get:

$$\sigma^{(1)}_{\varepsilon_2 \ell_2} = \frac{8\pi m^2 e^4}{3\,\varepsilon_1 \tau_2}(\ell_2+1)\int_{\sqrt{\frac{\ell_2+1}{m\varepsilon}}}^{\sqrt{\frac{\ell_2+1}{mI}}} \frac{r_2^2\, dr_2}{P(r_2)}\left(\ln\frac{P \cdot r_2}{\ell_2+1} + \frac{7}{4} - C - \ln 2\right), \tag{67}$$

and the dipole contribution really behaves as $\ln \varepsilon_1 / \varepsilon_1$ for $\varepsilon_1 \gg 1$.

## § 8. Numerical results for ionization

A number of calculations have been performed by using las (40–50) based on the local approximation (22–25) for all terms with $\ell \geqslant 1$, which will be further referred to as "local approximation", and using formulas (52–58), where the dipole term was treated more accurately, which we will call "nonlocal approximation". The Thomas-Fermi and Hermann-Skillman atomic potentials were used with little difference in the corresponding results.

Comparison of these results with calculations, made by classical-like formula (30) shows that the elimination of the monopole term makes the contribution from outer $p$- and especially from $d$-shells to the total cross-section much smaller. For Ag, for instance, the cross-section of ionization from 4d-shell falls from 313 Å$^2$ down to 3.3 Å$^2$ for $Ar$ (3p - shell) - from 15 down to 2.8 Å$^2$. The outer s-shells are less affected. In $Mg$ for instance, the maximum of the cross-section for 3s shell decreases from 6.6 Å$^2$ to 4.5 Å$^2$.

In Table 2 we compare the maximum values of the experimental cross-section with those calculated by our "local" (38), "non local" (52-58) and "high energy" (65-67) approximations.

Table 2

| Approximation | Ionization cross-section in Å$^2$ | | | | | |
|---|---|---|---|---|---|---|
| | Na3s | Na2p | Ne2s | Ne2p | Ag5s | Ag4d |
| $\sigma^{local}_{max}$ | 4.8 | 3.1 | 0.10 | 0.60 | 2.3 | 3.4 |
| $\sigma^{non\ local}_{max}$ | 7.5 | 0.49 | 0.16 | 0.90 | 2.8 | 3.9 |
| $\sigma^{high-energy}_{max}$ | 6.9 | 0.32 | 0.15 | 0.61 | 2.7 | 2.5 |
| $\sigma^{experim}_{max}$ | 7 | | 0.8 | | 4 | |

Such calculations have been performed for many atoms for which experiment is available. In all cases good agreement was observed. So we come to the conclusion that rather simple and suitable for numerical calculations semiclassical local approximation (38) gives reliable description of electron-atom ionization cross-sections.

As to the dipole term, one can see that it has comparatively small influence on the probability of ionization of s - shells. For instance for 4s - shell in Ca the dipole contribution increases the cross-section from 5.4 $\mathring{A}^2$ to 6.9 $\mathring{A}^2$ , for 4s shell in $Zn$ from 2.5 $\mathring{A}^2$ to 2.9 $\mathring{A}^2$. Its influence is much more pronounced for p - shells. For 4p shell in Kr the dipole term contributes 4.1 $\mathring{A}^2$ and the rest only 1.1 $\mathring{A}^2$, for 2p in $Ne$ - 0.5I $\mathring{A}^2$ against 0.09 from all other terms. Its contribution is equally important for $d$-shells. "Non-local" and "high energy" approximations are somewhat more satisfactory from the theoretical point of view, when compared to the simple local approximation (38) but, as Table 2 shows, they probably do not lead to big changes in final results.

At last we will say a few words about the relation between our calculations and purely classical calculations, based on the expression (3I). It is well known that in most cases such calculations give satisfactory description of the experiment, especially if the ionization potential of the shell is used for $\varepsilon_2$.

To be more exact, the results are good for those atoms where the main part of the total cross-section comes from outer $s$- or $p$-shells as is the case of practically all available experimental data. But for such atoms as Ag, Cu, Zn, which have d-shells with a very small ionization potential, classical calculations give too high cross-sections, and in the framework of purely classical considerations it is impossible to find the reason for such discrepancy. In Fig. 1 we illustrate typical results for the cases of Ar and Ag. More details on this subject can be found in [4].

From the point of view of the semiclassical theory developed above, the results of purely classical calculations are quite understandable. In such calculations three factors are involved. First, the use of the ionization potential for $\varepsilon_2$ makes the contribution of the second term in (3I) lower than it should be according to a more consistent value $\langle\varepsilon_2\rangle$. Further the use of classical formula (3I) corresponds to our asymptotic formula (30) which gives too low coefficients in the "centrifugal" term for s- and p- shells. On the other hand, in (30) we did not yet eliminate the contribution from the monopole term what led to the increase of the cross-section. For s- and partly for p - shells the first two factors usually are compensated by the third. Besides, the contribution from the first term in (3I) is comparatively large (especially for s - electrons). But for the shells with greater angular momentum the contribution from the monopole term greatly increases, and, unless it is eliminated, what cannot be done in the classical theory, the results become meaningless in some cases. In Figures 1 and 2 we show typical results of classical and semiclassical calculations for Ar and Ag.

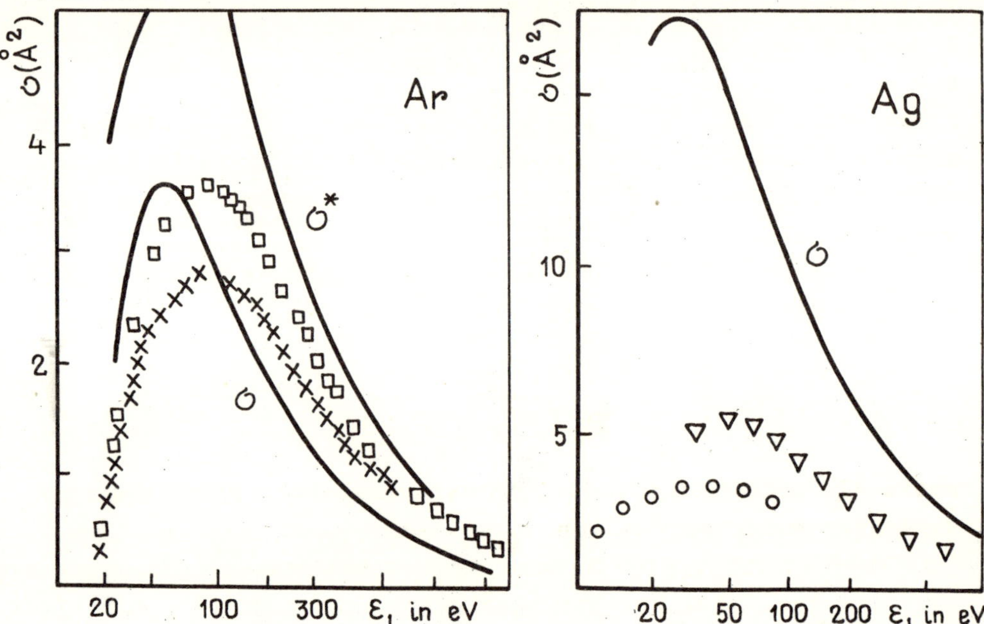


Fig. 1. Ionization of Ar and Ag in the classical binary approximation. $\sigma$ is calculated with $\varepsilon_2 = I$, $\sigma^*$ with $\langle\varepsilon_2\rangle$.

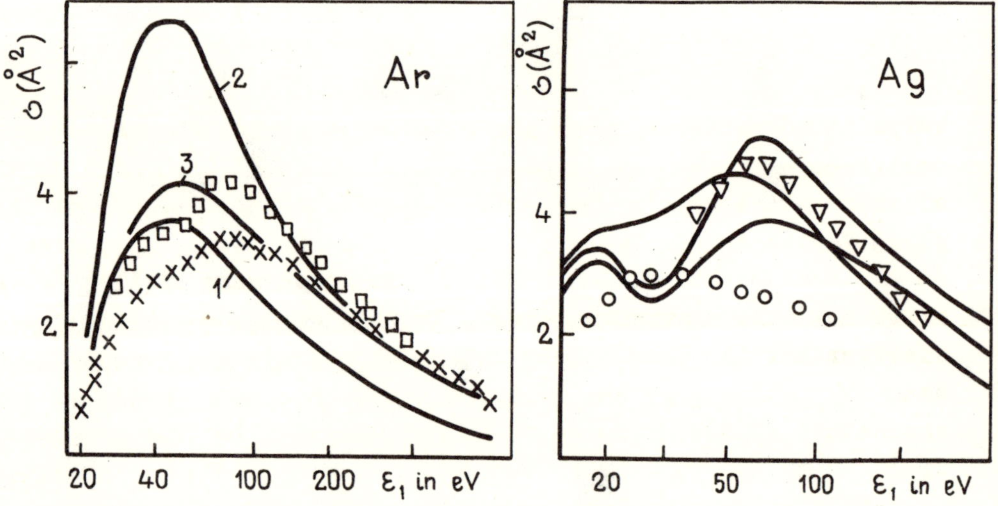


Fig. 2. Ionization of Ar and Ag in the semiclassical approximation. 1 -"local approximation"(38-50), 2 -"non-local approximation"(52,58), 3 -"High energy approximation" (65,67).

## § 9. Excitation

The approach described in previous sections can be applied to the problem of excitation as well. Only three obvious alterations are needed. The final state now has a definite energy $\varepsilon_2'$ and a definite angular momentum $l_2'$ So we have to drop the integration over $\varepsilon_2'$ and summation over $l_2'$. Keeping in mind that the final state now belongs to the discrete spectrum we have also to change the normalization factor $1/\pi$ of its wave function for $1/\tau_2$.

If we introduce these changes and restrict ourselves to the nonexchange transitions only, we obtain

$$\sigma_{\varepsilon_2 l_2 \to \varepsilon_2' l_2'} = \frac{\pi^3 e^4 (2l_2+1)(2l_2'+1)}{8\varepsilon_1 \varepsilon^3 \tau_2 \tau_2' r_2' p_2(r_2)} \left(1 + \frac{s_2}{m\varepsilon r_2^2}\right) \Psi \qquad (68)$$

where all notations are as before. Remembering also that our technique works only if the condition

$$(\varepsilon_2' - \varepsilon)(l_2' - l_2) > 0 \qquad (69)$$

is fulfilled we see that for excitation the monopole term never occurs and hence $\Psi = 1$.

So the characteristic feature of the expression (68) is a very simple energy dependence

$$\sigma = Const/\varepsilon_1 \qquad (70)$$

which is not realistic, at any rate for neutral atoms and small collision energies. It is not difficult to understand the nature of this behaviour. In the derivation of our formulas it was implied that both electrons, atomic and the colliding one, move in the same atomic field $U(r)$ in which many excited states exist near the continuum border. In this case the colliding electron has the possibility, when its energy $\varepsilon_1$ becomes less than $\varepsilon$, to occupy one of these levels near the continuum and the point $\varepsilon_1' = 0$ proves to be not singular for the cross-section.

This picture probably gives adequate description of the process of excitation of a sufficiently ionized atom. But in the problem of ionization of a neutral atom the colliding electron moves in quite a different field than the atomic one. The atomic field is almost screened for the free electron and there are no discrete states near the continuum.

We can take this into account introducing two different potentials $U_1(r)$ and $U_2(r)$. In this case our statement made after (25) that the argument of the step-function arising after integration over the parameters of the colliding electron in (2I) is always positive, becomes invalid, and some additional dependence of the cross-section on energy appears, which leads to more reasonable threshold behaviour. Introducing the mentioned $\theta$ -function in our formulas we arrive at (68) with a new $\Psi$ -factor, which can be represented as

$$\Psi = \Psi_\alpha - \Psi_\beta, \tag{7I}$$

where

$$\Psi_\alpha = \Psi(\xi_\alpha, \lambda_\alpha); \quad \xi_\alpha = \frac{m r_2^2 \varepsilon}{\alpha \sqrt{S_2}}; \quad \lambda = \frac{\sqrt{S_2}}{\alpha}; \tag{72}$$

$\Psi(\xi,\lambda)$ is given by (47) and

$$\alpha = \sqrt{2m}\cdot r_2\left(\sqrt{\tilde{\varepsilon}_1} + \sqrt{\tilde{\varepsilon}_1'}\right) \qquad \beta = \sqrt{2m}\cdot r_2\left(\sqrt{\tilde{\varepsilon}_1} - \sqrt{\tilde{\varepsilon}_1'}\right); \tag{73}$$

$$\tilde{\varepsilon}_1 = \varepsilon_1 + U_1(r_2) \qquad \tilde{\varepsilon}_1' = \varepsilon_1' + U_1(r_2); \tag{74}$$

$\Psi(\xi,\lambda)$ is non-trivial only if

$$\frac{1}{2\lambda} - 1 \leq \xi \leq \frac{1}{2\lambda} + 1. \tag{75}$$

Otherwise

$$\Psi = 1 \quad \text{if} \quad \xi > \frac{1}{2\lambda} + 1; \tag{76}$$

$$\Psi = 0 \quad \text{if} \quad \xi < \frac{1}{2\lambda} - 1.$$

Formulas (68-76) describe the cross-section for a given $U_1(r)$

Let us note that condition (75) for both values $\lambda_\alpha$ and $\lambda_\beta$ reduces to the same inequality

$$\tilde{\varepsilon}_1'/\varepsilon < S_2/(S_2' - S_2). \tag{77}$$

So for high energies for which $\tilde{\varepsilon}_1' > \varepsilon \cdot s_2/(s_2' - s_2)$ factor $\Psi = 1$ again and we again have the simple dependence on energy as in (70). But when $\varepsilon_1$ decreases we reach the region (77), where $\Psi$ depends on $\varepsilon_1$ in a complicated way. If we consider the field of the atom as completely screened for incoming electron and put $U_1 = 0$ we immediately see that for $\varepsilon_1' = 0$ the numbers $\alpha$ and $\beta$ in (73) become equal and the cross-section is zero at the threshold.

Formula (68) was obtained under the assumption that all angular momenta are large. But for comparison with experimental data available now we need the formulas for small $l_2$ and $l_2'$.

To derive them we start again with the general expression for excitation with given $l_2$ and $l_2'$:

$$\sigma^{(l)}_{\varepsilon_2 l_2 \to \varepsilon_2' l_2'} = \frac{\pi^3 m e^4}{\varepsilon_1 \varepsilon^2 \tau_2 \tau_2'} \frac{r_2}{P_2(r_2)} \frac{(2l_2+1)(2l_2'+1)|C^{l}_{l_2 l_2'}|^2}{(2l+1)^3} \times \tag{78}$$

$$\times \sum_{l_1, l_1'} |C^{l}_{l_1 l_1'}|^2 \frac{r_1}{P_1(r_1)} \frac{r_<^{2l}}{r_>^{2l+2}} (2l_1'+1)(2l_1+1)$$

If we restrict ourselves to practically most important cases of dipole and quadrupole transitions ( $l = 1, 2$ ), then

$$l_2' = l_2 + l; \qquad l_1' = l_1 - l$$

and summations over $l_1$ and $l_1'$ turn into a summation over $l_1$ only. Approximating as before the summation over $l_1$ by integration we obtain in the "non-local approximation":

$$\sigma^{(1)}_{\varepsilon_2 l_2 \to \varepsilon_2' l_2'} = \frac{\pi^3 m^3 e^4 (l_2+1) r_2}{3\varepsilon_1 \tau_2 \tau_2' P_2(r_2)} \int \frac{r_1^3 dr_1^2}{P_1(r_1)} \frac{r_<^2}{r_>^4} \tag{79}$$

where $r_2^2 = (l_2+1)/m\varepsilon$; $\quad r_1^2 = l_1/m\varepsilon \quad$ and

$$\sigma^{(2)}_{\varepsilon_2 l_2 \to \varepsilon_2' l_2'} = \frac{9\pi^3 m^3 e^4 (l_2+1)(l_2+2) r_2}{160\,\varepsilon_1 \tau_2 \tau_2' (2l_2+3) P_2(r_2)} \int \frac{r_1^3 dr_1^2}{P_1(r_1)} \frac{r_<^4}{r_>^6} \tag{80}$$

Here $r_2^2 = (2l_2+3)/m\varepsilon$ and $r_1^2 = (2l_1-1)/m\varepsilon$.

These formulas are similar to (53) for ionization.

We can move considerably farther with all these formulas, if we apply them to the specific case of the hydrogen-like ion with nuclear charge $Z$. In this case $U_2(r) = Ze^2/r$. As to the

colliding electron it is reasonable to put $U_1(r) = (Z-1)e^2/r$
Now the expressions (68,71) for the cross-section with large transfer of angular momentum give:

$$\sigma_{n_2 l_2 \to n_2' l_2'} = \frac{\pi a_0^2}{Z^4} \cdot \frac{F_{n_2 n_2'}^{l_2 l_2'}}{1+u} \cdot \Psi(u, Z), \tag{81}$$

where $u = \varepsilon'/\varepsilon$

$$F_{n_2 n_2'}^{l_2 l_2'} = \frac{2 n_2^5 n_2'^5}{(n_2'^2 - n_2^2)^{7/2}} \cdot \frac{s_2' + s_2}{s_2' - s_2} \sqrt{s_2'} \times \tag{82}$$

$$\times \left\{ s_2 n_2'^2 - s_2' n_2^2 + 2 n_2 n_2' \sqrt{(s_2' - s_2)(n_2'^2 - n_2^2)} \right\}^{-1/2},$$

and the region where $\Psi$ is non-trivial is:

$$u + \left(1 - \frac{1}{Z}\right) \sqrt{\frac{4 n_2^2 n_2'^2}{(n_2'^2 - n_2^2)(s_2' - s_2)}} < \frac{s_2}{s_2' - s_2} . \tag{83}$$

For $s-p$ transitions we obtain from (79):

$$\sigma_{s-p} = \frac{\pi a_0^2}{Z^4} \cdot \frac{I(u, Z)}{1+u} \cdot F_{n_2 n_2'}, \tag{84}$$

where

$$F_{n_2 n_2'} = \frac{16}{3} \cdot \frac{n_2^5 n_2'^5}{(n_2'^2 - n_2^2)^{7/2}} \left\{ \frac{n_2^2}{4} - \frac{9}{4} n_2'^2 + 2 n_2 n_2' \sqrt{2(n_2'^2 - n_2^2)} \right\}^{-1/2} \tag{85}$$

$$I(u, Z) = \int_1^\infty \frac{dr}{p(r)} + \int_{1/\sqrt{2}}^1 \frac{r^6 dr}{p(r)} \tag{86}$$

$$p^2(r) = 2m\left(u + \frac{a}{r} - \frac{1}{2}\left(r - \frac{1}{2r}\right)^2\right) \tag{87}$$

$$a = \left(1 - \frac{1}{Z}\right) \frac{n_2 n_2' \sqrt{2}}{(n_2'^2 - n_2^2)^{1/2}} . \tag{88}$$

Similar expressions can be obtained for $s-d$ and other transitions.

It is instructive to study the semiclassical selection rules which follow from the above expressions. They consist of the mentioned condition $(l_2' - l_2)(\varepsilon_2' - \varepsilon_2) > 0$
and the requirement

$$p_2(r) > 0 \tag{89}$$

which means that the atomic electron shoud be in the classically allowed region during the transition. Simple analysis showes that for the hydrogen-like ions Eq. (89) is fulfilled in a certain region of quantum numbers, which is qualitatively sketched in Fig. 3. For transitions from the levels with $n_2 = 1, 2, 3$ it is described in detail in Table 3.

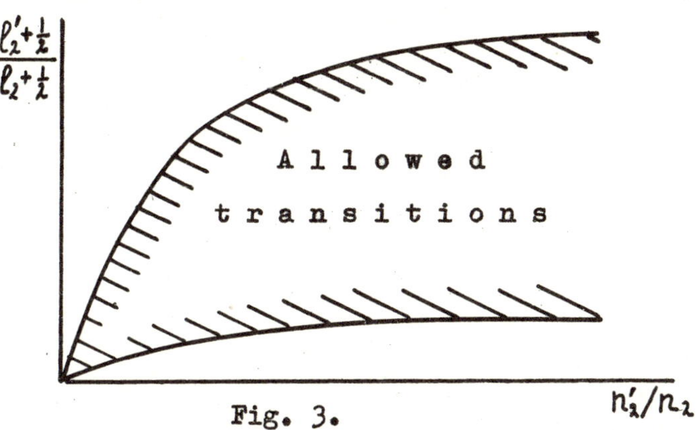


Fig. 3.

Table 3

Transitions allowed in the semiclassical approximation

| Initial state | | Final state | |
|---|---|---|---|
| $n_2 = 1$ | $l_2 = 0$ | $n_2' \geq 2$ | $l_2' = 1$ |
| $n_2 = 2$ | $l_2 = 0$ | $n_2' \geq 3$<br>$n_2' \geq 5$ | $l_2' = 1, 2$<br>$l_2' = 3$ |
| | $l_2 = 1$ | $n_2' \geq 3$<br>$n_2' \geq 7$ | $l_2' = 2$<br>$l_2' = 3$ |
| $n_2 = 3$ | $l_2 = 0$ | $n_2' \geq 4$<br>$n_2' \geq 8$ | $l_2' \leq 4$<br>$l_2' = 5$ |
| | $l_2 = 1$ | $n_2' \geq 4$<br>$n_2' \geq 10$ | $l_2' \leq 4$<br>$l_2' = 5$ |
| | $l_2 = 2$ | $n_2' \geq 4$<br>$n_2' \geq 6$ | $l_2' = 3$<br>$l_2' = 4$ |

Let us inspect more closely the expression (84) for $s-p$ transitions. The dependence on quantum numbers $n_2$ and $n_2'$ is given explicitly by (85). At first sight more complicated seems to be the dependence of the integral (86) on $Z$ and $u$. We have calculated this integral numerically as a function of $u$ and the parameter $a$ given by (88). The results are shown in Table 4.

Table 4

| Integral $I(a, u)$ | | | | | |
|---|---|---|---|---|---|
| $a$ \ $u$ | 0 | 0.125 | 0.5 | 1 | 2 |
| 0 | 0 | 0.56 | 0.53 | 1.06 | 1.18 |
| 0.4I | 0.73 | 0.80 | 0.94 | 1.05 | 1.16 |
| 0.82 | 0.8I | 0.85 | 0.95 | 1.04 | 1.15 |
| 1.63 | 0.86 | 0.89 | 0.96 | 1.03 | 1.13 |
| 2.83 | 0.90 | 0.9I | 0.96 | 1.03 | 1.11 |
| 4.7I | 0.93 | 0.94 | 0.97 | 1.02 | 1.10 |
| 9.73 | 0.96 | 0.96 | 0.99 | 1.02 | 1.07 |

One can see from this table that for $u \geqslant 0.5$ or $a \geqslant 1$ $I(a, u)$ depends on its arguments rather weakly. At the same time it follows from (88) that

$$a \geqslant \sqrt{2}\left(1 - \frac{1}{Z}\right) \cdot n_2 \qquad (90)$$

So $a \geqslant 1$ almost for all cases with the exception of hydrogen ($Z = 1$), where $a = 0$ for all transitions, and for $He^+$ ($Z = 2$) where the minimal value of $a$ is 0.7I.

In Table 5 we present a few examples of calculations of the cross-sections for transitions between comparatively low levels in neutral hydrogen made according to the expressions (8I) and (84) and to the conventional Born approximation for comparison.

Table 5

**Excitation cross-sections of s-p transitions in H (in units of $\pi a_0^2$)**

| $u$ | 1s-2p | | | 2s-3p | | | 3s-4p | | |
|---|---|---|---|---|---|---|---|---|---|
| | (81) | (84) | Born | (81) | (84) | Born | (81) | (84) | Born |
| 0 | 2.51 | 0 | 0 | 24.1 | 0 | 0 | 137 | 0 | 0 |
| 0.6 | 1.84 | 2.02 | 1.06 | 17.7 | 19.4 | 6.69 | 101 | 111 | 36.3 |
| 1.2 | 1.03 | 1.81 | 1.32 | 9.9 | 17.5 | 17.0 | 56.2 | 99.6 | 85.6 |
| 2.0 | 0.50 | 1.21 | 1.07 | 4.8 | 11.6 | 18.8 | 27.4 | 66.2 | 118 |
| 3.2 | 0.22 | 0.68 | 0.69 | 2.1 | 6.6 | 14.3 | 12.2 | 37.6 | 102 |
| 4.8 | 0.10 | 0.38 | 0.42 | 1.0 | 3.7 | 9.5 | 5.7 | 20.8 | 71.2 |

As the last example in fig. 4 we compare the cross-section for 1s-2p transition in $C^{5+}$ calculated using (84) with the results of Born-Coulomb [5] and distorted wave [6] methods.

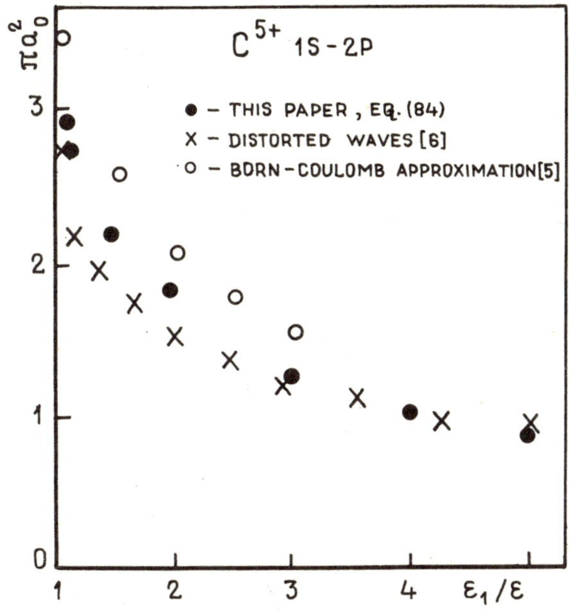

$C^{5+}$ 1S - 2P

● – THIS PAPER, EQ. (84)
X – DISTORTED WAVES [6]
O – BORN-COULOMB APPROXIMATION [5]

$\pi a_0^2$; 0, 1, 2, 3; $\varepsilon_1/\varepsilon$; 1, 2, 3, 4

Fig. 4.


## Summary

The results presented seem to be rather encouraging. Formula (84) is nothing but the semiclassical version of the Born-Coulomb approximation. However, in contrast to the latter, it is simple and easy to calculate. Thus, the semiclassical approach seems to be a useful instrument to treat a great variety of problems connected with the excitation and ionization of atoms and ions by electrons.

References

1. G.C. Mc. Coid, S.N. Milford, Phys. Rev. 130, 206, 1963.

2. A. Burgess, I.C. Percival, in "Advances in Atomic and Molecular Physics" 1968, v. 4, N.Y.-London, p.109-140.

3. L. Vriens, in "Case Studies in Atomic Collision Physics" 1969, v. I, N-Holland, Amsterdam, p. 335-398.

4. V.I. Ochkur, in "Atomic Collision Problems" 1975, v. I, Leningrad State University, p. 42-65.

5. M. Gailitis, Opt. and Spectr. 14, 465, 1963.

6. L.A.Vainstein, Opt. and Spectr. 11, 30I, 1961.

# SPECTROSCOPY OF MULTICHARGED IONS IN LABORATORY PLASMA

L.A.Vainstein*, V.A.Boiko*, E.Ya.Kononov**

* The Lebedev Physical Institute, USSR Academy of Sciences, Moscow

** The Institute for Spectroscopy, USSR Academy of Sciences, P/O Akademgorodok, Moscow District

## 1. Introduction

Spectra of multicharged ions are now widely investigated both in laboratory and astrophysical plasmas. The typical temperature values of such plasmas are from about 200 eV up to 2 keV and even higher (i.e. $(2\div20)\cdot10^6$K). The sources of laboratory plasma which were used in spectroscopy are laser produced plasmas, vacuum sparks, theta-pinch devices, tokamaks and some others.Data used in this report concern the first two sources. As for astrophysics, the solar corona is, in fact, a unique object, for which good line spectra are obtained in the X-ray region. The study of the line spectra of non-solar X-ray sources is extremely difficult due to small flux values. However, it seems very probable that the first such spectra are to be obtained with the large X-ray telescope of the USA orbiting satellite HEAO-B which is to be launched this autumn.

In the spectra of a hot plasma the main part of energy and, probably, of information falls onto the X-ray region between 1 and 20Å. This region contains, in particular, resonance lines of the H-like and He-like ions belonging to the elements from Ne to Ni, i.e. the ions with charges $Z \geqslant 10$. The valuable information seems to be obtained also from the EUV region between 100 and 1000 A. This review, however, does not cover the EUV-region spectra.

The main feature of spectra in the vicinity of resonance lines of multicharged ions is a satellite structure: every line is followed by a large group of satellites, often called dielectronic satellites because the principal (but not the only) mechanism of their formation is the dielectronic recombination.

Here we use the following notations:

DS - dielectronic satellite;

[H], [He] , etc - H-like, He-like, etc. ions;

Z - the spectroscopic symbol of an ion (Z=1 for a neutral atom);

$X_Z \equiv X^{(Z-1)+}$ - (Z-1) times ionized atom.

The DS's are one-electron transitions in the presence of an additional electron (or a group of electrons). These transitions differ from usual ones (which do not affect inner electrons) by the fact that their upper state is usually a vacancy in the inner-electron core. A general analysis of such transitions is beyond the scope of our review. We restrict ourselves to the case of [H] and [He] resonance-line satellites.

So we have:

| | Resonance lines | DS | | |
|---|---|---|---|---|
| [H] - | 1s-2p | 1snl-2pnl | - [He] | (1) |
| [He] - | $1s^2$-1s2p | $1s^2nl$-1s2pnl | - [Li] | |

We consider here multicharged ions ($Z \gg 1$) because satellite intensities relative to a resonance line intensity are rapidly increasing with Z. To the present time the DS's of ions up to Fe XXV have been studied and some data exist for even higher ionization stages.

Below we shall consider the satellites (1) only, except for some particular cases. For the [He] resonance line the DS's with a greater number of electrons are possible:

| | Resonance line | DS | | |
|---|---|---|---|---|
| [He] - | $1s^2$-1s2p | $1s^22snl$-1s2p2snl | - [Be] | |
| | | $1s^22pnl$-$1s2p^2nl$ | | |
| | | $1s^22s^22p^5$-$1s2s^22p^6$ | - [F] | (2) |

## 2. Two processes of satellite excitation

The main process of satellite excitation is that of dielectronic recombination (DR):

$$X_{Z+1}(\alpha_0) + e \longrightarrow X_Z(\gamma) \swarrow X_{Z+1}(\alpha_0) + e \quad \searrow X_Z(\gamma_1) + \hbar\omega \tag{3}$$

where $\gamma = [\alpha nlLS]$ ; $\alpha$ - states of the ion $X_Z$, e.g.

$$[He(1s^2)] + e \rightarrow [Li(1s2p(^{1,3}P)nl\,LS\,J)] \swarrow [He(1s^2)] + e \quad \searrow [Li(1s^2nl\,L_1S_1\,J_1)] + \hbar\omega \tag{3a}$$

Since the probability of electron capture is proportional to that of autoionization decay, the relative intensity of a DS is equal to:

$$i_d = \frac{I_s(\gamma)}{I_{res}} = \Phi_d(T_e)\cdot\frac{g_\gamma \Gamma(\gamma) A(\gamma,\gamma_1)}{A(\gamma) + \Gamma(\gamma)}, \tag{4}$$

$$A(\gamma) = \sum_{\gamma_1} A(\gamma,\gamma_1).$$

Here $\Gamma$ is the autoionization rate coefficient, A is the radiative transition probability, $g_\gamma$ is the statistical weight and $T_e$ is the electron temperature.

The factor $\Phi_d$ is, of course, proportional to $1/\langle v\sigma_{res}\rangle$. Using known analytical approximations and numerical calculations of $\langle v\sigma_{res}\rangle$ we get:

$$\Phi_d = C_d\,\delta\beta\, e^{\delta\beta} ; \quad \delta\beta = \frac{E_{Z+1}}{n^2kT_e} \tag{5}$$

$$Cd \approx 0.58\cdot10^{-16}\cdot n^2 \quad - [H] ;$$

$$Cd \approx 1.0\cdot10^{-16}\cdot n^2 \quad - [He] ,$$

where $E_{Z+1}$ is ionization energy of the ion $X_{Z+1}$.

In fact, $C_d$ is slightly dependent on $T_e$ and Z , but this dependence is very weak and can be neglected for estimates. We see that relative DS intensities are increasing with the decrease of temperature.

Radiative transition probabilities A are proportional to $Z^4$ and autoionization rate coefficients are practically independent on Z. Therefore, according to formula (4) $i_d$ increases as $Z^4$ up to $Z\sim20$ and thereafter is saturated.

Under these conditions intensities of satellite and resonance lines become of the same order. For illustration Fig.1 shows a satellite structure for [He] ions Mg XI and Sc XX /1/.

Apart from the DR process (3), generally speaking, there is a direct excitation of the electron shell 1s in the ion $X_Z$:

$$X_Z(\gamma_0) + e \rightarrow X_Z(\gamma) + e \quad \swarrow X_{Z+1} + e \qquad \searrow X_Z(\gamma_1) + \hbar\omega \tag{6}$$

For example, in the case of satellites to a [He] ion:

$$[\mathrm{Li}\ (1s^2 2s)] + e \rightarrow [\mathrm{Li}\ (1s2p2s)] + e \tag{6a}$$

In a laboratory plasma where the electron density is quite high, the states 2s and 2p of a [Li] ion are populated according their statistical weights. Hence, besides (6a) there is a process:

$$[\mathrm{Li}\ (1s^2 2p)] + e \rightarrow [\mathrm{Li}\ (1s2pnl)] + e \tag{6b}$$

i.e. practically all the satellites are included into the direct excitation process.

Obviously, for the satellites to a [H] ion the direct excitation process is not efficient, since it requires an excitation of two electrons.

Thus, there is an essential difference between satellites to [H] and [He] resonance lines: in the first case the only excitation process (DR) is possible, in the second case there are two processes. This circumstance is very important for plasma diagnostics.

In Fig.2 the satellite structure of the [H] ion Mg XII obtained from a laser-produced plasma is compared with the computed one /2/. As is seen, the DR process gives a sufficiently good description of the spectrum features. The satellites with $n = 2$ and $n = 3$ are distinguished clearly.

A similar comparison for [He] ions when only the DR process is taken into account shows "additional" satellites in experimental spectra (see below). These satellites proved to be essentially dependent on excitation mechanism.

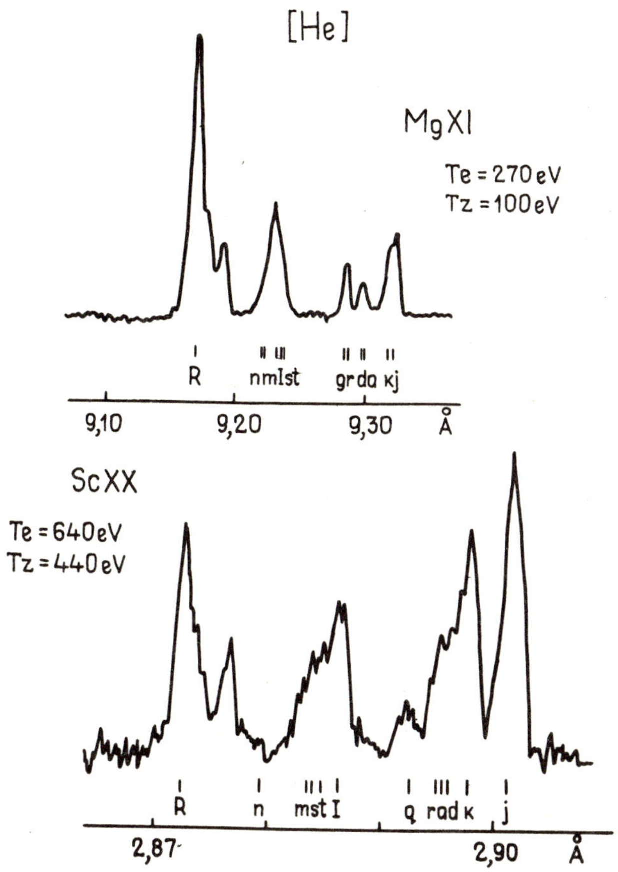


Fig. 1. The satellite structures of the [He] ions Mg XI and Sc XX in a laser-produced plasma. R and I are the resonance and intersystem lines respectively; n, m, ... j are DS in Gabriel's designations.

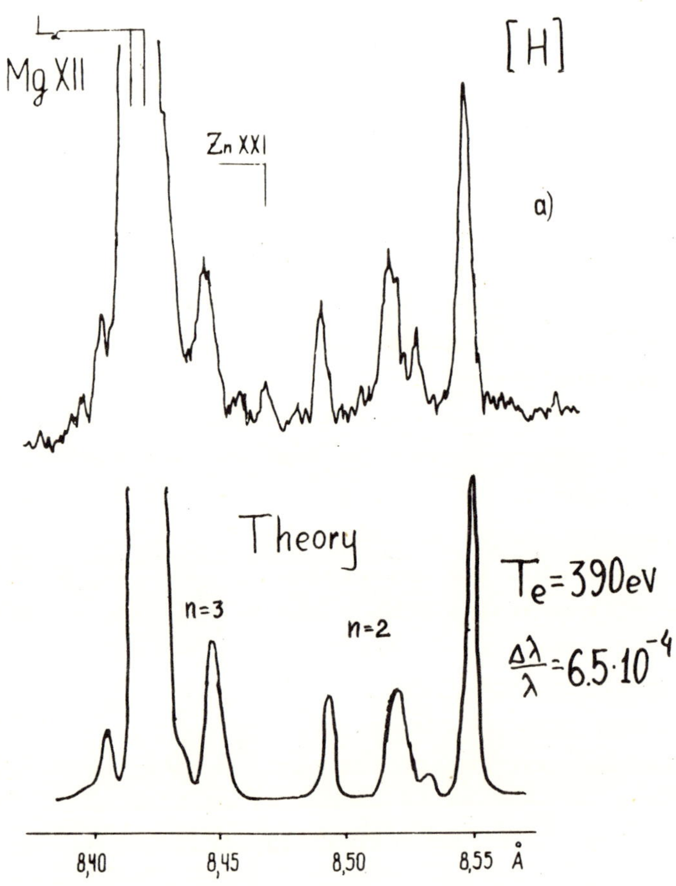


Fig. 2. The satellite structure of the [H] ion Mg XII observed in a laser-produced plasma in comparison with a calculated one (excitation by the DR process).

Relative intensity of a satellite due to the direct excitation process is:

$$i_c = \frac{N_Z}{N_{Z+1}} \cdot \frac{\langle v\sigma_s \rangle}{\langle v\sigma_{res} \rangle} \cdot \frac{A(\gamma,\gamma_1)}{A(\gamma)+\Gamma(\gamma)}, \tag{7}$$

where $N_Z$ and $N_{Z+1}$ are populations of ions $X_Z$ and $X_{Z+1}$, $\langle v\sigma_s \rangle$ is the full (over the initial state $\gamma_0$) excitation rate of a satellite. A comparison with (4) shows that the direct excitation process is especially important for satellites with low values of $\Gamma$. Let us write (7) in the form similar to (4):

$$i_c = \Phi_c(T_e)\,\frac{B_s \cdot A(\gamma,\gamma_1)}{A(\gamma)+\Gamma(\gamma)}, \tag{8}$$

where $B_s$ is determined by $\langle v\sigma_s \rangle$ and the factor $\Phi_c \sim N_Z/N_{Z+1}$ is dependent on the temperature and is practically the same for all satellites in a group. In the conditions of stationary ionization equilibrium $\Phi_c$ is proportional to $Z^4$. Hence, the role of the direct excitation process is small at low Z.

Numerical computations show that for the most of satellites the DR process is dominating, but some of them appear mainly due to the direct excitation of the ion $X_Z$. As examples of the first kind the satellites j and k can be considered and of the second kind - the satellites a and q (see Fig.1). We use here Gabriel's notations of satellites widely known at present /3/.

It is seen from formulae (4) and (8) that the DR process gives intense satellites when A and $\Gamma$ are high. The direct excitation process is essential at high A but low $\Gamma$ (the value of B being approximately proportional to A).

## 3. Comparison of two mechanisms. Steady-state and transient plasmas.

An essential feature of formulae (4) and (8) is the same (within the limits of each process) dependence $i(T_e)$ for all satellites having the given principal quantum number n. Hence, if the general picture of the intensity distribution in a group of satellites varies in experiments, it can be explained only by changes of the relative roles of two mechanisms. In accordance with this we should expect the more reproducible picture for [H] satellites.

The temperature dependence $\Phi_d(T_e)$ is given by the simple formula (5). Let us consider now the function $\Phi_c(T_e)$. If a steady-state ionization balance is valid for a plasma, then:

$$\Phi_c \sim \frac{N_z}{N_{z+1}} = \frac{\varkappa_v}{\langle v\sigma_z \rangle}(1+D); \qquad D = \frac{\varkappa_d}{\varkappa_v}, \tag{9}$$

where $\langle v\sigma_z \rangle$ is the ionization rate of the ion $X_Z$ by electron collisions and $\varkappa_v$, $\varkappa_d$ are the full radiative and dielectronic recombination rates respectively (into all levels of $X_Z$).

In the all cases of interest the three-particle recombination can be neglected at $Z \gg 1$. Using known analytical approximations for $\langle v\sigma_z \rangle$ and $\varkappa_v$ we get:

$$\Phi_c(T_e) = C_c(1+D)\,\delta\beta\, e^{\delta\beta}; \qquad C_c \approx \left(\frac{z+0.4}{49.1}\right)^4. \tag{10}$$

It is interesting that the temperature dependences $\Phi_c(T_e)$ and $\Phi_d(T_e)$ coincide at $D \ll 1$. Hence, the difference in the temperature dependences of two excitation processes of satellites is due to the DR. As is known:

$$\varkappa_d \sim \frac{1}{T_e}\exp(-\Delta E_{res}/kT_e); \quad \Delta E_{res} \approx \frac{3}{4}E_{z+1} \tag{11}$$

where $E_{res}$ is the excitation energy of the [He] resonance line. Therefore, at sufficiently low temperatures $D \ll 1$.

In Fig.3 the dependence of the ratio $N_Z/N_{Z+1}$ on the electron temperature $T_e$ is shown for several ions with and without account for the factor D. The DR results in appearance of a specific plateau (or even a maximum) in the temperature function $\Phi_c(T_e) \sim N_{Z+1}$ at $Z < 20$ and namely in the temperature range where a given ion emits its radiation.

All said above related to steady-state plasma conditions. Practical calculations show that the process of the DR is dominating in most cases. The direct excitation is important mainly for relatively weak satellites.

In a transient recombining plasma the ratio $N_Z/N_{Z+1}$ is even less than that for the steady-state case, so the contribution of the direct excitation process is smaller. Nevertheless, an increased level of plasma ionization can be observed

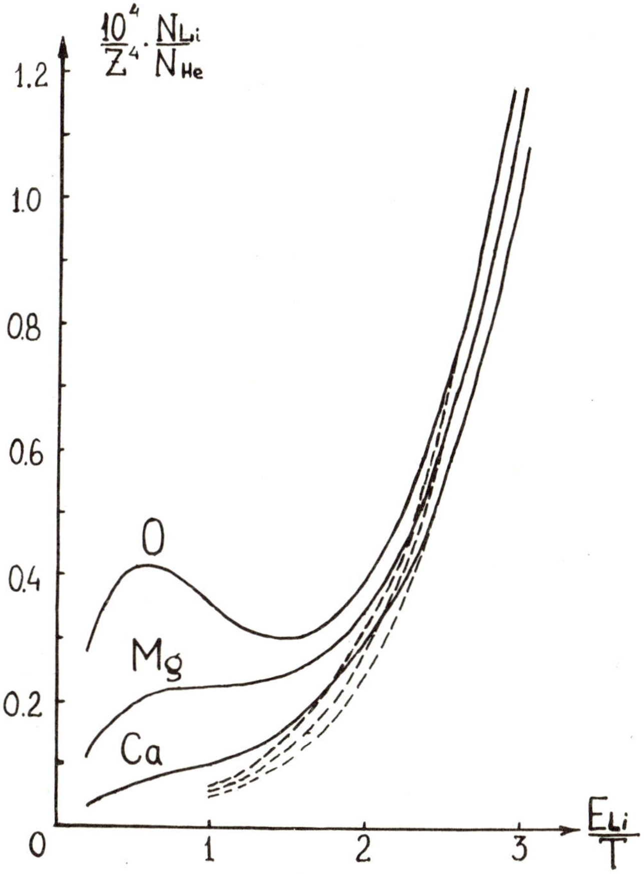


Fig. 3. Temperature dependences of N [Li] /N [He] calculated with account for the DR (solid line) and without it (dash line).

due to the absence of some satellites provided that a spectral resolution and a signal-to-noise ratio are sufficiently high.

The most interesting case, however, is with transient ionizing plasma when the ionization equilibrium is not reached. Here the population ratio $N_Z/N_{Z+1}$ of [Li] and [He] ions appears to be abnormally high. Therefore the value $i_c$ is here relatively higher, because the direct excitation process is based on [Li] ions whereas the DR - on [He] ions.

Fig.4 shows the satellite structures of the [He] ion of P, obtained in a laser produced plasma using a Nd-glass laser and a $CO_2$-laser /4/. Since the frequences of the heating laser radiation are different, in the first case the electron density is about $10^{21}$ $cm^{-3}$, and in the second - about $10^{19} cm^{-3}$ at nearly the same values of the plasma life-time $\sim 1$ ns and of electron temperature $\sim 0.3$ keV. Our estimations show that in the last case the ionization equilibrium is not reached. This appears in the spectrum (Fig.4) as a strong increase of the satellites a, d, q.

The intensity ratio of satellites having different contributions of each excitation mechanism allows to determine the relevant population ratio of ions [He] and [Li]. At rapid heating of a plasma it contains in sufficient quantitites not only [Li] ions but also the lower stages: [Be] , [B], etc; whichis clearly seen in Fig.4. Diagnostics of such ions can be fulfilled by using satellites of the type (2) in a similar way.

The deviation from the ionization equilibrium is often characterized by the temperature $T_z$ which is the electron temperature of such a steady-state plasma, where the ratio $N_Z/N_{Z+1}$ equals to the experimentally observed one. It should be mentioned, however, that this parameter is not very sucessful. As is seen in Fig.3 the ratio $N_Z/N_{Z+1}$ has a plateau exactly in the region of the greatest interest. Here the small errors in measurements or calculations result in large variations of $T_Z$.

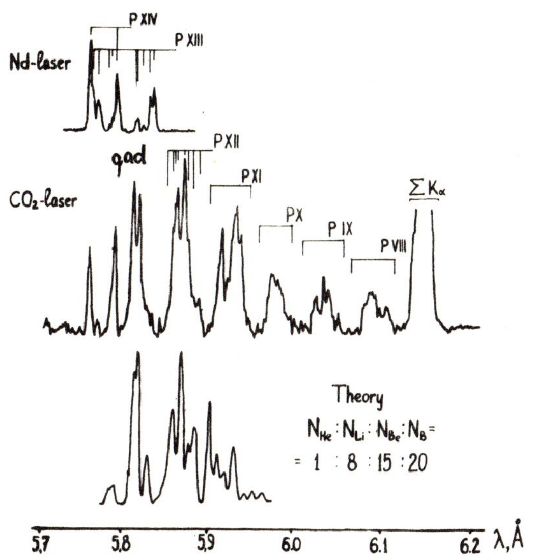


Fig. 4. The satellite structures of the [He] ion P XIV in the spectra of plasmas produced by the Nd-glass laser and the $CO_2$-laser.

## 4. Applications to plasma diagnostics

From all what is said above there are the following possibilities and/or peculiarities in applications of satellite spectra of multicharged ions to plasma diagnostics.

1. Intense satellites having purely dielectronic excitation are very convenient for determination of the electron temperature with the aid of formula (4). This method has two important advantages. Firstly, the resonance line and its satellites are in a narrow spectral region, and at the same time the intensity ratio is essentially temperature-dependent. Secondly, both the resonance line and the DS are excited from the same initial state.

2. Satellites with dominating direct excitation process allow to determine the population ratio N [Li] /N [He], i.e. the level of stationarity of plasma emitting these lines. N [Be] , N [B], etc can be determined in a similar way.

3. The formulae considered above are related to the case when a plasma has not very high density and the probability of a collision can be neglected comparing to that of radiative decay. With increasing density the collisional transitions between upper satellite levels $\gamma - \gamma_1$ become essential. In other words, an excitation is transferred from one satellite to another. It is easy to see that in these conditions the satellites with high A and low $\Gamma$ appear in spectra although they are not excited directly. Their intensity is proportional to the electron density $N_e$, that gives a possibility of measuring $N_e$ in the range $N_e \gtrsim 10^{20}$ cm$^{-3}$.

In Fig.5 the satellite structure of the Mg XII space-resolved spectra is shown. The electron density and therefore the intensity of the satellite marked by an arrow are decreasing with the increase of the distance from the target /5/.

This effect is especially important for [H] satellites. In the case of [He] satellites it is masked by the process of the direct excitation of [Li] ions, which also appears in the first place at low $\Gamma$ and high A.

It should be mentioned that non-stationarity effects are also indirectly connected with $N_e$: they are stronger at low

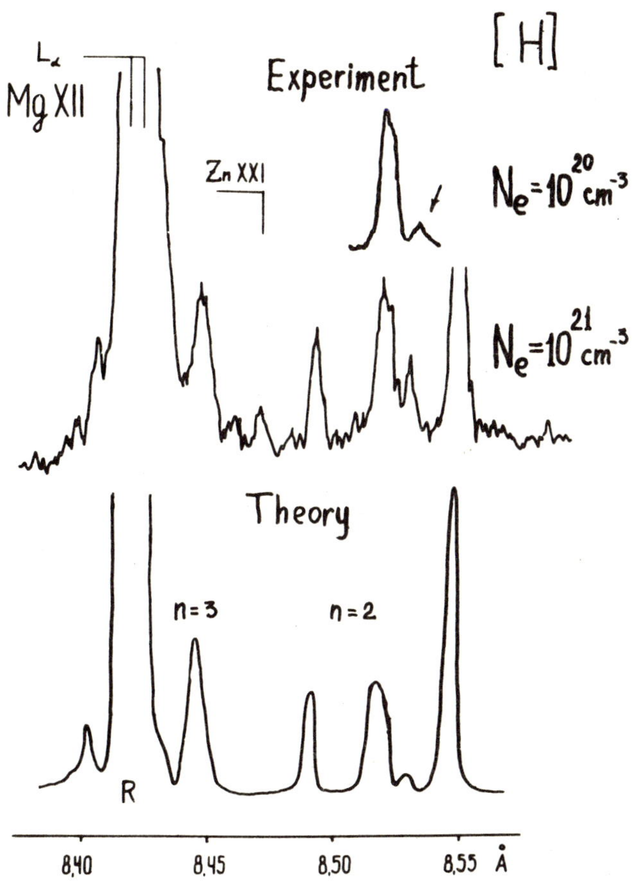


Fig. 5. The satellite structure of the [H] ion Mg XII observed at various distances from the target.

$N_e$ and can give estimates of $N_e$ in the range $N_e < 10^{20}$ $cm^{-3}$.

4. Plasma diagnostics with satellites is, unfortunately, not always easy due to rather dense distribution of lines. Hence a careful analysis is necessary based on various assumptions concerning line widths. An example of such analysis is shown in Fig.6: a spectrum in the vicinity of the Ca XIX resonance line is calculated using various line widths and various $T_Z$(i.e. values of $N_Z/N_{Z+1}$). An experimental spectrum is placed in the vicinity of the "optimally" fitted theoretical one /6/.

The situation which arises in a low-inductance vacuum spark - the "hottest" known plasma source - is still more complicated /7/. Some results of analysis of iron vacuum spark spectra are given in Fig.7. The line broadening is here more considerable than in a laser plasma and this disfigures the satellite structure. It is also seen in Fig.7 that the contributions of the DS and the direct excitation can be very different depending on the spark regimes.

5. The non-steady-state character of ionization observed on non-time-resolved spectrograms (a $CO_2$-laser plasma, a vacuum spark) suggests that the ratios $N_Z/N_{Z+1}$ vary to a large extent during the plasma life-time. An example is shown in Fig.8 where iron ion populations, computed theoretically, are plotted against the product of the density and the time $N_e\tau$, an initial jump of the temperature $T_e$ up to 5 keV being assumed at $\tau$=0 /8/. Line intensities observed experimentally are proportional to the relevant time-integrated value of $N_Z$. (Fig.9). A comparison of such calculations with experimental data brings into accordance observed relative intensities of satellites for [H] and [He] ions simultaneously, which was not possible when considering a simple steady-state model.

An analysis of plasma conditions in laser plasmas, vacuum sparks, tokamaks and solar flares shows that the values of $N_e\tau$ are close to $10^{12}$ $cm^{-3}$ sec (or even less). This suggests a non-steady-state solutions in the cases of highest ionization stages observed for heavy elements.

Time-resolved spectra of some H- and He-like ion reso-

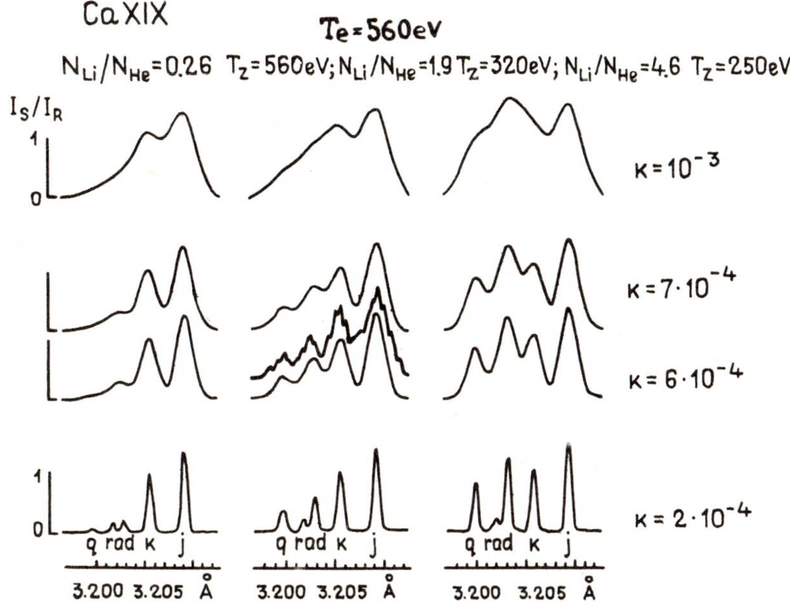


Fig. 6. The satellite structure of Ca XIX calculated for various relative Doppler widths k= $\Delta\lambda_D/\lambda$ and ratios $N_Z/N_{Z+1}$.

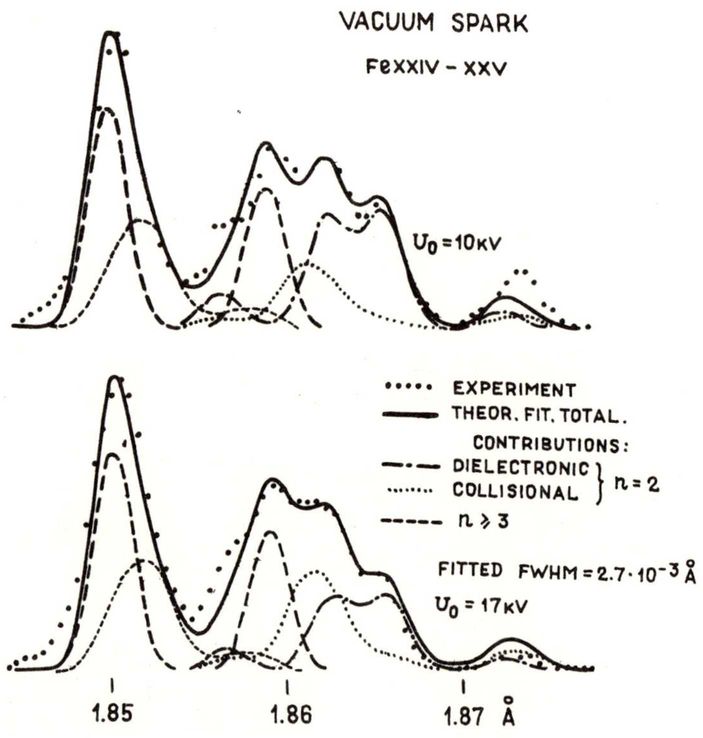


Fig. 7. The satellite structure of Fe XXV observed in a vacuum spark and calculated theoretically. The fitted contributions of the DR process (for n=2 and 3) as well as of the direct excitation are shown.

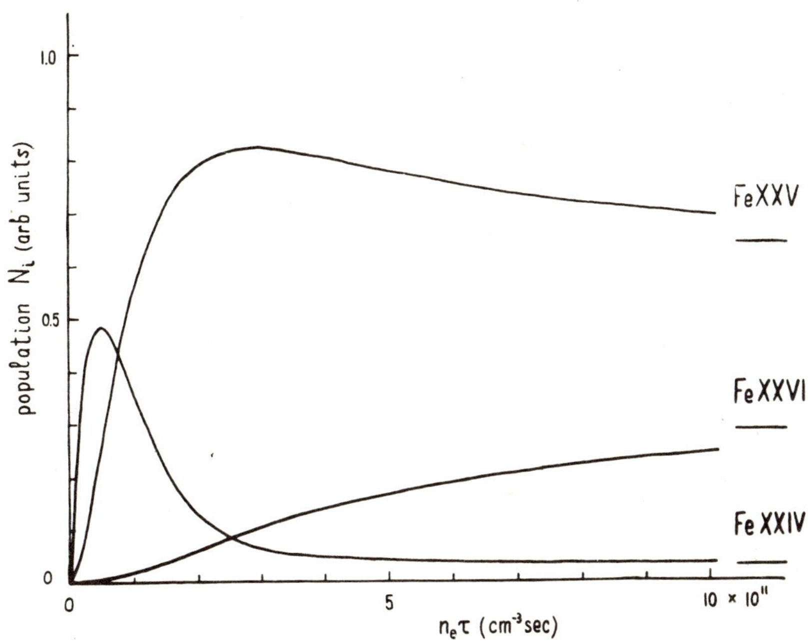


Fig. 8. Calculated relative densities of the [Li] , [He] and [H] iron ions as function of the product of electron density and time. An initial jump of $T_e$ up to 5 keV assumed.

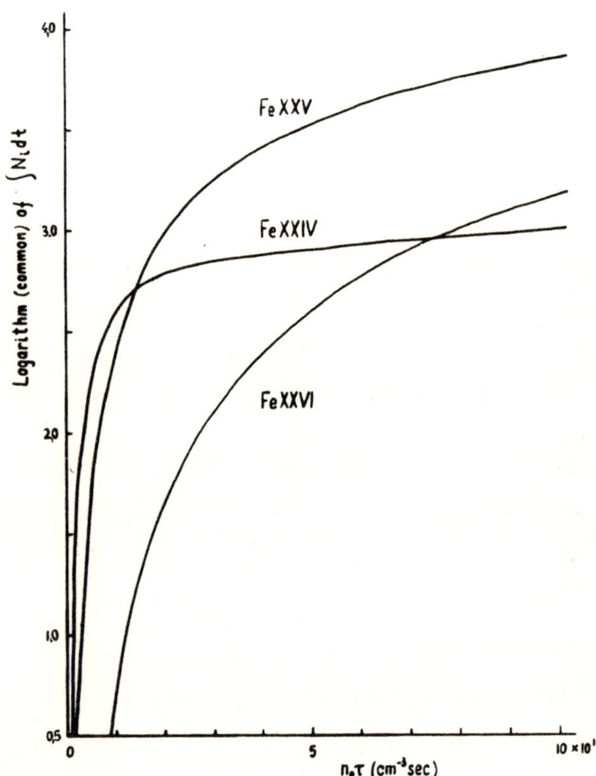


Fig. 9. Time integrals of the relative densities shown in Fig. 8 as function of $N_e\tau$.

nance lines and their satellites excited in the laser produced plasma were recently obtained at the Lebedev Physical Institute /9/.

The time resolution of the spectra obtained was about 1.8-2.0 nsec with the Nd-laser pulse duration about 5 nsec. Fig.10 shows the time dependence of the intensity of [H] and [He] ion resonance lines of Al together with Nd-laser pulse. The time dependences of electron temperature and density obtained from the relative intensities of dielectronic satellites ($T_e$) and of intercombination line ($N_e$) to the resonance line are shown in Fig.11.

The main feature of these experimental data is the marked difference between emission time duration of [He] and [H] ion resonance lines. The [He] line is emitted during the whole laser pulse; the time of emission of [H] line is shorter. The electron temperature is, apparently, slightly higher at first moments and comes to the value which is equal to that obtained from time-integrated spectra.

Further investigations of time-resolved spectra would be useful to understand the dynamic evolution of the plasma ionization.

6. Sometimes, in laser plasmas in particular, an analysis of plasma inhomogenity as well as optical opacity is essential. Such considerations can complicate the solution of interpretation problems.

## 5. Calculations of atomic data

All considered above shows that the satellite structure is a very useful instrument of diagnostics of a hot plasma both in laboratory and in astrophysics. For appropriate use of this instrument the knowledge of some atomic data is necessary such as wavelengths ($\lambda$), radiation and autoionization transition probabilities ($A, \Gamma$) and excitation cross-sections for resonance and satellite levels ($\sigma$). All these data are necessary for [H], [He], [Li] and other ions of high stages of ionization $Z \gtrsim 10$.

Values $\lambda$, $A$, $\Gamma$ for ions (the levels with $n = 2,3$) have been calculated using two methods: 1)the Hartree-Fock method with

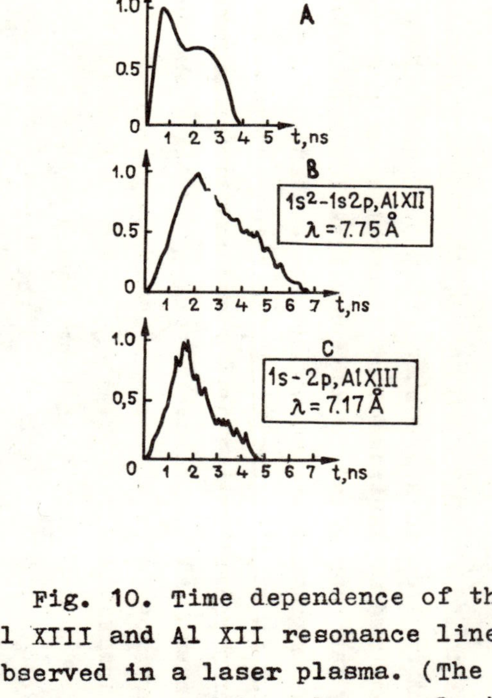


Fig. 10. Time dependence of the Al XIII and Al XII resonance lines observed in a laser plasma. (The curve A represents a laser pulse).

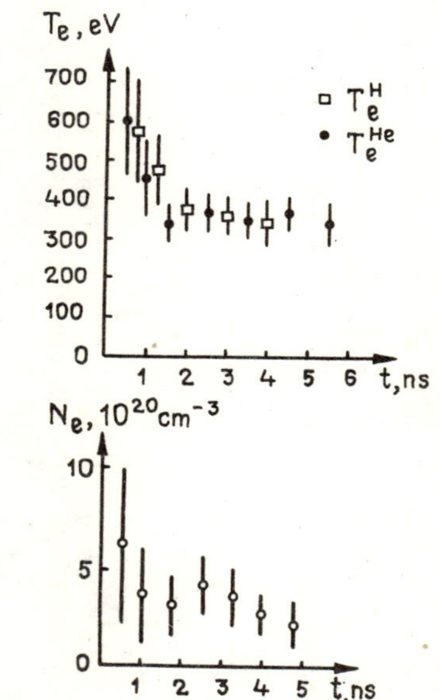


Fig. 11. Experimental values of electron temperature $T_e$ and density $N_e$ as function of time in a laser plasma.

the intermediate coupling scheme; 2) the perturbation theory method based on hydrogen-like functions, i.e. the 1/Z expansion (Z is here a nucleus charge). The first method was used by Jones, Gabriel and others in England and USA. The second was used in the USSR by Safronova et al. As a result the tables of $\lambda$, A, $\Gamma$ for ions in a wide range of Z have been published.

The situation is worse with cross-sections. Calculations were mainly based on an approximation without account for exchange. In these conditions:

$$\frac{\sigma_s}{\sigma_{res}} \approx \frac{A_s}{A_{res}} . \qquad (12)$$

However, the exchange strongly influences on $\sigma$ for some transitions and it should be taken into account. Such new calculations were recently carried out at the Lebedev Physicsl Institute.

Data obtained by different methods agree, in general, quite well. The largest deviations are for $\Gamma$ , probably due to using only the first-order term in the 1/Z expansion (without screening). On the other hand this method is the most convenient for systematic calculations, as it gives results for a whole isoelectronic sequence immediately.

Investigations of laboratory and astrophysical plasmas were extremely important for verification of the theoretical atomic data related to multicharged ions. At the present time an extensive material is collected and certain conclusions can be drawn.

It should be mentioned, that the accuracy of wavelength calculations turned out to be very high.Thus, for [He] and [Li] ions a relative error $\Delta\lambda/\lambda$ is of the order $10^{-4}$. This error is comparable with the experimental one and sometimes is even less than the last one: this is a situation quite rare in spectroscopy.

The accuracy of A, $\Gamma$ , $\sigma$ is not so easy to estimate, since experimentally observed intensities are essentially dependent on plasma parameters $T_e$, and $N_e$, on stationarity conditions, mechanisms and values of the line broadening. In any case, at present we have no evidence of large errors for A and $\Gamma$, at least when their values are not too small (as in the case of transitions allowed only due to admixture of other states).

## 6. Conclusion

Laboratory and astrophysical investigations of multicharged ions spectra brought into life in present situation new means for study of physical processes in hot plasmas. These are based on the analysis of satellite structures close to resonance lines observed in the X-ray region. These means require, however, thoroughful studies of various elementary processes related to populations of ionic energy levels. An excitation of multicharged ions spectra takes place in plasma regions with extremely high temperatures and, therefore, is connected with accumulation of energy. Since the rates of such accumulations are always finite, the transient phenomena in the ionization balance are important to the same extent as the ionization equilibrium situations. This is why the needs in detailed information concerning atomic constants such as transition probabilities and cross-sections of electron-ion interactions become more and more urgent.

## References


1. V.A.Boiko, A.Ya.Faenov, and S.A.Pikuz, J.Quant.Spectr.Radiat.Transfer. 19, 11 (1978)
2. V.A.Boiko, A.Ya.Faenov, S.A.Pikuz, and U.I.Safronova, Mon.Not.Roy.Astr.Soc. 181, 107 (1977)
3. A.H.Gabriel, Mon.Not.Roy.Astr.Soc. 160, 99 (1972)
4. V.A.Boiko, A.Yu.Chugunov, A.Ya.Faenov, T.G.Ivanova, I.A.Kholin, S.A.Pikuz, U.I.Safronova, A.M.Urnov, L.A.Vainstein, Mon.Not.Roy.Astr.Soc. (in press)
5. V.A.Boiko, S.A.Pikuz, A.Ya.Faenov. Kratkie soobshcheniya po fisike (Short Comm.Phys.) N 4, 38 (1977)
6. V.A.Boiko, S.A.Pikuz, A.Ya.Faenov, Kvantovaya elektronika. 5, 394 (1978)
7. E.Ya.Kononov, Physica Scripta, 17, 425 (1978)

8. E.Ya.Kononov, K.N.Koshelev, Yu.V.Sidelnikov. Fysika plasmy (Sov.J.Plasma Phys.) 3, 663 (1977)
9. Yu.S.Kasyanov, M.A.Mazing, V.K.Chevokin, A.P.Shevel'ko, Pis'ma JETF (Lett.JETP) 25, 373 (1977)

# MESIC ATOMIC AND MESIC MOLECULAR PROCESSES IN THE HYDROGEN ISOTOPE MIXTURES

L.I.Ponomarev

Joint Institute for Nuclear Research, Dubna, U S S R

## 1. Introduction

In recent years much attention has been focused on the physics of $\mu$ -atomic phenomena which occur during deceleration and stopping of muons in matter. The reasons are as follows. Experimentally: mesonic facilities are in operation now and relevant new experiments are carried out. Theoretically: only in recent years methods have been developed for the calculation of almost all $\mu$ -atomic processes with an accuracy necessary for reliable comparison with experiment.

From the theoretical viewpoint all the $\mu$ -atomic processes are particular cases of the three-body problem with the Coulomb interaction. Due to the smallness of mesic atoms (mesic atomic unit of length is $a_\mu = \hbar^2/m_\mu e^2 = 2.56.10^{-11}$ cm) one can neglect the influence of the electron shell of atoms in describing the mesic atomic processes. On the other hand, the $\mu$ -atomic physics can be treated as a unique laboratory for checking the most refined methods of calculation of different characteristics of the three-body system.

This review is devoted to the theoretical description of $\mu$ -atomic processes in hydrogen isotope mixtures ($H_2$, $D_2$ and $T_2$). It contains the results of the theoretical investigations performed in the Joint Institute for Nuclear Research from 1973 to 1978 by the Dubna group: M.P.Faifman, L.I.Ponomarev, I.V.Puzynin, T.P.Puzynina, L.N.Somov and S.I.Vinitsky.

The formation of mesic atoms $p\mu$ , $d\mu$ and $t\mu$ in those mixtures occurs during deceleration and stopping of $\mu^-$ -mesons, and then different processes take place, such as the elastic collisions of mesic atoms ($p\mu+p$ , $d\mu+p$ and so on), the isotope exchange processes ($p\mu+d \rightarrow d\mu+p$ ), the formation of mesic molecules ($d\mu+p \rightarrow pd\mu$ ), the transitions between the levels of the hyperfine structure of mesic atoms ( $p\mu(\uparrow\uparrow) + p \to p\mu(\uparrow\downarrow) + p$ ) and so on (see the reviews [1-2]).

For the description of the collision processes in the three-body system at the collision energies $\varepsilon < 1$ eV (where most of the $\mu$ -atomic processes occur) it is natural to use the adiabatic representation. The advantages and difficulties of this approach are rather well known[3]. However, in recent years most of these difficulties have been removed, which allowed one to determine many characteristics of the $\mu$ -mesic atomic processes more accurately. The main characteristics of the $\mu$ -mesic atoms are presented in Table I.

Table I

Main characteristics of mesic atoms*

| | $M_a$ | $-E_a$(eV) | $\Delta E$ (eV) | $\Delta E^{hfs}$(eV) |
|---|---|---|---|---|
| $m_\mu$ | 206.769 | - | - | - |
| $M_p$ | 1836.152 | 2528.437 | $\Delta E_{pd} = 134.705$ | 0.183 |
| $M_d$ | 3670.481 | 2663.142 | $\Delta E_{pt} = 182.745$ | 0.049 |
| $M_t$ | 5496.918 | 2711.182 | $\Delta E_{dt} = 48.040$ | 0.241 |

* The values indicated are calculated on the basis of the data in ref.[18]. The isotopic difference of the energy levels of the different mesonic atoms equals, e.g., $\Delta E_{pd} = E_p - E_d$ . The hyperfine splitting $\Delta E^{hfs}$ equals the energy difference of the ortho and para states of $p\mu$ and $d\mu$ atoms.

## 2. The adiabatic representation of the three-body problem

The Schrödinger equation for the three-particle system which is composed of two isotopes of hydrogen nuclei with masses $M_a \geqslant M_b$ and of the $\mu^-$-meson with mass $m_\mu$ in the coordinates $\vec{r}$ and $\vec{R}$ (see Fig.1) and in units $e = \hbar = 1$ = has the form[4-5]

$$(\hat{H} - E)\Psi(\vec{r}, \vec{R}) = 0$$

$$\hat{H} = \hat{T}_a + \hat{W}_a , \qquad \hat{W}_a = \hat{h}_a + \frac{1}{R} \tag{1}$$

$$\hat{T}_a = -(2M_0)^{-1}\left[\left(\nabla_{\vec{R}} + \frac{\varkappa}{2}\nabla_{\vec{r}}\right)^2 - \left(\frac{1+\varkappa}{2}\right)^2 \Delta_{\vec{r}}\right]$$

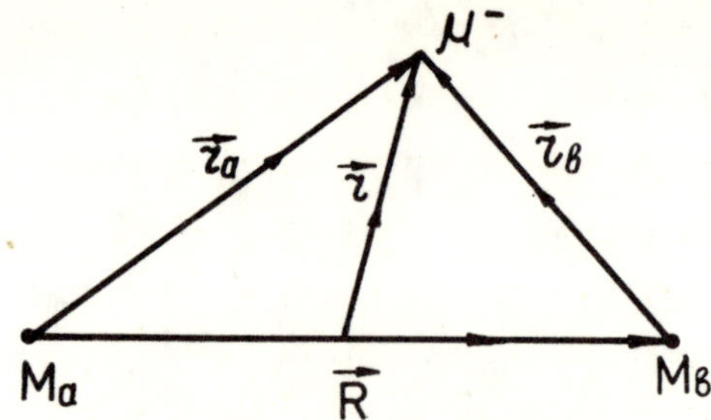


Fig. 1. Coordinates for the three-body problem.

where

$$\hat{h}_a = -(2m_a)^{-1}\Delta_{\vec{r}} - \frac{1}{r_a} - \frac{1}{r_b} \tag{2}$$

is the Hamiltonian of the two-center problem, which describes the motion of a negative charged particle in the Coulomb field of two positive charged centers separated by the distance R /6-7/. We introduce also the notation

$$m_a^{-1} = m_\mu^{-1} + M_a^{-1}, \quad M_0^{-1} = M_a^{-1} + M_b^{-1}; \quad \varkappa = \frac{M_b - M_a}{M_b + M_a} \tag{3}$$

The state of the three-body system is specified by the total angular momentum $J$, its projection $m_J$ and the total parity $\lambda = \pm(-)^J$. In the adiabatic representation the wave function of the three-body system is expanded over the adiabatic basis set, i.e. over the solutions of the two-center problem of quantum mechanics in the discrete and continuous spectrum, respectively

$$\hat{h}\,\Phi_{jm}(\vec{r};R) = E_{jm}(R)\,\Phi_{jm}(\vec{r};R) \tag{4a}$$

$$\hat{h}\,\Phi_{jm}(\vec{r};k,R) = \frac{k^2}{2}\Phi_{jm}(\vec{r};k,R) \tag{4b}$$

Here $\hat{h}$ is the Hamiltonian (2) of the two-center problem in units $m_a = 1$, and $j$ is the set of quantum numbers of the two-center problem /6-7/. In spheroidal coordinates

$$\xi = (r_a + r_b)/R, \quad \eta = (r_a - r_b)/R, \ \varphi \tag{5}$$

the function (4) can be represented as

$$\Phi_{jm}(\vec{r};R) = \varphi_{jm}(\xi,\eta;R)e^{im\varphi}(2\pi)^{-1/2} \times \begin{cases} (-)^m & \text{at } m>0 \\ 1 & \text{at } m<0 \end{cases} \tag{6}$$

In the discrete and continuous spectrum $j=\{n_1 q\}$, and $j=\{q\}$ respectively, where $n_1$ and $q$ are the numbers of zeroes of the wave functions $\Pi_{mn_1}(\xi;R)$, and $\Xi_{mq}(\eta;R)$ in the intervals $\xi\in(1,\infty)$ and $\eta\in$ (-1,1), respectively.

By introducing the functions

$$F^J_{mm_J}(\Phi,\theta,\varphi)=\left[4\pi(1+\delta_{0m})\right]^{-1/2}\left[(-)^m e^{im\varphi}D^J_{mm_J}(\Phi,\theta,0)+e^{-im\varphi}D^J_{-mm_J}(\Phi,\theta,0)\right] \quad (7)$$

the wave functions of the three-body system with the total positive parity $\lambda$ can be represented as

$$\Psi(\vec{r},\vec{R})=\sum_j\sum_m \varphi_{jm}(\xi,\eta;R)R^{-1}X^{Jm_J\lambda}_{jm}(R)F^J_{mm_J}(\Phi,\theta,\varphi), \quad (8)$$

where the summation includes the integration over the continuous spectrum of the two-center problem

$$\sum_j=\sum_{q=0}\left\{\sum_{n_1=0}+\int_0^\infty d\kappa\right\}. \quad (9)$$

Substituting expansion (8) into eqation (1) and averaging over the variables $\Phi,\theta,\varphi$ and the adiabatic basis set $\varphi_{jm}(\xi,\eta;R)$ we obtain the infinite system of ordinary integro-differential equations for the functions $X_{jm}$ = $X^{Jm_J\lambda}_{jm}(R)$ /5/

$$\left\{\frac{d^2}{dR^2}-\frac{J(J+1)-2m^2}{R^2}+2M\left[E-\frac{1}{R}-E_{im}(R)\right]\right\}X_{im}(R)= \quad (10)$$
$$=\sum_{jm'}U^J_{im,jm'}(R)X_{jm'}(R),$$

where $E_{im}(R)$ are the terms of the two-center problem (4a), and $U^J_{im,jm'}(R)$ are the effective potentials of the three-body problem in the adiabatic representation, i.e. the matrix elements of the operator (1) of the kinetic energy $\hat{T}_a$ over the wave functions of the two-center problem.

The properties of the adiabatic basis set $\Phi_{jm}(\vec{r};R)$ and $\Phi_{jm}(\vec{r};k,R)$ in the discrete and continuous spectrum, respectively, and the algorithms of its numerical calculation have been thoroughly studied in papers [/6-11/]. The effective potentials $U^J_{im,jm'}(R)$ connecting the states of the discrete spectrum of the two-center problem, and potentials $U^J_{im,qm'}(\kappa,R)$ connecting the states of the discrete and continuous spectrum have been calculated in papers [/9-12/] and [/13-14/], respectively. The developed algorithms allow one to calculate the terms $E_{jm}(R)$ and the potentials $U^{J}_{im,jm'}(R)$ with an accuracy of about $10^{-11}$, and potentials $U^{J}_{im,qm'}(\kappa,R)$ with an accuracy of about $10^{-5}$.

At present they are calculated for the sets $i=\{n_1\, q\, m\}$ and $j = \{n_1'\, q'\, m'\}$ with the quantum numbers $N=n_1+q+m+1 \leqslant 10, j=n_1'+q'+m'+1\leqslant 10$ in the interval of R=0.1(0.1)20(1)100 /12/ and for the sets $i=\{0\,0\,0\}$, $i=\{010\}$ and $j=\{\kappa q' m'\}$ with the quantum numbers $m'=0,1, q=0(1)7$ in the intervals $R=0.1\,(0.1)\,20$

and $\kappa=0.2\,(0.1)\,1.2\,(0.2)\,2\,(1)\,10$ /14/.

The system of ordinary integro-differential equations (10) is equivalent to the initial Schrödinger equation (1) and follows from it without any approximation. By solving this system one can calculate all the characteristics of the three-body problem, in particular the binding energy and cross sections of the scattering processes. The accuracy of the result obtained is restricted by the finiteness of the solved system of equations approximating the infinite system (10) and by the accuracy of the numerical algorithms of solving this system.

With the exception of several special cases, to obtain the cross sections and binding energies with an accuracy of about 10-30%, the two-level approximation of the adiabatic method is sufficient, i.e. it is sufficient to keep only two equations of the system (10). In this case the phase function method can effectively be used /15/. Based on this method, we have developed the stable algorithms for solving the scattering problems both for the case of two open channels /16/ and for the case of one open and one closed channel /17/.

As an important physical application of the above method we shall represent below the results of calculation of the cross sections and rates of different mesic atomic processes mentioned in the Introduction. The main characteristics of mesic atoms $p\mu, d\mu$ and $t\mu$ are presented in Table I. The general scheme of mesic atomic and mesic molecular processes in the mixture $H_2+D_2$ is given in Fig.2.

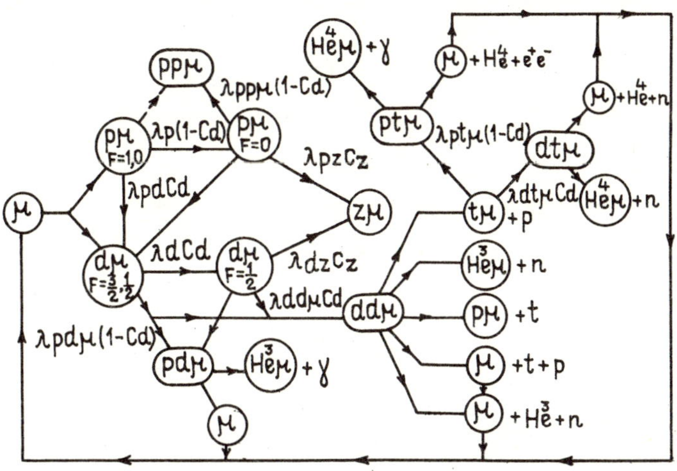


Fig.2. General sequence of mesic atomic and molecular processes in a hydrogen isotope mixture.

## 3. Elastic and Inelastic Collisions of Mesic Atoms of Hydrogen Isotopes

As an example let us consider the mesic atomic collisions in the hydrogen and deuterium mixture. At the collision energy $\varepsilon_1 > \Delta E_{pd}$ (see Fig.3a),i.e. when both the reaction channels are open, the following processes are possible:

$$
\begin{aligned}
&(\sigma_{11}) \quad d\mu + p \to d\mu + p \\
&(\sigma_{12}) \quad d\mu + p \to d + p\mu \\
&(\sigma_{21}) \quad p\mu + d \to p + d\mu \\
&(\sigma_{22}) \quad p\mu + d \to p\mu + d
\end{aligned}
\qquad (11)
$$

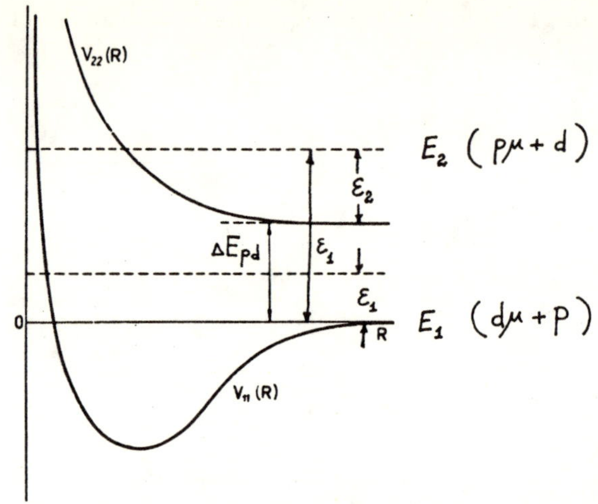


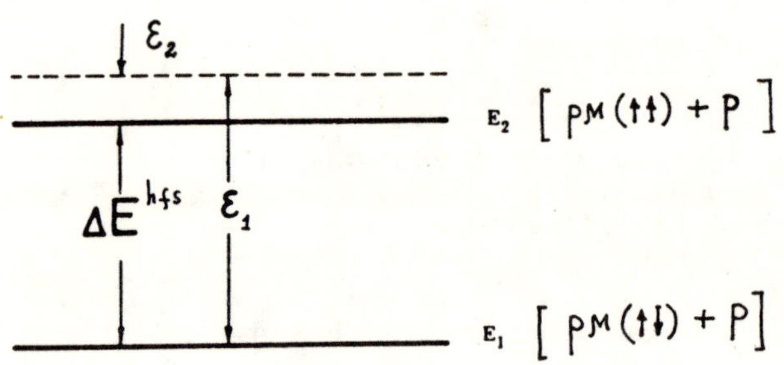


Figs.3a, 3b. Schematic behaviour of the diagonal potentials $U_{ii}(R)$ in the two-level approximation in equations (10): $E_i$ are the energies of the three-body system, $\varepsilon_i$ is the scattering energy in the channel $i$.

The cross sections of these processes at the collision energies $\varepsilon_2 \to 0, \varepsilon_1 \to \Delta E$ behave as follows:

$$\begin{aligned} &\sigma_{11} \sim \text{const}, \quad \sigma_{12} \sim (\varepsilon_2)^{1/2} \\ &\sigma_{21} \sim (\varepsilon_2)^{-1/2}, \quad \sigma_{22} \sim \text{const}. \end{aligned} \tag{12}$$

At the collision energy $\varepsilon_1 < \Delta E_{pd}$ only elastic scattering is possible

$$(\sigma_{11}) \qquad d\mu + p \rightarrow d\mu + p. \tag{13}$$

In the limit $\varepsilon_1 \rightarrow 0$ , $\sigma_{11} \rightarrow const.$

It is convenient to introduce the isotope exchange rates $\lambda$ and $\Lambda$ by formulae

$$\lambda = \sigma_{21} v \; cm^3 s^{-1} , \quad \Lambda = \lambda N_0 s^{-1}, \tag{I4}$$

where $v$ is the collision velocity and $N_0 = 4.25 \cdot 10^{22} cm^{-3}$ is the density of liquid hydrogen. From relations (12) there follows the limiting behaviour $\lambda_{pd} \rightarrow const$ as $\varepsilon_2 \rightarrow 0$ .

Table II represents the values $\sigma_{ij}$, $\bar{\sigma}_{11}$, $\lambda$ and $\Lambda$ for the systems $p+d$ , $d+t$ and $p+t$ at collision energy $\varepsilon = 0.04$ eV, i.e. at room temperature.

The energy dependence of the cross sections and the discussion of their peculiarities can be found in refs./16,17,19/.

Table II

Cross sections $\sigma_{ij}$ and $\bar{\sigma}_{11}$ and transfer rates $\lambda$ and $\Lambda$ in the mixture of hydrogen isotopes at the scattering energy $\varepsilon = 0.04$ eV

| | $p\mu+d$ | $p\mu+t$ | $d\mu+t$ |
|---|---|---|---|
| $\sigma_{11}, 10^{-19} cm^2$ | 1.8 | 2.0 | 1.6 |
| $\sigma_{12}, 10^{-22} cm^2$ | 3.3 | 1.2 | 0.15 |
| $\sigma_{21}, 10^{-18} cm^2$ | 1.2 | 0.56 | $1.8 . 10^{-2}$ |
| $\sigma_{22}, 10^{-20} cm^2$ | 6.8 | 3.3 | $3.4 . 10^{-2}$ |
| $\lambda$ , $10^{-13} cm^3 s^{-1}$ | 4.0 | 1.8 | $4.5 . 10^{-2}$ |
| $\Lambda$ , $10^{10} s^{-1}$ | 1.7 | 0.75 | $1.9 . 10^{-2}$ |
| $\bar{\sigma}_{11}, 10^{-20} cm^2$ | 0.57 | 1.8 | 14.5 |

## 4. Transitions Between H.F.S. Levels of Mesic Atoms

Since the value $\Delta E^{hfs}$ of the hyperfine splitting of the ground state of mesic atoms exceeds the average kinetic energy of mesic atoms at normal temperatures (see Table I), there are transitions between these levels in collisions of mesic atoms with nuclei. These processes are analogous to the isotope exchange reactions (11). For instance, in pure hydrogen for collision energies $\varepsilon_1 > \Delta E^{hfs}$ (see Fig. 3b) the following processes are possible:

$$\begin{aligned}
&(\sigma_{11}) \quad p\mu(\uparrow\downarrow) + p \longrightarrow p\mu(\uparrow\downarrow) + p,\\
&(\sigma_{12}) \quad p\mu(\uparrow\downarrow) + p \longrightarrow p\mu(\uparrow\uparrow) + p,\\
&(\sigma_{21}) \quad p\mu(\uparrow\uparrow) + p \longrightarrow p\mu(\uparrow\downarrow) + p,\\
&(\sigma_{22}) \quad p\mu(\uparrow\uparrow) + p \longrightarrow p\mu(\uparrow\uparrow) + p,
\end{aligned} \tag{15}$$

and at the energy $\varepsilon_1 < \Delta E^{hfs}$ the elastic scattering is possible

$$(\overline{\sigma_{11}}) \quad p\mu(\uparrow\downarrow) + p \longrightarrow p\mu(\uparrow\downarrow) + p. \tag{16}$$

Table III

Cross sections $\sigma_{ij}$ and $\overline{\sigma_{11}}$, and spin-flip rates $\lambda$ and $\Lambda$ at the scattering energy $\varepsilon = 0.04$ eV *

| | $p\mu + p$ | $d\mu + d$ | | $t\mu + t$ |
|---|---|---|---|---|
| | | $J = 1/2$ | $J = 2/3$ | |
| $\sigma_{11}, 10^{-19} cm^2$ | 1.0 | 1.5 | 2.3 | 0.33 |
| $\sigma_{12}, 10^{-19} cm^2$ | 5.8 | $4.9.10^{-2}$ | $3.0.10^{-2}$ | 0.35 |
| $\sigma_{21}, 10^{-19} cm^2$ | 27 | $9.6.10^{-2}$ | $6.0.10^{-2}$ | 2.7 |
| $\sigma_{22}, 10^{-19} cm^2$ | 27 | 1.9 | 1.2 | 7.8 |
| $\lambda$, $10^{-13} cm^3 s^{-1}$ | 3.9 | $2.9.10^{-2}$ | $1.8.10^{-2}$ | 0.21 |
| $\Lambda$, $10^9 s^{-1}$ | 17 | 0.12 | 0.78 | 0.89 |
| $\overline{\sigma_{11}}, 10^{-19} cm^2$ | 0.35 | 1.4 | 2.1 | 0.13 |

* For the system $d\mu + d$ there are two possibilities for the total angular momentum: $J = 3/2$ and $J = 1/2$. Taking into account the statistical

weights of the orthostates in both the states $J = 3/2$ and $J = 1/2$ we obtain the values

$\overline{\sigma}_{11}$ and $\Lambda$

$$\overline{\sigma}_{11} = \frac{1}{3}\overline{\sigma}_{11}(J=\tfrac{1}{2}) + \frac{2}{3}\overline{\sigma}_{11}(J=\tfrac{3}{2}) = 1.9 \cdot 10^{-19}\,\mathrm{cm}^2;$$

$$\lambda = \frac{1}{6}\lambda(J=\tfrac{1}{2}) + \frac{1}{3}\lambda(J=\tfrac{3}{2}) = 1.1 \cdot 10^{-15}\,\mathrm{cm}^3\,\mathrm{s}^{-1};$$

$$\Lambda = \lambda N_0 = 2.8 \cdot 10^8\,\mathrm{s}^{-1}.$$

The results of calculations of the cross sections of the processes of type (15) for the systems $p\mu + d$, $d\mu + d$ and $t\mu + t$ are given in Table III. In the limit $\varepsilon_i \to 0$ the cross sections $\sigma_{ij}$ can be expressed in terms of the scattering lengths $a_g$ and $a_u$ /20/. The recent calculations/21/ of the scattering lengths $a_g$ and $a_u$ are presented in Table IV. It should be noted, however, that the notion "scattering length" makes sense only in the energy region $\varepsilon \lesssim 10^{-2} - 10^{-3}$ eV /19/.

Table IV

Scattering lengths of the hydrogen mesic atoms (in units $a_\mu = 2.56 \cdot 10^{-11}$ cm)

| Process | $a_g$ | $a_u$ |
|---|---|---|
| $p\mu + p$ | -29.4 | 3.51 |
| $d\mu + d$ | 4.91 | 2.95 |
| $t\mu + t$ | -8.93 | 2.21 |

5. Energy Levels of Mesic Molecules

The accuracy of the adiabatic method in the two-level approximation is of the order of $(m_\mu/M_p)^2$. This is not enough for precision calculations of the energy levels of mesic molecules. To improve this accuracy it is necessary to solve the system of 30-40 differential equations. Such methods have been developed in refs./22-23/ on the basis of the continuous analog of the Newton method/24/. In combination with the perturbation theory these methods allowed one to calculate the energy levels of $\mu$ -mesic molecules with an absolute accuracy of about 0.1 eV.

All together there are 23 states of bound states of 6 mesic molecules of hydrogen isotopes. The calculated values of the binding energies $\varepsilon_{J\upsilon}$ of mesic molecules in the states with total moment $J$ and vibrational quantum number $\upsilon$ are presented in Table V. For more details see the papers [22-24]. The full list of the calculations and their comparison are given in papers [23] and review [2].

Table V

Binding energy $\varepsilon_{J\upsilon}$ (eV) of $\mu$ -molecules of the hydrogen isotopes*

| Angular momentum | J=0 | | J=1 | | J=2 | J=3 |
|---|---|---|---|---|---|---|
| Vibrational state | v=0 | v=1 | v=0 | v=1 | v=0 | v=0 |
| $pp\mu$ | 253.0 | - | 105.6 | - | - | - |
| $pd\mu$ | 221.5 | - | 96.3 | - | - | - |
| $pt\mu$ | 213.3 | - | 97.5 | - | - | - |
| $dd\mu$ | 325.0 | 35.6 | 226.3 | 2.0 | 85.6 | - |
| $dt\mu$ | 319.1 | 34.7 | 232.2 | 0.9 | 102.3 | - |
| $tt\mu$ | 362.9 | 83.7 | 288.9 | 44.9 | 172.0 | 47.7 |

* In the calculations we have used the mass values from Table I. The binding energy of $\mu$ -molecules composed of different nuclei is reckoned from the ground state of the mesic atom of the heavier isotope.

## 6. Rates of Nonresonant Formation of Mesic Molecules

The $\mu$ -molecule formation proceeds usually via the EI-electric dipole transition [1,25,26], (for some molecules the EO-transition is also essential [27,28]) from the scattering s-state of the mesic atom+nucleus system to the bound $p$ -state of the mesic molecule. The released binding energy of a mesic molecule is carried out by the Auger electron according to the scheme

$$d\mu + H_2 \rightarrow [(pd\mu)^+ pe]^+ + e. \qquad (17)$$

All the molecules are formed in the excited states, and then the molecules with different nuclei undergo transitions to the ground state with the rate $\sim 10^{11} s^{-1}$ /29/.

Table VI

Rates of nonresonant formation of $\mu$ -molecules

| | $J$ | $v$ | Formation rate $\lambda$, $10^6 s^{-1}$ |
|---|---|---|---|
| $pp\mu$ | 1 | 0 | 2.2 |
| $pd\mu$ | 1 | 0 | 5.9 |
| $pt\mu$ | 1 | 0 | 6.5 |
| $dd\mu$ | 0 | 1 | 0.02 |
| $dt\mu$ | 0 | 1 | 0.03 |
| $tt\mu$ | 1 | 1 | 3.0 |

The calculations performed according to this scheme (17)(see Table VI) are in fair agreement with experiment for the mesic molecules $pp\mu$ and $pd\mu$ (see Table VII).

Table VII

Comparison of the calculated and measured formation rates of $\mu$ -molecules $pp\mu$ and $pd\mu$

| References | $\lambda_{pp\mu}$, $10^6 s^{-1}$ | $\lambda_{pd\mu}$, $10^6 s^{-1}$ |
|---|---|---|
| Dzhelepov et al. (1962) /30/ | $1.5 \pm 0.6$ | - |
| Bleser et al. (1963) /31/ | $1.89 \pm 0.20$ | $5.8 \pm 0.3$ |
| Conforto et al. (1964) /32/ | $2.55 \pm 0.18$ | $6.82 \pm 0.25$ |
| Budyashov et al. (1968) /33/ | $2.74 \pm 0.25$ | - |
| Bystritsky et al. (1976) /34/ | $2.34 \pm 0.17$ | $5.53 \pm 0.16$ |
| Zeldovich and Gerstein (1960) /1/ | 2.6 | 1.3 |
| Cohen et al. (1960) /26/ | 3.9 | 3.0 |
| Ponomarev and Faifman (1976) /28/ | 2.2 | 5.9 |

Till recently there were serious discrepancies between the results of different measurements and the results of theoretical calculations of the formation rate of the $dd\mu$ -molecule (see Table VIII). The recent measurements of the Dzhelepov group /40/ have established definitely the temperature dependence of the formation rate of $dd\mu$ -molecules: it turned out that when the temperature of deuterium changes from -160 to +100°C, the rate increases by a factor of four (see Fig. 4a).

Table VIII

Comparison of the calculated and measured formation rates of the $dd\mu$ - molecules

| References | $\lambda_{dd\mu}$, $10^6 s^{-1}$ | $\varepsilon$ (eV) |
|---|---|---|
| Fetkovich et al. (1960) /35/ | 0.076±0.015 | 0.004 |
| Doede (1963) /36/ | 0.103±0.004 | 0.004 |
| Dzhelepov et al. (1966) /37/ | 0.66±0.19 | 0.042 |
| Bystritsky et al. (1974) /38/ | 0.73±0.07 | 0.046 |
| Zeldovich and Gerstein (1960) /1/ | 0.04 | - |
| Cohen et al. (1960) /26/ | 0.036 | - |
| Vinitsky et al. (1978) /39/ | 0.039 | 0.004 |
| - " - | 0.124 | 0.04 |

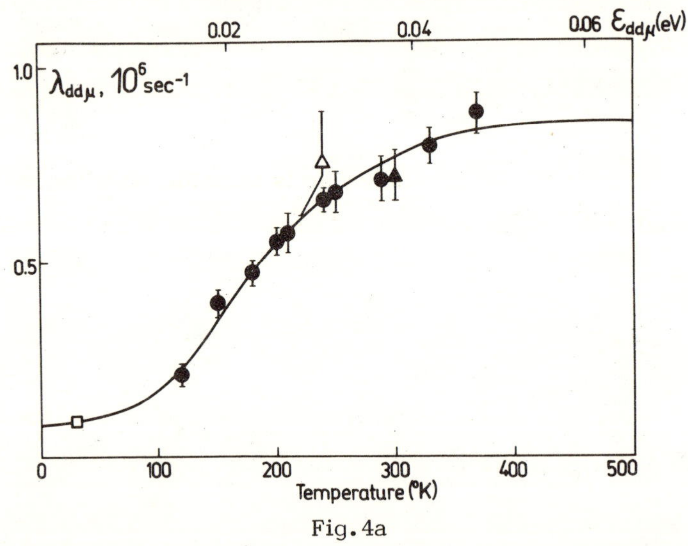


Fig. 4a

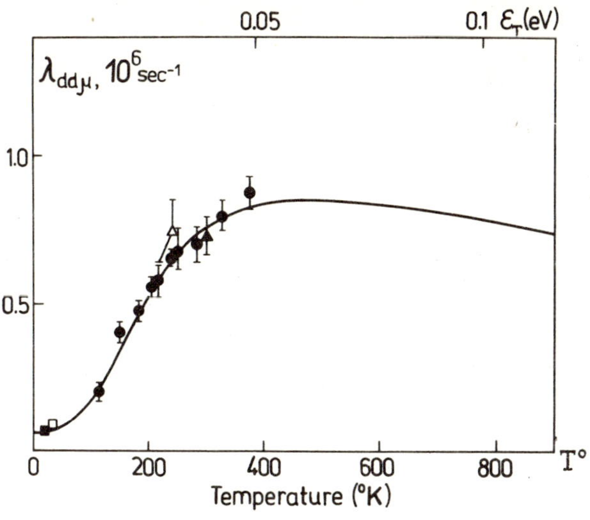


Fig. 4b

Resonance formation rates of the $dd\mu$ -molecules

a) the temperature dependence $\lambda_{dd\mu}(T)$ measured by different groups: ■ - Fetkovich et al.[35], □ - Doede[36], ▲ - Dzhelepov et al.[37], ▲ - Bystritsky et al.[38], ● - Bystritsky et al.[40];

b) the comparison of the measured values with the theoretical predictions[39].

## 7. Resonant Formation of the Mesic Molecules $dd\mu$ and $dt\mu$

For the explanation of the temperature dependence of the rate $\lambda_{dd\mu}$ Vesman[41] has considered the resonant mechanism of the $dd\mu$ -molecule formation. According to this mechanism if the $dd\mu$ -molecule has a weakly bound level with the binding energy $\varepsilon_{J\upsilon} \approx -2$ eV there is no electron emission during the formation of $dd\mu$ -molecules as in the reaction (17). Instead, the binding energy of the $dd\mu$ -molecule is transferred to the excitation of the vibrational states of the peculiar molecule $[(dd\mu)^+ d2e]$ in which the molecular ion $(dd\mu)^+$ plays the role of a nucleus

$$d\mu + D_2 \longrightarrow [(dd\mu)^+ d2e]. \qquad (18)$$

Such a resonance process occurs if the difference between the energy $E_\nu$ of the vibrational state of the excited molecule $[(dd\mu)^+ d2e]^*_\nu$ in the rotational state $K = 1$ and the energy $\overline{E_0}$ of the ground state of the molecule $D_2$ is a sum of the binding energy $\varepsilon_{11}$ of the $dd\mu$ -molecule and the kinetic energy $\varepsilon_0$ of the $d\mu$ -atom (see Figs.5 and 6).

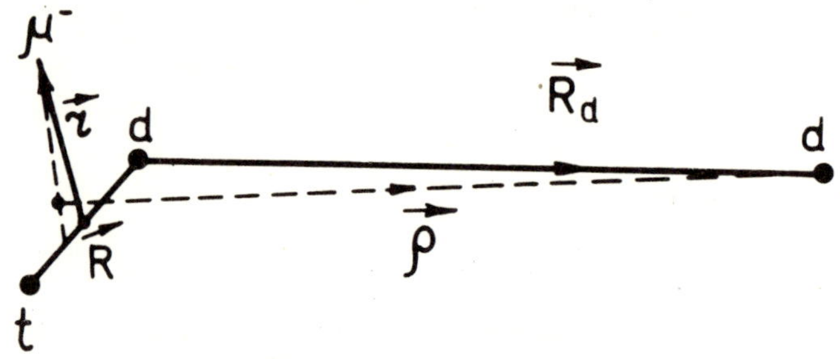


Fig.5. System of coordinates for the description of the resonant formation of $dt\mu$ -molecules: $\vec{r}$ and $\vec{R}$ are the coordinates of the $dt\mu$ -molecule; $R_d \approx \rho$ is the distance between nuclei of the $D_2$ -molecule.

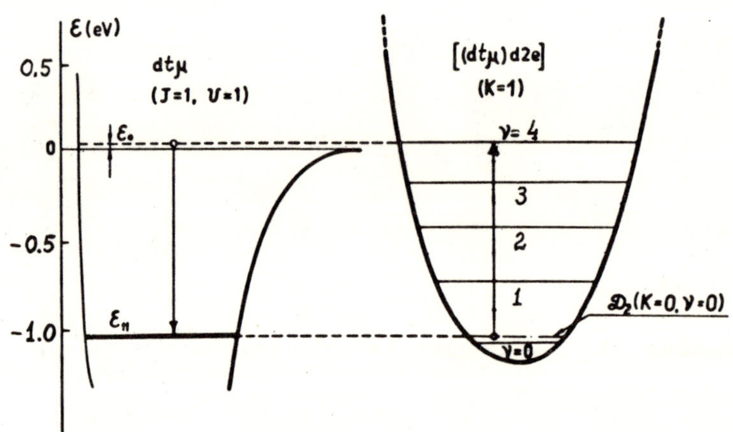


Fig.6. Schematic representation of the resonance formation of $dt\mu$ -molecules with transfer of energy to the excitation of the vibrational levels $\nu$ of the molecule $[(dt\mu)^+ d2e]$.

Of course, this mechanism is possible only due to the existence of the weakly bound state ( $J = 1$, $v = 1$) of the $dd\mu$ -molecule which was confirmed by the calculations (see Table V).

An analogous weakly bound state ( $J = 1$, $v = 1$) with the binding energy $\varepsilon_{Jv} \approx -1$ eV has been detected in the $dt\mu$ -molecule (see Table V). Hence it follows that the resonant mechanism should also take place in the $dt\mu$ -molecule formation. The calculated temperature dependence of the $dt\mu$ -molecule resonance formation rates has the form /39/

$$\lambda = \beta (N_0 a_0^3) \frac{8\pi^2}{3} \left(\frac{m_e}{m_\mu}\right)^5 \left(\frac{m_\mu}{m_a}\right)^3 |d_{fi}|^2 I_\nu^2 \, \gamma(\varepsilon_0, \varepsilon_T) \, \frac{m_e e^4}{\hbar^3} \, s^{-1}. \qquad (19)$$

Here: $\beta$ is the statistical factor which equals 1/3 for $dd\mu$ -molecule and 1 for $dt\mu$ -molecule; $N_0 = 4.25 \cdot 10^{22} cm^{-3}$ and $a_0$ is the Bohr radius; $m_e$ and $m_\mu$ are the masses of electron and muon; $m_a^{-1} = m_e^{-1} + M_a^{-1}$ and $M_a$ is the nuclear mass; $d_{fi}$ is the dipole matrix element calculated between the functions of the initial state (i) of system $d\mu + d$ (or $t\mu + d$) and the final state (f) ( $J = 1$, $v = 1$) of $dd\mu$- or $dt\mu$- molecules; $I_\nu$ is the matrix element of the dipole interaction, corresponding to the transition from the ground state ( $K = 0$, $\nu = 0$) of $D_2$ -molecule to the excited state ( $K = 1$, $\nu = 8$) or ( $K = 1$, $\nu = 4$) of the molecule $[(dd\mu)^+ d2e]$ or $[(dt\mu)^+ d2e]$ ; the function $\gamma(\varepsilon_0, \varepsilon_T) = (27\varepsilon_0/2\pi)^{1/2} \varepsilon_T^{-3/2} \exp\{-3\varepsilon_0/2\varepsilon_T\}$ is the Maxwell distribution, where $\varepsilon_T = \frac{3}{2} kT$ is the average kinetic energy of mesic atoms $d\mu$ (or $t\mu$ ) at the temperature $T_0$ and $\varepsilon_0$ is the collision energy.

The dependence (19) can be expressed in the form

$$\lambda = \lambda_1 + \lambda_2 \varepsilon_T + \lambda_3 \varepsilon_0^{1/2} \varepsilon_T^{-3/2} \exp\{-3\varepsilon_0/2\varepsilon_T\}, \qquad (20)$$

where constants $\lambda_i$ are presented in Table IX. The maximal value $\lambda^{max}$ is achieved at the collision energy $\varepsilon_T = \varepsilon_0 = \varepsilon_{11} + E_\nu + \overline{E}_0$.

As can be seen from Fig. 4, the calculated (19) and measured /40/ dependences of the formation rate of $dd\mu$ -molecules are in fair agreement.

The value $\lambda_{dt\mu}$ of the $dt\mu$ -molecule resonant formation rate

$$t\mu + D_2 \longrightarrow [(dt\mu)^+ d2e]_\nu^* \qquad (21)$$

**Table IX**

Main characteristics of the resonant formation of $\mu$ -molecules*

| Values | $dd\mu$ | $dt\mu$ |
|---|---|---|
| $\nu$ | 8 | 4 |
| $E_\nu$ eV | -2.307 | -3.395 |
| $\bar{E}_0$ eV | -4.556 | -4.556 |
| $\bar{\varepsilon}_{11}$ eV | -2.2 | -1.1 |
| $\varepsilon_0$ eV | 0.053 | 0.07 |
| $T_0^\circ$ K | 400 | 500 |
| $\lambda_3, 10^6$ s$^{-1}$ eV | 0.15 | 29.4 |
| $\lambda_2$, s$^{-1}$ | 0.40 | 0.033 |
| $\lambda_3$, s$^{-1}$ eV$^{-1}$ | 2.3 | 13.8 |
| $\lambda^{max}$, $10^6$ s$^{-1}$ | 0.80 | 95 |

* The indicated values of the binding energy $\bar{\varepsilon}_{11}$ of $dd\mu$ and $dt\mu$ -molecules are recalculated taking into account the experimental measurements /40/ of the resonant temperature $T_0$ (see ref. /39/); the values $\lambda^{max}$ are calculated at the temperature $T_0$ and density $N_o$ = $4.25 \cdot 10^{22}$ cm$^{-3}$ of the liquid hydrogen.

was calculated by the same scheme as the $dd\mu$ -molecule formation rate and equals /39/: $\lambda^{max}_{dt\mu} \approx 10^8$ s$^{-1}$.

## 8. Catalysis of the nuclear fusion reactions by $\mu^-$ -mesons

The synthesis reaction of the hydrogen and deuterium nuclei proceeding through the intermediate state of $pd\mu$ -molecule in the chain

$$d\mu + p \xrightarrow{\lambda_{pd\mu}} pd\mu \xrightarrow{\lambda_f} \begin{cases} \text{I} \nearrow \mu\,^3He + \gamma \\ \text{II} \searrow \,^3He + \mu^- \end{cases} + 5.4\ \text{MeV} \qquad (22)$$

was suggested by Frank[42] and ten years later was observed by Alvarez et al.[43]. This process is known as the catalysis of nuclear fusion in "cold" hydrogen by $\mu^-$-mesons, and has been intensively studied theoretically[26,27,44,45] to understand the possibility of practical application in energy production. This process is effective if one muon can cause sequentially several fusion reactions. For this purpose it is necessary to satisfy at least three conditions

$$\lambda_{pd\mu} \gg \lambda_0 , \ \lambda_f \gg \lambda_0 , \ W_s \ll 1, \qquad (23)$$

where $\lambda_0 = 0.455 \cdot 10^6 \ s^{-1}$ is the decay rate of the free muon $\lambda_{pd\mu}$ is the $pd\mu$-molecule formation rate, $\lambda_f$ is the nuclear fusion reaction rate in the $pd\mu$-molecule and $W_s$ is the "sticking" probability of muons to $^3He$ which is equal to the ratio of the reaction probabilities (I) and (II).

Table X

Characteristics of the $\mu$-catalysis of the nuclear fusion in mesic molecules*

| Process | $\lambda_m, 10^6 s^{-1}$ | $\lambda_f, 10^6 s^{-1}$ | $W_s$ |
|---|---|---|---|
| $d\mu + p \xrightarrow{\lambda_m} pd\mu \xrightarrow{\lambda_f} \mu\,^3He + \gamma + 5.4$ MeV | 5.9 | 0.3 | 1 |
| $d\mu + d \rightarrow dd\mu \rightarrow \mu\,^4He + n + 3.3$ MeV | 0.8 | $10^5$ | 0.13 |
| $t\mu + p \rightarrow pt\mu \rightarrow \mu\,^4He + \gamma + 20$ MeV | 6.5 | 10 | 1 |
| $t\mu + d \rightarrow dt\mu \rightarrow \mu\,^4He + n + 17.6$ MeV | 100 | $10^6$ | $10^{-2}$ |
| $t\mu + t \rightarrow tt\mu \rightarrow \mu\,^4He + 2n + 10$ MeV | 3.0 | - | $\sim 1$ |

* The maximal rate of the $\mu$-mesic molecule formation is given at density $N = 4.25 \cdot 10^{22} \ cm^{-3}$ of the liquid hydrogen.

As can be seen from Table X, the conditions (23) are fulfilled for the mixture of deuterium and tritium only. It means that in the mixture $D_2 + T_2$ one $\mu^-$ meson can catalyze $\sim$ 100 nuclear fusions in the chain of processes

$$t\mu + d \xrightarrow{\lambda_{dt\mu}} dt\mu \xrightarrow{\lambda_f} \left\{ \begin{array}{l} \mu\,^4He + n \\ ^4He + n + \mu \end{array} \right. + 17.6 \ MeV \qquad (24)$$

and release $\sim 2$ GeV of energy, i.e. $\sim 20$ times larger than the rest muon mass[46,47].

## 9. The Vacuum Polarization in $\mu$-mesic Molecules

Because of the critical dependence of the resonance formation rates of $dd\mu$ and $dt\mu$-molecules on their binding energy $\varepsilon_{J,v}$ it is necessary to take into account various corrections which can shift the calculated values. The most important correction is the vacuum polarization correction. The calculated values of the vacuum polarization corrections to the $\mu$-mesic molecule energy levels are presented in Table XI[48].

Table XI

The vacuum polarization corrections to the energy levels of the hydrogen isotope mesic molecules

| | J=0 | | J=1 | | J=2 | J=3 |
|---|---|---|---|---|---|---|
| | v=0 | v=1 | v=0 | v=1 | v=0 | v=0 |
| $pp\mu$ | -0.285 | - | -0.064 | - | - | - |
| $pd\mu$ | -0.290 | - | -0.096 | - | - | - |
| $pt\mu$ | -0.325 | - | -0.124 | - | - | - |
| $dd\mu$ | -0.397 | -0.030 | -0.227 | 0.008 | -0.016 | - |
| $dt\mu$ | -0.428 | -0.056 | -0.267 | -0.003 | -0.058 | - |
| $tt\mu$ | -0.479 | -0.097 | -0.329 | -0.034 | -0.130 | 0.044 |

## 10. Unsolved Problems

The two-level approximation well describes almost all processes of elastic and inelastic scattering of mesic atoms. However, in some cases it fails to describe, e.g., elastic scattering (16) of $p\mu$ atoms in the lower state of the hyperfine structure. In this case both the results of different experiments and theoretical calculations differ by three orders of magnitude (see Table XII). The reason for the experimental discrepancies is not yet clear. The discrepancy between different theoretical results seems to be a consequence of the resonance character of this process, which is manifested in the extreme sensitivity of the cross sections $\overline{\sigma_{11}}$ for the process to the shape of the effective potentials $U_{ij}(R)$ in the system of equations (10)[21].

Table XII

Elastic cross section $\overline{\sigma_{11}}$ of $p\mu$ atoms in the lower h.f.s. state

| References | $\overline{\sigma_{11}} \cdot 10^{-21}$ cm$^2$ |
|---|---|
| Dzhelepov et al. (1965)[49] | 167±30 |
| Albergi Quaranta et al. (1967)[50] | 7.6±0.7 |
| Zeldovich and Gerstein (1960)[1] | 1.2 |
| Cohen et al. (1960)[26] | 8.2 |
| Matveenko and Ponomarev (1970)[19] | 2.5 |
| Matveenko et al. (1975)[16] | 0.23 |
| Ponomarev et al.[21] | ~ 30 |

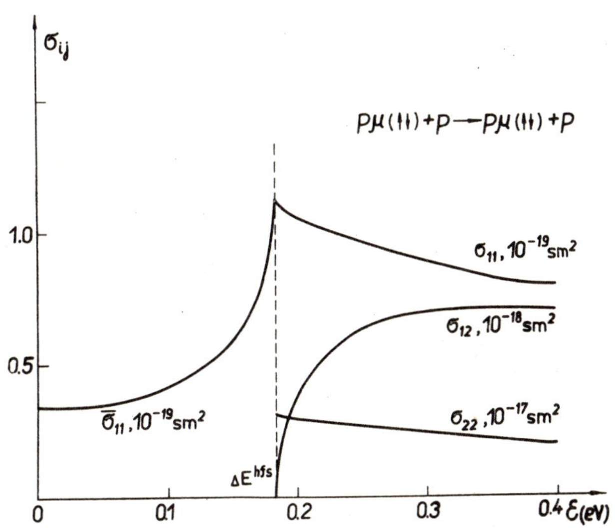


Fig.7. Cross sections $\sigma_{ij}$ and $\overline{\sigma_{11}}$ for the processes (15) and (16).

The disagreement between the calculated rates $\Lambda_d$ of the spin-flip process in the reaction

$$d\mu(\uparrow\uparrow) + d \longrightarrow d\mu(\uparrow\downarrow) + d \qquad (25)$$

$\Lambda_d = 0.7\cdot 10^8\ s^{-1}$ (Gerstein[20])

$\Lambda_d = 2.8\cdot 10^8\ s^{-1}$ (see Table III)

is perhaps due to the smallness of the value $\Delta E^{hfs} = 0.049$ eV which is compared with the thermal energy of mesic atoms.

Finally, it is not yet clear what mechanism can provide the spin-flip reaction

$$d\mu(\uparrow\uparrow) + p \longrightarrow d\mu(\uparrow\downarrow) + p \qquad (26)$$

which was assumed in the paper[51]. (See, however, the objections[40]).

One possibility is the following. The differences of energies $\Delta E_{LL'}$ between the neighbouring rotational levels $L$ and $L' = L + 1$ of the $H_2$-molecule are compared with the value of the hyperfine splitting $\Delta E^{hfs}$ of $d\mu$-atom. Under these conditions the resonance spin-flip reaction can occur in the mesic atom $d\mu$ with the simultaneous transition $L$ $L' = L + 1$ in the $H_2$-molecule, analogous to that described above in section 8 devoted to the calculation of the resonance formation rates of the molecules $dd\mu$ and $dt\mu$.

The accuracy of the calculations of the $\mu$-molecule energy levels is now $\sim 0.1$ eV. For the description of the resonant formation of the $dd\mu$ and $dt\mu$-molecules it is necessary to improve these calculations up to $\sim 0.01$ eV.

## 11. Conclusion

The method of the calculation presented here allows one to obtain all the essential characteristics of the mesic atomic and molecular processes in the hydrogen isotope mixtures. In a few cases it is necessary to improve the method. These results demonstrate the fruitfullness and possibilities of the adiabatic method.

It is very desirable to check experimentally some predictions of calculations, especially the rate of the resonant formation of the $dt\mu$-molecules.

## References


1. Ya.B.Zeldovich and S.S.Gerstein, Usp.Fiz.Nauk., 71, 581 (1960). English transl.Sov.Phys. - Usp. 3, 593 (1961).
2. S.S.Gerstein and L.I.Ponomarev, "Mesomolecular Processes induced by $\mu^-$ and $\pi^-$ mesons" in "Muon Physics" v.III, Eds. V.Hughes and C.S.Wu, Academic Press, New York and London (1975).
3. D.R.Bates, H.S.W.Massey and A.L.Stewart, Proc.Roy.Soc. A216, 437 (1953)
   D.R.Bates and Mc Carroll, Proc.Roy.Soc. A245, 175 (1958);
   J.C.Y.Chen, V.H.Ponce and K.M.Watson, J.Phys. B6, 965 (1973).
4. M.Born, Gott.Nachr. 1 (1951);
   G.Hunter, B.F.Gray and H.O.Pritchard, J.Chem.Phys. 45, 3806 (1966);
   A.M.Halpern, Phys.Rev. 14, 186(1969);
   R.T.Pack and J.O.Hirschfelder, J.Chem.Phys. 52, 521(1970).
5. S.I.Vinitsky and L.I.Ponomarev, Yadernaja Fizika, 20, 576 (1974). English transl. Sov.Phys.Nucl.Phys. 20, 310 (1975);
   S.I.Vinitsy and L.I.Ponomarev, Preprint JINR P4-11332, Dubna (1978).
6. D.R.Bates and R.H.Reid, in "Advances in Atomic and Molecular Physics", vol.IV Academic Press, New York and London (1968);
   J.D.Power, Phil.Trans.Roy.Soc. London, A272, 663 (1973).
7. I.V.Komarov, L.I.Ponomarev and S.Yu.Slavjanov, "Spheroidal and Coulomb Spheroidal Functions" (in Russian), Nauka, Moscow (1976).
8. D.R.Bates, Kathleen Ledsham and A.L.Stewart, Phil.Trans. A246, 215 (1953).
9. L.I.Ponomarev and T.P.Puzynina, Preprint JINR P - 5040, Dubna (1970); Zh.Eksp.Teor.Fiz. 52, 1273 (1967).
   English trans. Sov.Phys. - JETP, 25, 846 (1967).
10. L.I.Ponomarev and L.N.Somov, J.Comput.Phys. 20, 183 (1976).
11. N.F.Truskova, Communications JINR P11-10207, P11-10218, Dubna (1977), P2-11269, Dubna (1978).
12. L.I.Ponomarev and T.P.Puzynina, Preprint JINR P4-3405, Dubna (1967);
    L.I.Ponomarev, T.P.Puzynina and N.F.Truskova, Preprint JINR P4-11185, Dubna (1978).

13. M.P.Faifman, L.I.Ponomarev and S.I.Vinitsky, J. Phys. B9, 2255 (1976).

14. L.I.Ponomarev, T.P.Puzynina and L.N.Somov, J.Phys. B10, 1335 (1977).

15. V.V.Bibikov, "The phase function method in quantum mechanics", (in Russian), Nauka, Moscow (1968);
F.Calogero, "Variable Phase Approach to Potential Scattering", Academic Press, New York (1967).

16. A.V.Matveenko, L.I.Ponomarev and M.P.Faifman, Zh.Eksper. Teor. Fiz. 68, 437 (1975); English transl. Sov.Phys. - JETP, 41, 212 (1975).

17. M.P.Faifman, Yadern.Fiz. 26, 434 (1977).

18. E.R.Cohen and B.N.Taylor, J. Phys. and Chem. Ref. Data, 2, 663 (1973).

19. A.V.Matveenko and L.I.Ponomarev, Zh.Eksper.Teor.Fiz. 58, 1640 (1970). English transl. Sov.Phys. - JETP, 31, 880 (1970);
A.V.Matveenko and L.I.Ponomarev, Zh.Eksp.Teor.Fiz. 59, 1593 (1970). English transl. Sov.Phys. - JETP, 32, 871 (1971).

20. S.S.Gerstein, Zh.Eksp.Teor.Fiz. 34, 463 (1958); English transl. Sov.Phys. - JETP, 7, 318 (1958);
S.S.Gerstein, Zh.Eksp.Teor.Fiz. 40, 698 (1961); Engl. transl. Sov. Phys. - JETP, 13, 488 (1961).

21. L.I.Ponomarev, L.N.Somov and M.P.Faifman, Preprint JINR P4-11446, Dubna (1977).

22. S.I.Vinitsky and L.I.Ponomarev, Zh. Eksp. Teor. Fiz. 72, 1670 (1977).

23. S.I.Vinitsky, L.I.Ponomarev, I.V.Puzynin, T.P.Puzynina, L.N.Somov, Preprint JINR P4-10336, Dubna (1976); "Mesons in Matter", Proceedings of the Int. Symposium on mesonic chemistry and mesonic molecular processes in matter, Dubna, 7-10 June (1977).

24. L.I.Ponomarev, I.V.Puzynin and T.P.Puzynina, J. Comp. Phys. 13, 1 (1973).

25. V.B.Belyaev, S.S.Gerstein, B.N.Zakhar'ev and S.P.Lomnev, Zh. Eksp.Teor.Fiz. 37, 1652 (1959); Engl.transl. Sov.Phys. - JETP, 10, 1171 (1960).

26. S.Cohen, D.L.Judd and R.J.Riddel, Phys. Rev. 119, 384 (1960).

27. Ya.B.Zeldovich, Dokl.Akad.Nauk SSSR, 95, 493 (1954).

28. L.I.Ponomarev and M.P.Fairman, Zh.Eksp.Teor.Fiz. 71, 1689 (1976). Engl.transl. Sov.Phys. - JETP, 44, (1976).

29. L.I.Ponomarev and M.P.Faifman, Communications JINR P4-10635, Dubna (1977).

30. V.P.Dzhelepov, P.F.Ermolov, E.A.Kushnirenko, V.I.Moskalev, S.S.Gerstein, Zh.Eksp.Teor.Fiz. 42, 439 (1962); English transl. Sov.Phys. - JETP, 15, 306 (1962).

31. E.Bleser, L.Lederman, J.Rosen, J.Rothberg, E.Zavattini, Phys. Rev. 132, 2679 (1963).

32. G.Conforto, C.Rubbia, E.Zavattini, S.Focardi, Nuovo Cimento, 33, 1001 (1964).

33. Yu.G.Budyashov, P.F.Ermolov, V.G.Zinov, A.D.Konin, A.I.Mukhin and K.O.Oganesyan, Preprint JINR P15 - 3964, Dubna (1968).

34. V.M.Bystritsky, V.P.Dzhelepov, V.I.Petrukhin, A.I.Rudenko, V.M.Suvorov, V.V.Filchenkov, H.Hemnitz, N.N.Khovanskii, B.A.Khomenko, Zh.Eksp.Teor.Fiz. 70, 1167 (1976). English transl. Sov.Phys. - JETP, 43, 606 (1976).

35. J.G.Fetkovich, T.H.Fields, G.B.Yodh and M.Derrick, Phys.Rev. Lett. 4, 570 (1960).

36. J.H.Doede, Phys.Rev. 132, 1782 (1963).

37. V.P.Dzhelepov, P.F.Yermolov, V.I.Moskalev and V.V.Filchenkov, Zh.Eksp.Teor.Fiz. 50, 1235 (1966). English transl. Sov.Phys. - JETP, 23, 820 (1966).

38. V.M.Bystritsky, V.P.Dzhelepov, K.O.Oganesyan, M.N.Omeljanenko, S.Yu.Porokhovoy, A.I.Rudenko and V.V.Filchenkov, Zh.Eksp. Teor.Fiz. 66, 61 (1974). English transl. Sov. Phys. - JETP, 39, 27 (1976).

39. S.I.Vinitsky, L.I.Ponomarev, I.V.Puzynin, T.P.Puzynina, L.N.Somov and M.P.Faifman. Zh.Eksp.Teor.Fiz. 74, 849 (1978).

40. V.M.Bystritsky, V.P.Dzhelepov, V.I.Petrukhin, A.I.Rudenko, L.N.Somov, V.M.Suvorov, V.V.Filchenkov, G.Chemnitz, B.A. Khomenko, N.N.Khobansky and D.Horvath. Proceedings of the International Symposium on Mesonic Chemistry and Mesic Molecular Processes in Matter, Dubna, 7-10 June (1977); The talk at the VII Int. Conf. on High Energy Phys. and Nucl. Str. 28 August - 2 September 1977, Zurich, Switzerland.

41. E.Vesman, Zh.Eksp.Teor.Fiz. Pisma 5, 113 (1967). English transl. Sov.Phys. - JETP, Letters, 5, 91 (1967).

42. F.C.Frank, Nature, 160, 525 (1947).

43. L.W.Alvarez, H.Brander, F.S.Crowford Jr., J.A.Crowford, P.Falk - Vairant, M.L.Good, J.D.Gow, A.H.Rosenfeld, F.Solmitz, M.L.Stevenson, H.K.Ticho and R.D.Tripp. Phys.Rev. 105, 1127 (1957).

44. J.D.Jackson, Phys.Rev. 106, 330 (1957).

45. B.P.Carter, Phys.Rev. 141, 863 (1966).

46. L.I.Ponomarev, The talk at the VII Int.Conf. on High Energy Physics and Nuclear Structure, 28 August - September 1977, Zurich, Switzerland.

47. S.S.Gerstein and L.I.Ponomarev, Phys. Lett. 72B, 80 (1977).

48. V.S.Melezhik and L.I.Ponomarev, Preprint JINR P4-11186, Dubna (1978).

49. V.P.Dzhelepov, P.F.Yermolov and V.V.Filchenkov, Zh.Eksp. Teor.Fiz. 49, 393 (1965). English transl. Sov.Phys. - JETP, 22, 275 (1966).

50. A.Alberigi Quaranta, A.Bertin, G.Matone, F.Palmonari, A.Placci, P.Dalpiatz, G.Torelli, E.Zavattini, Nuovo Cimento, 47B, 72 (1967).

51. A.Bertin, A.Vitale, A.Placci and E.Zavattini, Phys.Rev. 80, 3774 (1973).

# DETERMINATION OF THE LAMB SHIFT (H, n=2) BY THE "ATOMIC INTERFEROMETER" METHOD

Yu.L.Sokolov

I.V.Kurchatov Institute of Atomic Energy, Moscow

Studies of the interference of atomic states intensively developed in the 60's occupied a great deal of attention up to now. Special importance of such investigations is that the finest properties of the atomic states (e.g. of H-atom) may, in principle, manifest themselves in some new aspects, since the interference pattern is extremely sensitive to the characteristics of its components. It seems possible that the negative results of the experiments which have been made to search the applicability limits of quantum electrodynamics may be due to the fact that there exists some kind of phenomena, though small, but playing an important role, which fails to come into view when using the modern experimental techniques.

A few years ago we proposed a method ("atomic interferometer method") that allowed to observe the stationary interference pattern for a long time with the phase shift arbitrarily changed, thus noticeably improving the measurement accuracy. Since only one paper covering this work was published[1], and thus it remained practically unknown, I shall permit myself to begin with a brief description of the method mentioned above.

The interference of atomic states can be observed, in principle, with the aid of device, similar in main outline to a standard two-channel optical interferometer (e.g. Michelson's interferometer). The latter can be considered as a system consisting of three regions separated by "active" zones changing the states of photons entering in them. If a light beam passing through an interferometer, crossing the first active zone is split into two components, then the photons transit into the superposition of the states corresponding to two channels of the interferometer. In this case in the third region (i.c.

after passing through the second active zone) the interference of the two-phase shifted light beam components resulted from the states associatted with the first and second channels will be observed.

The operation principle of one of the atomic interferometer constructed according to such a scheme can be explained in the following way (Fig. 1).

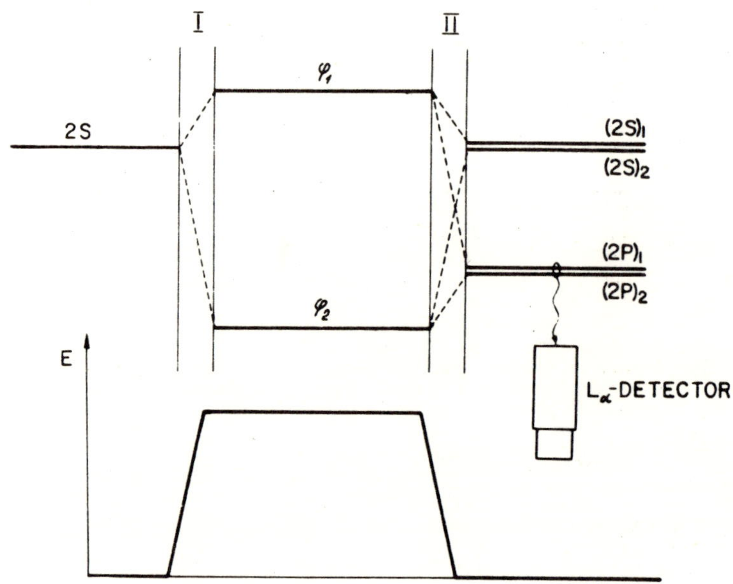


Fig. 1 - Schematic diagram of the atomic interferometer with the electric field varying non-adiabatically within the active zones I and II.

Suppose a beam of metastable $2s_{1/2}$ hydrogen atoms passing subsequently through two spatially separated zones I and II. When being inside these zones, the atoms are subject to a perturbation that enables their transition into other states, e.g. $2p_{1/2}$ and $2p_{1/2}$. A perturbing factor of this kind may be an electric field changing nonadiabatically within each zone.

The nonadiabaticity criterion is that the transit frequency $\omega = v/d$ should be higher than or of the order of the Lamb frequency (for the $2s_{1/2}$ - $2p_{1/2}$ transition) or of the fine structure frequency (for the 2s -2p transition). Here $v$ is the atom velocity and $d$ is the length at which the field rises or decreases. To simplify the picture, the further analysis will be perfomed on the exsample of the two-level $2s_{1/2}$-$2p_{1/2}$ system, which is justified when the fields are not too strong.

It follows from the above considerations that in the simplest version of the interferometer the boundaries I and II of the field localized in the prescribed region can act as the active zones. Then, when crossing boundary I, the beam atoms are subject to perturbing effect of the rising field and go over into the superposition of the eigenstates $\varphi_1$ and $\varphi_2$ with enegies $\mathcal{E}_1$ and $\mathcal{E}_2$ respectively, determined by the strength of the field E. At the boundary II where the field decreases to zero, beam components arise representing both the $2s_{1/2}$ and $2p_{1/2}$ states, each of the terms $\varphi_1$ and $\varphi_2$ forming the pair of such states as $\varphi_1 \rightarrow (2s)_1+(2p)_1$ and $\varphi_2 \rightarrow (2s)_2+(2p)_2$. Thus, in the field-free region adjacent to the boundary II the atom state will be represented by the superposition of four components: $(2s)_1$, $(2s)_2$, $(2p)_1$ and $(2p)_2$.

Outside the field the amplitudes of the $2s_{1/2}$ and $2p_{1/2}$ eigenstates will be defined by the transition amplitudes and phase difference between the components of each pair $(2s)_1$-$(2s)_2$ and $(2p)_1$-$(2p)_2$, which depends on the time of flight in the field and on the transition frequency between the $\varphi_1$ and $\varphi_2$ terms splitted by the electric field. The magnitude of such a splitting is entirely determined by the strength of the field E. Thus, at its monotonous variation, periodical (occuring in counterphase) intensity oscillations of the $H_{2s}$ and $H_{2p}$ atom fluxes will be observed due to interference of $(2s)_1$-$(2s)_2$ as well as $(2p)_1$-$(2p)_2$ waves arising on boundary II. A similar situation will take place when the time of flight T, i.e. the distance between the field boundaries is changed.

It should be emphasized that the interference of the 2s- or 2p-components of the beam can be only observed if the variation of the field on both boundaries meets the non-adiabaticity condition. The interference pattern of the $(2p)_1$-$(2p)_2$ components can be recorded measuring the intensity of the short-lived 2p-part of the beam having passed through the interferometer. Thus, the detector placed behind the second active zone (i.e. in the field-free region) must count quanta corresponding to the one-photon transition 2p-1s, i.e. the resonant line of the Lyman series ($\lambda$=1216 Å). One can also observe the interference of the $(2s)_1$-$(2s)_2$ components occuring in counterphase, for which purpose the beam should be passed through the additional field "quenching" the $2s_{1/2}$-state.

To simplify the analysis, it is reasonable to divide the field strength range into the region of "normal" and the region of "strong" fields. The normal fields should be taken to mean the fields for which the condition $x = \langle d \rangle E/\pi\hbar\delta \sim 1$ is satisfied, i.e. the Stark shift of the $2s_{1/2}$ and $2p_{1/2}$ levels, caused by them, proves to be of the same order as the Lamb shift (here $\langle d \rangle$ is the matrix element of the $2s_{1/2}$-$2p_{1/2}$ transition, E is the field strength and $\delta$ is the Lamb shift). In the region of the normal fields the Lamb shift effect manifests itself most clearly, while the presence of the $2p_{3/2}$-level does but weakly, which allows to reduce the problem to considering the two-level $2s_{1/2}$-$2p_{1/2}$ system. In this case the effect of the $2p_{3/2}$-level can be taken into account by small corrections.

In the two-level case and under assumption of sudden termination of the field on the boundaries, the yield of the $H_{2p}$-atoms is proportional to the quantity I:

$$
\begin{aligned}
I = \Big\{ & C_1 \frac{x_1^2}{1+x_1^2}\Big[ch\frac{s_1 T}{2\tau\sqrt{p_1+x_1^2}} - \cos 2\pi T\left(\delta+\tfrac{2}{3}\nu\right)\sqrt{1+x_1^2}\Big] + \\
& + C_2 \frac{x_2^2}{1+x_1^2}\Big[ch\frac{s_2 T}{2\tau\sqrt{p_2+x_2^2}} - \cos 2\pi T\left(\delta-\tfrac{10}{3}\nu\right)\sqrt{1+x_2^2}\Big] + \\
& + C_3 \frac{x_3^2}{1+x_3^3}\Big[ch\frac{s_3 T}{2\tau\sqrt{p_3+x_3^3}} - \cos 2\pi T\left(\delta+2\nu\right)\sqrt{1+x_3^2}\Big]\Big\} e^{-\frac{T}{2\tau}(1+\kappa x^2)}
\end{aligned}
\qquad (1)
$$

where $\tau$ is the lifetime of the $H_{2p}$-atom; T is the time of flight in the interferometer field; $\delta$ is the Lamb shift; $\nu$ is the hyperfine splitting frequency; $c_i$ , $s_i$ and $p_i$ are the constants; $\kappa = \delta/2\nu_1$ ; $\nu_1$ =9911MHz is the $(2p_{3/2}$-$2s_{1/2})$-splitting frequency. The factor $\kappa x^2$ is the $2p_{3/2}$-level effect correction.

In the $2s_{1/2}$-$2p_{1/2}$ system there are transitions between the s and p hyperfine structure sublevels with total angular momentum projections 1,0 and -1 (Fig.2). The energy differences $2s_{1/2}(F_z=1)$--$2p_{1/2}(F_z=1)$ and $2s_{1/2}(F_z=-1)$-$2p_{1/2}(F_z=-1)$ coincide, so that the summation in (1) is carried out for three components of the hyperfine splitting with the following values of $x_i$:

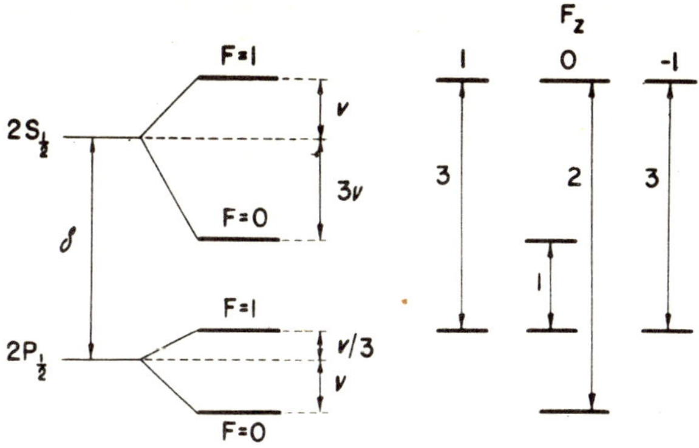


Fig.2 - Hyperfine structure of the 2s and 2p levels of the hydrogen atom.

$$x_1 = x/(1+\frac{2}{3}\frac{\nu}{\delta})$$

$$x_2 = x/(1-\frac{10}{3}\frac{\nu}{\delta})$$

$$x_3 = x/(1+2\frac{\nu}{\delta})$$

It follows from the formula (1) that the dependence of the $H_{2p}$-atom yield on E and T found experimentally allows to determine the values of $\delta$ and $\nu$. It should be noted however, that the method capacities can not be fully realized in the case of the simple interferometer described, since the values of $\delta$ and $\nu$ are found by an approximated formula, which is equivalent to introduction of a certain systematic error. A detailed theoretical analysis by V.P.Yakovlev has shown that the limitations in the accuracy resulted mainly from the uncertainty of the field E characteristics on the boundaries and a complicated atom behaviour in the field can be entirely eliminated by modification of the interferometer electrodes geometry. The most reasonable proves to be a system consisting of two independent flat capacitors separated by a variable distance L defining the flight time (Fig.3). Since the field on the lenth L is equal to zero, the eigenstates in this region will be $2s_{1/2}$, $2p_{1/2}$ and $2p_{3/2}$. In the case of such a "dual" interferometer the problem of $I_{2p}(L)$-dependence

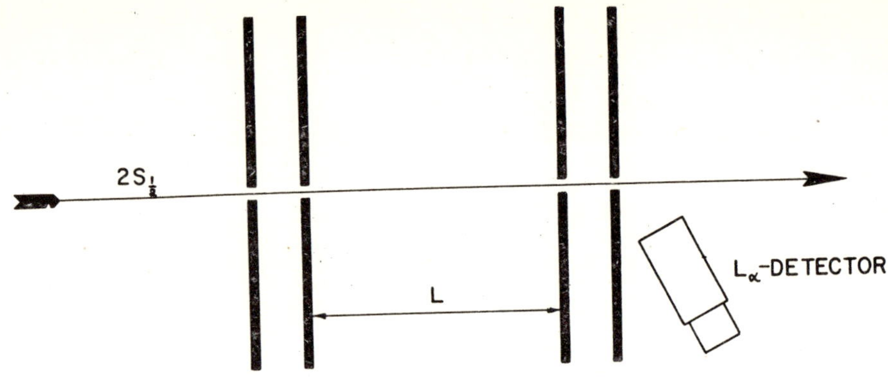


Fig.3 - Diagram of the "dual" atomic interferometer.

determination admits an accurate solution taking into account the effect of the $2p_{3/2}$-level at all the values of the field E, which makes it also possible to determine the value of the fine structure constant $\alpha$. A further essential simplification of the experimental data reduction is achieved by removing the $2s_{1/2}$-state component with the total momentum F=1 from the beam. In this case the effect observed will be only due to the transition $2s_{1/2}$(F=0, $F_z$=0) - $2p_{1/2}$(F=1, $F_z$=0) which reduces to minimum the number of unknown parameters when fitting the theoretical curve to the experimental one.

Fig.4 shows the diagram of the "Pamir" unit on which the phenomenon of the atomic interference was studied.

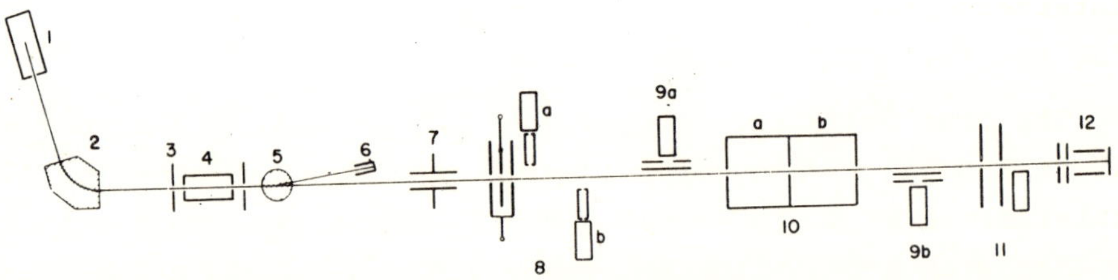


Fig.4 - Diagram of "Pamir" unit (notations in the text).

Protons with the energies of about 20 kev emerging from the ion source 1 were then passed through the velocity analyser consisting of magnet 2 and slit diaphragm 3. The diaphragm width and the distance between the diaphragm and the magnet were chosen so that the spread of the velocities of the protons passed through the analyser did not exceed 5 ev.

Neutral hydrogen atoms were produced in the charge exchange cell 4. The mixed beam passed through a weak magnetic field 5 deflecting the proton component by an angle about $0.2^{\circ}$ (the proton current was measured by Faraday cap 6). Flat condenser 7 was designed for "quenching" the $2s_{1/2}$-atoms when determining the background magnitude. The H-atom velocities were measured using system 8 which was essentially the simple interferometer described above with a fixed distance between the plates. The $2p_{1/2}$-atoms produced by non-adiabatic field were registered by the detector a placed in a fixed position and detector b which could be moved along the beam trajectory in a range of 1.5 cm. From the analysis of the exponential curves of the 2p-atoms decay their velocity was found with an error not exceeding $2\cdot10^{-4}$. The system for measuring the atom velocity, when inoperable, was removed as a whole from the beam trajectory.

To monitor the $2s_{1/2}$-atom flux a part of the beam was passed through the pickup 10a consisting of a flat capacitor where "quenching" $2s_{1/2}$-atoms occured and a L -quanta detector. To eliminate the $2s_{1/2}$-state component with total momentum F=1, the beam was transmitted through the rf-fields 9a and 9b with frequencies 1147 and 1087 MHz (in preliminary experiments one rf-field was used with frequency 1100 MHz). In that way only component 1 (see Fig.2) with the frequency of 909 MHz was present in the beam. Its intensity was monitored with the pickup 10b analogous to 10a. The remained part of the beam passed via interferometer 11 and then was trapped by the end meter 12 recording the full neutral atom current.

The beam trajectory length from the ion source 1 to the meter 12 is about 6 m; the neutral beam atom length from the cell 4 to interferometer 11 is 400 cm. Such a length was chosen so that the states with n=3-7 which are densely populated and at the same time short-lived, had time to radiate, since their transitions to the ground 1s-state produce $L_{\alpha}$ -quantum background.

In the experiments under consideration the interferometers with both a transverse and longitudinal (relative to the atom trajectory ) fields were used. In the interferometers of the latter type the distance between the electrodes could be changed, which allowed to obtain both $I_{2p}(E)_{T=const.}$ and $I_{2p}(T)_{E=const.}$ relations. Fig.5 is the diagram of the interferometer with transverse field; its electrodes could be switched in such a way that the field variation on the boundary was either adiabatic or non--adaiabatic. Fig.6 shows the interferometer with longitudinal

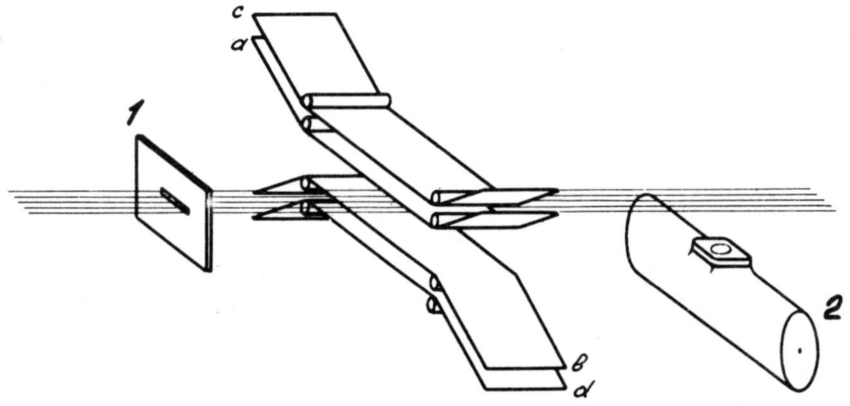


Fig.5 - Interferometer with the transverse field. 1-collimator; 2-L -quantum detector; a,b,c and d - ribbon electrodes.

field together with 2s -atom monitor. The diagram of the "dual" unterferometer is given in Fig.3.

Figures 7 and 8 show typical interference curves of the 2p --state of the hydrogen atom, obtained using the interferometer presented in Fig.6, with the rf-field switched off.

Fig.9 shows the interference pattern corresponding to the transition 2s $(F=0,F_z=0)$-2p $(F=1,F_z=0)$ obtained in passing through the rf-field. The curves of such a type were used in calibration experiments for determination of the Lamb shift $\delta$ and the hyperfine splitting frequency $\nu$ , since both these values

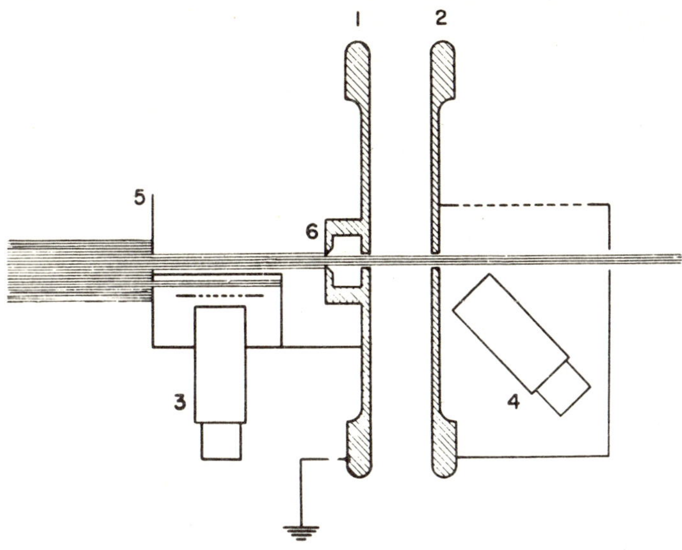


Fig.6 - Interferometer with the longitudinal field and a 2s - -atom flux monitor; 1 and 2 are interferometer electrodes; 3 is the monitor detector; 4 is the interferometer detector; 5 and 6 are diaphragms.

enter expression (1). The thoretical curve was fitted to the experimental points by variation of parameters $\delta$ and $\nu$. Treatment of three curves $I_{2p}(T)$ for E=250 v/cm gave the following values:

1. $\delta$= 1057.931 MHz   $\nu$= 44.390 MHz
2. $\delta$= 1057.935   $\nu$= 44.396
3. $\delta$= 1057.929   $\nu$= 44.393

There values, homogenous enough as they are, differ, nevertheless, from the theoretical and experimental Lamb shift values known at present time:

$\delta_{theo}$ = 1057.912(11) MHz (G.W.Erickson[2])
$\delta_{theo}$ = 1057.864(14) MHz (P.J.Mohr[3])
$\delta_{exp}$ = 1057.893(20) MHz (S.R.Lundeen and F.M.Pipkin[4])
$\delta_{exp}$ = 1057.862(20) MHz (D.A.Andrews and G.Newton[5])

This can be accounted for primarily by the fact that, as has been noted, in the case of the simple interferometer $\delta$ was found using an approximate formula. Other possible reasons for such a

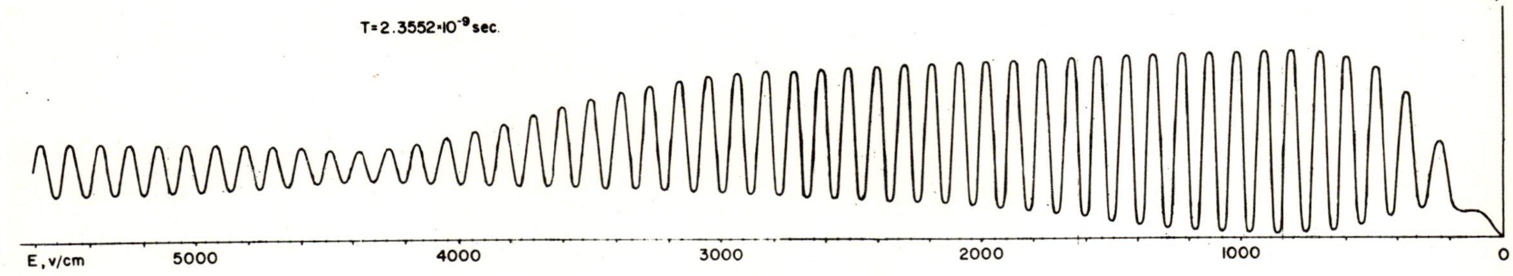


Fig.7 - Curve of $2p_{1/2}$-state interference at T=const.

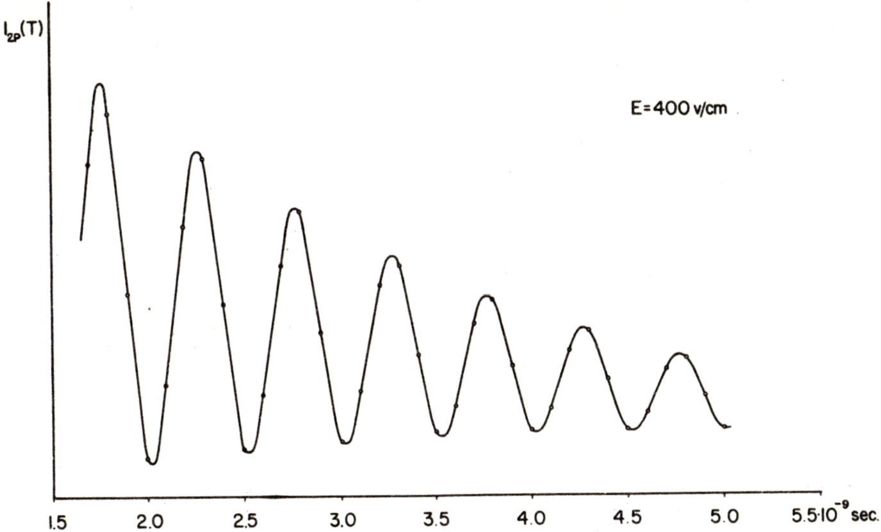


Fig.8 - Curve of $2p_{1/2}$-state interference for E=400 v/cm.

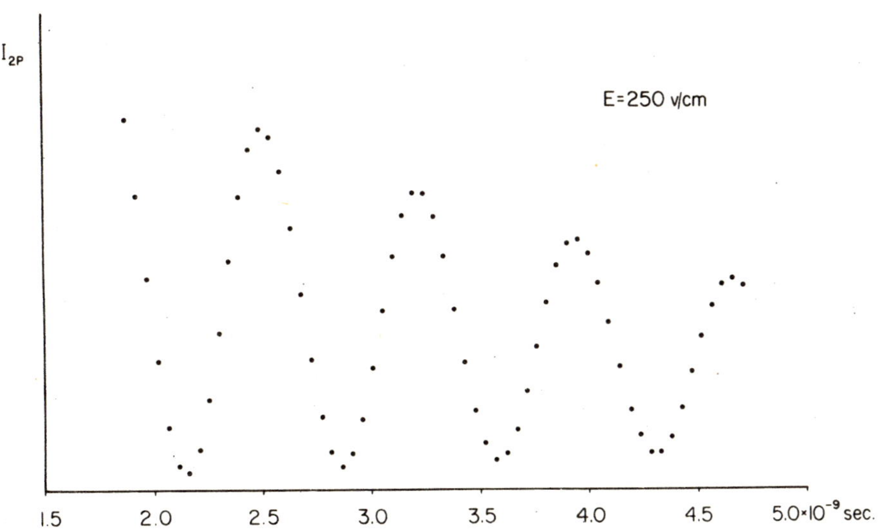


Fig.9 - Interference pattern corresponding to the transition $2s_{1/2}(F=0,F_z=0)-2p_{1/2}(F=1,F_z=0)$.

discrepancy will be discussed later. It was difficult to estimate the veritable statistical error in this case because of a small number of measurements.

In order to evaluate the actual accuracy of the method its resolving power must be determined, for example, relative to the Lamb shift. It follows from $x = \kappa \frac{E}{\delta}$ that change in $\delta$ required for such an estimation can be imitated by changing the field strength E. For this purpose the relation $I_{2p}(T)_{E-\Delta E}$ was determined for various values of $\Delta E$. As a result sets of numbers were obtained representing the interference curves $I_{2p}(T)_{E=const.}$. Then it was found at which minimum $\Delta E$ value the difference between $I_{2p}(T)_{Eo}$ and $I_{2p}(T)_{Eo+\Delta E}$ remained statistically significant (for example, with 99% reliability) and is not the result of a random spread of numbers. Thus it was found that the method proposed has a high sensibility to change in $\delta$ and can rather easily provide measurement of this value with an error not exceeding in any case 10 ppm.

It follows from the above considerations that for determination of $\delta$ and $\nu$ it is reasonable to consider (in the region of normal fields) the interference curves having a great number of oscillations (i.e. corresponding to the flight time of the order of $7.5 \cdot 10^{-9}$ sec) which are extremely sensitive to the smallest variations of $\delta$ and $\nu$ parameters.

Before the accurate measurement of $\delta$ and $\nu$ was performed with the aid of the "dual" interferometer, it became necessary to investigate in details the effect of the setup operating conditions on the population of the hyperfine structure sublevels of 2s-state. For this purpose the interference curves $I_{2p}(E)$ were taken, with the rf-system switched off. Thus, each of the curves represented, as has been mentioned, the sum of three components corresponding to the transitions between $2s_{1/2}$ and $2p_{1/2}$ states:

$$I = \sum_{1}^{3} c_i I_i \tag{2}$$

The treatment of such curves consisted in fitting of coefficients $C_1$, $C_2$ and $C_3$ at the given values of T, $\delta$ and $\nu$, i.e. the population of each of the hyperfine structure transitions was determined. In another version of the treatment four parameters $C_1$, $C_2$, $C_3$ and T were found.

However, quite an unexpected phenomenon was observed at this point, which changed the whole course of the investigation.

At small flight times ($T \sim 2.5 \cdot 10^{-9}$sec) agreement between the theoretical and experimental curves $I_{2p}(E)$ was good enough within the range of the field strength from 0 to 600 v/cm (Fig.10). However, at large flight times ($\sim 7.5 \cdot 10^{-9}$sec) such a curve fitting was found to be impossible.

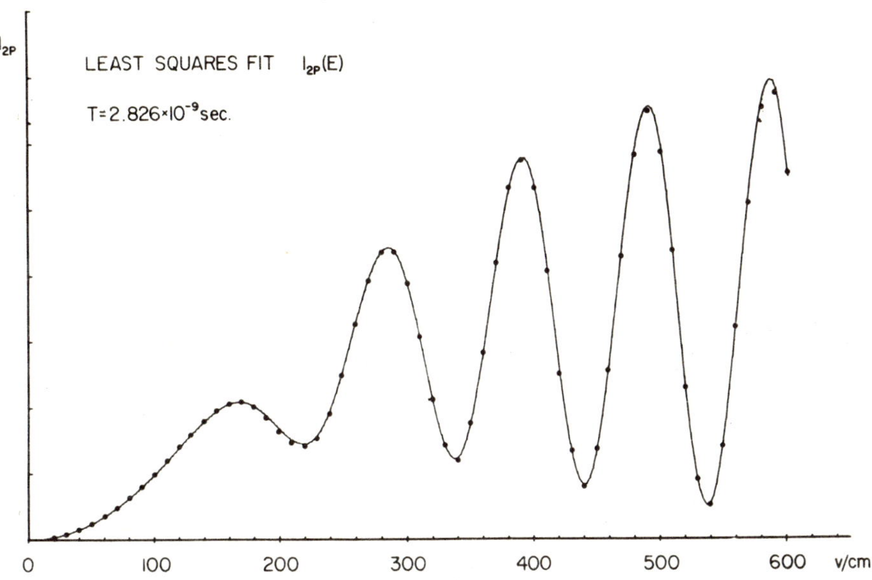


Fig.10 - Fitting of the theoretical curve to the experimental interference pattern.

As has been pointed out, the intensity of the 2p-component of the beam having passed through the interferometer is determined as:

$$I = \frac{x^2}{1+x^2}\left(ch\frac{sT}{2\tau\sqrt{p+x^2}} - \cos 2\pi T\delta\sqrt{1+x^2}\right)e^{-\frac{T}{2\tau}(1+\kappa x^2)} \qquad (3)$$

Thus, the structure of the interference curve appears to be the folowing: the term with the hyperbolic cosine defines its mean line while one with the trigonometric cosine represents oscillations of the 2p-atom flux intensity relative to this mean line. The former, as the field E strength increases, tends to a certain asymptotic value which is reached at the E value of the order 750 v/cm for all flight times. The difference between the theoretical and experimental curves consists mainly in the fact

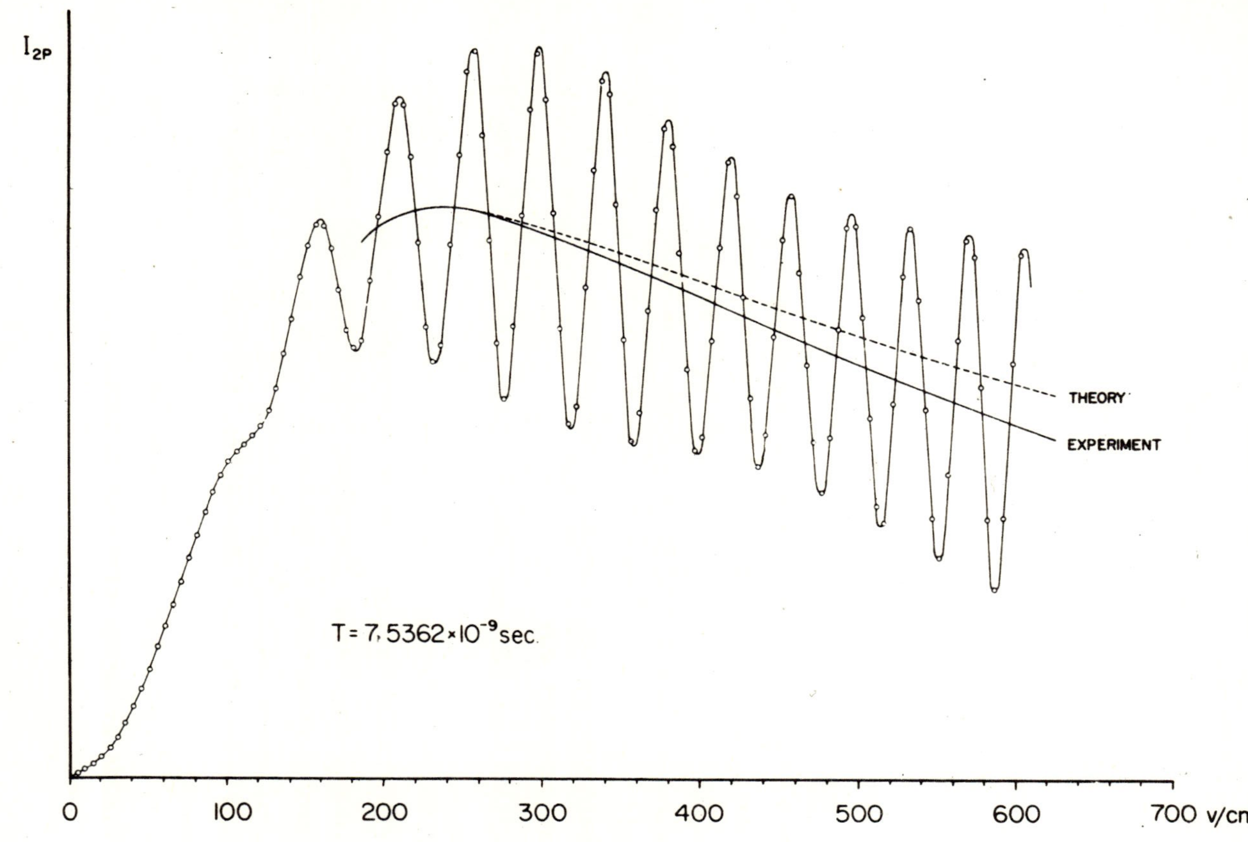


Fig.11 - Experimental interference curve at $T=7.5362\cdot10^{-9}$ sec and its theoretical mean line.

that in the latter the mean line, at large T and with increase in E, approaches abscissa more closely than it follows from formula (1) (Fig.11).

This phenomenon can be represented in another form. It follows from (3) that for strong fields ($x \gg 1$), when Lamb shift effect is negligible, the position of the mean line is defined as follows:

$$\ln \bar{I} = a - \gamma_{eff} T/2 \qquad (4)$$

where $\gamma_{eff} = \gamma(1+15x^2)$ is the decay constant determined taking into account the $2p_{3/2}$-level influence.

Relation (4) is shown in Fig.12. As it seen at large flight times a deviation of the experimental points from the theoretical curve is observed. This deviation lies beyound the limits of the measurement and calculation errors. The above circumstance means that the rate of de-excitation of atoms in the interferometer field during the time T is higher than this is defined by the decay constant $\gamma_{eff}$.

Investigations perfomed revealed no errors or any features of the experiment which could cause occurence of such an effect. The theoretical analysis likewise fails to give any suitable explanation. Nevertheless, one should take into account the above effect in all the cases, for example, in the analysis of the $I_{2p}(T)$ curves attemting to define $\delta$, $\nu$ and $\alpha$. We have considered several different assumptions concerning the nature of the phenomenon described, which require, however, a thorough experimental verification. For this purpose measurements using the "dual" interferometer have been planned in the region of strong fields at large times of flight.

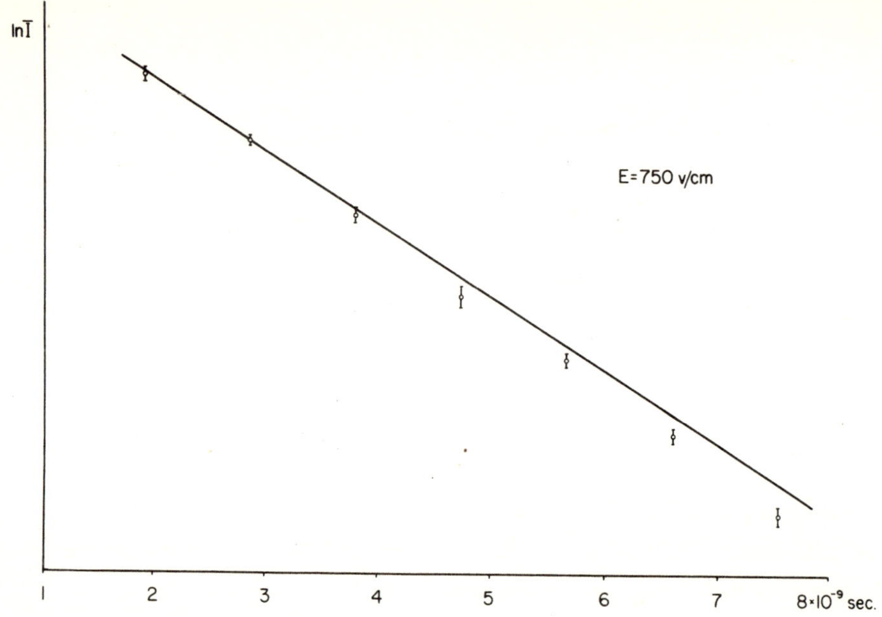


Fig.12 - Plot of $\ln T = a - \gamma_{eff} T/2$ relation. —— theory, ♀ - experiment.

The author is deeply grateful to V.M.Galitskii, V.P.Yakovlev, B.B.Kadomtcev and I.N.Golovin for constant help and numerous discussions, and V.V.Chashchin for taking part in the measurements.

## REFERENCES

1. Yu.L.Sokolov, Zh.Eksp.Teor.Fiz. 63, 461 (1972)
2. G.W.Erickson, Phys.Rev.Lett. 27, 780 (1971)
3. P.J.Mohr, Phys.Rev.Lett. 34, 1050 (1975)
4. S.R.Lundeen and F.M.Pipkin, Phys.Rev.Lett. 34, 1368 (1975)
5. D.A.Andrews and G.Newton, Phys.Rev.Lett. 37, 1254 (1976)

# SPECTROSCOPY OF NEGATIVE IONS

John S. Risley

Department of Physics
North Carolina State University
Raleigh, North Carolina 27650

## I. INTRODUCTION

This paper is concerned with the spectroscopy of negative ions - ions which have either a permanent or temporary excess of negative charge. Negative ions which retain their extra electron for an indefinite period of time are considered stable: additional energy is required to remove the electron. Negative ions which hold on to their extra electron for $10^{-6}$ seconds are metastable: this period of time is long enough for the ion to be useful for a laboratory experiment. Negative ions which last $10^{-12}$ sec. are a noticeable entity decaying predominately by electron emission. Shorter lived species, $10^{-15}$ sec., having a fleeting moment of existence, can be readily observed as a broad resonance in electron scattering measurements. An atom which is brushed by an electron as it passes by in $10^{-17}$ sec. is not granted the distinction of being called a negative ion.

The periodic table in Fig. 1 shows that most elements are capable of forming a stable negative ion.[1] The excited states of negative ions fall into two classes; those lying below the detachment limit corresponding to the neutral atom plus a free electron, and those lying in the continuum which are energetically allowed to decay via an ejection of an electron, a process which is called autodetachment. An example of the possible excited states of a negative ion is shown in Fig. 2. For $C^-$ the ground state lies 1.268 eV below the $^3P$ state of C plus a free electron with no kinetic energy. A second, bound excited state of $C^-$ has been firmly established[2] lying just 35 meV below the detachment limit. Higher lying states of $C^-$ which can autodetach when excited have been predicted by Matese[3] and one transition, indicated in Fig. 2, has been detected by Lee and Edwards.[4]

In general, few stable negative ions have bound excited states, but apparently all atoms have autodetaching states. Many of the known autodetaching states have also been observed as resonances in electron-atom scattering experiments.

In this paper we will center our discussion on the nature of the autodetaching states of negative ions. For a discussion of the bound states

| 1 H 0.7542 | | | | | | | | | | | | | | | | | 2 He <0 |
|---|---|---|---|---|---|---|---|---|---|---|---|---|---|---|---|---|---|
| 3 Li 0.620 | 4 Be <0 | | | | | | | | | | | 5 B 0.28 | 6 C 1.268 | 7 N ≤ 0 | 8 O 1.462 | 9 F 3.399 | 10 Ne <0 |
| 11 Na 0.546 | 12 Mg <0 | | | | | | | | | | | 13 Al 0.46 | 14 Si 1.385 | 15 P 0.743 | 16 S 2.0772 | 17 Cl 3.615 | 18 Ar <0 |
| 19 K 0.5012 | 20 Ca <0 | 21 Sc <0 | 22 Ti 0.2 | 23 V 0.5 | 24 Cr 0.66 | 25 Mn <0 | 26 Fe 0.25 | 27 Co 0.7 | 28 Ni 1.15 | 29 Cu 1.226 | 30 Zn <0 | 31 Ga 0.3 | 32 Ge 1.2 | 33 As 0.80 | 34 Se 2.0206 | 35 Br 3.364 | 36 Kr <0 |
| 37 Rb 0.4860 | 38 Sr <0 | 39 Y ≈0 | 40 Zr 0.5 | 41 Nb 1.0 | 42 Mo 1.0 | 43 Tc 0.7 | 44 Ru 1.1 | 45 Rh 1.2 | 46 Pd 0.6 | 47 Ag 1.303 | 48 Cd <0 | 49 In 0.3 | 50 Sn 1.25 | 51 Sb 1.05 | 52 Te 1.9708 | 53 I 3.061 | 54 Xe <0 |
| 55 Cs 0.4715 | 56 Ba <0 | 57 La 0.5 | 72 Hf <0 | 73 Ta 0.6 | 74 W 0.6 | 75 Re 0.15 | 76 Os 1.1 | 77 Ir 1.6 | 78 Pt 2.128 | 79 Au 2.3086 | 80 Hg <0 | 81 Tl 0.3 | 82 Pb 1.1 | 83 Pi 1.1 | 84 Po 1.9 | 85 At 2.8 | 86 Rn <0 |

Fig. 1. Periodic table of the elements listing the electron affinity in eV, see Ref. 1.

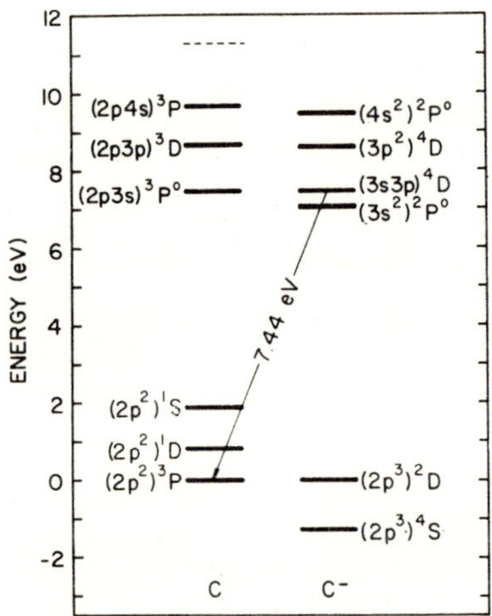


Fig. 2. The excited states of C and $C^-$ and an observed autodetaching transition.[14]

the reader is referred to Hotop and Lineberger.[1] Because $H^-$ is the simplest negative ion, we start with it emphasizing recent work published since the review on the structure and collision processes involving $H^-$ by Risley[5] in 1975. Examples of the properties of other negative ions follow the discussion on $H^-$.

## II. NEGATIVE HYDROGEN ION

The simplicity of $H^-$ has attracted the attention of both theorists and experimentalists in the hope of fully understanding its structure. An energy level diagram with some observed transitions is shown in Fig. 3.

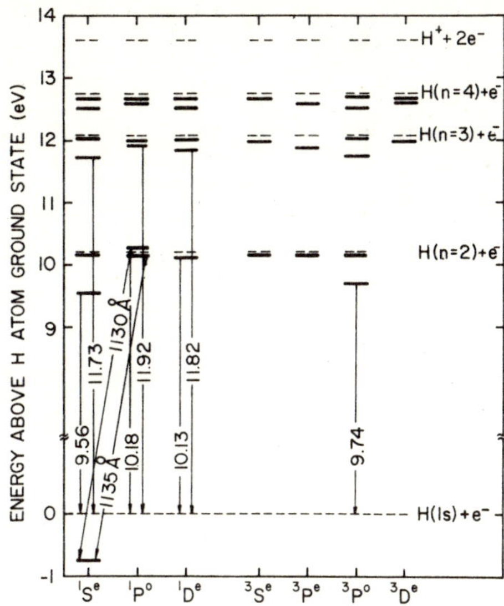


Fig. 3. Energy level diagram of $H^-$.

During the past several years progress has centered on precision theoretical calculations, a new model for the energy levels of $H^-$, high resolution electron spectroscopy, photoemission of $H^-$ in arcs and the photodetachment of $H^-$.

## II.A. Theoretical Advances

### II.A.1. Precision calculations.

A summary of the precision calculations[6-14] for the energy and width of the lowest $^1S$ autodetaching state of $H^-$ is listed in Table I and plotted chronologically in Fig. 4. Except for two cases, the latest results all agree to within 1 meV of one another or to within 2% of the natural width of the state. The value of the energy denoted "best" refers to comments by Ho, Bhatia and Temkin on the reliability of their calculation.[14]

The large width associated with the state, 0.047 eV, makes it unlikely that an experiment can be designed which will yield the state energy to sufficient precision to distinguish between the various theoretical methods. In general, one concludes that, with care, there are several theoretical approaches which can be used to accurately predict the energy of the autodetaching states of $H^-$. It remains to be determined just how well suited each method is for calculating autodetaching states of more complicated negative ions.

### II.A.2. Hyperspherical representation.

A new model, yielding a unusual graphic description of $H^-$, has come

Table I. Precision calculations of the lowest $^1S$ autodetaching state of $H^-$.

| E(eV) | Γ(eV) | Method | Author |
|---|---|---|---|
| 9.555 | 0.048 | Scattering, 3 state close coupling with correlation | Burke (1968)[6] |
| 9.5522 | 0.047 | Kohn variational | Shimamura (1971)[7] |
| 9.5517 | 0.0474 | Complex coordinates | Bardsley and Junker (1972)[8] |
| 9.5490 | 0.0411 | Feshbach technique | Chung and Chen (1972)[9] |
| 9.5518 | | Complex rotation | Doolen, Nuttall and Stagat (1974)[10] |
| 9.5518 | 0.0458 | Stabilization | Bhatia (1974,1977)[11,14] |
| 9.5512 | 0.0476 | QHQ and polarized orbital | Bhatia and Temkin (1975)[12] |
| 9.55176 | 0.04722 | Uncorrelated Kohn-Feshbach | Chung and Chen (1976)[13] |
| 9.55215 | 0.04714 | Optical potential formalism | Ho, Bhatia and Temkin (1977)[14] |

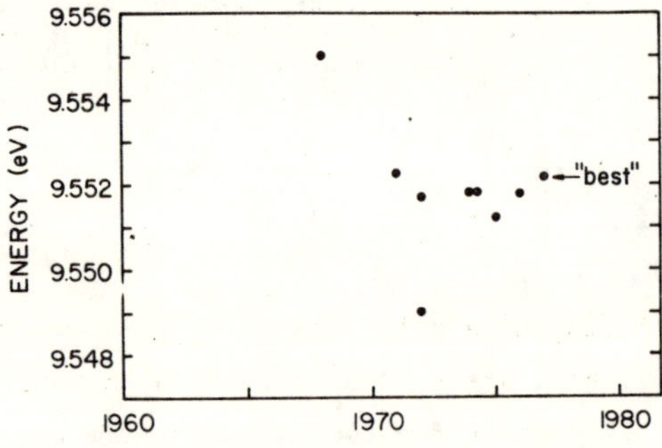


Fig. 4. Energy of the $^1S$ state of $H^-$.

from a search for the important characteristics of the effects of correlation in 2 electron systems, i.e. angular correlations, radial correlations and exchange.[15-19] In this method, the 2-electron Schrödinger equation is solved directly by looking for quasiseparability when the wave function is written in hyperspherical coordinates.

In this coordinate system the distances of the two electrons from the nucleus $r_1$ and $r_2$ are replaced by the hyperradius $R = (r_1^2 + r_2^2)^{\frac{1}{2}}$ and a hyperangle $\alpha = \tan^{-1}(r_2/r_1)$. The angle $\alpha$ and the usual polar coordinates $(\theta_1,\phi_1)$ and $(\theta_2,\phi_2)$ of the two electrons are represented collectively by

$\Omega = \{\alpha,\theta_1,\phi_1,\theta_2,\phi_2\}$. The angular term $\Omega$ identifies the angular orientation of the two electrons where the hyperangle $\alpha$ is a measure of the relative magnitudes of the electron positions. Coordinate R is a measure of the size of the negative hydrogen ion.

Because the angular coordinates are confined to finite ranges, the kinetic energy or velocity associated with the angular motion is larger than that associated with the radial motion[15] producing an effective potential for the radial motion. Since the resulting equation is reminiscent of that used for solving the diatomic molecular problem, the wave function is expanded as $\psi(R,\Omega) = \Sigma_\mu F_\mu(R)\Phi_\mu(R,\Omega)$ where $\Phi_\mu(R,\Omega)$ is similar to the electronic wave function at a fixed internuclear distance R and $F_\mu(R)$ is similar to the nuclear wave function in the diatomic molecular case. To make the problem tractable a Born-Oppenheimer type separation is used in which the couplings between the radial and angular motions are neglected.[19]

The angular wave function $\Phi_\mu(R,\Omega)$ gives rise to a potential energy $U_\mu(R)$: an energy which when plotted as a function of R allows one to make cogent statements about the structure of $H^-$. Once $U_\mu(R)$ is known, evergy levels of $H^-$ can be calculated as vibrational levels in the potential well. An example of the potential energies for the n=2 level of hydrogen can be seen in Fig. 5. The short horizontal lines on the

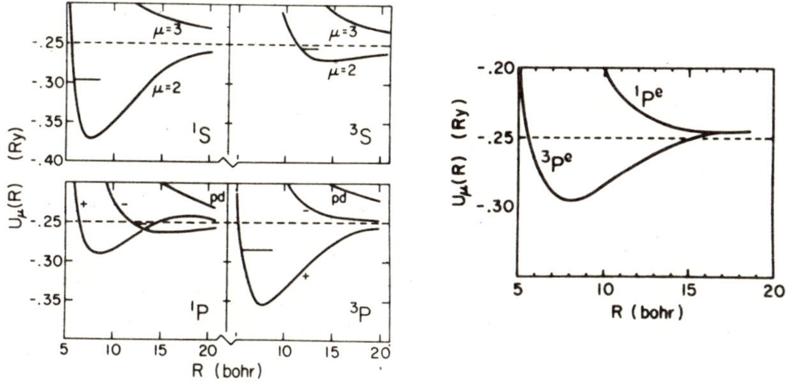


Fig. 5. Potential curves of $H^-$ converging to the n=2 level of the hydrogen atom.[19]

left side of the potential curves indicate the energy of the excited states, e.g. for the $^1S$ series the lowest excited state for $\mu = 2$ is slightly less than -0.30 Ry. A cursory inspection of the curves can often determine which set of quantum numbers corresponds to excited states, the range of possible energy levels and the spatial extent of the excited ion.

From these curves several important properties for $H^-$ were discovered: except for the ground state there is no other bound state associated with the n=1 limit of H, there is no other shape resonance associated with the n=2 threshold of H except the $^1P^o$ shape resonance, and the lowest $^1P^o$ Feshbach state is better designated as 2s3p - 2p3s than as the lowest member of the "+" series, i.e. 2snp + 2pns.[18,19] Although Lin[19] has calculated the energies for the first few members of the $^1S$, $^3S$, $^1P$ and $^3P$ states of $H^-$ below the n=2 limit, the results are not particularly precise since his method does not include level shifts due to the interaction with the continuum. His model is best suited for predicting the existence or nonexistence of excited states giving us insight into the underlying reasons.

## II.B. Experimental Observations

Excited states of $H^-$ have been detected in four distinct types of experiments:

a. photoemission

$$e^- + H \rightarrow H^{-*} \rightarrow H^- + h\nu$$

b. photodetachment

$$h\nu + H^- \rightarrow H^{-*} \rightarrow H + e^-$$

c. electron scattering

$$e^- + H \rightarrow H^{-*} \rightarrow H + e^-$$

d. collisional excitation

$$H^- + A \rightarrow H^{-*} + A \rightarrow H + e^- + A$$

Experiments in which photons are either detected or are shined on an $H^-$ beam have the advantage, in general, of much higher instrumental resolution. However, such observations are limited to studies of the $^1P$ series or states which mix with the $^1P$ states when external fields are applied. Electron scattering experiments have exhibited fairly high resolution ~ 20 meV and yielded absolute energies for the position and widths of several states. However, resonances which are extremely narrow, less than a few meV, cannot be observed. The collisional excitation technique, in which the ejected electrons are energy analyzed, has had the poorest resolution to date, but this failing is partly offset by the ability to control the excitation of some states over others by using different collisional velocities, target atoms and observation angles. Additionally, because of reaction kinematics, one can shift the laboratory energy of a line associate with the decay of an excited state over a fairly large range. For example, the laboratory energy of the $^1S$ line can be adjusted from below 1 eV to over 30 eV for an $H^-$ collision energy of 10 keV simply by changing the observation angle. This procedure allows one to place the line close to other lines in the ejected electron spectrum for accurate calibration purposes.

II.B.1. Electron scattering experiments.

Recently, a variety of detection modes have been used in high resolution electron scattering experiments on atomic hydrogen to study $H^-$ resonances. Preliminary results with 20 meV resolution have been reported by Williams for the elastic differential scattering cross section at $30^{\circ}$ and $90^{\circ}$ shown in Fig. 6.[20] For the first time one can clearly observe the strong angular dependence of the lowest $^1S$ and $^3P$ resonances.

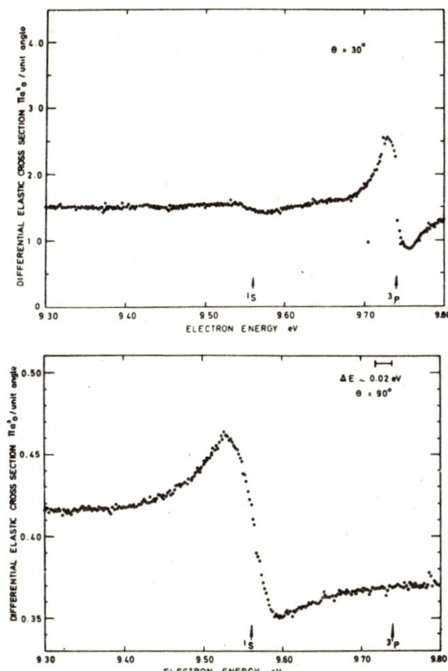


Fig. 6. The elastic differential scattering cross section for electrons on atomic hydrogen at $30^{\circ}$ and $90^{\circ}$, see Ref. 20.

By observing the Lyman alpha radiation from the decay of the 2P and 2S states of atomic hydrogen excited via electron impact, resonant structure can be seen in the excitation cross section. Results from the work of Williams[21] is shown in Fig. 7 along with other experimental measurements and theoretical calculations. The $^1P^o$ shape resonance just above the n=2 level dominates the threshold region while $H^-$ states below the n=3 and n=4 levels can be clearly discerned in the 11 to 13 eV electron energy range.

Resonances below the n=3 level also appear in the 10.2 eV energy loss spectra of electrons scattered from atomic hydrogen,[21] see Fig. 8. The angular distributions obtained by this energy-loss technique help to identify the angular momentum and spin of the $H^-$ states.

Complimentary measurements using electron transmission spectroscopy on atomic hydrogen by Spence[22] have yielded precise results for the lowest

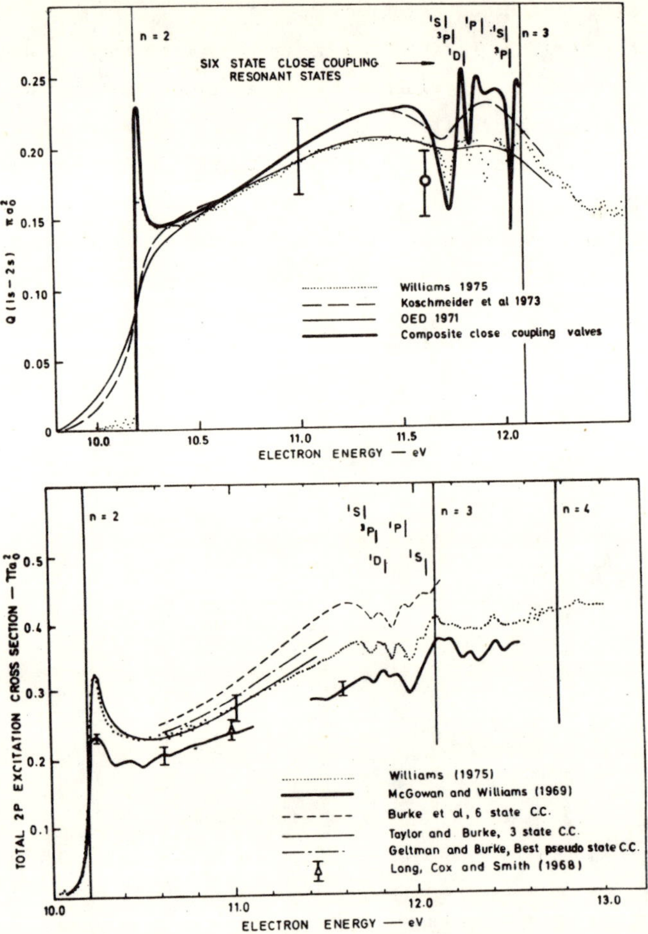


Fig. 7. The 2S state and 2P state total excitation cross sections as a function of incident electron energy.[21]

$^1D$ state of $H^-$ below the n=3 level. Figure 9 shows the weak $^1D$ resonance superimposed in a band of molecular $H_2^-$ structures and the strong n=2 resonances at lower energy.

A detailed comparison of all the experimental results[21-24] with the "best" theoretical calculations[25] for $H^-$ states below the n=3 level can be made using Table II. It is apparent that some discrepancies exist between the measurements of the McGowan et al.[24] and Williams.[21] At present a full interpretation of the high resolution data of Williams[21] has not been made. Because of the small energy separation between the resonances, comparable to the electron energy resolution, and because of the uncertainty in the electron energy scale, no attempt was made to deconvolute the apparatus function from the observed profiles or to fit the profiles to theoretical line shapes which would yield the true resonance energy. Such a procedure

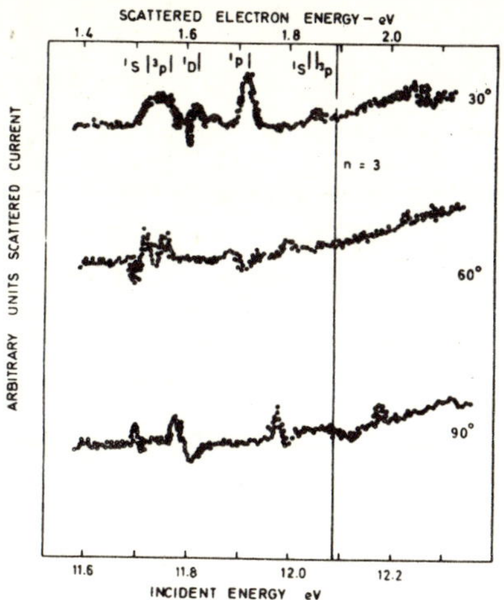


Fig. 8. The 10.2 eV energy-loss spectra for electrons scattered on atomic hydrogen.[21]

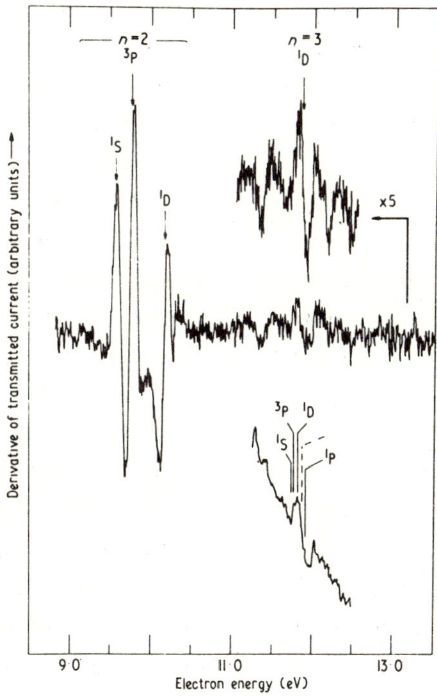


Fig. 9. Derivative of the electron current transmitted through atomic hydrogen.[22]

is necessary in electron scattering experiments if accurate comparison is to be made with theory.

Table II. Measurement of $H^-$ states below the n = 3 level. State energies derived from electron scattering resonances in the laboratory reference frame have been corrected to the center of mass frame.

| | Experiment | | | | Theory |
|---|---|---|---|---|---|
| State | Williams (1976)[21] | Spence (1975)[22] | Risley, et al. (1974)[23] | McGowan, et al. (1969)[24] | Burke, et al. (1967)[25] |
| $^1S$ | 11.73(8) | | | 11.64(3) | 11.727 |
| $^3P$ | | | | 11.76(2) | 11.759 |
| $^1D$ | 11.84(8) | 11.85(3) | 11.86(4) | 11.88(2) | 11.813 |
| $^1P$ | 11.93(8) | | | | 11.910 |
| $^1S$ | 12.04(8) | | | | 12.031 |

II.B.2. Photoemission experiments using hydrogen arc plasmas.

A hydrogen plasma generated in a wall-stabilized steady-state arc at atmospheric pressure has been used in attempts to observe the $H^-$ shape resonance at 1130 Å.[26,27] The shape resonance lies slightly above the n=2 level, see Fig. 3.

The effect of the shape resonance on the photoabsorption cross section of $H^-$ has been discussed by Macek.[28] His calculations show that the peak in the cross section should be about 25 times greater than the continuous free-bound cross section. However, the 1130-Å region is dominated by the Ly-α wing and thus only a small, 50% structure is expected.[26]

The shape of the feature depends on the magnitude of the oscillator strength, f-value, and to a lessor extent on the width. In the first experiment by Ott et al.[26] no obvious indication of the shape resonance was observed, see Fig. 10. The lower curve shows the experimental data of Ott et al.[26] The dashed peak indicates the lineshape expected based on an oscillator strength of 0.044 as predicted by Macek,[28] with no microfield Stark broadening. The upper curve from Wendoloski and Reinhardt[29] is a calculation using an oscillator strength of 0.024 from Broad and Reinhardt,[30] dc Stark broadening, an average over a Holtzmark distribution of microfields appropriate to an LTE plasma at 15,000 K, an addition of a simulated molecular background and a convolution of the spectrum with a 10% random signal to noise ratio. Thus, the negative results of Ott et al.[26] can be explained.

In a second experiment in which Behringer and Thoma[27] used essentially the same experimental procedure as Ott et al.[26] the $H^-$ shape resonance is clearly indicated at the predicted wavelength. The recorded spectra, see

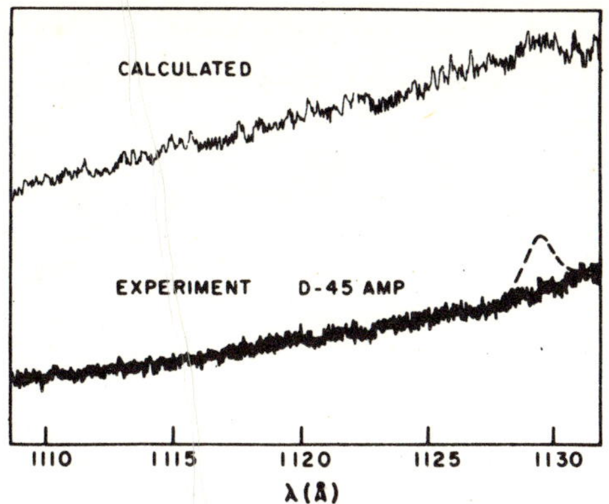


Fig. 10. Spectra in the vicinity of the predicted shape resonance of $H^-$. The experimental data is from Ott et al.[26] and the calculated spectra from Wendoloski and Reinhardt.[29]

Fig. 11, demonstrate that Macek's theory[28] overestimates the strength of the resonance and confirm the reduced oscillator strength of Broad and Reinhardt[30] but with half the peak height and twice the width - a possible indication for pressure broadening of the resonance.

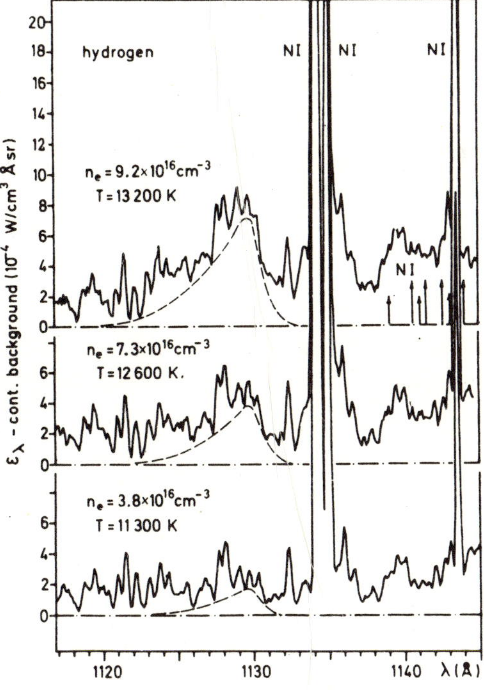


Fig. 11. Spectra in the vicinity of the $H^-$ shape resonance as measured by Behringer and Thoma.[27]

Listed in Table III are the theoretical predictions for the $^1P^o$ of $H^-$.

Table III. Theoretical predictions for the $^1P^o$ shape resonance of $H^-$ measured above the n = 2 level of 10.19892 eV.

| E(eV) | Γ(eV) | f | Method | Author |
|---|---|---|---|---|
| 0.01803 | 0.0151 | 0.044 | 3 state close coupling with correlation | Macek and Burke (1967)[31] Macek (1967)[28] |
| 0.018 | | | group-theoretical | Herrick and Sinanoglu (1975)[32] |
| 0.032 | 0.028 | | hyperspherical coordinates | Lin (1975)[18] |
| 0.018 | 0.015 | 0.024 | pseudo state close coupling | Broad and Reinhardt (1976)[30] |
| 0.0176 | 0.0141 | 0.015 | complex coordinates | Wendoloski and Reinhardt (1978)[29] |

Except for Lin's calculation[18] the energy and width of the state are well-determined but serious disagreement exists for the f-value or oscillator strength. It is interesting to note that in Lin's model the natural lifetime of the shape resonance is governed by tunneling through the angular momentum barrier. The presence of an electric field will indeed lower the potential barrier causing the resonance to broaden and to increase the transition rate, but additionally it will mix in other angular momentum states making the potential curve complex and allowing detachment into open channels with a further increase in width and transition rate.[29]

### II.B.3. Photodetachment of $H^-$.

The photodetachment of $H^-$ has recently been studied in the 11-eV region using a unique experimental setup. No tuneable lasers or intense light sources exist in this wave length region - (1130 Å). To overcome this difficulty Bryant et al.[33] directed light from a nitrogen laser 3370 Å to intersect a beam of 800-MeV $H^-$ ions obtained at the Los Alamos Meson Physics Facility. By varying the angle of intersection between the two beams the Doppler-shifted energy of the photon as seen by the $H^-$ ion was tuned through the important 11-eV region.

An example of the photodetachment cross section is shown in Fig. 12. The experimentally measured cross section was normalized to the low-energy continuum as predicted by Broad and Reinhardt[30] and the energy scale was shifted by 33 meV, but not dilated, in order to align the low-energy peak with the Feshbach resonance.[33] The resolution of the apparatus was about 10 meV, thus the narrow Feshbach resonances are artificially broadened.

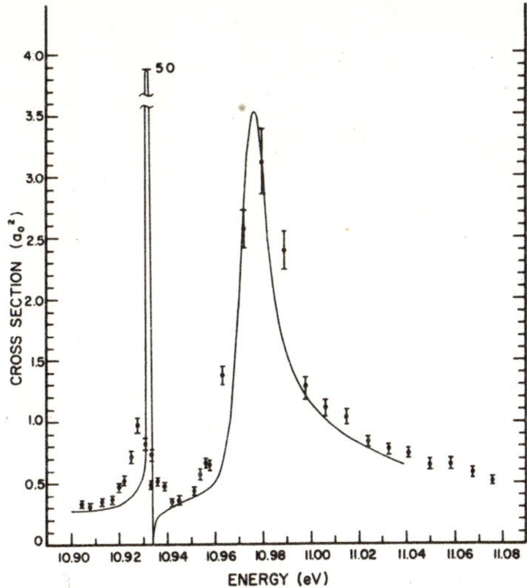


Fig. 12. Photodetachment cross section of $H^-$. Solid curve is theory from Broad and Reinhardt,[30] and data points from the experiment of Bryant et al.[33]

Although the prediction of Broad and Reinhardt[30] yields only one Feshbach resonance, more accurate calculations predict several additional states, see for example the photodetachment cross section as determined by Ajmera and Chung.[34] Good agreement is found in comparing experiment with the theory of Broad and Reinhardt folded with a 10-meV Gaussian resolution function.[35]

In a subsequent experiment the New Mexico team investigated the lowest $^1P^o$ Feshbach state and the shape resonance under the influence of a large variable electric field.[35] By applying a magnetic field at the location of the intersection of the laser beam and the $H^-$ ion beam, relativistic kinematics of the 800 MeV $H^-$ ions transforms the pure magnetic field so that an ion in its rest frame experiences a magnetic field of about twice the laboratory strength and a huge electric field. For example, a nominal 1200 Gauss magnetic field in the laboratory is transformed to an effective 2200 Gauss field and a 560 kV/cm electric field. The interaction energy between the ion and electric field is about 100 times that for the ion and magnetic field, thus the observed effects are assumed to be due to the electric field.[35]

Some results of the experiment are shown in Fig. 13. Figure 13 (a) exhibits the best resolution obtained so far, 5.5 meV FWHM. In the sequence from 12 kV/cm to 130 kV/cm, the single Feshbach state is seen to decline in height, split into two components, with a third component barely visible at a lower energy for an electric field strength of 130 kV/cm. For larger electric fields, e.g. 220 kV/cm, the Feshbach lines

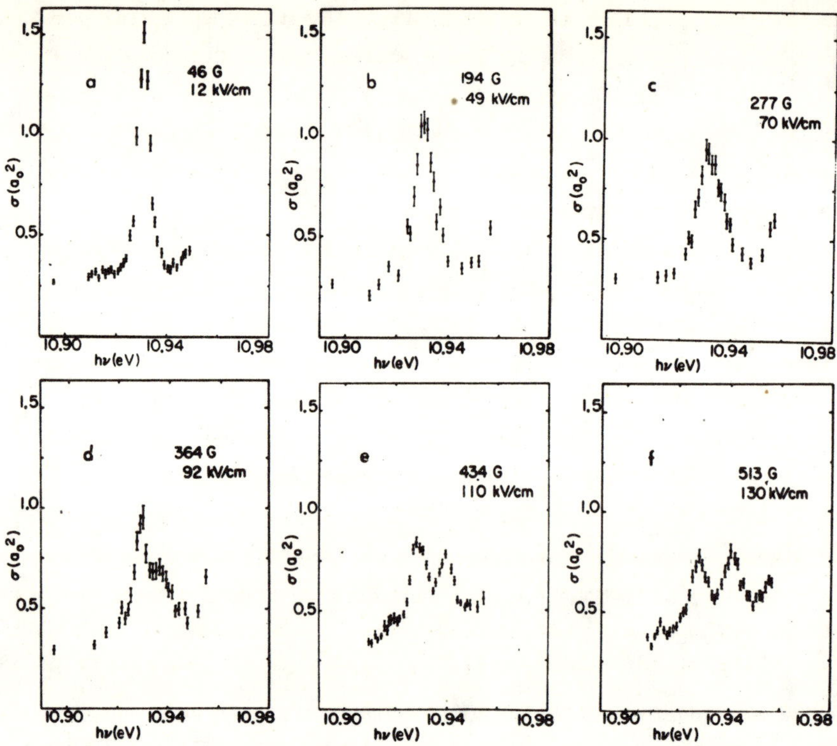


Fig. 13. Behavior of the Feshbach states of $H^-$ as a function of applied fields.[35]

disappear completely, and at 550 kV/cm the shape resonance has noticeably decreased in height and in area and broaden from 21 to 28 meV.[35]

The experimental observation of the broadening of the shape resonance compares favorably with a theoretical calculation of Wendoloski and Reinhardt in which they predict that the width of the $^1P^o$ shape resonance increases from 14.1 meV for zero field to 26.1 meV for an applied electric field of 380 kV/cm.[29]

An estimate for the electric field in which one might expect the $^1P^o$ shape resonance to be reduced significantly can be made using Lin's model assuming that tunnelling out through the potential barrier is the primary decay mechanism. Lin's calculation[18] shows that the barrier has a maximum height of 65 meV at a distance of 10 Å. This leads to an estimate of 600 kV/cm for the electric field at which the structure is expected to be affected, in agreement with the results of Gram et al.[35,36]

The effect of an electric field on the Feshbach states of $H^-$ below the n=2 level has been studied using the stabilization method by Callaway and Rau.[36] They find that the electric field couples a close lying $^1S^e$ state with the $^1P^o$ state leading to an approximately linear Stark effect and accounts for experimental result of Gram[35] that the splitting between the two main peaks is linear with the electric field. Only the m = 0 component of the $^1P^o$ state mixes with the $^1S^e$; the m = ±1 components are affected only in second order and decrease slowly with increasing field.[36]

Because the explanation of the behavior of the states relies strongly on the existence of near degeneracy between the $^1P^o$ state and the second $^1S^e$ state, more detailed information is needed on the separation between these states. Burke and Taylor's original calculations for these two states showed that they are separated by only 1 meV and that the width of the $^1S^e$ state is 2.2 meV which is larger than the separation.[6] Thus one expects a linear Stark effect because of the accidental degeneracy. Recently Callaway undertook a precise scattering calculation employing an algebraic variational method.[37] His results using 7 hydrogenic states and 11 pseudostates confirm those of Burke and Taylor[6] and show that the energy of the $^1S^e$ state is 10.16966 eV with a width of 5.6 meV and the energy of the $^1P^e$ state is 10.170269 eV with a width of $3.7 \times 10^{-2}$ meV. The separation between the two states of 0.61 meV is roughly 10 times smaller than the width of the $^1S^e$ state, indicating that an extremely small electric field will cause a strong mixing of the states.

The $^1P^o$ Feshbach state should disappear when the additional Stark energy lifts it above the n=2 level. Using results from Lin's model[18] of $H^-$, Callaway and Rau[36] estimate that the critical field-induced ionization should occur at roughly 170 kV/cm. The data of Gram et al.[35] are consistent with this estimate showing the resonances at 130 kV/cm but not at 220 kV/cm.

## III. OTHER NEGATIVE IONS

Referring to Fig. 1 one finds that the number of stable bound atomic negative ions is large. Recognizing that even unstable negative ions have numerous excited state level, it is clear that the spectroscopy of negative ions has barely begun. Figure 2 shows some of the types of excited states which are possible. For $C^-$ only one autodetaching transition has been observed leaving intriguing questions as to how to excite other states and observe the resulting transitions.

Excited states of negative ions have been reviewed most recently by Edwards[38] (particularly the halogen ions) and as electron resonances by Schulz[39] (primarily rare gases and alkali ions) and Golden[40] (simple atoms and diatomic molecules).

### III.B. The $He^-$ Problem

The next simplest negative ion after $H^-$ is the negative He ion. In contradistinction to $H^-$, experiments involving He are easier; whereas theoretical calculations are more difficult. However, much progress in both areas has been made in the past decade. A summary of the results for the lowest $^2S$ excited state of $He^-$ is listed in Table IV.[41]

Two problems are apparent:

(1) Since 1970 except for one measurement, the position of the resonance has been determined experimentally to be 19.35 eV to within an accuracy of 20 meV. However, since 1970 different theoretical approaches predict a mean value of 19.38 eV with no result lying more than 18 meV above or below this value. At present, it is not possible to reconcile the discrepancy between experiment and theory.

(2) A more puzzling problem concerns the width of the $^2S$ state of $He^-$. Except for one measurement, all transmission experiments give a width of about 12 meV; whereas, differential scattering measurements give a width of about 8 to 9 meV. The latest theoretical calculations yield a width of 12 meV. No satisfactory explanation is yet available. In an experiment random errors and unknown systematic effects invariably broaden a spectral line, not contract it, especially if the spectral feature has been recorded numerous times using a repetitious scanning procedure.

## III.C. Collisional Excitation Technique

During the past few years, Edwards and coworkers have used the collisional excitation technique to observe the decay of autodetaching states of $H^-$,[23] $C^-$,[4] $O^-$,[42,43] $F^-$,[44] $Cl^-$,[45] $Br^-$,[46] and $I^-$.[47]

In the collisional-excitation method the negative ion to be studied is produced in an ion source (duoplasmatron), accelerated through several thousand volts, mass analyzed and directed into a gas cell. Electrons leaving the collision region are energy analyzed at one particular angle with respect to the ion beam. The electron spectra consist of lines superimposed on a continuous background. Each line corresponds to a transition from an excited state of the negative ion to a lower state of the neutral atom. Energy analysis and angular distributions of the lines allow one to identify and determine the energy of excited states of the negative ion. Because the electrons are ejected from a moving ion, energy analysis must include reaction kinematics to transform laboratory energies into rest frame energies.[23]

To properly identify each spectral feature one varies the collision energy, observation angle and target gas. Since spectra from multi-electron negative ions contain lines both from excited states of the negative ion and of the neutral atom, Edwards and co-workers also found it necessary to neutralize part of the negative ion beam, apply a deflection voltage and allow only neutral atoms to enter the collision region;

Table IV. Comparison of experiment and theory for the lowest $^2S$ resonance of $He^-$, ref. 41.

| Energy (eV) | Width (meV) | Method | Reference |
|---|---|---|---|
| | | *Experiment* | |
| 19.3 | 17.5 ± 2.5 | Transmission | Andrick and Ehrhardt (1966) |
| 19.3 | 12 | Transmission | Andrick and Ehrhardt (1968) |
| 19.3 | 12 | Transmission | Ehrhardt et al. (1968) |
| | 8 | Differential scattering | Gibson and Dolder (1969) |
| 19.30 ± 0.01 | 8 ± 2 | Transmission | Golden and Zecca (1970) |
| 19.34 ± 0.02 | | Transmission | Sanche and Schulz (1972) |
| 19.35 ± 0.02 | | Differential scattering | Mazeau et al. (1972) |
| 19.367 ± 0.009 | 9 ± 1 | Differential scattering | Cvejanovic et al. (1974) |
| 19.35 ± 0.02 | 13 | Transmission | Golden et al. (1974) |
| | | *Theory* | |
| 19.45 ± 0.15 | 19 ± 11 | Variational | Kwok and Mandl (1965) |
| 19.33 | 39 | Close coupling | Burke et al. (1966) |
| 19.3 | | Stabilization | Ellezer and Pan (1970) |
| 19.368 | | Stabilization | Weiss and Krauss (1970) |
| 19.368 | 14 | Quasi-projector operators | Temkin et al. (1972) |
| 19.4 | 15 | Variational | Sinfailam and Nesbet (1972) |
| 19.38 | 11.5 | Close coupling | Ormonde and Golden (1973) |
| 19.398 | 12 | Complex coordinate | Bain et al. (1974) |
| 19.387 | 12.1 | Complex coordinate | Junker and Huang (1978) |
| 19.386 | 11.73 | Correlated wave function | Junker (1978) |

thereby studying the autoionizing lines of the neutral atom without interference from autodetaching lines from the negative ion. A particularly striking example is seen in Fig. 14 for $Br^-$ on Ar and He and Br on He.[46] No satisfactory explanation is available for why only

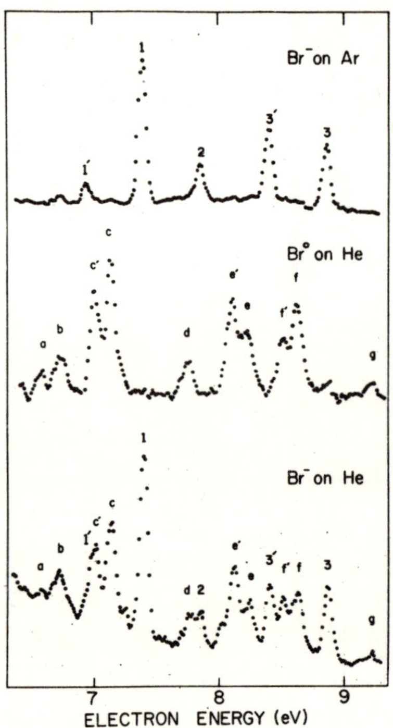


Fig. 14. Electron spectra produced by 2 keV collisions of $Br^-$ on Ar, Br on He, and $Br^-$ on He.[46]

autodetaching lines are observed for $Br^-$ on Ar whereas for $Br^-$ on He both autodetaching lines of $Br^-$ and autoionizing lines of Br are seen.

Table V lists all the autodetaching lines which have been observed using the collisional excitation technique.[38] In general, the autodetaching states can be classified as doubly-excited states with dominant configurations of (positive core)$n\ell^2$ or (positive core)$n\ell n\ell$ with the most intense line usually being the lowest available $ns^2$.[38] A notable except is the $2s2p^6$ subshell excitation of $O^-$.[48] Excellent agreement between experiment and published theory exists for $C^-$,[3] $O^-$,[48] and $Cl^-$.[49] The results listed in Table V come from only a small range of collision energies, a few target gases, and just a fraction of the total number of

Table V. Autodetaching states of negative ions measured using the collisional excitation method, ref. 38.

| Ion | Energy (eV) | Ion | Energy (eV) |
|---|---|---|---|
| $H^-$ | 9.59 ± 0.03 | | 9.15 ± 0.05 |
| | 9.76 ± 0.03 | | 9.97 ± 0.04 |
| $C^-$ | 7.44 ± 0.07 | | 12.09 ± 0.06 |
| $O^-$ | 9.50 ± 0.02 | $Br^-$ | 7.39 ± 0.06 |
| | 10.11 ± 0.02 | | 7.84 ± 0.06 |
| | 10.87 ± 0.02 | | 8.85 ± 0.06 |
| | 12.12 ± 0.02 | $I^-$ | 6.41 ± 0.06 |
| | 13.71 ± 0.02 | | 6.75 ± 0.06 |
| $F^-$ | 14.85 ± 0.04 | | 7.15 ± 0.06 |
| $Cl^-$ | 8.53 ± 0.05 | | 8.06 ± 0.06 |

possible negative ions shown in Fig. 1. It is apparent we are at the threshold of the quest to discover many excited states of negative ions.

## III.D. Branching Ratios

If an autodetaching state of a negative ion lies above an excited state of the neutral atom, then the state can decay either to the ground state ejecting an electron with a large kinetic energy or to the excited state ejecting an electron with a small kinetic energy. The branching ratio or relative transition probability for the two decay channels can be determined from the ratio of the intensities of the two lines.

### III.D.1. $H^-$.

Macek and Burke have calculated the relative transition probabilities from several $H^-$ states below the n=3 level.[31] They found that it is 10 to 50 times more probable that low energy electrons, 0.5 to 1.8 eV, are ejected than high energy electrons, 11.7 to 12.0 eV. Risley et al.[23] observed a structure lying just below the n=3 level which was 40 times less intense than lines below n=2. Attempts to observe the low energy lines with the apparatus failed.

### III.D.2. $Kr^-$ and $Xe^-$.

Branching ratios involving more complex negative ions are important in understanding the electron spectra produced in collisions of $H^-$ on the heavy rare gas atoms.[50] Figure 15 shows autodetaching transitions from the $^2P_{3/2}$ and the $^2P_{1/2}$ states of $Ar^-$, $Kr^-$ and $Xe^-$. The excited states were created in an electron transfer process with the $H^-$ ion - a powerful

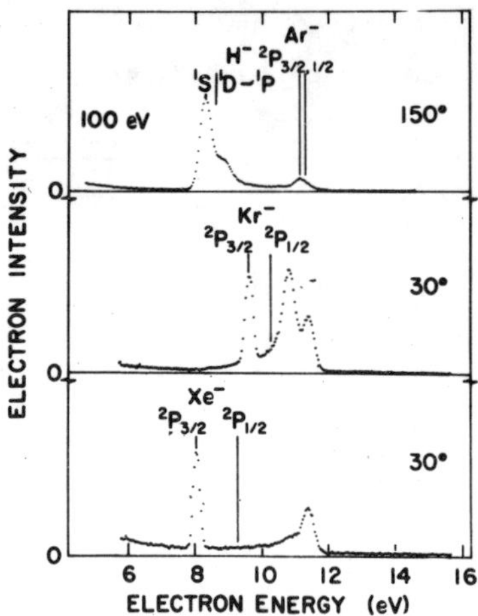


Fig. 15. Electron spectra for 100 eV $H^-$ on Ar, Kr and Xe.[50]

technique for studing states of negative ions which do not have a stable ground state.

Measurements for $H^-$ on Ar show that the $^2P_{3/2}$ state of Ar is populated roughly two times more often than the $^2P_{1/2}$ state in accordance with the statistical weights. If we assume that statistical population holds for the two states of $Kr^-$ and $Xe^-$, then by measuring both the widths and the relative intensities of each line, one can determine the total transition probabilities and the relative transition probabilities, respectively. However, the low resolution scans in Fig. 15 show only a line corresponding to the lowest state of $Kr^-$ and $Xe^-$, i.e. a transition from the $^2P_{3/2}$ state. Figure 16 exhibits the behavior of the $^2P_{3/2}$ line of Kr and $Xe^-$ as a function of $H^-$ collision energy. Additional measurements at other observation angles and with high resolution are needed in order to fully understand the missing $^2P_{1/2}$ transition.

III.D.3. $N_2^-$.

An example in which branching ratios play an important role is the excitation and decay of vibrational autodetaching states of $N_2^-$ formed in collisions of $H^-$ on $N_2$.[51] Figure 17 shows a significant perturbation in the energy distribution of electrons between 1 and 4 eV. The spacings between each line above about 1.9 eV are roughly the same as the spacings between the vibrational levels of $N_2^-$ (0.24 eV), and below 1.9 eV they are roughly the same as the $N_2$ levels (0.29 eV).

Figure 18 shows a possible electron transfer reaction involving $H^-$ with $N_2$ using a molecular model for $N_2^-$. In this example $N_2$ in the ground state is excited to $\nu^- = 3$ of $N_2^-$. The $N_2^-(3)$ state can decay to each

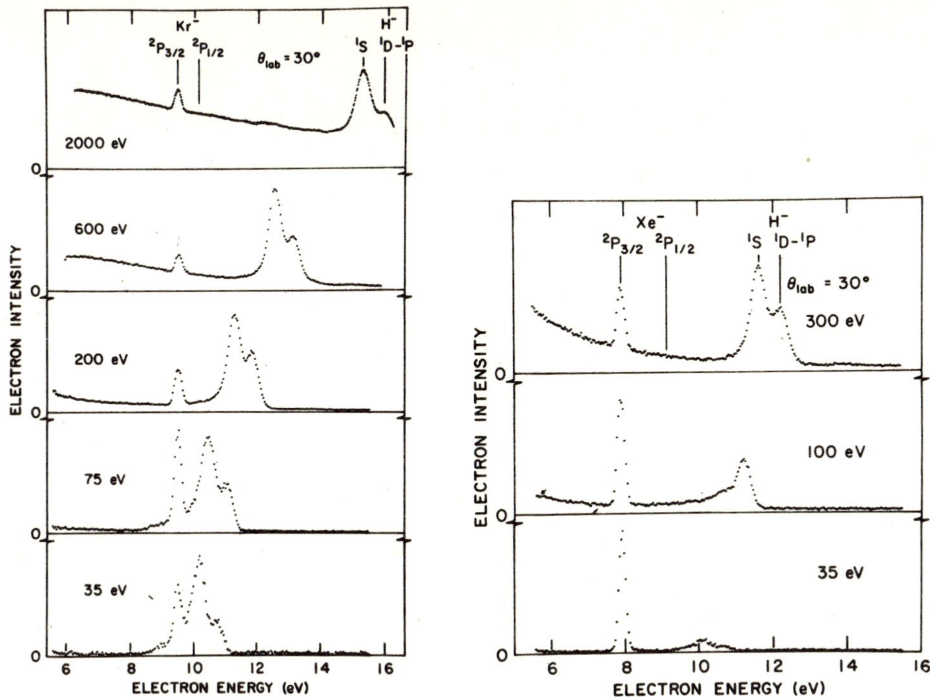


Fig. 16. Electron spectra for $H^-$ on Kr and Xe.[50]

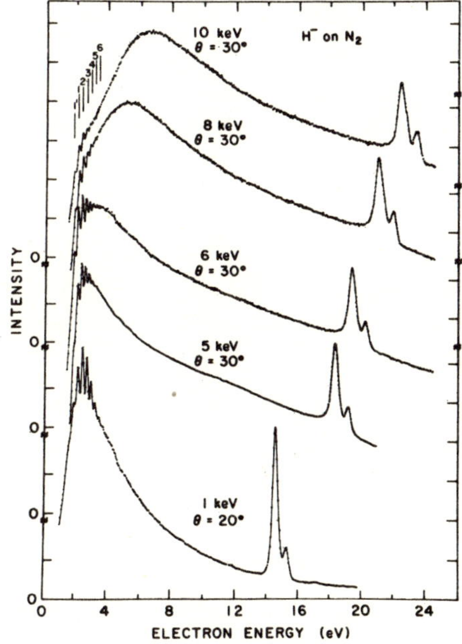


Fig. 17. Energy distribution of secondary electrons produced in collisions of $H^-$ with $N_2$.[51]

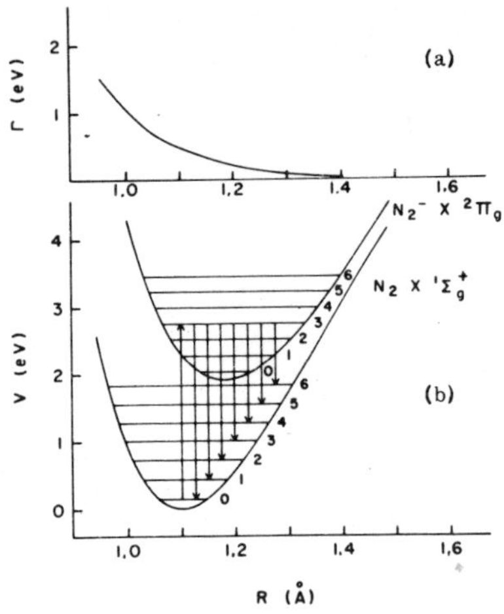


Fig. 18. Potential energy curves for the $N_2$ - $N_2^-$ system showing one excitation transition to $\nu^-$ = 3 and subsequent decay channels.[51]

energetically allowed vibrational state of $N_2(\nu)$ with a certain branching ratio. If decay occurs to the ground state, an electron with a kinetic energy of 2.62 eV is emitted. If decay occurs to an excited vibrational level of $N_2(\nu)$, then an electron is emitted with a kinetic energy less than 2.62 eV. A series of lines will result with a spacing equal to the vibrational levels of $N_2$. The lowest kinetic energy of the autodetaching electron, in this example, is less than 0.13 eV for a decay to $\nu$ = 9 of $N_2$.

On the other hand, if each $N_2^-$ state decayed only to ground state of $N_2$, then a series of lines would result with a spacing equal to the vibrational levels of $N_2^-$, about 0.24 eV. The energy of the lowest lying line would be 1.89 eV.

To test various excitation and decay possibilities, a model was developed for the expected electron energy distribution. The model had to exhibit the general features of the ejected electron spectra:

1) well defined structure exists above 3.0 eV,
2) the oscillations are more pronounced above 2.0 eV than below,
3) the relative intensity of each oscillation exhibits little change as the $H^-$ collision energy is varied by two orders of magnitude, and
4) the relative separation between each line remains constant regardless of observation angle or collision energy (75 eV to 10,000 eV).

Thus, the model must give rise to ejection of high energy electrons with respect to the $N_2$ - $N_2^-$ system. In addition, because of lack of the usual characteristics of interference effects, that is, rapid variation of the spectral features with collision velocity or observation angle, the relative phase difference in the amplitudes for excited each $N_2^-$ state may not be important (in contradistinction to electron scattering). In other words, we observe, experimentally, an averaging of the relative phase in exciting the upper $N_2^-$ states.

With these ideas as a basis, an expression for the electron intensity, as observed with our analyzer, can be written as being proportional to

$$\sum_{\nu^-=0}^{6} \sum_{\nu=0}^{6} \sigma(\nu^-)\ \gamma(\nu^-,\nu)\ E\ L(E)$$

where $\sigma(\nu^-)$ is the cross section for exciting the $\nu^-$ state of $N_2^-$ from the ground state $N_2$, $\gamma(\nu^-,\nu)$ is the branching ratio of the upper $N_2^-(\nu^-)$ state to the lower $N_2(\nu)$ state, E which is equal to the kinetic energy of the electron accounts for the transmission factor of our electrostatic analyzer at high energies, and L(E) is the usual Lorentzian line profile.

$$L(E) = \frac{\Gamma^2/4}{[(E - \Delta E(\nu^-,\nu)]^2 + \Gamma^2/4}$$

where $\Delta E(\nu^-,\nu)$ is the difference between the $N_2^-(\nu^-)$ energy level and the $N_2(\nu)$ level. $\Gamma$ is the effective or average FWHM for the $N_2^-$ state.

Because the relative intensities of each line did not change with collision velocity and because the energy transfer was small compared with the collision energy, it was assumed that electron transfer to the $N_2^-$ state could be described in terms of a fast or sudden transition. Therefore the Franck-Condon principle was invoked to yield

$$\sigma(\nu^-) = |a(\nu^-)|^2 \quad |\int\psi_\nu\text{-}\psi_0\ dR|^2$$

where $a(\nu^-)$, the transition amplitude for all the electrons from the initial $H^- + N_2$ state to the final $H + N_2^-$ state, is assumed not to depend strongly on $\nu^-$ and set equal to a constant value. $|\int\psi_\nu\text{-}\psi_0 dR|^2$ is the Franck-Condon factor FCF or overlap integral squared of the nuclear vibrational wavefunctions for $N_2^-(\nu^-)$ and $N_2(0)$.

The important factor missing for the model is the branching ratios $\gamma(\nu^-,\nu)$. Various possibilities were explored using conventional molecular principles, see Ref. 51.

Comparison of the model calculations with experiment reveals that the Franck-Condon principle seems to apply for excitation of the $N_2^-$ states but not for the decay, see Fig. 19. In fact, if the current spectra are not attenuated below 2 eV, then we are led to the unexpected conclusion that $N_2^-$ decays primarily to the ground state of $N_2$. Further work both theoretically and experimentally is warranted to understand the correct branching ratios for $N_2^-$.

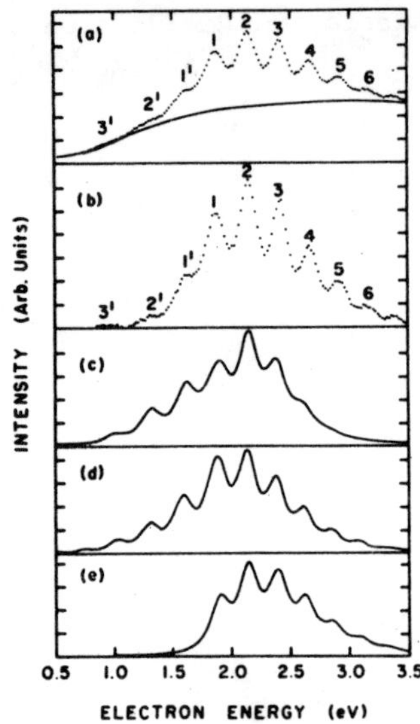


Fig. 19. Electron energy distribution from $N_2^-$: (a) experiment, 4-keV $H^-$ on $N_2$ observed at 150°; (b) experiment, background subtracted from (a); (c) model, branching ratio proportional to $\Delta E^{5/2}$ x FCF; (d) model, branching ratio proportional to $\Delta E^{5/2}$; (e) model, branching ratio equal to $\delta_{\nu 0}$.[51]

## IV. REFERENCES


1. H. Hotop and W.C. Lineberger, J. Phys. Chem. Ref. Data 4, 539-576 (1975).
2. V.A. Oparin, R.N. Il'in, T.T. Serenkov, E.S. Solov'ev, and N.V. Fedorenko, JETP Lett. 13, 249 (1971).
3. J.J. Matese, Phys. Rev. A 10, 454-457 (1974).
4. N.Lee and A.K. Edwards, Phys. Rev. A 11, 1768-1770 (1975).
5. J.S. Risley in *Atomic Physics 4*, ed. G. zu Putlitz, E.W. Weber, and A. Winnacker, (Plenum Press, New York, 1975), pp. 487-528.
6. P.G. Burke, Advan. At. Mol. Phys. 4, 173 (1968).
7. Shimamura, J. Phys. Soc. Jap. 31, 852 (1971).
8. J.N. Bardsley and B.R. Jonker, J. Phys. B 5, L178-L180 (1972).
9. K.T. Chung and J.C.Y. Chen, Phys. Rev. A 6, 686-693 (1972).
10. G.D. Doolen, J. Nuttall and R.W. Stagat, Phys. Rev. A 10, 1612-1615 (1974).
11. A.K. Bhatia, Phys. Rev. A 9, 9-11 (1974).
12. A.K. Bhatia and A. Temkin, Phys. Rev. A 11, 2018-2024 (1975).

13. K.T. Chung and J.C.Y. Chen, Phys. Rev. A 13, 1655-1656 (1976).
14. Y.K. Ho, A.K. Bhatia and A. Temkin, Phys. Rev. A 15, 1423-1429 (1977).
15. U. Fano and C.D. Lin, in *Atomic Physics 4*, eds. G. zu Putlitz, E.W. Weber, and A. Winnacker (Plenum Press, New York, 1975), pp. 47-70.
16. C.D. Lin, Phys. Rev. A 10, 1986-2001 (1974).
17. C.D. Lin, Phys. Rev. A 12, 493-497 (1975).
18. C.D. Lin, Phys. Rev. Letters 35, 1150-1153 (1975).
19. C.D. Lin, Phys. Rev. A 14, 30-35 (1976).
20. J.F. Williams, in *Electronic and Atomic Collisions, Abstracts of Papers for the Ninth ICPEAC,* ed. J.S. Risley and R. Geballe (University of Washington Press, Seattle 1975), pp. 684-685.
21. J.F. Williams, J. Phys. B 9, 1519-1527 (1976) and in *The Physics of Electronic and Atomic Collisions, Invited Lectures, Review Papers and Progress Reports of the Ninth ICPEAC,* ed. J.S. Risley and R. Geballe (University of Washington Press, Seattle, 1976) pp. 139-150.
22. D. Spence, J. Phys. B 8, L42-L45 (1975).
23. J.S. Risley, A.K. Edwards and R. Geballe, Phys. Rev. A 9, 1115-1129 (1974).
24. J.W. McGowan, J.F. Williams and E.K. Curley, Phys. Rev. 180, 132 (1969).
25. P.G. Burke, S. Ormonde, and W. Whitaker, Proc. Phys. Soc. (London) 92, 319 (1967).
26. W.R. Ott, J. Slater, J. Cooper and G. Gieres, Phys. Rev. A 12, 2009-2016 (1975).
27. K. Behringer and P. Thoma, Phys. Rev. A 17, 1408-1415 (1978).
28. J. Macek, Proc. Phys. Soc. 92, 365-369 (1967).
29. J.J. Wendoloski and W.P. Reinhardt, Phys. Rev. A 17, 195-200 (1978).
30. J.T. Broad and W.P. Reinhardt, Phys. Rev. A 14, 2159-2173 (1976).
31. J. Macek and P.G. Burke, Proc. Phys. Soc. (London) 92, 351 (1967).
32. D.R. Herrick and O. Sinanoglu, Phys. Rev. A 11, 97-110 (1975).
33. H.C. Bryant, B.D. Dieterle, J. Donahue, H. Sharifian, H. Tootoonchi, D.M. Wolfe, P.A.M. Gram and M.A. Yates-Williams, Phys. Rev. Letters 38, 228-230 (1977).
34. M.P. Ajmera and Kwong T. Chung, Phys. Rev. A 12, 475-479 (1975).
35. P.A.M. Gram, J.C. Pratt, M.A. Yates-Williams, H.C. Bryant, J. Donahue, H. Sharifian, and H. Tootoonchi, Phys. Rev. Letters 40, 107-111 (1978).
36. J. Callaway and A.R.P. Rau, J. Phys. B. letter to be published.
37. J. Callaway, to be published.
38. A.K. Edwards in *The Physics of Electronic and Atomic Collisions, Invited Lectures, Review Papers, and Progress Reports of the IXth ICPEAC,* ed. J.S. Risley and R. Geballe (University of Washington Press, Seattle) 1976, pp. 790-802.
39. G.J. Schulz, Rev. Mod. Phys. 45, 378-422 (1973).

40. D.E. Golden, in *Advances in Atomic and Molecular Physics*, eds. D.R. Bates and B. Bederson, to be published.

41. B.R. Junker and C.L. Huang, Phys. Rev. A., to be published, Aug., 1978.

42. A.K. Edwards, J.S. Risley and R. Geballe, Phys. Rev. A 3, 583-586 (1971).

43. A.K. Edwards and D.L. Cunningham, Phys. Rev. A 8, 168-173 (1973).

44. A.K. Edwards and D.L. Cunningham, Phys. Rev. A 9, 1011-1012 (1974).

45. D.L. Cunningham and A.K. Edwards, Phys. Rev. A 8, 2960-2964 (1973).

46. A.K. Edwards and D.L. Cunningham, Phys. Rev. A 10, 448-450 (1974).

47. D.L. Cunningham and A.K. Edwards, Phys. Rev. Letters 32, 915-917 (1974).

48. J.J. Matese, S P. Rountree, and R.J.W. Henry, Phys. Rev. A 7, 846-850 (1973).

49. J.J. Matese, S.P. Rountree, and R.J.W. Henry, Phys. Rev. A 8, 2965-2967 (1973).

50. Risley, unpublished.

51. J.S. Risley, Phys. Rev. A 16, 2346-2351 (1977).

# THRESHOLD BEHAVIOUR OF ELEMENTARY ATOMIC PROCESSES

M.Gailitis

Latvian SSR Academy of Sciences

Institute of Physics Riga-Salaspils USSR

## I. Introduction

Two basic aspects to be treated here are: first, the threshold behaviour of e-H scattering and photodetachment of $H^-$ ion, secondly, the effect of an electric field on the resonances in He and helium like ions.

Attention will be directed toward the characteristic peculiarities caused by the participation of two electrons with one nucleus in these reactions. Such phenomena are pertinent only to the three-particles case.

Our studies on the threshold behaviour of e-H scattering have been under way since 1963 [1,2]. By now the majority of predictions have been confirmed by the measurements. Although oscillations in cross sections above the thresholds remain yet unobserved.

The first experimental studies concerning electric field influence upon the double-excited states of $H^-$ and Sr have been published this year [3,4]. As for He and heliumlike ions no data are available yet.

### 1.1. The threshold laws

When the forces between the particles are short-range we have a comparatively simple threshold behaviour. Above the threshold the new state excitation cross section follows Wigner's law

$$\sigma_{if} \sim k_f^{2l_f+1} \tag{1}$$

where $k_f$ and $l_f$ are the momentum and angular momentum of an outgoing particle. Below the threshold other cross sections may have some resonances at such energies when an outgoing particle is remaining for some time bound with the target.

The long-range character of the Coulomb forces is of crucial importance for the electron-atom scattering. We have some cases when excitation cross sections above the threshold tend to zero slower than in (1), with numerous resonances below and close to the threshold [5-10].

2. Electron-hydrogen scattering

For e-H scattering the degeneracies of the hydrogen atom are of major importance. The levels with different parity have equal energies and the excited atom can have a nonzero dipole momentum. Interaction between the atom and outgoing electron is proportional to $r^{-2}$ . A similar dependence relates to the centrifugal energy. This leads to the changing power in (1) which now also depends on the dipole terms. At small total angular momentum L the dipole attraction exceeds the centrifugal repulsion and the power tends to zero: the excitation cross section becomes finite at the threshold.

2.1. The multichannel effective range theory for e-H scattering

Our discussion will be based on the matrix relationship [1,2]

$$S=e^{\frac{i\pi\ell}{2}}A\,e^{-\frac{i\pi\lambda}{2}}\left[1+k^{\lambda+1/2}\frac{2i}{M-(tg\,\pi\lambda+i)k^{2\lambda+1}}k^{\lambda+1/2}\right]e^{-\frac{i\pi\lambda}{2}}A^{-1}e^{\frac{i\pi\ell}{2}}. \tag{2}$$

It describes the S-matrix behaviour via the following matrices: k and $\ell$ - diagonal matrices of momentum and angular momentum for scattered electron, A - the orthogonal matrix which diagonolizes the sum of centrifugal terms $\ell(\ell+1)r^{-2}$ and dipole terms $\alpha_{ij}\, r^{-2}$ connecting the degenerate channels

$$\Lambda=\ell(\ell+1)+\alpha, \tag{3}$$

$$A^{-1}\Lambda A=a=\lambda(\lambda+1). \tag{4}$$

Dimensions of matrices in (2-4) are equal to the number of open channels above the threshold under investigation. Diagonal matrix $\lambda$ can be expressed through eigenvalues of matrix $\Lambda$

$$\lambda+1/2=\sqrt{a+1/4}\,. \tag{5}$$

Matrices $\Lambda$ and A are cell-diagonal. Each cell corresponds to subspace with constant L, total parity P, atoms principal quantum number N and gives finite number of eigenvalues $a$ and $\lambda$ (5). Relation (2) represents the generalization of the effective range theory formula

$$k^{2\ell+1}\,ctg\,\delta_\ell=-\frac{1}{a}+r_e k^2 \tag{6}$$

for a multichannel case taking into account also the dipole interaction between degenerate channels. Matrix M in Eq. (2) represents the right hand side of Eq. (6). It has a smooth energy dependence.

2.2. Excitation cross sections behaviour

The threshold behaviour is determined through factor $\kappa^{\lambda+1/2}$ in (2). It differs from that of the short-range case by the replacement of $\lambda$ for $\ell$. On each threshold opens the transitions to atomic states with new N. We use N for threshold numeration. At sufficiently large L all $\lambda$ are real for given N and $|\kappa_f^{\lambda_{min}+1/2}|^2$ is predominant in the threshold behaviour. Here $\lambda_{min}$ is the smallest of $\lambda$ for the threshold under investigation.

At small L we have some complex $\lambda$ with $\mathrm{Re}\lambda = -1/2$ and $|\kappa_f^{\lambda+1/2}|^2 = 1$. It leads to the finiteness of cross sections on the thresholds. This has been experimentally tested by Chamberlain et al.[11].

To describe the photodetachment of $H^-$ ion we have to introduce one additional channel with $H^-$ ion plus photon $h\nu$. Its coupling with the e-H scattering channels can be calculated in the first approximation of the perturbation theory. Thus we can obtain the (2) type formulae with L=1 and spin S=0. The new M-matrix has one additional row and a column for $H^-+h\nu$ channel. All other elements remain equal to those of the pure e-H scattering. Therefore the $H^-$ ion photodetachment cross sections are also finite at all the thresholds with an excited H atom.

2.3. Resonances spectrum

The dominator in (2) is responsible for resonances below the threshold occuring at energies with a small determinant of the dominator. The resonances distribution under the threshold is sensible to $\lambda$ values. Some separate resonances may occur for the thresholds and L with all real $\lambda$. For complex $\lambda$ terms

$$\kappa^{2\lambda+1} = e^{2i\,\mathrm{Im}\lambda\cdot\ln\kappa} \tag{7}$$

oscillate. They take equal values when $E_t - E$ differs by factor $\exp(2\pi/\mathrm{Im}\lambda)$. Thus the families of the resonant states below the thresholds are established. For each $\lambda$ we have the following type of distribution

$$E_t - E_n = (E_t - E_1)\left(e^{-\frac{2\pi}{\mathrm{Im}\lambda}}\right)^{n-1}. \tag{8}$$

Slight deviations from law (8) are possible owing to some smooth changes of M matrix with energy and to the presence of terms $k^{2\lambda+1}$ with different $\lambda$ . So, it is clear that the situation depends on the presence of complex $\lambda$ , i.e. on the existence of eigenvalues $a<-1/4$ .

2.4. Eigenvalues of $\Lambda$

In the present section the nuclear charge $Z$ will not be fixed. The eigenvalues of $\Lambda$ can be obtained separately for each of its cells in subspaces with constant L, P and N. The number of eigenvalues in a cell with $P=(-1)^L$ is $N(N+1)/2$, for $L \geqslant N-1$ and $(L+1)(N-L/2)$ for $L<N-1$. At $P=(-1)^{L+1}$ the numbers are $N(N-1)/2$ and $L(N-(L+1)/2)$ correspondingly.

According to Seaton[12] for N=2 the eigenvalues are given by quadratic equation. At $P=(-1)^L$ we have

$$a_{1,2}=L(L+1)+1\pm\sqrt{(2L+1)^2+36/Z^2} \qquad (9)$$

and also $a_3=L(L+1)$ if $L\geqslant 1$.

At $P=(-1)^{L+1}$ $a_1=L(L+1)$ if $L\geqslant 1$.

We find that $a_2<-1/4$ if $Z=1, P=(-1)^L$ and $L\leqslant 2$ . It is interesting to note that we have eigenvalue $L(L+1)$ for both parities. The coincidence is not haphazard. It has been recently demonstrated by Herrick[13,14], Nikitin and Ostrovsky[15] that for arbitrary Z and N the eigenvalues of $\Lambda$ are degenerated, i.e. all the eigenvalues of $\Lambda$ at $P=(-1)^{L+1}$ are also the eigenvalues at $P=(-1)^L$ . They have constructed the following operator

$$\mathcal{B}=\vec{A}_2\vec{\ell}+\frac{3N}{2Z}\frac{\vec{r}\vec{\ell}_2}{r} \qquad (10)$$

with commutation relations

$$[\mathcal{B},\vec{L}]=[\mathcal{B},\Lambda]=\mathcal{B}P+P\mathcal{B}=0. \qquad (11)$$

Index 2 denotes the variables of an atomic electron. The variables of a scattered electron have no index.

$$\vec{A}_2=\frac{1}{\sqrt{-2E_2}}\left\{\vec{p}_2(\vec{r}_2\vec{p}_2)-\vec{r}_2\left(p_2^2-\frac{Z}{r_2}\right)\right\} \qquad (12)$$

is the Runge-Lenz vektor. Taking into account identity

$$(N\ell_2'|\vec{r}_2|N\ell_2)=-3N(N\ell_2'|\vec{A}_2|N\ell_2)/(2Z)$$

we can write (3) in the following form

$$\Lambda = \vec{l}^{\,2} - \frac{3N}{Z}\vec{A}_2\frac{\vec{r}}{r} \qquad (13)$$

and derive relations (11) responsible for the degeneration of $\Lambda$ eigenvalues through the parity.

For N=3 the eigenvalues also have the algebraic formulae. They are the solutions of cubic equations [16].

Numerical computation remains possible at any N. Approximate expressions can also be used. In case of small angular momentum the first term in (3) is smaller than the second one and can be treated as perturbation. In a zero-order approximation the $\Lambda$ eigenvalues are equal to the matrix $\alpha$ eigenvalues. Since $\alpha$ in fact describes the linear Stark effect caused in the atom by the electric field of a scattered electron the eigenvalues of $\alpha$ are

$$a^{(0)}_{KT} = -3NK/Z. \qquad (14)$$

We can verify (14) by the following expressions for elements

$$\alpha_{ij} = 2(-1)^{L+1}(N l_{2i}|r_2|N l_{2j})\sqrt{(2l_i+1)(2l_j+1)(2l_{2i}+1)(2l_{2j}+1)} \times$$

$$\times \begin{pmatrix} l_i & 1 & l_j \\ 0 & 0 & 0 \end{pmatrix} \begin{pmatrix} l_{2i} & 1 & l_{2j} \\ 0 & 0 & 0 \end{pmatrix} \begin{Bmatrix} l_{2i} & 1 & l_{2j} \\ l_j & L & l_i \end{Bmatrix} \qquad (15)$$

and

$$A^{(0)}_{l l_2, KT} = (l_2 L T -T | l 0)\left(\frac{N-|T|-1}{2}, \frac{N+|T|-1}{2}, \frac{K}{2}, -\frac{K}{2} \,\middle|\, l_2 0\right) \qquad (16)$$

with $(-1)^{l+l_2} = P$

for the eigenvectors due to eigenvalues (14). The eigenvectors are characterized by two integer quantum numbers K and T . Their values are restricted to two conditions:

(i) $N-1-|K| - |T|$ must be even, greater or equal to zero,

(ii) $T=0$ only if $P=(-1)^L$ .

We can interpret K as the difference between the both parabolic quantum numbers and T as the atoms angular momentum projection on the direction from the nucleus to the scattered electron.

Spectrum (14) is equidistant and degenerated according to T, P and L. Being a zero approximation, eigenvalues (14) are simultaneously the lower bounds for $\Lambda$ eigenvalues. Indeed operator $\ell(\ell+1)$ is positive and restricted

$$\ell_{min}(\ell_{min}+1) \leqslant \ell(\ell+1) \leqslant (L+N-1)(L+N),$$
$$\ell_{min} = max(0, L+1-N). \qquad (17)$$

Therefore all eigenvalues of $\Lambda$ are placed above corresponding $a^{(0)}_{KT}$, the difference being less than (L+N-1)(L+N). The lowest eigenvalue for hydrogen is within the limits

$$\ell_{min}(\ell_{min}+1) - 3N(N-1) < a_{min} < -2N(N-1) + L(2N-1+L) \qquad (18)$$

It is less than $-1/4$ at least for all

$$L < (N-1/2)(\sqrt{3}(1-(2N-1)^{-2})^{1/2}-1) \approx (N-1/2)(\sqrt{3}-1). \qquad (19)$$

On the other hand, it is definitely exceeding $-1/4$

if $L > (N-1/2)(\sqrt{3}(1-(2N-1)^{-2})^{1/2}+1)-1 \approx (N-1/2)(\sqrt{3}+1)-1.$ (20)

For all the thresholds with $N \gg 2$ we have partial waves with finite cross sections. Their number increases with N and is approximately proportional to 2N-1 .

Taking $\ell(\ell+1)$ as perturbation according to $\alpha$ Herrick[13] obtained in the second approximation

$$a^{(2)}_{KT} = -3KN/Z + L(L+1) + (N^2-1-K^2-3T^2)/2 -$$
$$-KZ(8L(L+1)+N^2-1-K^2-15T^2)/(12N). \qquad (21)$$

At small L the values of (21) are close to the numerical results. For large L it will be more grounds to use the perturbation theory in the opposite direction, i.e. to take $\ell(\ell+1)$ as zero approximation and $\alpha$ as perturbation. So, Nikitin and Ostrovsky[15] have derived

$$a_{n'n''} = L(L+1) - 2\omega(n'+n'') + (n'+n'')^2 + O\left(\frac{N}{L^2 Z}\right),$$
$$\omega = \sqrt{\left(L+\tfrac{1}{2}\right)^2 + \left(\frac{3N}{2Z}\right)^2}. \tag{22}$$

Their quantum numbers $n'$ and $n''$ vary independently within the $\mp(N-1)/2$ limits taking correspondingly only the integer or half-integer values. They are connected with quantum numbers in (21)

$$K = n' + n'', \quad T = n' - n''. \tag{23}$$

In addition to the perturbation theory the semiclassical approximation has been also used in [15] obtaining (22) (since $L \gg 1$). Therefore (22) have an additional degeneration - the eigenvalues depend only on K but not on T.

2.5. Cross sections oscillations above the threshold

The presence of imaginary $\lambda + \tfrac{1}{2}$ in Eq. (2) causes the series of resonances (8) below the threshold and the cross sections finitness above the threshold - the phenomena observed experimentally as well as oscillations above the threshold which are not yet identified. These oscillations are due to the phase change of terms (7).

We should mention that it is the form of oscillations (a periodical dependence of the logarithm of energy distance to the threshold) that makes observation difficult. The measurements should be carried out in a large interval and especially very close to the threshold. It requires the high resolution of energy. Besides, at the first threshold N=2 two additional difficulties arise. First, above the threshold there is an $^1P^0$ shape resonance with $\lambda = L = 1$ in no way connected with oscillations under discussion, though this resonance gives large peak in the background. Secondly, we have only one imaginary $\lambda + \tfrac{1}{2}$ for each $L \leqslant 2$. Therefore in total cross sections oscillations are caused only by the dominator in (2). In differential cross sections the oscillations are expected to be more expressed owing to the changing interference among the partial waves due to the phase variations of the numerator in (2). Oscillations are also expected to be relatively weak for elastic scattering on excited states and for transitions between excited states with equal N. In these processes the first term in

(2) is most important, it is very slowly tending to a unity with L increasing. Many partial waves participate in the scattering and the cross sections are large. We have oscillations only for the first partial waves their contribution being relatively small.

2.6. The angular dependence of elastic scattering on excited states

It also follows from (2) that elastic scattering on excited states has a sharp forward peak whereas the differential cross sections has a logarithmic singularity[17]

$$d\sigma_{ii} = C_i k^{-2} \left| \ln(\sin \vartheta/2) + \mathcal{D}_i + O(\vartheta) \right|^2 d\Omega. \qquad (24)$$

Coefficients $C_i$ are dependent on state i and independent of a scattered electron velocity. For example, $C_{2s}=364, C_{2p}=81$ (if the magnetic quantum number for 2p state on the initial direction of a scattered electron is $\pm 1$). The second terms $\mathcal{D}_i$ have a weak energy dependence.

Formulae (24) can be obtained owing to the fact that the singularities of cross sections of small $\vartheta$ are determined by the S matrix behaviour at large L described by the first term of (2). In it we substitute the expressions for $\lambda$ and A at $L \gg 1$ and obtain the first terms for the S-matrix asymptotic expansion in powers of $1/L$. Summation over L in expansion for scattering amplitude yields (24). Fig. 1. illustrates the angular behaviour for N=2. We see that at $\vartheta \lesssim 8^\circ$ the results of (24) are close to the numerical. Dependence (24) also follows from the second Born approximation[18].

2.7. The number of resonances

Formulae (2) and law (8) can be applied only if the energies are considerably further from the threshold than the relativistic splitting $\Delta\varepsilon = (N-1)/(2N^4(137)^2)$ which we do not take into account in (2)

$$|E - E_t| \gg (L+1)\Delta\varepsilon. \qquad (25)$$

Closer to the threshold the resonance series is ending, whereas the excitation cross sections tend to zero. From (8) and (25) we find that the number of resonances in each series is finite

$$n_{max} \approx 1 + \frac{\operatorname{Im}\lambda}{2\pi} \ln \frac{E_t - E_1}{(L+1)\Delta\varepsilon}. \qquad (26)$$

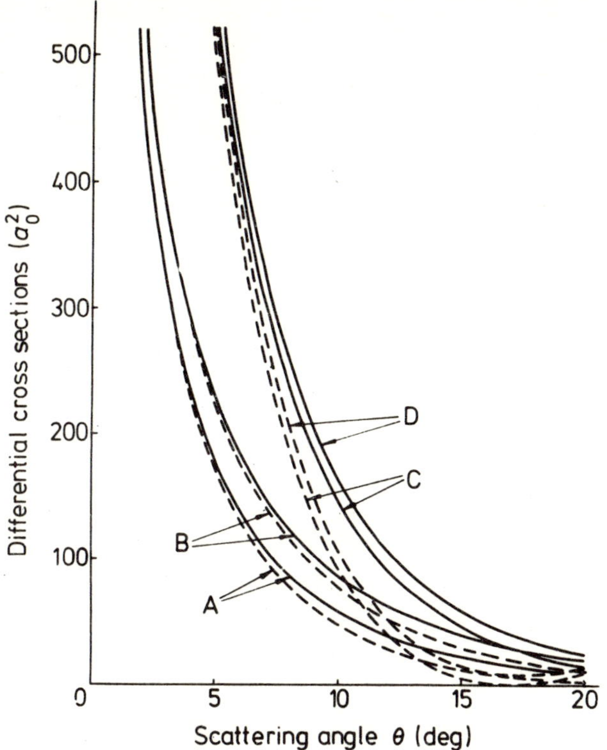


Figure 1. Angular dependence of elastic scattering cross sections at $k^2$ =0.09. Full curves, numerical results; broken curves, equation (24). A, triplet scattering from 2p ; B , singlet scattering from 2p; C, triplet scattering from 2s ; D, singlet scattering from 2s states.

The number of experimentally resolved resonances $n_{exp}$ is compared to (26) in Table 1. The values of $E_1$ are taken from reviews by Risley [10] (N=2) and Callaway [19] (N=3).

Due to the large values of $e^{\frac{2\pi}{3m\lambda}}$ the predicted number of resonances $n_{max}$ in each series is not large, though it is essentially exceeding the observed ones. Only in $^1S^e$ series at N=2 two resonances are resolved. The measured ratio $(E_t - E_1)/(E_t - E_2) = 22$. For the rest of the series only the first resonance is observed yet. At N=3 we have two series of P and D resonances. In Table 1 it is assumed that the lowest resonances calculated for N=3 are related to eigenvalue $a$ with quantum numbers K,T=2,0 . For $^1P^o$ resonance at 11.90eV the validity of such assumption is in fact unclear.

Table 1. Series of resonances below thresholds N = 2,3

| N | $^{2S+1}L^P$ | KT | $a$ | $\exp\frac{2\pi}{\mathrm{Im}\,\lambda}$ | $E_1$(eV) | $n_{max}$ | $n_{exp}$ |
|---|---|---|---|---|---|---|---|
| 2 | $^1S^e$ | 10 | -5.0828 | 17.429 | 9.56 | 4 | 2 |
| | $^3S^e$ | 10 | -5.0828 | 17.429 | 10.15 | 3 | |
| | $^3P^e$ | 01 | 2.0000 | $\infty$ | 10.19 | 1 | |
| | $^1P^0$ | 10 | -3.7082 | 29.334 | 10.17 | 2 | 1 |
| | $^3P^0$ | 10 | -3.7082 | 29.334 | 9.74 | 3 | 1 |
| | $^1D^e$ | 10 | -0.8102 | 4422.2 | 10.13 | 1 | 1 |
| | $^3D^e$ | 10 | -0.8102 | 4422.2 | | | |
| 3 | $^1S^e$ | 20 | -16.199 | 4.8225 | 11.73 | 6 | 1 |
| | $^3S^e$ | 20 | -16.199 | 4.8225 | 12.00 | 5 | |
| | $^1P^0$ | 20 | -14.897 | 5.164 | 11.90 | 5 | 1 |
| | | 11 | -5.220 | 16.752 | | | |
| | $^3P^0$ | 20 | -14.897 | 5.164 | 11.76 | 6 | 1 |
| | | 11 | -5.220 | 16.752 | | | |
| | $^1D^e$ | 20 | -12.249 | 6.134 | 11.81 | 5 | 1 |
| | | 11 | -2.300 | 80.552 | | | |
| | $^3D^e$ | 20 | -12.249 | 6.134 | 12.00 | 4 | |
| | | 11 | -2.300 | 80.552 | | | |
| | $^1F^0$ | 20 | -8.171 | 9.323 | 12.07 | 3 | |
| | $^3F^0$ | 20 | -8.171 | 9.323 | 11.93 | 4 | |
| | G | 20 | -2.56 | 62.43 | | | |

The number of oscillations above the threshold is expected to have the same order as the number of resonances below the threshold.

Complex $\lambda$ is not the only cause of the resonances. Some individual resonances can also appear for real $\lambda$. Though they are unconnected with the (8) type series. For instance, $^3P^e$ level with energy 0.00965 eV below the threshold N=2[20-22]. This level has $\lambda$ =L=1 and the autoionisation is going through spin-orbital interaction.

2.8. The electric field linear influence on $^1P^0$ resonance

It seems interesting to note that the calculated energy 10.17 eV for the second $^1S^e$ resonance at N=2 with an accuracy to several thousands eV coinsides with the energy of the first $^1P^0$ resonance. The reasons of such a close coincidence remain yet unexplained though the energy difference is of the same order as the calculations accuracy [23]. Such close energies and the opposite parities lead to the linear Stark effect observed by Gram et al.[3]. In the frequency dependence of the photodetachment cross section of $H^-$ the $^1P^0$ resonance is splitting into two. The splitting is approximately proportional to the field strength and is close to the calculated one[23].

3. The electric field linear influence on the resonance energies in helium-like ions

The eigenvalues $\lambda$ replace angular momentum $\ell$ also in the multichannel effective range theory[24] describing two electron cases at nuclear charge $Z \gg 2$: the electron scattering on the one-electron ions and photodetachment of He and helium-like ions. However, the consequences are considerably less essential as compared with e-H scattering. Now the Coulomb field is of crucial importance. As for all other ions the excitation cross sections are finite at the threshold and the positions of resonances below the thresholds are described by the Rydberg formula. The differences appear for quantum defects in it. For scattering on hydrogen-like ions the quantum defect is considerably exceeding the usual one. For series with imaginary $\lambda$ the quantum defect can take various values because it is determined by the behaviour of the wave function inside the ion. For series with real $\lambda$ the quantum defect is close to $\lambda$. With increasing $\lambda$, the difference between the quantum defect and $\lambda$ tends to zero. The energies of resonances

$$E_{\lambda n} = -\frac{Z^2}{2N^2} - \frac{(Z-1)^2}{2n^2}, \qquad (27)$$

$$n = n_r + \lambda + 1, \qquad n_r \gg 0.$$

It is important that $n_r$ is always integer but $n$ and $\lambda$ may be non-integer. The degeneration of $\lambda$ with different parities causes the degeneration of energies (27). Nikitin and Ostrovsky [15] noted that it generally leads to the possibility of the linear Stark effect. Though its experimental observation is unknown. Therefore below we calculate its magnitude.

3.1. Dipole approximation

The calculations will be carried out in the so-called "dipole approximation"[12], where two conditions are assumed to be fulfilled:

i) a contribution of the inner electron to the total wave function can be written as a linear combination of the hydrogen-like functions with one total quantum number N, though with different angular momenta $l_2$ and their projections;

ii) in the interaction between electrons it is sufficient to include only the dipole term:

$$\frac{1}{|\vec{r}-\vec{r}_2|} \approx \frac{1}{r} + \frac{\vec{r}\,\vec{r}_2}{r^3} \qquad (28)$$

this is possible if only the space-part $r \gg r_2$ is significant.

The dipole approximation was also used deriving Eq. (2), however, only for the investigation of the scattered electron motion outside the atom. It will be used below over all the space. This is rather a rough assumption therefore the smaller effects, such as the exchange, the finite width and additional shifts of resonances, the spin-orbital interaction, will be neglected.

The wave function can be written in the form

$$\Psi(\vec{r},\vec{r}_2) = \frac{1}{r r_2} \sum_{l l_2} F_{l l_2}(r)\, R_{N l_2 Z}(r_2)\, Y_{l l_2 LM} \qquad (29)$$

with

$$Y_{l l_2 LM} = \sum_{m m_2} Y_{lm}(\vec{r})\, Y_{l_2 m_2}(\vec{r}_2)\,(l\, l_2\, m\, m_2 | LM) \qquad (30)$$

and the Coulomb function $R_{NL_2Z}$ - for the inner electron. In the dipole approximation we have equations for F in matrix form

$$\left(\frac{d^2}{dr^2} - \frac{\Lambda}{r^2} + \left(2E + \frac{Z^2}{N^2}\right) + \frac{2(Z-1)}{r}\right) F(r) = 0. \qquad (31)$$

It differs from the Coulomb equation only in the substitution of centrifugal term $l(l+1)$ for matrix $\Lambda$. The solutions of (31) are the products of matrix A columns and the Coulomb function

$$\Delta E = \varepsilon M((n^2 - \lambda(\lambda+1)/3) S_1 - S_2) \tag{36}$$

where n is given by (27), the coefficients $S_1$ and $S_2$ depend on Z, N, L and $\lambda$. Their values for Z=2 and N=3,4 are presented in Table 2.

Table 2. Coefficients in (36) at Z=2

| N | L | $\lambda$ | $S_1$ | $S_2$ |
|---|---|---|---|---|
| 3 | 1 | 2.5289 | 0.4421 | 1.0579 |
| | 2 | 3.2867 | 0.1020 | 0.8623 |
| | | 1.7957 | 0.3199 | 1.3090 |
| | | 0.5680 | 0.2752 | 0.4465 |
| | 3 | 4.1714 | 0.0403 | 0.6524 |
| | | 2.9121 | 0.1300 | 1.0816 |
| | | 1.7990 | 0.1127 | 0.3915 |
| | 4 | 5.1116 | 0.0206 | 0.5148 |
| | | 3.9544 | 0.0652 | 0.8954 |
| | | 2.8791 | 0.0539 | 0.3585 |
| 4 | 1 | 3.9019 | 0.3708 | 1.4101 |
| | | 2.3212 | 0.5259 | 3.0817 |
| | 2 | 4.5654 | 0.0828 | 1.2512 |
| | | 3.3623 | 0.2684 | 1.8159 |
| | | 2.6390 | 0.2542 | 0.3992 |
| | | 0.7631 | 0.3816 | 0.6691 |
| | 3 | 5.3773 | 0.0334 | 0.9401 |
| | | 4.2463 | 0.1087 | 1.5456 |
| | | 3.3484 | 0.0979 | 0.5137 |
| | | 2.5898 | 0.2167 | 1.9626 |
| | | 1.4676 | 0.2535 | 0.8013 |
| | 4 | 6.2662 | 0.0177 | 0.7340 |
| | | 5.1711 | 0.0555 | 1.2779 |
| | | 4.2026 | 0.0458 | 0.5039 |
| | | 3.7852 | 0.1152 | 1.6707 |
| | | 2.6972 | 0.1231 | 0.8372 |
| | | 1.1898 | 0.0816 | 0.3626 |

$$F_{\ell\ell_2}(r) = A_{\ell\ell_2,\lambda p}\, R_{n\lambda Z-1}(r). \tag{32}$$

Values $\lambda$ and Z-1 are substituted in the Coulomb functions instead of the angular momentum and the charge. Energy E and value $n$ are given by (27). In the cases of complex $\lambda$ the dipole approximation over all the space is unacceptable. These cases will not be considered below.

The effect of the electric field essentially differs for the resonances at the threshold N=2 and at other thresholds $N \geqslant 3$.

3.2. The Stark effect at $N \geqslant 3$

At $N \geqslant 3$ all $\lambda$ are noninteger and dependent on L. Between them $N(N-1)/2$ if $L \geqslant N-1$ or $L(N-(L+1)/2)$ if $L < N-1$ are degenerated with respect to parity. In the electric field these resonances are splitting

$$E = E_{\lambda n} \pm \Delta E. \tag{33}$$

The splitting is proportional to the dipole matrix element between the states with equal $\lambda$ and different parity

$$\Delta E = \mathcal{E}(\lambda^e | z + z_2 | \lambda^o), \tag{34}$$

$$(\lambda^e | z + z_2 | \lambda^o) = \frac{3}{2}(-1)^M (2L+1) \begin{pmatrix} 1 & L & L \\ 0 & -M & M \end{pmatrix} \sum_{\ell'\ell_2'\ell\ell_2} A_{\ell'\ell_2',\lambda^e} A_{\ell\ell_2,\lambda^o} \times$$
$$\times\Bigg[(-1)^{\ell_2}\delta_{\ell_2'\ell_2}\big((2\ell'+1)(2\ell+1)\big)^{1/2}\begin{pmatrix} \ell' & 1 & \ell \\ 0 & 0 & 0 \end{pmatrix}\begin{Bmatrix} 1 & L & L \\ \ell_2 & \ell & \ell' \end{Bmatrix}\frac{n^2-\lambda(\lambda+1)/3}{Z-1} -$$
$$-(-1)^{\ell}\delta_{\ell'\ell}\big((2\ell_2'+1)(2\ell_2+1)\big)^{1/2}\begin{pmatrix} \ell_2 & 1 & \ell \\ 0 & 0 & 0 \end{pmatrix}\begin{Bmatrix} 1 & L & L \\ \ell & \ell_2 & \ell_2' \end{Bmatrix}\frac{N\sqrt{N^2-\max(\ell_2,\ell_2')^2}}{Z}\Bigg]. \tag{35}$$

The dependence of $\Delta E$ on M is expressed via the common factor $(-1)^M(2L+1)\begin{pmatrix} 1 & L & L \\ 0 & -M & M \end{pmatrix} = M(-1)^{L+1}\sqrt{\frac{2L+1}{L(L+1)}}$ which follows from the Wigner-Eckart theorem. The electric field removes the degeneration on the parity as well as on $|M|$ and splits the resonance into 2L+1 equidistant components. The contributions of external and inner electrons give the first and the second terms in (34, 35). In each series only the factor $n^2-\lambda(\lambda+1)/3$ at the first term depends on the number of resonances $n_r$. Therefore

### 3.3. The Stark effect at N=2

At N=2 and $L \geqslant 1$ the picture is more complicated. Now only $\lambda = L$ is degenerated to the parity. But it is an integer, therefore energies (27) have an additional degeneration over L, for L changing within the limits $\max(1,M) \leqslant L \leqslant n-1$. The splitting equals to $\mathcal{E}$ multiplied by the eigenvalues of symmetric matrix with the order $2(n-\max(1,M))$ and with three nonzero elements in each row and each column. Apart from (35) it has elements undiagonal to $\lambda$:

$$(L'^e|z+z_2|L^o) = (-1)^M((2L'+1)(2L+1))^{1/2}\begin{pmatrix} 1 & L' & L \\ 0 & -M & M \end{pmatrix} \times$$

$$\times \sum_{l'l l_2' l_2} \Big\{ (-1)^{l_2}\delta_{l_2' l_2}((2l'+1)(2l+1))^{1/2}\begin{pmatrix} l' & 1 & l \\ 0 & 0 & 0 \end{pmatrix}\begin{Bmatrix} 1 & L' & L \\ l_2 & l & l' \end{Bmatrix}(nL'|r|nL)$$

$$- \frac{3\sqrt{3}}{Z}(-1)^{l}\delta_{l'l}\,\delta_{|l_2'-l_2|,1}\begin{Bmatrix} 1 & L' & L \\ l & l_2 & l_2' \end{Bmatrix}(L'|L)\Big\} A_{l'l_2',L'^e} A_{l l_2,L^o}$$

with $L'=L\pm 1$, $(nL-1|r|nL) = -\dfrac{3n\sqrt{n^2-L^2}}{2(Z-1)}$, $(L-1|L) = -\dfrac{\sqrt{n^2-L^2}}{n}$.

The spectrum of splitting shown in Fig.2 and 3 depends on Z, n and M. The splitting of hydrogen-like ion with nuclear charge Z-1 is presented for comparison. It is seen that the splitting has the same order of magnitude as that for the one electron ion. Although this spectrum is not equidistant and dependent on |M|. Therefore the full picture has more lines and is more complicated.

### 3.4. Region of Application

Our results are applicable only in the middle intensity fields. At weak $\mathcal{E}$ the errors of the dipole approximation should be taken into account. The resonances are not in fact exactly degenerated to the parity. Our results are applicable only when the splitting by the electric field is sufficiently exceeding the existing splitting without the field. On the other hand, at very large $\mathcal{E}$ it is necessary to take into account increasing resonances widths due to an additional ionization in the electric field. With $\mathcal{E}$ increasing it leads to the complete disappearance of the resonance.

The present problem as well as the $^1P^0$ resonance behaviour in $H^-$ are examples of the cases where the influence of the external electric field leads to a linear splitting in the circumstances when we cannot say that the movement of an outer electron occurrs in a pure Coulomb field and is independent of the inner electron movement.

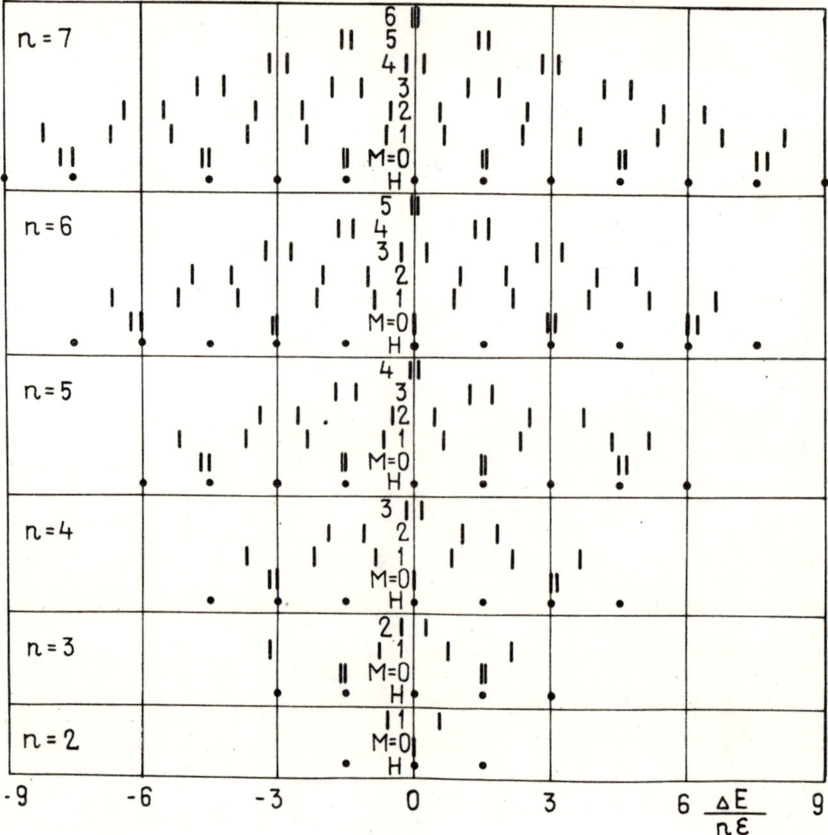


Figure 2. The He double-excited states energies with N=2 splitting linear to external electric field $\mathcal{E}$. Dots - hydrogen atom data.

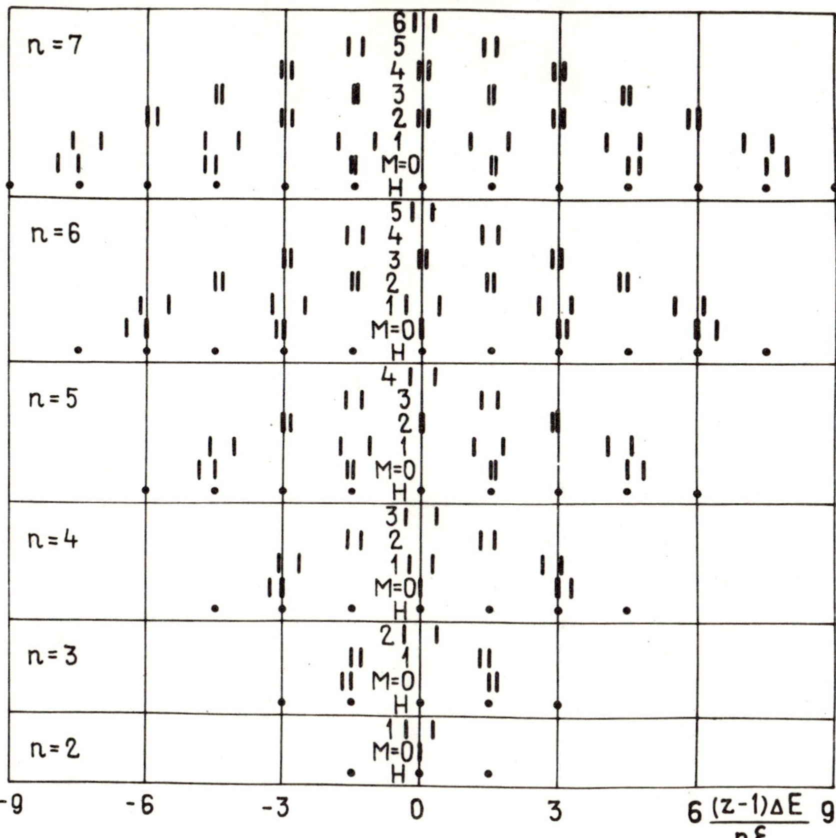


Figure 3. The $B^{3+}$ double-excited states energies with N=2 splitting linear to external electric field $\mathcal{E}$. Dots - $Be^{3+}$ data.

## REFERENCES

1. M.Gailitis, R.Damburg, Zh.Eksperim.i Teor.Fiz.,44,1644(1963) (English transl.Soviet Phys.JETP.17,1107(1963)).
2. M.Gailitis, R.Damburg,Proc.Phys.Soc.,82,192(1963).
3. P.A.M.Gram,J.C.Pratt,M.A.Yates-Williams,H.C.Bryant,J.Donahue, H.Sharifian,H.Tooboonhi,Phys.Rev.Lett.,40,107(1978).
4. R.R.Freeman,G.C.Bjorklund,Phys.Rev.Lett.,40,118(1978).
5. P.G.Burke,Advances in Atomic and Molecular Physics,4,173(1968).
6. U.Fano,J.W.Cooper,Rev.Mod.Phys.,40,441(1968).
7. U.Fano,Atomic Physics,1,209(1969).
8. M.J.Seaton, Atomic Physics,1,295(1969).
9. G.J.Schulz,Rev.Mod.Phys.,45,378(1973).
10. J.S.Risley,Atomic Physics, 4, 487(1975).
11. G.E.Chamberlain,S.J.Smith,D.W.O.Heddle,Phys.Rev.,Lett.,12,647 (1964).
12. M.J.Seaton, Proc.Phys.Soc.,77,174(1961).
13. D.R.Herrick,Phys.Rev., A12,413(1975).
14. D.R.Herrick,Phys.Rev., A12,1(1978).
15. S.I.Nikitin,V.N.Ostrovsky,J.Phys.,B11,1681(1978),Abstr.Pap. X ICPEAC,688,Paris(1977),Izv.Akad.Nauk SSSR,Ser.Fiz.,41,2468(1977).
16. S.I.Nikitin,V.N.Ostrovsky,Vestnik LGU(in Russian),No.4,31(1977).
17. M.Gailitis,J.Phys.,B11,L279(1978).
18. C.J.Joachain,K.N.Winters,L.Cartiaux,E.M.Mendez-Moreno,J.Phys., B10,1277(1977).
19. J.Callaway,Physics Reports,(1978).
20. E.Holøien,Physica Norvegica,1,53(1961-62),J.Chem.Phys.,33,301(1960).
21. G.W.Drake,Phys.Rev.Lett.,24,126(1970).
22. A.K.Bhatia,Phys.Rev.,A2,1667(1970).
23. J.Callaway,A.R.P.Rau,J.Phys.,B11,L289(1978).
24. M.Gailitis,Zh.Eksperim.i Teor.Fiz.,44,1974(1963) (English translation: Soviet Phys.JETP 17,1328(1964)).

# QUASI-RESONANCE PROCESSES IN ATOMIC COLLISIONS

E.E. Nikitin

Institute of Chemical Physics Academy of Sciences, USSR

B.M. Smirnov

Kurchatov Institute of Atomic Energy, USSR

Quasi-resonance processes in atomic collisions correspond to low or virtually zero changes in electronic energy of colliding partners. This often results in large cross sections comparable to or even larger than the gas-kinetic ones, which makes these processes important in phenomena occuring in excited or ionized gases under different conditions. In slow collisions when the relative velocity of colliding particles is small as compared to the velocities of outermost electrons, the theoretical approach is usually based on the quasi-molecular approximation, that is on the notion of electronic states of a diatom with fixed interatomic distance.

There are two important peculiarities of quasi-resonance processes making them readily tractable. First, not many molecular states are coupled by non-adiabatic interaction which is proportional to the velocity. Second, the coupling is localized at large interatomic distances where the molecular states are still not very different from the parent atomic states. Thus, it appears to be possible to reliably estimate the cross sections of many processes using information only on the free states of colliding atoms.

This approach in the theory of atomic collisions which is mainly due, to a great extent, to contributions of Soviet physicists and is described in books [I-6] is sometimes referred to as the asymptotic theory of atomic collision. The main idea of the method is to use the ratio "atomic radius to coupling distance" as a small parameter in the expansion of a cross section. Thus one can control the approximation which is of course an advantage over model or semiempirical calculations. Moreover, once the necessary calculations are done, the asymptotic theory usually provides final results in a comparatively simple form(analytical or semianalytical), thus making it possible to correlate a cross section of any particular process with the parameters of colliding particles and the form of the trajectory adopted to describe their relative motion.

As the scattering matrix (and transition probability) is expressed via parameters entering into the molecular adiabatic terms, and as we need to know these terms at large interatomic distances $R$ , the discussion will be focused first on the interatomic interaction at large $R$ . Similar information is of use also for other than scattering problems, e.g. for calculating transport coefficients at room and higher temperatures or estimating the bonding and related parameters of condensed media consisting of weakly interacting particles.

Calculation of molecular terms at large destances constitutes a part of the general problem of the molecular structure theory. The direct, variational method which is in use now requires the best computers. This is of course some disadvantage when one needs knowledge of the potential curves and of the coupling matrix elements over a large range of interatomic distances. However, neither the complexity of such calculations nor the high cost of their accomplishment are the main obstacles to systematic use of variational methods in collisions with small energy change. The most essential shortcoming stems from the fact that with R increasing the accuracy of the variational method diminishes for a fixed basis set of electronic functions, since the interaction energy becomes a very small fraction part of a diatom total energy.

An alternative approach would be the use of the perturbation theory taking the free atomic states as zero order functions. This method is known [7] to yield first- and second-order interaction energy, provided multipole expansion is used and the overlap is neglected. The interaction energy is represented by a sum in reciprocal powers of $R$, the leading term $R^{-n}$ depending on the electronic and charge state of a diatom. Simple examples would be n = 4 for the interaction of an ion with an atom in the S-state (interaction of a point charge with the induced dipole), and n = 6 for the interaction of atoms in S-states (the van-der-Waals interaction). This approach has been developed even at the early stage of quantum mechanics.[8-14]

The main shortcoming of the multipole perturbation theory is that it fails even at distances at which the long range interaction energy $U$ is still very low compared to the characteristic atomic excitation energy $E_1$. The only way to overcome this, is to take into account the overlap.

In attempting the Heitler-London approach [7] it will be found that the condition of small first order correction to the zero order wave function in the region of overlap, i.e. at the exponential tail of atomic functions, is not fulfilled. This immediately shows that calculation of overlap integrals, and also of resonance or exchange integrals with unperturbed atomic functions yields incorrect results for the exponentially decaying (often called short-range or exchange) part of interatomic interaction.

The problem thus arises - how to take into account the deformation of electronic clouds at large $R$, at which the motion of electrons is classically forbidden. It is rather small, but essential for exchange interaction. The solution of this problem, giving asymptotically correct (at the limit of large $R$) values for the exchange interaction, often in analytical form represents the main result of the asymptotic method of calculating the short-range interatomic interaction [1, 4].

Fig. 1 shows three basic types of exchange interaction. Two of these involve actual transition of electron from one classically allowed region to another. The solution of the corresponding wave equation is demonstrated qualitatively in Fig. 2. For illustration we give here the expression for an exchange splitting between gerade and ungerade terms of symmetrical

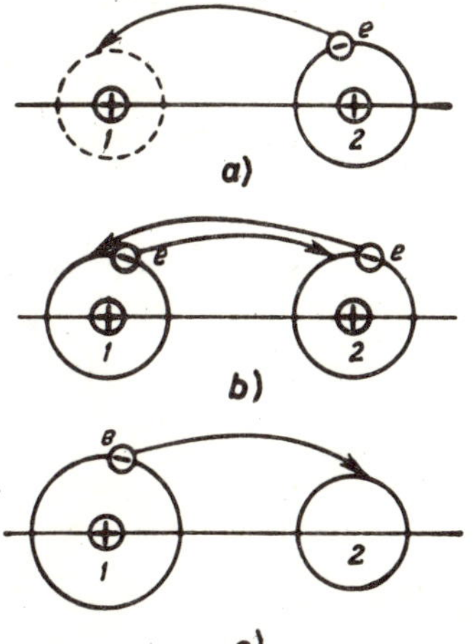


Fig. 1. Three basic types of short range interactions:
a) Resonant interaction due to tunneling of an electron from one potential well to another.
b) Exchange interaction.
c) Tunneling of a bound electron to the repulsive or weakly attractive foreign core.

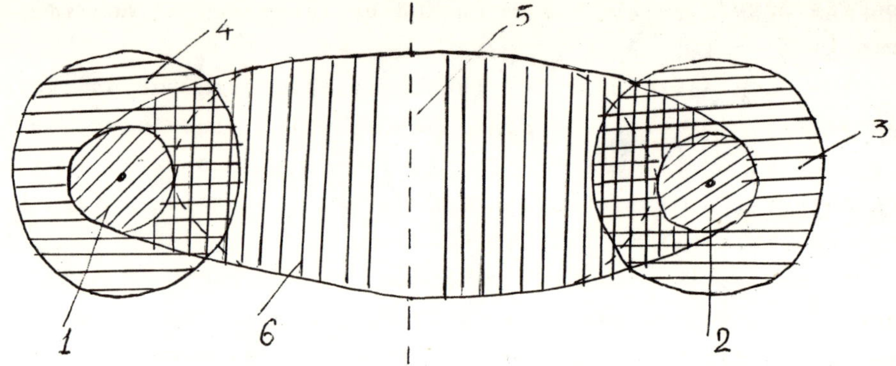


Fig. 2. Schematic illustration of the two-electron two-core problem. The shaded areas correspond to:

I, 2 - inner regions close to cores;

3, 4 - asymptotic regions outside the classical radius of valence electron orbit;

5 - region where the WKB one-dimensional approximation is valid;

6 - region mostly contributing to overlap and the exchange integrals.

The ratio of six-dimensional volumes corresponding to regions I and 5 is of the order $a_0^2/R^2$, which gives the accuracy of the asymptotic approach.

diatom consisting of ion and a parent atom [I5] :

$$\Delta \equiv \mathcal{E}_g - \mathcal{E}_u = \frac{n(J+\frac{1}{2})}{2s+1} \left| G_{SL}^{s\ell} \right|^2 \sum_{\mu} (\ell_e \mu; \ell M_\ell | L, M_L + \mu) \cdot (\ell_e \mu; \ell, M_L - \mu | L M_L) \cdot \Delta_\mu \qquad (1)$$

Here $J$ is the total electronic angular momentum of a diatom, $L, M$ are the atomic orbital momentum and its projection on the molecular axis, $S$ is the atomic spin; $\ell, M_\ell, s$ are similar quantum numbers for atomic core and $\ell_e, \mu, \frac{1}{2}$ are those for the valence electron. The coupling between momenta in the system of n valence electrons is described by the Clebsch-Gordan coefficients $(j_1 m_1; j_2 m_2 | j m)$ and by geneological coefficients $G_{SL}^{sl}$.

Now, $\Delta_\mu$ represents the resonance splitting for one valence electron having a projection of the angular momentum $l_e$ [16]:

$$\Delta_\mu = A^2 \frac{(2l_e+1)(l_e+|\mu|)!}{(l_e-|\mu|)!\,|\mu|!\,\gamma^{|\mu|}} R^{\frac{2}{\gamma}-1-|\mu|} \exp\left(-R\gamma-\frac{1}{\gamma}\right) \quad (2)$$

Here $\frac{\gamma^2}{2}$ is the binding energy of an electron (in atomic units) and $A$ is the asymptotic coefficient describing the unperturbed tail of atomic function.

We reiterate here an important point: the problem is reduced to one-electronic (or two-electronic in case of exchange), and its solution is asymptotically correct.

Note an essential feature of this approach: the first and second-order multipole long-range interactions can be added to the short range interaction, and the spin-orbital coupling can be calculated as in the absence of perturbation of atomic wave functions. This simplifying picture is due to separation of the configuration space of electrons into three parts: one close to the nuclei, unperturbed and contributing to the spin-orbital coupling, the second perturbed by admixture of other states at intermediate distances from nuclei and contributing to long-range interaction, and the third perturbed by tunneling (and in case of exchange also by interelectron repulsion in the tunneling region) and responsible for the short range interaction.

This approach allows to set the energy matrix, and diagonalization of the latter yields molecular terms. For instance, for a noble gas ion in the state $^2P$ and its parent atom in the state $^1S$ the matrix in the Russel-Saunders approximation is shown in Table I. [17]

Table I.

Block of the Energy Matrix for a Diatom Consisting of an Ion $^2P$ and its Parent Atom $^1S$.

| Quantum numbers | $M_L=0,\ m_s=\frac{1}{2}$ | $M_L=1,\ m_s=-\frac{1}{2}$ | $M_L=1,\ m_s=\frac{1}{2}$ |
|---|---|---|---|
| $M_L=0,\ m_s=\frac{1}{2}$ | $E_0+u_0\pm\frac{1}{2}\Delta_0$ | $-\frac{\sqrt{2}}{3}\varepsilon_T$ | 0 |
| $M_L=1,\ m_s=-\frac{1}{2}$ | $-\frac{\sqrt{2}}{3}\varepsilon_T$ | $E_0+u_1-\frac{\varepsilon_T}{3}\pm\frac{1}{2}\Delta_1$ | 0 |
| $M_L=1,\ m_s=\frac{1}{2}$ | 0 | 0 | $E_0+u_1+\frac{\varepsilon_T}{3}\pm\frac{1}{2}\Delta_1$ |

Here $E_0$ is the energy at $R\to\infty$ without spin-orbital coupling, $u_M$ is the long-range interaction for the spinless molecular state with M-projection, $\Delta_M$ is the resonance splitting; plus or minus signs correspond to gerade or ungerade molecular state and $\delta$ is the fine-structure splitting in the ion.

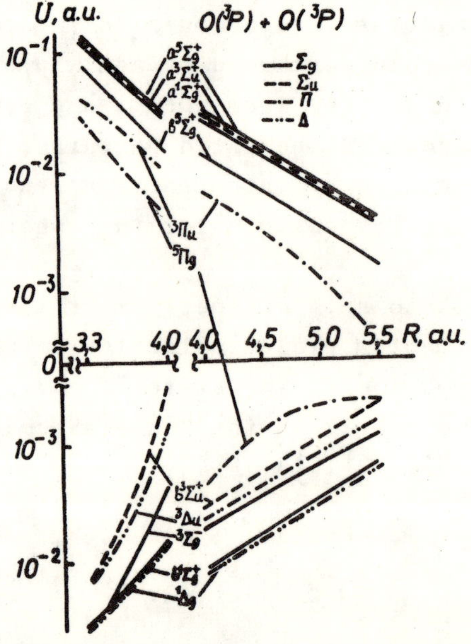


Fig 3. Adiabatic terms of the diatom $O(^3P)+O(^3P)$
The left part is for variational calculation, [I8]
the right part for the asymptotic method. [I9]

Fig 3 gives an example of calculated terms for the case of two oxygen atoms in the ground state $^3P$ [19]. Here the spin-orbital coupling is so small, that it is undiscernable within the given scale. Asymptotic terms are seen to be quite accurate except one term for which severe compensation of different contributions occurs (term $\Pi_g$). Note that at the matching region interaction energie are about 1/20 of the excitation energy to the next atomic state

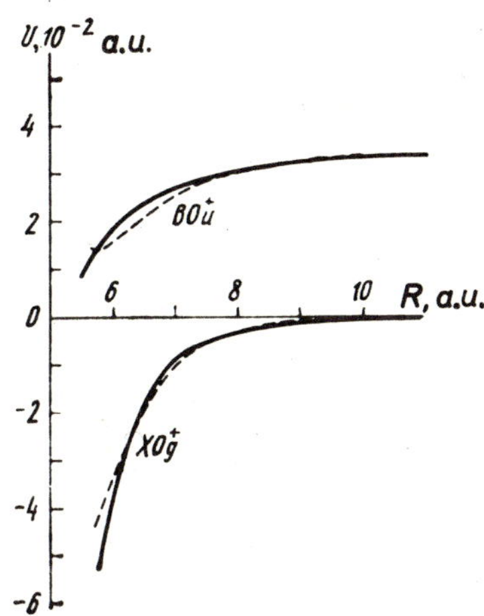


Fig 4. Attractive terms for $I(^2P_{3/2}) + I(^2P_{3/2})$ (lower curve) and $I(^2P_{1/2}) + I(^2P_{3/2})$ (upper curve). Full lines correspond to asymptotic calculation, broken lines to the experimental data.

Fig. 4 shows a different case with large spin-orbital interaction ($I_2$ molecule). The asymptotic terms (full lines) are seen to be very close to those obtained from experiment (broken lines). [20]

For cases referred above there is an extended region over which the short range interaction is dominant. Even for alkali diatoms in the resonance excited state where the long-range interaction is strongest for neutral systems (decreasing as $R^{-3}$) exchange interaction was found [19] to be important for the intramultiplet mixing with the energy change as small as 100 $cm^{-1}$.

We briefly formulate now the conditions under which the asymptotic method is applicable - large distances $R$ and small difference in the binding energies of valence electrons at each Coulomb well:

$$R \gg \frac{1}{\gamma}, \quad R \gg \frac{1}{\gamma^2}, \quad \gamma = \frac{\gamma_1 - \gamma_2}{2} \tag{3}$$

$$R\,|\gamma_1 - \gamma_2| \ll 1. \tag{4}$$

Under these conditions the overlap is actually determined by the extended region close to the molecular axis and sufficiently separated from the cores.

Consider now the transitions between quasi-resonance states in slow collisions. The efficiency of coupling between molecular states depends on the value of the local Massey parameter $\xi$ which is the ratio of the characteristic collision time $a/v$ to the time of electronic transition $\hbar/\Delta\varepsilon$

$$\xi = \frac{\Delta\varepsilon \cdot a}{\hbar v}. \tag{5}$$

Here $\Delta\varepsilon$ is the energy difference between molecular terms, $v$ is the relative velocity and $a$ is the extension of the coupling region. If $\xi \gg 1$ the transition probability will be proportional to $\exp(-\xi)$. However we shall focus attention of the case $\xi \ll 1$, when transitions are rather effective.

Under quasi-classical conditions, when the be Broglie wavelength $\frac{\hbar}{\mu v}$ is small compared to the atomic radius $R_0$, a strong coupling regime ($\xi \ll 1$) means that the transferred energy is small compared with the kinetic energy $E$

$$\frac{\Delta\varepsilon}{E} \sim \xi \frac{\lambda}{a} \ll 1 \tag{6}$$

even though usually $R_0/a \gtrsim 1$. If, moreover, $\Delta\varepsilon$ is of the order of molecular energies, at interatomic distances of interest, i.e.

$$U(R) \ll E, \tag{7}$$

the scattering equations for quasi-resonance channels reduce to the semiclassical form appropriate for the rectilinear trajectory. To this approximation, the time dependence of the wave function is

$$\Psi(t) = \sum_n a_n \Psi_n \exp\left(-\frac{i}{\hbar}\int^t \varepsilon_n dt'\right), \quad (8)$$

where the summation is over all molecular states correlating at $R\to\infty$ with initial and final atomic states. It is tacitly assumed that coupling with "far-lying" states can be neglected. Equations for the coefficients are of the form

$$i\hbar \dot{a}_n = \sum_{n'} C_{nn'}(t)\, a_{n'}(t) \exp\left[-\frac{i}{\hbar}\int^t (\varepsilon_n - \varepsilon_{n'}) dt'\right] \quad (9)$$

with dynamical coupling matrix elements $C_{nn'}$. Integration of equation (9) along a straight path directly gives the scattering matrix which can be used to calculate the low-angle differential cross-section.

The advent of computers makes the problem of solving equations like (9) a perely technical one. Nevertheless, approximate and physically transparent methods still play an important role in helping to trace intimate details of the collision mechanism. One method uses transformation of the electronic basis in order to achieve the maximal localisation of the coupling. If this is done, the general solution can be constructed by the matching of piece-wise solutions over limiting interatomic intervals of R. Within each interval the multichannel problem reduces to several independent two-or three-channel ones [22].

A two-channel problem occurs most frequently. It is described by $\Psi$ written as a superposition of two molecular adiabatic states $\Psi_1$ and $\Psi_2$

$$\Psi(t) = a_1(t)\Psi_1(t)\, e^{-\frac{i}{\hbar}\int^t \varepsilon_1 dt'} + a_2(t)\Psi_2(t)\, e^{-\frac{i}{\hbar}\int^t \varepsilon_2 dt'} \quad (10)$$

with equation (9) reduced to

$$\begin{aligned} i\hbar a_1 &= C_{12} \exp\left[\frac{i}{\hbar}\int^t (\varepsilon_1 - \varepsilon_2) dt'\right] a_2 \\ i\hbar a_2 &= C_{21} \exp\left[\frac{i}{\hbar}\int^t (\varepsilon_2 - \varepsilon_1) dt'\right] a_1, \end{aligned} \quad (11)$$

and $C_{12} = \langle \Psi_1 | -i\hbar \frac{\partial}{\partial t} | \Psi_2 \rangle$

The decoupling of equations in (II) correspond to the simplest process, such as resonant charge exchange. For this case

$$\Psi(t) = \frac{(\Psi_1+\Psi_2)}{2} e^{-\frac{i}{\hbar}\int^t \varepsilon_g dt'} + \frac{(\Psi_1-\Psi_2)}{2} e^{-\frac{i}{\hbar}\int^t \varepsilon_u dt'}, \qquad (I2)$$

where $\Psi_1$ and $\Psi_2$ are atomic functions describing localisation of the electron on one or other core. The transition probability, which can be easily got from (I2), is [2I]

$$W = |\langle \Psi(\infty) | \Psi_2 \rangle|^2 = \sin^2 \int_{-\infty}^{+\infty} \frac{\Delta dt}{2\hbar}, \qquad (I3)$$

where $\Delta = \varepsilon_g - \varepsilon_u$

At present there are several two-state models [22], which can be solved analytically. To make a simple presentation of these, we define $a_1 a_1^* - a_2 a_2^*$, $a_1 a_2^*$ and $a_1^* a_2$ as spherical components of a unit vector $\vec{m}$ . Then the two complex equations (9) reduce to one vector equation for precessing

$$\vec{m} = \frac{1}{\hbar} [\vec{\nu} \vec{m}] \qquad (I4)$$

with $\vec{\nu}$ defined as

$$\nu_x = 2 \operatorname{Re} H_{12}, \ \nu_y = H_{11} - H_{22}, \ \nu_z = 2 \operatorname{Im} H_{12}. \qquad (I5)$$

If $H_{I2}$ is real, as is mostly the case, the tip of the vector draws a curve in the x-y plane, the length of $\vec{\nu}$ being the adiabatic splitting $\Delta$ and the rate of rotation corresponding to dynamic coupling.

Initially, before reaching the coupling region, $\vec{\nu}$ and $\vec{m}$ are oriented along the $y$-axis. Then the coupling region is reached, $\vec{\nu}$ rotates, and $\vec{m}$ follows $\vec{\nu}$ , precessing about some direction different from $\vec{m}$ . After passing the coupling region, $\vec{\nu}$ tends to a constant vector $\vec{\nu}(\infty)$, and $\vec{m}$ precesses about $\vec{\nu}(\infty)$ The angle of precession $\beta$ characterises the transition probability, namely,

$$P_{12} = \frac{1}{2}(1 - \cos\beta) = \frac{1}{2}\left[1 - \frac{\vec{m}(\infty)\vec{\nu}(\infty)}{m(\infty)}\right]. \qquad (I6)$$

The change in the adiabatic basis after transition through the coupling region is reflected in this gyroscopic model as a non-zero angle between vectors $\vec{m}$ and $\vec{\nu}$ . Fig. 5 shows the trajectories $\vec{m}=\vec{m}(t)$ for which analytical solutions of Eq. (I4) are known [8,23-30]. Recently some of the two-state models have been generalized for complex terms, thus describing non-adiabatic coupling between quasi-stationary decaying states [3I, 32].

Now, if the transition matrix N for each localized non-adiabaticity region is known, the total S matrix is built by multiplication of N matrices and of diagonal A matrices that describe adiabatic evolution of a diatom between coupling regions. In this way several two-state models are used to construct a multichannel matrix.

In case of one coupling region centered at $R = R_p$ we have

$$S = A^{>} N A^{<} N A^{>} \qquad (I7)$$

where $A^{>}$ and $A^{<}$ are diagonal matrices containing phase factors for motion at $R > R_p$ and $R < R_p$ and N is the transition matrix at $R = R_p$. The transition probability $\mathcal{P}_{12}$ is expressed via the one-way transition probability $P_{I2}$ and the Stueckelberg phase $\Phi$ which contains also an important dynamic correction $\varphi$ to the adiabatic action [22, 33]:

$$\mathcal{P}_{12} = |S_{12}|^2 = 4 P_{12}(1 - P_{12}) \sin^2(\Phi + \varphi) \qquad (I8)$$

More complicated expressions will be found for a multichannel problem, but all these - provided the Stueckelberg phases are not too small- can be interpreted in terms of severel classical paths channelling the appropriate amplitudes.

Let us discuss now what can be learned about the mechanism of quasi-resonance processes if we use simultaneously both the asymptotic method of calculation of molecular terms and the model approach to non-adiabatic coupling. Once the accuracy of both is known, we can estimate also the error in the total (or differential) cross section. The typical quasi-resonance processes in atomic collisions are listed in Table II.

The resonance charge exchange is the simplest process which is successfully treated by the method described. Here, even at energy of I kev, transfer of an electron occurs at distances about ten times larger than the radius of the electron orbit. The main error in the total cross section comes from the asymptotic coefficient A in eq. (2).

Table II.

Quasi-resonance processes

| Process | Designation |
|---|---|
| I. Resonant charge transfer. | $A^+ + A \rightarrow A + A^+$ |
| 2. Non resonant charge transfer. | $A^+ + B \rightarrow A + B^+$ |
| 3. Excitation transfer. | $A^* + B \rightarrow A + B^*$ |
| 4. Spin-exchange and transitions between hyperfine states. | $A\uparrow + B\downarrow \rightarrow A\downarrow + B\uparrow$ (Arrow shows the direction of electronic spin) |
| 5. Depolarization and multipole relaxation. | $A(M) + B \rightarrow A(M') + B$ (M is the projection of electronic angular momentum on a space-fixed axis.) |
| 6. Intramultiplet mixing. | $A(J) + B \rightarrow A(J') + B$ (J is the total electronic angular momentum). |
| 7. Penning ionisation and associative ionisation. | $A^* + B \rightarrow A + B^+ + e$; $A^* + B \rightarrow AB^+ + e$ |
| 8. Electron detachment | $A^- + B \rightarrow A + B + e$; $A^- + B \rightarrow AB + e$ |
| 9. Mutual neutralisation of positive and negative ions. | $A^- + B^+ \rightarrow A + B^*$ |
| IO. Collisional rediative deactivation of metastables. | $A^* + B \rightarrow A + B + h\nu$ |

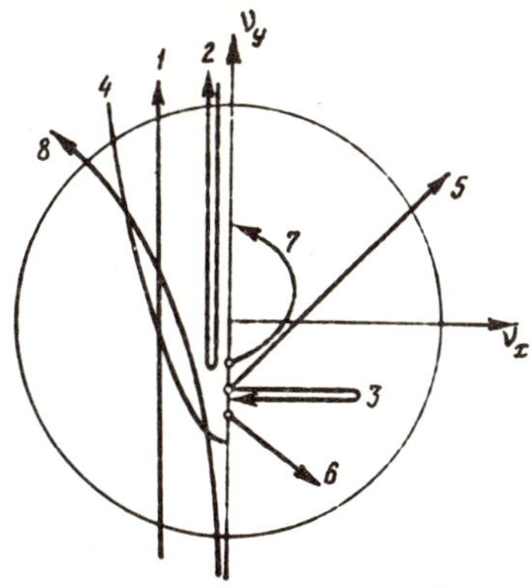


Fig. 5. Motions of the vector $\vec{\nu}$ in cases for which analytical solution of the two-state problem are known:

I - is the original Landau-Zener-Stueckelberg model [8]; 2 - the extended Landau-Zener model accounting for the effect of the turning point (Bykhovskii, Nikitin, Ovchinnikova [23] ); 3 - the Rosen-Zener-Demkov model [24,25]; 4 - the Vitlina - Chaplik model [26]; 6, 7 - the Demkov - Kunicke hypergeometric models [28]; 8 - the Nikitin - Bykhovskii - Ovchinnikova linear-exponential model [29, 30].

The typical uncertainties of A, $\Delta A/A$ = 0.I - 0.3, result in a relative error in the cross section 3 - 10%. Very often the accompanying effects, such as depolarization, transitions between fine structure levels, coupling to other states at small interatomic distances and oscillations due to failure of the random phase approximations contribute less than the error mentioned above. Experimental data and asymptotic calculations for charge transfer in $Rb$-$Rb^+$ and $Xe$-$Xe^+$ collisions are compared in Figs. 6 and 7.

For non-resonant charge exchange the situation is similar to resonance processes in case $\xi \ll 1$ (see (5) ), where the cross section decreases with increasing energy. However, for $\xi > 1$ the cross section increases with energy (see Figs. 8, 9) which is in line with the theory.

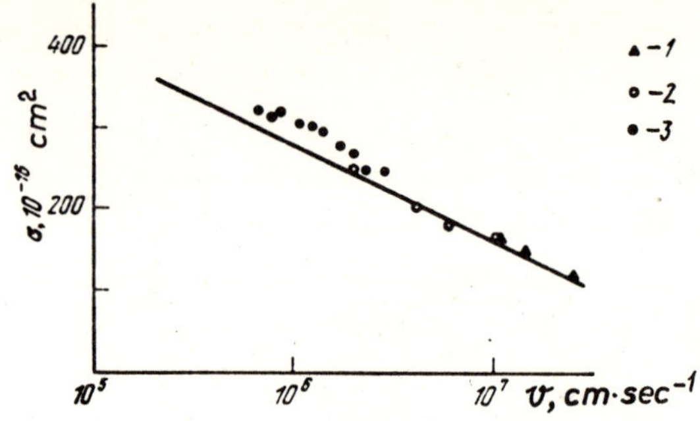


Fig. 6. Total cross section for resonant charge transfer in collisions $Rb^+ + Rb$. Full curve corresponds to the asymptotic approach. [34] Experimental data I- [35], 2 - [36], 3 - [37].

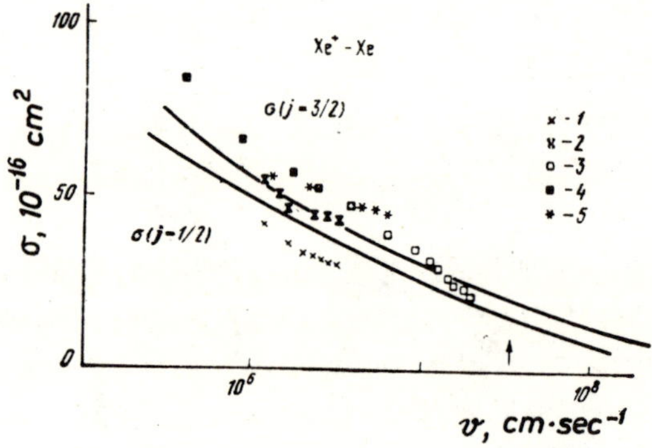


Fig. 7. Total cross section for resonant charge transfer in $Xe^+$ - Xe collisions. Full curves correspond to the asymptotic approach [38] for the total ion moment I/2 and 3/2. The experimental data: I - [39] ; 2 - [40]; 3 - [41]; 4 - [42]; 5 - [43].

The spin-exchange is similar to the resonant charge transfer and is brought about by dephasing during the motion of a diatom on singlet and triplet potentials. The cross sections of these processes have been calculated by the asymptotic method (see table III).

As for excitation transfer, two types of it - with and without electron exchange - can be distinguished. It is quite evident that for the first case the short-range interaction is decisive as e.g. in the process of excitation transfer between He($2^3S$) and He($I^IS$). The cross section can be estimated as for the resonant charge transfer, provided the critical distance corresponds to that for an unexcited electron.

A more delicate situation is observed for excitation transfer which can in principle proceed without electron exchange. A most vivid example of this is given by collision of the ground state atom with a similar atom in the resonant state (i.e. in the first optically allowed excited state).[68-76]. In this case the long range dipole-dipole interaction is operative, so that for resonance channels the splitting between gerade and ungerade molecular terms of the same angular symmetry is $\Delta \sim d^2/R^3$ , where $d$ is the dipole moment matrix element between ground and excited states. The excitation transfer cross section for this case is of the order

$$\sigma \sim d^2/\hbar v \qquad (19)$$

which is about $I0^{-I2}$ $cm^2$ for thermal energy. This large cross section markedly effects the shape of the resonance emission line.

For quasi-resonance channels corresponding to excitation transfer with a change of the fine-structure state the situation is different in that dipole-dipole interaction is not always necessarily responsible for the process, and the exchange interaction is coming into play. Let us consider intramultiplet transitions in an excited alkali metal $M^{*}(^2P)$ colliding with a similar ground-state atom $M(^2S)$. If only long-range interaction is taken into account, the pattern of molecular exhibits crossing of two states of different angular symmetry (Fig. I0). This crossing was shown to be responsible for the intermultiplet transitions $^2P_{3/2} \longrightarrow {}^2P_{I/2}$ in Na ( $\Delta\varepsilon$ is rather small, $\Delta\varepsilon$ = I7 $cm^{-I}$). This mechanism predicts for arbitrary $\Delta\varepsilon$ the cross-section of the order [75]:

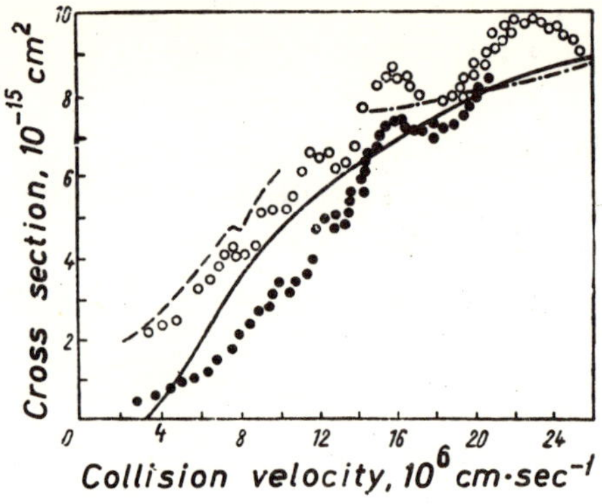


Fig. 8. Total cross section for non-resonant charge transfer in collisions $Rb^+ + Cs$ and $Cs^+ + Rb$. Full curve corresponds to the asymptotic calculation.[16] Experimental data: $Rb^+ + Cs \rightarrow Rb + Cs^+$, – – – [44], o – [45], —·— [46]; $Rb + Cs^+ \rightarrow Rb^+ + Cs$, ● – [45]

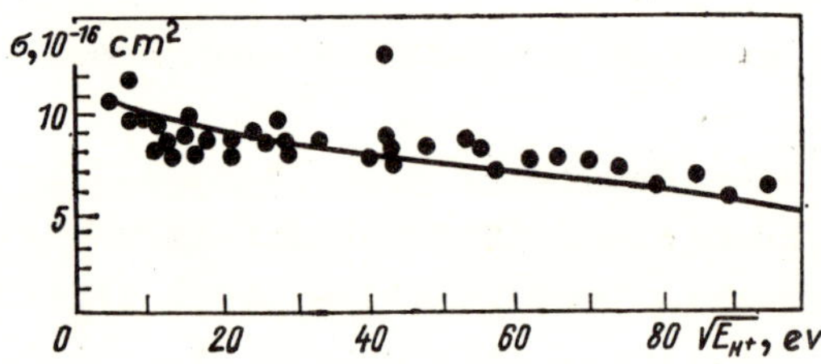


Fig. 9. The cross section of the process $H^+ + O \rightarrow H + O^+$. Full curve – the calculation [16], the points – experiment [47].

Table III.

| The colliding particles | The spin exchange cross section $10^{-15}$ cm$^2$ | | The small parameter of asymptotic theory $1/2\gamma R_0$ |
|---|---|---|---|
| | Theory | Experiment | |
| H – H | 2.0 | 2.4 [48-54] | 0.074 |
| Na – Na | 11 | 10 [55,56] | 0.051 |
| K – K | 15 | 21 [56,57] | 0.048 |
| Rb – Rb | 17 | 19 [54,58,63] | 0.046 |
| Rb – Cs | 18 | 24 [62-66] | 0.045 |
| Cs – Cs | 19 | 22 [56,62,66,67] | 0.045 |

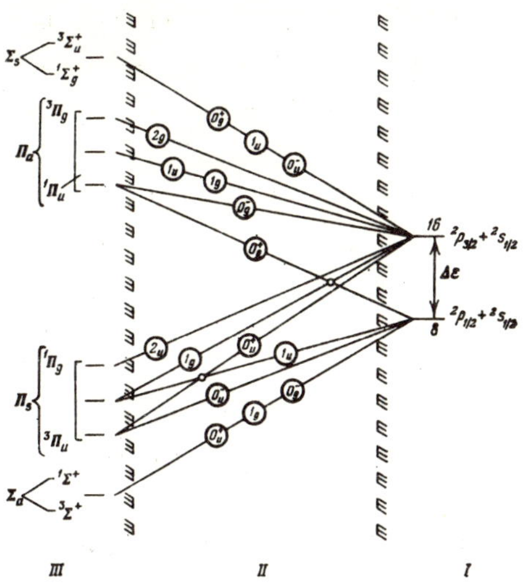


Fig. IO. Long-range correlation diagram of molecular terms for alkali diatom $M(^2P_j) + M(^2S_{I/2})$. [77]

$$\sigma \sim \frac{\Delta R}{R_o} R_o^2 \sim \frac{\hbar v\, d^{2/3}}{\Delta \varepsilon^{4/3}} \qquad (20)$$

where $R_o$ is the transition distance, $\Delta R$ is the width of the transition range. This formula gives cross-sections which are compared to the experimental values for heavier alkalies (K, Rb, Cs). This implies the existence of other coupling regions located presumably at shorter distances. Hence the importance of exchange forces, whether the asymptotic approach will suffice remains to be seen. At any rate this approach will be useful in correlating ab initio terms at intermediate distances calculated variationally with long-range tails of potentials. Recent study of the mechanism of collisional excitation of alkalies provides an example of such as application.

In conclusion, we emphasize the essential features of asymptotic approach in the theory of quasi-resonance atomic collision. First, transitions between molecular terms are supposed to occur at distances $R_o$ outside the classical radius of the electron orbit. This allows to calculate molecular terms and the dynamical coupling between them analytically, or at least in terms of simple numerical functions. Second, the multichannel scattering problem is simplified by explicit accounting only for coupling of a small number of channels in several regions. Third, in each region the coupling problem is solved in terms of simple models, and the multichannel scattering matrix is constructed by matching pieces of solution within different regions.

It is hoped that this approach in the theory of atomic collisions may be applied to other fields such as inelastic and reactive molecular collisions [78, 84], calculation of transport coefficients [79-8I] and estimation of interaction in molecular crystals.[82, 83]

## References


1. Б.М. Смирнов. Атомные столкновения и элементарные процессы в плазме, М., Атомиздат, 1968.
2. Е.Е. Никитин. Теория элементарных атомно-молекулярных процессов, М., "Химия", 1970.
3. Е.Е. Никитин. Теория элементарных атомно-молекулярных реакций. Новосибирск, Изд-во Новосиб. Ун-та, ч. I, 1971; ч. 2, 1974.
4. Б.М. Смирнов. Асимптотические методы в теории атомных столкновений. М., Атомиздат. 1973.
5. Ю.Н. Демков, В.Н. Островский. Метод потенциалов нулевого радиуса в атомной физике, Л., Изд-во Ленингр. Ун-та, 1975.
6. И.В. Комаров, Л.И. Пономарев, Ю.С. Славянов, Сфероидальные и куловонские сфероидальные волновые функции, М.,Наука,1976
7. H.F. Schaeffer, The Electronic Structure of Atoms and Molecules, A survey of Rigonos Quantum Mechanical Results, Reading, Mass, Addition-Wesley, 1972.
8. Л.Д.Ландау, Е.М. Лифшиц. Квантовая механика,М.,Наука,1974.
9. J.C. Slater, Phys. Rev. 32, 349 (1928).
10. F. London, Zs. Phys. Chem. BII 222 (1930).
11. E. Hylleraas, Zs. Phys. 71, 739 (1931).
12. J.C. Slater, C.G. Kirkwood. Phys. Rev. 37, 682 (1931).
13. J.G. Kirkwood, J. Phys. Zs. 33, 39, (1932).
14. H. Bethe, Quantenmechanik der Ein-und Zwei-Elektronen-problem. Handb. Phys. 1933
15. Е.Л. Думан, Б.М. Смирнов, ЖТФ 40, 91 (1970).
16. Б.М. Смирнов. Теплоф.выс.темп. 3, 429 (1966).
17. Е.Д.Думан, Б.М.Смирнов. Оптика и спектр. 32, 448 (1972).
18. H.F. Schaefer, F.E. Harris, J. Chem. Phys. 48, 1918 (1968).
19. А.И. Резников, С.Я. Уманский. Теор. и экспер. хим. 8, 56 (1972).
20. A.I. Voronin, E.P. Gordeev, S.Ja. Umanski, Chem. Phys. 23, 524 (1953).
21. О.Б. Фирсов ЖЭТФ, 21, 1001 (1951).

22. Е.Е. Никитин, С.Я. Уманский. Неадиабатические переходы при медленных атомных столкновениях. Атомиздат, М., I978.
23. В.К. Быховский, Е.Е. Никитин, М.Я. Овчинникова, ЖЭТФ, 9750 (I964).
24. N. Rozen, C. Zener, Phys. Rev. 40, 502 (I932).
25. Ю.Н. Демков, ЖЭТФ, 45, I95 (I963).
26. Р.З. Витлина, А.В. Чаплик, М.В. Энтин, ЖЭТФ, 67, I667 (I974).
27. Е.Е. Никитин. Опт. и спектр. I3, 76I (I962).
28. Ю.Н. Демков, М. Кунике, Вестн. Ленингр. Ун-та, № I6, 39 (I969).
29. В.К. Быховский, Е.Е. Никитин. ЖЭТФ 48, I499, I965.
30. М.Я. Овчинникова, ЖЭТФ, 64, I29 (I973).
3I. А.З. Девдариани, В.Н. Островский, Ю.Н. Себякин, ЖЭТФ 7I, 909 (I976).
32. В.А. Базылев, Н.К. Жеваго, М.И. Чибисов, ЖЭТФ 7I, I286 (I976).
33. Г.В. Дубровский, ЖЭТФ, 58, I075 (I970).
34. Б.М. Смирнов. ЖЭТФ 57, 5I8, I964.
35. Д.В. Чкуасели. У.Д. Николайшвили, А.И. Гулдамишвили, ЖЭТФ, 30, 8I7 (I960); Изв. АН СССР, сер физ,24,920 (I960).
36. J. Perel. R.H. Vernon, H. Delay, Phys. Rev. AI38, 937 (I965).
37. W.R. Gentry, Y. Lee, B.H. Mahan, Phys. Rev. AI38, 937 (I965).
38. Е.Л. Думан, Б.М. Смирнов. ТВТ, I2, 502 (I974).
39. J.A. Dillon et al., J. Chem. Phys. 23, 776 (I955).
40. S.N. Grosh, N.F. Sheridan, Ind. J. Phys. 3I, 337 (I957).
4I. И.Г. Флакс, Е.С. Соловьев, ЖТФ, 28, 599 (I958).
42. Р.М. Кушнир, Б.М. Палюх, Л.А. Сена, Изв. АН СССР, сер. физ. 23, I007 (I959).
43. R.C. Amme, P.O. Hagsjaa, Phys. Rev. I65, 63, I965.
44. L.L. Marino, In book: "Atomic collision processes" Amsterdam I964. p. 823.
45. J. Perel, R.H. Vernon, H. Daley, Phys.Rev. I38A, 937, I965.
46. D.V. Chkuaceli, A.I. Gouldamachvili, U.D. Nicoleychvili, Proc. 6 Inter. Conf. on Ioniz. Phen. in Gases. Paris. I963, p. 475.

47. R.F. Stebbings. A.C. Smith, C.J. Ehrhardt, Geophys. Res. 69, 2349, I964.
48. J.P. Wittle, R.H. Dicke, Phys. Rev. I03, 620 (I956).
49. A.F. Hildengrandt, F.B. Booth, C.A. Barth, J. Chem. Phys. 3I, I73, (I959).
50. R.M. Mazo, ib id. 34, I69 (I96I).
5I. R.L. Bender, Phys. Rev. I32, 2I54 (I963).
52. H. Hellwig, ibid. I66, 4 (I968).
53. C. Audoin, Phys. Lett. A28, 372 (I968).
54. N.P. Asselt, J.G. Maas, Chem. Phys. I6, 8I (I976).
55. L.W. Anderson, A.T. Ramsey, Phys. Rev. I32, 7I2 (I963).
56. N.W. Ressler, R.H. Sands, T.E. Start, ibid. I84, I02 (I969).
57. F.J. Grossetete, M.J. Brossel, C.R.Ac. Sci. 254, 3829 (I967).
58. M. Desainfuscien, C. Audoin, Phys. Rev. AI3, 2070 (I976).
59. J. Carver, in: Proc. of Ann-Arbor Conference of Optical Pumping, I959, p. 29.
60. P. Davidovits, N. Knable, Bull. Am. Phys. Soc. 8, 352 (I963).
6I. S.M. Jahrett, Phys. Rev. AI33, III (I964).
62. H.W. Moos, R.H. Sands, ibid. AI35, 59I.
63. M. Elbel, H. Wieder, Phys. Lett. I8, 276 (I965).
64. H.M. Gibbs. R.J. Hull, Phys. Rev. I53, I32 (I967).
65. J. Vanier, ibid. I68, I29 (I968).
66. M.A. Bouchiat, F.J. Grossetete, Phys. Lett. 27, 353 (I966).
67. K. Ernst, F. Strumia, Phys. Rev. I70, 48 (I968).
68. M. Mori, T. Watanabe, K. Katsuura, J. Phys. Soc. Japan I9, 380, I504 (I964).
69. T. Watanabe, Phys. Rev. IA30, I375 (I965).
70. А.И. Вайнштейн, В.М. Галицкий, Препринт НГУ, Новосиб. I965.
7I. А.И. Вайнштейн, В.М. Галицкий, в кн.: Вопросы теории атомных столкновений, под. ред. Ю.А. Вдовина, М., Атомиздат, I970, с. 39.
72. A.J. Omont, J. Phys. 26, 26 (I965).

73. Ю.А. Вдовин, В.М. Галицкий, В.В. Якимец, цит. в 68 с. 3.
74. Ю.А. Вдовин, Н.А. Добродеев. ЖЭТФ 55, I047 (I969).
75. Ю.А. Вдовин, В.М. Галицкий, Н.А. Добродеев. ЖЭТФ 56, I344 (I969).
76. Ю.А. Вдовин, В.М. Галицкий. ЖЭТФ 69, I03 (I975).
77. E.I. Dashevskaya, A.I. Voronin, E.E. Nikitin, Can. J. Phys. 47, I237 (I969).
78. E.E. Nikitin, Adv. Chem. Phys. 28, 3I7 (I975).
79. Л.А. Палкина, Б.М. Смирнов, М.И. Чибисов, ЖЭТФ, 56, 340 (I969).
80. Б.М. Смирнов, М.И. Чибисов, ТВТ, 9, 5I3 (I974).
81. А.В. Елецкий, Л.А. Палкина, Б.М. Смирнов. Явления перехода в слабоионизованной плазме, М., Атомиздат,I975.
82. Е.Л. Думан, Б.М. Смирнов. Опт. и спектр. 29, 425 (I970).
83. Б.М. Смирнов, Г.В. Шляпников, УФН, I20, 69I (I976).
84. Е.Е. Никитин, Усп. хим. 43, I905 (I974).

# CORRELATED TRANSITIONS IN ATOM WITH TWO INNER-SHELL VACANCIES

V.V.Afrosimov, A.P.Shergin

A.F.Ioffe Physical-Technical Institute,
Academy of Sciences of the USSR,
Leningrad, USSR

## I. Introduction

It is known, that collisions between many-electron atoms represent an effective means for the creation of two or more vacancies in the inner electron shells. Due to the quasimolecular mechanism of the interaction the cross-sections for the creation of two vacancies can reach up to $10^{-19}$- $10^{-17}cm^2$, i.e. the value of the geometrical cross-section of an inner shell. The main channels of the two-vacancy state decay are the conventional Auger and radiative transitions, when the filling of each of the vacancies occurs independently and is accompanied by the ejection of an Auger electron or photon. A study of the energy spectra of Auger electrons and X-ray photons created in atomic collisions resulted in a discovery of decay channels of another type, namely, of correlated transitions in an atom with two inner-shell vacancies.

In a correlated transition, the two inner-shell vacancies are filled by two outer-shell electrons with the corresponding energy carried off by one Auger electron or photon. Evidently, the energy of the Auger electron or photon created in a correlated transition exceeds by about a factor of two that of electrons or photons formed in ordinary transitions. As an example Fig.I presents possible Auger decay channels for a state with two K-vacancies with a conventional transition shown for comparison. The radiative and Auger transitions can be conveniently denoted with the shell symbol and the actual number of the initial and final vacancies. Naturally, correlated transitions can also occur with two vacancies in other inner shells of many-electron atoms (L, M, N-shells, etc.).

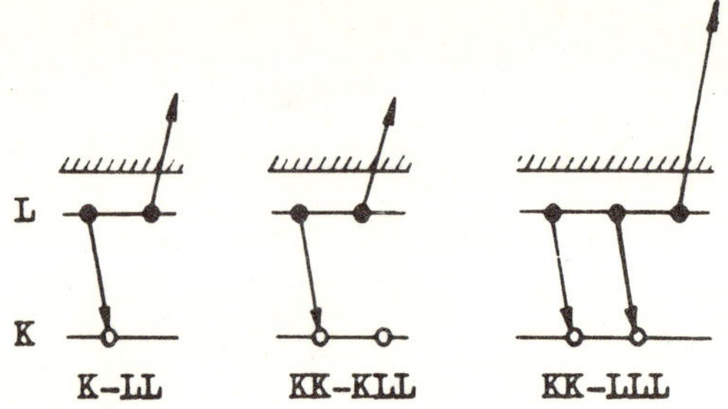


Fig.I. Diagram of Auger transitions for the case of one and two K-vacancies. K-LL - conventional transition; KK-KLL - hypersatellite transition (decay of one of K-vacancies in the presence of another); KK-LLL - three-electron transition (correlated decay of two K-vacancies involving ejection of one Auger electron).

Correlated transitions in atoms with two vacancies which were first observed about three years ago, have now been thoroughly studied experimentally [I-I2]. The correlated transitions, mainly radiative, were also adequately described in the frame of the shake-off model and many-body atom theory [I3-23].

2. Experiment

The decay of atomic states with inner-shell vacancies is accompanied by ejection of electrons and photons, and can be studied by measuring their energy spectra. If the transition scheme is known, it is possible to estimate the energies of Auger electrons $E_e$ and characteristic photons $E_x$ at the peaks in the spectra. The transition energies were compared with the theoretical values and, as shown in section 4, found to be in a good agreement in all cases.

Fig.2 taken from review [II], shows the spectra of electrons formed in collisions $Ar^+$-Ar, $Cl^+$-Ar and $S^+$-Ar. The strong peaks at I75, I40 and I20 eV in the cases $Ar^+$-Ar, $Cl^+$-Ar and $S^+$-Ar are due to conventional Auger transitions $L_{2,3}$-MM in Ar, Cl and S ions, respectively. The peaks at approximately twice the electron energies in the same spectra, 445, 370 and 305 eV are connected with the three-electron transitions $L_{2,3}L_{2,3}$-MMM.

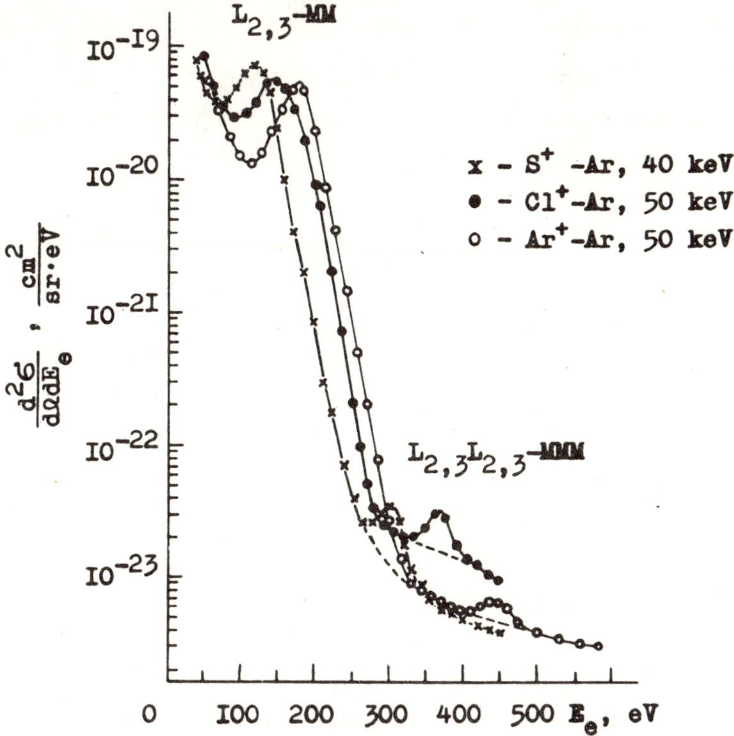


Fig.2. Energy spectra of electrons formed in collisions of $S^+$, $Cl^+$ and $Ar^+$ ions with Ar atoms.

When interpreting these transitions, one should consider the ratio of decay probabilities of the states with two inner vacancies into different channels. If $\sigma(KK)$ and $\sigma(LL)$ are the cross-sections for the formation of two K- or L-vacancies in an ion, then the ratios:

$$\gamma = \frac{\sigma_A(KK\text{-}LLL)}{\sigma(KK)} \text{ or } \frac{\sigma_A(LL\text{-}MMM)}{\sigma(LL)}, \text{ and } \beta = \frac{\sigma_x(KK\text{-}LL)}{\sigma(KK)} \text{ or } \frac{\sigma_x(LL\text{-}MM)}{\sigma(LL)}$$

characterize the relative probabilities of decay of the states with two K- or L-vacancies by the three-electron Auger transition ( $\gamma$ ) or radiative two-electron-one-photon transition ( $\beta$ ).

Information on the cross-section for formation of two vacancies in one particle $\sigma(KK)$ and $\sigma(LL)$ in a collision can be obtained from inelastic energy loss experiments or by calculation. In high resolution experiments studying K-radiation it was possible to reveal lines, corresponding to hypersatellite transitions, and to determine the intensity ratio of correlated and hypersatellite transitions $\alpha = \frac{\sigma_x(KK-LL)}{\sigma_x(KK-KL)}$. Evidently, $\alpha$ should not depend on the probability for the two vacancy formation, since for correlated and hypersatellite transitions the initial is the state with two vacancies. The latter ratio allows a direct comparison with the theory to be made because the probabilities of the hypersatellite and correlated transitions in an atom with two K-vacancies can be calculated.

To illustrate experimental information which can be derived by studying radiative transitions, consider the radiation spectrum in the case of Fe-Fe, 40 MeV collision obtained by Stoller et al. [6] (Fig.3).

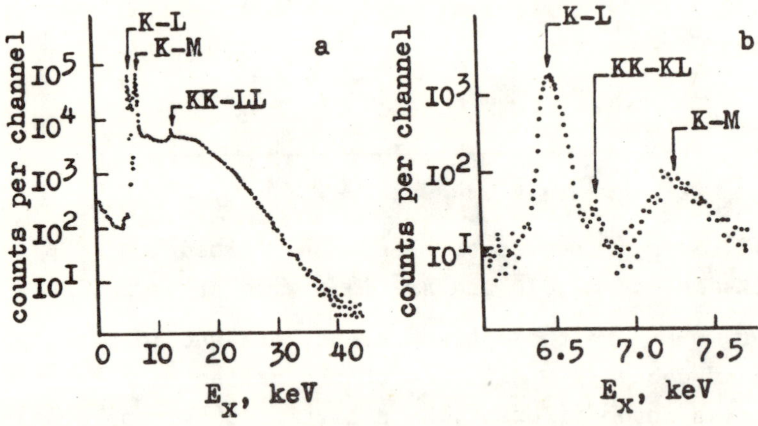


Fig.3. X-ray spectra for Fe-Fe collisions at incident energy $E_0$= 40 MeV. a) Note peaks corresponding to K-L and K-M transitions and to two-electron-one-photon transition KK-LL; b) the hypersatellite KK-KL transition peak is resolved in a crystal spectrometer.

Fig.3a shows a spectrum obtained with a high sensitivity Si(Li)-detector. Besides the peaks at $E_x$=6.5 and 7.3 keV corresponding to the conventional K-L and K-M transitions, one can see at $E_x$=13 keV a KK-LL peak related to a single-photon decay of two K-vacancies in Fe. This experiment allows us to determine the cross-section ratio for correlated and conventional transitions, $\sigma_x(KK-LL)/\sigma_x(K-L) = 3.0 \cdot 10^{-6}$. It should be noted that the intensity of K-L transition in the spectrum of Fig.3a is strongly attenuated by an Al filter. Fig.3b reproduces a part of the spectrum in the region of K-L and K-M transitions measured with a LiF-crystal spectrometer of a high resolution (40 eV). The purpose of this experiment was to select the hypersatellite KK-KL transition. The ratios obtained from these two measurements, $\sigma_x(KK-LL)/\sigma_x(K-L)$ and $\sigma_x(KK-KL)/\sigma_x(K-L)$, permits one to derive $\alpha=2.45 \cdot 10^{-4}$, which can be compared with theory.

It was determined experimentally, that the fraction of correlated transitions in the decay of two-vacancy states is $10^{-4}-10^{-6}$. Because of this, quantitative study of correlated transitions presents a serious problem and requires high sensitivity techniques. Such techniques have been developed recently [1,2,3,8,10].

The existence of correlated transitions was particularly well established when turning from symmetric to asymmetric atomic collisions [1]. For instance, in collisions $Cl^+$-Ar and $Ar^+$-Ar the cross-sections for three-electron Auger transitions $\sigma_A$(LL-MMM) in the Cl and Ar ions is found to differ approximately by an order of magnitude while the cross-sections for conventional transitions $\sigma_A$(L-MM) remain close to one another. This may be attributed to the fact that in asymmetric $Cl^+$-Ar collisions both L-vacancies are formed almost exclusively in Cl-ions, whereas in symmetric $Ar^+$-Ar collisions the vacancies can be shared between the collision partners.

Up to now correlated transitions have been studied in the decay of states with vacancies in the K and L-shells. Three-electron Auger transitions involving $L_{2,3}$-vacancies were studied in the Si, S, Cl and Ar atoms [1,4], and with K-vacancies, in the C and N atoms [9] in the incident ion energy range $E_0$=10-50 keV. Data on three-electron transition to $L_{2,3}$-vacancies in Ar in the $N^+$-Ar collisions were also obtained by Stolterfoht et al.[10]. Agreement between the results of these experiments is quite satisfactory. Single-photon decay of two K-vacancies over a wide range of atomic number Z (in Al, Ca, Fe and Ni atoms) was studied by Wölfli et al.[2,5,6] in collisions at $E_0$=25-40 MeV. Radiative transitions to K-vacancies for the light atoms N, O and Ne were studied by Hoogkamer et al.[3,12] in collisions at energies ranging from 200 to 600 keV. Precise measurements of the energies of KK-LL transitions for atoms with Z=12-26 were carried out by Knudson et al.[7]. Radiative $L_{2,3}L_{2,3}$-MM transitions for the same atoms S, Cl and Ar as in the experiment on three-electron Auger transitions [I,4] were investigated by Afrosimov et al.[8] which permitted one in the same experiment to establish the role of both the Auger and radiative correlated transitions.

3. Theoretical Models

Observation of correlated transitions in atoms with two vacancies stimulated quantitative theoretical consideration of this phenomenon. The probabilities of correlated transitions were estimated theoretically within the framework of the shake-off model and many-body theory of the atom. According to the shake-off model, correlated transitions occur in two stages. For instance, in a three-electron Auger transition (LL-MMM) one of the L-vacancies is filled as a result of a conventional Auger transition, with the corresponding energy transferred to an outer M-electrons. In the changed field another outer M-electron is shaken off on to the second L-vacancy with a transfer of additional energy to the ejected electron. The three-electron Auger transition probability in the shake-off approximation is [I6]

$$W_A(LL\text{-}MMM)=2\pi\left|\langle \Psi_f \mid V \mid \Psi_i\rangle\right|^2 \simeq W_A(L\text{-}MM)\cdot W_S$$

where $\Psi_i$ and $\Psi_f$ - wavefunctions of initial and final states, $V=\sum_{k<m} I/r_{km}$ - electron-electron interaction operator, $W_S=\left|\langle \varphi_f \mid \varphi_i\rangle\right|^2$ - probability of the shake-down process which is equal to the overlap integral between initial and final wavefunctions of the electron involved. Within the framework of the shake-off model $\gamma \sim W_S$, since

$$\gamma = \frac{\sigma_A(LL\text{-}MMM)}{\sigma(LL)} \simeq \frac{W_A(LL\text{-}MMM)}{2W_A(L\text{-}MM)} = \frac{W_S}{2} .$$

The probability of single-photon decay of K-vacancies by the shake-off model was calculated by Aberg et al.[I4] and Gavrila and Hansen [I5,20]. The single-particle wavefunctions of the initial and final states used in the shake-off model are non-orthogonal because the screening in the initial and final states is different. This results in a non-zero overlap integral. Aberg et al.[I4] carried out calculations of radiative KK-LL transitions with Hartree-Fock wavefunctions for a number of atoms from Mg to Ni. They showed that

$$\alpha = \frac{\sigma_x(KK\text{-}LL)}{\sigma_x(KK\text{-}KL)} \sim \left|\langle (Is)_f \mid (2s)_i\rangle\right|^2 \sim Z^{-2}$$

i.e. the relative role of electron correlations in radiative decay channels of states with two inner-shell vacancies decreases with the increase of atomic number. Gavrila and Hansen[15,20] calculated the probability of transitions to K-vacancies for atoms with Z = 7-28 with filled outer shells and in the presence of 2p-vacancies. As shown in [20], $\alpha$ is practically insensitive to the degree of outer-shell ionization. This result can be easily understood, since both the conventional and correlated radiative transitions involve only one 2p-electron.

In the refs.[13,17-19,21-23] correlated transitions, mainly radiative, were considered from the viewpoint of many-body theory of the atom. The transition probabilities were calculated in the first order of perturbation theory in interelectron interaction. The probability of two-electron-one-photon transition is defined as $W_x = \frac{16}{3} \alpha^3 \omega^3 |M|^2$, where $\alpha$ is the fine structure constant, $\omega$ - transition energy in atomic units, M - the matrix element determined by the corresponding number of Feinman diagrams. The diagrams included in calculations of the single-photon decay probability of two vacancies in refs. [18,21] are shown in Fig.4.

It is of interest to consider two particular cases. As shown in [21], at the energy of the virtual electron (diagrams b in Fig.4) $E_k = E_{f1} + E_{f2} - E_{in}$, where $E_i$ and $E_f$ are the initial and final vacancy energies, the two-electron-one-photon transition can be considered as occurring in two stages: Auger decay of one vacancy followed by recombination of the ejected Auger electron with the filling of the other vacancy. Of interest is also the second case (diagrams a). If the total energy of two inner vacancies $E_{i1}$ and $E_{i2}$ is close to that of vacancy formation in a deeper electron shell $E_k$ (i.e. the condition $E_k + E_{fn} \simeq E_{i1} + E_{i2}$ is satisfied), single-photon decay can be also considered as a two-stage process. Initially, the vacancies $i_1$ and $i_2$ transfer to k and $f_n$, with a subsequent decay of the intermediate vacancy k. The decay probability of a state with k-and $f_n$-vacancies can evidently be quite high, because it is determined by the probability of the conventional transition. It should be noted, that a transfer from a state with two inner vacancies to that with a vacancy in a deeper shell was observed earlier experimentally [24].

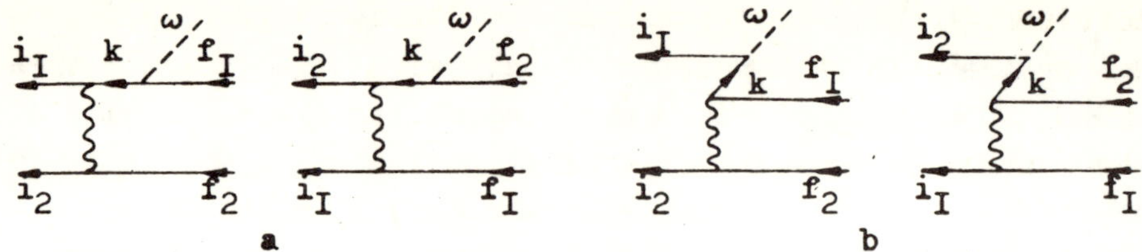


Fig.4. Feynman diagrams of single-photon decay of the state with two inner-shell vacancies: a) direct interaction of vacancies; b) interaction via intermediate virtual Auger process.

The most accurate calculations of the two-electron-one-photon transition probability for many-electron atoms were carried out with Hartree-Fock wavefunctions [13,18,21]. The choice of objects for the calculations was determined by the possibility of comparison with experimental data [3,6,8]. Kelly [13] calculated the two-electron-one-photon transition probability for the Fe ion with two K-vacancies. Amusia et al.[18,21] calculated the probabilities of single-photon decay of K-vacancies in Ne, $L_I$ and $L_{2,3}$-vacancies in Ar, and $N_{4,5}$-vacancies in Xe.

Simpler calculations of the single-photon decay probabilities of two K-vacancies can be carried out with Coulomb wavefunctions. Khristenko [17] calculated the probabilities of single-photon decay of two K-vacancies for transitions of the type $nln'l'\,^1P \rightarrow 1s^2\,^1S$ in helium-like ions (Z=13-28). The electrostatic interaction between two electrons $e^2/r_{12}$ was considered as a perturbation. Safronova and Senashenko[19,22] determined the probabilities of KK-LL transitions in ions with two, three and more electrons. A similar method of calculation was used in a recent paper by Ivanov et al.[23] for determination of the decay probability via another channel, i.e. three-electron Auger transitions to K-vacancies in ions with three electrons. Calculations of correlated Auger transitions are essentially more complicated than those of radiative ones. A comparable contribution to the total three-electron transition probability, as shown in [23], comes from processes described by 9 Feynman diagrams.

An analysis of the calculations carried out in the framework of the many-body theory shows [18,21] that under some conditions a correlated transition can be considered as a two-stage process. In this case the description of correlated transitions is particularly straightforward, and the transition probabilities can be estimated relatively simply. Note, however, that in the general case correlated transitions should not be considered as occurring in stages. A comparison between the two methods of theoretical description of correlated transitions using the shake-off model and many-body theory shows[22] that the shake-off approximation includes the contribution of only a part of the mechanisms. Thus, as it could be anticipated, consideration of correlated transitions from the viewpoint of the many-body theory is more comprehensive than using the shake-off model. At the same time, the shake-off model has the advantage of the calculations of correlated transitions being comparatively simple.

In the following sections a comparison of the available experimental and theoretical results will be presented.

## 4. Dependence of the Correlated Transition Energy on the State of the Outer Shell

To calculate the energies of correlated transitions, one should know the energies of the initial and final ion configurations involved in the transition. It is known, that in a collision simultaneously with the formation of inner-shell vacancies, ionization of the outer shells takes place. Thus, according to an experimental study by Fastrup et al.[25], Afrosimov et al.[26] of inelastic energy losses in the collisions $S^+$, $Cl^+$ and $Ar^+$-Ar (incident ion energy $E_0$=IO-IOO keV) accompanied by the formation of $L_{2,3}$-vacancies in S,Cl and Ar ions one or two electrons are removed, on the average, from the outer shells of given ions.

The energies of various ion configurations with one or two $L_{2,3}$-vacancies were calculated using the solutions of nonrelativistic Hartree-Fock equations by Larkins for Ar [27], Savukynas et al. for Si, S and Cl [28]. In the Table taken from a review by Shergin and Gordeev [II] are presented calculated electron energies $E_e$ for transitions $(2p)^{-2} \rightarrow (3p)^{-3}$ in the case of S, Cl and Ar, and $(2p)^{-2} \rightarrow (3s)^{-I}(3p)^{-2}$ in the case of Si for various degrees of ionization in the initial state. As seen from the Table, the calculated energies for $m \simeq I$ agree well with experimental data.

Table. Energies of electrons $E_e$ formed in three-electron Auger transitions to $L_{2,3}$-vacancies in ions with various numbers of vacancies m in outer M-shell.

| Ion | Initial ion configuration | $E_e$, ev (Theory) m = 0 | m = I | m = 2 | m = 3 | $E_e$, ev (Exper.) |
|---|---|---|---|---|---|---|
| Ar | $2p^4 3s^2 3p^{6-m}$ | 457.6 | 449.4 | 440.7 | 432.5 | 445 $Ar^+$-Ar, 50 kev |
| Cl | $2p^4 3s^2 3p^{5-m}$ | 382.9 | 368.3 | 358.8 | | 370 $Cl^+$-Ar, 50 kev |
| S | $2p^4 3s^2 3p^{4-m}$ | 3I0.6 | 300.9 | | | 305 $S^+$ -Ar, 40 kev |
| Si | $2p^4 3s^2 3p^{2-m}$ | I87.8 | | | | I95 $N_2^+$-$SiH_4$,50 kev |

Experiment and theory for radiative transitions are compared in Fig.5 taken from the paper by Knudson et al.[7], where the experimental and calculated values of the difference of photon energies in two-electron-one-photon transitions $E_x$(KK-LL) and double photon energy in a conventional transition $2E_x$(K-L) are plotted as functions of atomic number.

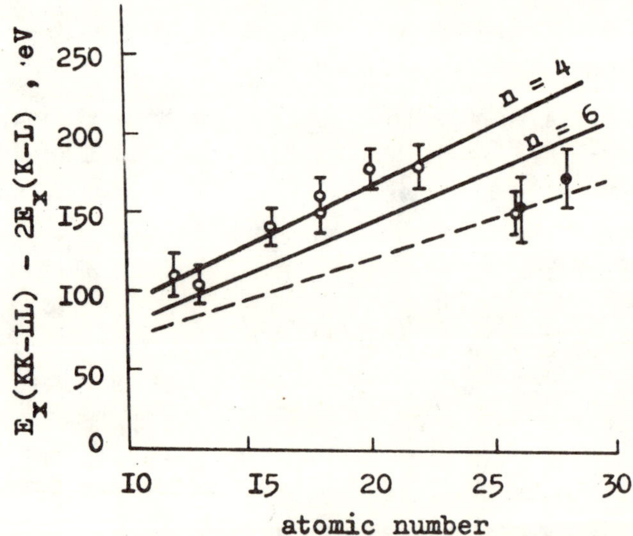


Fig.5. The difference between the photon energy in a single-photon decay of two K-vacancies $E_x$(KK-LL) and twice the photon energy in a conventional transition $2E_x$(K-L) vs. atomic number.

Experiment: o - Knudson et al.[7], $E_0$=3.0-3.5 MeV; • - Wölfli et al.[2,5], $E_0$=40 MeV; ▬ - Hartree-Fock calculation for transitions to K-vacancies in atoms with filled(n=6) and doubly-ionized (n=4) L-shell (--- shift with multiplet splitting taken into account for n=6).

Circles denote experimental data. Straight lines show the Hartree-Fock calculations for transitions $2s^2 2p^n \to Is^2 2s2p^{n-I}$ in the case of a filled L-shell (n=6) and in the presence of two L-vacancies (n=4). The energies were calculated assuming a statistical population of different multiplet states in the initial configuration. Fig.5 shows that in the interval $I2 \leq Z \leq 22$ the experimental values of the difference $E_x$(KK-LL)-2$E_x$(K-L) correspond to transitions in ions with two vacancies in the L-shell. Only the values for Fe and Ni (Z=26 and 28) are not in agreement with the above dependence.

## 5. Relative Probabilities of Three-Electron Auger Transitions

To determine relative probabilities of filling of two inner-shell vacancies in three-electron Auger transitions, it is necessary to compare the experimental cross-sections $\sigma_A$(LL-MMM) and $\sigma_A$(KK-LLL) with the cross-sections for formation of two vacancies in one particle $\sigma$(LL) and $\sigma$(KK). The cross-sections for the formation of two $L_{2,3}$-vacancies in the Si, S, Cl and Ar ions were determined[I,4,2,3] on the basis of the preceding study of scattering and inelastic energy losses in collisions involving the above mentioned atoms [25,26]. Thus, in the case of the L-shell the necessary information is available. For the K-shells the situation proves to be more complicated. At present we have no experimental data which could be used to derive the cross-sections $\sigma$(KK). The values of $\gamma$ for the case of transitions to K-vacancies in the ions C and N were determined [9] from the experimental ratios $\sigma_A$(KK-LLL)/$\sigma_A$(K-LL) and ratios $\sigma_A$(K-LL)/$\sigma$(KK), calculated using the data by Taulbjerg et al.[29] and Macek and Briggs [30]. As is known, in collisions of light particles ($Z \leq I0$) K-vacancies are formed as a result of rotational coupling between the $2p\pi_x$ and $2p\sigma$ orbitals at small internuclear distances. Taulbjerg et al.[29] used a scaling procedure for the calculation of the probability of $2p\sigma - 2p\pi_x$ transition and obtained an universal dependence of transition probability on the impact parameter and collision velocity for different combinations of interacting particles. In the paper by Macek and Briggs [30] the probability of finding 0, I and 2 vacancies on the $2p_x$ orbital was calculated as a function of the number of 2p-vacancies in atoms before collision.

Fig.6 shows the relative probability of three-electron transition $\gamma$ to K and L-vacancies vs. the number of outer electrons n. It is seen that the probabilities of transitions to vacancies in different shells are close and lie within $10^{-4}$-$10^{-3}$. It is evident, that $\gamma$ for the same shell should depend on the number of outer electrons (for $n \leq 2$, three-electron Auger transition is generally not possible). The observed dependence $\gamma(n)$ was explained in [4] on the basis of a very simple statistical consideration. Assuming that filling of an inner vacancy by any outer electron is equiprobable, the three-electron transition probability should be proportional to the number of combinations of n three at a time, i.e. $C_n^3$, and the conventional transition probability, of n two at a time, i.e. $C_n^2$. Thus, the expression for $\gamma$ is: $\gamma \sim \frac{C_n^3}{C_n^2} = \frac{n-2}{3}$. As seen from Fig.6, the experimental data for the K and L-shells agree with the relationship $\gamma \sim n-2$ (the dashed lines in Fig.6)

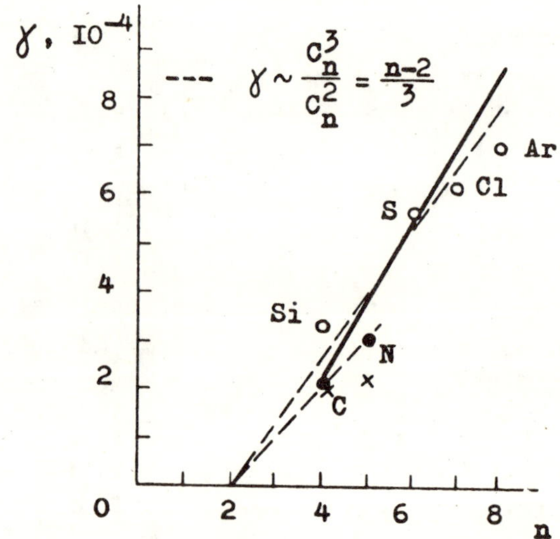


Fig.6. The relative probability of three-electron Auger transitions as a function of the number of outer shell electrons. o and • - experimental results by Afrosimov et al.[1,4,9] for LL-MMM and KK-LLL transitions, accordingly; — - calculation of $\gamma$ for LL-MMM transitions by Kishinevsky et al.[16] using shake-off model; x - the calculation of $\gamma$ for KK-LLL transitions by Ivanov et al.[23] using many-body theory.

Shown in Fig.6 by a solid line are the values for LL-MMM transitions calculated by Kishinevsky et al. [16] using the shake-off model involving single-electron wavefunctions. One can see that the calculated $\gamma$'s agree with experiment both in the magnitudes and in the dependence on the number of outer electrons. The experimental values of $\gamma$ for KK-LLL transitions, as seen from Fig.6, are also in a good agreement with the theoretical values calculated in [23] in the first order of perturbation theory in interelectron interaction.

## 6. Probabilities of Single-Photon Decay of Two Vacancies

For a comparison of experiment with theory in the case of KK-LL transitions one can use the values $\beta = \sigma_x(KK\text{-}LL)/\sigma(KK)$, $W_x(KK\text{-}LL)$ and $\alpha = \sigma_x(KK\text{-}LL)/\sigma_x(KK\text{-}KL)$. The values $\beta(KK\text{-}LL)$ and $W_x(KK\text{-}LL)$ can be determined only using data from different experimental and theoretical studies. This, naturally, reduces reliability of conclusions derived from the comparison. When comparing theory with experiment, it is reasonable to use the ratio $\alpha$ (or, as it is done in most papers, the reciprocal ratio $\alpha^{-1}$), since only this value was independently calculated theoretically and measured experimentally. For the comparison of various theoretical models, it is convenient to use directly the values of the probabilities $W_x(KK\text{-}LL)$.

Fig.7 shows experimental and theoretical dependences of $\alpha^{-1}$ on Z. One can remind that $\alpha^{-1}$ is nothing else than the ratio of probabilities of radiative decay of a state with two vacancies via the hypersatellite (single-electron) and correlated (two-electron) transitions. The probabilities $W_x(KK\text{-}LL)$ as a function of Z are given in Fig.8. Here are also shown the probabilities derived from the experimentally obtained values of $\alpha$ using the approximate expression

$$W_x(KK\text{-}LL) = \alpha \cdot W_x(KK\text{-}KL) \simeq \alpha \cdot 2W_x(K\text{-}L)\left[\frac{E_x(KK\text{-}KL)}{E_x(K\text{-}L)}\right]^3$$

including corrections for the difference in energy between the hypersatellite and conventional transitions. The factor 2 reflects the increase of decay probability due to the existence of two vacancies.

From the analysis of data presented in Fig.7 and 8 two main conclusions follow. Firstly, as seen from Fig.7, there is a satisfactory agreement between the values of $\alpha^{-1}$, **measured and calculated using different models (the discrepancy does not exceed the factor of 2-3). It is seen also that** the dependence of $\alpha^{-I}$ on Z as $Z^2$, which is predicted in many theoretical papers, agrees with the experimental data of Stoller et al.[6]. The results of calculations using the many-body theory fit best of all to experiment. This could be anticipated because it is this model that takes into account the interelectron correlations resulting in the two-electron-one-photon transition most accurately.

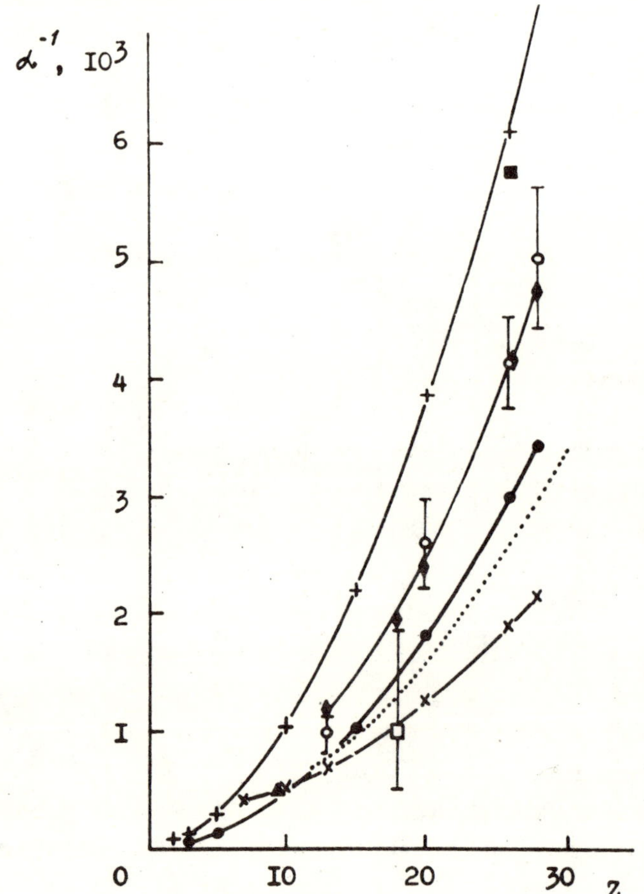


Fig.7. The ratio of probabilities of radiative decay of the state with two K-vacancies via hypersatellite and correlated transitions $\alpha^{-I}$ as a function of atomic number. Experiment: o – Stoller et al.[6]; □ – Knudson et al.[7]; calculation: ▲ – Amusia et al.[I8]; ■ – Kelly[I3]; ♦ – Khristenko[I7]; ●,+ – Safronova and Senashenko [I9] (transitions $2s2p\ ^{I}P \rightarrow Is^2\ ^{I}S$ and $2s^22p\ ^2P \rightarrow Is^22s\ ^2S$, respectively);.... – Aberg et al.[I4]; x – Gavrila and Hansen[I5].

Secondly, comparing the models one can see that there is a discrepancy of about a factor of 3 between $\alpha^{-I}$ calculated for Fe by Aberg et al.[I4], Gavrila and Hansen [I5] using the shake-off model, and Kelly [I3] using the many-body theory. As one can see from Fig.8, this difference of $\alpha^{-I}$ is due completely to the difference in $W_x$(KK-LL). The calculated value $W_x$(KK-LL) for Ne from [I5] also exceeds that obtained in [I8] using the many-body theory by 2.5 times. Hence, the incomplete inclusion of correlations in the model used in [I4,I5,20] results in overvalued KK-LL transition probability. The existing difference can be attributed to not including the contribution from some mechanisms in the shake-off model. As was shown in papers considering correlated transitions from the viewpoint of the many-body theory, the contributions of some mechanisms quite frequently can be comparable and can cancel.

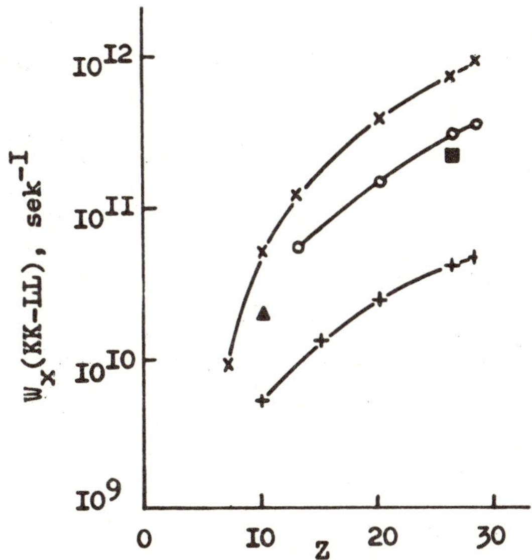


Fig.8. The dependence of single-photon decay probability of two K-vacancies on atomic number Z. Calculation: x - Gavrila and Hansen [I5]; ▲ - Amusia et al.[I8]; ■ - Kelly[I3]; + - Safronova and Senashenko [I9] (transition $2s^2 2p\ ^2P \to Is^2 2s\ ^2S$); Experiment: o - Stoller et al.[6].

As seen from Fig.7, the best agreement with experiment was obtained by Khristenko [I7] for the transitions $2s2p\ ^IP \to Is^2\ ^IS$ in helium-like ions. However, as it was mentioned by Gavrila and Hansen [20], in the initial state of heluim-like ions two multiplets - $2s2p\ ^IP$ and $2s2p\ ^3P$ can be populated, whereas only $^IP$ is able to decay into the state $Is^2\ ^IS$. Thus, when taking into account possible population of two initial multiplet states there will appear a discrepancy between calculations[17] and experiment.

Concluding the comparison one can note that for the LL-MM transitions studied experimentally by Afrosimov et al.[8], only the calculation made by Amusia et al.[18,21] is available. The experimental $(1.7 \div 4.3)\cdot 10^{-6}$ and calculated $(1.8\cdot 10^{-6})$ values of the relative probability $\beta = \sigma_x(LL\text{-}MM)/\sigma(LL)$ of single-photon decay of two $L_{2,3}$-vacancies in Ar agree well.

The relative role of the two-electron-one-photon transition and three-electron Auger transition in the decay of two inner vacancies in an atom can be characterized by the value of "fluorescence yield" $\omega_{xx}$ for the correlated transitions. Thus, in the case of decay of two L-vacancies $\omega_{xx}$ is determined by the expression $\omega_{xx} = \sigma_x(LL\text{-}MM)/[\sigma_x(LL\text{-}MM)+\sigma_A(LL\text{-}MMM)]$. As was shown in [8], $\omega_{xx}$ should be close to the value of fluorescence yield for conventional transitions $\omega_x$. Indeed, according to the shake-off model:

$$\omega_{xx} \simeq \frac{W_x(LL\text{-}MM)}{W_A(LL\text{-}MMM)} = \frac{W_x(L\text{-}M)\cdot W_s}{W_A(L\text{-}MM)\cdot W_s} \simeq \omega_x .$$

Fig.9 shows the values of $\omega_{xx}$ obtained in [8] for transitions to $L_{2,3}$-vacancies in the S, Cl and Ar atoms.

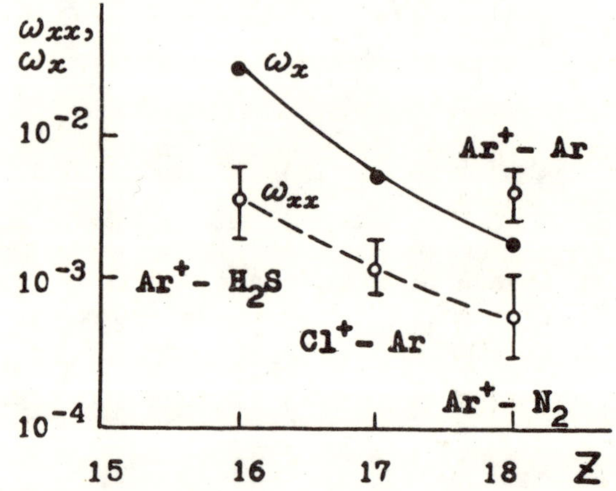


Fig.9. Fluorescence yield for correlated ( $\omega_{xx}$) and conventional ( $\omega_x$) transitions vs. atomic number Z. $\omega_{xx}$ - Afrosimov et al.[8]; $\omega_x$ - Fortner [31].

For comparison, there are also presented the values of $\omega_x$ for conventional transitions taken from [31]. As one can see, the values of $\omega_{xx}$ and $\omega_x$ are indeed close to one another. Estimates of $\omega_{xx}$ for the case of decay of two K-vacancies in

N [9] give the value $\sim 10^{-2}$ which is also close to $\omega_x$ for conventional transitions. With the above mentioned experimental estimate, $10^{-2}$, agrees the value $\omega_{xx}=0.023$, obtained by calculations of correlated transitions in the framework of the many-body theory [23].

7. Conclusion

1. Correlated radiative and Auger transitions in atoms with two inner-shell vacancies were studied both experimentally and theoretically. The probabilities of decay of states with two K or L-vacancies by three-electron Auger transition are about $10^{-4}$-$10^{-3}$ of the total decay probability and by the two-electron-one-photon transition, about $10^{-6}$-$10^{-5}$ of it.

2. The shake-off model and many-body theory of the atom are used for a theoretical description of correlated transitions in the decay of two inner vacancies. A good agreement is obtained between experiment and theory. The many-body theory provides a more complete description of correlated transitions.

3. The results of the study of correlated transitions allow to interpret correctly the spectra of Auger electrons and photons which is necessary for the analysis of composition of various materials. When interpreting a spectrum, one should take into account the possible contribution of correlated transitions. Otherwise, in the spectra of electrons or photons the lines corresponding to these transitions may be wrongly attributed to the presence of traces of impurities in the material under study.

References


1. V.V.Afrosimov, Yu.S.Gordeev, A.N.Zinoviev, D.H.Rasulov, A.P.Shergin. Abstr.IX ICPEAC, Seattle, 1975, v.2, p.1068; Pisma Zh.Eksp.Teor.Fiz., 21, 535, 1975.

2. W.Wölfli, Ch.Stoller, G.Bonani, M.Suter, M.Stöckli. Phys.Rev.Lett., 35, 656, 1975.

3. T.P.Hoogkamer, P.Woerlee, F.W.Saris, M.Gavrila. J.Phys.B, 9, L145, 1976.

4. V.V.Afrosimov. Proc. 2nd Int. Conf. on Inner Shell Ionization Phenomena, Invited Papers, Freiburg, 1976, p.258.

5. W.Wölfli, Ch.Stoller, G.Bonani, M.Stöckli, M.Suter. Proc. 2nd Int. Conf. on Inner Shell Ionization Phenomena,

Invited Papers, Freiburg, 1976, p.272.

6. Ch.Stoller, W.Wölfli, G.Bonani, M.Stöckli, M.Suter. Phys. Lett., 58A, 18, 1976.

7. A.R.Knudson, K.W.Hill, P.G.Burkhalter, D.J.Nagel. Phys.Rev.Lett., 37, 679, 1976.

8. V.V.Afrosimov, Yu.S.Gordeev, V.M.Dukelsky, A.N.Zinoviev, A.P.Shergin. Pisma Zh.Eksp.Teor.Fiz., 24, 273, 1976; Abstr. X ICPEAC, Paris, 1977, v.I, p.182.

9. V.V.Afrosimov, A.P.Ahmedov, Yu.S.Gordeev, A.N.Zinoviev, A.P.Shergin. Abstr. X ICPEAC, Paris, 1977, v.I, p.184; Pisma Zh.Techn.Fiz., 3, 971, 1977.

10. N.Stolterfoht, D.Schneider, D.Brandt. Abstr. X ICPEAC, Paris, 1977, v.2, p.902.

11. A.P.Shergin, Yu.S.Gordeev, Proc.X ICPEAC, Invited Lectures and Progress Reports, Paris, 1977, p.377.

12. T.P.Hoogkamer, P.H.Woerlee, F.W.Saris. J.Phys.B, 1978, (in press).

13. H.P.Kelly. Proc. 2nd Int. Conf. on Inner Shell Ionization Phenomena, Invited Papers, Freiburg, 1976, p.198.

14. T.Aberg, K.A.Jamison, P.Richard. Phys.Rev.Lett., 37, 63, 1976.

15. M.Gavrila, J.E.Hansen. Phys.Lett., 58A, 158, 1976.

16. L.M.Kishinevsky, V.I.Matveev, E.S.Parilis. Pisma Zh.Techn.Fiz., 2, 710, 1976.

17. S.V.Khristenko. Phys.Lett., 59A, 202, 1976.

18. M.Ya.Amusia, I.S.Lee, A.N.Zinoviev. Phys.Lett., 60A, 300, 1977.

19. U.I.Safronova, V.S.Senashenko. J.Phys.B, 10, L271, 1977; Abstr. X ICPEAC, Paris, 1977, v.I, p.194.

20. M.Gavrila, J.E.Hansen. Abstr. X ICPEAC, Paris, 1977, v.I, p.192; J.Phys.B, 11, 1353, 1978.

21. M.Ya.Amusia, I.S.Lee, Zh.Eksp.Teor.Fiz., 73, 430, 1977.

22. U.I.Safronova, V.S.Senashenko. Izv.Akad.Nauk SSSR (Ser.Fiz.), 41, 2610, 1977.

23. L.N.Ivanov, U.I.Safronova, V.S.Senashenko, D.S.Viktorov. J.Phys.B, 11, L175, 1978.

24. V.V.Afrosimov, Yu.S.Gordeev, A.N.Zinoviev, D.H.Rasulov, A.P.Shergin. Abstr. IX ICPEAC, Seattle, I975, v.2, p.I066.

25. B.Fastrup, G.Herman, K.J.Smith. Phys.Rev., A3, I59I, I97I.

26. V.V.Afrosimov, Yu.S.Gordeev, A.M.Polyansky, A.P.Shergin. Zh.Eksp.Teor.Fiz., 63, 799, I972.

27. E.P.Larkins. J.Phys.B, 4, I, I97I.

28. A.Savukynas, S.Šhadžiuviene, J.Glembockis. Lit.Fiz.Sbornik, I7, 3I5, I977.

29. K.Taulbjerg, J.S.Briggs, J.Vaaben, J.Phys.B, 9, I35I, I976,

30. J.H.Macek, J.S.Briggs. J.Phys.B, 6, 84I, I973.

3I. R.J.Fortner. Phys.Rev., AI0, 22I8, I974.

# OPTICAL PUMPING INVESTIGATION OF ATOMIC INTERACTIONS

R.A.Zhitnikov

A.F.Ioffe Physico-Technical Institute, USSR Academy of Sciences, Leningrad, I9402I, USSR

Optical orientation of atoms (optical pumping), discovered by A.Kastler in I949, is a transfer of the angular momentum to an ensemble of atoms as a result of their absorption by circularly polarized resonance radiation and subsequent transition to the ground state. This orientation of atomic momenta can be detected and investigated by means of measuring variation in absorption or scattering of resonance pumping light. Nonpolarized light can also be sometimes used for optical pumping with resulting alignment of atomic momenta /I/. Many important scientific and practical results have been obtained with the aid of optical pumping /I/. One application has been the study of a number of atomic interactions. Depolarizing collisions of alkali metal atoms in the ground and excited states with atoms of noble gases and molecules of hydrogen and nitrogen, Van der Waals and exchange interactions involved in these collisions as well as spin exchange in collisions between like and unlike paramagnetic atoms and between atoms and electrons have been investigated /I/.

This paper describes recent results obtained by means of optical pumping in atomic interactions involving helium atoms in the $2^3S_I$ metastable state. Ionization (Penning) interaction, metastability exchange and interaction of metastable helium atoms with hydrogen molecules have been investigated. This work has been based on optical orientation of atoms in gas discharge plasma where apart from various atoms and molecules in the ground state, a high concentration of metastable helium atoms can also be obtained. The discussion will be mainly concerned with the results obtained at the A.F.Ioffe Physico-Technical Institute, USSR Academy of Sciences, Leningrad.

## Penning interaction of atoms

In 1966, an investigation of optical orientation of $He^4$ atoms in the $2^3S_1$ state in a gas discharge helium plasma led to the discovery that space orientation of spin moments of these neutral atoms affected plasma electrical conductivity and helium emission intensity /2/. Later /3, 4/, this phenomenon was explained along the following lines. The excitation energy and ionization potential of a He atom in the $2^3S_1$ state are 19.8eV and 4.7eV, respectively. When two $2^3S_1$ atoms collide in plasma, one of them can be ionized at the expense of the excitation energy of the other; the ionization process can be described as follows:

$$He(2^3S_1) + He(2^3S_1) \longrightarrow He(1^1S_0) + He^+(1^2S_{1/2}) + e^- \quad (1)$$

If it is assumed, as in /5/, that the total spin is conserved in the reaction, the ionization probability appears to depend on the relative spin orientation of colliding atoms. For example, if the colliding He atoms in the $2^3S_1$ state have the same spin orientation and their total spin equal to 2 (measured in $\hbar$ units) is in excess of the total spin of reaction products, the maximum value of which cannot be more than 1, and if the total spin is assumed to be conserved, then the ionization process described by (1) will be forbidden, collisions will be elastic and the states of metastable atoms will remain unchanged. However, if the spin orientations of colliding metastable atoms are different, reaction (1) will be allowed.

An alternative explanation is also possible. When a He atom is in the $2^3S_1$ state, its electrons occupy different electron shells (1s2s configuration) and have their spins oriented in the same direction, thus forming the atom's spin equal to 1. The mechanism of the ionization process (1) is, then, that there is a transition of the 2s-electron of the one atom onto the 1s-shell of the other, either directly or through the 2s--shell of the latter atom. The excitation energy equal to 19.8eV is sufficient for the 2s-electron of the latter atom to escape. As a result, a helium atom in the ground state, a helium ion and a free electron are produced. Now, if the total spins of colliding $2^3S_I$ helium atoms are oriented in the same direction and, as a consequence, the spins of all four electrons are parallel, the ionization process (I) cannot occur because on Pauli exclusion principle two electrons with parallel spins never share the same s-shell. The condition assumed in this case is that the electron spin orientations of two $2^3S_I$ atoms undergo no changes during the collision. Thus it appears that the present consideration has led us a conclusion which is identical to that of a more general though somewhat phenomenological consideration based on the requirement that the total spin of atoms involved in ionization collisions should be conserved.

When He atoms in the $2^3S_I$ state are optically oriented in plasma, with atomic spin polarization or alignment produced depending on whether or not the light is circularly polarized, there is a significant growth in the portion of parallel spin orientation collisions and a corresponding reduction in the occurrence of ionization of type (I) and in the free electron yield leading to lowered electron density, electrical conductivity and glow in plasma. If optical orientation of metastable helium atoms is destroyed by magnetic resonance, a technique widely used in optical pumping, changes in both helium plasma conductivity and glow should occur. An observation of this effect in /2, 4/ has provided evidence of the conservation of total spin in the ionization process (I).

Spin polarization of free electrons in helium plasma with $2^3S_I$ helium atoms oriented by circularly polarized light has been observed and investigated in /5, 6/ and was attributed to the conservation of total spin in (I).

An experimental determination of the degree of total spin conservation in /7/ confirmed within I0% the existence of I00% conservation of total spin in (I). An essentially arbitrary assumption was made in /7/ that the probability of Penning ionization did not depend on whether the total spin of colliding particles was equal to 0 or I (for the total spin equal to 2 the ionization probability, provided total spin is conserved, is equal to 0). It was shown, however, in /8/ that an account taken of the difference in the ionization probability for zero and unity total spin could not essentially alter the quantitative estimation given in /7/, nor did it throw any doubt on the conservation of total spin in the reaction (I).

High frequency and probe measurements of electron density in helium plasma performed in /7/,/9/ confirmed the evidence obtained in /2/ of the decreasing effect of optical orientatation of $2^3S_I$ atoms on electron density.

The question of total spin conservation in Penning collisions between different atoms is of considerable interest. For example, total spin conservation should cause the relative spin orientation of colliding atoms to affect free electron yield not only in collisions between metastable helium atoms, but also in collisions between $2^3S_I$ helium atoms and alkali metal atoms. It was supposed in /I0/ that in the latter collisions the density of Penning electrons should change, provided the atoms of both substances are polarized by optical pumping.

Experimental evidence of this effect should provide new possibilities in the study of spin dependent Penning collisions; one of them would be an independent spin orientation of the two colliding atoms, which cannot be achieved in collisions between metastable helium atoms.

The aim of /II/ was to observe and investigate experimentally the dependence of Penning ionization probability on relative spin orientation in collisions between metastable He and alkali metal (rubidium) atoms. That this dependence can exist follows from a consideration of the process:

$$He(2^3S_I) + Rb(5^2S_{I/2}) \longrightarrow He(I^IS_o) + Rb^+(4^IS_o) + e^- \qquad (2)$$

When total spin is conserved this reaction is forbidden if the spins of colliding atoms are parallel because the total spin of the atoms will be 3/2, in excess of that of reaction products, equal to I/2. As a result, the collision is of elastic type and no ionization occurs. If the relative orientation of the colliding helium and rubidium atoms changes and their spins become antiparallel, the total spin of the colliding atoms will be I/2, thus making ionization reaction (2) allowed and consequently leading to an increased free electron yield detectable from a rise in plasma electron density.

Ref. II was devoted to the study of the presence and behaviour of electron density change in helium-rubidium plasma due to simultaneous optical orientation of $2^3S_I$ helium atoms and those of rubidium in the ground $5^2S_{I/2}$ state. The electron density change was detected from changed plasma conductivity.

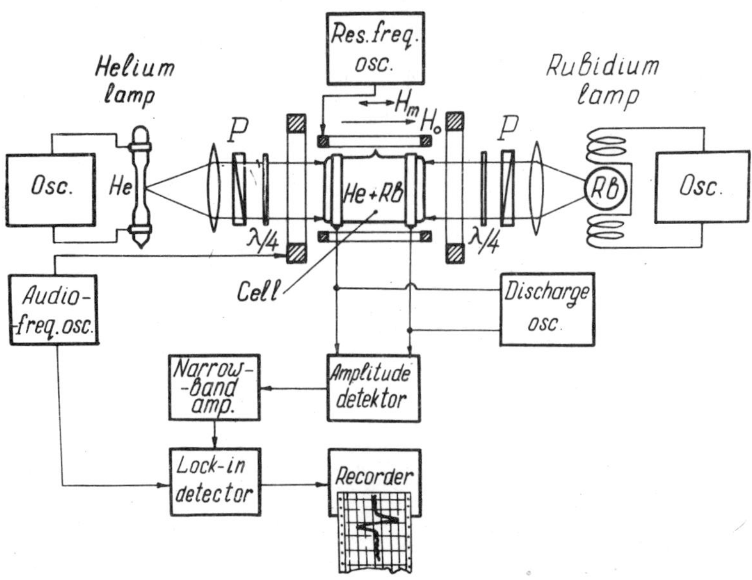


FIG. I. Diagram of experimental setup.

Fig.I shows diagrammatically a setup designed for observation of plasma conductivity change induced by optical orientation of helium and alkali metal atoms. Metastable helium atoms are produced by an electrode-less high frequency discharge excited in an absorption cell filled with rubidium vapour and helium. Two beams of circularly polarized light are incident on the cell from two opposite directions; one of them is emitted by a helium lamp and serves to orient the metastable helium atoms, while the other is produced by a rubidium lamp to orient the rubidium atoms. Depending on the sense of circular polarization of these beams, helium and rubidium atom spins can be oriented either parallel or anti-parallel to each other. The cell is placed in a constant magnetic field $H_o$ and an arrangement consisting of two coil systems and two rf oscillators is used to produce rf magnetic fields in the cell, which can excite magnetic resonance transitions either only in helium or rubidium atoms, or in both. (For simplicity, Fig.I shows only one coil system and one oscillator.)

To detect the variation in plasma conductivity caused by magnetic resonance, the magnetic field $H_o$ is modulated by an audio-frequency oscillator and the coil system. The modulation frequency component is separated by an amplitude detector from the voltage on the cell electrodes and is fed through a narrow band amplifier to a lock-in detector whose output voltage registered by an automatic recorder is found proportional to the derivative of plasma conductivity variation. Also, the constant field $H_0$ is slowly swept over the values satisfying the magnetic resonance conditions.

As compared to experiments on Penning collisions between helium atoms, this technique offers new possibilities. If optical orientation is performed in pure helium plasma, all colliding atoms are either oriented in the same direction or have their orientation destroyed. In the case of helium-alkali metal plasma the spins of collision partners, i.e. helium and alkali metal atoms, can be oriented independently of each other either in the same or opposite direction. The magnetic resonance destruction of their respective orientations can also be performed independently and one can detect the resulting variation in plasma conductivity. It is also possible to make magnetic resonance switch on or switch off the ionization process for parallel and anti-parallel original orientations, respectively, as well as to separate the ionization collisions between helium-helium and helium-alkali metal atoms.

Fig.2 shows the derivatives of plasma conductivity variation signals due to magnetic resonance for rubidium (Fig.2a) and for helium (Fig.2b) atoms. In Fig.2a which summarizes the results on resonance excitation of rubidium atoms alone, signal I corresponds to parallel orientation of rubidium and helium atoms, while signal 2 to their opposite orientation. It can be seen that though the plasma conductivity changes due to rubidium atom optical orientation destruction are equal in size in both cases, they are of opposite sign. This is evidence of an increased number of ionization collisions between helium and rubidium atoms in the first case and a decreased number of the same collisions in the second case. Fig.2b demonstrates the results obtained for resonance excitation of helium atoms alone. Here signal I corresponds to parallel orientation of helium and rubidium atoms, while signal 3 to opposite orientation; signal 2 has been obtained in the absence of rubidium lamp emission, that is, in the absence of rubidium atom orientation. Signal 2 is exclusively due to ionization

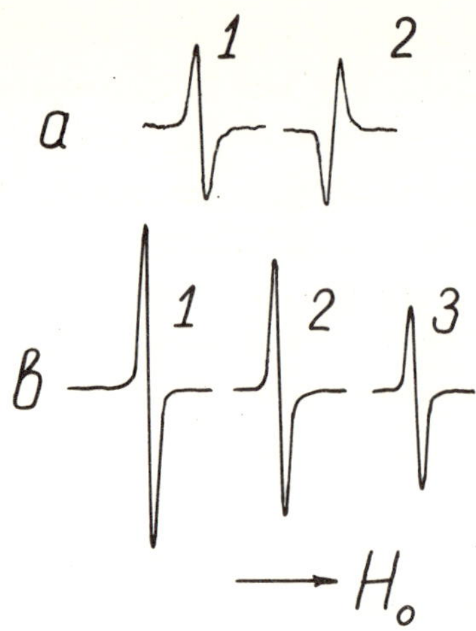


FIG.2. Signals representing the variation in the electrical conductivity of plasma: (a) magnetic resonance in the ground $5^2S_{1/2}$ state of $Rb^{85}$ atoms ($f_o$ = 290 kHz); (b) Magnetic resonance in the metastable $2^3S_1$ state of $He^4$ ($f_o$ = 1760 kHz).

collisions between metastable helium atoms because non-oriented rubidium atoms do not contribute to it. Signal I includes the effect of ionization collisions between rubidium and helium atoms added to signal 2, while in signal 3 the same effect is subtracted from signal 2.

Experiments on simultaneous excitation of magnetic resonance in optically oriented helium and alkali metal atoms open up a new and interesting possibility. The fact is that under magnetic resonance the total moment of optically oriented atoms, being now somewhat smaller, turns normal to the constant magnetic field $H_o$ and starts precessing at the Larmor frequency. The time averaged orientation of atoms which becomes zero was described above as vanishing or "destroyed". But it does not actually disappear despite its time-averaged value being equal to zero. This raises the question if the precessing transverse component of atomic momenta orientation has an effect on plasma conductivity. The only orientation component revealed in the experiments described above has been the one constant in time and parallel to the field $H_o$. Optical orientation of helium and alkali metal atoms permits experiments to observe the effect of the transverse precessing orientation component on plasma conductivity. If magnetic resonance is simultaneously excited in optically oriented helium and rubidium atoms, their total moment transverse components precessing at different frequencies will alternate their movement to and away of each other, thus changing their relative orientation from parallel to anti-parallel and visa versa. Parallel and anti-parallel orientations of these components will recur at the frequency equal to the difference frequency of precession of both atoms. If the electron yield due to ionization collisions of these atoms is to vary with the angle between the transverse components of their orientations, plasma conductivity must oscillate at a frequency equal to the difference frequency of magnetic resonance of helium and rubidium atoms.

To observe this effect simultaneous magnetic resonance for rubidium and helium atoms was excited in the setup shown in Fig.I. Neither the field $H_o$ was modulated, nor the audio-frequency oscillator or synchronous detector used. The narrow-band amplifier was tuned to a frequency equal to the difference frequency of magnetic resonance for helium and rubidium atoms, and directly connected to the cell electrodes. The amplifier output was fed via an amplitude detector to an automatic recorder.

A signal obtained as described above is presented in Fig.3. It shows the variation in plasma conductivity at the frequency equal to the difference frequency of magnetic resonance of helium and rubidium atoms. The magnetic field $H_o$ is slowly growing over the value $H_p$ indicating to the presence of magnetic resonance for both helium and rubidium atoms. Ample evidence

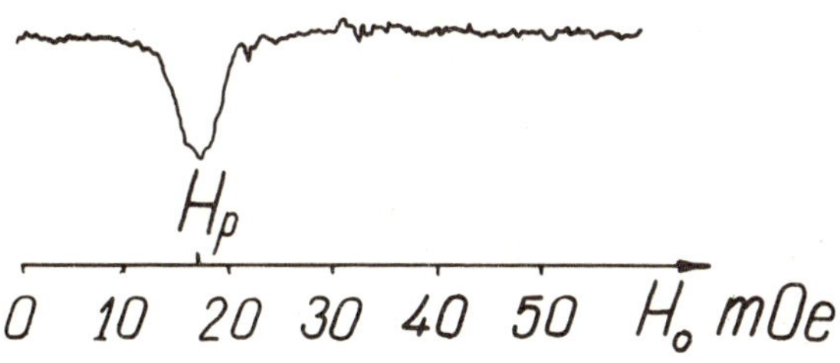


FIG. 3. Signal showing the variation in electrical conductivity of plasma for the difference frequency $f = f_o(He^*) - f_o(Rb^{85})$ in the case of simultaneous magnetic resonance of the $Rb^{85}$ and $He^*$ atoms . $f_o(He^*) = 48.4$ kHz, $f_o(Rb^{85}) = 8.I$ kHz . $He^*$ is the $He(2^3S_I)$.

is thus provided that the transverse component does change plasma conductivity. It may be safely assumed that this should also be true for a pure helium plasma and could be observed, though with less ease, for example, when optically oriented $2^3S_I$ helium atoms are under magnetic resonance.

It follows that our experimental results provide indication of the plasma conductivity dependence on the orientation of both longitudinal and transverse components of spin moments of helium and rubidium atoms. This means that Penning ionization probability should also depend on relative orientation of the total spin moments of colliding atoms, and that the sum of both atoms total moments should be conserved in Penning reactions. This would seem a natural conclusion if one considers that under the present experimental conditions the period of Larmor precession of atomic spins in an external magnetic field by far exceeds the time of a Penning collision, being about $10^{-6}$ seconds as compared to $\sim 10^{-12}$ seconds for the latter. Since the Penning ionization cross section $\sigma$ is a value averaged over all possible collisions between rubidium and metastable helium atoms, it has to be expressed in terms of macroscopic spin moment values averaged over the ensembles of atoms involved:

$$\sigma = \sigma ( \langle \vec{S}_{He} \rangle , \langle \vec{S}_{Rb} \rangle ) \qquad (3)$$

The experimental signals obtained for electron density variation can be regarded as proportional to variation in $\sigma$; consequently, they would be indicative of the nature of the relationship between $\sigma$ and the macroscopic spin moments $\langle \vec{S}_{He} \rangle$ and $\langle \vec{S}_{Rb} \rangle$ /II/.

The features of the dependence of the observed electron density variation signals on macroscopic moments $\langle \vec{S}_{He} \rangle$ and $\langle \vec{S}_{Rb} \rangle$ are as follows:

I. Signals change sign with reversion of $\langle \vec{S}_{He} \rangle$ or $\langle \vec{S}_{Rb} \rangle$ ; 2. Signals depend on the modulus $\langle \vec{S}_{He} \rangle$ and $\langle \vec{S}_{Rb} \rangle$ ; 3. Change of the angle between $\langle \vec{S}_{He} \rangle$ and $\langle \vec{S}_{Rb} \rangle$ at the rate $\omega$ causes the appearance of signals at the frequency $\omega$ . Since in (3) the scalar value $\sigma$ is a function of the two vectors $\langle \vec{S}_{He} \rangle$ and $\langle \vec{S}_{Rb} \rangle$ , the value $\sigma$ can be supposed to show linear dependence of the scalar product of these vectors:

$$\sigma = A + B \langle \overrightarrow{S_{He}} \rangle \cdot \langle \overrightarrow{S_{Rb}} \rangle , \qquad (4)$$

where A and B are independent of spin moments.

The experimental evidence available at present is insufficient for a quantitative verification of equation (4). It should be noted however, that this equation is in good agreement with all the above features of electron density variation signals obtained experimentally. On the assumption of total spin conservation, an equation of the form (4) has been derived theoretically in /8/ for the cross section of Penning reaction involving $2^3S_I$ metastable He and Rb atoms.

## Metastability exchange

When one of the two atoms of the same element is in the ground state while the other is in the metastable state, their collision may be followed by an exchange of states, that is, a metastability exchange occurs. The cross section $\sigma$ of this process is very great and for He its value reaches $8 \cdot 10^{-16} cm^2$ at $300^oK$ /I/. The process is very important for helium whose discharge metastable triplet $2^3S_I$ ortostate has the life time of about $10^{-3}$sec and is usually densely populated ( $10^{10}$ ÷ $10^{11}cm^{-3}$). Knowledge of the value $\sigma$ is highly desirable because it would yield information on the interaction potential V(r) of the two He atoms, one being in the ground $I^1S_o$ state and the other in the $2^3S_I$ metastable state.

However, on the principle of indistinguishability, $He^4$ resists all attempts at direct observation of this process. This becomes possible in the case of $He^3$ isotope whose non-zero nuclear spin and nuclear magnetic moment permit employment of optical orientation and magnetic resonance of $2^3S_I$ helium atoms to study the metastability exchange and to measure its cross section. The fact is that the atoms resulting from the metastability exchange between an optically oriented $2^3S_I$ metastable $He^3$ atom and a $He^3$ atom in the ground state are essentially different from the original ones: the metastable atom may now have no nuclear spin orientation, while the one in the ground state will. Two consequences follow: I. Atoms in the ground state have their nuclear moments oriented, with the values of nuclear polarization reaching as high as 20 to 60% corresponding to equilibrium magnitization in the field equal to $10^9$ Oersted. This effect is of wide use in many current applications. 2. The line width of magnetic resonance in the metastable state is much greater for $He^3$ than for $He^4$ and for $He^3$ is almost totally due to metastability exchange.

Prior to /12/ the metastability exchange cross section was thought to be independent of whether the colliding atoms were of the same isotope, say, $He^3$ or $He^4$, or of different isotopes /13/ because nuclear moment was not associated with metastability exchange. Indeed, as the hyperfine interaction is small the collision time is too short for the nuclear moment to change its orientation. The experiments carried out in /13/ confirmed the equality of the cross sections $\sigma$ ($He^3$ - $He^3$) and $\sigma$ ($He^3$ - $He^4$).

Optical orientation and magnetic resonance of $2^3S_1$ metastable helium atoms in a mixture of $He^3$ and $He^4$ isotopes have been studied in /13/. The measurements were performed at room temperature. The results gave evidence that within 10% of experimental error $\sigma$ ($He^3$ - $He^3$) = $\sigma$ ($He^3$ - $He^4$). The suggestion made in this work was that the cross sections under consideration should indeed be equal because the nuclear moments of the two colliding atoms did not change their orientation and had no effect on the result of the collision as $\tau \Delta W \ll \frac{h}{2\pi}$ where $\tau$ is the collision time, $\Delta W$ is the hyperfine structure splitting and $h$ is Plank's constant.

It has been pointed out in /12/, however, that, firstly, the supposition of the necessary equality of cross sections for collisions between atoms of the same and different isotopes is not evident as the energy of the metastable state shows isotopic shift and the metastability exchange for collisions of atoms of different isotopes may not be completely resonant. Secondly, the experimental evidence of the equality of these cross sections has been obtained for room temperature only. At lower temperatures the part played by isotopic shift in metastability exchange should grow in importance. We have, therefore, undertaken an investigation of optical orientation and magnetic resonance of metastable helium atoms in a mixture of $He^3$ and $He^4$ isotopes at liquid nitrogen temperature. This experiment provided the first evidence of inequality between the cross sections $\sigma(He^3 - He^3)$ and $\sigma(He^3 - He^4)$ /I2/. The qualitative explanation given in this work related the observed effect to the presence of the resonance defect $\Delta E$ occurring in metastability exchange between atoms of different isotopes and, consequently, to the non-resonance character of this process. Here $\Delta E = 6cm^{-I}$ is the difference excitation energy for $2^3S_I$ state of $He^3$ and $He^4$ atoms (see Fig.4). This resonance defect is considerable at low temperatures because in this case $\tau_{exc} \cdot \Delta E \sim \frac{h}{2}$ , where $\tau_{exc}$ is the metastability exchange time growing quite large as thermal velocities of atoms decrease.

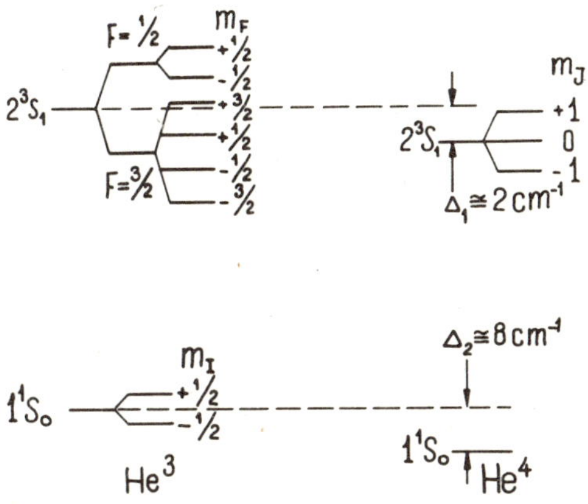

FIG. 4. Scheme of energy levels of the $He^3$ and $He^4$ atoms.

In /I2/ the cross sections $\sigma_I = \sigma(He^3 - He^3)$ and $\sigma_2 = \sigma(He^3 - He^4)$ measured at 77°K were deduced from the line width of magnetic resonance for $2^3S_I$ state of optically oriented atoms in a mixture of $He^3 - He^4$ isotopes. It was found that at the liquid nitrogen temperature these cross sections were significantly different ( $\sigma_I / \sigma_2 = 0.6$).

Owing to the presence of a potential barrier in the formation of a $He(I^IS_0) \cdot He(2^3S_I)$ molecule which is intermediate in metastability exchange /I5/, the cross section $\sigma$ of this process as measured for $He^3$ isotope was found to be strongly dependent of the gas temperature /I6/,/I7/. It was of considerable interest, therefore, to determine the cross sections of metastability exchange between atoms of different isotopes for various temperatures.

The aim of /I4/ was to investigate the process of metastability exchange between atoms in a mixture of $He^3$ and $He^4$ isotopes at various temperatures ranging from $77^o$ to $300^oK$, to determine the dependence of cross sections $\sigma_I$ and $\sigma_2$ on temperature, and also to study the nature of this dependence. The experiment was based on the well-known technique of optical orientation and magnetic resonance of metastable helium atoms /I/,/I6/. A weak high-frequency gas discharge excited in the absorption cell filled with a mixture of $He^3$ and $He^4$ atoms caused transition of a portion of these atoms into metastable state, the density of helium atoms in the ground $I^IS_o$ state being of the order of $IO^{I6}cm^{-3}$ while in the $2^3S_I$ state about $IO^{IO}$ - $IO^{II}cm^{-3}$. The helium atoms in the $2^3S_I$ state were optically oriented by a resonance line of $\lambda$ = IO830 Å ($2^3S_I \longrightarrow 2^3P$ transition), for which purpose the cell was irradiated by circularly polarized light from a helium lamp. Optically oriented $2^3S_I$ helium atoms were in a weak magnetic field and by means of a rf field magnetic resonance was excited in them and detected through observation of variation in the pumping light absorption. The metastability exchange cross sections $\sigma_I = \sigma(He^3 - He^3)$ and $\sigma_2 = \sigma(He^3 - He^4)$ were determined from the dependence of a magnetic resonance line width of optically oriented $He^3$ and $He^4$ atoms in the $2^3S_I$ state on the percentage of $He^3$ in the mixture of $He^3$ - $He^4$ isotopes at constant total gas pressure. The contribution of metastability exchange to magnetic resonance line width was due to collisions of the following three types:

$$He^{3*} + He^3 \longrightarrow He^3 + He^{3*},$$
$$He^{3*} + He^4 \longrightarrow He^3 + He^{4*},$$
$$He^{4*} + He^3 \longrightarrow He^4 + He^{3*}$$

(the atoms in the $2^3S_I$ state are marked with asterisk).

An examination of the above processes shows /I4/ that their contribution to magnetic resonance line width is given by the expressions:

$$(\Delta\omega)_{3/2} = \frac{4}{9}\alpha N \bar{v}_I \sigma_I + (I - \alpha) N \bar{v}_2 \sigma_2 \qquad (5)$$

$$(\Delta\omega)_{I/2} = \frac{7}{9}\alpha N \bar{v}_I \sigma_I + (I - \alpha) N \bar{v}_2 \sigma_2 \qquad (6)$$

$$(\Delta\omega)_{He^4} = \alpha N \bar{v}_2 \sigma_2 \qquad (7)$$

where $(\Delta\omega)_{3/2}$ and $(\Delta\omega)_{I/2}$ are magnetic resonance line half widths for $He^3$ in the metastable $2^3S_I$ state with F = 3/2 and F = I/2, respectively; $(\Delta\omega)_{He^4}$ is magnetic resonance half width for $He^4$ in the $2^3S_I$ state, N is helium atom density in the absorption cell, $\alpha = \frac{N_{He^3}}{N}$ is the fraction of $He^3$ atoms in the helium isotope mixture, $\bar{v}_I$ and $\bar{v}_2$ are mean relative velocities of $He^3 - He^3$ and $He^3 - He^4$ atoms, respectively; $\sigma_I = \sigma(He^3 - He^3)$ and $\sigma_2 = \sigma(He^3 - He^4)$. Coefficients 4/9 and 7/9 allow for the incomplete destruction of orientations in F = 3/2 and F = I/2 states for collisions between $He^3$ atoms /I3/,/I8/. Expressions (5) - (7) were used in our calculations of metastability exchange cross sections.

We used absorption cells filled with a mixture of $He^3$ and $He^4$ isotopes of varying relative proportion but at constant total pressure of 0.4 Torr at 300°K. Partial pressure values for $He^3$ were 0.05, 0.I, 0.2, 0.3 and 0.35 Torr. The measurement error of the pressure value was about 8% total pressure.

The recording of the magnetic resonance signal derivatives was made at various temperatures in the range from 77° to 300°K. For this purpose the absorption cell was placed in a Dewar vessel through which nitrogen vapour was passed. Temperature was controled by changing the flux of nitrogen vapour passing in the vessel and was measured by a thermo-couple. Temperature stability for a run was 3°K.

Fig.5 shows the dependence of magnetic resonance line widths on $He^3$ concentration for T=I25°K and T=300°K. Data shown in Fig.5 were corrected for rf broadening, modulation effects, inhomogeneity of the constant magnetic field $H_0$ and the effects of discharge and diffusion of metastable atoms towards the absorption cell walls.

Shown in Fig.6a are the values of $\sigma_I$ and $\sigma_2$ as determined for six temperature values within 77 - 300°K range, the error being within 8%. It can be seen that as temperature increases from 77°K to 300°K, the absolute values of $\sigma_I$ and $\sigma_2$ also show monotonous increase. Two regions of faster increase in these values can be noticed, namely those of 77-I20° and 220-300°K. At T=300°K the cross sections $\sigma_I$ and $\sigma_2$ can be regarded as equal within the experimental error.

Variation in relative difference $\Delta\sigma/\sigma_I = (\sigma_I - \sigma_2)/\sigma_I$ with temperature is shown in Fig.6b. It can be seen here

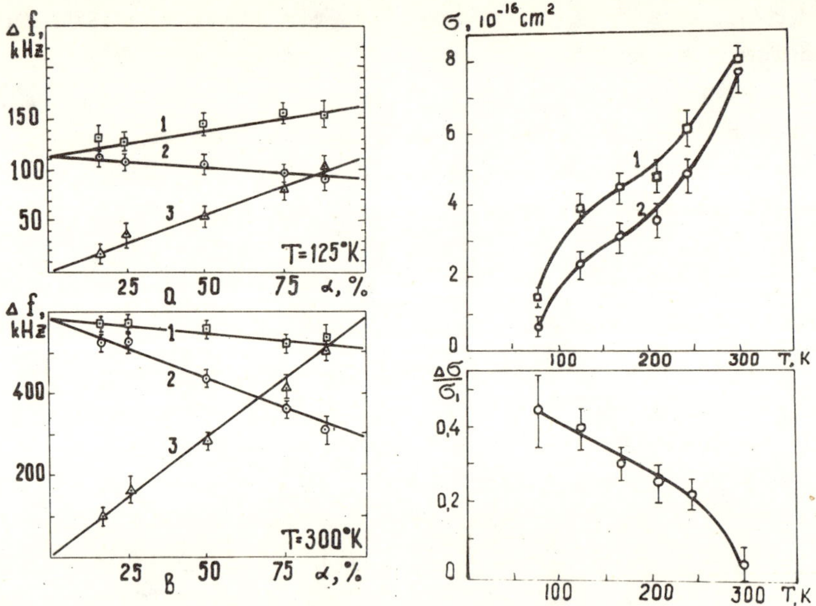


FIG. 5. Dependence of magnetic resonance line widths of optically oriented helium atoms in a mixture of $He^3$ - $He^4$ isotopes on percentage content $\alpha$ of $He^3$ at a total pressure of 0.4 Torr; I - resonance line widths for F = I/2, $2^3S_I$ state of $He^3$; 2 - resonance line widths of the $2^3S_I$ state of $He^3$ with F = 3/2; 3 - resonance line widths of the $2^3S_I$ state of $He^4$: a) T = I25°K, b) T = 300°K.

FIG. 6. Temperature dependence of the cross sections of metastability exchange: a) cross sections: I) $\sigma_I = \sigma(He^3 - He^3)$ 2) $\sigma_2 = \sigma(He^3 - He^4)$; b) relative difference of the cross sections $\Delta\sigma/\sigma_I = (\sigma_I - \sigma_2)/\sigma_I$.

that the value $\Delta\sigma/\sigma_I$ changes much slower in the temperature range of I00° to 200°K than in the range of 200° to 300°K.

Thus, the metastability exchange cross sections $\sigma_I$ and $\sigma_2$ differ significantly throughout the entire range of temperatures studied, with the exception of a small region near T = 300°K. This experimental fact can be explained by considering the interaction of two helium atoms, one of which is in the $I^IS_0$ ground state and the other in the $2^3S_I$ metastable

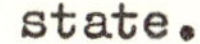
state.

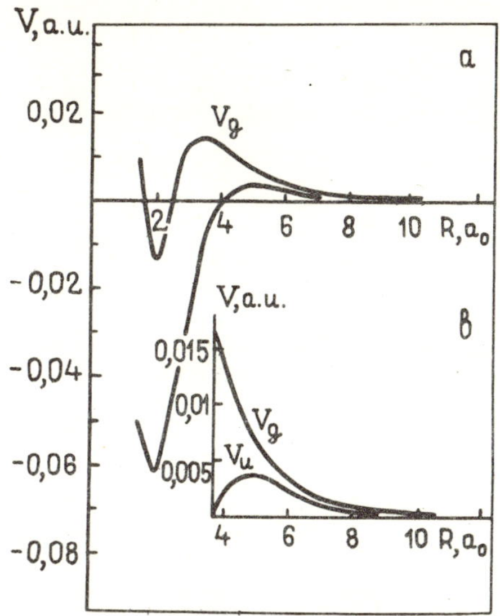


FIG. 7. Potential energies $V_g$ and $V_u$ of the symmetric and antisymmetric states of $He(I^1S_0) \cdot He(2^3S_I)$ (from the data of Ref. I9); a) with the use of I2 basis functions; b) with the use of 20 basis functions.

This interaction is described by potentials $V_g$ and $V_u$ which correspond to symmetric and antisymmetric states of a helium molecule $He_2^*$ $He(I^1S_0) \cdot He(2^3S_I)$ /I5/,/I6/. Fig.7 shows the potentials $V_g$ and $V_u$ obtained in Ref. I9. As can be seen, these potentials differ appreciably at $r < I0\ a_0$ where $a_0$ is the Bohr's radius. The metastability exchange cross section is determined by the difference value of these potentials averaged over the effective time of helium atom interaction and multiplied by the interaction time, i.e. by the time over which the two atoms are in the region of noticeable difference between $V_g$ and $V_u$ potentials. As temperature rises, the region of effective interaction and, hence, the mean value of potential difference $\Delta V = V_g - V_u$ also significantly rises (see Fig.7b). This leads to a sharp growth in interaction cross section values $\sigma_I$ and $\sigma_2$ within the temperature range 77° - - I20°K. However, when the two colliding atoms enter the region of a faster growing potential difference ($r < 8a_0$), the effective interaction time begins to fall leading to a slower rate of increase in the metastability exchange cross section for the temperature range between I20° and 220°K. Since the growth rate of potential $V_u$ decreases upon further increase in temperature ($T > 220°K$), the increase of the potential difference $\Delta V$ is so significant that a much faster growth rate for cross sections $\sigma_I$ and $\sigma_2$ is observed. As the potential barrier value for $V_u$ is about 0.IeV /I9/, then at $T \leqslant 300^oK$ only a very small fraction of helium atoms would have kinetic energy high enough for them to penetrate the barrier and form a stable $He_2^*$ molecule. Therefore, within this temperature range the dependence $\sigma(T)$ is far from saturation, a possibility of which for high temperatures was pointed out in /I6/.

Of the cross section values $\sigma_I$ measured so far the reliable ones seem to have been given in Ref.I7 for I5 - II5$^o$K temperature range and in Ref.I3 for 300$^o$K. The values of $\sigma_I$ reported in Ref.I7 for the 77 - II5$^o$K range are slightly lower than the present values, but the measurements in this temperature range involve large experimental error, and so the agreement between the results can be considered quite satisfactory. For example, $\sigma_I = (2.37 \pm 0.6I) \cdot 10^{-16}$ for $T = 113.8^oK$ given in /I7/ closely fits the experimental curve I of Fig.7a of the present work. The value of $\sigma_I = (7.6 \pm 0.4) \cdot 10^{-16} cm^2$ cited in Ref.I3 matches our value of $\sigma_I$ given above.

For resonance transfer of excitation energy in metastability exchange between helium atoms, the cross section value $\sigma_2 = \sigma(He^3 - He^4)$ would have to be slightly higher for the same temperature than $\sigma_I = \sigma(He^3 - He^3)$ because the former system, having a larger reduced mass, spends more time in the region of effective interaction. However, the difference in the excitation energy for the $2^3S_I$ state of $He^3$ and $He^4$ isotopes (see Fig.4), which is mainly due to the difference in mass of $He^3$ and $He^4$ atoms, results in an essentially nonresonant character of metastability exchange between atoms of different helium isotopes. As a consequence of this, the cross section $\sigma_2 = \sigma(He^3 - He^4)$ comes to be smaller than the cross section $\sigma_I = \sigma(He^3 - He^3)$.

The value of relative difference in cross sections $\sigma_I$ and $\sigma_2$ is to a great extent determined by the time which the system of two, $He^3$ and $He^4$, atoms spend in the region of effective interaction. Since the increase of temperature from 77$^o$K to 300$^o$K, as has been estimated from the results of /I9/, halves this time, the values of $\sigma_I$ and $\sigma_2$ must be close to each other near the room temperature. This is in agreement with the behaviour of the experimental curves shown in Fig.6 (a and b) as well as with results obtained in Ref.I3 which indicate to the coincidence of $\sigma_I$ and $\sigma_2$ values at $T = 300^oK$

within I0% of experimental error.

There is a considerable number of papers dealing with the calculation of the interaction potentials $V_g$ and $V_u$ or their determination from various experiments such as elastic scattering, diffusion, optical orientation of atoms (cf. Refs I5, I9 - 23). For the long range interaction region ($r > 7a_0$), the potentials obtained in various investigations are in remarkable disagreement. Knowledge of temperature dependences for $\sigma_I$ and $\sigma_2$ can help one choose satisfactory potentials. The question arises, however, whether an unambiguous choice is possible. The following remarks could be made in this connection.

As noted above, the value $\sigma_I$ is determined by the integral $\int \Delta V dt$ over the region of effective interaction of atoms. In the first approximation, $\Delta V \sim \exp(-\beta r)$ and the interaction region is determined by the behaviour of the potential curves $V_g$ and $V_u$ when $r > r_{min}$, where $r_{min}$ is the distance of closest approach of two helium atoms, determined by their kinetic energy and collision parameter. It can be concluded, therefore, that the interaction potentials $V_g$ and $V_u$ whether they have steep /2I/,/22/ or smooth /I9/,/23/ falloff, should result in $\sigma_I$ values that would match, provided the divergence between $V_g$ and $V_u$ for steep potentials is larger than that for smooth ones. The value proposed in /22/ is $\beta = 2.23$, while the results presented in /I9/ could be approximated by $\beta = I.I$. Our preliminary estimation shows that while the potentials are so different in the long range interaction region, they are in satisfactory agreement with the temperature dependence $\sigma_I$ shown in Fig.6a (curve I). This means that restriction to cross sections $\sigma_I$ would not allow us to choose the potentials $V_g$ and $V_u$ having a suitable slope in the region where $r > 7a_0$.

It should be noted, however, that a correct choice of inter action potentials becomes possible if the $\sigma_2$ temperature dependence is included in the analysis because the value of $\sigma_2$ is very sensitive to the time spent by the two helium atom system in the effective interaction region, i.e. to the steepness of the potential curves $V_g$ and $V_u$ in the region of their falloff. Needless to say, the choice of a potential that agrees best with $\sigma_2$ temperature dependence can be made only after the problem of nonresonant excitation transfer in the $He^3$-

$-He^4$ system is solved, which should be the subject of a separate theoretical investigation. At present we can only say that the smoother potentials such as reported in /I9/,/20/, /23/ might perhaps show better agreement with the experimental curve for $\sigma_2$ presented in Fig.6a.

While there are theoretical calculations /22/,/23/ relating the cross section $\sigma_I = \sigma(He^3 - He^3)$ and its temperature dependence to potentials $V_g$ and $V_u$, there are none as yet for the interaction between $He^3$ and $He^4$ and the cross section $\sigma_2 = \sigma(He^3 - He^4)$. When these are available, it should be possible to determine more accurately the potentials $V_g$ and $V_u$ of interactions between $He(I^IS_o)$ and $He(2^3S_I)$ atoms from the $\sigma_I$ and $\sigma_2$ values and their temperature dependences as obtained in the present work.

The excitation of magnetic resonance results in the coherent precession of spins in a constant magnetic field $H_o$.

In the experiments on optical orientation of atoms, coherence circulation between the ground and excited states leads to a well-known light shift in the frequency of magnetic resonance for the ground state due to the existence of real optical transitions /I/. The value of this shift is relatively small, much smaller than the magnetic resonance line width because the usual light intensity is not so great and the excitation time for an atim is short compared to the time when it is in the ground state.

Metastability exchange in helium provides a unique situation when real transitions produce considerable magnetic resonance frequency shifts which are easy to measure and investigate. In the presence of metastability exchange an atom is excited from its ground state to a metastable one and then in time $\tau$ comes back again to its ground state. For a $He^3$ isotope these transitions result in coherence transfer from the ground to the metastable state and back. Since the rate of precession in the metastable state is considerably higher (the giromagnetic ratio difference between the $2^3S_I$ and $I^IS_o$ states for $He^3$ atoms being about three orders of magnitude), the increase in resonance frequency in the ground state becomes comparable with the resonance line width. Nuclear magnetic resonance frequency shifts in the ground state of $He^3$ atoms due to metastability exchange have been observed and investigated in great detail both experimentally and theoretically in /I/,/25/.

Metastability exchange collisions should also involve coherence transfer of metastable state from one atom to the other. Under optical orientation and magnetic resonance of the $2^3S_I$ $He^3$ and $He^4$ atoms in a mixture of these isotopes, a shift of resonance frequency of metastable atoms should arise due to metastability exchange between atoms of different isotopes. The transfer of excitation energy in a mixture of isotopes proceeds somewhat differently as compared to a single-isotope system. Thus, in isotope mixtures the transfer partly occurs via atoms of the second isotope. The difference in the giromagnetic ratio values of $He^3$ and $He^4$ in the $2^3S_I$ state and, consequently, the difference in the precession frequency for different isotopes in this state should cause the magnetic resonance frequency shifts for metastable atoms as a result of coherence conservation in metastability exchange. It is of interest, therefore, to know the relationship between the values of resonance frequency shifts and metastability exchange cross sections for the atoms of the same ( $\sigma_I$) and different ( $\sigma_2$) helium isotopes.

The effect of coherence transfer in metastability exchange on magnetic resonance frequencies of $He^3$ and $He^4$ metastable $2^3S_I$ atoms in a mixture of $He^3$ - $He^4$ isotopes was studied in /26/. The values of resonance frequencies for various sublevels of the $2^3S_I$ state of $He^3$ and $He^4$ atoms were derived from the equations for the evolution of transverse orientations of various sublevels for metastable state. Simple analytical expressions for resonance frequency shifts in $He^3$ sublevels with $F = I/2$ and $F = 3/2$ and in $He^4$ metastable state can be derived if the condition $\omega_m \tau \gg 1$ is fulfilled, where $\omega_m$ is the resonance frequency of $He^4$ $2^3S_I$ state, $I/\tau$ is the probability of the metastable exchange.

The values of frequency shifts grow as temperature increases, giving a dependence of the form $\delta\omega \sim T^{7/2}$, or as atomic density in the absorption cell grows or as the constant magnetic field value decreases. The magnitudes of shifts are square functions of the resonance metastability exchange cross section $\sigma_I = \sigma\,(He^3 - He^3)$, while the relationship between various shifts is determined by the $He^3$ partial pre sure and the cross section ratio $\sigma_I/\sigma_2$, where $\sigma_2$ is the cross section of nonresonant metastability exchange between $He^3$ and $He^4$ atoms. Resonance frequency shifts should be large reaching several tens of kHz and be comparable to resonance line widths. This means that by measuring these shifts one can obtain quite precise values of cross sections $\sigma_I$ and $\sigma_2$.

The experimental determination of magnetic resonance frequency shifts for metastable $He^3$ and $He^4$ atoms optically oriented in a mixture of $He^3$ - $He^4$ isotopes, as a function of the fraction of $He^3$ isotope ( $\alpha$ ) at constant total pressure of the mixture, was performed following the above procedure, but the magnetic resonance frequency rather than the resonance line width was measured.

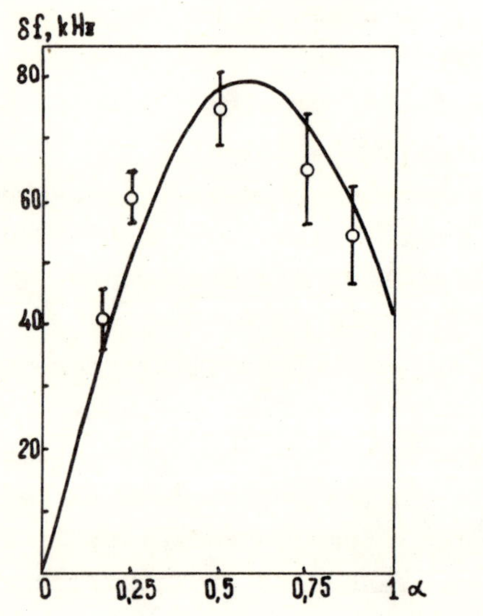


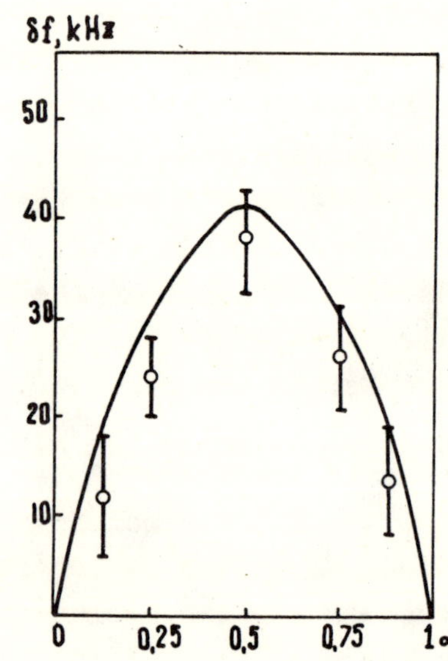


FIG. 8. Dependence of frequency shift of helium metastable atoms in a $He^3$ - $He^4$ mixture at T = 300°K on partial pressure of $He^3$ in a field $H_0$ = I.I78 Oe for $He^3$ sublevel F = 3/2; the solid curve represents a theoretical calculation.

FIG. 9. The same as the previous figure but for $He^4$ in the $2^3S_I$ state.

Figs 8 and 9 show the experimental values of magnetic resonance frequency shifts for F = 3/2 sublevel of $He^3$ atoms in the $2^3S_I$ state and for $2^3S_I$ $He^4$ atoms at different partial pressures of $He^3$ in the $He^3$ - $He^4$ mixture. Here the solid curves show the theoretical dependence derived from equations given in /26/ for $\sigma_I = \sigma_2 = 7.6 \cdot 10^{-16} cm^2$ and pressure of 0.4 Torr (T = 300°K). It can be seen that the agreement between theory and experiment is very good.

Measuring resonance frequency shifts can also permit determination of metastability exchange cross sections $\sigma_I$ and $\sigma_2$ from experimental data on optical orientation and magnetic resonance of helium $2^3S_I$ atoms. As a rule, these values are derived from magnetic resonance line widths, as above. However, at low temperatures helium metastability exchange cross sections exhibit a sharp decrease and magnetic resonance line width is determined not only by metastability exchange but also, to a large extent, by relaxation on the absorption cell walls, inhomogeneity of constant and rf magnetic fields, various processes occurring in gas discharge and so on. It appears that information obtained at low temperatures on $\sigma_I$ and $\sigma_2$ from magnetic resonance frequency shifts is very valuable, and sometimes about the only reliable one. Then, to obtain considerable frequency shifts, it is necessary that magnetic fields be low and helium atom densities in the absorption cell be as high as possible ($N \sim 10^{17} cm^{-3}$).

## Interaction between $2^3S_I$ metastable He atoms and hydrogen molecules

Interaction between molecular hydrogen and helium atoms has received a great deal of attention in recent time. Since $He+H_2$ system contains only a few electrons, it permits accurate calculation using various methods. The calculations made for helium atoms in the metastable state $He^*$ have been reported in /27/,/28/. Because of a high excitation energy of $He^*$ atoms reaching 19.8eV and 20.6eV for $2^3S_I$ and $2^1S_0$ states, respectively, the quazi-molecule $He^* - H_2$ formed during the collision is unstable to autoionization. The following reactions are possible in this case:

$$He^* + H_2 \rightarrow \begin{cases} He + H_2^+ + e^- & \text{(Penning ionization)} \\ HeH_2^+ + e^- & \text{(associative ionization)} \\ HeH^+ + H + e^- & \text{(rearrangement ionization)} \\ He + H + H^+ + e^- & \text{(dissociative ionization)} \end{cases} \qquad (8)$$

A theoretical calculation of the total ionization cross section which is the sum of the cross sections of all the reaction channels (8) has been given in /27/,/28/.

So far the experimental investigation of reaction (8) has been carried out by means of examination of afterglow of the reaction products /29/ or using molecular beams /30/. A list of total cross sections obtained experimentally for ionization of hydrogen by helium in the $2^3S_1$ metastable state at $300^oK$ is given in /29/. According to /29/, this cross section $\sigma_{H_2}$ = $(1.3 \pm 0.4) \cdot 10^{-16} cm^2$, while according to /30/, it is $(3.4 \pm 0.7) \cdot 10^{-16} cm^2$. This discrepancy is accountable by a difficulty in separating between singlet and triplet reaction channels, i.e. the reactions (8) with $2^3S_1$ and $2^1S_o$ atoms, respectively, as well as by a necessity to estimate the extent to which the particle velocity distribution differs from Maxwellian.

The first investigation of reaction (8) using optical orientation of helium atoms in the $2^3S_1$ state was made in /31/. This method permitted a complete separation of the triplet channel of reaction (8) and provided experimental data at low energies of colliding particles. The experiment was carried out at temperatures ranging from $77^o$ to $300^oK$.

The work was based on the well-known technique of optical orientation of helium atoms in the $2^3S_1$ state. A weak gas discharge excited in the absorption cell filled with a mixture of $He^4$ at 0.36 Torr and $H_2$ at 0.019 Torr at $300^oK$ resulted in a transition of a fraction of helium atoms to metastable state. Helium atoms in the $2^3S_1$ state were optically oriented by circularly polarized light with the wave length $\lambda$= 10830 Å ($2^3S_1 \longrightarrow 2^3P$ transition). Magnetic resonance of optically oriented helium atoms in the $2^3S_1$ state was observed. Addition of hydrogen led to the destruction of the metastable state of helium atoms due to reaction (8). This caused the broadening of the $2^3S_1$ helium resonance line. To separate the contribution of reaction (8) to the resonance line width, comparison was made with the line width measured in pure helium under the same experimental conditions. Ionization cross section $\sigma_{H_2}$ was derived from the equation:

$$\pi \Delta f = N_{H_2} \cdot v \cdot \sigma \qquad (9)$$

where $\Delta f$ is the contribution of reaction (8) to the resonance line width, v is the mean relative velocity for the He-$H_2$ system.

It should be noted that with hydrogen additions the destruction of helium metastable state may also be due to collisions with H, $H_2^+$, $H_3^+$, $HeH^+$ and so on.

To estimate experimentally the contribution of these collisions to destruction of helium metastable state, resonance line widths were measured at various discharge intensities. Since at low discharge intensity the number of H, $H_2^+$, $H_3^+$, $HeH^+$ increases with intensity, the line width for helium atoms should vary with the discharge intensity if the part played by the above particles is important. No such variation was observed, however, and the conclusion was made that compared to $H_2$ these particles are unimportant in the destruction of helium atom metastability state.

It appears, therefore, that the molecular hydrogen ionization cross section practically coincides with that of helium metastability destruction induced by hydrogen additions.

The table below lists the ionization cross sections $\sigma_{H_2}$ derived experimentally using equation (9). The experimental errors in the cross section values are due to errors in measuring hydrogen pressure and resonance line width and are equal to ±15%.

Table. Total cross section for ionization of a $H_2$ molecule by helium $2^3S_1$ atoms at various temperatures

| T,K | $\sigma_{H_2}$, $10^{-16} cm^2$ |
|---|---|
| 77 | 0.69 ± 0.10 |
| 162 | 1.11 ± 0.16 |
| 220 | 1.33 ± 0.21 |
| 300 | 1.53 ± 0.23 |

A sharp temperature dependence of ionization cross sections will be noticed: as temperature goes down from 300° to 77°K the cross section more than halves. This is evidence of the repulsive character of the potential of interaction between $2^3S_1$ He and $H_2$ at all distances of approach investigated for colliding particles. One practical advantage is that helium atoms can be optically oriented at high concentrations of admixed hydrogen by lowering the temperature of the absorption cell. For example, optical orientation and magnetic resonance of helium atoms was observed at 77°K when the absorption cell contained 0.36 Torr of $He^4$ and 0.034 Torr of $H_2$.

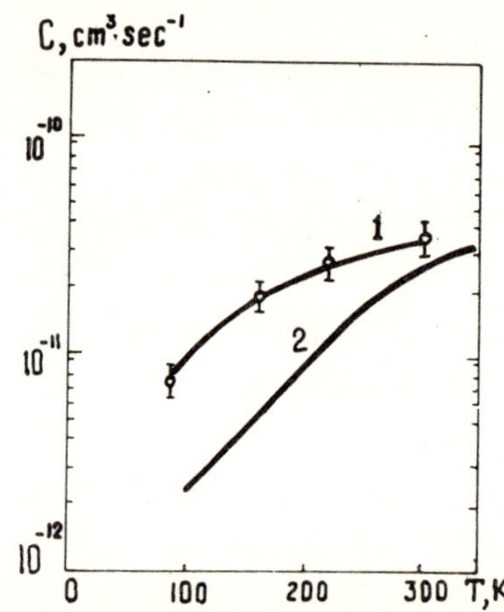


FIG. IO. Temperature dependence of the reaction rate constant for ionization of $H_2$ molecules by metastable helium; curve I - experimental data; curve 2 - theoretical calculation in Ref.28

Fig.IO shows the rate constant $C = \sigma_{H_2} \cdot v$ for reaction (8) experimentally obtained in the present work (curve I) and theoretically calculated in /28/ (curve 2). The course of the experimental dependence is in agreement with theoretical calculation. Quantitative discrepancy between curves I and 2 is accounted for by the approximate nature of the model on which the calculations in /28/ have been based.

Thus, the method of optical orientation has been successfully used in the investigation of the ionization of a hydrogen molecule $H_2$ by $2^3S_I$ helium atoms alone and in the study of this process at low energies of colliding particles.

## References


I. W.Happer, Rev. Mod. Phys. 44, I69 (I972)

2. B.N.Sevast'yanov and R.A.Zhitnikov, Zh. Eksp. Teor. Phys. 56, I508 (I969) /Sov. Phys. JETP, 29, 809 (I969)/

3. R.A.Zhitnikov, Usp. Fiz. Nauk I04, I68 (I97I) /Sov. Phys. Usp. I4, 359 (I97I)/

4. R.A.Zhitnikov, E.V.Blinov and L.S.Vlasenko, Zh. Eksp. Teor. Fiz. 64, 98 (I973) /Sov. Phys. JETP 37, 53 (I973)/

5. M.V.McCusker, L.L.Hatfield and G.K.Walters, Phys. Rev. Lett 22, 8I7 (I969)

6. M.V.McCusker, L.L.Hatfield and G.K.Walters, Phys. Rev. A 5, I77 (I972)

7. J.C.Hill, L.L.Hatfield, N.D.Stockwell and G.K.Walters, Phys. Rev. A5, I89 (I973)
8. A.I.Okunevich, Zh. Eksp. Teor. Fiz. 70, 899 (I976)
9. L.D.Schearer and L.A.Riseberg, Phys. Lett. 33A, 325 (I970)
I0. L.D.Schearer, Phys. Rev. I7I, 8I (I968)
II. S.P.Dmitriev, R.A.Zhitnikov and A.I.Okunevich, Zh. Eksp. Teor. Fiz. 70, 69 (I976) /Sov. Phys. JETP, 43, 35 (I976)/
I2. R.A.Zhitnikov, V.A.Kartoshkin, G.V.Klement'ev and L.V.Usacheva, Pis'ma Zh. Eksp. Teor. Phys. 22, 293 (I975) / JETP Lett. 22, I36 (I975)/
I3. J.Dupont-Roc, M.Leduc and F.Laloe, Phys. Rev. Lett. 27, 467 (I97I)
I4. R.A.Zhitnikov, V.A.Kartoshkin, G.V.Klement'ev and L.V.Usacheva, Zh. Eksp. Teor. Phys. 7I, I76I (I976) /Sov. Phys. JETP, 44, 924 (I976)/
I5. R.A.Buckingham and D.Dalgarno, Proc. Roy. Soc.(London) A2I3, 327 (I952)
I6. F.D.Colegrove, L.D.Schearer and G.K.Walters, Phys. Rev. AI35, 353 (I964)
I7. S.D.Rosner and F.M.Pipkin, Phys. Rev. A5, I909 (I972)
I8. J.Dupont-Roc, M.Leduc and F.Laloe, J. de Phys. 34, 96I (I973)
I9. H.J.Kolker and H.G.Michels, J. Chem. Phys. 50, I762 (I969)
20. B.M.Smirnov, Zh. Eksp. Teor. Phys. 53, 305 (I967) /Sov. Phys. JETP, 26, 204 (I968)/
2I. P.L.Pakhomov and I.Ya.Fugol', Dokl. Akad. Nauk SSSR, I79, 8I3 (I968) /Sov. Phys. Dokl. I3, I7 (I968)/
22. A.P.Hickman and N.F.Lane, Phys. Rev. AI0, 444 (I974)
23. B.M.Smirnov. Asimptoticheskie metody v teorii atomnykh stolknovenii (Asymptotic methods in the theory of atomic collisions) (Nauka, Moscow, I973)
24. J.R.Barrat and C.J.Cohen-Tannoudji, J. de Phys. et Rad. 7, 329, 443 (I96I)
25. A.Donszelmann, Physica, 56, I38 (I97I)
26. R.A.Zhitnikov, V.A.Kartoshkin and G.V.Klement'ev, Zh. Eksp. Teor. Phys. 73, I738 (I977)
27. A.P.Hickman, A.D.Isaakson and W.H.Miller, J. Chem. Phys. 66, I492 (I977)
28. J.S.Cohen and N.F.Lane, J. Chem. Phys. 66, 586 (I977)

29. W.Lindinger, A.L.Schmeltekopf and F.C.Fehsenfeld, J. Chem. Phys. 6I, 2890 (I974)

30. J.C.Howard, J.P.Riola, R.D.Rundel and R.F.Stebbings, J. Physics, B6, I09 (I973)

3I. R.A.Zhitnikov, V.A.Kartoshkin and G.V.Klement'ev, Pis'ma Zh. Eksp. Teor. Phys. 26, 65I (I977)

# THE ROLE OF ATOMIC AND MOLECULAR PROCESSES IN THE COMPUTER MODELING OF GAS LASERS

Kenneth Smith and S.A. Roberts

Department of Computer Studies
The University
Leeds, England

## ABSTRACT

The first section of this paper is pedogogical in introducing the reader to the basic physics of gas lasers. The principal objective of the paper is to draw the attention of the atomic and molecular physics community to the cross sections currently being used in models of the $CO_2$ and CO lasers. This is accomplished in Secs. 2 and 3. A subsidiary objective is to present a sketch of the processes involved in the exotic lasers, excimers and mercury monohalides, which have received considerable attention over the last few years.

CONTENTS

1. Basic Physics
   1.1. Electromagnetic Waves
   1.2. Energy Levels

2. $CO_2$ Lasers
   2.1. TLASER
   2.2. BOLTZ

3. CO Lasers
   3.1. BOLTZ
   3.2. COLASE

4. Bound-Free Lasers
   4.1. Rare Gas Excimers
   4.2. Rare-Gas-Halide Lasers
   4.3. Boltzmann Transport Equation

5. Mercury Monohalide Lasers

6. References

## 1. BASIC PHYSICS

### 1.1. Electromagnetic Waves. [1]

The properties of electromagnetic waves in regard to their interaction with matter are strongly dependent on the wavelength $\lambda$ (centimeters per cycle). The eye responds only to waves extending roughly from $\lambda = 3.8 \times 10^{-5}$ cm (violet) to $7.8 \times 10^{-5}$ cm (red). Radiations for which $\lambda < 3.8 \times 10^{-5}$ cm are called ultraviolet down to about $\lambda \sim 100 \times 10^{-8}$ cm, below which they are called X-rays.

Radiations for which $\lambda > 7.8 \times 10^{-5}$ cm belong to the infrared. For $\lambda < 10\ \mu$ or $10^{-3}$ cm the radiations are said to be in the near infrared, and for $\lambda > 10\mu$ they are said to be in the far infrared. The boundary at $10\mu$ is arbitrary and not always used in exactly the same way. Radiations whose wave-lengths are of the order of millimeters and centimeters are called microwaves and those of still longer wavelengths are called radio waves, or Hertzian waves.

In Fig.(1) we present a schematic of the radiation spectrum of some important gas lasers from the long-wave-length $CO_2$ laser at $1.06 \times 10^4$ nm down to the UV wavelength of KrCl at $2.22 \times 10^2$ nm.

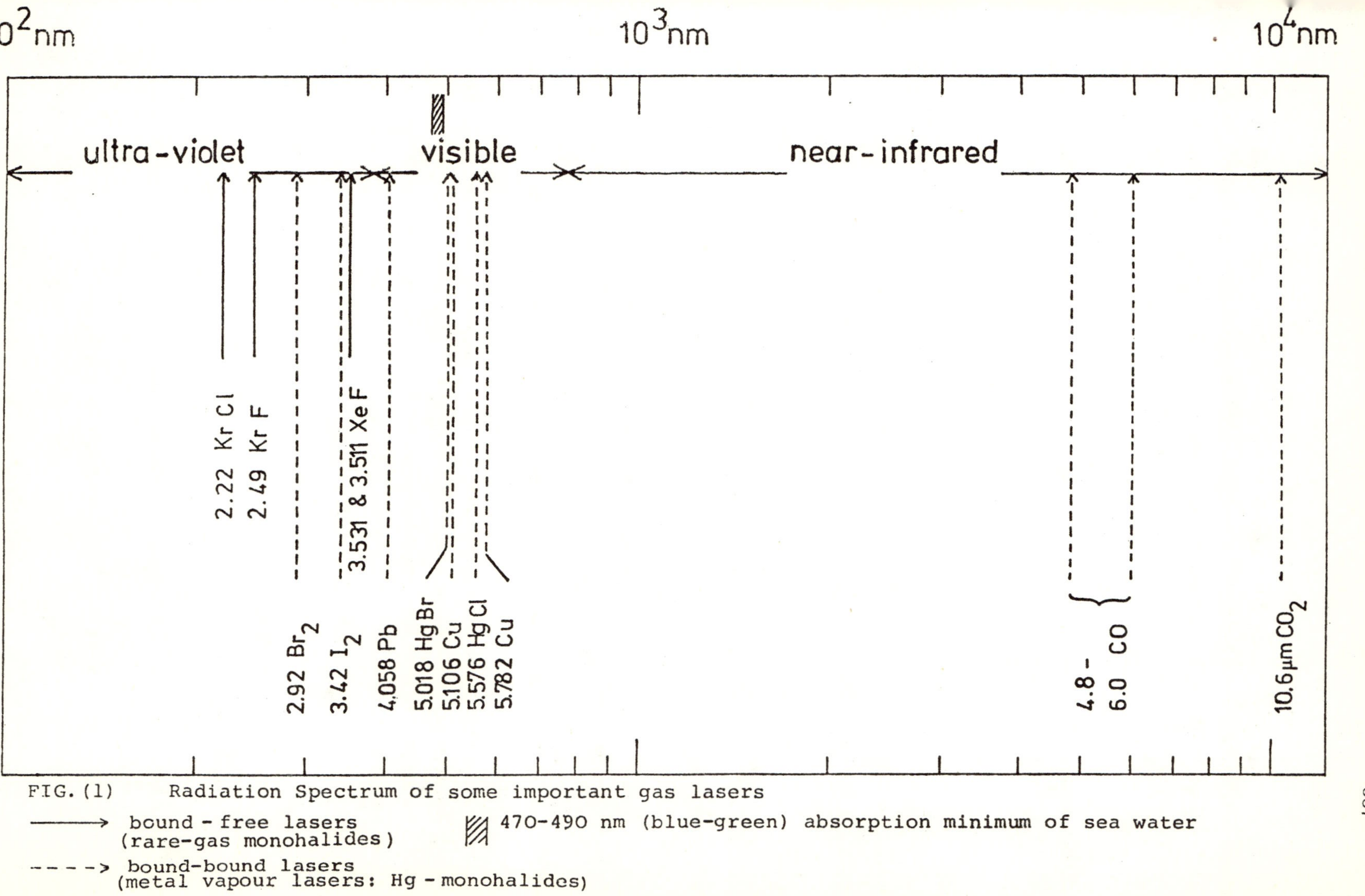


FIG. (1) Radiation Spectrum of some important gas lasers

⟶ bound - free lasers (rare-gas monohalides)

- - - -> bound-bound lasers (metal vapour lasers: Hg - monohalides)

▨ 470-490 nm (blue-green) absorption minimum of sea water

## 1.2 ENERGY LEVELS

Electromagnetic waves are generated in $CO_2$ by pumping the ground state by electron excitation and by V-V transfer in $N_2$ - $CO_2$ collisions. In Fig. (2) we present a diagram of the lower vibrational levels of the electronic ground state of $CO_2$ together with their associated rotational manifolds. The populations of the rotational substates are represented by the lengths of their lines. Dipole transitions of $CO_2$ satisfy the selection rule $\Delta J = \pm 1$, where $\Delta J$ is the change in rotational quantum number corresponding to the observed R and P branches of the emission spectrum. The lines are designated P (J) or R (J) according to whether J + 1, or J - 1, is the upper laser level with lower laser level J.

In Fig. (3) we present a diagram of the energy transfer processes in a CO-$N_2$ laser in which there has been a resurgence of interest in recent months. The excited states of CO are pumped by electron excitation, V-V energy transfer and V-T energy transfer. We see from Fig. (3) that the CO system can lase on several lines simultaneously.

In the last five years or so considerable effort has been devoted to rare-gas monohalides for which the lower state of the laser is dissociative, that is the electromagnetic radiation is emitted as a consequence of bound-free transitions. More recently, attention is being focussed on mercury monohalides which emit radiation in bound-bound transitions. We will now recall the usual notation for electronic transitions, see Herzberg [2].

If $\underset{\sim}{L}$ is the electronic orbital angular momentum of a molecule, then a precession of $\underset{\sim}{L}$ takes place about the internuclear axis with constant component $M_L$ which can take only the values

$$M_L = L, L-1, L-2, \ldots, -L \tag{1.1}$$

In diatomic molecules states differing only in the sign of $M_L$ are degenerate, and electronic states of diatomic molecules are classified according to the value of $|M_L|$

$$\Lambda = |M_L| \tag{1.2}$$

The corresponding angular momentum vector $\underline{\Lambda}$ represents the component of the electronic orbital angular momentum along the internuclear axis. Its magnitude is $\Lambda$ $(h/2\pi)$.

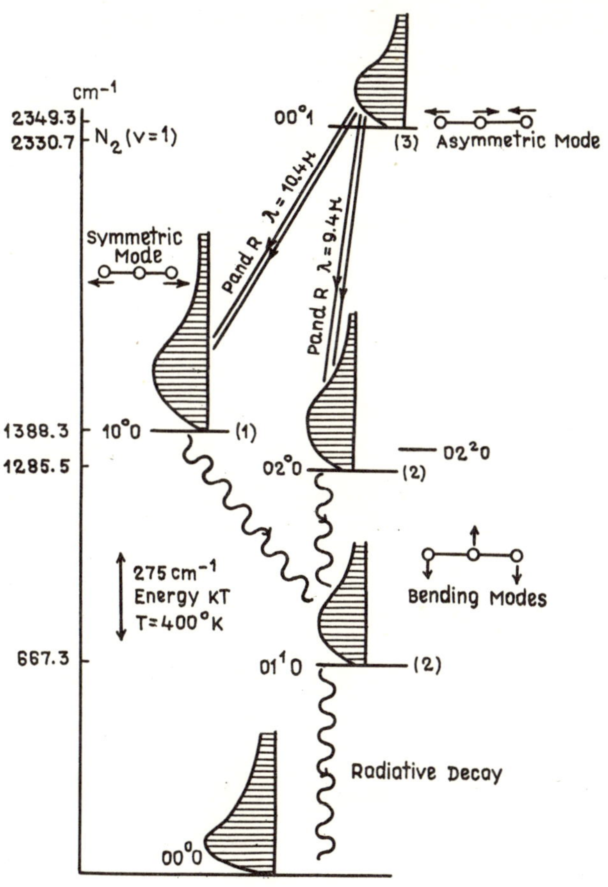


Fig. 2. Lower vibrational levels of the electronic ground state of $CO_2$. Vibrational levels are designated by $v_1$, $v_2^l$, $v_3$, where $l=v_2$, $v_2-2$, ..., 0 for even $v_2$, and $l=v_2$, $v_2-2$, ..., 1 for odd $v_2$. The populations of the rotational substates of the vibrational states are indicated by the lengths of the lines specifying the rotational states. Dipole transitions of $CO_2$ satisfy $\Delta J=\pm 1$, where $\Delta J$ is the change in rotational quantum number, corresponding to the observed R and P branches of the emission spectrum. The lines are designated P(J) or R(J), where J is the rotational quantum number of the lower level.

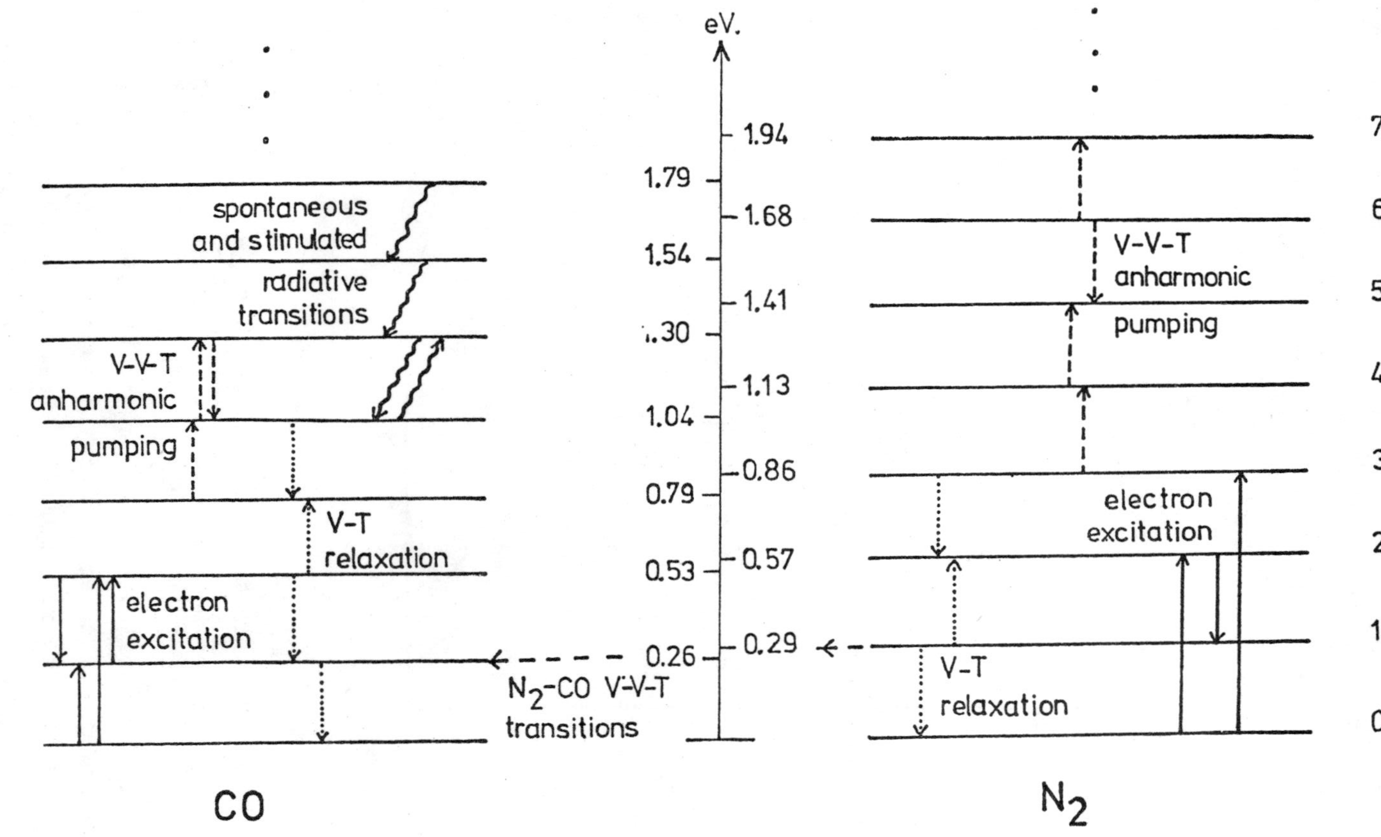


Figure 3. Schematic diagram of the energy transfer processes in a $CO-N_2$ laser

According to Eq. (1.1), for a given value of L, the quantum number $\Lambda$ can take the values

$$\Lambda = 0,1,2,\ldots,L \tag{1.3}$$

Thus in the molecule for each value of L there are L+1 distinct states with different energy. However, often the value of L cannot be given at all, since the corresponding angular momentum L is not defined.

According as $\Lambda = 0,1,2,3,\ldots,$ the corresponding molecular state is designated a $\Sigma,\Pi,\Delta,\Phi,\ldots,$ state, analogous to the mode of designation for atoms. Greek letters are used throughout in the designation of molecular quantities referring to components of electronic angular momenta, while the corresponding italic letters used for atoms refer to the electronic angular momenta themselves.

$\Pi,\Delta,\Phi,\ldots$ states are doubly degenerate since $M_L$ can have the two values $+\Lambda$ and $-\Lambda$ (see above); $\Sigma$ states are non-degenerate.

In designating a given electronic transition, the upper state is always written first and then the lower. $^1\Pi-{}^1\Sigma$ thus means a transition for which the upper state is $^1\Pi$ and the lower $^1\Sigma$; $^1\Sigma-{}^1\Pi$ means a transition for which $^1\Sigma$ is the upper state and $^1\Pi$ is the lower state. Sometimes an arrow is used to indicate whether the transition under consideration is observed in emission or absorption:
$^1\Pi\rightarrow{}^1\Sigma$ means a transition from $^1\Pi$ to $^1\Sigma$ in emission, and $^1\Pi\leftarrow{}^1\Sigma$ indicates a transition in absorption from the lower $^1\Sigma$ state to the upper $^1\Pi$ state. Naturally, the validity of a selection rule is independent of which state is the upper and which is the lower. This independence is indicated by a double arrow $\leftrightarrow$; for example, $\Pi\leftrightarrow\Sigma^+$ means that $\Pi-\Sigma^+$, as well as $\Sigma^+-\Pi$, is possible.

If several electronic states of a molecule are known, some of which may be of the same type, they are distinguished by a letter X,A,B...., a,b,..., in front of the term symbol or sometimes by one or more asterisks added to it. Thus we describe transitions by symbols such as $A^1\Pi-X^1\Sigma$ or $B^1\Sigma-X^1\Sigma$ or $B^1\Sigma-X^1\Sigma$ (or $\Sigma^{**}-\Sigma^*$) and so on.

For short reference, in conformity with a fairly well established custom, the ground state is referred to as X, the excited states of the same multiplicity as A,B,C,..., those of different multiplicity as a,b,c,.... The only exception to this rule is $N_2$ where the custom of designating the excited singlet states a,b,... and the excited triplet states A,B,... appeared too well established to make a change. The order a,b,c,...A,B,C,... is usually the order of increasing energy. But in some cases, particularly for $H_2$ and $He_2$, exceptions were made in order to be in agreement with earlier designations.

In Fig. (4), see Herzberg and Fig. (5) see Bhaumik [3] we illustrate potential energy curves for possible bound-bound and bound-free transitions.

## 2 $CO_2$ LASERS

### 2.1 TLASER

TLASER, see Davies et al. [4] is a computer program which predicts the output power pulse shape and gain profile from a model of an electrically excited $CO_2$ gas laser, including the effects of dissociation and a variable ambient gas temperature.

In Fig. (6) we present a schematic of the collision processes taken into account in TLASER. Continuous, arrowed lines denote V-V transitions; dotted, arrowed lines denote V-T energy transfer processes; and dashed arrowed lines denote electron excitation processes. In this figure we have introduced the relaxation time, $\tau$, which is directly related to the rate coefficient, $k_{if}$, see Ross et al. [5], which itself is defined in terms of the cross section, $\sigma_{if}(u)$ $cm^2$, for the collision between X and Y, initially in state "i", finally in state "f", that is

$$k_{if}\ (X,Y) \equiv \int\int u\ \sigma_{if}\ (u)\ f_X(\underset{\sim}{v}_X) f_Y(\underset{\sim}{v}_Y) d\underset{\sim}{v}_X d\underset{\sim}{v}_Y \qquad (2.1)$$

where $f_X\ (\underset{\sim}{v}_X)$ is the velocity distribution for particle X and

$$u = |\underset{\sim}{v}_X - \underset{\sim}{v}_Y| \qquad (2.2)$$

is the relative speed of X and Y.

In TLASER we have collisions between heavy particles, where, whether justified or not, the velocity distribution is assumed to be Maxwellian, that is

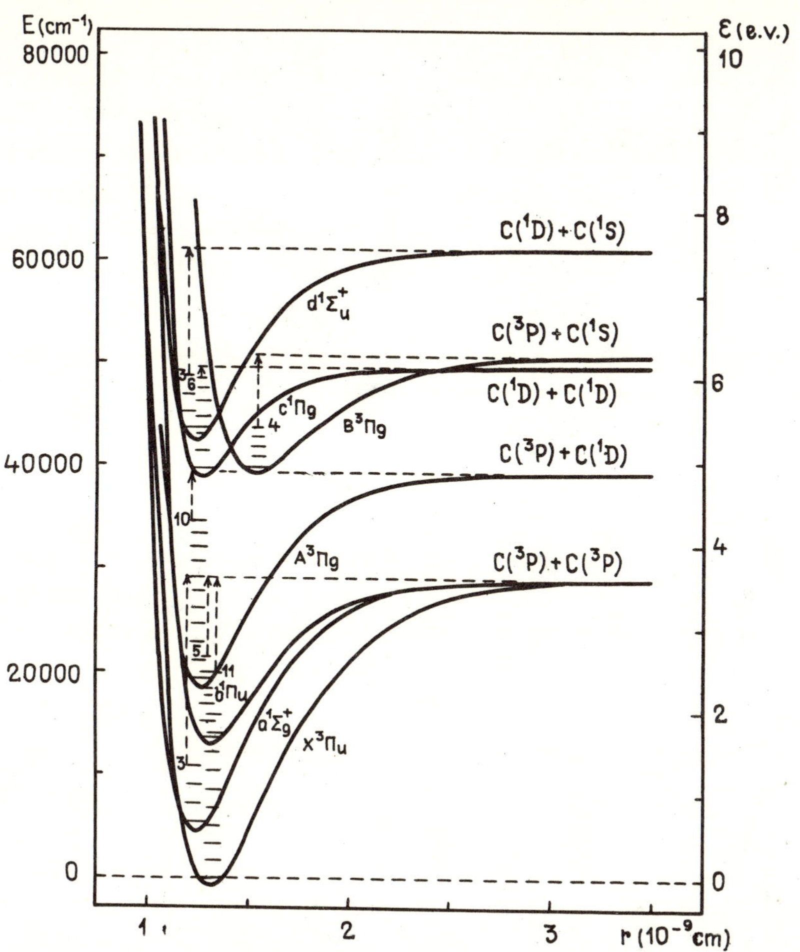


Fig. 4. Potential Curves of the Observed Electronic States of the $C_2$ Molecule. The relative position of singlet and triplet levels is not definitely known. It has been assumed here that the excitation energy of the $a^1\Sigma_g^+$ state is 5300cm$^{-1}$ The dissociation energy, also, is not definitely established. The value D =3.6 e.v. has been assumed here.

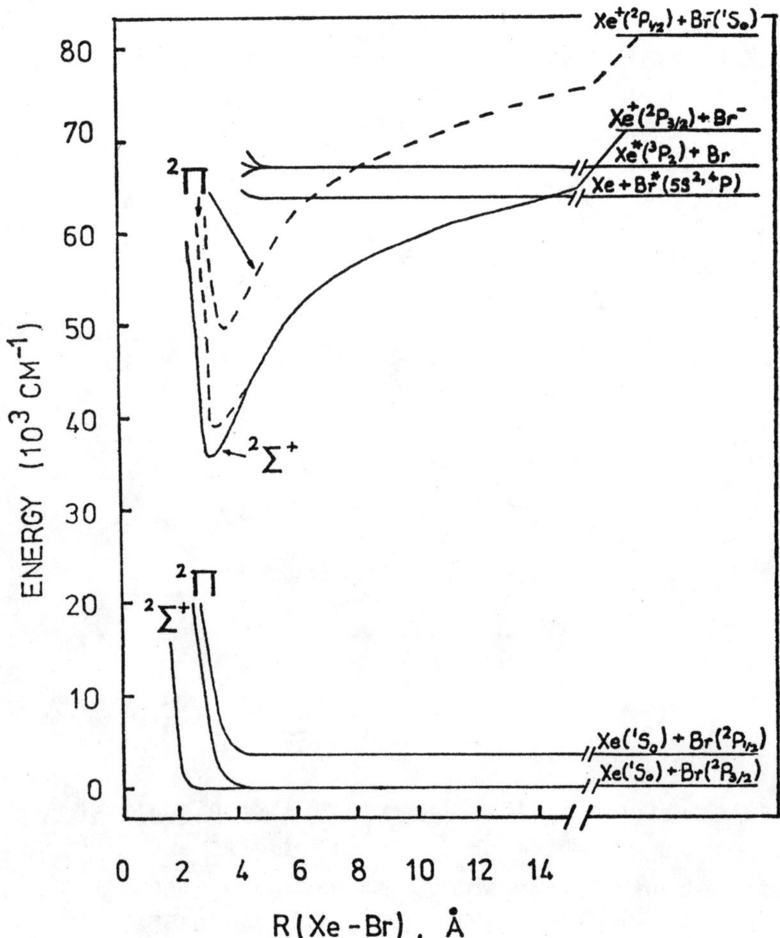


Fig.5

POTENTIAL-ENERGY CURVE for xenon bromide is similar to those of other excimers with repulsive ground states. Curve was supplie by D.W. Setser of Kansas State University.

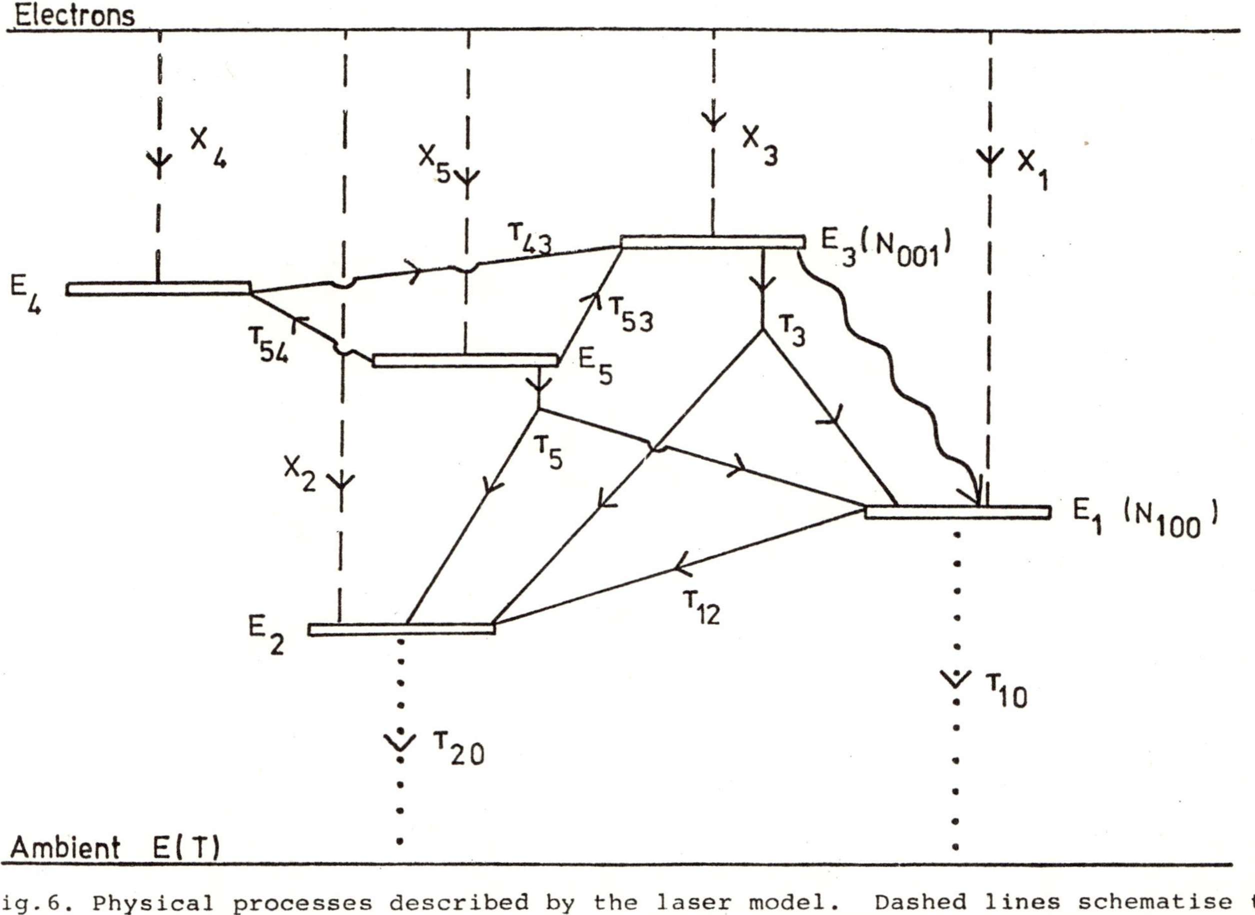


Fig.6. Physical processes described by the laser model. Dashed lines schematise the electron excitation of vibrational modes, solid lines represent vibrational energy exchange between modes, and dotted lines refer to the relaxation of vibrationally excited molecules to the ground state. $E_1$, $E_2$, $E_3$ are respectively the symmetric, bending, and asymmetric modes of $CO_2$, and $E_4$ and $E_5$ are the vibrational modes of $N_2$ and CO.

$$f(\nu) = (m/2\pi k_B T)^{3/2} \quad e^{-(m\nu^2/2k_B T)}, \ cm^{-3} \ sec^3 \qquad (2.3)$$

When one of the collision partners is an electron the Boltzmann transport equation is solved for $f_e$, see Sec. (2.2)

In Table (1) we present the empirical temperature-dependent equations for the heavy particle rate coefficients. These rates were assembled from the literature in 1974, whether or not these are the best available values today is why we present them explicitly to this conference as TLASER is in frequent production. What we have not done with TLASER is to use it as a code to test the sensitivity of laser output power to variations in rate coefficients.

## 2.2 BOLTZ

In Fig. (6) we have shown the electron excitation rates $X_i$ which are defined, according to Smith and Thomson [6] by

$$X_i \ (E/N,T) = \sum_j \alpha_{ji} \ (\frac{2e}{m})^{\frac{1}{2}} \int_0^\infty \ Q^i{}_j(u) \ f(E/N,T,u) \ du \qquad (2.4)$$

where T is the ambient temperature, E/N Volts $cm^2$ is the applied electric field divided by the total particle number density,while $Q^i{}_j$ $cm^2$ is the velocity-dependent electron molecule cross section for excitation of the jth vibrational level from the ground state of species i. The coefficients $\alpha_{ji}$ is the number of $h\nu_i$ quanta given to state j of species i.

Atomic and molecular collision cross sections are required for $Q^i{}_j(u)$ in Eq. (2.4) and in the Boltzmann transport equation for f(E/N,T,u), where the momentum transfer cross sections $Q^i{}_m$ and the cross sections for superelastic collisions, $Q^i{}_{-j}$, as well as $Q^i{}_j$ are required. These various cross sections are shown in figures (7) to (9).

# 3. CO LASERS

## 3.1 BOLTZ

In an electrically excited CO laser, approximately the first eight vibrational levels of CO are excited from the ground vibrational state, by electron impact. These processes are shown by the solid arrowed lines in Fig. (3). The rates for excitation are defined by equation (2.3) of section (2.2). The electron energy distribution f (E/N, T, u) is obtained from the code BOLTZ, the cross sections for vibrational excitation by electron impact are given in Figs.(10a and b).

Table 1
Arguments of RATES (T5,T20,T10,T43,T12,T3012,T63,T64,T612,T11)and temperature-dependent equations for corresponding rates $k_{ij}$

| Argument | Description | Rates ($cm^3 sec^{-1}$) | |
|---|---|---|---|
| T5 | Ambient temp,T | $k^{CO_2}_{010;000}$ | $4.6 \times 10^{-10} \exp(-77/T^{1/3})$ |
| T20 | $\tau_{20}$ | $k^{N_2}_{010;000}$ | $9.6 \times 10^{-11} \exp(-77/T^{1/3})$ |
| | | $k^{He}_{010;000}$ | $8.1 \times 10^{-11} \exp(-45/T^{1/3})$ |
| | | $k^{CO}_{010;000}$ | $6.82 \times 10^{-8} \exp(-77/T^{1/3})$ |
| T10 | $\tau_{10}$ | $4.5 \times \tau_{20}$ | |
| T43 | $\tau_{43}$ | $k^{N_2-CO_2}_{1,000;0,001}$ | largest of $1.71 \times 10^{-6} \exp(-175.3/T^{1/3})$, $6.07 \times 10^{-14} \exp(15.3/T^{1/3})$ |
| T12 | $\tau_{12}$ | $k^{CO_2}_{020;100}$ | $8.65 \times 10^{-15} \times T^{3/2}$ |
| | | $k^{N_2}_{020;100}$ | $3.68 \times 10^{-16} \times T^{3/2}$ |
| | | $k^{He}_{020;100}$ | $4.23 \times 10^{-17} \times T^{3/2}$ |
| | | $k^{CO}_{020;100}$ | $3.68 \times 10^{-16} \times T^{3/2}$ |
| T3012 | $\tau_3$ | $k^{CO_2}_{001;110}$ | $9.6 \times 10^{23} \times T^{-5.89} \times F(T)$ |
| | | $k^{CO_2}_{001;110}$ | $6.87 \times 10^{23} \times T^{-5.89} \times F(T)$ |
| | | $k^{He}_{001;110}$ | $2.43 \times 10^{23} \times T^{-5.89} \times F(T)$ |
| | | $k^{CO}_{001;110}$ | $6.87 \times 10^{23} \times T^{-5.89} \times F(T)$ |
| T63 | $\tau_{53}$ | $k^{CO-CO_2}_{1,000;0,001}$ | $1.56 \times 10^{-11} \exp(-30.1/T^{1/3})$ |
| T64 | $\tau_{54}$ | $k^{CO-N_2}_{10,01}$ | largest of $1.78 \times 10^{-6} \exp(-210/T^{1/3})$, $6.98 \times 10^{-13} \exp(25.6/T^{1/3})$ |
| T612 | $\tau_5$ | $k^{CO-CO_2}_{1,000;0,110}$ | $5.96 \times 10^{-22} \times T^{-5.86} \times F(T)$ |
| T11 | $\tau_1$ | $k^{CO_2-CO_2}_{100,000;000,100}$ | $10^n \times k^{CO_2}_{020;100}$ |

$F(T)=\exp\{-4223/T - 672.7/T^{1/3} + 2683/T^{2/3}\}$, $n \geq 0$ signifying that $k^{CO_2}_{020;100}$ is a lower limit to $k^{CO_2-CO_2}_{100,000;000,100}$

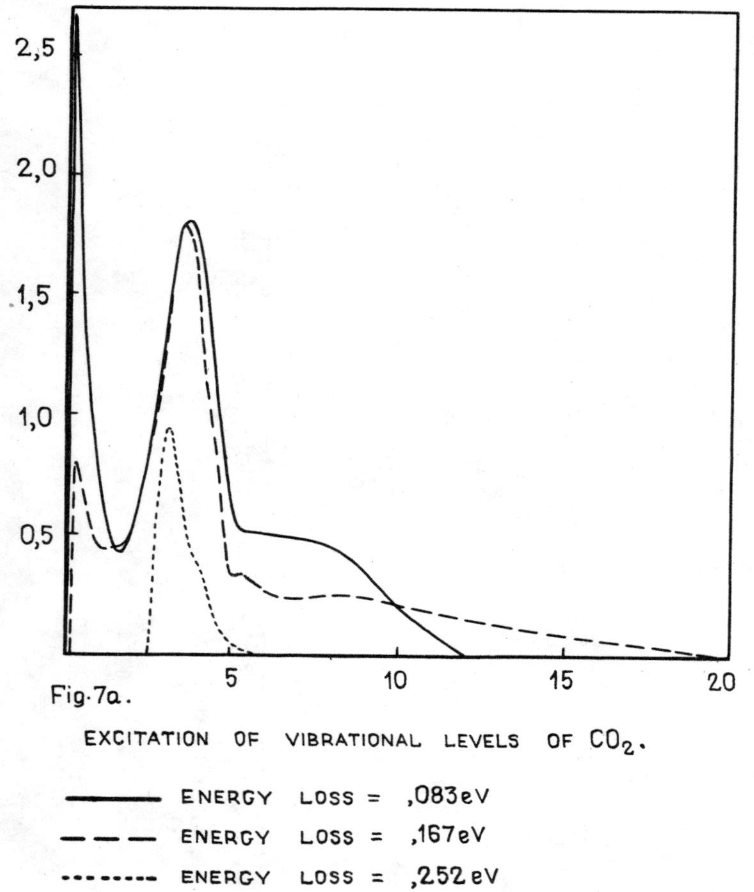


Fig.7a.

EXCITATION OF VIBRATIONAL LEVELS OF $CO_2$.

——— ENERGY LOSS = ,083 eV
— — — ENERGY LOSS = ,167 eV
-------- ENERGY LOSS = ,252 eV

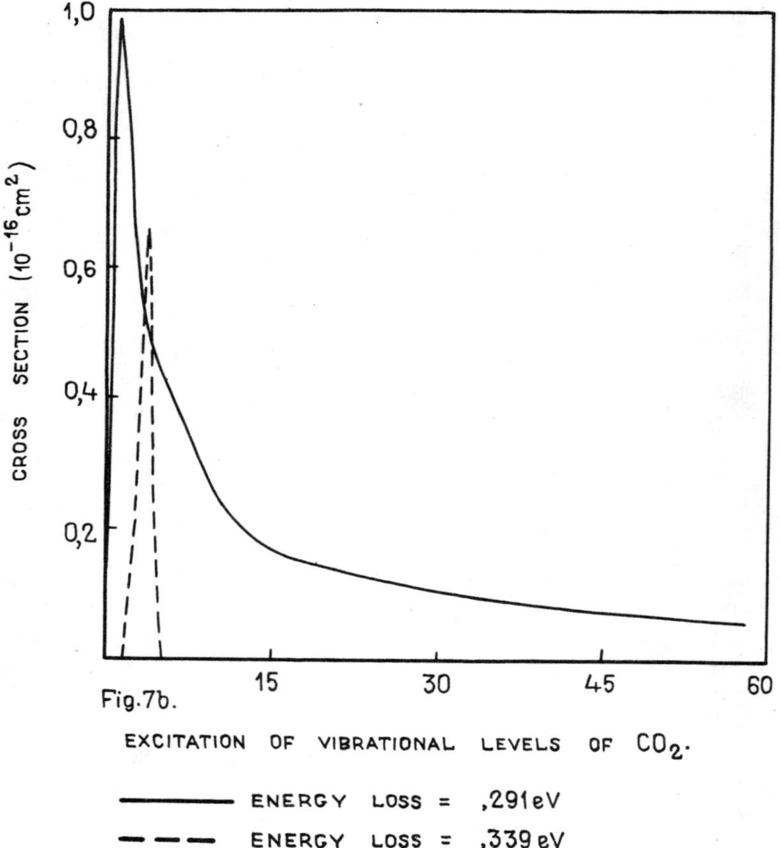


Fig.7b.

EXCITATION OF VIBRATIONAL LEVELS OF $CO_2$.

——— ENERGY LOSS = ,291 eV
— — — ENERGY LOSS = ,339 eV

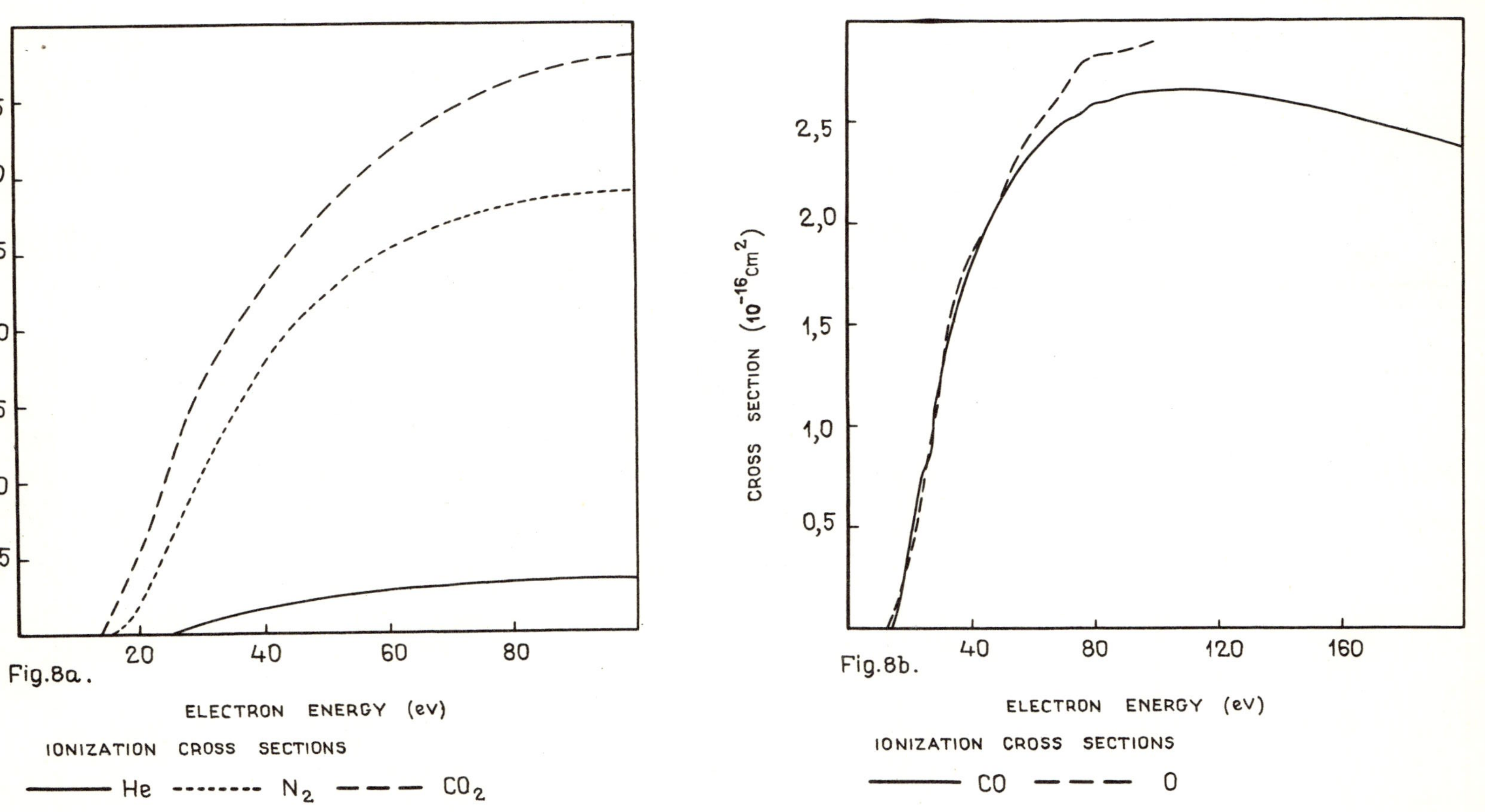


Fig.8a.

IONIZATION CROSS SECTIONS

—— He ------ $N_2$ — — — $CO_2$

Fig.8b.

IONIZATION CROSS SECTIONS

—— CO — — — O

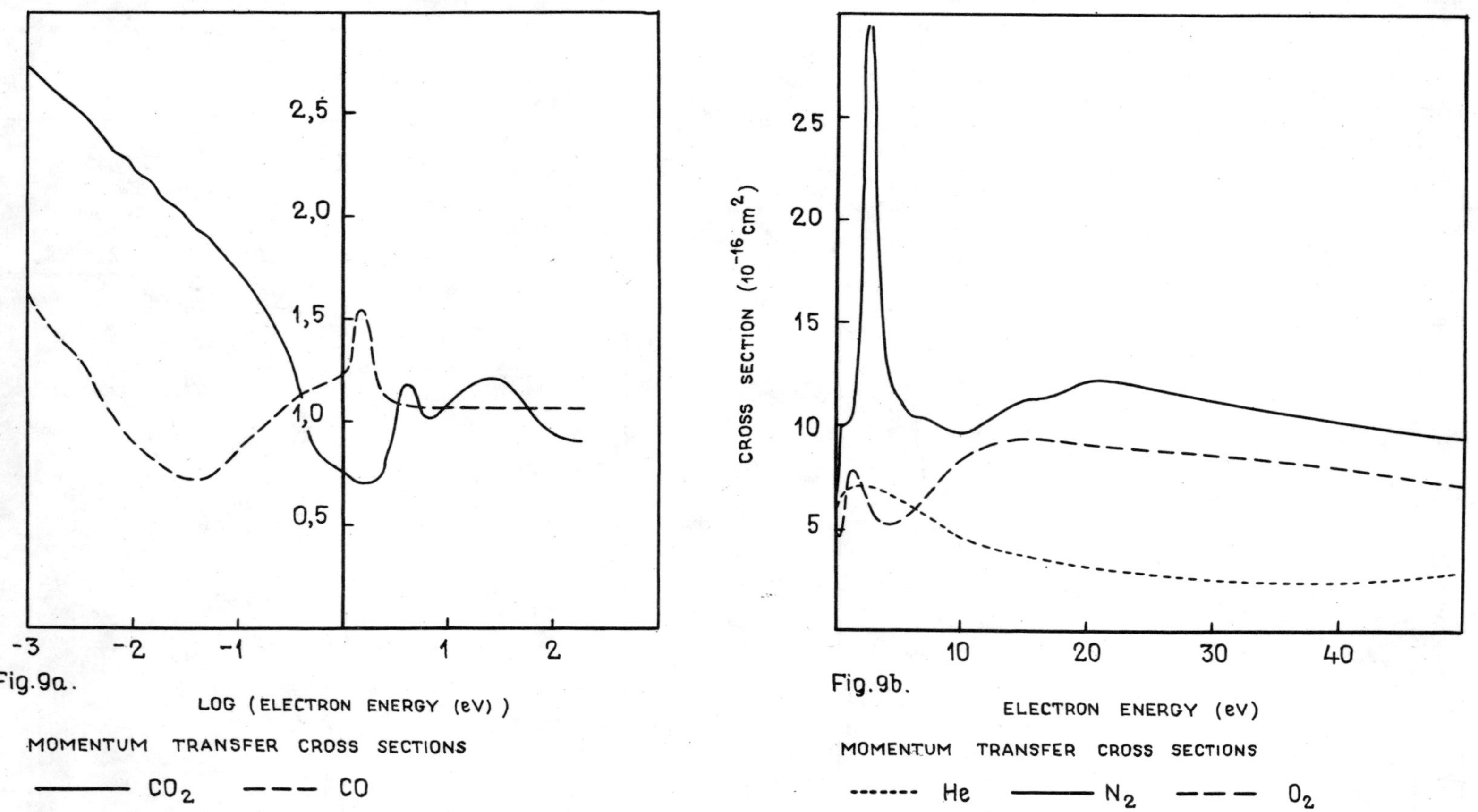


Fig.9a. MOMENTUM TRANSFER CROSS SECTIONS
—— $CO_2$ — — — CO

Fig.9b. MOMENTUM TRANSFER CROSS SECTIONS
------ He —— $N_2$ — — — $O_2$

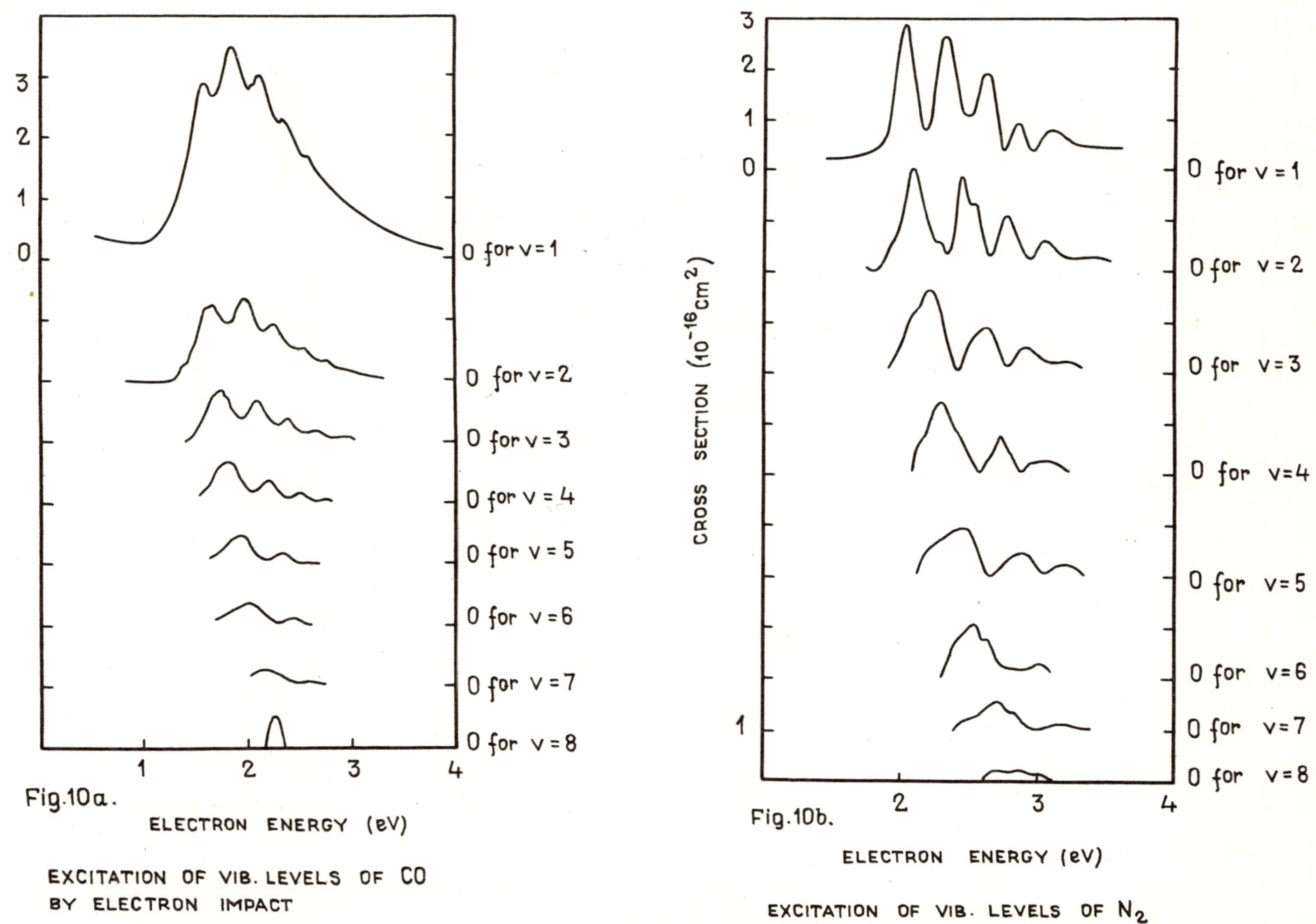


Fig.10a. EXCITATION OF VIB. LEVELS OF CO BY ELECTRON IMPACT

Fig.10b. EXCITATION OF VIB. LEVELS OF $N_2$ BY ELECTRON IMPACT

The efficiency of a CO laser may be increased by the addition of $N_2$ which, because of its slow V-T relaxion rate, acts as an energy reservoir. With $N_2$ included in the model, the code BOLTZ is required also to calculate the rates of excitation of the first eight vibrational levels of $N_2$.

## 3.2 COLASE

The computer program COLASE [7] predicts the build-up of gain and the output power from an electrically excited CO gas laser, which may include partial pressures of $N_2$, He and Ar. The various kinetic processes which are taken into account are shown in Fig. (3). Dotted lines denote V-T transitions, solid arrowed lines, electron excitation, dashed lines denote V-V transitions and wavy lines denote radiative transitions.

Because of the anharmonicity of the CO and $N_2$ molecules, the V-V transitions are not exactly on resonance, the energy difference being supplied by or transferred to the translational energy. For this reason we denote these transitions as V-V-T transitions.

The effect of this anharmonicity is that energy is "pumped" up the vibrational ladder of CO and $N_2$ allowing lasing to occur simultaneously from several levels of CO, as indicated in Fig. (3).

Our knowledge of V-V-T and V-T rate coefficients, defined by Eq.(2.1) of section (2.1), is incomplete from experimental measurement. We have computed these rate coefficients from parameterised analytic formulae, with the parameters adjusted to give the best fit to what experimental data is available.

The analytic formulae used are based on those given by Schwartz *et al.* [8] with modifications due to Keck and Carrier [9] and also Shin [10]. For the V-V-T rate coefficients an extra term is added, given by Sharma and Brau [11], to account for long-range interactions. The V-T rate coefficients are shown in Figs.(11a and 11b), the V-V-T rate coefficients in Figs.(12a and 12b).

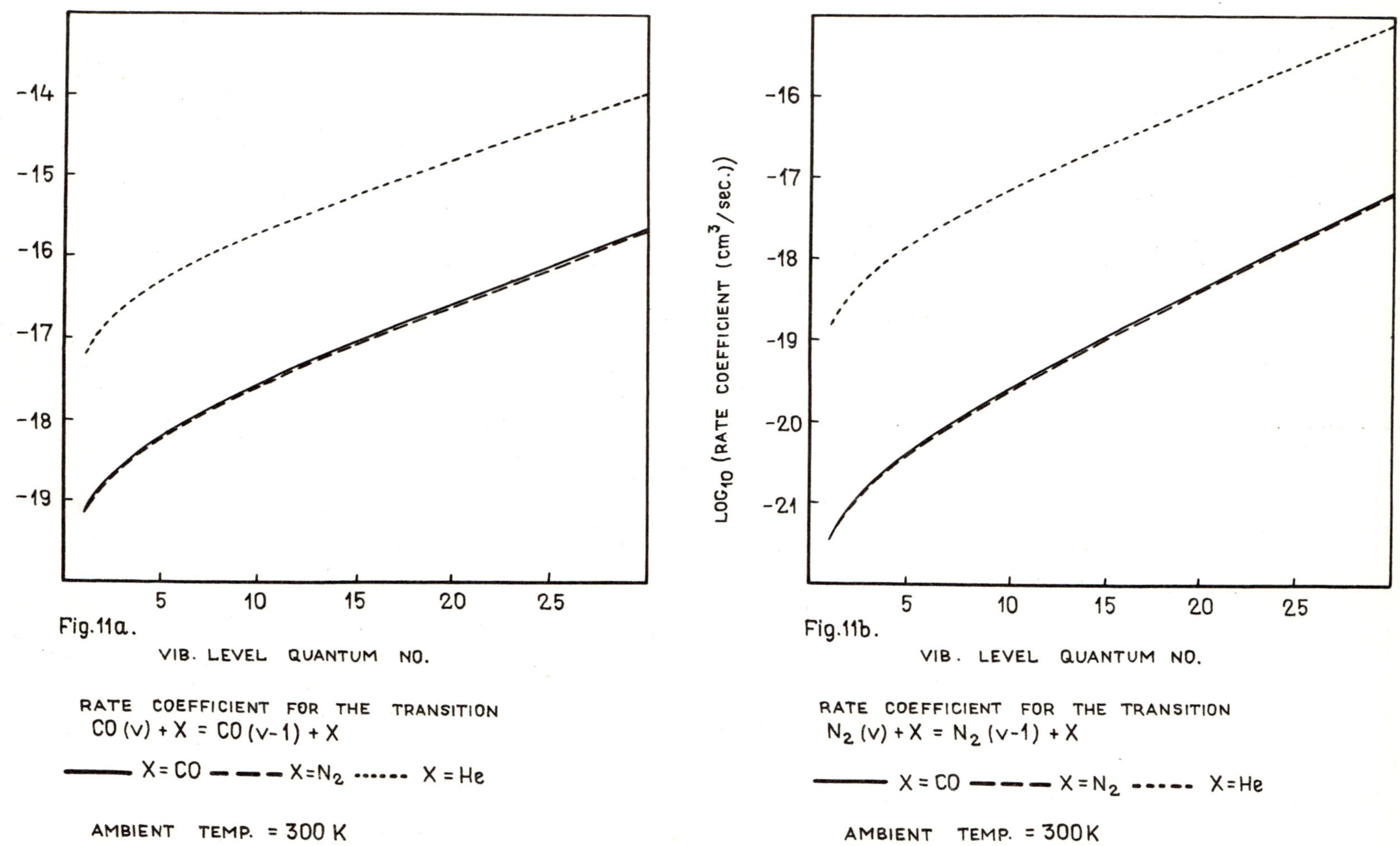


Fig. 11a.

RATE COEFFICIENT FOR THE TRANSITION
$CO\,(v) + X = CO\,(v-1) + X$

—— X = CO — — — X = $N_2$ ······ X = He

AMBIENT TEMP. = 300 K

Fig. 11b.

RATE COEFFICIENT FOR THE TRANSITION
$N_2\,(v) + X = N_2\,(v-1) + X$

—— X = CO — — — X = $N_2$ ----- X = He

AMBIENT TEMP. = 300 K

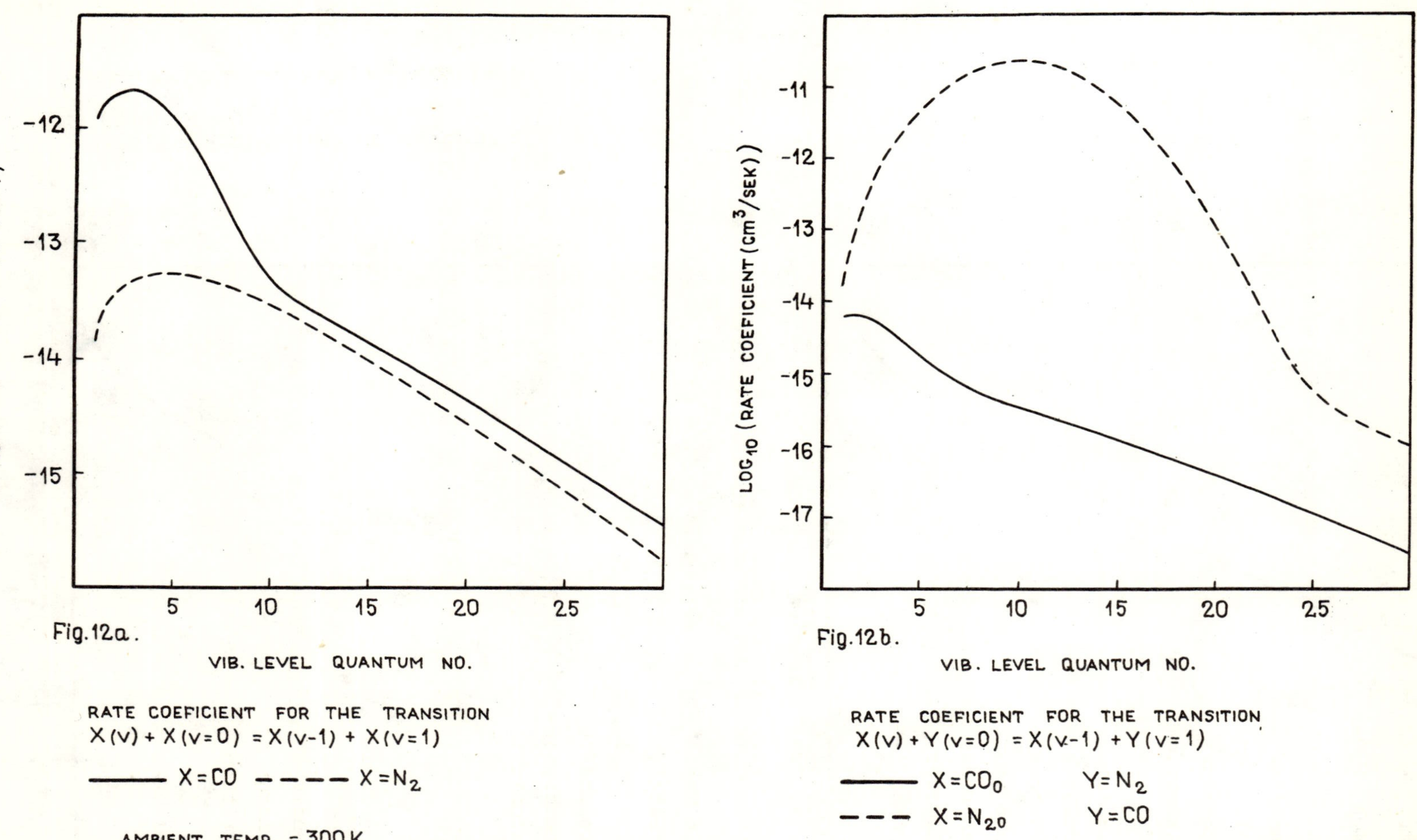
LOG10 (RATE COEFICIENT (cm3/sek))
-12
-13
-14
-15
5
10
15
20
25
VIB. LEVEL QUANTUM NO.
RATE COEFICIENT FOR THE TRANSITION
X(v) + X(v=0) = X(v-1) + X(v=1)
X = CO
X = N2
AMBIENT TEMP. = 300 K
LOG10 (RATE COEFICIENT (cm3/sek))
-11
-12
-13
-14
-15
-16
-17
5
10
15
20
25
VIB. LEVEL QUANTUM NO.
RATE COEFICIENT FOR THE TRANSITION
X(v) + Y(v=0) = X(v-1) + Y(v=1)
X = CO0 Y = N2
X = N20 Y = CO

Fig.12a.

Fig.12b.

# 4. BOUND-FREE LASERS

## 4.1 Rare Gas Excimers

An excimer is a molecule which only exists in the excited state and falls apart when it reaches the ground state after emission. Such depopulation of the ground state prevents terminal state bottlenecking, a major problem in many other lasers.

First reports of stimulated emission making use of a radiative transition between a bound and a repulsive molecular state came from N. G. Basov and colleagues in 1971. Theoretical descriptions of these systems have lagged, unfortunately, the experimental progress, owing largely to the enormous complexity of the atomic and molecular processes. However, Werner *et al*. [12] have formulated a model which takes into account

A. relativistic electron beam pumping;

B. kinetic reactions;

C. energy balance.

In Table (2) we reproduce Werner *et al*.'s numerical values for reaction parameters for the rare gas excimers of Xe and Kr.

Since the majority of excimer systems are pumped by relativistic e-beams, then it is important to understand the mechanisms involved in this system of excitation. The atomic ion so formed may be lost by several precesses:

i) collisional-radiative recombination at low pressures, and at high pressures.

ii) $X^+ + X + X \rightarrow X_2^+ + X$

where the third body is necessary to remove the binding energy of the molecule. The rate for such a reaction is written in terms of the three-body formation coefficient $\beta^+$.

Once formed, the diatomic ion may participate in two major loss processes

TABLE 2

Numerical Values of Important Reaction Parameters Used in the Computational Model

| Reaction | Coefficient | Value (Xe) | Value (Kr) |
|---|---|---|---|
| $X^+ + 2X \rightarrow X_2^+ + X$ | $\beta^+$ ($cm^6$ $sec^{-1}$) | $3.5 \times 10^{-31}$ | $2.7 \times 10^{-31}$ |
| $X_2^+ + 2X \rightarrow X_3^+ + X$ | $\beta_3^+$ ($cm^6$ $sec^{-1}$) | $9.0 \times 10^{-32}$ | $5.0 \times 10^{-32}$ |
| $X_2^+ + e \rightarrow X^{**} + X$ | $\alpha$ ($cm^3$ $sec^{-1}$) † | $1.4 \times 10^{-6}$ | $1.2 \times 10^{-6}$ |
| $X_3^+ + e \rightarrow X^{**} + 2X$ | $\alpha_3$ ($cm^3$ $sec^{-1}$) | $9.5 \times 10^{-5}$ | $2.4 \times 10^{-5}$ |
| $X^*(s(3/2,1)) + 2X \rightarrow X_2^*\ (^1\Sigma) + X$ | $\beta_1$ ($cm^6$ $sec^{-1}$) | $1.7 \times 10^{-32}$ | $2.9 \times 10^{-32}$ |
| $X^*(s(3/2,2)) + 2X \rightarrow X_2\ (^3\Sigma) + X$ | $\beta_3$ ($cm^6$ $sec^{-1}$) | $3.00 \times 10^{-32}$ | $5.3 \times 10^{-32}$ |
| $X_2^* + e \rightarrow X_2^+ + 2e$ | $\langle Qv \rangle$ ($cm^3$ $sec^{-1}$) | $5 \times 10^{-9}$ | $3 \times 10^{-9}$ |
| $X_2^* + X_2^* \rightarrow X_2^+ + 2X + e$ | $\langle Qv \rangle$ ($cm^3$ $sec^{-1}$) | $8 \times 10^{-11}$ | $3 \times 10^{-11}$ |
| $X_2^*(^1\Sigma) \rightarrow 2X + h\nu$ | $\tau$ (sec) | $6.2 \times 10^{-9}$ | $5.8 \times 10^{-9}$ |
| $X_2^*(^3\Sigma) \rightarrow 2X + h\nu$ | $\tau$ (sec) | $1 \times 10^{-7}$ | $6.5 \times 10^{-7}$ |
| $X_2^*(^1\Sigma) + e \rightarrow X_2^*(^3\Sigma) + e$ | $\langle Qv \rangle$ ($cm^3$ $sec^{-1}$) | $2.3 \times 10^{-7}$ | $1.5 \times 10^{-7}$ |
| $X_2^*(^1\Sigma) + X \rightarrow X_2^*(^3\Sigma) + X$ | $\langle Qv \rangle$ ($cm^3$ $sec^{-1}$) | $4.3 \times 10^{-13}$ | $2.8 \times 10^{-13}$ |
| $X_2^* + h\nu \rightarrow X_2^* + e$ | $Q$ ($cm^2$) | $1.9 \times 10^{-18}$ | $8.0 \times 10^{-19}$ |

† - $kT = kT = 0.025$ eV

direct electron capture followed by dissociation, i.e. dissociative recombination

$$e + X_2^+ \rightarrow (X_2^*)_{\text{unstable}} \rightarrow X + X^*$$

cluster-ion formation via three-body collisions with ground state atoms

$$X_2^+ + X + X \rightarrow X_3^+ + X$$

Cluster ions may also be lost by dissociative recombination.

The excited atomic species may also participate in destructive processes of the form

iii) $e + X^* \rightarrow X^+ + 2e$

iv) $X^* + X^* \rightarrow X^+ + X + e$

the latter being Penning ionization, which has not been measured in the heavier rare gases!

Once formed, the two molecular states $^1\Sigma_u^+$ and $^3\Sigma_u^+$ may decay via several processes, of primary importance is radiation, either by spontaneous or stimulated emission, for which a knowledge of the radiative lifetimes is necessary. Both excimers may undergo destructive processes of the form

$$X_2^* + e \rightarrow X_2^* + 2e$$

$$X_2^* + X_2^* \rightarrow X_2^+ + 2X + e$$

In addition to the above excited state loss processes, the kinetic model of Werner *et al*. incorporates superelastic electron collisional deactivation of the excited atomic and molecular states.

An important loss mechanism of the excimer states during laser oscillation involves photoionization by a UV photon. This process reduces the net optical gain of the system. Calculations of these cross sections have been performed by Lorents [13] using a quantum defect method and are thought to be accurate to within a factor of two!

Two reactions play a key role in determining relative populations

$$e + X_2^* (^1\Sigma) \rightleftarrows e + X_2^* (^3\Sigma)$$

$$M + X_2^* (^1\Sigma) \rightleftarrows M + X_2^* (^3\Sigma)$$

where M designates a heavy atomic collision partner. Collisional mixing by heavy particles is critical in determining the relative populations of the singlet and triplet states, yet there is not sufficient evidence to know precisely what should go into a model.

## 4.2 Rare-Gas-Halide Lasers

The earlier excimers resulted in lasers whose over-all efficiency was limited by the ratio of the stimulated - emission cross section to that of photoionization. In the latter half of 1975 many excimer lasers were discovered which did not have this difficulty. These lasers, emitting in the ultra violet include

Xe F , Xe Br , Xe Cl and Kr F

The KrF has the highest efficiency because several processes are favourable.

$$e + (Ar\ ,\ Kr\ ,\ NF_3) \rightarrow Ar^+\ ,\ Ar^*\ \text{(predominantly)}$$

$$Ar^+ \xrightarrow{\text{kinetic processes}} Ar^*$$

$$Ar^* + Kr \rightarrow Ar + Kr^* \qquad \text{(100\% efficient)}$$

$$Kr^* + (NF_3, F_2) \rightarrow Kr\,F^*$$

The high efficiency of these energy transfer processes means that nearly every Ar* , whose formation requires about 20eV of e-beam energy, results in a 5eV KrF laser photon giving a theoretical efficiency of almost 25% , consistent with observations.

Rokni *et al*. [14] have given a detailed discussion of rare gas fluoride lasers, while the most comprehensive description of the KrF is given by Cohn and Lacina [15]. In Fig. (13) the dominant formation kinetics are shown schematically.

The different measured quenching rate constants of Kr F* have been collected together by Rokni *et al* and are given in Table (3).

There are three main methods for pumping the rare-gas monohalides:

a) pure electron-beam pumping;
b) electron-beam controlled discharge pumping;
c) UV preionized (avalanche) discharge pumping;

In methods (b) and (c) most of the pump power is

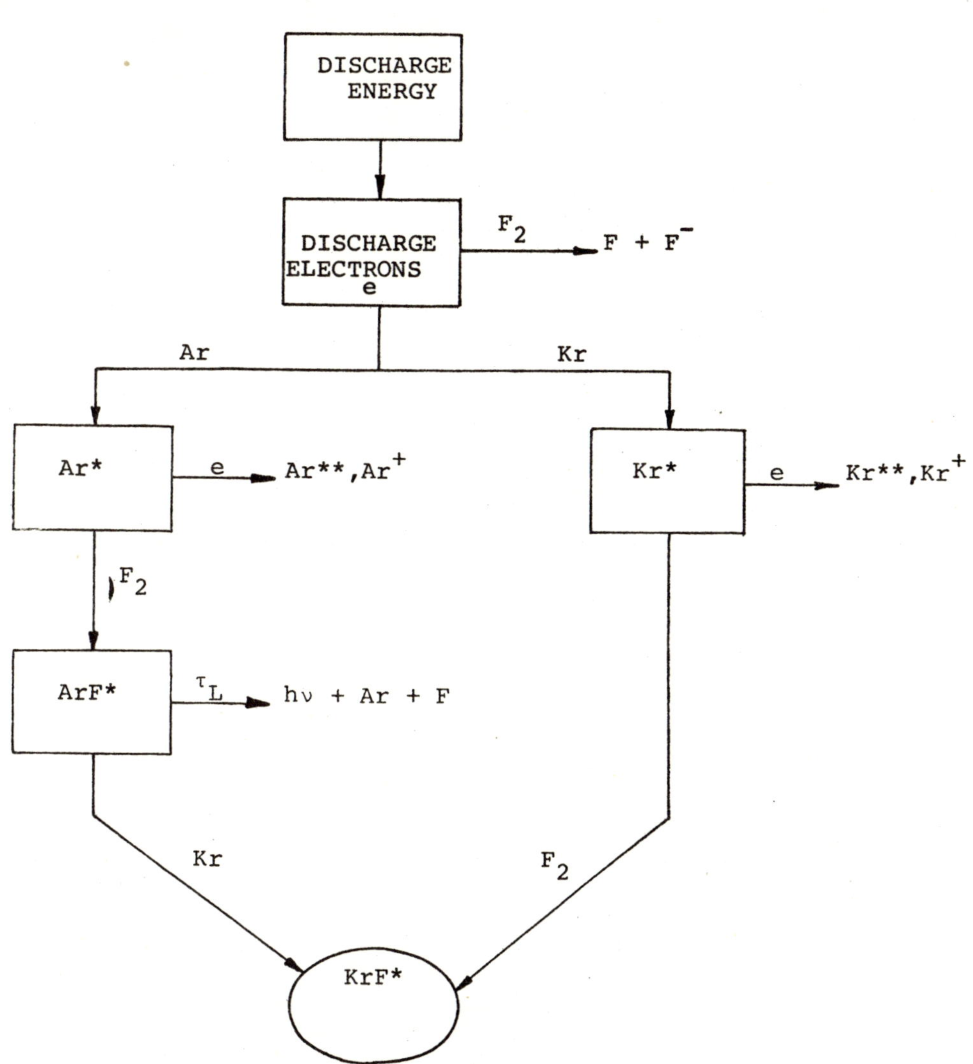
DISCHARGE ENERGY
DISCHARGE ELECTRONS e
$F_2$
F + $F^-$
Ar
Kr
Ar*
e
Ar**, $Ar^+$
Kr*
e
Kr**, $Kr^+$
$F_2$
ArF*
$\tau_L$
hν + Ar + F
Kr
$F_2$
KrF*

Fig.13

TABLE 3 DOMINANT QUENCHING PROCESSES OF $KrF^*$

| REACTION | $\kappa\tau_R$ ($KrF^*$) | $\kappa$ ($\tau_R$ = 6.5 nsec) |
|---|---|---|
| $KrF^* + F_2 \rightarrow$ PRODUCTS | $5 \times 10^{-18}$ $cm^3$ | $7.8 \times 10^{-10}$ $cm^3$ $sec^{-1}$ |
| $KrF^* + 2Kr \rightarrow Kr_2F^* + Kr$ | $4.4 \times 10^{-39}$ $cm^6$ | $6.7 \times 10^{-31}$ $cm^6$ $sec^{-1}$ |
| $KrF^* + Kr \rightarrow$ PRODUCTS | $\leq 1.1 \times 10^{-20}$ $cm^3$ | |
| $KrF^* + Kr + Ar \rightarrow Kr_2F^* + Ar$ | $4.2 \times 10^{-39}$ $cm^6$ | $6.5 \times 10^{-31}$ $cm^6$ $sec^{-1}$ |
| $KrF^* + 2Ar \rightarrow$ PRODUCTS | $4.6 \times 10^{-40}$ $cm^6$ | $7 \times 10^{-32}$ $cm^6$ $sec^{-1}$ |

provided by the discharge. The discharge physics is strongly affected by electron impact excitation and ionization of the rare gas metastables. To model these effects, Rokni <u>et al</u> have considered the Krypton metastable as rubidium and the argon metastable as potassium. Some of their electron impact cross sections are shown in Fig. (14). These cross sections are input into a code, similar to BOLTZ, which solves the Boltzmann electron transport equation. Clearly it is not satisfactory to be guessing these cross sections in this way. One of the few measurements of these cross sections has been reported by Fel'tsan [16].
In Table (4) we present a list of the important rate constants used in the analysis of Cohn and Lacina [15]. The theoretical model of these authors begins with the solution of the electron transport Boltzmann equation which requires cross sections for

elastic and inelastic electron-neutral collisions;
momentum transfer with recoil;
electron-electron scattering;
superelastic collisions of electrons with excited states.

The model also includes external circuit equations to describe the discharge current and voltage. These calculations are coupled to the master equation for the population densities of the electrons, ions, excited states and excimers-seventy reactions were included! These authors also used a code to translate symbolic reactions into rate equations, as done by Roberts, see Ref [17], in his plasma chemistry model of the $CO_2$ laser.

### 4.3 Boltzmann Transport Equation

The electron velocity distribution function $f(\underset{\sim}{v}, \underset{\sim}{r}, t)$ for a multicomponent gas at temperature T in an external electrical field $\underset{\sim}{E}$ satisfies

$$\left[ \frac{\partial}{\partial t} + \underset{\sim}{v} \cdot \underset{\sim}{\nabla}_r - \frac{e}{m} \underset{\sim}{E}(\underset{\sim}{r}, t) \cdot \underset{\sim}{\nabla}_v \right] f(\underset{\sim}{v}, \underset{\sim}{r}, t) = \left. \frac{\delta f}{\delta t} \right|_c$$

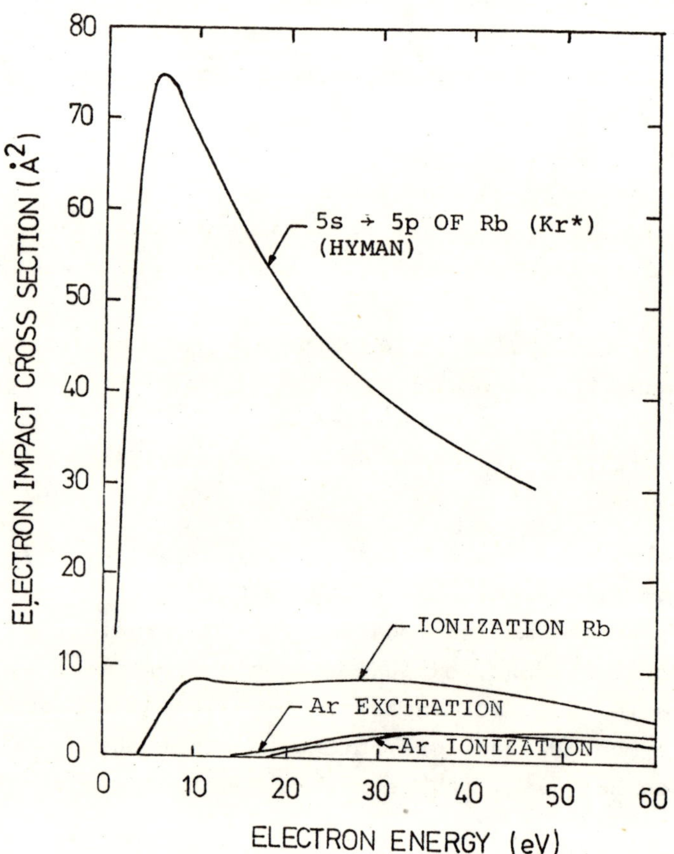
80
70
60
50
40
30
20
10
0
ELECTRON IMPACT CROSS SECTION (Å$^2$)
5s → 5p OF Rb (Kr*)
(HYMAN)
IONIZATION Rb
Ar EXCITATION
Ar IONIZATION
0
10
20
30
40
50
60
ELECTRON ENERGY (eV)

Fig.14

Table 4. Important Kinetic Processes and Rate Constants Assumed for the Present Analysis.

| Reaction | ($s^{-1}$, $cm^3s^{-1}$, $cm^6s^{-1}$, or $cm^2$) |
|---|---|
| $(Ar,Kr) + e \rightarrow (Ar,Kr)^* + e$ | Calc. |
| $(Ar,Kr) + e \rightarrow (Ar,Kr)^+ + 2e$ | Calc. |
| $(Ar,Kr) + \vec{e} \rightarrow (Ar,Kr)^* + \vec{e}$ | (1.2,2.3)(-18) |
| $(Ar,Kr) + \vec{e} \rightarrow (Ar,Kr)^+ + e + \vec{e}$ | (4.3,8.0)(-18) |
| $(Ar,Kr)^* + e \rightarrow (Ar,Kr)^+ + 2e$ | Calc. |
| $F_2 + e \rightarrow F + F^-$ | Calc. |
| $Ar^+ + Ar + M \rightarrow Ar_2^+ + M$ | 2.0(-31) |
| $Ar^+ + Kr + M \rightarrow ArKr^+ + M$ | 2.5(-31) |
| $Kr^+ + Kr + M \rightarrow Kr_2^+ + M$ | 2.4(-31) |
| $Kr^+ + Ar + M \rightarrow ArKr^+ + M$ | 1.0(-31) |
| $Ar_2^+ + Kr \rightarrow Kr^+ + Ar + Ar$ | 7.5(-10) |
| $ArKr^+ + Kr \rightarrow Kr_2^+ + Ar$ | 3.2(-10) |
| $(Ar,Kr)_2^+ + e \rightarrow (Ar,Kr)^* + (Ar,Kr)$ | Calc. |
| $(Ar,Kr)_2^+ + F^- \rightarrow (Ar,Kr)F^* + (Ar,Kr)$ | 5.0(-7) |
| $\rightarrow (Ar,Kr)_2F^*$ | 5.0(-7) |
| $(Ar,Kr)^+ + F^- \rightarrow (Ar,Kr)F^*$ | 1.0(-6) |
| $Kr^* + F_2 \rightarrow KrF^* + F$ | 7.2(-10) |
| $Ar^* + F_2 \rightarrow ArF^* + F$ | 7.5(-10) |
| $KrF^* + Ar + M \rightarrow ArKrF^* + M$ | 6.0(-32) |
| $KrF^* + Kr + M \rightarrow Kr_2F^* + M$ | 5.0(-31) |
| $ArF^* + Ar + M \rightarrow Ar_2F^* + M$ | 4.0(-31) |
| $ArF^* + Kr \rightarrow KrF^* + Ar$ | 1.5(-10) |
| $Kr_2^* + F \rightarrow KrF^* + Kr$ | 3.0(-10) |
| $Kr_2^* + F_2 \rightarrow Kr_2F^* + F$ | 3.0(-10) |
| $Kr_2F^* + F_2 \rightarrow 2Kr + F + F_2$ | 1.0(-9) |
| $ArKrF^* + Kr \rightarrow Kr_2F^* + Ar$ | 1.0(-10) |
| $Ar_2F^* \rightarrow Ar + Ar + F$ | 2.0(8) |
| $Kr_2F^* \rightarrow Kr + Kr + F$ | 6.7(7) |
| $KrF^* \rightarrow Kr + F + h\nu$ | 1.6(8) |
| $KrF^* + h\nu \rightarrow Kr + F + h\nu$ | 2.0(-16) |
| $F_2 + h \rightarrow F + F$ | 1.5(-20) |
| $F^- + h \rightarrow F + e$ | 5.4(-18) |
| $Ar_2^+ + h\nu \rightarrow Ar + Ar^+$ | 1.5(-17) |
| $Kr_2^+ + h\nu \rightarrow Kr + Kr^+$ | 1.0(-18) |

Note: Number in ( ) denotes exponent of 10. e and $\vec{e}$ denote secondary and high energy beam electrons.

where the right-hand side collision terms includes the effects of elastic and inelastic scattering processes. It is usual to neglect the first and second terms and solve the equation by expanding $f(\underset{\sim}{v})$ in Legendre Polynomials

$$f(\underset{\sim}{v}) \simeq f_o(v) + \frac{\underset{\sim}{v}}{v}\underset{\sim}{f}_1 \quad (v) \quad + \quad (\frac{3\underset{\sim}{v}\underset{\sim}{v}}{v^2} - \hat{1}) : \hat{f}_2 \; (v) + \ldots$$

where the hat denotes second rank tensors (dyadics). Usually, a two-term approximation is taken. The justification is related to assumptions about the degree of anisotropy to be expected for the velocity distribution. If the electric field is small enough that the directed speed of the electrons, as measured by their drift velocity, is much less than their random thermal velocities, then the small first-order perturbation.

$$\underset{\sim}{v} \cdot \underset{\sim}{f}_1$$

from an isotropic distribution, $f_o$, should be good enough. For high values of E/N, characteristic of self-sustaining discharges, Bailey [18] has speculated that the $f_2$ term may be comparable to the $f_1$ term, although no analysis has included these terms until the work of Shkarofsky [19].

## 5. MERCURY MONOHALIDE LASERS

Unfortunately, the UV wavelength of the rare-gas-halide lasers are unsuitable for many applications and so UV to visible conversion schemes are sometimes necessary, see Djeu and Burnham [20]. The mercury-halides are attractive candidates in the visible.

A year ago, Parks [21] reported a new high-power visible laser operating on the

$$B\,\Sigma^+_{\frac{1}{2}} \rightarrow X^2\,\Sigma^+_{\frac{1}{2}}$$

transition of Hg Cl at 5576 Å (v' = 0 → v" = 22). This bound-bound laser transition originates on the first excited state of Hg Cl, which is predominantly ionic in nature correlating with the separated ions

$$Hg^+\,(^2S_{\frac{1}{2}}) + Cl^-\,(^1S_o)$$

This ionic character of the upper laser level provides the opportunity to use the highly efficient formation processes important in lasers such as KrF, for example

$$Hg^{*} + CCl_4 \rightarrow HgCl^{*} + CCl_3$$

which has a reactive cross section of 34Å$^2$ and a branching ratio into the upper laser level of near unity.

Parks [22] has also reported laser action on the corresponding band of HgBr at 5018 Å which he achieved by e-beam excitation of a mixture of Hg, Ar, and a halogen bearing hydrocarbon.

Laser action on the B $^2\Sigma^+ \rightarrow$ X $^2\Sigma^+$ transition of the Hg Br radical by photodissociating Hg $Br_2$ in the vapour phase has been reported by Schimiltschek _et al_. [22]. The parent molecule dissociates and recombines according to

$$Hg\,Br_2 + h\nu_{pump} \rightarrow Hg\,Br\,(B^2\Sigma^+) + Br$$

$$Hg\,Br\,(B^2\Sigma^+) \rightarrow Hg\,Br(X^2\Sigma^+) + h\nu_{fluor.}$$

$$Hg\,Br\,(X^2\Sigma^+) + Br + M \rightarrow Hg\,Br_2 + M.$$

Further work on electron-beam mercury monohalides has been carried out by Eden [24] and Witney [25], any theoretical model of this laser will require accurate rates and cross sections for the following system of reactions.

$$e + Ar \rightarrow Ar^+ + 2e$$

$$Ar^+ + 2Ar \rightarrow Ar_2^+ + Ar$$

$$Ar_2^+ + Xe \rightarrow Xe^+ + 2Ar$$

$$Xe^+ + Xe + M \rightarrow Xe_2^+ + M \qquad (M = Ar, Xe)$$

$$Xe_2^+ + Hg \rightarrow Hg^+ + 2Xe$$

$$CCl_4 + e \rightarrow Cl^- + CCl_3$$

and

$$Hg^+ + Cl^- + M \rightarrow HgCl\,(B^2\Sigma) + M\,.$$

## REFERENCES


1. E. U. Condon, in *Handbook of Physics*, edited by E. U. Condon and H. Odishaw (McGraw Hill, New York, 1967), 2nd Edition, p.6-3.

2. G. Herzberg, *Molecular Spectra and Molecular Structure, I. Spectra of Diatomic Molecules* (van Nostrand, New Jersey, 1967), 2nd Edition, p.212

3. M. L. Bhaumik, Laser Focus, 12, 54 (1976).

4. A. R. Davies, K. Smith and R. M. Thomson, Comp. Phys. Comm. 10, 117 (1975).

5. J. Ross, J. C. Light and K. E. Schuler, in *Kinetic Processes in Gases and Plasmas*, edited by A. R. Hochstim (Academic Press, New York, 1969), p.281.

6. K. Smith and R. M. Thomson, *Computer Modeling of Gas Lasers* (Plenum, New York, 1978), p.96

7. A. R. Davies, S. A. Roberts, K. Smith and R. M. Thomson, Report No. 105, Department of Computer Studies, University of Leeds, November (1977).

8. R. N. Schwartz, Z. I. Slawsky, K. F. Herzfeld, J. Chem. Phys. 20, 1591 (1952)

9. J. Keck and G. Carrier, J. Chem. Phys. 43, 2284 (1965)

10. H. K. Shin, J. Chem. Phys. 42, 59 (1965)

11. R. D. Sharma and C. A. Brau, J. Chem. Phys. 50, 924 (1969)

12. C. W. Werner, E. V. George, P. W. Hoff and C. K. Rhodes, IEEE J. Quantum Electron. QE-13, 769 (1977).

13. D. C. Lorents, Physica 83C, 19(1976), and a more recent report from SRI, *Kinetic Processes in Rare Gas Halide Lasers*.

14. M. Rokni, J. A. Mangano, J. H. Jacob and J. C. Hsia *Rare Gas Fluoride Lasers*, Avco Everett Res.Lab., (1978)

15. D. B. Cohn and W. B. Lacina, *High Power Short Wavelength Laser Development*, Northrop Corp. NRTC-77-43R (1977)

16. P. V. Fel'tsan, Ukr. Fiz. Zh. 12, 1425 (1967)

17. S. A. Roberts, K. Smith and R. M. Thomson, Report No. 92 Centre for Computer Studies, University of Leeds (1976)

18. W. F. Bailey, Paper DD5 presented at the 25th Gaseous Electronics Conf., London, Ontario, Canada, October (1972)

19. I. P. Shkarofsky, *Study of Electrostatic Probes in Non-Equilibrium Plasmas*, Wright-Patterson AFB, TR75-0228 June (1975).

20. N. Djeu and R. Burnham, Appl. Phys. Letts. 30, 473 (1977)

21. J.H. Parks, Appl. Phys. Letts. 31, 192 (1977)

22. J.H. Parks, Appl. Phys. Letts. 31, 297 (1977)

23. E.J. Schimitschek, J.E. Celto and J.A. Trias, Appl. Phys. Letts. 31, 608 (1977)

24. J.G. Eden, Appl. Phys. Letts. 31, 448 (1977)

25. W.T. Witney, Appl. Phys. Letts. 32, 239 (1978)

# MOLECULAR FORMATION AND DESTRUCTION IN THE INTERSTELLAR MEDIUM

H. VAN REGEMORTER, A. GIUSTI SUZOR, E. ROUEFF

Observatoire de MEUDON - 92190 MEUDON

Abstract

A brief review is given of the most important processes of formation and destruction of diatomic molecules in gaz phase reactions : the radiative association, the photodissociation, the predissociation, the inverse predissociation and the dissociative recombination.

## Introduction

The study of interstellar molecules in the interstellar clouds, which can be considered as the sites of star formation, implies a practical cooperation between atomic physicists, quantum chemists and astrophysicists. Thanks to many recent observations by radiotelescopes or at optical ultraviolet wavelengths (in particular with satellite Copernicus) the astrophysicists are prepared to explain the physical conditions (temperature, density, chemical composition, thermal and ionization balance) as well as the evolution of these gravitationnally collapsing interstellar clouds.

In brief, one can consider two different kinds of clouds, the diffuse clouds, in which the temperature is of the order of 100 K and the densities of the order of 10 to 100 $cm^{-3}$, and the dense clouds, where the temperature is below 100 K and the densities are between $10^3$ and $10^6$ $cm^{-3}$. The essential difference between them is the presence of visible, UV galactic radiation (below the H ionization limit at 13,6 eV) inside the diffuse clouds, which can ionize most of the atoms and dissociate the molecules. In the dense clouds, only the high energy cosmic rays can penetrate and the constituants are mainly molecules. Hydrogen is by far the most abundant element (90 %).

From studies of interstellar extinction of starlight the presence of solid particles has been proved. These dust grains play an important rôle in the formation of interstellar molecules. It has been confirmed that two hydrogen atoms are converted into a $H_2$ molecule at every collision in presence of an interstellar grain.

In the present paper, we shall concentrate on physical processes for reactions in the gas phase, excluding all aspects relevant to surface and solid state physics. We shall also exclude all molecular excitation processes by collision which are necessary to obtain the thermal balance of these regions which are very far of being in thermodynamical equilibrium.

The specific conditions of the interstellar clouds, very low temperature and very low density, can explain at the same time the interest and the difficulties of the problem of molecular formation and destruction.

From the experimental point of view essentially no experiment has been made below 300 K and it is very dangerous to extrapolate the rate coefficient when some reaction channels are the most important above or below a given temperature. Some reactions are not detectable in the laboratory, like the radiative association, the excess of energy being absorbed by a third particle of the plasma or by the wall of the cell. If it is true to say that the theoretical approach is of considerable importance for explaining these processes, one has to have in mind that the most elaborate and expensive calculations of potential curves and surfaces for molecular complexes give energies at best to 0,05 eV. This incertitude can lead to considerable errors for the cross sections or the rate coefficients at a temperature of 100 K corresponding to a collision energy of 0,01 eV !

A few very good review articles by [1]W.D. WATSON (1976), [2]A. DALGARNO (1975) [3]A. DALGARNO and J. BLACK (1976) have shown that considerable progress have been made recently for explaining the formation and the destruction of interstellar molecules.

In the following, we shall consider briefly a few elementary processes involving diatomic molecules, like the radiative association, the photodissociation and predissociation, the dissociative recombination. Chemical or reactive reactions will not be considered although they are very important in the cloudy medium : this field of study will certainly develop in the near future both in the laboratory and within accurate theoretical treatments.

## I - Direct Radiative Association

At the very low densities of the interstellar clouds - from 10 to $10^6$ $cm^{-3}$ - only two body association can occur. The excess of energy cannot be absorbed by the surrounding particles but can only appear as a photon emission. This is the process of radiative association

$$X + Y \longrightarrow (XY)_i \longrightarrow XY + h\nu$$

in which two particles approach along a particular attractive excited state $(XY)_i$ which can emit a photon leaving the molecule in a bound state XY (see figure I).

The probability of emission during the time of collision can be written in the form

$$P = \int_{-\infty}^{+\infty} g\, A\, dt \simeq g\, A\, (R_o)\, T$$

where g is the probability of populating the excited state $(XY)_i$ with

$(o < g < 1)$

A $(R_o)$ is the transition probability at the internuclear distance $R_o$ corresponding to the minimum of the ground potential curve. A is of the order of $10^{+7}\ s^{-1}$ for a dipolar transition, and T is the collision time of the order of $10^{-14}$ s.

This gives a probability of the order of $10^{-7}$ and a reaction rate $\alpha = (v\sigma)$ with $\sigma = 2\pi \int P(b)\, b\, db$ of the order of $10^{-16}\ cm^3 s^{-1}$.

From a semi-classical point of view([4]BATES, 1951), this reaction rate has the form

$$\alpha = \bar{v}\ g\ A\ (R_o)\ T\ \pi b_c^2$$

where $\pi b_c^2$ is the capture cross section, $b_c$ being the orbiting parameter such that for a classical collision with $b > b_c$ the relative particle cannot surmount the centrifugal barrier and enter the emission region unless tunelling occurs.

Astrophysical interest in radiative association has centered mainly on the $C^+$ + H collisions, in order to explain the $CH^+$ abundance and the $CH/CH^+$ abundance ratio.

In the case of $CH^+$, $^1\Pi \to {}^1\Sigma$ is the strong transition involved in the radiative association. As one sees on the correlation diagram of figure 2, the èxcited state $^1\Pi$ is not correlated to the initial state $C^+ (P_{1/2})$ + H (IS). The $^1\Pi$ level can only be populated through non adiabatic collisions.

A semi-classical treatment has been developped by [5]GIUSTI et al. (1976) which takes account of such collisional effects as well as tunnelling effects through the centrifugal barrier. In the equivalent quantal treatment, the rate can be expressed as $\alpha = g\,\bar{v}\,\sigma_{em}$ where the emission cross section $\sigma_{em} = \sum_\ell (2\ell + 1)\,\sigma_\ell$ is given in terms of partial cross section $\sigma_\ell$ proportionnal to

$$\sum_{v'} \nu_{v'E} \left| \int P_{v'\ell}(R)\, M(R)\, F_{E\ell}(R)\, dR \right|^2$$

M (R) is the electronic transition moment, $P_{v'\ell}$ is the vibrationnal final radial wave function, $F_{E\ell}$ the free initial wave function of the nuclear motion, $\nu_{v'E}$ is the photon frequency.

In the quantum calculations, peaks appear in the emission cross section which are due to shape resonances, i.e. the formation of quasi bound states in the well of the excited state effective potential. These resonance effects give a small increase of the rate coefficient, and for $T < 150$ K the quantum results exceed the semi-classical ones by at most 25 %. The rate at 100° K, 17 $10^{-18}\ cm^3\ s^{-1}$, is too weak to explain the discrepancy between observed and theoretically deduced abundances of the interstellar $CH^+$. (see ABGRALL et al. (1976).

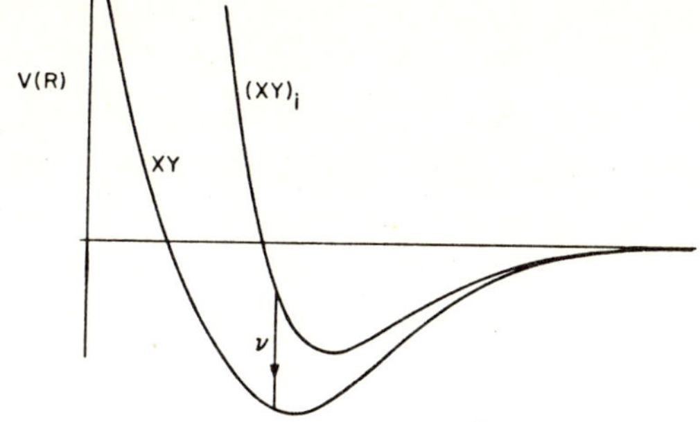


Figure 1. Radiative association of X and Y which approach along $(XY)_i$ and radiate to XY.

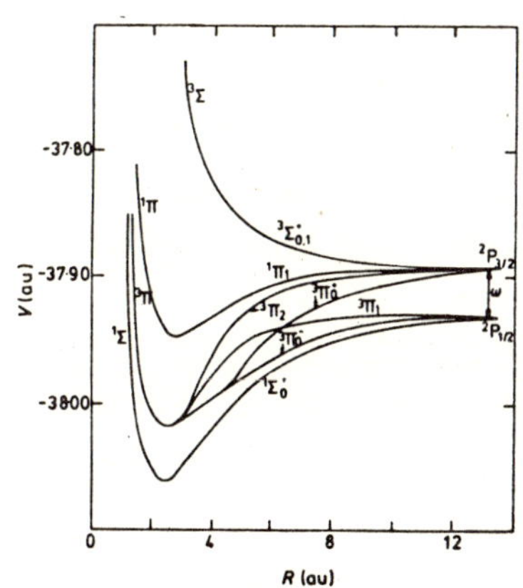


Figure 2. Correlation of atomic and molecular states in the $C^+$ + H collision. The transition involved in the radiative association is $^1\Pi \rightarrow {}^1\Sigma$ . Fine structure splitting $\omega$ is exaggerated. (from GIUSTI-SUZOR et al., 1976)[5]

Additional natural barrier may exist in the excited potential curve as in the case of state $B^2\Sigma$ of CH ([7]BAIN and BARDSLEY, 1972). This small barrier of 500 $cm^{-1}$ is higher than the mean kinetic energy at 100 K (70 $cm^{-1}$). Small incertitudes on the potential curve and on the dissociation energy can lead to serious errors in the cross sections.

The calculation of the direct radiative association process for polyatomic molecules is very difficult and involves handling the nuclear motions classically along an excited state surface potential. [8]HERBST and DELOS (1976) have studied the formation of $CH_2^+$ in its first excited $^2B_1$ state from low energy collisions of $C^+ + H_2$ for which a rate of the order of $10^{-14}$ $cm^3$ $s^{-1}$ at 90° K has been found.

Some other indirect processes of radiative association can be considered as the inverse of predissociation processes. Therefore, we shall first consider the question of molecular destruction by photodissociation.

## II - Molecular destruction by photodissociation

Photodissociation is the most effective process of molecular destruction in the diffuse clouds, in which all photons with wavelengths larger than the Lyman limit of 912 Å are present.

Direct photodissociation by absorption of a photon by a molecule in its ground state, either to a repulsive electronic state or to the vibrational continuum of an attractive electronic state, is most of the time a weak process. Following [9]ALLISON and DALGARNO (1971) this can be understood from the sum rule of oscillator strengths : most part of the photon absorption is going into discrete states, the bound bound oscillator strengths being much stronger than the strength of a bound free transition to vibrational continuum. But many dissociation channels are, in fact, unknown and there are few reliable data on direct photodissociation of interstellar molecules. If new experimental techniques have been developped recently, mostly for the measurement of photodissociation of positive molecular ions (see [10] J. DURUP, 1977), a comprehensive review on photodissociation of polyatomic molecules by [11]W.M. GELBART (1977) shows that in this case, because of the many degrees of freedom involved, theoretical work is still in its infancy.

Predissociation is also present as a process of photodestruction of interstellar molecules. We shall consider two different modes of predissociation before discussing the spontaneous radiative dissociation which is the process of destruction of $H_2$ and HD in the interstellar clouds. Predissociation by rotation - through a potential centrifugal barrier - in diatomic molecules has been reviewed recently by [12] M.S. CHILD (1974).

### A. Predissociation involving a dissociative electronic state

The process is illustrated in figure 3. Absorption takes place into an attractive excited state coupled to the continuum of the dissociating state. Absorption into all vibrational levels with $v > 1$ may be followed by molecular dissociation.

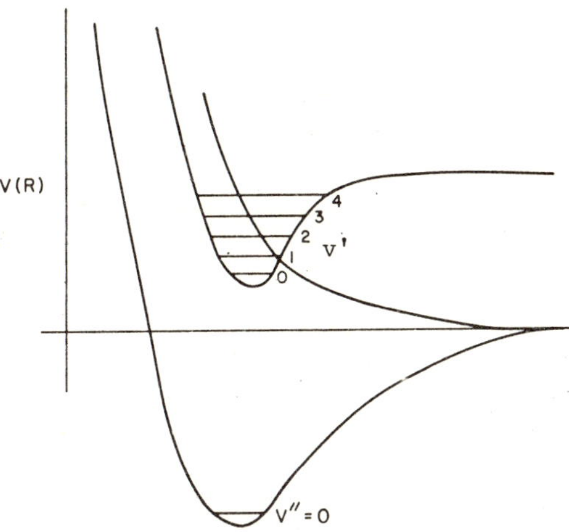


Figure 3. Predissociation following absorption into $v' \geq 1$.

The total life time t of the discrete level embedded in the continuum is given by

$$t^{-1} = A^r + A^d$$

where the radiative transition probability $A^r$ corresponds to the photon spontaneous reemission and where the predissociation transition probability may be due to different couplings (spin orbit, coupling between the electron motion and the nuclear motion).

In the common case, when $A^d \gg A^r$, all absorption is followed by predissociation which is more rapid than photon reemission, and the predissociation rate is simply given in terms of discrete band oscillator strengths

$$f\ (v\ v') = \frac{4}{3}\ \Delta E \left| \langle \chi_v (R) \mid D (R) \mid \chi_{v'} \rangle \right|^2 \quad \text{in a.u.}$$

where D (R) is the transition moment in function of the internuclear distance, $\chi_v$ (R) and $\chi_{v'}$ (R) are the initial and final vibrational wave functions.

When the predissociation is not very strong ($A^d$ of the order or smaller than $A^r \sim 10^8$, the channel corresponding to the photon reemission competes with the predissociation channel and both $A^d$ and $A^r$ must be known from theory for calculating the predissociation rate.

From precise experimental determination of life times t using the high frequency deflection technique, and theoretical determination of the radiative life time $A_r^{-1}$, [14]BRZOZOWSKI et al. (1978) and [15]SMITH et al. (1976) have recently obtained good estimation of predissociation in OH, NO and NH.

B. Predissociation without any repulsive state

Predissociation can occur without any repulsive electronic state when the vibrational level of an excited attractive state is above the dissociation limit of the ground state, as one sees on figure 4.

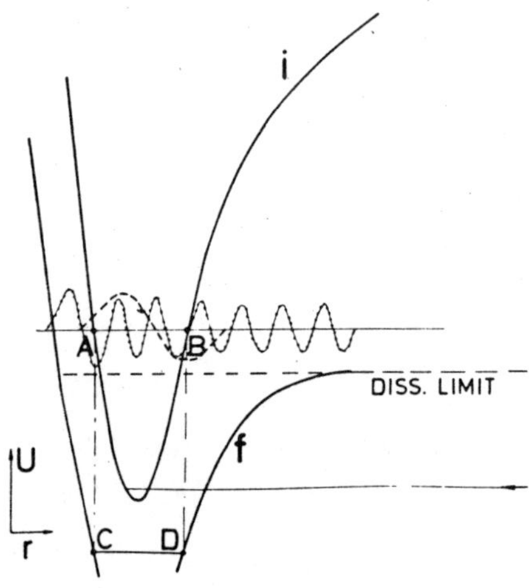


Figure 4. Weak predissociation of a state i via direct vibrational interaction with the continuum states of a lower attractive state f.

Such an interaction between two stable non crossing states, which can be due to spin rotational coupling, is expected to be weaker than the process involving a repulsive crossing state. It has been experimentally found in the CH radical by [16]BRZOZOWSKI et al. (1976) after its theoretical prediction by [17]JULIENNE and KRAUSS (1972).

Its interest is due to the importance of the inverse process which is a formation mechanism, as we shall see below. The predissociation probability $A^d$ being small, all molecular formation in the excited state is followed by photon emission and the formation of a stable molecule in the ground state.

C. The spontaneous radiative dissociation

$H_2$ or HD molecules cannot be destroyed by predissociation in the diffuse clouds, the photon energy being below 13.60 eV, the Lyman limit. As one sees from the level diagram of $H_2$ in figure 5, the excitation of free nuclear excited states from the ground state is energetically impossible. Only can the vibrational levels of $^1\Pi_u$ and $^1\Sigma_u$ be excited and these excitations can be followed by spontaneous transitions to free nuclear levels of the ground state $^1\Sigma_g^+$.

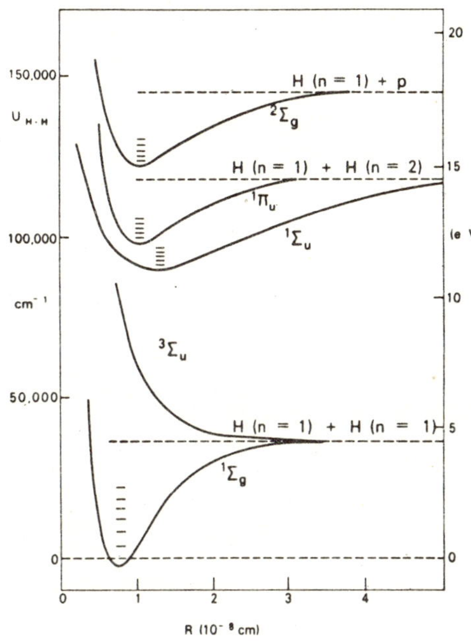


Figure 5. Electronic energy for low lying states of molecular hydrogen vs internuclear separation.

[18]DALGARNO and STEPHENS (1970) have shown that 23 % of the Lyman excitation lead to dissociation through $^1\Sigma_u^+ - {}^1\Sigma_g^+$ transitions.

Being due to the line absorption, this process decreases when the optical depth increases inside the cloud. HD being much less abundant than $H_2$ the photodissociation of the former is much more efficient by a factor of $10^4$ than for the last. The recent observations by satellite Copernicus give an abundance ratio of $^{HD}/H_2$ of the order of $10^{-6}$, a typical ratio in most of the interstellar clouds containing $H_2$. Knowing from observation the ratio $D/_H \sim 10^{-5}$, one can deduce that the formation mechanisms must be different for $H_2$ and HD, which is built by gas phase reactions, involving cosmic protons (see WATSON, 1976).

## III - Molecular formation by inverse predissociation

Inverse predissociation is similar to radiative association because the stabilization of the molecule is also achieved by emission of a photon.

The direct radiative association is not possible at thermal energies when no excited attractive state, from which a dipolar transition to the ground state is allowed, is connected to the initial state of the two atom system.

This is the case in OH of great astrophysical interest, shown in figure 6 where the repulsive state, i.e. the $^4\Sigma^-$ states crosses the excited intermediate state $A^2\Sigma^+$, an attractive state different from the state in which the atoms collide. Coupling between these two states are due to spin orbit interaction or to the breakdown of the Born Oppenheimer approximation. The excited $A\ ^2\Sigma^+$ state then radiates to the bound state and a stable molecule is formed.

This process

$$X + Y \rightarrow (X\,Y)_i \rightarrow (X\,Y)_m \rightarrow X\,Y + h\nu$$

where the quasi stationnary state $(X\,Y)_m$ is coupled to the continuum of the initial state $(X\,Y)_i$ , is the inverse of the predissociation process.

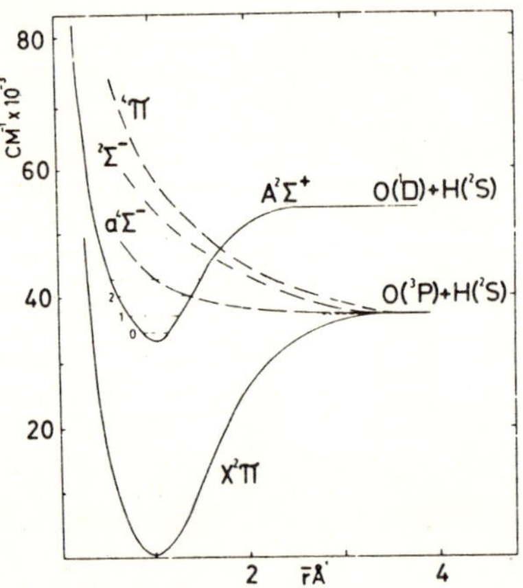


Figure 6.

Potential curves in OH (from [14] BRZOZOWSKI et al.(1978)).

The rate can be conveniently written in terms of the predissociation width $\Gamma^d$ and the spontaneous radiative width $\Gamma^r$ of all the quasi stationnary levels situated above the crossing point ($v > 2$ for $A^2\Sigma^+$ of OH, in figure 6) which are coupled to the free channel of the system (the two atoms approaching each other along the potential curve $^4\Sigma^-$).

From the theoretical point of view, this coupling gives rise to Fesbach type or "excited core" resonances. The predissociation width $\Gamma^d$ can be calculated using the configuration interaction theory of [19] FANO (1961).

Theoretical calculations have been done by [20] JULIENNE et al., 1973. A value of $\alpha$ (at 100° K) $\simeq$ 5 $10^{-20}$ $cm^3\ s^{-1}$ has been found for OH. From experimental determination of predissociation life times, a similar rate has been obtained recently by [14] BRZOZOWSKI et al. (1978). At low temperature this rate is very sensitive to an accurate determination of the ground state energy.

We have seen above that weak predissociation ($\Gamma^d < \Gamma^r$) can also occur without the presence of any <u>repulsive</u> state (see II.B). The inverse process may lead to important rates of molecular formation. At T = 100 K rates of $\alpha = 2.10^{-18}\ cm^3\ s^{-1}$ for CH and of $\alpha = 4.10^{-19}\ cm^3\ s^{-1}$ for OH have been found by ERMAN's group (see references 14 and 16).

In the case of OH, shown in figure 6, this additional inverse predissociation is due to the coupling between the vibrational continuum of the ground state $X\,^2\Pi$ and all the vibrational levels corresponding to the excited state $^2\Sigma^+$ which lie above the ground state dissociation limit.

## IV - Dissociative recombination

Electrons coming from the photoionization of atomic species with ionization potentials less than 13.6 eV (such as $C^+$) are also present in the interstellar space. Reactions of molecular ions with electrons, called recombination processes, are important because the involved rates are high. Two different channels are present, one being the radiative (or dielectronic) recombination and the other the dissociative recombination which is much more efficient for molecular ions when allowed.

In the dissociative recombination

$$A\,B^{+} \;+\; e \longrightarrow A\,B'' \longrightarrow A + B$$

the electron is captured into a repulsive state A B" of the neutral molecule which may dissociate, giving A + B, or autoionize giving back $A\,B^{+} + e$ (see figure 7).

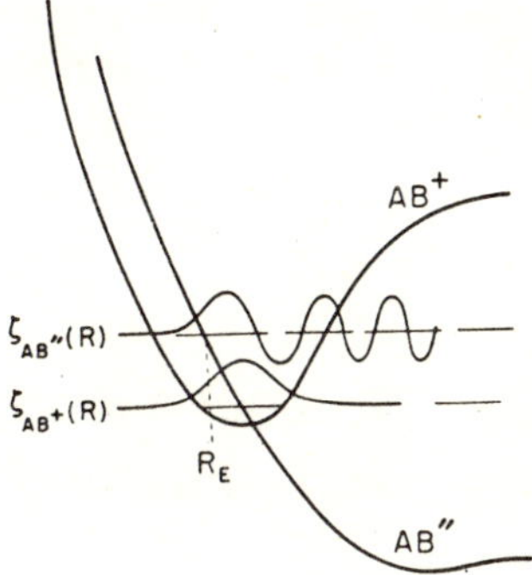


Figure 7. Hypothetical potential energy curves and associated wave functions for one of the states of the molecular ion $AB^{+}$ and of the unstable molecule AB" involved in the direct dissociative recombination process. (from [25] BARDSLEY and BIONDI (1970)).

The dissociative recombination is a resonance process which competes with the resonance scattering where the final channels are of the form $A\,B^{+} + e$ where $A\,B^{+}$ can be in its ground state (elastic) or in an excited state (inelastic scattering).

As well known in scattering physics, if $A\,B^{+}$ is replaced by a neutral molecule, the "dissociative attachment" is a special case of electron-molecule scattering where the intermediate resonance state dissociates into $A + B^{-}$ before it can decay by electron ejection into the continuum.[21] G.J. SCHULZ (1973) and[22] R. HALL (1976) and[23] (1978) have given excellent reviews on recent experiments concerning dissociative resonant states in electron-molecule scattering.

From the theoretical point of view, the exact formulation of the problem must include the coupling between the different final channels. Recent progress in a priori techniques in electron molecule scattering will give the possibility of including the dissociative final channel but no calculation has been done in this way up to now.

Following[24] BARDSLEY (1968) and[25] BARDSLEY and BIONDI (1970), who were the first to study the dissociative recombination, the cross section for this process can be approximated by the product of an electron capture cross section $\sigma_{cap}$ and a "survival factor" $S$ which is the probability that this state will decay by dissociation rather than by autoionization. Estimation of the last can be given in function of the autoionization width of the resonance state A B" assuming classical concepts of motion for the nuclei.

In a more accurate quantal treatment, the cross section for dissociative recombination is proportional to the outgoing flux of atoms

$$\lim_{R \to \infty} | \xi_{AB''} (R) |^2$$

where $\xi_{AB''}$ is the nuclear wave function of the dissociative state perturbed by the electronic continuum.

Assuming the Born Oppenheimer approximation, using the quasistationnary state formalism or the equivalent projection operator technique[27] O. MALLEY (1966),[24] BARDSLEY (1968) and[26] BOTTCHER (1974) are able to calculate $\xi$, taking account of both the electron capture process and the possibility of electron reemission through autoionization. When the last is weak, the cross section is just given by the capture cross section

$$\sigma_{cap} \text{ prop } | \langle \xi_{AB^+} (R) | V(R) | \xi_{AB''} (R) \rangle |^2$$

very sensitive to the Franck Condon overlap factor between $\xi_{AB''}$ and the discrete vibrational wave function $\xi_{AB^+}$ of the initial ion. This overlap is maximum when the ion curve $AB^+$ crosses the repulsive one near the turning point $R_t$ of the free motion in potential A B". V (R) is the electronic potential

$$\langle \Psi_{AB^+} (r R) \varphi_\varepsilon | H_{el} | \Psi_{AB''} (r R) \rangle$$

where $H_{el}$ is the electronic part of the molecule hamiltonian, the $\Psi$'s are the electronic wave functions of the molecular states involved and $\varphi_\varepsilon$ the wave function of the free incoming electron.

Difficulties appear at each step of the calculation. They have been discussed by[28] GIUSTI et al. (1977a) (1977b) in view of solving the problem of dissociative recombination of $CH^+$. (see also [41] RASEEV et al. (1978)).

Firstly, the relevant interatomic potential is very sensitive to the number of orbitals included in the basis set.

Secondly, the incoming electron cannot be represented by a Coulomb wave but $\varphi_\varepsilon$ must be calculated accurately or eventually using the quantum defect method.

Thirdly, as underlined by[24] BARDSLEY (1968), an indirect dissociative recombination may occur in which the incident electron is temporarily trapped into an excited vibrational level related to a Rydberg state, through an electronic-vibration coupling, i.e. a breakdown of the Born Oppenheimer approximation. This causes resonances in the cross section which were found for $H_2^+$ by the high resolution merged beam experiment of[30] Mc GOWAN et al. (1976) and may enhance the recombination rate at low temperature considerably, even after the Maxwell averaging. We must mention the semi-empirical quantum defect method of[31] LEE (1978) who calculates simultaneously the rate for direct and indirect process in $NO^+$ using all the spectroscopic data available for NO.

The extension of the theory to polyatomic ions is highly desirable. One can mention the theoretical attempt of[32] HERBST (1978) for $H_2CN^+$, $H_3O^+$, $NH_4^+$ and $CH_3^+$, all ions of astrophysical interest. On the experimental side, dissociative recombination can be studied by different ways. Amongst those, the merged beam technique developped by Mc GOWAN (see his paper in this volume) for electron-molecular ions collisions is particularly interesting because it can reproduce the very low energies found in the interstellar medium. Results have been obtained for $H_2^+$ [30](Mc GOWAN et al, 1976), $H_3^+$ [33](CAUDANO et al., 1975) and $CH^+$ [34](MITCHELL and Mc GOWAN, 1978).

## V - Chemical equilibrium in interstellar clouds

To conclude this brief review, it is interesting to take the point of view of the astrophysicist who has to build a self consistent model of the chemical composition and of the thermal balance of the clouds, taking account of the transfer of UV radiation at the surface of the cloud.

Assuming a given uniform density $n(H) = 10^4$ and a given temperature, let us say T = 30 K and taking account of the most important cooling processes due to collisional excitation by $H_2$ of $C^+$, C and molecular CO, it is necessary to know the chemical reaction scheme for the CO formation which is given in figure 8.

Only gas phase reactions are considered apart from the formation of $H_2$, CH an OH on grains in the work of[35] CLAVEL et al. (1978). The heating of the gas is mainly due to the cosmic ray ionization $H_2 + CR \rightarrow H_2^+$. Radiative association between $C^+$ and $H_2$ is taken into account to form CH. Chemical reaction involving the "carbon hydride" and the "oxygen hydride" cycles are important and contribute to the cooling.

The assumed gas temperature must be consistent with the temperature deduced from the CO rotational line observations by radiotelescopes. This kind of analysis has been so far applied to diffuse clouds by[36] GLASSGOLD and LANGER (1974),[37] OPPENHEIMER and DALGARNO (1975),[38] BARSUHN and WALMSLEY (1977), BLACK and DALGARNO (1977), and to dense clouds by[39] KUDO (1973) and[35] CLAVEL et al. (1978).

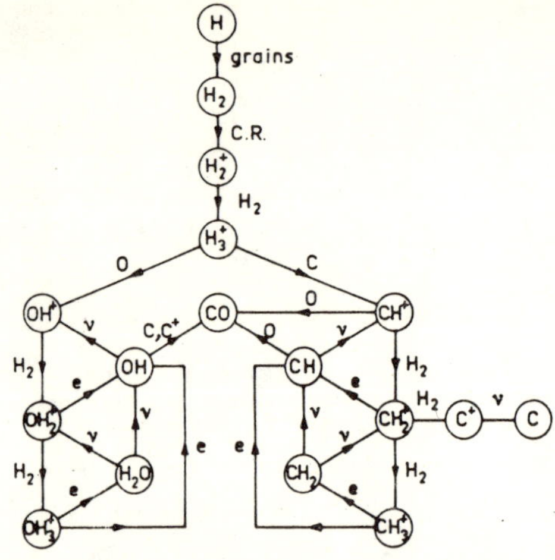


Figure 8. Chemical reaction scheme for the CO formation through the "carbon hydride" and "oxygen hydride" cycles.
(from[35] CLAVEL et al., 1978)

Although the "theoretical" chemical schemes suffer from many uncertainties, they permit to reproduce almost coherently the observed and theoretical abundances, and populations of the rotational levels of molecular hydrogen. Even more,[13] BLACK and DALGARNO (1977) predicted some unobserved species in the ζ Ophiuchi cloud. On these bases,[40] CHAFFEE and LUTZ (1978) discovered very recently the $C_2$ molecule in this particular cloud.

If considerable progress has been made recently towards a quantitative understanding of the formation and destruction processes for interstellar molecules, and if semi quantitative agreement between predicted and observed abundances can be achieved for small species, accurate tests are too rare to justify definite conclusions. This is mainly due to the poor actual knowledge of the physical conditions of the interstellar matter and of the processes of formation of large molecules (larger than diatomic) in gas phase reactions or in surface reactions on dust grains.

REFERENCES


[1] W.D. WATSON, Rev. Mod. Phys., 48, 315 (1976).

[2] A. DALGARNO, in Atomic and Molecular Processes in Astrophysics (1975) published by the Geneva Observatory.

[3] A. DALGARNO and J. BLACK, Rep. Prog. Phys. 39, 573 (1976).

[4] D.R. BATES, Mon. Not. R. Astron. Soc. 3, 303 (1951).

[5] A. GIUSTI-SUZOR, E. ROUEFF and H. VAN REGEMORTER, J. Phys. B Atom Molec. Phys. 9, 1021 (1976).

[6] H. ABGRALL, A. GIUSTI-SUZOR and E. ROUEFF, Astrophys. J 207, L69 (1976)

[7] R.A. BAIN and J.N. BARDSLEY, J. Phys. B 5, 277 (1972).

[8] E. HERBST and J.B. DELOS, Chem. Phys. Let. 42, 54 (1976).

[9] A.C. ALLISON and A. DALGARNO, J. Chem. Phys. 55, 4342 (1971)

[10] J. DURUP, in Etats atomiques et moléculaires couplés à un continuum, Colloque 273, C.N.R.S. (1977).

[11] W.M. GELBART, Ann. Rev. Phys. Chem. 28, 323 (1977).

[12] M.S. CHILD, in Molecular Spectroscopy Vol. 2, Ed. The Chemical Society, London (1974), p. 466

[13] J. BLACK and A. DALGARNO, Astrophys. J. Suppl. 34, 405 (1977).

[14] J. BRZOZOWSKI, P. ERMAN and M. LYYRA, Phys. Scripta 17, 507 (1978) and 14, 290 (1976).

[15] W.H. SMITH, J. BRZOZOWSKI, P. ERMAN, J. Chemical Phys. 64, 4628 (1976).

[16] J. BRZOZOWSKI, P. BUNKER, N. ELANDER, P. ERMAN, Astrophys. J. 207, 414 (1976).

[17] P. JULIENNE and M. KRAUSS in Molecules in the galactic environment, ed. M.A. Gordon and L.E. Snyder. Wiley (New York) 1973.

[18] A. DALGARNO and T.L. STEPHENS, Astrophys. Let. 160, 107 (1970).

[19] U. FANO, Phys. Rev. 124, 1866 (1961).

[20] P.S. JULIENNE, M. KRAUSS and B. DONN, Astrophys. J. 170, 65 (1971).

[21] G.J. SCHULZ, Rev. Mod. Phys. 45, 423 (1973).

[22] R. HALL in Physics of ionized gases. Dubrovnik Summer School, ed. B. Navinsek, Jugoslavia (1976).

[23] R. HALL in Electronic and Atomic Collisions. Proceedings of the Xth I.C.P.E.A.C., ed. G. Watel, North Holland p. 25 (1978).

[24] J.N. BARDSLEY , J. Phys. B 2, 365 (1968).

[25] J.N. BARDSLEY and M.A. BIONDI, Adv. Atom. and Mol. Phys. 6, 1 (1970).

[26] C. BOTTCHER, Proc. Roy. Soc. London A 340, 301 (1974).

[27] T.F. O'MALLEY, Phys. Rev. 162, 98 (1967).

[28] A. GIUSTI-SUZOR and H. LEFEBVRE-BRION, Astrophys. J 214, L101 (1977).

[29] A. GIUSTI-SUZOR, H. LEFEBVRE-BRION and E. ROUEFF, Xth I.C.P.E.A.C., Abstract of papers. Paris 1977.

[30] J.W. Mc GOWAN, R. CAUDANO and J. KAYSER, Phys. Rev. Let. 36, 1447 (1976).

[31] C.M. LEE, Phys. Rev. A 16, 109 (1977)

[32] E. HERBST, Astrophys. J. 222, 508 (1978).

[33] R. CAUDANO, S.F.J. WILK, J.W. Mc GOWAN, Xth I.C.P.E.A.C., Abstract of papers, 389 (1975).

[34] J.B.A. MITTCHEL and J.W. Mc GOWAN, Astrophys. J 222, L77 (1977).

[35] J. CLAVEL, Y.P. VIALA and N. BEL, Astron. Astrophys. 65, 435 (1978)

[36] A.E. GLASSGOLD and W.D. LANGER, Astrophys. J 193, 73 (1974).

[37] M. OPPENHEIMER and A. DALGARNO, Astrophys. J. 200, 419 (1975).

[38] J. BARSUHN and C.M. WALMSLEY, Astron. Astrophys. 54, 345 (1977).

[39] A. KUDO, Sci. Rep. Tohoku Univ. 56, 152 (1973).

[40] F.H. CHAFFEE and B.L. LUTZ, Ap. J. 221, L91 (1978).

[41] G. RASEEV, A. GIUSTI-SUZOR and H. LEFEBVRE-BRION, J. Phys. B 11, 2735 (1978).

# LASER - PHOTOCHEMICAL DISSOCIATION OF SMALL MOLECULES: CALCULATION OF ISOTOPE EFFECTS WITH RESPECT TO DISSOCIATION PROBABILITIES

H. Johansen, K. Johst

Zentralinstitut für Isotopen- und Strahlenforschung der Akademie der Wissenschaften der DDR, 705 Leipzig, Permoserstr. 15

## I. Introduction

An understanding of dissociation processes of polyatomic molecules is important in such fields as photochemistry, mass spectroscopy, and chemical reaction theory. The present report investigates the isotope effect on the photodissociation of a polyatomic molecule. Considerable interest is being focused upon the problem of laser isotope separation because of its technological importance as well as its inherent scientific interest.

New light sources, such as laser and synchroton radiation, makes it now possible to experimentally study photodissociation in detail. The vibrational states of the products resulting from photodissociation may be studied as a function of the properties of the incident radiation.

There are a number of different types of dissociation processes that a molecule may undergo, depending upon the energy transferred to the polyatomic molecule. If a high frequency photon is absorbed by a molecule that has no quasibound levels in that energy region, the molecule dissociates directly into continuum states. The continuum states may be labelled by the electronic structure, the vibrational quantum numbers, the relative kinetic energy, and the relative angular momentum of the fragments.

Dissociation may also take place via an intermediate excited (metastable) level which is degenerate with a continuum level or many continua. The term predissociation is given to such a process.

Some recent attempts at laser isotope separation involve a multiphoton photofragmentation process that utilizes infrared lasers to excite the vibrational levels of molecules. An anharmonic oscillator is driven by a pulsed, intense laser field.

Collision energy transfer in thermal unimolecular reactions leads to noticeable isotope effects with respect to the specific rate constants $k(E)$ for dissociation processes (at excitation energies $E$ ), generally small however in comparison with isotope effects occuring in photodissociation processes.

The ideas presented in the following and the application to simple reactions indicate the utility of photodissociation to practical isotope separation problems. Conversely, the study of the isotopic dependence of photodissociation processes provides a valuable tool for probing the structure of the potential surfaces.

## II. Direct dissociation

The theoretical formulation of polyatomic dissociation processes is expressed using simple scattering theory. In most molecules the reaction coordinate, a bond vibration, is not a normal coordinate of the initial electronic state of the molecule, rather it is a linear combination of normal modes. Band and Freed have provided the foundation for describing dissociation of polyatomic molecules in terms of a dynamical theory in which the proper normal modes of the polyatomic and of the decay fragments are fully incorporated [1,2].

### A. Theory of molecular direct dissociation

The full wavefunction is taken in the form

$$\psi(x,Q)=\psi_i(x,Q)\chi_i(Q)+\psi_f(x,Q')\chi_f(Q'), \qquad \text{(II.1)}$$

where $x$ represents the set of electronic coordinates, and the nuclear normal coordinates of the initial state are represented collectively by $Q$. $Q'$ are the set of normal coordinates of the fragment states together with the reaction coordinate.
The full Hamiltonian may be equivalently expressed in terms of coordinates appropriate to either nuclear arrangement,

$$\mathcal{H}(x,Q) = T(Q) + H(x,Q) \qquad \text{(II.2)}$$

$$= T(Q') + H(x,Q'), \tag{II.3}$$

where $T$ represents the nuclear kinetic energy operator. The wavefunctions $\psi_i(x,Q)$ and $\psi_f(x,Q')$ may be taken to be eigenfunctions of the full electronic Hamiltonian $H(x,Q)$ or $H(x,Q')$, respectively. Some part of $H(x,Q)$, say $H^{(1)}$ (the matter-radiation interaction Hamiltonian), is responsible for the coupling between $\psi_i(x,Q)$ and $\psi_f(x,Q')$:

$$H(x,Q) = H^{(0)}(x,Q) + H^{(1)}(x,Q), \tag{II.4}$$

$$H(x,Q') = H^{(0)}(x,Q') + H^{(1)}(x,Q'). \tag{II.5}$$

The transition matrix element involves a multidimensional Franck - Condon overlap integral

$$T_{fi} = \iint\cdots\int \chi_{E,L,M}(Q_1')\chi_{n_2'}(Q_2')\cdots\chi_{n_p'}(Q_p')X_{fi}(\{Q\})$$

$$\chi_{n_1}(Q_1)\chi_{n_2}(Q_2)\cdots\chi_{n_p}(Q_p)\,dQ_1'\cdots dQ_p', \tag{II.6}$$

where $X_{fi}(\{Q\}) = \langle\psi_f(x,Q')|H^{(1)}(x,Q')|\psi_i(x,Q)\rangle_x$

is the electronic matrix element between the initial and final electronic states, $Q_1'$ is the dissociation (reaction) coordinate, and $\chi_{E,L,M}(Q_1')$ is the wavefunction describing the relative motion between the fragments which have asymptotic relative kinetic energy $E$ and relative angular momentum $L$ with $z$ projection $M$. $\{Q_2',\cdots,Q_p'\}$ are the normal modes of the fragments, which are in vibrational and rotational quantum states $\{n_2',\cdots,n_p'\}$; $\{Q_1,\cdots,Q_p\}$ are the normal modes of the initial bound electronic state. The set $\{Q'\}$ is a function of $\{Q\}$. For the case of a direct photodissociation process from a ground electronic molecular state the $X_{fi}(Q) = -(1/2m)\langle\psi_f(x,Q')|\sum_i p_i\cdot A(x_i)|\psi_i(x,Q)\rangle_x$ represents the coupling of the matter to radiation. For optically allowed transitions, $X_{fi}(Q)$ is assumed to be only weakly dependent upon the nuclear coordinates and may be taken out of the integration over $\{Q'\}$. For allowed direct photodissociation, different isotopic molecules have the same values of $\chi_{fi}(Q)$ since the electronic structure of the molecules is identical.

The full transition amplitude is given by a sum of products of Franck - Condon factors and final state amplitudes for the half collision [1, 3, 4]

$$\tau_{fi} = \sum_{\bar{f}} S_{f\bar{f}} \, T_{\bar{f}i} \, , \tag{II.7}$$

where the sum extends over all possible intermediate states $\bar{f}$ for the nuclear degrees of freedom on the repulsive electronic state potential surface. The half-collision transition matrix, $S$ , for the final state interactions couples vibrational and rotational states on the repulsive potential energy surface characterized by the electronic channel in question.

B. The dissociation of formaldehyde

The formaldehyde molecule is described in the framework of a Urey - Bradley force field approximately only by three vibrational normal modes. The normal coordinates $Q_i \; (i = 1, 2, 3)$ are approximated by symmetry coordinates $S_i$ .

$$\begin{aligned}
Q_1 &\approx S_1 = \Delta r_3 && \text{(C-O)-stretching} \\
Q_2 &\approx S_2 = \tfrac{1}{\sqrt{6}}\left(2\Delta q_4 - \Delta q_1 - \Delta q_2\right) && \text{HCH - bending} \\
Q_3 &\approx S_3 = \tfrac{1}{\sqrt{2}}\left(\Delta q_1 - \Delta q_2\right) && \text{(C-H)-stretching} \\
Q_4 &\approx S_4 = \tfrac{1}{\sqrt{3}}\left(\Delta q_4 + \Delta q_1 + \Delta q_2\right) \equiv 0
\end{aligned} \tag{II.8}$$

$\chi_i(Q)$ is then given by $\chi_{n_1}(Q_1)\,\chi_{n_2}(Q_2)\,\chi_{n_3}(Q_3)$ where $n_1, n_2, n_3$ are the vibrational quantum numbers of the modes. The final state $\chi_f(Q')$ consists of two diatomic molecules in states $\chi_{n_2'}(Q_2')$ , $\chi_{n_3'}(Q_3')$ describing the diatomics vibrational states, together with the wave-function $\chi_E(Q_1')$ for the relative motion of the molecule

CO with respect to the molecule $H_2$.
We consider the coplanar dissociation of the $H_2CO$-molecule. The additional (initial) bending and (final) relative rotational degrees of freedom slightly complicate the analysis but add no new conceptual problems. Furthermore in many cases the inclusion of these degrees of freedom does not produce substantial changes in the stretching vibrational distribution in the fragments. For coplanar dissociation of the $H_2CO$-molecule, there is a linear relationship between the variables $\{Q\}$ and $\{Q'\}$,

$$\begin{bmatrix} Q_1 \\ Q_2 \\ \vdots \\ Q_p \end{bmatrix} = [C] \begin{bmatrix} Q_1' - a_1 \\ Q_2' - a_2 \\ \vdots \\ Q_p' - a_p \end{bmatrix} , \qquad \text{(II.9)}$$

where the matrix $C$ and the constants $a_1, \cdots, a_p$ are given in terms of the masses, equilibrium distances, and normal frequencies of the initial state of the molecule, the final one for the fragments and the geometry of the saddle point of the potential energy surface for dissociation.
The ground state $S_0$ potential surface for the dissociation of formaldehyde $H_2CO \longrightarrow H_2 + CO$, calculated by Jaffe and Morokuma with the ab initio MCSCF method, has been used [5]. Figure 1 is a potential surface plot showing energy contours as a function of $D'$ and $R_{HH}$ with $R_{CO}$, $\phi$ and $\Theta$ fixed at the values given in Table I. The geometry of the saddle point in Table I was described in terms of the coordinates shown in Fig. 2. Motion along the reaction coordinate near the crest of the barrier involves almost exclusively changes in $D'$ and $R_{HH}$.
Substituting for the variables $\{Q\}$ in the initial state vibrational wavefunctions and employing a harmonic approximation for the molecular and fragment vibrations, the integration over $\{Q_2', \cdots, Q_p'\}$ reduce to a multidimensional Gaussian integration which can analytically be performed and yields a linear combination of harmonic oscillator wavefunctions in the

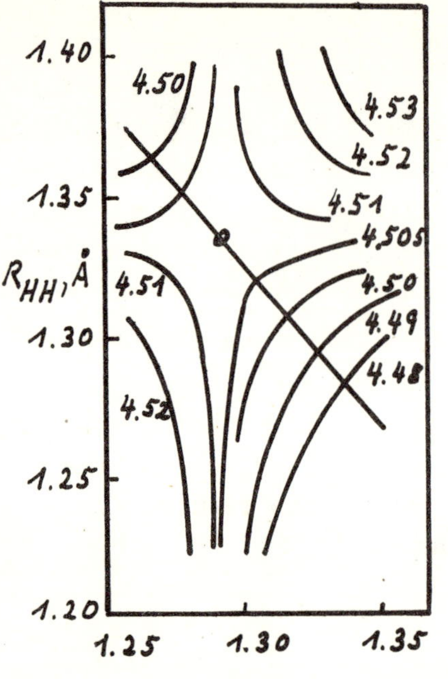


Fig. 1. Potential energy surface (in e V) as a function of $R_{HH}$ and $D'$ near the saddle point. The saddle point is located at ⊙ . The dissociation pathway is also shown.

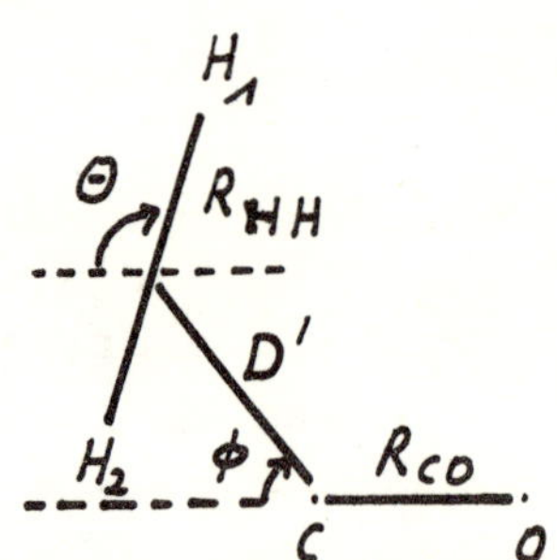


Fig. 2. The definition of the coordinate system used for planar $H_2CO$.

TABLE I. (α) Geometry of the $H_2CO$ ground state; (β) geometry and energy of the saddle point for $H_2CO \rightarrow H_2 + CO$ in $C_s$ symmetry in the ground state.

| (α) | (β) | |
|---|---|---|
| $r_{CO}$ = 1.24 Å | $D'$ = 1.293 Å; | $\phi$ = 49° |
| $r_{CH}$ = 1.084 Å | $R_{HH}$ = 1.336 Å; | Θ = 108° |
| ∢HCH = 118° | $R_{CO}$ = 1.176 Å; | $\Delta E^{\odot}$ = 4.55 eV |

variable $Q_1'$ . The Franck - Condon transition amplitude is then exactly given by the sum of one-dimensional integrals

$$T_{fi} = X_{fi} \int X_E(Q_1') \left[ \sum_n^{n_{max}} b_n H_n(Q_1') e^{-\gamma(Q_1' - r_0)^2} \right] dQ_1', \quad \text{(II.10)}$$

where $n_{max} = \sum_{i=1}^{p} n_i + \sum_{i=2}^{p} n_i'$ .

$H_n(Q_1') e^{-\gamma(Q_1' - r_0)^2}$ are the unnormalized harmonic oscillator wavefunctions in an effective oscillator along the reaction coordinates. The coefficients $b_n$ and the parameters $\gamma$ and $r_0$ contain all information about the molecule and fragment vibrational quantum numbers and normal modes. The bottom of the effective harmonic well at $Q_1' = r_0$ and the force constant $k = \mu\gamma^2$ (where $\mu$ is the reduced mass between the fragments) are functions of $C, a_1, \cdots, a_p$ and therefore determined by the parameters of the initial electronic state of the molecule, the final one for the fragments and the geometry of the saddle point of the potential energy surface. This analysis is applied here for the photodissociation of

$H_2{}^{12}C^{16}O$, $D_2{}^{12}C^{16}O$, $H_2{}^{13}C^{16}O$ and $H_2{}^{12}C^{18}O$ (ground state $S_0$ potential surface),

$$H_2{}^{12}C^{16}O + h\nu \longrightarrow H_2 + {}^{12}C^{16}O, \quad \text{(II.11a)}$$

$$D_2{}^{12}C^{16}O + h\nu \longrightarrow D_2 + {}^{12}C^{16}O, \quad \text{(II.11b)}$$

$$H_2{}^{13}C^{16}O + h\nu \longrightarrow H_2 + {}^{13}C^{16}O, \quad \text{(II.11c)}$$

$$H_2{}^{12}C^{18}O + h\nu \longrightarrow H_2 + {}^{12}C^{18}O. \quad \text{(II.11d)}$$

## C. Isotope Effects

Given the normal modes of $H_2{}^{12}C^{16}O, D_2{}^{12}C^{16}O, H_2{}^{13}C^{16}O, H_2{}^{12}C^{18}O$ and $H_2$, $D_2$, ${}^{12}C^{16}O$, ${}^{13}C^{16}O$ and ${}^{12}C^{18}O$ the effective oscillator for the reaction coordinate can readily be evaluated. These are displayed in Fig. 3 as curves (a) and (b).

The repulsive $H_2$ --- CO curve $V(Q_1') = V_0 \exp(-DQ_1')$ [ (e) in Fig. 3 ] has been fit to the potential energy surface (Fig. 1). The largest contribution to integrals in (II.10) comes from the small region about the classical turning point $(R_E)$ on the repulsive potential. Thus, this potential is linearized about the turning point $(R_E)$ for each value of $E$.

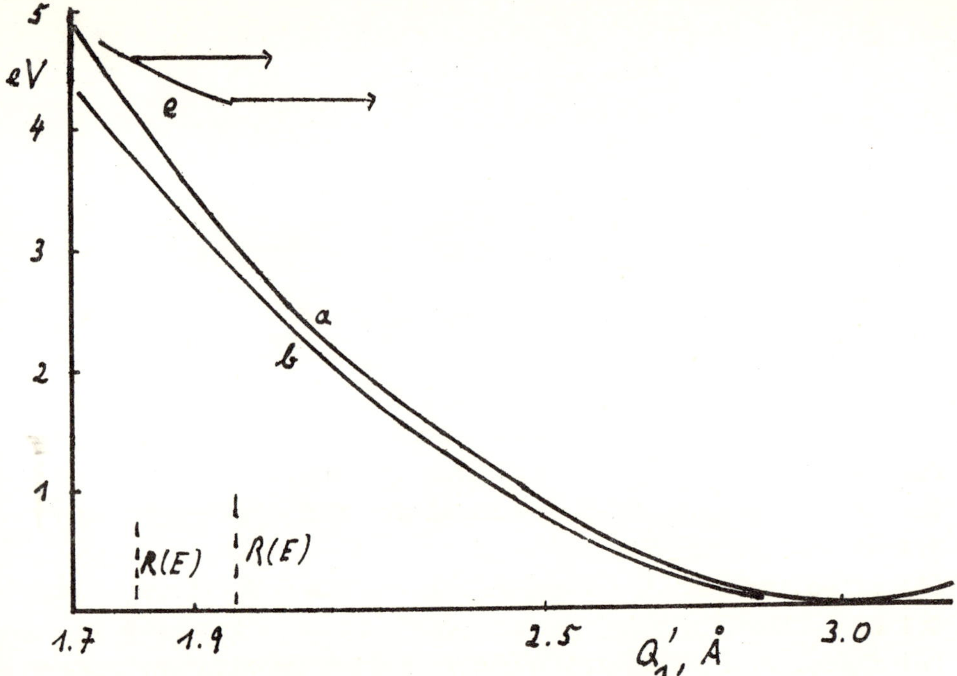


Fig. 3. Potential curves: (a) effective oscillator potential for $H_2CO + h\nu \longrightarrow H_2 + CO$ [$k$ = 0,86 mdyn/Å; $r_0$ = 3,065 Å;] (b) effective oscillator potential for $D_2CO + h\nu \rightarrow D_2 + CO$ [$k$ = 0,76 mdyn/Å; $r_0$ = 3,118 Å;] (e) fitted repulsive electronic potential between the fragments as a function of reaction coordinate $Q_1'$.

Given this approximation, (II.10) can be evaluated analytically [1].

Before discussing the resulting dissociation probabilities of formaldehyde and the isotope effects, we explain the problem once more in the simpler case of collinear dissociation of HCN and DCN,

$$HCN + h\nu \longrightarrow H(1s) + CN(B^2\Sigma^+), \qquad \text{(II.12a)}$$

$$DCN + h\nu \longrightarrow D(1s) + CN(B^2\Sigma^+), \qquad \text{(II.12b)}$$

treated by Band and Freed [1,2] (Fig. 4).

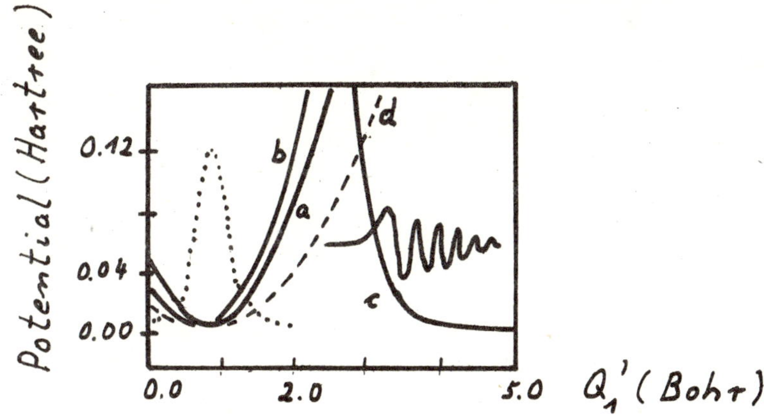


Fig. 4. Potential curves and wavefunctions:
(a) effective oscillator potential for $HCN + h\nu \longrightarrow H(1s) + CN(B^2\Sigma^+)$,
(b) effective oscillator potential for $DCN + h\nu \longrightarrow D(1s) + CN(B^2\Sigma^+)$,
(c) fitted repulsive electronic potential between the fragments as a function of reaction coordinate $Q_1'$,
(d) oscillator potential for C-H in HCN.
Also shown are the ground state wavefunctions in effective oscillator potentials and the continuum wavefunction for the relative motion on the repulsive potential for $H(1s) + CN(B^2\Sigma^+)$.

It is clear from the figure that the bound-continuum overlap-integral corresponds to a tunnelinglike process. The Franck-Condon transition amplitude $T_{fi}$ varies as $exp[-(\mu_1)^{1/2} V_b d]$ with $V_b$ a barriere height and $d$ a barriere width. The isotopic ratio of transition amplitudes is of the form $exp[(\mu_2^{1/2} - \mu_1^{1/2}) V_b d]$, where $\mu_1$ and $\mu_2$ are the reduced masses between the fragments (different isotopes). Large percentage isotope mass change and a maximization of the product $V_b d$ lead to very large isotope effects. A compromise can be accomplished by choosing the exciting wavelength so the isotope effect is large, but the over-all dissociation rate of the light isotope is sufficient.
As the wavelength of the incident photon is diminished, $V_b d$ becomes smaller before increasing again.

In the case of the formaldehyde molecule the turning point $(R_E)$ on the repulsive potential is far to the left of the effective oscillator well (Fig. 3), but the isotopic ratio of transition amplitudes $T_{fi}$ is of the same form as in the case of HCN. At first however an inverted vibrational population of the fragments is expected. But in such a situation (Fig. 3) there is also a high probability that the final state interaction mechanism (equ. II.7) may change this initial inversion population. In Table II an impression is given about the magnitudes of the transition amplitudes $T_{fi}$ and the isotopic ratios of these amplitudes. Final state interactions may be treated with some sophisticated methods, advanced by Band and Freed [1].

TABLE II. $T_{fi}$ and isotopic ratios $T/T^*$ for $H_2CO$

| $h\nu$ [eV] | 4.428 | 4.528 | 4.628 |
|---|---|---|---|
| $T_{fi}(n_i=0;\, n'_{C-O}=1;\, n'_{HH}=0)/T_{fi}(n_i=n'_i=0)$ | | | |
| | $1.55\times10^3$ | $0.93\times10^3$ | $0.70\times10^3$ |
| $T_{fi}(n_i=0;\, n'_{C-O}=1,2,3,4;\, n'_{HH}=0)/T_{fi}(n_i=n'_i=0)$ | | | |
| $h\nu = 4.428$ [eV] | $n'_{C-O}$: 2<br>$0.62\times10^7$ | 3<br>$0.42\times10^8$ | 4<br>$0.63\times10^{14}$ |
| $T_{fi}(n_i=0;\, n'_{C-O}=0;\, n'_{HH}=1,2,3,4)/T_{fi}(n_i=n'_i=0)$ | | | |
| $n'_{HH}$: 1<br>$0.34\times10^3$ | 2<br>$0.12\times10^6$ | 3<br>$0.29\times10^8$ | 4<br>$0.12\times10^{12}$ |
| $T_{fi}(n_1=n_2=0;\, n_3=n_{C-H}=1,2,3,4;\, n'_i=0)/T_{fi}(n_i=n'_i=0)$ | | | |
| $n_3$: 1<br>$0.41\times10^4$ | 2<br>$0.11\times10^8$ | 3<br>$0.50\times10^{11}$ | 4<br>$0.27\times10^{15}$ |
| $h\nu = 4.428$ [eV] | $T_{fi}(n_i=0;\, n'_{C-O}=4;\, n'_{HH}=0)$ | | |
| $T_{H_2{}^{12}C^{16}O}/T_{H_2{}^{12}C^{18}O}$ : $1.9\times10^4$ | | $T_{H_2{}^{12}C^{16}O}/T_{H_2{}^{13}C^{16}O}$ : $2.6\times10^{-3}$ | |
| $T_{fi}(n_i=0;\, n'_{C-O}=0;\, n'_{HH}=1)$ | | | |
| $T_{H_2CO}/T_{D_2CO}$ : $3\times10^{-12}$ | | | |

## III. Multiphoton Photodissociation

A quantum mechanical model for multiphoton photofragmentation in intense laser fields of an "isolated", collision-free molecule on the electronic ground state potential surface is considered, advanced by Mukamel and Jortner [6,7]. Isotopic selectivity of the process, anharmonicity effects, power-broadening phenomena, intramolecular nonreactive vibrational relaxation and nonradiative decomposition, i.e., predissociation are incorporated. Multiphoton photofragmentation is considered in terms of a truncated anharmonic oscillator driven by a pulsed, intense laser field. The effective - hamiltonian formalism is applied to derive expressions for the photofragmentation yield and for the isotopic separation factor.

### A. Quantum mechanical model for multiphoton photofragmentation

The model rests on the following assumptions:

(a) The molecule is optically pumped with the laser frequency $\hbar\omega$, being near-resonance with a certain vibrational mode ( $\nu_3$ in the case of $SF_6$). The molecular levels constitute a truncated anharmonic oscillator, being represented by the energy levels

$$E_v = \omega_0 v - \omega_0 x v^2, \quad v = 0, 1, \cdots, \bar{v}. \qquad \text{(III.1)}$$

$\omega_0$ is the oscillator frequency and $\omega_0 x$ the anharmonicity constant.

(b) Three energy regions can be distinguished (Fig. 5):
Region I. The density $\rho_B(E)$ of background states corresponding to other vibrational modes is low. Dynamical Stark level shifts are neglected.
Region II. The levels of the vibrational ladder are quasidegenerate with a background manifold of vibrational states which correspond to other modes. $V_A(E)$ is the anharmonic coupling matrix element.
The condition for strong coupling

$$V_A(E)\,\rho_B(E) \gg 1 \qquad \text{(III.2)}$$

Fig. 5. Molecular energy level scheme for a multiphoton fragmentation.

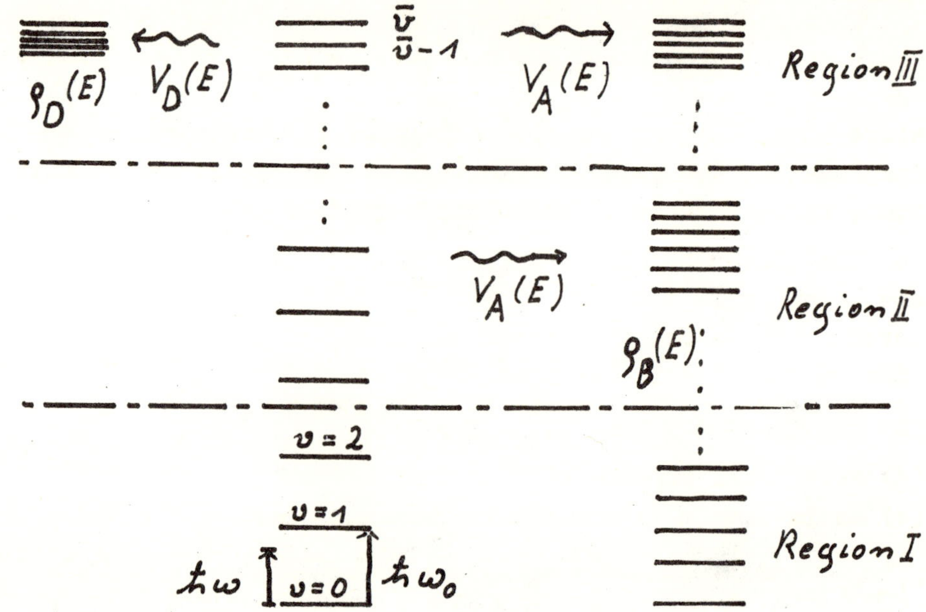

$\bar{v}$ $\bar{v}-1$ $\rho_D(E)$ $V_D(E)$ $V_A(E)$ Region III

$V_A(E)$ Region II $\rho_B(E)$

$v=2$ $v=1$ $v=0$ $\hbar\omega$ $\hbar\omega_0$ Region I

marks the onset of region II. When the level spacing $[\rho_B(E)]^{-1}$ in region II exceeds the width of these levels, $\gamma_B$ (which originates from infrared decay in the case of the isolated molecule), i.e.,

$$\gamma_B(E)\rho_B(E) < 1 , \qquad \text{(III.3)}$$

and when relation (III.2) is obeyed the intensity of the $v \longrightarrow v+1$ transition will be spread over the energy range $W_B(E) \approx 2\pi\,|V_A(E)|^2\rho_B(E)$.
No intramolecular vibrational relaxation (IVR) occurs. When

$$\gamma_B(E)\rho_B(E) > 1 , \qquad \text{(III.4)}$$

overlap of the background levels is exhibited, and the fanifold acts as an dissipative continuum for IVR. The IVR rate from a level $|v\rangle$ located at $\approx E$ is then

$$\Gamma_v^{IVR} \approx 2\pi\,|V_A(E)|^2\rho_B(E) . \qquad \text{(III.5)}$$

Region III. The high energy reactive region where a dissociative channel opens up. Each $|v\rangle$ state in that region will be characterized by a decay width $\Gamma_v^D$,

$$\Gamma_v^D = 2\pi\,|V_D(E)|^2 \rho_D(E), \tag{III.6}$$

where $V_D(E)$ and $\rho_D(E)$ (Fig. 5) correspond to the coupling and the density of states in the dissociative continuum, respectively, at the energy $E \approx E_v$.

(c) The higher levels in region II and the levels in region III are characterized by IVR widths $\Gamma_v^{IVR}$. To the highest $\bar{v}th$ level a predissociative level width $\Gamma_{\bar{v}}^D$ is assigned; the total width of this level is $\Gamma_{\bar{v}} = \Gamma_{\bar{v}}^D + \Gamma_{\bar{v}}^{IVR}$.
The model assumes that $\Gamma_{\bar{v}}^D >> \Gamma_{\bar{v}}^{IVR}$.

(d) Rotational effects will be disregarded.

(e) An intense electromagnetic field is tuned on the time scale $0 \lesssim t \lesssim T$. The field is specified in terms of a state $|n\rangle$ containing $n$ photons of frequency $\omega$ in a single mode. The off-resonance energy for the 0 → 1 molecular transition is

$$\Delta = \omega_0 - \omega . \tag{III.7}$$

(f) The zero - order states of the entire system $|v,n\rangle$ are characterized by the energy levels

$$E(v,n) = n\omega + v\omega_0 - \omega_0 x v^2 . \tag{III.8}$$

The model considers a group of near - resonant "dressed" molecular states corresponding to the sequence $|v, n-v\rangle$, i.e.,

$$E(v, n-v) = n\omega + \Delta v - \omega_0 x v^2, \; v = 0,1,\dots,\bar{v}. \tag{III.9}$$

(g) Radiation - matter interaction matrix elements $H_{int}$ are taken only between adjacent levels corresponding to a harmonic oscillator

$$\langle v'n'|H_{int}|vn\rangle = \mu E\,\delta_{n',n\pm 1} \otimes \tag{III.10}$$

$$[(v+1)^{1/2}\delta_{v',v+1} + v^{1/2}\delta_{v,v-1}],$$

where $\mu$ is the dipole matrix element for the $0 \rightarrow 1$ transition, $E$ corresponds to the amplitude of the electromagnetic field.

(h) Spontaneous infrared emission $v \longrightarrow v-1$ will be neglected.

B. The probability for photofragmentation

The time evolution of a system of discrete molecular states $|0,n\rangle, |1,n-1\rangle, \cdots, |\bar{v}, n-\bar{v}\rangle$, some of which are coupled to an intramolecular IVR quasicontinuum and to a dissociative continuum can be treated by the effective hamiltonian formalism [8 - 11]. The effective hamiltonian, $H_{eff}$, for the problem is

$$H_{eff} = \begin{pmatrix} \text{region I} & & \text{region II} & & \text{region III} \\ E_0 & \mu E & & & \\ \mu E & E_1 & \mu E\sqrt{2} & & \\ & & \ddots & & \\ & \mu E\sqrt{v} & E_v - \frac{1}{2} i \Gamma_v^{IVR} & \mu E\sqrt{v+1} & \\ & & & \ddots & \\ & & & & E_{\bar{v}} - \frac{1}{2} i \Gamma_{\bar{v}} \end{pmatrix}$$

for the time $0 \leq t \leq T$. (III.11)

The probability amplitude for the population of the dissociative state $|\bar{v}, n-\bar{v}\rangle$ at time $t$ is

$$C_{\bar{v}}(t) = \langle 0, n | U(t,0) | 0, n \rangle$$

$$= -\frac{1}{2\pi i} \int \langle 0, n | (E - H_{eff})^{-1} | 0, n \rangle e^{-iEt} dE, \quad \text{(III.12)}$$

where $|0,n\rangle$ is the initial state of the system and $U(t,0)$ is the time-evolution operator. This amplitude can be expressed in terms of the complex eigenvalues $\Lambda_j = \varepsilon_j - \frac{1}{2} i \gamma_j$ and the (non - orthogonal) eigenvectors $\{|j\rangle\}$ of the effective hamiltonian, resulting in

$$\begin{aligned} C_{\bar{v}}(t) &= 0, & t<0; \\ &= \sum_j \langle \bar{v}|j\rangle\langle 0|j\rangle \exp(-i\varepsilon_j t - \tfrac{1}{2}\gamma_j t), & 0\leq t\leq T; \\ &= C_{\bar{v}}(T)\exp(-iE_{\bar{v}}t - \tfrac{1}{2}\Gamma_{\bar{v}}t), & t>T. \end{aligned} \tag{III.13}$$

The probability for photofragmentation, $P_D(t)$, is finally given by

$$\begin{aligned} P_D(t) &= \Gamma_{\bar{v}}^D \int_0^t |C_{\bar{v}}(\tau)|^2 d\tau, \quad 0\leq t\leq T; \\ &= P_D(T) + \exp[-\Gamma_{\bar{v}}^D(t-T)]\,|C_{\bar{v}}(T)|^2, \quad t>T. \end{aligned} \tag{III.14}$$

C. Numerical simulation of multiphoton dissociation of the $SF_6$ molecule

Numerical simulations were conducted using the following energetic parameters: $\bar{v} = 9$, $\omega_0 x = 1\,cm^{-1}$, $\Delta$ = 0, 10, 20, 30, 100 cm$^{-1}$. For the decay widths we have assumed that IVR is slow relative to the decay width. $\Gamma_v^{IVR}$ was varied in the range 0 - 0,5 x 10$^{-3}$ cm$^{-1}$ for the relevant levels in region II and in region III, while $\Gamma_{\bar{v}}^D$ was varied in the range 10$^{-5}$ - 10$^{-3}$ cm$^{-1}$. The pulse duration was taken as $T = 5000 (1/cm^{-1})$, while the Rabi frequency $\omega_R = \mu E/\hbar$ was chosen in the range $\omega_R$ = 1, 5, 10 cm$^{-1}$ corresponding to experimentally accesible intense pulses. For a laser power of 1 GW/cm$^2$ and a transition dipole $\mu \approx 0.10\,D$ we have $\mu E = 10\,cm^{-1}$. Typical numerical results which provide information concerning field effects, off-resonance energy and IVR effects and isotope effects are shown in Tables IV, V.

Table III

Energetic data for $SF_6$ pumped by an pulsed $CO_2$ laser

| Parameter | | |
|---|---|---|
| (S - F)-stretching frequency | $\hbar\omega_0(\nu_3)$ | $948\ cm^{-1}$ |
| laser frequency | $\hbar\omega$ | $944-960\ cm^{-1}$ |
| dissociation energy | $D_e$ | $24000\ cm^{-1}$ |
| anharmonicity | $\omega_0 x$ | $1-10\ cm^{-1}$ |
| number of vibrational states in $\nu_3$ | $M$ | $\approx 50$ |
| isotope shift $^{32}S/^{34}S$ | $S$ | $17\ cm^{-1}$ |
| Rabi frequency | $\mu E$ | $1-10\ cm^{-1}$ |
| pulse duration | $T$ | $100\ ns$ |
| spread of P, Q, R branches | | $10\ cm^{-1}$ |

Numerical simulations were performed to probe the effects of $\Gamma_{\bar{v}}^{D}$ and $\Gamma_v^{IVR}(v = 5, \cdots, 8;\ \Gamma_{0-4} = 0)$ in dependence on $\omega_R, \Delta$ and $S$ (Tables IV, V). At low fields $P_D$ increases rapidly with $\omega_R$ and for higher values of $\omega_R$ saturation occurs. The $P_D$ versus $\Delta$ dependences are characterized by a maximum at the off - resonance energy

$$\bar{\Delta} \approx \omega_0 x\ \bar{v}\ , \tag{III.15}$$

which exhibits only a weak dependence on $\omega_R$ . The most effective photofragmentation process will be induced at off - resonance energies satisfying eq. (III.15). A comparison of Table IV with Table V indicates that vibrational isotope shifts of a few wavenumbers, $S$ , (e.g. $S = 17\ cm^{-1}$ for $^{32}SF_6 - {}^{34}SF_6$) induce a high isotopic selectivity of the photofragmentation process.

D. Anharmonic coupling $V_A(E)$

For the numerical simulations of multiphoton dissociation physically feasible energetic parameters $\Gamma^{IVR}$ are needed. An ansatz for the calculation of the (high-order) anharmonic coupling $V_A(E)$ in the simple case of a triatomic linear molecule is demonstrated. The ansatz may be explained by the potentials portrayed in Fig. 4. The effective oscillator

TABLE IV. Dissociation probabilities $P_D$ for $^{32}SF_6$ ($P_D = 0.268 - 12 \wedge 0.268 \times 10^{-12}$)

| $\omega_R$ [$cm^{-1}$] | $\Delta$ [$cm^{-1}$] | | | | | | |
|---|---|---|---|---|---|---|---|
| | $\Gamma_9^D$ : | $10^{-5}$ | $10^{-4}$ | $10^{-3}$ | $10^{-3}$ | $10^{-3}$ | $10^{-3}$ |
| | $\Gamma_8^{IVR}$ : | 0 | 0 | 0 | $10^{-4}$ | $0.5 \times 10^{-3}$ | $0.75 \times 10^{-3}$ |
| | $\Gamma_7$ : | 0 | 0 | 0 | $10^{-5}$ | $10^{-4}$ | $0.5 \times 10^{-3}$ |
| | $\Gamma_6$ : | 0 | 0 | 0 | $10^{-5}$ | $10^{-4}$ | $0.5 \times 10^{-3}$ |
| | $\Gamma_5$ : | 0 | 0 | 0 | $10^{-5}$ | $10^{-4}$ | $10^{-4}$ |
| | | $P_D$ | | | | | |
| 1 | 0 | 0.268 −12 | 0.268 −11 | 0.268 −10 | 0.267 −10 | 0.260 −10 | 0.256 −10 |
| | 10 | 0.182 −5 | 0.178 −4 | 0.144 −3 | 0.130 −3 | 0.901 −4 | 0.680 −4 |
| | 20 | 0.119 −14 | 0.119 −13 | 0.118 −12 | 0.115 −12 | 0.942 −13 | 0.736 −13 |
| | 30 | 0.324 −20 | 0.324 −19 | 0.324 −18 | 0.318 −18 | 0.271 −18 | 0.222 −18 |
| 5 | 0 | 0.127 −2 | 0.127 −1 | 0.126 | 0.124 | 0.111 | 0.870 −1 |
| | 10 | 0.170 −2 | 0.163 −1 | 0.116 | 0.111 | 0.924 1 | 0.731 −1 |
| | 20 | 0.720 −5 | 0.719 −4 | 0.711 −3 | 0.701 −3 | 0.635 −3 | 0.551 −3 |
| | 30 | 0.334 −8 | 0.333 −7 | 0.330 −6 | 0.323 −6 | 0.279 −6 | 0.224 −6 |

TABLE IV.

| | | | | | | | | | | | | | |
|---|---|---|---|---|---|---|---|---|---|---|---|---|---|
| 10 | 0 | 0.816 | −3 | 0.792 | −2 | 0.607 | −1 | 0.588 | −1 | 0.497 | −1 | 0.399 | −1 |
| | 10 | 0.414 | −2 | 0.405 | −1 | 0.330 | | 0.320 | | 0.278 | | 0.236 | |
| | 20 | 0.452 | −3 | 0.443 | −2 | 0.369 | −1 | 0.358 | −1 | 0.311 | −1 | 0.262 | −1 |
| | 30 | 0.153 | −4 | 0.153 | −3 | 0.152 | −2 | 0.150 | −2 | 0.136 | −2 | 0.117 | −2 |
| | 40 | 0.282 | −6 | 0.282 | −5 | 0.279 | −4 | 0.274 | −4 | 0.243 | −4 | 0.200 | −4 |
| 30 | 0 | 0.684 | −3 | 0.663 | −2 | 0.492 | −1 | 0.483 | −1 | 0.443 | −1 | 0.364 | −1 |
| | 10 | 0.385 | −2 | 0.376 | −1 | 0.302 | | 0.292 | | 0.248 | | 0.200 | |
| | 20 | 0.665 | −2 | 0.646 | −1 | 0.491 | | 0.485 | | 0.447 | | 0.388 | |
| | 30 | 0.466 | −2 | 0.455 | −1 | 0.373 | | 0.370 | | 0.346 | | 0.306 | |
| | 40 | 0.116 | −2 | 0.113 | −1 | 0.939 | −1 | 0.924 | −1 | 0.847 | −1 | 0.747 | −1 |

TABLE V. Dissociation probabilities $P_D$ for $^{34}SF_6$ ($P_D = 0.856 - 22 \wedge 0.856 \times 10^{-22}$)

| $\omega_R$ [$cm^{-1}$] | $\Delta$ [$cm^{-1}$] | | | | | | |
|---|---|---|---|---|---|---|---|
| | | $\Gamma_9^D$ : $10^{-5}$ | $10^{-4}$ | $10^{-3}$ | $10^{-3}$ | $10^{-3}$ | $10^{-3}$ |
| | | $\Gamma_8^{IVR}$ : 0 | 0 | 0 | $10^{-4}$ | $0.5 \times 10^{-3}$ | $0.75 \times 10^{-3}$ |
| | | $\Gamma_7$ : 0 | 0 | 0 | $10^{-5}$ | $10^{-4}$ | $0.5 \times 10^{-3}$ |
| | | $\Gamma_6$ : 0 | 0 | 0 | $10^{-5}$ | $10^{-4}$ | $0.5 \times 10^{-3}$ |
| | | $\Gamma_5$ : 0 | 0 | 0 | $10^{-5}$ | $10^{-4}$ | $10^{-4}$ |
| | | $P_D$ | | | | | |
| 1 | -17 | 0.856 -22 | 0.856 -21 | 0.856 -20 | 0.848 -20 | 0.791 -20 | 0.757 -20 |
| | -7 | 0.861 -18 | 0.861 -17 | 0.861 -16 | 0.854 -16 | 0.807 -16 | 0.781 -16 |
| | 3 | 0.168 -8 | 0.168 -7 | 0.168 -6 | 0.165 -6 | 0.145 -6 | 0.132 -6 |
| | 13 | 0.849 -9 | 0.848 -8 | 0.837 -7 | 0.835 -7 | 0.814 -7 | 0.798 -7 |
| 5 | -17 | 0.176 -9 | 0.176 -8 | 0.176 -7 | 0.174 -7 | 0.161 -7 | 0.149 -7 |
| | -7 | 0.716 -6 | 0.716 -5 | 0.716 -4 | 0.709 -4 | 0.658 -4 | 0.599 -4 |
| | 3 | 0.731 -2 | 0.726 -1 | 0.685 | 0.659 | 0.546 | 0.427 |
| | 13 | 0.258 -2 | 0.257 -1 | 0.252 | 0.252 | 0.249 | 0.247 |

TABLE V.

| | | | | | | | | | | | | | |
|---|---|---|---|---|---|---|---|---|---|---|---|---|---|
| 10 | -17 | 0.932 | -5 | 0.932 | -4 | 0.929 | -3 | 0.918 | -3 | 0.839 | -3 | 0.731 | -3 |
| | -7 | 0.245 | -2 | 0.244 | -1 | 0.241 | | 0.237 | | 0.210 | | 0.167 | |
| | 3 | 0.345 | -2 | 0.331 | -1 | 0.232 | | 0.227 | | 0.202 | | 0.163 | |
| | 13 | 0.424 | -2 | 0.414 | -1 | 0.333 | | 0.323 | | 0.282 | | 0.233 | |
| | 23 | 0.196 | -3 | 0.195 | -2 | 0.183 | -1 | 0.180 | -1 | 0.161 | -1 | 0.135 | -1 |
| 30 | -17 | 0.155 | -2 | 0.151 | -1 | 0.122 | | 0.119 | | 0.104 | | 0.822 | -1 |
| | -7 | 0.118 | -2 | 0.112 | -1 | 0.656 | -1 | 0.632 | -1 | 0.517 | -1 | 0.362 | -1 |
| | 3 | 0.956 | -3 | 0.934 | -2 | 0.761 | -1 | 0.726 | -1 | 0.587 | -1 | 0.440 | -1 |
| | 13 | 0.458 | -2 | 0.448 | -1 | 0.368 | | 0.348 | | 0.269 | | 0.193 | |
| | 23 | 0.582 | -2 | 0.564 | -1 | 0.421 | | 0.414 | | 0.376 | | 0.310 | |

potential in the variable $Q_1'$ (dissociation coordinate) is identified with the exact potential of the molecule including (high - order) anharmonic coupling. The oscillator potential for C-H in HCN is identified with the potential of the oscillator driven by a intense laser field. The transition matrix element $V_A(E)$ then involves a Franck-Condon overlap integral given by the sum of one-dimensional integrals

$$V_A(E) = \int \chi_{C-H}(Q_1')\left(V_a(Q_1') - V_d(Q_1')\right)\left[\sum_m^{m_{max}} b_m H_m(Q_1') \otimes e^{-\gamma(Q_1' - r_0)^2}\right] dQ_1' \qquad \text{(III.16)}$$

IV. Thermal unimolecular reactions

Specific rate constants k(E) (dissociation probabilities) for dissociation processes at different excitation energies E have been calculated by the use of transition state theory. Activated complexes have been localized by the minimum density of states criterion. Especially isotope effects have been calculated. The isotope effects are very small in comparison with the isotope effects due to photodissociation processes, but are typical of isotope effects in the case of total thermalization, which may occur in multiphoton photodissociation processes.

A. Theory of specific rate constants

The specific rate constants $k(E)$ are determined in the framework of transition state theory (activated complex theory) by the expression [12] :

$$k(E) = \frac{W^{\neq}(E - E_0)}{h\,\rho(E)} \quad . \qquad \text{(IV.1)}$$

The excess energy $E - E_0$ ( $E_0$ : dissociation energy) is spread over the number of activated complex quantum background states of the molecule (excluding the reaction coordinate $q$ ) $W^{\neq}(E - E_0)$. $\rho(E)$ is the state density, the number of quantum states of the molecule in the interval $(E, E + dE)$ divided by $dE$ , for the molecule before the activated complex. The position of the activated complex,

$q^{\neq} = q^{\neq}(E)$, is determined by the minimum state density principle [13,14]:

$$\frac{\partial \varrho(E,q)}{\partial q} = 0 . \qquad \text{(IV.2)}$$

$\varrho(E,q)$ is the local state density along the reaction (dissociation) coordinate. The state density $\varrho(E)$ in the denominator of the expression (IV.1) is given by the integral

$$\varrho(E) = \int_{q_{min}(E)}^{q^{\neq}} \varrho(E,q)\,dq . \qquad \text{(IV.3)}$$

( $q_{min}(E)$ is the minimal accessible value of $q$ at the excitation energy $E$ .) $\varrho(E,q)$ may be evaluated for small molecules in a semiclassical manner in a simple way, and $W^{\neq}$ may be determined by simple counting. The potential surface for dissociation is determined in the surroundings of the position of the activated complex, $q^{\neq}$ , by the BEBO - method [15] .

B. Isotope effects

Isotope effects with respect to the specific rate constants (dissociation probabilities) for dissociation processes at different excitation energies $E$ have been calculated for the molecules $NO_2$, $SO_2$ and $CO_2$. The ratios $k(E)/k^{*}(E)$ ( $*$ : isotopic substituted molecule) are portrayed in Figs. 6 - 10 [16] . Also the influence of the total angular momentum state of the dissociating molecule is shown. The quantum character of the dissociating systems is demonstrated by the oscillations of the ratios $k(E)/k^{*}(E)$ and by the occurence of inverse isotope effects. The magnitudes of these ratios are typical for multiphoton photodissociation processes with total intramolecular thermalization.

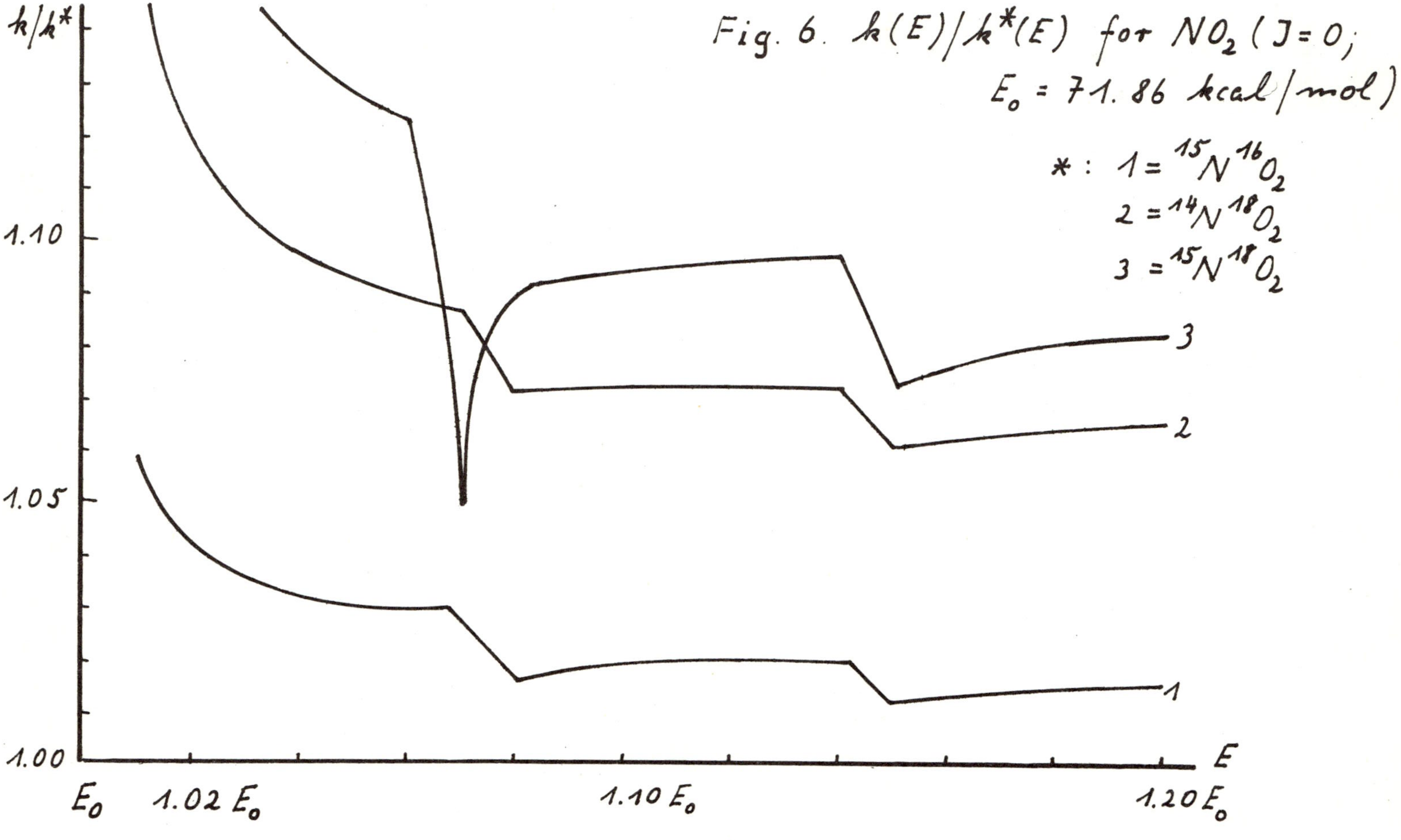


Fig. 6. $k(E)/k^*(E)$ for $NO_2$ ($J = 0$; $E_0 = 71.86$ kcal/mol)

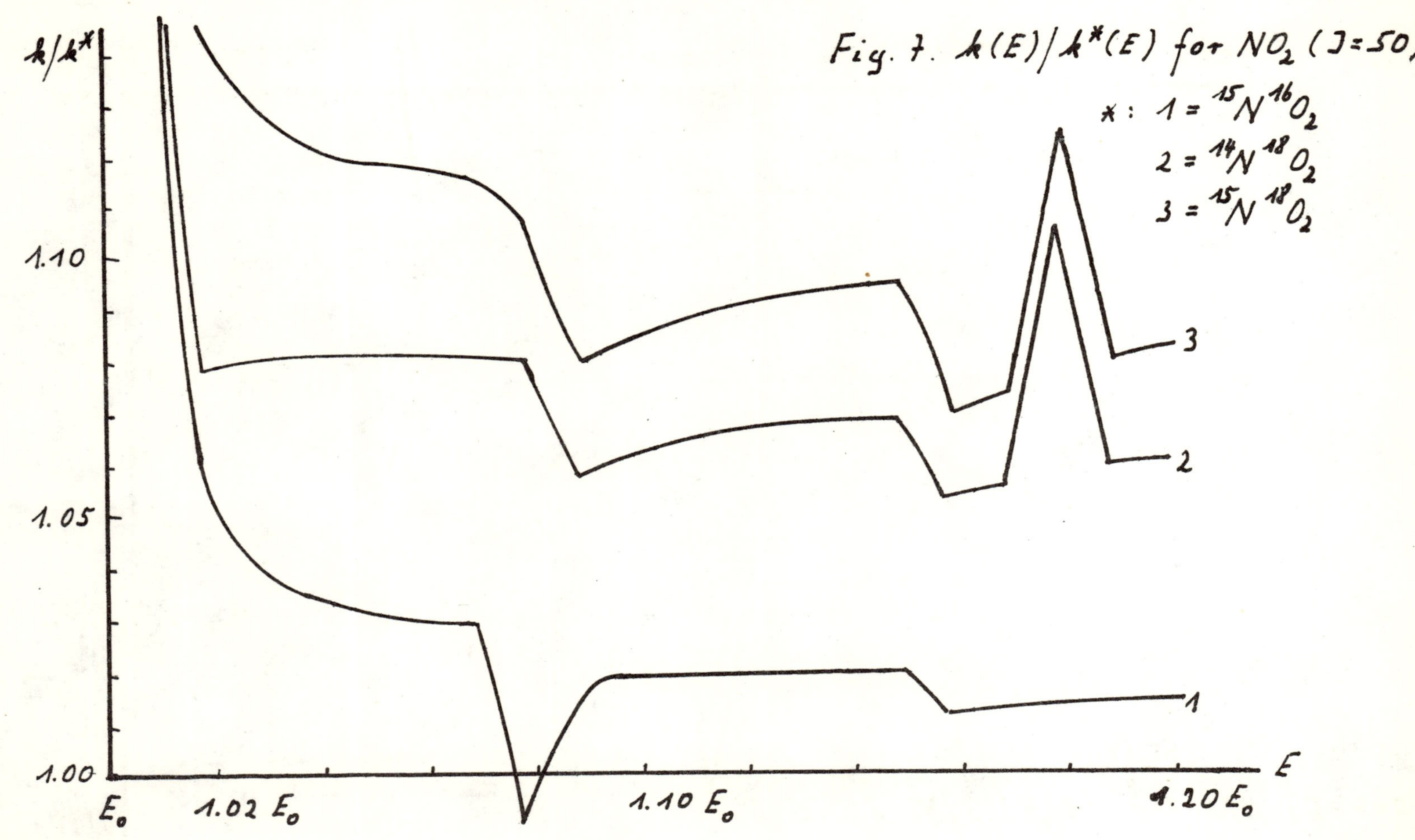


Fig. 7. $k(E)/k^*(E)$ for $NO_2$ $(J=50)$

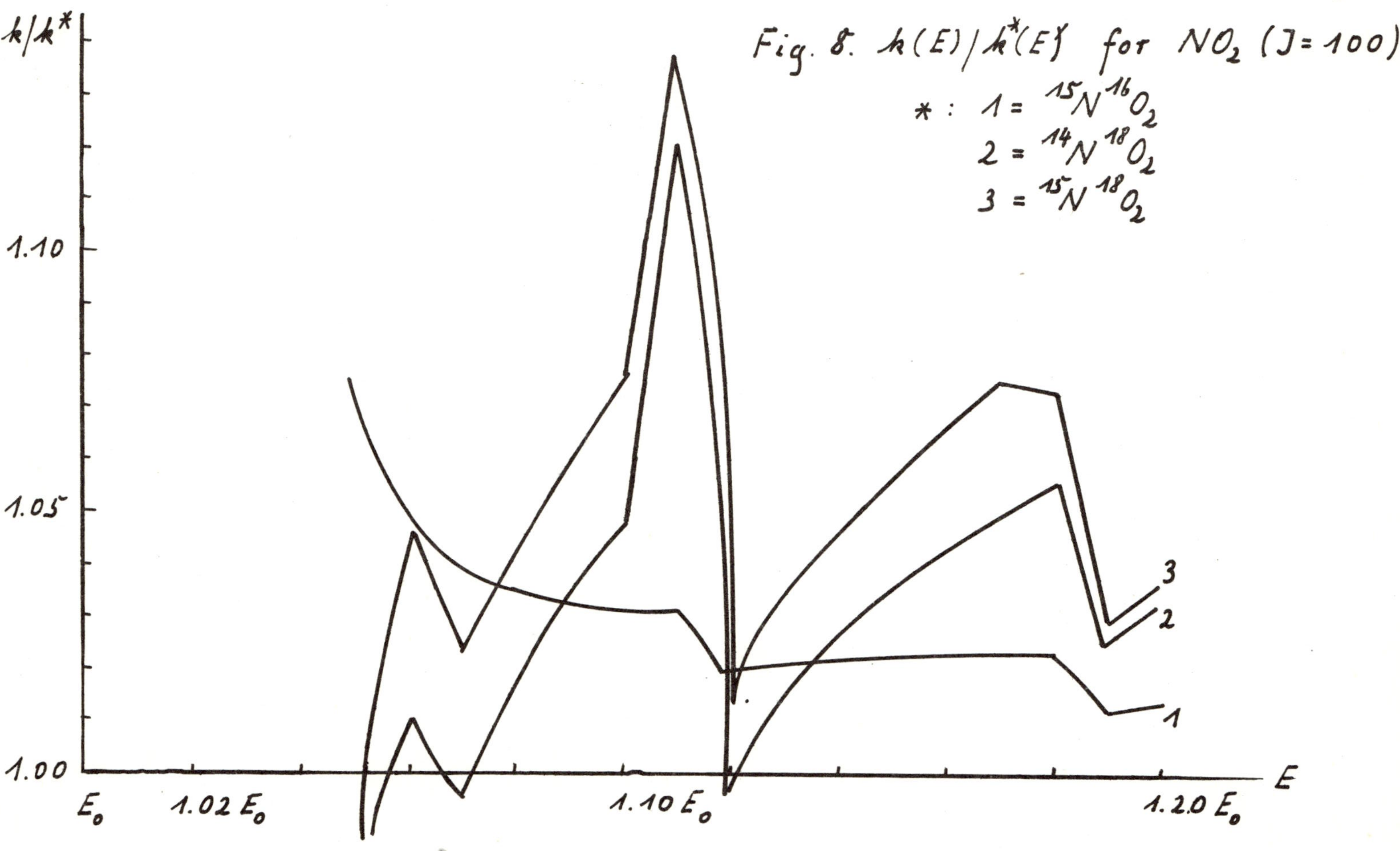


Fig. 8. $k(E)/k^*(E)$ for $NO_2$ ($J = 100$)

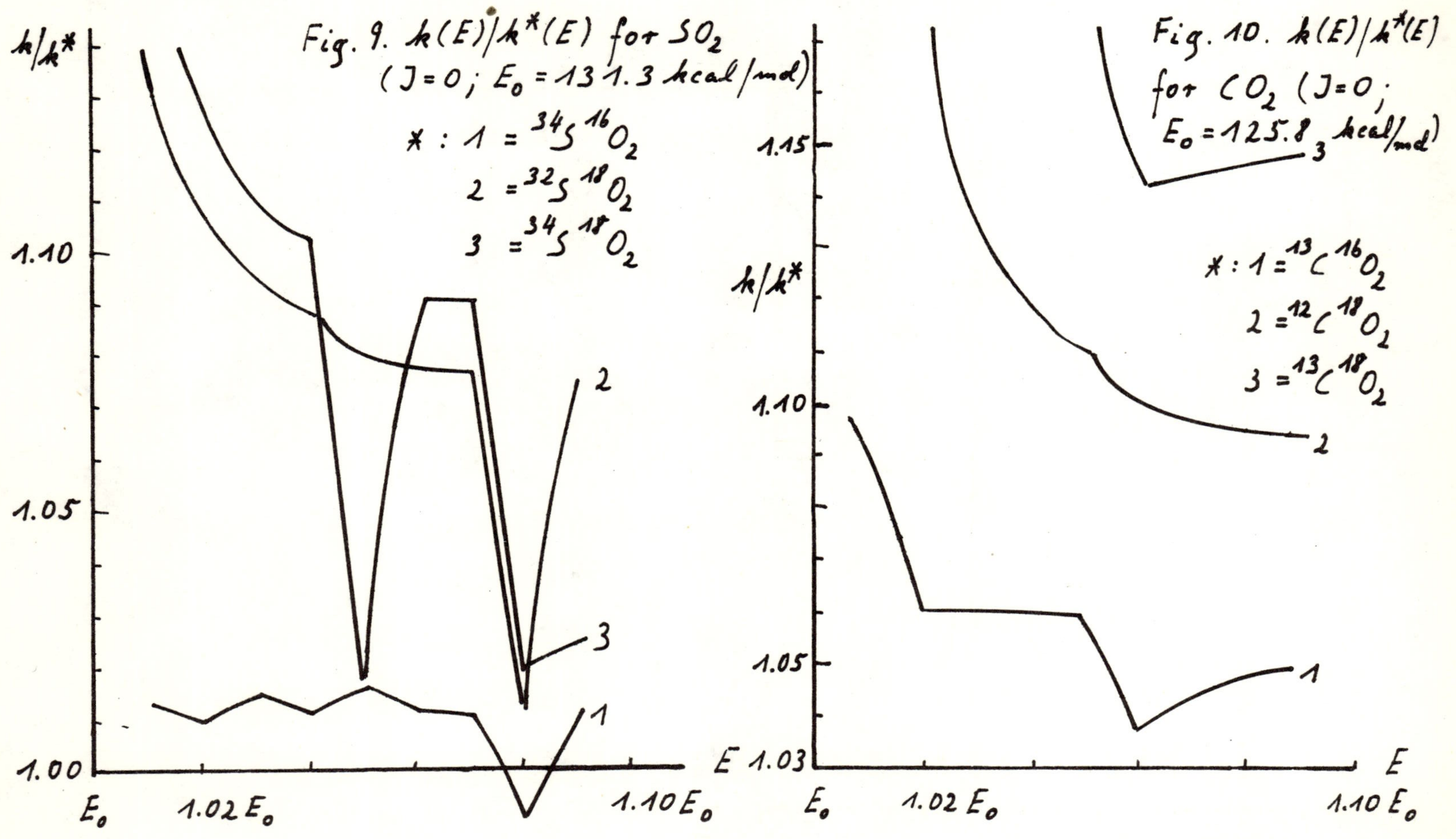


Fig. 9. $k(E)/k^*(E)$ for $SO_2$ ($J=0$; $E_0 = 131.3$ kcal/mol)
$*$: $1 = {}^{34}S^{16}O_2$
$2 = {}^{32}S^{18}O_2$
$3 = {}^{34}S^{18}O_2$

Fig. 10. $k(E)/k^*(E)$ for $CO_2$ ($J=0$; $E_0 = 125.8$ kcal/mol)
$*$: $1 = {}^{13}C^{16}O_2$
$2 = {}^{12}C^{18}O_2$
$3 = {}^{13}C^{18}O_2$

References


[1] Y.B.Band and K.F.Freed, J.Chem.Phys, 63, 3382 (1975).

[2] Y.B.Band and K.F.Freed, J.Chem.Phys. 63, 4479 (1975).

[3] M.Shapiro and R.D.Levine, Chem.Phys.Lett. 5, 499 (1970).

[4] H.Gebelein and J.Jortner, Theor.Chim.Acta 25, 143 (1972).

[5] R.L.Jaffe and K.Morokuma, J.Chem.Phys. 64, 4881 (1976).

[6] S.Mukamel and J.Jortner, Chem.Phys,lett. 40, 150 (1976).

[7] S.Mukamel and J.Jortner, J.Chem.Phys. 65, 5204 (1976).

[8] L.Mower, Phys.Rev. 142, 799 (1966).

[9] C.Cohen, Tannoudji, in: Cargese Lectures in Physics, Vol.2 (Gordon and Beach, London, 1967).

[10] S.Haroche, Ann.Phys. (Paris) 6, 189 (1971).

[11] J.Jortner and S.Mukamel, in: The world of quantum chemistry, eds. R.Daudel and B.Pullman (Reidel, Dordrecht, 1973).

[12] H.Gaedtke und J.Troe, Ber.Bunsenges.physik.Chem. 77, 24 (1972).

[13] D.L.Bunker and M.Pattengill, J.Chem.Phys. 48, 772(1968).

[14] W.H.Wong and R.A.Marcus, J.Chem.Phys. 55, 5625 (1971).

[15] H.S.Johnston, Gas Phase Reaction Rate Theory, Ronald Press, New York,1966.

[16] H. Johansen, K.Johst, Isotopenpraxis (DDR) 14 (1978).

# DECAY OF ATOMIC POLARIZATION MOMENTS

M.I.Dyakonov, V.I.Perel

A.F.Ioffe Physico-Technical Institute
of the USSR Academy of Sciences, Leningrad, USSR

The state of a free atom is described by the velocity of its motion, its internal energy and angular momentum. All these quantities may change in the act of a collision. For a long time the object of the collision theory was almost entirely confined to studying the changes in the velocity and energy of an atom. However, in connection with the ideas and methods of optical pumping, radiofrequency and laser spectroscopy a new field of the collision theory is intensely developing. It is concerned with processes in which the angular momentum of an atom changes.

## Description of Relaxation in Terms of Polarization Moments

The main question was, what precisely are the characteristics of a collision which determine the changes in the polarization properties of an ensemble of atoms. Adequate means of description of these properties are provided by the density matrix $f_{mm'}$. In the simplest case, when atoms on a single energy level with the value of the total angular momentum j are studied, the subscripts m and m' correspond to different projections of the angular momentum on the quantization axis.

In each collision the density matrix changes by the value of $\Delta f_{mm'}$ which depends on the parameters of the collision and is linearly related to the density matrix before the collision. Accordingly, the variation of the density matrix in time due to collisions may be described by the equation

$$\frac{df_{mm'}}{dt} = -\sum_{m_1 m_1'} \Gamma^{m_1 m_1'}_{m m'} f_{m_1 m_1'} . \qquad (1)$$

The right-hand side of Eq. (1) represents obviously the quantity $\Delta f_{mm'}$ summed over all collisions occurring during a unit of time. Eq. (1) serves as a starting point for most of the work in the field considered, although, strictly speaking, it is not correct, since it does not take into account the velocity dependence of the changes in the density matrix (see below). There is a case, however, when this equation is true: the depolarization of heavy atoms in a gas of light atoms. In other cases Eq. (1) may serve as a reasonable approximation, if one is not interested in the velocity dependence of the density matrix.

The relaxation matrix $\hat{\Gamma}$ provides the description of the depolarizing action of collisions. However, even for the simplest case j = 1 this matrix has generally 81 elements. The solution of the problem requires obviously that the matrix $\hat{\Gamma}$ be diagonalized. This may be done in a general way if the relaxation process posesses rotational symmetry. The density matrix fmm' transforms according to a reducible representation of the rotation group. It is natural, therefore, to split it in irreducible parts which obviously should decay independently. Such an expansion may be done by using the irreducible tensor operators $\hat{T}^{\varkappa}_{q}$

$$\left(T^{\varkappa}_{q}\right)_{m'm} = (-1)^{j-m'} \begin{pmatrix} j & \varkappa & j \\ -m' & q & m \end{pmatrix} (2\varkappa+1)^{1/2}. \tag{2}$$

The density matrix may be presented in the form

$$\hat{f} = \sum_{\varkappa q} f^{\varkappa}_{q} \hat{T}^{\varkappa}_{q} \quad ; \quad f^{\varkappa}_{q} = Sp\left(\hat{f}\hat{T}^{\varkappa}_{q}\right). \tag{3}$$

The coefficients of the expansion $f^{\varkappa}_{q}$ are called polarization moments and transform according to irreducible representations of the rotation group of rank $\varkappa$ $(0 \leq \varkappa \leq 2j, -\varkappa \leq q \leq \varkappa)$. The rotational symmetry of relaxation processes allows one to write down, instead of Eq. (1), the following equation describing the decay of the polarization moments

$$\frac{df^{\varkappa}_{q}}{dt} = -\gamma_{\varkappa} f^{\varkappa}_{q}, \tag{4}$$

where

$$\delta_{\varkappa\varkappa_1}\delta_{qq_1}\gamma_{\varkappa} = \sum_{\substack{mm' \\ m_1 m_1'}} \Gamma^{m_1 m_1'}_{mm'} \left(T^{\varkappa}_{q}\right)_{mm'} \left(T^{\varkappa_1}_{q_1}\right)_{m_1 m_1'}. \tag{5}$$

Thus the influence of collisions on the atomic polarization characteristics is determined by the set of relaxation constants $\gamma_{\varkappa}$. The polarization moments have a clear physical meaning: moments with $\varkappa = 0, 1, 2$ in particular, describe, respectively, the population of the level, orientation and alignment.

The density matrix expansion in terms of irreducible tensor operators was introduced by Fano /1/. The polarization moments (also called "irreducible components of the density matrix" or "state multipoles") were widely used in nuclear physics. Application of this formalism to problems of optical pumping and relaxation of the atomic density matrix was begun in the works of the present authors /2-4/ and of Omont /5/. The relaxation constants $\gamma_{\varkappa}$ were first introduced in /2/, where relaxation due to radiation trapping was considered. The methods of calculating these constants for collisional relaxation were formulated in papers /4,5/. The existence of different relaxation constants for orientation ( $\gamma_1$ ) and alignment ( $\gamma_2$ ) was first demonstrated experimentally by Omont /6/ and Saloman and Happer /7/.

The polarization moments formalism proved to be very useful for the description of interaction of atoms with external magnetic fields as well as with radiation in process of resonant fluorescence /3/, and also for the description of polarization phenomena in gas lasers /8/. This formalism provides a straightforward description of hyperfine structure effects, taking into account the fact that broadband excitation and collisions directly influence only the electronic state of the atom but not the nuclear spin /9-12/. An exhaustive review of applications of the polarization moments formalism to problems of optical pumping was given recently by Omont /13/.*) See also Chaika's book /14/ and earlier reviews /15,16/.

## Velocity Dependence of Relaxation and Generalized Polarization Moments

We will now discuss the effects due to the dependence of the density matrix on the velocity $\vec{v}$. The most general form of the equation describing the density matrix relaxation is

$$\frac{\partial f_{mm'}(\vec{v})}{\partial t} = - \sum_{m_1 m_1'} \int d^3\vec{v}_1\, K_{mm'}^{m_1 m_1'}(\vec{v},\vec{v}_1)\, f_{m_1 m_1'}(\vec{v}_1) \,. \qquad (6)$$

This integral equation takes into account the possible changes in the velocity of an atom during depolarizing collisions. However, the cross-sections of depolarizing collisions are often much larger than the gas-kinetic cross-sections, so that the velocity changes are small. Accordingly the kernel K has

---

*) We follow this review in the definition of the operators $\hat{T}_q^{\varkappa}$ and the polarization moments $f_q^{\varkappa}$ in Eqs. (2), (3). In our papers /2-4, 8, 9/ the normalization was different.

a sharp maximum for $\vec{v}_1 = \vec{v}$. If the velocity dependence of $\hat{f}(\vec{v})$ is smooth enough, Eq. (16) may be simplified

$$\frac{\partial f_{mm'}(\vec{v})}{\partial t} = -\sum_{m_1 m_1'} \Gamma_{mm'}^{m_1 m_1'}(\vec{v})\, f_{m_1 m_1'}(\vec{v}), \tag{7}$$

where

$$\Gamma_{mm'}^{m_1 m_1'}(\vec{v}) = \int d^3\vec{v}_1\, K_{mm'}^{m_1 m_1'}(\vec{v}, \vec{v}_1). \tag{8}$$

Eq. (7) has a form analogous to Eq. (1), with the essential difference that now the matrix $\hat{\Gamma}$ depends on vector $\vec{v}$. This dependence arises because for a moving atom the collision processes are non-isotropic even if the velocity distribution of atoms-perturbers is isotropic. This circumstance does not allow to diagonalize $\hat{\Gamma}$ by means of Eq. (3) and to obtain independent equations (4) for the polarization moments.

A complete utilization of the rotational symmetry of the scattering medium may be achieved by expanding the density matrix in terms of generalized tensor operators $\hat{\Psi}_M^{J\varkappa\ell}$ which depend on the direction of the velocity

$$\hat{\Psi}_M^{J\varkappa\ell}(\vec{n}) = \sum_{q\mu} (\varkappa \ell J M | \varkappa q \ell \mu)\, \hat{T}_q^{\varkappa}\, Y_{\ell\mu}(\vec{n}), \tag{9}$$

where $Y_{\ell\mu}(\vec{n})$ are spherical functions, $\vec{n}$ is the unit vector along $\vec{v}$. Eq. (9) corresponds to coupling of 'angular momentum' $\varkappa$ and $\ell$ with the aid of Clebsch-Gordan coefficients, $J$ and $M$ playing the role of the total angular momentum and its projection on the quantization axis. The operators $\hat{\Psi}$ are ortogonal and form a complete set:

$$Sp \int d\Omega \left( \hat{\Psi}_M^{J\varkappa\ell +} \hat{\Psi}_{M_1}^{J_1 \varkappa_1 \ell_1} \right) = \delta_{JJ_1} \delta_{MM_1} \delta_{\varkappa\varkappa_1} \delta_{\ell\ell_1}; \tag{10}$$
$$\sum_{\substack{\varkappa\ell \\ JM}} \left( \hat{\Psi}_M^{J\varkappa\ell +}(\vec{n}) \right)_{mm'} \left( \hat{\Psi}_M^{J\varkappa\ell}(\vec{n}_1) \right)_{m_1 m_1'} = \delta_{mm_1} \delta_{m'm_1'}\, \delta(\vec{n}-\vec{n}_1).$$

The integration is over the polar angles of vector $\vec{n}$.

Let us introduce the generalized polarization moments $f_M^{J\varkappa\ell}$ by the following formulae which generalize Eq. (3)

$$\hat{f}(\vec{v}) = \sum_{\varkappa\ell JM} f_M^{J\varkappa\ell}\, \hat{\Psi}_M^{J\varkappa\ell}; \qquad f_M^{J\varkappa\ell} = Sp \int d\Omega \left( \hat{f}\, \hat{\Psi}_M^{J\varkappa\ell +} \right). \tag{11}$$

The quantities $f_M^{J\varkappa\ell}$ (depending on the modulus of the velocity) transform accordingly to irreducible representations of the rotation group of rank $J$ ($|\varkappa-\ell| \le J \le \varkappa+\ell$, $-J \le M \le J$, $0 \le \varkappa \le 2j$, $\ell = 0, 1, 2 \ldots$).

The density matrix of the entire atomic ensemble may be obtained by integrating $\hat{f}(\vec{v})$ over the velocity. Only terms with $\ell = 0$ contribute, the number $J = \varkappa$ characterizing the polarization moment of the whole atomic ensemble. The generalized polarization moments with $\ell \neq 0$ and $\varkappa \neq 0$ describe the so-called "hidden" orientation, alignment etc.

The relaxation equation (6) may be written in the generalized polarization moments representation in the form

$$\frac{\partial f_M^{J\varkappa\ell}(v)}{\partial t} = - \int_0^\infty v_1^2 dv_1 \sum_{\varkappa_1 \ell_1} K_{\varkappa\ell}^{\varkappa_1 \ell_1}(J; v, v_1)\, f_M^{J\varkappa_1\ell_1}(v_1), \tag{12}$$

where the kernel K in the generalized polarization moments representation may be easily expressed in terms of $K_{mm'}^{m_1 m_1'}(\vec{v}, \vec{v}_1)$ by use of Eqs. (10), (11).

It is essential that due to the rotational symmetry of the scattering media Eqs. (12) for different values of $J$ and $M$ are independent, and that only the moduli of velocities $v$ and $v_1$ enter. However, for a given $J$ one has, in general, to solve a system of equations, unlike the case when a velocity-independent relaxation matrix $\Gamma$ may be introduced. The number of these equations (that is, the number of different pairs $\varkappa, \ell$ for a given $J$ ) is, generally speaking, equal to $(2j+1)^2$ . Further simplifications arise if symmetry properties with respect to inversion and time reversal are used.

The necessity of taking into account the correlation between the polarization of an atom and its velocity was first pointed out by Kazantsev /17/, who derived an equation in the form of Eq. (6) for resonant collisions. The notion of hidden alignment was introduced by Chaika /18,14/, who investigated effects related to this phenomenon in gas discharge. Generalized polarization moments were introduced by Rogova and one of the authors /19/ in connection with the problem of atomic depolarization due to resonant trapping. It was possible in this case to obtain an exact solution of the system of integral equations (12).

## Generalized Helical Moments

Now we introduce another method of description of the simultaneous relaxation of the atomic velocity and polarization distribution, founded on the use of the helical basis. This method also taking full account of the rotational symmetry of the scattering medium, has some advantages over the method described in the previous section.

In the helical basis one takes the direction of the particle velocity as the quantization axis, and the states are labelled by the projection $\lambda$ of the angular momentum on this direction (helicity) /21/. The density matrix in the helical basis $g_{\lambda\lambda'}$ is related to the density matrix in the laboratory frame $f_{mm'}$ by the equation

$$f_{mm'}(\vec{v}) = \sum_{\lambda\lambda'} \mathcal{D}^{j}_{m\lambda}(\omega)\, \mathcal{D}^{j*}_{m'\lambda'}(\omega)\, g_{\lambda\lambda'}(\omega), \qquad (13)$$

where $\mathcal{D}^{j}_{m\lambda}$ are the rotation matrices,*) $\omega$ is the rotation of the z-axis of the laboratory frame towards the direction of the velocity $\vec{v}$. The rotation is defined by Euler angles $\varphi$, $\theta$, $\psi$, $\theta$ and $\varphi$ being obviously the polar angles of $\vec{v}$. The angle $\psi$ is arbitrary, corresponding to the arbitrariness of the direction of the x-axis in the new frame. The left-hand side of Eq. (13) not depending on $\psi$, the matrix $g_{\lambda\lambda'}(\omega)$ should be proportional to $\exp[i(\lambda-\lambda')\psi]$.

It is natural to expand $g_{\lambda\lambda'}(\omega)$ in terms of irreducible tensor operators acting in the helical basis

$$g_{\lambda\lambda'}(\omega) = \sum_{\varkappa q} g^{\varkappa}_{q}(\omega)\, (\hat{T}^{\varkappa}_{q})_{\lambda\lambda'}. \qquad (14)$$

The "helical moments" $g^{\varkappa}_{q}$ are in fact the polarization moments in the frame where the z-axis coincides with $\vec{v}$. It is convenient to expand the quantities $g^{\varkappa}_{q}(\omega)$ in terms of rotation matrices

$$g^{\varkappa}_{q}(\omega) = \sum_{JM} f^{J\varkappa}_{Mq} \left(\frac{2J+1}{4\pi}\right)^{1/2} \mathcal{D}^{J*}_{Mq}(\omega). \qquad (15)$$

Thus finally we obtain the following expansion of the density matrix $\hat{f}(\vec{v})$

$$\hat{f}(\vec{v}) = \sum_{JM\varkappa q} f^{J\varkappa}_{Mq}\, \hat{\Phi}^{J\varkappa}_{Mq}(\vec{n}); \quad f^{J\varkappa}_{Mq} = Sp \int d\Omega\, \hat{f}\, \hat{\Phi}^{J\varkappa +}_{Mq}(\vec{n}). \qquad (16)$$

Using formulae (13), (14) and doing the summation over $\lambda$ and $\lambda'$ we find an explicit expression for the operators $\hat{\Phi}$ which depends, in fact, only upon the velocity polar angles and $\varphi$

$$\hat{\Phi}^{J\varkappa}_{Mq}(\vec{n}) = \left(\frac{2J+1}{4\pi}\right)^{1/2} \mathcal{D}^{J*}_{Mq}(\omega) \sum_{q'} \hat{T}^{\varkappa}_{q'}\, \mathcal{D}^{\varkappa}_{q'q}(\omega). \qquad (17)$$

We shall call the operators $\hat{\Phi}$ and the coefficients $f^{J\varkappa}_{Mq}$ 'generalized helical operators' and 'generalized helical moments respectively. (The latter may of course depend on the

---

*) We follow Edmonds /20/ in the definition of the D-matrices.

modulus of the velocity). The operators $\hat{\Phi}$, as well as $\hat{\Psi}$ transform under rotations according to irreducible representations of rank $J$ and have properties analogous to those given by Eq. (10)

$$Sp\int d\Omega\left(\hat{\Phi}^{J\varkappa +}_{Mq}\ \hat{\Phi}^{J_1\varkappa_1}_{M_1q_1}\right) = \delta_{JJ_1}\delta_{\varkappa\varkappa_1}\delta_{MM_1}\delta_{qq_1};$$

$$\sum_{J\varkappa Mq}\left(\Phi^{J\varkappa +}_{Mq}(\vec{n})\right)_{mm'}\left(\Phi^{J\varkappa}_{Mq}(\vec{n}_1)\right)_{m_1'm_1} = \delta_{mm_1}\delta_{m'm_1'}\delta(\vec{n}-\vec{n}_1). \tag{18}$$

The moments $f^{J\varkappa\ell}_M$ and $f^{J\varkappa}_{Mq}$ are related via a unitary matrix U:

$$f^{J\varkappa\ell}_M = \sum_q U^\ell_q(J\varkappa)\, f^{J\varkappa}_{Mq}, \tag{19}$$

where

$$U^\ell_q\,\delta_{JJ'}\delta_{\varkappa\varkappa'}\delta_{MM'} = Sp\int d\Omega\ \hat{\Psi}^{J\varkappa\ell +}_M\ \hat{\Phi}^{J'\varkappa'}_{M'q'}. \tag{20}$$

A direct calculation gives

$$U^\ell_q(J\varkappa) = (-1)^{J-q}(2\ell+1)^{1/2}\begin{pmatrix} J & \ell & \varkappa \\ -q & 0 & q\end{pmatrix}. \tag{21}$$

## Relaxation Equation for Generalized Helical Moments

The equation describing the decay of the helical moments $g^\varkappa_q(\omega)$ has the form

$$\frac{\partial g^\varkappa_q(\omega)}{\partial t} = -\int_0^\infty v_1^2 dv_1 \int\frac{d\omega_1}{2\pi}\sum_{\varkappa_1 q_1} K^{\varkappa\varkappa_1}_{q\,q_1}(v, v_1;\omega_{\vec{v}\vec{v}_1})\, g^{\varkappa_1}_{q_1}(\omega_1). \tag{22}$$

The integration over $d\omega_1$ icludes integration over Euler's angles $\varphi_1$, $\theta_1$, $\psi_1$ defining a rotation from the laboratory frame to a coordinate system with the z-axis along the velocity $\vec{v}_1$. The integrand actually does not depend on $\psi_1$, and integration over this angle is added for the sake of convenience. The kernel $K^{\varkappa\varkappa_1}_{q\,q_1}$ in the helical moments representation depends only upon the moduli of the velocities $v$ and $v_1$ and upon the parameters defining the rotation $\omega_{\vec{v}\vec{v}_1} = \omega^{-1}\omega_1$ of the coordinate system related to $\vec{v}_1$ towards the system related to $\vec{v}$. This is a consequence of the isotropy of the scattering medium.

The calculation of the kernel K on a microscopic basis may be done in a natural way in the representation of helical moments. However, the equation of relaxation is more conveniently written in terms of the generalized helical moments which depend only upon the absolute value of the velocity. Using expansion (15), we obtain

$$\frac{\partial f^{J\varkappa}_{Mq}(v)}{\partial t} = -\int_0^\infty v_1^2 dv_1 \sum_{\varkappa_1 q_1} K^{\varkappa\varkappa_1}_{qq_1}(J, v, v_1)\, f^{J\varkappa_1}_{Mq_1}(v_1). \tag{23}$$

The new kernel is the coefficient in the expansion of the kernel of Eq. (22) in terms of the rotation matrices

$$K^{\varkappa\varkappa_1}_{qq_1}(J; v, v_1) = \int \frac{d\omega_{\bar v \bar v_1}}{2\pi} K^{\varkappa\varkappa_1}_{qq_1}(v, v_1; \omega_{\bar v \bar v_1}) \mathcal{D}^{J*}_{qq_1}(\omega_{\bar v \bar v_1}). \tag{24}$$

If the velocity changes accompanying depolarizing collisions are small and may be neglected, the following approximation is readily obtained from Eq. (24)

$$v_1^2 K^{\varkappa\varkappa_1}_{qq_1}(J; v, v_1) = \delta(v - v_1)\, \delta_{qq_1}\, \Gamma^{\varkappa\varkappa_1}_q(v). \tag{25}$$

Then Eq. (23) takes the form

$$\frac{\partial f^{J\varkappa}_{Mq}(v)}{\partial t} = -\sum_{\varkappa_1} \Gamma^{\varkappa\varkappa_1}_q(v)\, f^{J\varkappa_1}_{Mq}(v). \tag{26}$$

Thus in the case considered the equations for different generalized helical moments are separate not only for different $J$ and $M$ but also for different $q$. The relaxation matrix not depending on the values of $J$ and $M$, Eq. (25) may be written directly for the helical moments $g^{\varkappa}_q$.

The customarily used relaxation constants $\gamma_\varkappa$ entering Eq. (4) are obtained by substitution of the relaxation matrix $\Gamma^{m_1 m_1'}_{m m'}$ by its value averaged over the Maxwellian distribution. They are expressed via the quantities (26) as follows

$$\gamma_\varkappa = \frac{1}{2\varkappa+1} \sum_q \int \Gamma^{\varkappa\varkappa_1}_q(v)\, f_0(v)\, d^3 v, \tag{27}$$

where $f_0(v)$ **is the Maxwellian distribution normalized to unity.**

If it is necessary to take into account the velocity changes the system of equations (23) as well as the system (12) contains for each, generally speaking, equations. The generalized helical moments representation gives the advantage of a simple transition to the extreme case (24); in addition, the kernel of Eq. (23) is more directly related to the microscopic characteristics of the collisions.

Depolarizing collisions of an atomic beam with a fixed velocity were first considered by Rebane /22/ who in fact calculated the relaxation matrix $\Gamma^{\varkappa\varkappa_1}_q$ for j = 1. Effects due to coupling of different polarization moments in the course of collisional relaxation were studied in detail for various experimental conditions by Lombardi /23/ (see also Omont's review /13/).

## Calculation of the Relaxation Matrix in the Impact Parameter Approximation

We shall consider now the relation between the relaxation matrix $\Gamma$ and the microscopic characteristics of an individual act of collision. We shall confine ourselves in this section to the case when velocity changes during collisions may be neglected, and the trajectories of the colliding atoms considered as rectilinear. In other words, we assume that the impact parameter approximation is valid.

The main characteristic of a collision in this approximation is the transition amplitude $A_m^{m_1}$ equal to the amplitude of state $m$ after the collision, under the condition that before the collision the atom was in state $m_1$. Obviously, the amplitudes $A_m^{m_1}$ depend on the vector of the relative velocity $\vec{w}$ and on the vector $\vec{\rho}$, where $\rho$ is the impact parameter.

The change in the density matrix $\Delta f_{mm'}$ during one collision may be expressed via the transition amplitudes:

$$\Delta f_{mm'}(\vec{u},\vec{w},\vec{\rho}) = \sum_{m_1 m_1'} \left[ A_m^{m_1} (A_{m'}^{m_1'})^* - \delta_{mm_1}\delta_{m'm_1'} \right] f_{m_1 m_1'}(\vec{v}). \quad (28)$$

Summing expression (28) over all collisions occurring in a unit of time, we obtain the relaxation matrix

$$\Gamma_{m_1 m_1'}^{mm'}(\vec{v}) = n \int \mathcal{F}(u)\, d^3\vec{u} \int d^2\vec{\rho}\, w \left[ \delta_{mm_1}\delta_{m'm_1'} - A_m^{m_1}(A_{m'}^{m_1'})^* \right], \quad (29)$$

where $n$ is the concentration of atoms-perturbers, $\mathcal{F}(u)$ is their velocity distribution normalized to unity and assumed to be isotropic.

The transition amplitudes are naturally calculated in the helical basis with the z-axis directed along the relative velocity $\vec{w}$, and the x-axis along vector $\vec{\rho}$. The transition amplitudes $a_\mu^{\mu_1}$ calculated using this basis depend only on the absolute value of $w$ and the impact parameter $\rho$.

Let us introduce a matrix

$$\tilde{\Gamma}_s^{\varkappa\varkappa_1}(w) = 2\pi w \int_0^\infty \rho\, d\rho \sum_{\mu\mu'\mu_1\mu_1'} (T_s^{\varkappa})_{\mu\mu'} \left[ \delta_{\mu\mu_1}\delta_{\mu'\mu_1'} - a_\mu^{\mu_1} (a_{\mu'}^{\mu_1'})^* \right] (T_s^{\varkappa_1})_{\mu_1\mu_1'}, \quad (30)$$

which gives a full description of collisions with a fixed relative velocity. It may be shown that the relaxation matrix entering Eq. (25) for the generalized helical moments is related to the matrix $\tilde{\Gamma}$ in the following way:

$$\Gamma_q^{\varkappa\varkappa_1}(v) = \sum_s \int_0^\infty w^2 dw\, I_{qs}^{\varkappa\varkappa_1}(v,w)\,\tilde{\Gamma}_s^{\varkappa\varkappa_1}(w). \tag{31}$$

Here we have introduced a kinematical factor

$$I_{qs}^{\varkappa\varkappa_1}(v,w) = \int d\Omega_{\vec{w}}\, \mathcal{F}(\vec{v}-\vec{w})\, \mathcal{D}_{qs}^{\varkappa}(\omega_{\vec{w}\vec{v}})\, \mathcal{D}_{qs}^{\varkappa_1 *}(\omega_{\vec{w}\vec{v}}), \tag{32}$$

where integration is performed over the polar angles $\alpha$ and $\beta$ of vector $\vec{w}$. The rotation $\omega_{\vec{w}\vec{v}}$ of the coordinate system related to $\vec{w}$ to the system related to $\vec{v}$ is defined by Euler angles $\alpha$, $\beta$, $\gamma$. Actually, the integrand is independent of $\alpha$ and $\gamma$, so that integration over $\alpha$ gives $2\pi$ and only an integral over the angle $\beta$ between $\vec{v}$ and $\vec{w}$ remains. (The distribution function $\mathcal{F}$ depends only on $|\vec{v}-\vec{w}| = (v^2+w^2-2vw\cos\beta)^{1/2}$).

Consider expression (32) in extreme cases. If the velocity $v$ is much less than the thermal velocity of perturbers, we obtain

$$I_{qs}^{\varkappa\varkappa_1}(v,w) = \frac{4\pi\,\mathcal{F}(w)}{2\varkappa+1}\,\delta_{\varkappa\varkappa_1}. \tag{33}$$

In this case the equation describing the relaxation is reduced to the simple Eq. (4), the constants $\gamma_\varkappa$ defined by the formula analogous to Eq. (27):

$$\gamma_\varkappa = \frac{1}{2\varkappa+1}\sum_q \int \tilde{\Gamma}_q^{\varkappa\varkappa_1}(w)\,\mathcal{F}(w)\, d^3w. \tag{34}$$

In the opposite extreme case when the velocity $v$ is large, we have $w^2 I_{qs}^{\varkappa\varkappa_1} = \delta_{qs}\,\delta(v-w)$, so that $\Gamma_q^{\varkappa\varkappa_1}(v) = \tilde{\Gamma}_q^{\varkappa\varkappa_1}(v)$.

In the impact parameter approximation the transition amplitudes $a_\mu^{\mu_1}$ entering formula (30) are found by solving the system of equations

$$i\hbar\frac{d a_\mu(t)}{dt} = \sum_{\mu'} V_{\mu\mu'}(\vec{\rho}+\vec{w}t)\, a_{\mu'}(t) \tag{35}$$

subject to initial conditions $a_\mu(-\infty) = \delta_{\mu\mu_1}$. Then $a_\mu^{\mu_1} = a_\mu(\infty)$. Here $V_{\mu\mu'}(\vec{r})$ is the matrix element of the operator describing the interaction of the atom with a perturber. It should be noted that if the interaction is such that $V \sim r^{-n}$, the amplitudes $a_\mu^{\mu_1}$ depend on $\rho$ and $w$ only via the combination $w\rho^{n-1}$. In this case the dependence of $\tilde{\Gamma}_q^{\varkappa\varkappa_1}$ on $w$ may be established: $\tilde{\Gamma}_q^{\varkappa\varkappa_1} \sim w^{\frac{n-3}{n-1}}$.

The transition amplitudes $a_\mu^{\mu_1}(\rho, w)$ are the result of the microscopic consideration of the collision act in the impact parameter approximation. Though these quantities are calculated assuming the trajectories to be rectilinear one can see from the following section that the same amplitudes define the kernel of the integral Eq. (6) in case when the scattering angles are small.

## Kernel of Integral Relaxation Equation for Small Scattering Angles

If the cross-section for depolarizing collisions is large compared to the gas-kinetic cross-section, the scattering accompanying the depolarization occurs mostly on small angles. However, the related changes in the velocity distribution function of atoms may in some cases be of special interest. Such is, e.g., the problem of collisional broadening of narrow holes in the Doppler contour of a spectral line produced by monochromatic radiation.

It is well known /24/ that the eikonal approximation allows one to calculate the scattering amplitudes for small scattering angles by considering rectilinear classical trajectories. This method may be easily generalized for the case of degenerate atomic states (see also Ref. /25/).

In the eikonal approximation the expression for an element of the S-matrix corresponding to a transition from the state $\mu_1 \vec{q}_1$ to the state $\mu \vec{q}$ ( $\vec{q}_1$ and $\vec{q}$ being the relative momenta of two particles before and after the collision) has the form

$$S_{\mu\vec{q}}^{\mu_1\vec{q}_1} = L^{-2}\,\delta_{q_1 q_z} \int d^2\vec{\rho}\;\; a_{\mu}^{\mu_1}(\rho, w_1)\, e^{-i(\mu-\mu_1)\psi}\, e^{-\frac{i}{\hbar}(\vec{q}-\vec{q}_1)\vec{\rho}} \tag{36}$$

Here $q_z$ denotes the projection of vector $\vec{q}$ on the direction of the initial relative momentum $\vec{q}_1$, $\vec{w}_1$ is the relative velocity before the collision, vector $\vec{\rho}$ is perpendicular to $\vec{w}_1$, $L$ is the linear dimension of the normalizing volume. The transition amptitudes $a_{\mu}^{\mu_1}$ are defined by solving Eqs. (35) in the impact parameter approximation. The appearance of the $\delta$-symbol in Eq. (36) is related to the fact that the change of the longitudinal component of the momentum for small-angle scattering is a quantity of second order in the scattering angle.

A complete description of the result of collisions with a fixed value of the initial relative velocity $\vec{w}_1$ is given by an integral operator with the kernel

$$\tilde{K}_{ss_1}^{\varkappa\varkappa_1}(\vec{w}_1, \Delta\vec{w}) = \left(\frac{M}{2\pi\hbar}\right)^2 \delta(v_z - v_{1z}) \int d^2\vec{\rho}\, d^2\vec{\rho}\,'\; G_{ss_1}^{\varkappa\varkappa_1}(\vec{\rho}, \vec{\rho}\,', w_1) \times \tag{37}$$

$$\times \exp\left\{-i\frac{M}{\hbar}(\vec{v}-\vec{v}_1)(\vec{\rho}-\vec{\rho}\,')\right\},$$

where

$$G_{ss_1}^{\varkappa\varkappa_1}(\vec{\rho},\vec{\rho}\,',w_1) = w_1 \sum_{\mu\mu'\mu_1\mu_1'} \left[\delta_{\mu\mu_1}\delta_{\mu'\mu_1'} - a_{\mu}^{\mu_1}(\rho,w_1)\,a_{\mu'}^{\mu_1'*}(\rho' w_1)\right] \times$$
$$\times \left(T_s^{\varkappa}\right)_{\mu\mu'} \left(T_{s_1}^{\varkappa_1}\right)_{\mu_1\mu_1'} \exp\left[-i(\mu-\mu_1)\psi + i(\mu'-\mu_1')\psi'\right]. \tag{38}$$

Here $M$ is the atomic mass, $v_{12}$ and $v_2$ are the projections of the atomic velocities before and after the collision on the initial relative velocity $\vec{w}_1$, $\vec{\rho}$ and $\vec{\rho}\,'$ are vectors perpendicular to $\vec{w}_1$, $\psi$ and $\psi'$ are the angles of these vectors with respect to vector $\Delta\vec{w}$ (also perpendicular to $\vec{w}_1$). Obviously

$$\Delta\vec{w} = \vec{w} - \vec{w}_1 = \frac{M+M_0}{M_0 M}(\vec{q}-\vec{q}_1) = \frac{M+M_0}{M_0}(\vec{v}-\vec{v}_1),$$

where $M_0$ is the mass of the perturbing atom. The kernel (37) is written down in the helical moments representation in a coordinate system with the z-axis along $\vec{w}_1$ and the x-axis along $\Delta\vec{w}$. The kernel depends only on the absolute values of vectors $\vec{w}_1$ and $\Delta\vec{w}$ and the angle between them.

The kernel of the integral equation for the helical moments (22) is expressed via the kernel (37) by the following formula analogous to Eq. (31)

$$K_{q\,q_1}^{\varkappa\varkappa_1}(v,v_1;\omega_{\bar{v}\bar{v}_1}) = \sum_{ss_1}\int d^3w_1\, J_{qs,\,q_1s_1}^{\varkappa\,\varkappa_1}\,\tilde{K}_{s\,s_1}^{\varkappa\varkappa_1}(\vec{w}_1,\Delta\vec{w}), \tag{39}$$

where a new kinematical factor is introduced

$$J_{qs,\,q_1s_1}^{\varkappa\,\varkappa_1} = \mathcal{F}(|\vec{v}_1-\vec{w}_1|)\,\mathcal{D}_{qs}^{\varkappa}(\omega_{\bar{w}_1\bar{v}})\,\mathcal{D}_{q_1s_1}^{\varkappa_1 *}(\omega_{\bar{w}_1\bar{v}_1}). \tag{40}$$

Eqs. (37)-(40) are rather cumbersome. They give, however, a complete solution of the problem, and allow us to calculate the kernel of the relaxation equation (22) via the transition amplitudes $a_{\mu}^{\mu_1}$ obtained in the impact parameter approximation.

In the case of small-angle scattering which we have considered above, one might simplify the solution of the integral equation (22) by a transition to the Fokker-Plank equation. Such an equation as applied to depolarizing collisions was recently derived and investigated by Dymnikov /26/.

## References


1. U.Fano. Rev. Mod. Phys., 29, 74 (1957).
2. M.I.Dyakonov, V.I.Perel. ZhETF, 47, 1483 (1964) /Sov.Phys. JETP, 20, 997 (1965)/.
3. M.I.Dyakonov ZhETF, 47, 2213 (1964) /Sov. Phys. JETP, 20, 1484 (1965)/.
4. M.I.Dyakonov, V.I.Perel. ZhETF, 48, 345 (1965) /Sov. Phys. JETP, 21, 227 (1965)/.
5. A.Omont. J. Phys. 26, 26 (1965).
6. A.Omont. C.R. Acad. Sci. 260, 3331 (1965)
7. E.B.Saloman, W.Happer. Phys. Rev. 144, 7 (1966); 160, 23 (1967).
8. M.I.Dyakonov, V.I.Perel. Optika i spektroskopiya, 20, 472 (1966) /Opt. Spectr. 20, 257 (1966)/.
9. M.I.Dyakonov. Optika i spektroskopiya, 19, 662 (1965).
10. A.Omont. J. Phys. 26, 576 (1965).
11. J.P.Faroux. Thesis, Paris (1969).
12. A.I.Okunevich, V.I.Perel. ZhETF, 58, 666 (1970) /Sov.Phys. JETP, 31, 356 (1970).
13. A.Omont. Progr. in Quantum Electronics, 5, 69 (1977).
14. M.P.Chaika. Interferentsiya vyrozhdennykh atomnykh sostoyanij. Ed. Leningrad State University, Leningrad (1975).
15. W.Happer. Rev. Mod. Phys. 44, 169 (1972).
16. U.Fano, J.H.Macek. Rev. Mod. Phys. 45, 553 (1973).
17. A.P.Kazantsev. ZhETF, 51, 1751 (1966).
18. H.Kallas, M.P.Chaika. Optika i spektroskopiya 27,694(1969).
19. V.I.Perel,I.V.Rogova. ZhETF 61, 1814 (1971); 65,1012(1973).
20. A.R.Edmonds. Angular Momentum in Quantum Mechanics. Princeton University Press (1957).
21. V.B.Beresteckij, E.M.Lifshits, L.P.Pitaevskij. Relyativistskaya kvantovaya teoriya, I. "Nauka", Moscow (1968).
22. V.N.Rebane. Optika i spektroskopiya, 24, 163, 296 (1968).
23. M.Lombardi. J. Phys. 30, 631, 789 (1969).
24. L.D.Landau, E.M.Lifshits. Kvantovaya mekhanika. Fizmatgiz, Moscow (1963).
25. M.R.Flannery, K.J.McCann. In: Electron and Photon Interactions with Atoms. Ed. by H.Kleinpoppen, M.R.C.McDowell. Plenum Press, New York, London, p. 275 (1976).
26. V.D.Dymnikov. Integral stolknovenij dlya nediagonalnoj matritsy plotnosti. Preprint N°584 of Ioffe Physico-Technical Institute. Leningrad (1978).

# ALIGNMENT OF EXCITED ATOMS IN A GAS DISCHARGE

M.P. Chaika
Leningrad State University, U.S.S.R.

Both the Hanle effect and the phenomenon of level crossing have become well known. The Hanle effect, i.e. depolarization of resonant radiation by external magnetic field, was described 55 years ago, while the phenomenon of level crossing which is actually an expansion of the Hanle effect was described nearly 20 years ago. In these twenty years the level crossing method has become widely used alongside with the other spectroscopy methods of atom investigation. The major modifications of the method can be listed as follows:

1. Excitation of atomic vapours by polarized resonant light and investigation of polarization dependence or resonant scattering intensity upon the value of the applied magnetic field. This method may be called classical.

2. Replacement of optical excitation by electronic impact. The method is not widely applied due to the complexity of signal treatment resulting from cascade transitions.

3. Application of an atomic beam replacing the vapour cell, which allows to increase considerably the number of elements that can be studied by this method.

4. Use of the "auto-alignment" model in a gas discharge, i.e. polarization of spontaneous scattering in a gas discharge, which allows to apply the method for the study of highly excited states.

This latter method will be now examined more closely. Atomic radiation is polarized if the excited state is not symmetrical spherically, which occurs whenever the process of excitation is anisotropic, and the structure of atoms in this state need not be of spherical symmetry (it is sufficient to have $J \geqslant 1/2$ for orientation and $J \geqslant 1$ for alignment). The most characteristic example is the excitation caused by a polarized light beam. The circularly polarized light brings about a macroscopic moment to the ensemble of excited atoms (the excited state is oriented), which is described by a vector, while the linearly polarized light brings about a macroscopic electric quadruple moment (the excited state is aligned) described by a tensor.

In order to obtain the alignment the light need not be polarized, it is sufficient for the light to have a certain direction; moreover, the alignment is possible even if the intensity of light is different along different directions. Such conditions take place in a gas discharge, since due to a limited volume the light intensity in any point within the light source is not the same in different directions. The light propagating within the source is partially absorbed by the atoms whose excited states become aligned while their radiation becomes partially polarized. When the magnetic field is applied, the radiation becomes depolarized, which leads to a change in intensity that is called a signal of alignment.

It is obvious that the radiation diffusion must necessarily lead to the alignment of atom states if $J > 1$ near the boundaries of a radiating volume, whatever its origin, in particular the statement is valid for cosmic bodies as well. The phenomenon is actually observed in the solar corona and solar prominences, and astrophysicists use the data of polarization degree and direction in order to define more accurately the solar magnetic field.

Radiation diffusion appears to be the most acceptable explanation and the most probable - though not the single - cause for generation of alignment in a gas discharge. The movement of electrons has the same symmetry as the light beam; directed electronic impact also results in alignment. Thus, experiments on a hollow cathode tube and the analysis of signals of hollow cathode discharge emission prove that it is the electronic impact that is responsible for the polarization of spontaneous emission /1/.

We found the signals of "auto-alignment" in a gas discharge in 1969 when studying the influence of laser generation upon the polarization of spontaneous radiation of an active medium /2/. At the same time we discovered similar signals, of the same shape and value, on the lines beginning at the levels with $J = 0$, though theoretically these levels do not allow of any alignment since their origin implies spherical symmetry.

Explanation for these abnormal signals was found two years later /3/. The explanation is based on the origination of "hidden" alignment and its influence upon the population of levels.

The hidden alignment is an alignment of a sub-ensemble of atoms with an isolated direction of heat motion, when the axis of alignment coincides with the direction of motion while its value depends on the atoms velocity. Fig.1 illustrates the process of generation of the alignment: in the system of coordinates connected with the moving atom, the radiation of resonant frequency directed along, and against, the direction of the atom motion is of a smaller intensity than the radiation propagated perpendicular to the direction of the atom motion, which is explained by the Doppler effect. Therefore, such ensemble of atoms is excited in anisotropic manner mainly due to the "perpendicular" light, and, as a result, the ensemble is aligned along the velocity vector. The alignment disappears by integrating over all directions of atomic motion, therefore we call it "hidden" alignment. But this alignment causes a change in the emission of radiation.

It is not unreasonable to expect that this hidden alignment can be observed directly in experiments on absorption. In order to observe the hidden alignment influencing the absorption, let us divide the ensemble of atoms into three groups according to the directions of their velocity vectors. It is sufficient to study the absorption of these three groups of atoms moving, respectively, in three orthogonal directions, i.e. for the first group the direction coincides with that of the light beam whose absorption is under observation, for the second group - with a magnetic field which is applied later in the experiment, and for the third group the direction should be perpendicular to each of the other two. Let us take that the electric vector of the linearly polarized light beam coincides with the direction of the third group of atoms. Besides, one more condition must be satisfied: the light must be monochromatic, or, at any rate, its absorption spectrum must be much narrower than the width of the absorption line.

Let us discuss the light absorption by each of the three groups of atoms. In this problem the absorbing atoms are aligned in respect to the directions of their motions; therefore, their absorption coefficients depend both on the direction of light beam and that of light polarization. The dependence of absorption coefficient upon these two directions can be shown in the form of an ellipsoid as is generally done in crystallo-optics. This is illustrated by Fig.2, where ellipsoids are replaced by cylinders as it is easier to show graphically the direction of axes of a cylinder; the replacement makes the figure more illustrative, while the quantitative differences caused by the replacement are not essential for a qualitative study. The additional diagram in Fig.2 shows the correlation between the sizes of the cylinders and the absorption coefficient at different directions of polarization. When polarization (the electrical vector) is directed along the arrow, the absorption coefficient is proportional to the length of the arrow, while the value of the length is proportional to the respective value of the cylinder cross-section (see Fig.2).

Thus, when polarization is vertical, the light is mainly absorbed by the first group of atoms, since the direction of polarization coincides with the axis of cylinder; in this situation the other two groups of atoms absorb less light. When polarization is orthogonal to the figure, the atoms in the first and third groups absorb little light, while the second group absorbs it actively. It should be noted here that the atoms of the first and second groups move perpendicularly to the light beam direction, thus absorbing light within frequencies near the centre of the Doppler contour of absorption, and it is the third group of atoms that is responsible for absorption at the edges of the line. Let us now apply magnetic field in the manner shown in Fig.2. Due to precession of atomic dipoles in the magnetic field, cylinders "2" and "3" will rotate and the diagram of the absorption coefficient will assume the shape of a low and wide cylinder, with a vertical axis and with a diameter greater than that of a "tall" cylinder but smaller than its length. For the first group of atoms the diagram of absorption coefficient will remain unaltered.

Let us now discuss the changes in absorption that will occur when magnetic field is applied. There will be obviously no changes in the absorption when polarization is vertical (the first group of atoms will continue to absorb light actively, while in groups 2 and 3 the absorption is weak). Of particular interest is the study of absorption in the case when polarization is perpendicular to the drawing. For the second and third groups of atoms the absorption will now be proportional to the diameters of the "wide" cylinders. With magnetic field applied, in the second group the absorption is greater, and in the third group smaller than in the absence of magnetic field. It should be noted again that the second group absorbs light within the central frequencies of the line, while the third group absorbs light at the edges of the line. Consequently, in this experiment when magnetic field is applied the observer will register an increase of absorption in the centre of the Doppler line and a decrease of absorption at the edges of the line.

Calculation for the model which takes into account only the precession of dipoles caused by magnetic field gives a dependence of the change in absorption at any light frequency, which can be written down in the usual Lorentzian form of

$$K = A_\nu + \frac{B_\nu}{1 + \frac{4\omega^2}{\Gamma^2}} \qquad (1)$$

where the magnitude of frequency influences only $B_\nu$, i.e. the value and the sign of the coefficient of proportionality.

The actual picture, however, is much more complicated. Apart from the "one-stage" mechanism described above, there are "multi-stage" processes which affect both the absorption and the spontaneous emission. One of such processes is diffusion of light. The example described above proves that the hidden alignment causes a change in the contour of the absorption coefficient for the light beam of a certain polarization. The same effect is observed for spontaneous emission: its contour differs from the Doppler form. The change in the contour of the line, which is introduced by application of magnetic field, results in a change in the average number of re-absorption of photons in a volume, and, consequently, it causes changes in the lifetime and magnitude of population. It should be noted that this case is different from the one discussed above; we are now considering photons emitted by the state with hidden alignment and absorbed by the normal state of atoms, which we take to be symmetrical spherically.

It seems to be a rather hopeless task to try to observe changes in the spectral contour of emission of a whole radiating volume, since the emission is influenced upon not only by the hidden alignment but also by the absorption in the volume which, in its turn, also depends on the hidden alignment. Integral intensity of emission need not change with changes in diffusion; in fact, if, apart from the radiative decay into the normal state, there are no other channels of irrevocable destruction of the state under consideration, then the intensity cannot depend on diffusion; it can only be determined by the power of primary excitation. However, if other channels of destruction exist the hidden alignment will affect the intensity of resonant radiation, and its dependence on the applied magnetic field will be determined by a law which can be written down in the form of

$$\bar{I} = \bar{I}_0 + \bar{I}'\left(\frac{1}{1+\frac{\omega^2}{\Gamma^2}} + \frac{1}{1+\frac{4\omega^2}{\Gamma^2}}\right) \qquad (2)$$

(higher orders of $\left(1+\frac{\omega^2}{\Gamma^2}\right)$ and $\left(1+\frac{4\omega^2}{\Gamma^2}\right)$ are omitted).

Let us now discuss another process, i.e. a multi-stage excitation when one can observe spontaneous emission from higher excited states (Fig.3). There are two mechanisms which relate the intensity of this radiation with the hidden alignment of the first excited state. The first mechanism is easy to explain: through radiation diffusion the hidden alignment influences the population of the first excited state "1", and, consequently, the number of acts of transition into state "2" per unit of time. It is obviously the same law (2) that determines the changes in intensity of radiation emitted by state "2" due to magnetic field.

The other mechanism is more difficult to explain, and it only refers to the optical multi-stage excitation. Similar to the $|0\rangle \rightarrow |1\rangle$ transition, in the case of $|1\rangle \rightarrow |2\rangle$

transition - in the system of axes related to the moving atom - the light of resonant frequency is not isotropic, i.e. it is more intensive in the direction perpendicular to that of the moving atom than in any other direction. Besides, the absorbing state is aligned along the direction of the motion. The average number of acts of absorption per unit of time can be written down in terms of polarization moments as

$$N \approx \sum_{q\,æ} f(æ, q)\, \rho_q^{æ}\, \mathcal{F}_q^{æ}$$

where $\rho$ is tensor of alignment of the absorbing state
$\mathcal{F}$ is tensor describing the anisotropy of exciting light
$f$ means coefficients depending on $q$ and $æ$ only.

Since the axes of tensors $\rho$ and $\mathcal{F}$ coincide, and since they are both unit axial, then q=0 throughout the expression and

$$N \approx f_0^{(0)} \rho_0^0 \mathcal{F}_0^0 + f_0^{(2)} \rho_0^2 \mathcal{F}_0^2$$

The second term differs from zero if the absorbing state is aligned; the term disappears if the absorbing state cannot be aligned, i.e, if it is symmetrical spherically, irrespective of the characteristics of the radiating light. It also disappears when radiation is isotropic (in the system of axes related to the atom), irrespective of the characteristics of the absorbing state. The magnetic field destroys partially the alignment of atomic state $\rho$ , thus causing the change of atomic stream from state "1" into state "2". The intensity of spontaneous emission by state "2" is also changed. It is remarkable that the dependence on the magnetic field has the same form (2) again.

The signals of hidden alignment have been observed experimentally /2,4/. The experimental set-up is shown in Fig.4. The magnitude of magnetic field was modulated, and the alternating part of photocurrent was recorded by phase-sensitive detection. Fig.5 represents the shape of the signal, i.e. the dependence of the increase in intensity upon the increase of tension in the magnetic field, $\Delta I/\Delta H$ .

On line $\lambda$ =607,4 nm of Neon, which corresponds to the transition $J=0 - J=1$ , the signal has the same sign, irrespective of the directions of the magnetic field or the characteristics of polarization. This is in perfect agreement with the above discussed model which relates the changes in population in state "2" with the destruction of hidden alignment; state "2" has $J$ =0 and possesses no other polarization moments apart from population.

The mere presence of the signal on the line cannot be considered as an exhaustive experimental proof of the existence of a hidden alignment. If macroscopic alignment is present in state "1", then its destruction under the influence of magnetic field, with the diffusion of radiation, can bring about a change in the rate of population of state "1" and, consequently, in state "2" as well. The decisive factors are the shape and the width of the signal. Macroscopic alignment gives a signal of Lorentzian shape, with the width equal to $2\Gamma$ . The shape of the signal of hidden alignment is more complicated and is described by formula (2).

In the investigation of atomic magnetic resonance the method of phase-sensitive detection is widely used, as well as observation of the signal in the form $\Delta I/\Delta H$ ; various methods have been developed for finding the parameters (mainly the width) of the primary function, i.e. of $I=\Phi(H)$ with $\Delta H$ approaching zero. All these methods have been developed for the Lorentzian form of the signal and have proved to be unsuitable for our experimental curves; the results depend too greatly on the values of $\Delta H$ used for extrapolation. Neither could the correct value of $\Gamma$ be estimated from a simple extrapolation of the width of the signal towards its boundary value when $\Delta H \to 0$. Hence, the signal cannot have the Lorentzian shape.

This year we have resumed the study of alignment signals at the level $^3P_1$ of Neon. The method of signal recording has been improved by the use of the principle of accumulation, which allowed to get a nondistorted signal of $I=\Phi(H)$ on the same line of Neon ($\lambda$ =607,4 nm) /5/. Calculation by formula (2) gave a value of $\Gamma$ =6,87 MHZ. At the same time we received a usual Hanle signal on the resonant line of Ne $^3P_1 - {}^1S_0$ $\lambda$ =74,3 nm, with optical excitation presence, i.e. in the classical set-up of the experiment. The signal was of the Lorentzian shape, with the width corresponding to $\Gamma$ =7,08 MHz. The good agreement of these two values is a sufficient proof of the existence of hidden alignment and of its influence upon the intensity of spectral lines.

Signals of hidden alignment on the same state $^3P_1$ of Neon were observed in 1973 in Novosibirsk, when the absorption of monochromatic radiation $\lambda$ =632,8 nm by a gas discharge in a magnetic field was measured, at various polarizations /6/. The authors applied formula (2) for calculation and estimated a value of $\Gamma$ =6,3±1 MHz.

Fig.1.

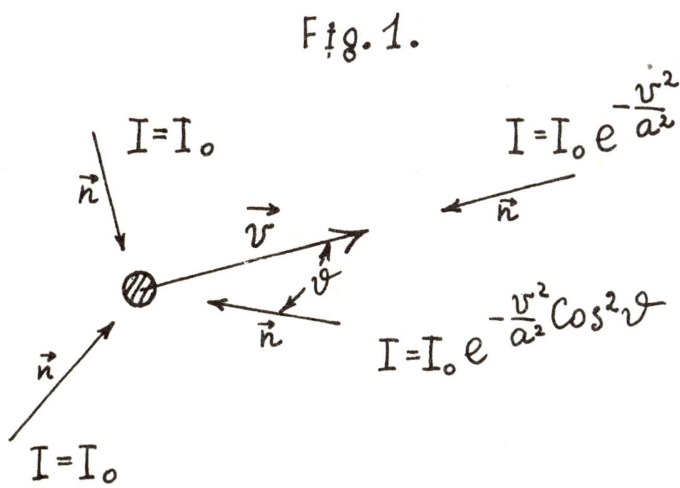

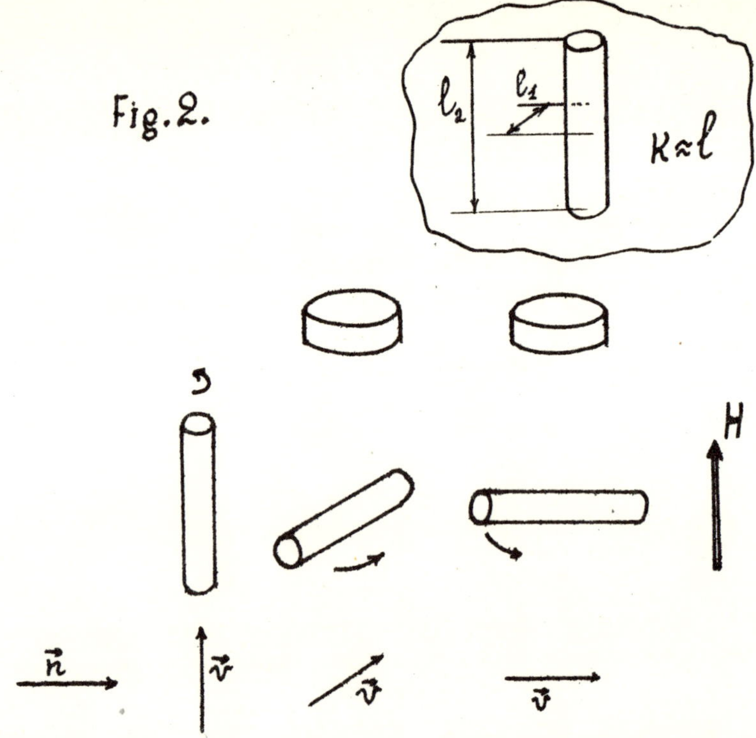


Fig.2.

Fig.3.

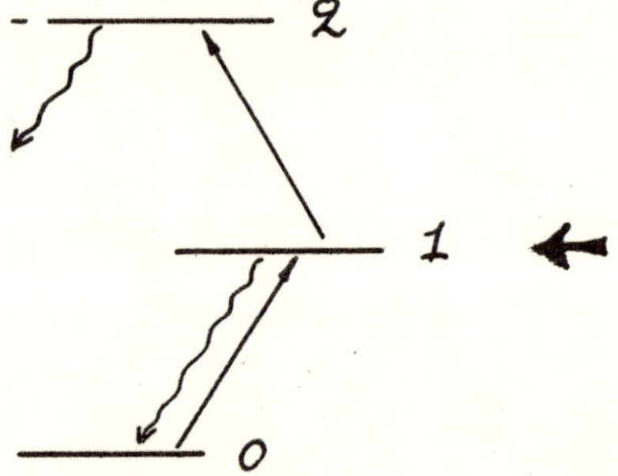

Fig. 4.

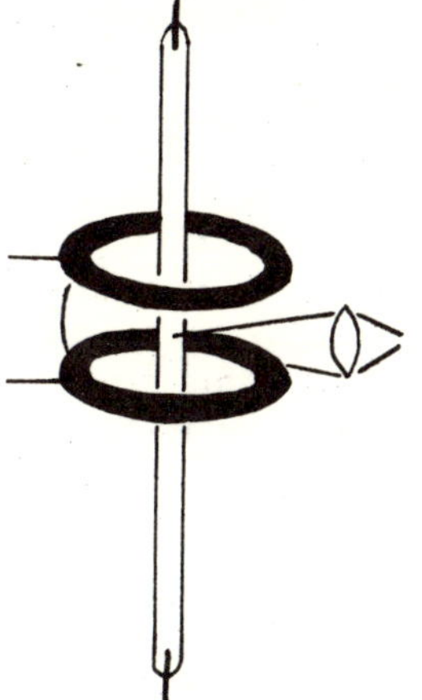

Fig. 5.

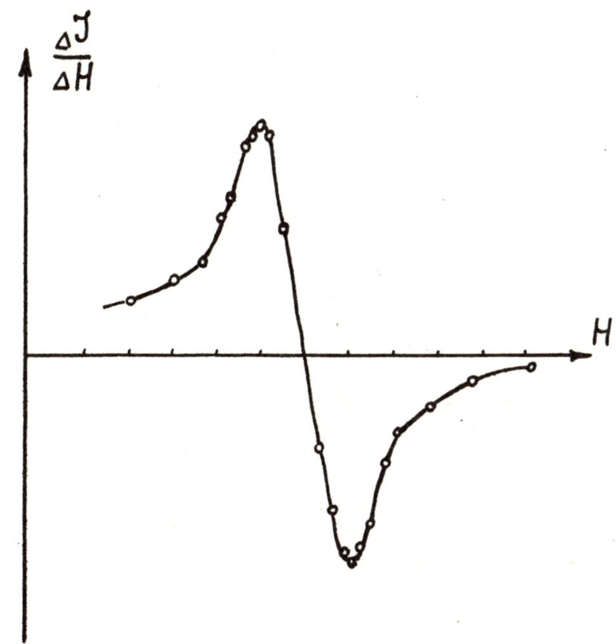

References

1. D. Zhechev, M. Chaika, Opt. i.Spektr., 43, 590, 1977; ibid, in press.
2. H. Kallas, M. Chaika, Opt. i Spektr., 27, 694, 1969.
3. M. Chaika, Opt. i Spektr., 30, 822, 1971; ibid., 31, 67, 1971; ibid., 31, 513, 1971.
4. S. Kasanzev, M. Chaika, Opt. i Spektr., 31, 510, 1971.
5. E. Mishchenko, S. Kasanzev, P. Telbisov, M.Chaika, Vestnik LGU, dep., No. 1343-78.
6. Im Gkhek-de, E. Saprikin, A. Shalagin, Opt. i Spektr., 35, 202, 1973.

# APPLICATIONS OF ANTICROSSING SPECTROSCOPY

H.-J. Beyer and H. Kleinpoppen†

Institute of Atomic Physics,
University of Stirling,
Stirling, Scotland

## 1. Introduction

During the last twenty years both the level crossing and the anticrossing methods have been used extensively to study the atomic and molecular structure and properties of atoms and molecules in external fields. The level crossing technique in fact goes back to 1924 when Hanle[1] discovered the depolarisation of resonance fluorescence light in a weak magnetic field (Hanle effect). At the time this could be explained as the precession in the magnetic field of a damped linear oscillator.[2] With progress in the development of quantum mechanics, the polarisation of resonance fluorescence light was related to the coherent superposition of the degenerate excited states and the depolarisation was readily explained by the removal of the degeneracy in the magnetic field.[3] Looking at it the other way round, a resonance type signal of the polarisation or the intensity of the resonance fluorescence light is observed when the magnetic substates converge (cross) at zero magnetic field, and for this reason the Hanle effect is also referred to as zero field level crossing. The magnetic field required to remove the degeneracy is directly connected with the natural width (i.e. lifetime) of the states, and Hanle signals have, therefore, provided the basis for many lifetime measurements of excited states.

In 1959 Colegrove et al[4],[5] discovered that level crossing signals can also be observed when suitable states cross at non-zero magnetic field thus making it possible to carry out simple and precise investigations of fine and hyperfine structure intervals. The coherence conditions for the

† temporary address: Fakultät für Physik, Universität Bielefeld, 48 Bielefeld, Postfach 8640, W. Germany.

excitation and the decay restrict the level crossing method to states with the same parity.

This restriction does not apply to a different type of crossing signal, observed for the first time by Eck et al[6] during the course of a level crossing study when the coherence conditions were not fulfilled. Here the states were coupled near the crossing by a perturbation which causes the states to repel each other so that the actual crossing is removed. For this reason the signals were called anticrossings. Since perturbations between states can be introduced in many different ways, anticrossing signals can be observed between a large variety of states.

The level crossing technique has been well covered by review papers,[7]-[13] and detailed discussions of applications of the anticrossing method have also appeared.[9][12][14][19]

## 2. The Crossing Signal

The level crossing and the anticrossing techniques have in common that the excited states are created by resonance light or collisional excitation and that a change of the intensity or polarisation of the subsequent decay light (to the original or another state) is recorded when the external (usually magnetic) field is varied through the crossing of the states. Both effects are connected with the fact that the wavefunctions of the states cannot be distinguished at the crossing. In the case of the level crossing signal the "interaction" at the crossover is inherent in the coherence condition for excitation and decay while the anticrossing signal relies on an explicit internal or external perturbation. Indeed, a common theory can be worked out, and this results in some interesting border line cases of signals with both level crossing and anticrossing properties. The theory may be based on the density matrix formalism[16] or on the time dependent Schrödinger equation.[17] The latter was used by Wieder and Eck[17] to obtain a very general equation for the intensity, S, of the resonance fluorescence light as a function of the (frequency) separation, $\Delta\nu$, in the vicinity of a crossing. The results are also valid for other excitation processes (e.g. electron impact excitation).

$$
\begin{aligned}
S = {} & \underbrace{(1/\gamma_a)\Sigma\{|f_a|^2|g_a|^2\}}_{1} + \underbrace{(1/\gamma_b)\Sigma\{|f_b|^2|g_b|^2\}}_{2} \\
& + \underbrace{(\gamma_a\gamma_b/\bar{\gamma}D)\Sigma\{f_a f_b^* g_a g_b^* + f_a^* f_b g_a^* g_b\}}_{3} \\
& - \underbrace{(i\gamma_a\gamma_b\Delta\nu/\bar{\gamma}^2 D)\Sigma\{f_a f_b^* g_a g_b^* - f_a^* f_b g_a^* g_b\}}_{4} \\
& - \underbrace{(2|V|^2\gamma_a\gamma_b/\bar{\gamma}D)\Sigma\{fg\}}_{5} \\
& + \underbrace{(2/\bar{\gamma}D)\Sigma\{(V^* f_a f_b^* + V f_a^* f_b)(V g_a g_b^* + V^* g_a^* g_b)\}}_{6} \\
& + \underbrace{(\Delta\nu\gamma_a\gamma_b/\bar{\gamma}^2 D)\Sigma\{f(V g_a g_b^* + V^* g_a^* g_b) + g(V^* f_a f_b^* + V f_a^* f_b)\}}_{7} \\
& + \underbrace{(i\gamma_a\gamma_b/\bar{\gamma}D)\Sigma\{f(V g_a g_b^* - V^* g_a^* g_b) + g(V^* f_a f_b^* - V f_a^* f_b)\}}_{8}
\end{aligned}
\tag{1}
$$

with $D = \gamma_a\gamma_b + |2V|^2 + (\gamma_a\gamma_b/\bar{\gamma}^2)\Delta\nu^2$,

$\bar{\gamma} = \frac{1}{2}(\gamma_a+\gamma_b)$,

$f = (|f_a|^2/\gamma_a) - (|f_b|^2/\gamma_b)$,

$g = (|g_a|^2/\gamma_a) - (|g_b|^2/\gamma_b)$,

$\gamma_a = 1/(2\pi\tau_a) \qquad \gamma_b = 1/(2\pi\tau_b) \qquad \tau$ lifetime

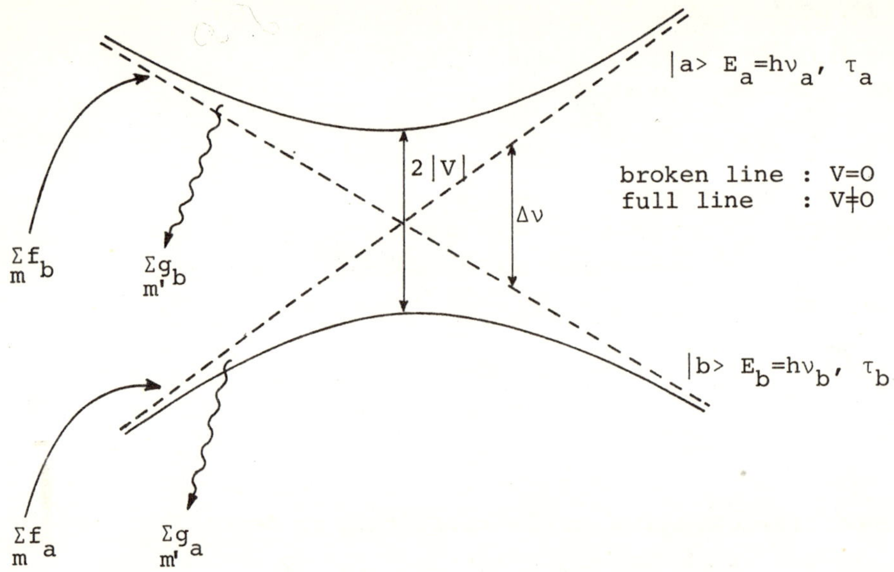


Fig. 1: Level crossing - Anticrossing.

The nomenclature used is explained in Fig. 1. V is the perturbation between states |a> and |b> and gives rise to a minimum separation of the mixed states of 2V. $f_a$, $f_b$, $g_a$, $g_b$ are abbreviations for appropriate electric dipole matrix elements including polarisation and instrumental effects and the sums are to be extended over all relevant initial and final states m and m' respectively. In the case of electron impact excitation $\Sigma|f_a|^2$ and $\Sigma|f_b|^2$ may be replaced by the excitation rates $r_a$ and $r_b$ respectively.

Terms 1 and 2 represent the background intensity of light scattered from states |a> and |b> independent of their energies.

Terms 3 and 4 describe the level crossing signal. It is an interference effect which results in a spatial redistribution of the radiation when sweeping through the crossing and thus requires coherence for both the excitation and decay part. The selection rules for a given experimental arrangement derive from the coherence condition: Both states must be reached from a common ground or lower state and the lower state selected for the decay must also be reached from both crossing states. In particular this reduces the level crossings to states with the same parity.

No perturbation is required, and in the common case of V=0, the width of the signal is determined only by the natural width of the two crossing states.

Level crossing signals (i.e. signals based only on the coherence conditions) may still be observed if the two states anticross ($V \neq 0$), since even then both branches are still a superposition of both states. However, in this case the width of the signal is increased by a contribution related to V.

Terms 5 to 8 describe various types of anticrossing signals, i.e. signals which require the perturbation $V \neq 0$. Term 5 is usually called the pure anticrossing signal since it requires only $V \neq 0$. Terms 6 to 8 on the other hand require $V \neq 0$ and at the same time coherence in the excitation or decay (7,8) or both (6) and, therefore exhibit properties of the level crossing as well as of the anticrossing signals.

The pure anticrossing signal (term 5) represents an absorption Lorentzian shape centered at the point of closest approach of the substates. Its width (FWHM) increases steadily with V+

$$2B = (\gamma_a + \gamma_b) \quad (1 + |2V|^2/\gamma_a\gamma_b)^{\frac{1}{2}} \qquad (2)$$

The signal amplitude also increases with V, but reaches a saturation value for large V.

f is equivalent to the population difference between the crossing states outside the crossing, and no anticrossing signal can be detected for f=0. g is a measure of the degree to which the selected decay is able to discriminate between crossing states. In contrast to the level crossing signals, which are at their best if excitation and decay are evenly divided among the crossing states, the anticrossing signals are strongest if only one state is excited and the decay of only one state is observed. They have this in common with

---

\+ The width of anticrossing signals for states with differing lifetimes and especially the influence of an additional perturbation on the signal width has recently been discussed in more detail by Dohnalik.[(18),(19)]

radio frequency signals.

The selection rules for pure anticrossing signals are determined by the perturbation V which may be inherent in the atomic or molecular system under investigation (internal perturbation) or may be introduced by an external (e.g. electric) field (external perturbation) or may be a combination of both. Using an external perturbation has the advantage over an internal perturbation that the strength of the perturbation and therefore the degree of saturation and the width of the anticrossing signal can be chosen for optimum conditions. On the other hand the width of the anticrossing signals with internal coupling provides information about the strength of the internal coupling (assuming that the natural lifetimes are known).

## 3. Experimental Method

For anticrossing signals the only restriction to the excitation process is that the states must have different populations. This condition can be satisfied by resonance light excitation and by particle impact excitation including beam foil and charge exchange processes on fast beams. Electron impact excitation is most commonly used, and a typical experimental arrangement is shown in Fig. 2. A vacuum system, containing the electron gun and of the order of $10^{-3}$ to $10^{-2}$ Torr of the gas to be studied, is placed in a magnetic field parallel to the electron beam. The appropriate spectral line is selected by an interference filter or a monochromator and its intensity is measured by the photomultiplier and recorded as a function of the magnetic field. It is often advantageous to introduce a polariser into the light path or even to record the anticrossing signal as the change of the polarisation which can be accomplished by a rotating polariser with synchronous lock-in detection. If necessary, a static electric field can be applied (at right angles to both the magnetic field and the direction of light detection) through the Stark plates marked S.

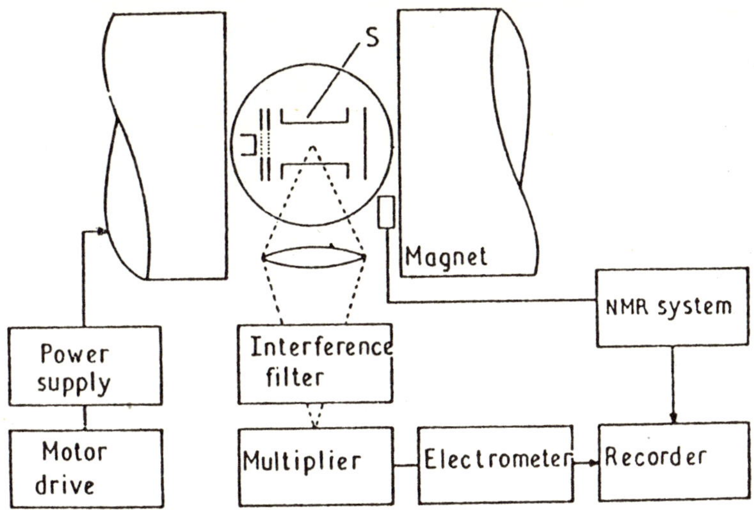


Fig. 2: Typical experimental arrangement used in anticrossing experiments based on electron excitation.

## 4. Applications of Anticrossing Spectroscopy

The discussion in the preceeding sections has already given an indication as to what properties and effects may be investigated with the help of anticrossing signals.

First and foremost comes the measurement of fine and hyperfine structure intervals of atomic and molecular states. These are derived directly from the magnetic field values of the signal centres, assuming that the Zeeman effect is known (→ 4.1).

Secondly, it is possible to investigate the differential Stark effect of the crossing states by studying the influence of a static electric field on the signal positions and on the signal width (→ 4.2).

Thirdly, if the fine or hyperfine structure interval is known or if several signals can be observed which have to form a consistent set, it may be possible to investigate higher order corrections to the Zeeman effect (e.g. quadratic Zeeman effect) (→ 4.3).

Fourthly, anticrossing signals based on internal coupling permit the measurement of the strength of the underlying perturbation from the width of the signal if the lifetimes of the states are known. Anticrossing signals based on external coupling could lead to a measurement of the combined lifetime of the states by extrapolation to zero perturbation (→ 4.4).

Fifthly, the amplitudes of the anticrossing signals, and in particular the saturation amplitudes, should in favourable cases allow to derive relative cross sections and possibly excitation functions of the relevant substates (→ 4.5).

In the following we shall give a more detailed discussion of these applications taking some typical examples of anticrossing signals of atomic states. Emphasis is put on typical and more recent investigations. A more comprehensive review of earlier results has been given elsewhere(15) by the present authors and molecular anticrossings are included in the review by Miller and Freund.(12)

### 4.1 Atomic fine and hyperfine structure

#### 4.1.1 Li. $\Delta L=0$. Internal magnetic hyperfine coupling

The first investigation of an anticrossing signal was carried out by Eck et al(6) and Wieder and Eck(17) in the course of a study of the fine and hyperfine structure of the $2^2P$ states of Li. A classical resonance fluorescence arrangement was used to investigate the crossing of the $2^2P_{3/2}(m_J = -3/2)$ and $2^2P_{1/2}$ $(m_J = -1/2)$ substates at approximately 4800G. Including the hyperfine structure, four closely spaced $\Delta m_F=1$ level crossing signals were expected between the dashed lines in Fig. 3 in the case of $^7$Li. These are marked by full circles. The situation is changed drastically by off-diagonal matrix elements of the magnetic hyperfine interaction. This internal perturbation couples substates with $\Delta m_F=0$ and results in the full lines in Fig. 3 and gives rise to three $\Delta m_F=0$ anticrossings as indicated by squares. Two substates without partner for $\Delta m_F=0$ coupling are unaffected and only two of the $\Delta m_F=1$ level crossings remain

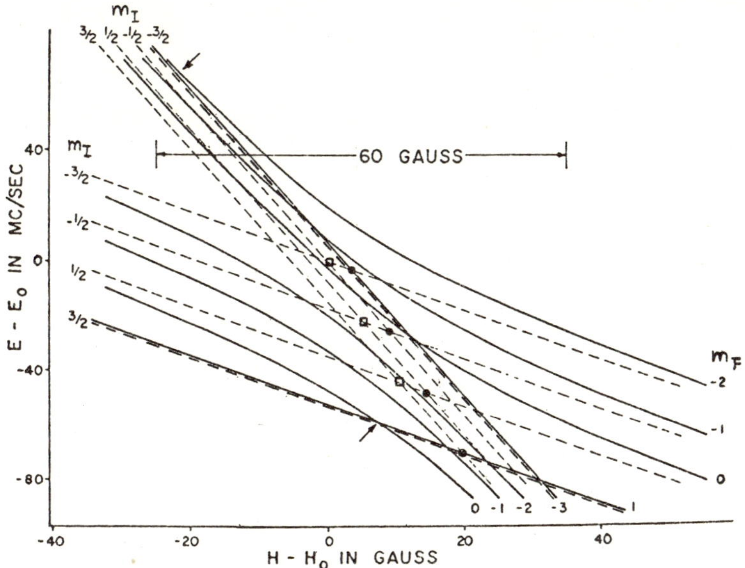


Fig. 3: The eight hyperfine levels involved in the fine-structure crossing $2^2P_{3/2}(m_J = -3/2)$ - $2^2P_{1/2}(m_J = -1/2)$ of $^7Li$. To show more clearly the details of the crossing the spacing of the four $m_I$ levels with $m_J = -3/2$ has been increased by a factor of approximately 2.5. The spacing of the $m_I$ levels with $m_J = -1/2$ is correct as shown.

(now marked by arrows) shifted from their original positions. The three anticrossing signals cannot be separated in the experiment which thus resulted in a single signal ∿35G wide from which the fine structure interval was derived.

### 4.1.2 H, He$^+$. ΔL ≥ 1. External electric field coupling

The perturbation most widely used to convert crossings into anticrossings is a homogeneous static electric field, F. This results in electric dipole coupling of states differing in L by 1 with the perturbation energy

$$V_{ab} = \vec{F} \cdot \langle a|e\vec{r}|b\rangle \quad . \tag{3}$$

Such electric field induced anticrossing signals were employed in precision measurements of the $2^2S_{1/2}$ - $2^2P_{1/2}$ Lamb shift intervals of H and D[(20)-(22)] and in investigations of some other fine structure intervals of H[(23)] and He$^+$[(24)(25)(26)]

Details may be found in a recent review of fine structure measurements.[27]

A homogeneous static electric field can also couple states with $\Delta L>1$ by a multistep dipole coupling process which is analogous to multiphoton transitions. Such higher order coupling is easily accomplished in fine structure systems with near degenerate states of different L, like hydrogenic states and Rydberg states of other atoms, where intermediate states are available in near resonant conditions (i.e. with near zero frequency intervals).

Very favourable conditions for the observation of higher order anticrossing signals are offered by the n=4 fine structure system of $He^+$, and various measurements have been reported by Eck and Huff,[28] Beyer and Kleinpoppen[29] and Billy et al.[26] Fig. 4 shows the Zeeman splitting of this system. Crossings which have been converted into anticrossings by an electric field perpendicular to the magnetic field and have provided anticrossing signals are marked. A set of third order signals S-F is reproduced in Fig. 5. It shows the typical increase in width and amplitude when the coupling, i.e. the electric field is increased. It will be noted that the signals are fairly narrow compared with the more common S-P radio frequency or anticrossing signals. This is a result of the relatively long lifetime of the F state compared with the P state and gives hope that the centre of the anticrossing signal (and therefore the corresponding fine structure interval) can ultimately be determined with good accuracy. A detailed discussion of these signals and of results may be found in earlier review papers.[14][15]

Similar measurements were carried out by Beyer and Kleinpoppen on n=5 of $He^+$[30][31] and by Glass-Maujean on n=3[32] and n=4[33] of H. The final results of these experiments on H were recently published by Glass-Maujean et al.[34] In this experiment the excited atoms were created by electron impact on $H_2$ in a dissociation/excitation process and no external electric field was applied since the motional electric field experienced by the atoms in the magnetic field provided sufficient state mixing. The difference signal of $\sigma$ and $\pi$-light intensity is measured through a rotating linear polariser in

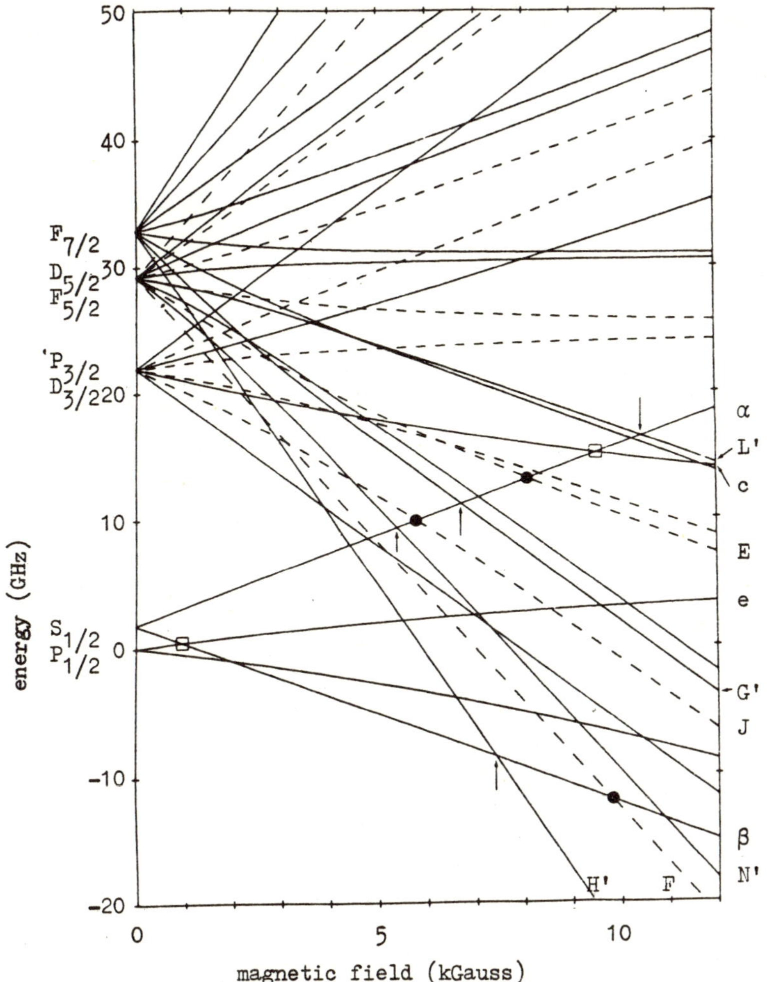


Fig. 4: Zeeman effect of the fine structure system n=4 of $He^+$. The D substates are shown as dashed lines, the others as solid lines. Anticrossings occurring between substates from S and substates from P,D, and F (on application of an electric field perpendicular to the magnetic field) are marked by squares, circles, and arrows, respectively.

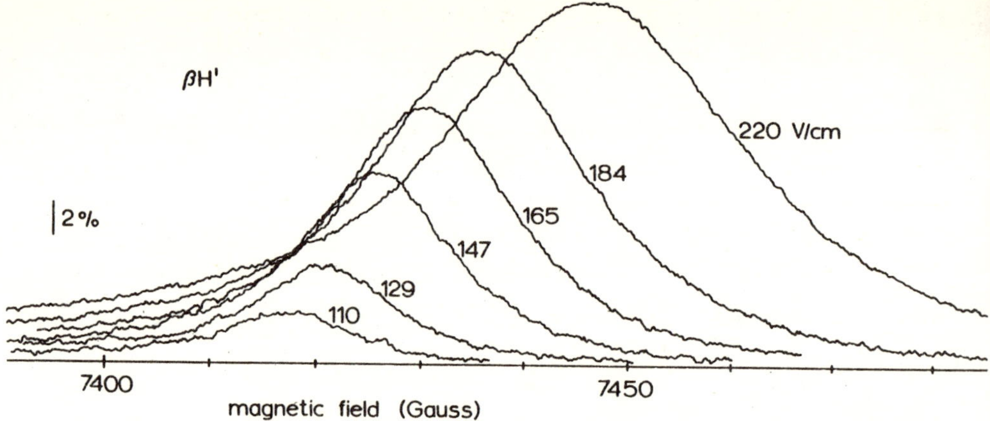


Fig. 5: Superposition of several recordings of the anti-crossing $S_{1/2}(m_J = -1/2) - F_{7/2}(m_J = -7/2)$ in n=4 of $He^+$ showing clearly the influence of the electric field on amplitude, width, and position of the anticrossing signal. The bar indicates a change of approximately 2% of the total light intensity.

conjunction with a lock-in amplifier. The signals are complicated by the presence of the hyperfine structure (absent in $He^+$) and are, therefore, doublepeaked. This is borne out by the recording of a second order S-D signal shown in Fig. 6. The

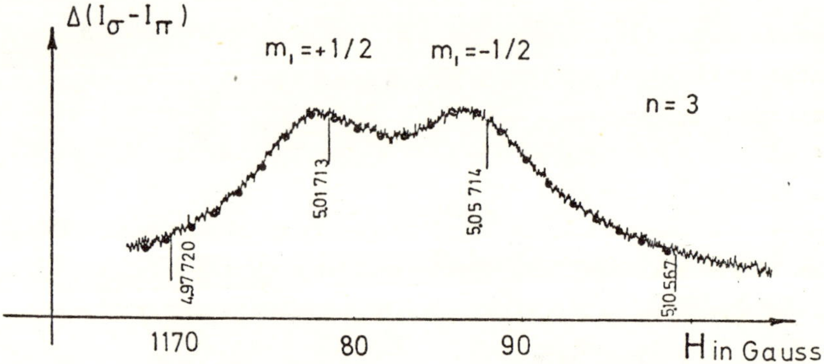


Fig. 6: (From Ref. 34) Recorder trace and computed points of the αE anticrossing signal $3^2S_{1/2}(m_J = 1/2) - 3^2D_{5/2}(m_J = -3/2)$ of hydrogen taken as the difference between the intensities of σ and π light in the $H_\alpha$ line. The vertical bars are magnetic field calibration marks.

resulting fine structure intervals are collected in Table 1. Other intervals were derived from these values.[34]

| Interval | Experiment[a] | Theory[b] |
|---|---|---|
| $3^2S_{1/2} - 3^2P_{1/2}$ | 314.9 ± 0.9 | 314.898 ± 0.003 |
| $3^2D_{3/2} - 3^2S_{1/2}$ | 2930.0 ± 2.3 | 2929.859 ± 0.003 |
| $3^2D_{5/2} - 3^2P_{3/2}$ | 1078.0 ± 1.1 | 1078.0059 ± 0.0007 |
| $3^2D_{5/2} - 3^2S_{1/2}$ | 4013.75 ± 0.60 | 4013.197 ± 0.003 |
| $4^2D_{3/2} - 4^2S_{1/2}$ | 1235.0 ± 2.1 | 1235.756 ± 0.001 |
| $4^2D_{5/2} - 4^2S_{1/2}$ | 1693.0 ± 0.4 | 1692.790 ± 0.001 |

a) Glass-Maujean et al,(Ref. (34)).

b) G.W. Erickson, private communication, Jan. 1977. (Compare footnote a of Table III in (Ref. (27)).

Table 1: Fine structure intervals in H, derived from electric field induced anticrossing signals (in MHz).

### 4.1.3 $Mg^+$. $\Delta L=0$. External electric field coupling

Second order coupling by a static electric field (as discussed in 4.1.2) may, in two steps of $\Delta L=1$, also couple states with $\Delta L=0$ and thus permit the observation of $\Delta L=0$ anticrossing signals. Attempts to detect $\Delta L=0$ signals in n=4 of $He^+$ have so far failed[35] but a successful study on the fine structure of $^2D$ and $^2F$ states of $Mg^+$ has just been reported by Andersen et al.[36] In this experiment a fast beam (50 - 500 keV) of $^{24}Mg^+$ ions was excited in an He or $H_2$ filled gas cell placed in the field of an electromagnet (magnetic field perpendicular to the beam axis). A monochromator in conjunction with a linear polariser was employed to detect the decay light at right angles to the magnetic field and at $6^o$ or $90^o$ with respect to the beam axis. When the motional electric field experienced by the fast ions was compensated by an external electric field, weak $\Delta m_J = 2$ level crossing signals could be detected between $^2D_{3/2}$ and $^2D_{5/2}$ substates. The Zeeman splitting of these states and the crossing positions are shown in Fig. 7. The crossings are converted to

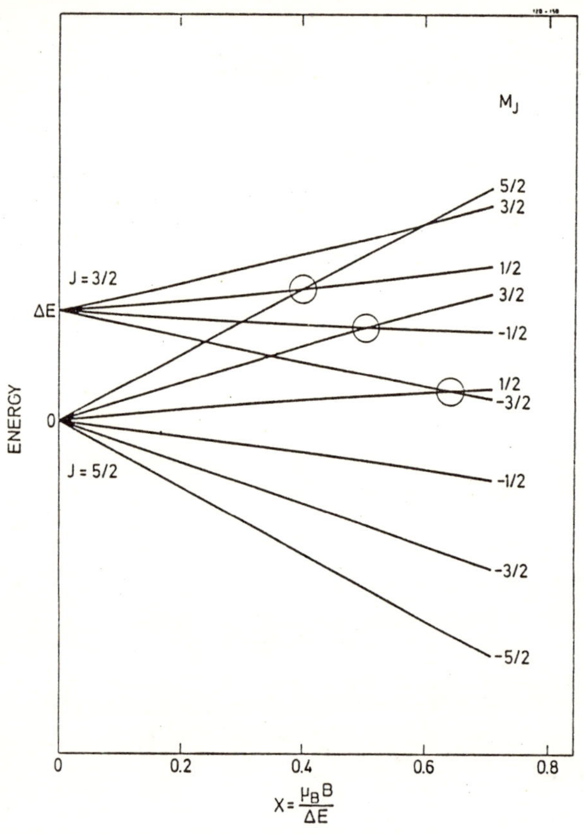


Fig. 7: (from Ref. (36). Energy values and Zeeman splitting of the inverted $^2D_{5/2}$ and $^2D_{3/2}$ terms of $Mg^+$. The $\Delta m_J = 2$ level crossings are indicated. These are converted to anticrossings by a static electric field perpendicular to the magnetic field.

anticrossings by second order coupling through $^2F$ states of the same n, if the motional electric field is left uncompensated. In this case anticrossing signals were found to dominate over the level crossing signals by a large margin. Fig. 8 shows a recorder trace of the lowest field anticrossing signal for n=5 using σ light and the decay $^2D_{5/2,3/2} \rightarrow {}^2P_{3/2}$ at 90° with respect to the beam axis. The predominant dispersion shape of the signal indicates that the pure anticrossing signal (term 5 of Eq. 1) which should have absorption shape, plays only a minor role. The major part of the signal is provided by term 7 of Eq. 1 by way of interference effects in the decay (both states can decay to common lower $^2P_{3/2}$ states and coherence is induced by the electric field coupling). Indeed, the signal shape changes drastically if the interference part is suppressed by either selecting π light or decay to $^2P_{1/2}$ states. The sign of the Stark shift of the signal positions could be used to

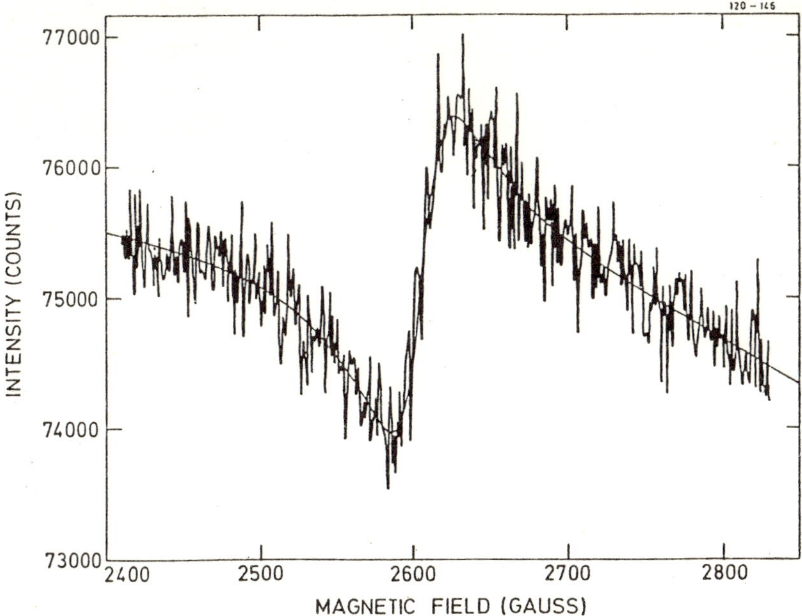


Fig. 8: (from Ref. (36)). Anticrossing signal $5^2D_{3/2}(m_J = 1/2) - 5^2D_{5/2}(m_J = 5/2)$ in $Mg^+$. σ light. Electric field strength ∿4.6 kV/cm. The dispersion shape indicates that interference terms play a major role in the observed signal.

verify the inverted fine structure $^2D_{3/2} - ^2D_{5/2}$ as shown in Fig. 7. Similar measurements were done on $^2F_{5/2} - ^2F_{7/2}$ crossings, where the fine structure is normal. Results are shown in Table 2.

### 4.1.4 He. ΔL=0, ΔS=1. Internal spin orbit coupling

States may also be coupled by off-diagonal matrix elements of the spin orbit interaction, and this permits the observation of anticrossing signals between states of different spin systems. In this way it was possible to measure the $n^1D - n^3D$ intervals (n=3,....,20) in helium.(37)-(41) The selection rules for this coupling are ΔL=0, ΔS=1, ΔM=0 ($M = m_L+m_S$), and the Zeeman diagram in Fig. 9 for n=10 of He shows the anticrossings complying with these selection rules as full circles.

| Interval | measured[a] (MHz) | theory (MHz) | difference \|th\| - \|exp\| (%) |
|---|---|---|---|
| $5^2D_{5/2}$ - $5^2D_{3/2}$ | -9084 ± 18 | -9860[b] | 8.0 |
| $6^2D_{5/2}$ - $6^2D_{3/2}$ | -5546 ± 15 | -6030[b] | 8.0 |
| $7^2D_{5/2}$ - $7^2D_{3/2}$ | -3565 ± 15 | -3930[b] | 9.2 |
| $4^2F_{7/2}$ - $4^2F_{5/2}$ | 3530 ± 5 | 3652[c] | 3.4 |
| $5^2F_{7/2}$ - $5^2F_{5/2}$ | 1757 ± 10 | 1869[c] | 6.0 |
| $6^2F_{7/2}$ - $6^2F_{5/2}$ | 1001 ± 10 | 1082[c] | 7.4 |

a) Andersen et al, Ref. (36).

b) Holmgren et al, Ref. (65) and Mårtenssen, private communication (1978) to Andersen et al, Ref. (36).

c) Theoretical hydrogenic fine structure splittings, see Andersen et al, Ref. (36).

Table 2: Fine structure intervals in $Mg^+$.

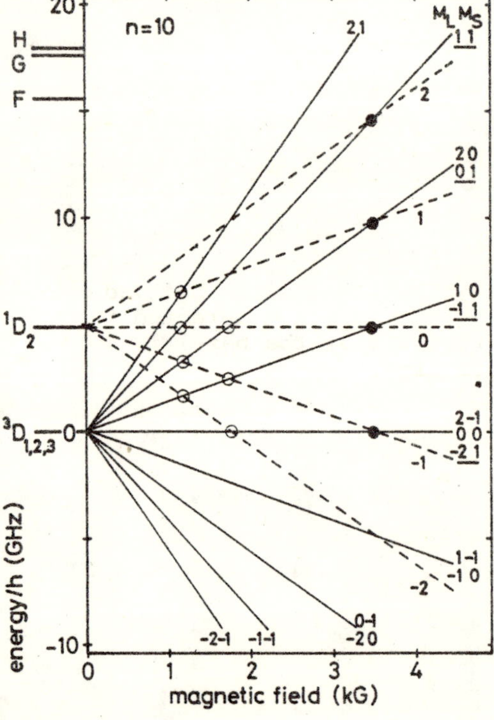


Fig. 9: Zeeman splitting of the 10D states of helium (the fine structure splitting of the $^3D$ states is too small to be seen on this scale). The selection rules for singlet-triplet anticrossings induced by internal spin-orbit coupling provide four nearly degenerate crossings marked by full circles. The crossings marked by open circles may also be converted into anticrossings if, in addition, an external electric field is applied perpendicular to the magnetic field (→ discussion in 4.2.2).

Since the populations of singlet and triplet states from electron impact are very different and since singlet and triplet decay lines can readily be isolated, strong anticrossing signals are recorded both in singlet and (with opposite sign) in triplet decay. This is shown in Fig. 10. Unfortunately, the four signals as indicated in Fig. 10 are nearly degenerate and cannot be resolved as a result of the broadening by the strong and fixed internal interaction. Nevertheless, the accuracy of the $^1D$ - $^3D$ intervals measured in this way by far exceeds that of earlier optical investigations.[59]

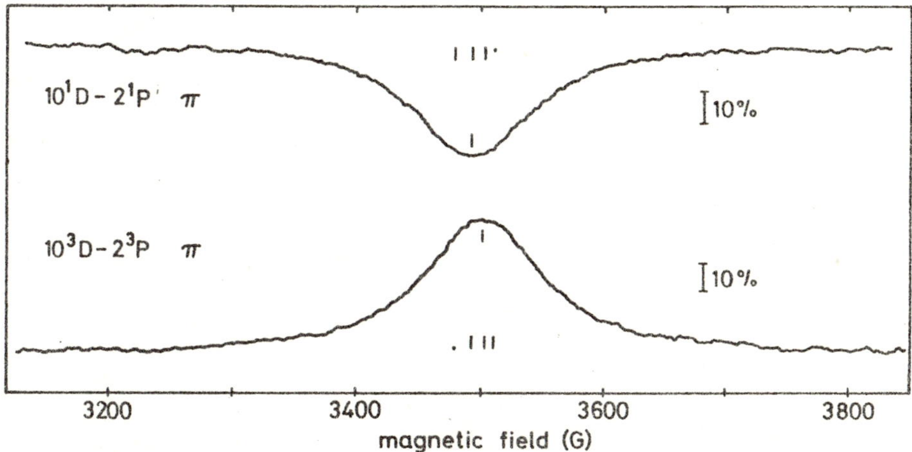


Fig. 10: Recorder traces of the singlet-triplet anticrossing signal in n=10 of He observing singlet decay (upper trace) and triplet decay (lower trace), both in π-polarisation. The theoretical positions of the four anticrossings are indicated near the respective baselines using lines for components contributing to the observed signal and dots for the components not detected in π-light. The bars indicate 10% of the baseline intensity of the observed line.

## 4.2 Stark effect measurements

The energy of magnetic substates is affected by external static electric fields and the differential Stark effect of the crossing states can, therefore, be investigated from the shift of the centre of anticrossing signals. The electric field may be applied in addition to the internal coupling of the states or it may be required to induce the anticrossing signals. In the latter case, and in particular for higher order electric field induced anticrossings, the necessary extrapolation of the crossing positions to zero electric field is equivalent to a measurement of the Stark effect.

Alternatively, the width of electric field induced anticrossing signals may also be used to extract information about the Stark effect.

### 4.2.1 The non-linear Stark effect in hydrogenic systems

Optical investigations of the Stark effect of H are restricted to the high field region where the Stark shifts are large compared with the fine structure. Under this condition the well-known linear Stark effect of H is observed. However, at low electric field, where the Stark shifts are small compared with the fine structure, a quadratic Stark effect is expected in H as it is the usual case in other atomic systems. Because of the smallness of the hydrogenic fine structure the quadratic Stark effect could not be verified until fairly recently when it was investigated by H maser,(42)(43) quantum beat,(44)-(48) anticrossing(49) and level crossing(50) studies.

The shift of the third order anticrossing signal S-F in n=4 of $He^+$ is obvious from Fig. 5 in 4.1.2 and may serve as example. In Fig. 11 the centre position of the same signal is plotted as a function of the squared electric field. The measured results (full circles) within their error bars lie on a straight line, thus confirming the quadratic nature of the Stark effect. The fitted slope of the measured results is in agreement with the theoretical dependence (solid line), obtained from diagonalisation of the energy matrix of the n=4 fine structure system in combined electric and magnetic fields. Satisfactory agreement was obtained for all signals investigated(49) and more accurate measurements are under way.(66)

### 4.2.2 He

In non-hydrogenic systems a measurement of the Stark effect provides information on the wavefunction of the states. In many cases the Stark effect is fairly accurately predicted on the basis of hydrogenic wavefunctions. Excited states of He,

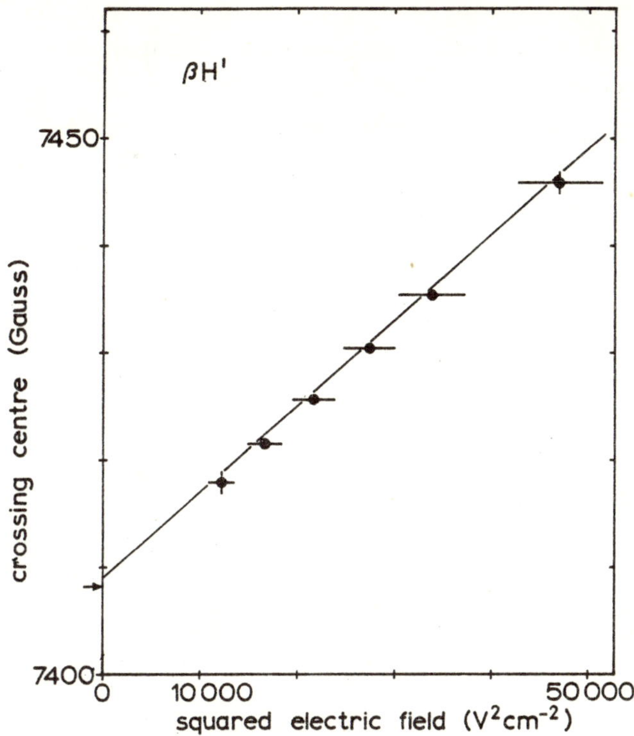


Fig. 11: Positions measured for the anti-crossing signal $4^2S_{1/2}(m_J = -1/2)$ - $4^2F_{7/2}(m_J = -7/2)$ in $He^+$ as a function of the squared electric field. The line represents the theoretical dependence. The measured points extrapolate to the zero electric field crossing position indicated by the arrow.

in particular those with not too small L values, would be expected to fall into this group, and a measurement on the $n^1D$ and $n^3D$ states (n=5-7)[(51)] confirms that. The $^1D$ - $^3D$ anticrossing signals described in 4.1.4 are not particularly suitable for such a measurement since they are a superposition of several components with differing Stark shifts. However, by applying a static electric field perpendicular to the magnetic field it is possible also to convert the crossings marked in Fig. 9 with open circles into anticrossings (by a combined electric field-spin orbit coupling process). The width of these signals is much less than that of the pure $^1D$ - $^3D$ signals of 4.1.4 and the components can be fully resolved and their Stark effect measured. Preliminary results[51] showed agreement with the theoretical expectation based on hydrogen wavefunctions within the fairly large experimental error limits of 10% to 15%. Further measurements so far indicate that the agreement is much closer, possibly of the order of 1%.[52]

### 4.2.3 Signal width, Cs

Khvostenko[53] proposed to determine the Stark effect from the width of crossing signals which increases if an electric field is applied in such a way that the crossing states are coupled and converted into an anticrossing (→ Eq. 1). He estimated that this would permit a measurement of the Stark effect at much lower electric field strength than that required to shift the crossing signal by a noticeable amount.

An experimental study was carried out on the $7^2P_{3/2}$ hyperfine crossings of $^{133}$Cs. Several $\Delta m_F = 2$ level crossing signals were observed in a resonance fluorescence arrangement. Applying a static electric field parallel to the magnetic field (which does not couple states with $\Delta m_F = 2$) resulted in a significant shift of the crossing positions[54] but only for $F > 4$ kV/cm. If an electric field is applied perpendicular to the magnetic field, the crossing states are coupled, and a significant increase in the width was observed for electric fields of 0.8 to 2 kV/cm from which the Stark constant was derived.[55]

However, the corresponding set of signals shown in Fig. 12 does not change amplitude or general shape with increasing electric field perturbation. This indicates that the perturbation does not contribute to the signal other than by increasing its width. Thus, the signal appears to be governed entirely by terms 3 and 4 of Eq. 1, and in the sense of our definition in Section 2, would remain a pure level crossing signal even though the crossing is actually converted into an anticrossing. This has, of course, no effect on the measurements taken, since the width is the same for all terms of Eq. 1.

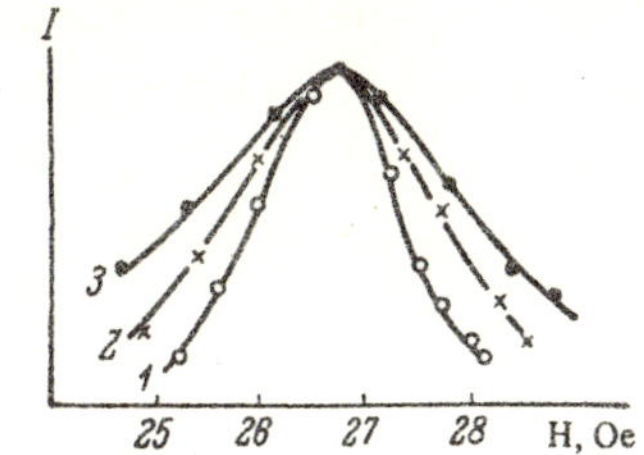


1–E = 0, 2–E = 1.6, 3–E = 2 kV/cm.

Fig. 12: (from Ref. (55)). $\Delta m_F=2$ level crossing signals observed in the hyperfine structure of the $7^2P_{3/2}$ state of $^{133}Cs$. Application of an electric field perpendicular to the magnetic field results in second order coupling of the crossing states which leads to the removal of the actual crossing and an increased signal width.

### 4.3 Quadratic Zeeman effect

Any influence of the quadratic Zeeman effect can normally be neglected when investigating radio frequency or crossing signals of low n atomic states using ordinary electromagnets. However, the sensitivity increases quickly with n, and even at low n, corrections may be necessary if strong and narrow signals allow a very precise location of the signal centres. In most cases appropriate corrections would be calculated from the theory[(56)-(58)] and, if noticeable, applied to the measured results. For the calculation of the quadratic Zeeman effect it is common to distinguish an isotropic part, described by the constant $\chi_i$, and an anisotropic part, described by the constant $\chi_a$.[(60,61)] The isotropic part is a function of n and L and shifts the whole set of states $|nL\rangle$ while the anisotropic part also depends on $m_L$ and results in specific shifts of the various magnetic substates.

Contributions of the anisotropic quadratic Zeeman effect are in part responsible for the lifting of the degeneracy of the four $^1D$ - $^3D$ anticrossing components in He (→ 4.1.4), and appropriate corrections were applied since the superposition of the components could not be unravelled.

Fairly noticeable (mainly isotropic Zeeman effect)

corrections had to be applied to the electric field induced anticrossing signals between $^{1,3}D$ and high L states in He.[61]

An actual measurement of the anisotropic Zeeman effect at low magnetic field was reported by Miller and Freund[60],[62] who studied the relativistic fine structure of the $4^3P_{0,1,2}$ and $5^3P_{0,1,2}$ states of He using radio frequency transitions. The results at different magnetic field values were found to be consistent only in conjunction with a certain value for $\chi_a$. The measured values of $\chi_a$ were about 20% larger than expected for hydrogenic wavefunctions and it was pointed out that further investigations would be necessary to confirm this effect.

Such a study is possible on the D states of He making use of the electric field induced $^1D$ - $^3D$ anticrossing signals mentioned in 4.2.2. As is shown by the open circles in Fig. 9 seven anticrossing signals can be detected and, after extrapolation to zero electric field and application of the linear Zeeman effect, this leads to seven raw fine structure values determined by the common fine structure interval $^1D$ - $^3D_{average}$ and specific contributions related to $\chi_a$ and the constants A and b describing the relativistic fine structure of the $^3D$ state. A least squares fit of the four constants to the experimental results thus provides a value for $\chi_a$. The values of $\chi_a$ obtained from a preliminary investigation of such electric field induced anticrossing signals $n^1D$ - $n^3D$ (n=5-7) in He[51] are not sufficiently accurate to merit comparison with theory, but considerable improvements of the accuracy appear to be possible, and further measurements are in progress.[52]

## 4.4 Coupling energy and lifetimes

The width of an anticrossing signal cannot usually be determined with the same accuracy as the position, and it is also more likely to be disturbed by effects like collision broadening or the distribution of motional electric fields (through Stark shift). Keeping these restrictions in mind, useful results can nevertheless be derived from a measurement of the signal width which, according to Eq. 2 in Section 2, depends on the lifetimes of the states and the interaction energy. Thus, if the lifetimes of the states are known, the signal width leads to the value of the interaction energy and vice versa.

For signals induced by an external perturbation (e.g. a static electric field), not too much attention is normally paid to the interaction energy unless one wants to verify the coupling process as for instance in the case of the higher order electric field induced signals in $He^+$ (see Refs. (14), (15) for a discussion). In other cases a determination of the interaction energy may be used to derive Stark constants and to gain information about the underlying wavefunctions without having to apply fields strong enough to shift the signals (compare 4.2.3).

The determination of the interaction energy is particularly useful if the coupling is provided by an internal perturbation which may not otherwise be accessible to measurement. This was already noticed by Wieder and Eck[(17)] in their original anticrossing study on Li (→ 4.1.1). After having extracted the width of single signal components from the unresolved superposition of signals the matrix elements responsible for the anticrossing signals were derived (using known lifetimes of the states). Both for $^6$Li and $^7$Li they were found to be in excellent agreement with the theoretical magnetic dipole matrix elements expected to be responsible for the anticrossing signals.

In a similar way and with corresponding problems of the superposition of unresolved signals, the off-diagonal spin-orbit interaction parameter, $\alpha$, was determined from the width of $^1$D - $^3$D anticrossing signals in He[(37)-(39)] (→ 4.1.4). $\alpha$ is directly related to the off-diagonal matrix elements of the Breit operator $H_3$ in the Breit-Bethe approximation (→ discussion in Ref. (64)). The experimental values of $\alpha$ for the nD states of He (n=3-8) were found to be in good agreement with the theory.

Lifetime measurements could conveniently be carried out on electric field induced anticrossing signals which can be extrapolated to zero electric field (V=0) where the width depends only on the lifetime. However, these signals occur mostly between states having different L values and thus different lifetimes so that only the combined lifetimes could be obtained. To our knowledge no serious lifetime studies have been carried out on the basis of anticrossing signals.

## 4.5 Excitation cross sections

According to Eq. 1 in Section 2, the amplitude of the pure anticrossing signal (term 5) is proportional to the steady state population difference outside the crossing region. In particular the saturation amplitude ($V \to \infty$) is independent of V:

$$S_\infty \propto \frac{\gamma_a \gamma_b}{\gamma_a + \gamma_b} \left( {}^{r}a/\gamma_a - {}^{r}b/\gamma_b \right)$$

Thus it should be possible to deduce differential excitation cross sections for the interacting states.

A tentative study with this aim was carried out[63] following the investigation of electric field induced anti-crossing signals in n=4 of $He^+$ (→ 4.1.2) where the states were excited by electron impact. Since anticrossing signals S-P, S-D and S-F were observed it was hoped to combine these to extract in the first instance relative cross sections S:P:D:F. The matter is complicated by the many magnetic substates of the fine structure system which all contribute to the detected light while only two are involved in each anticrossing. On the one hand the resultant background light may be much stronger than the signals and affected by non resonant Stark mixing; on the other hand even the population difference between the crossing states (outside the crossing region) may be altered by non-resonant Stark mixing with these substates. This will affect the signal amplitudes, in particular by reducing the steady state population of the long lived S states (even a small amount of S-P mixing leads to a noticeable decrease of the S state lifetime). Thus a calculation was made to establish the actual degree of saturation (in terms of the zero electric field populations) corresponding to the relative signal amplitudes measured at certain values of the electric field. The calculation was based on the eigenvector components of the crossing states obtained from matrix diagonalisation of the full fine structure matrix in combined magnetic and electric fields. The results for the near saturation electric fields used in the experiment indicate that the measured relative amplitudes are less than the undisturbed amplitudes by about 10 - 30% for S-F and by as much as 50% for S-D. The measured relative signal amplitudes were corrected for this reduction as well as for polarisation effects, changes in the light intensity with the electric field and background light not related to the n=4 states. Fairly consistent values of the corrected saturation amplitudes were obtained for each group of signals (e.g. S-F, S-D) but no sensible set of relative cross sections could be derived from these values. Further investigations of both the theoretical and experimental aspects are required.

## REFERENCES


1. W. Hanle, Z. Physik 30, 93-105 (1924).
2. W. Hanle, Erg. Ex. Nat. 4, 214-232 (1925).
3. G. Breit, Rev. Mod. Phys. 5, 91 (1933).
4. F.D. Colegrove, P.A. Franken, R.R. Lewis and R.H. Sands, Phys. Rev. Lett. 3, 420 (1959).
5. P.A. Franken, Phys. Rev. 121, 508 (1961).
6. T.G. Eck, L.L. Foldy and H. Wieder, Phys. Rev. Lett. 10, 239 (1963).
7. G. zu Putlitz, Atomic Physics 1 (Plenum Press), 227-264 (1969).
8. W. Hanle and R. Pepperl, Physikal. Blätter 27, 19-27 (1971).
9. V.G. Pokozan'ev and G.V. Skrotskii, Soviet Physics Uspekhi 15, 452-470 (1973).
10. A. Kastler, Physikal. Blätter 30, 394-404 (1974).
11. N.I. Kalitejewski and M. Tschaika, Atomic Physics 4 (Plenum Press), 19-45 (1975).
12. T.A. Miller and R.S. Freund, Advances in Magnetic Resonance, Vol. 9, Ed. J.S. Waugh (Academic Press, New York), pp.50-189 (1977).
13. W. Happer and R. Gupta, Progress in Atomic Spectroscopy, Ed. W. Hanle and H. Kleinpoppen (Plenum Press, New York), pp. 391-462 (1978).
14. H.-J. Beyer and H. Kleinpoppen, Int. J. Quantum Chem. Symp. 11, 271-287 (1977).
15. H.-J. Beyer and H. Kleinpoppen, Progress in Atomic Spectroscopy, Ed. W. Hanle and H. Kleinpoppen (Plenum Press, New York), pp. 607-637 (1978).
16. M. Glass-Maujean and J.P. Descoubes, J. Phys. B11, 413-19 (1978) and Opt. Comm. 4, 345-51 (1972).
17. H. Wieder and T.G. Eck, Phys. Rev. 153, 103 (1967).
18. T. Dohnalik, private communication, July 1977, Dec. 1977.

19. T. Dohnalik, Acta Phys. Polonica A53, 619-632 (1978).

20. R.T. Robiscoe, Phys. Rev. 138, A22-34 (1965) and Phys. Rev. 168, 4-11 (1968).

21. R.T. Robiscoe and T.W. Shyn, Phys. Rev. Lett. 24, 559-562 (1970).

22. B.L. Cosens, Phys. Rev. 173, 49-55 (1968).

23. M. Glass-Maujean and J.P. Descoubes, C.R. Acad. Sci. (Paris), 273, B721-724 (1971).

24. M. Baumann, A. Eibofner, Phys. Lett. 33A, 409-410 (1970).

25. A. Eibofner, Z. Phys. 249, 58-72 (1971).

26. N. Billy, C. Lhuillier and J.P. Faroux, J. Physique Lett. (Paris) 38, L429-434 (1977).

27. H.-J. Beyer, Progress in Atomic Spectroscopy, Ed. W. Hanle and H. Kleinpoppen (Plenum Press, New York), pp. 529-605 (1978).

28. T.G. Eck and R.J. Huff in "Beam Foil Spectroscopy", Ed. S. Bashkin (Gordon and Breach, New York, 1968) pp. 193-202 and Phys. Rev. Lett. 22, 319-321 (1969).

29. H.-J. Beyer and H. Kleinpoppen, J. Phys. B4, L129-32 (1971) and J. Phys. B5, L12-L15 (1972).

30. H.-J. Beyer, H. Kleinpoppen and J.M. Woolsey, Phys. Rev. Lett. 28, 263-5 (1972).

31. H.-J. Beyer and H. Kleinpoppen, J. Phys. B8, 2449-55 (1975).

32. M. Glass-Maujean, Opt. Comm. 8, 260-262 (1973).

33. M. Glass-Maujean, Thesis, Université de Paris (1974), unpublished.

34. M. Glass-Maujean, L. Julien and T. Dohnalik, J. Phys. B11, 421-430 (1978).

35. H.-J. Beyer and K.-J. Kollath, unpublished.

36. T. Andersen, S. Isaksen, D.B. Iversen and P.S. Ramanujam, accepted for publication in Phys. Rev. A, Sept. 1978.

37. T.A. Miller, R.S. Freund, F. Tsai, T.J. Cook and B.R. Zegarski, Phys. Rev. A9, 2474-84 (1974).

38. T.A. Miller, R.S. Freund and B.R. Zegarski, Phys. Rev. A11, 753-7 (1975).

39. J. Derouard, R. Jost, M. Lombardi, T.A. Miller and R.S. Freund, Phys. Rev. A14, 1025-1035 (1976).

40. H.-J. Beyer and K.-J. Kollath, J. Phys. B8, L326-30 (1975).

41. H.-J. Beyer and K.-J. Kollath, J. Phys. B9, L185-88 (1976).

42. E.N. Fortson, D. Kleppner and N.F. Ramsey, Phys. Rev. Lett. 13, 22-23 (1964).

43. P.C. Gibbons and N.F. Ramsey, Phys. Rev. A5, 73-78 (1972).

44. I.A. Sellin, C.D. Moak, P.M. Griffin and J.A. Biggerstaff, Phys. Rev. 188, 217-221 (1969).

45. H.J. Andrä, Phys. Rev. A2, 2200-2207 (1970).

46. E.H. Pinnington, H.G. Berry, J. Desesquelles and J.L. Subtil, Nucl. Instrum. Meth. 110, 315-320 (1973).

47. A. van Wijngaarden, E. Goh, G.W.F. Drake and P.S. Farago, J. Phys. B9, 2017-2025 (1976).

48. J. Bourgey, A. Denis and J. Désesquelles, J. Physique (Paris), 38, 1229-36 (1977).

49. H-J. Beyer, H. Kleinpoppen and J.M. Woolsey, J. Phys. B6, 1849-55 (1973).

50. K-J. Kollath and H. Kleinpoppen, Phys. Rev. A10, 1519-21 (1974).

51. H-J. Beyer and K-J. Kollath, J. Phys. B10, L5-L9 (1977).

52. H-J. Beyer, to be published.

53. G. Khvostenko, Opt. Spectrosc. 26, 352-3 (1968).

54. G. Khvostenko and M. Chaika, Opt. Spectrosc. 25, 450-51 (1968).

55. G. Khvostenko, V.I. Khutorshchikov and M.P. Chaika, Opt. Spectrosc. 36, 475 (1974).

56. L.I. Schiff and H. Snyder, Phys. Rev. 55, 59-63 (1939).

57. H.A. Bethe and E.E. Salpeter, "Quantum Mechanics of One- and Two- Electron Atoms" (Berlin, Springer Verlag) (1957).

58. R.H. Garstang, Rep. Prog. Phys. 40, 105-154 (1977).

59. W.C. Martin, Phys. Chem.Ref. Data 2, 257-266 (1973).

60. T.A. Miller and R.S. Freund, Phys. Rev. A4, 81 (1971).

61. H-J. Beyer and K-J. Kollath, J. Phys. B11, 979-991 (1978).

62. T.A. Miller and R.S. Freund, Phys. Rev. A5, 588-591 (1972).

63. H-J. Beyer, unpublished.

64. K.B. MacAdam and W.H. Wing, Phys. Rev. A12, 1474 (1975).

65. L. Holmgren, I. Lindgren, J. Morrison and A.M. Mårtensson, Z. Physik A 276, 179 (1976).

66. G. Tepehan, H-J. Beyer and H. Kleinpoppen, to be published.

# RESONANT INTERACTION OF ATOMS WITH INTENSE RADIATION FIELDS

M.L.Ter-Mikaelian, M.A.Sarkissian

Institute for Physical Research, Armenian Academy of Sciences 378410, Ashtarak-2, Armenia, USSR

Phenomena, occuring near the resonance has always attracted physicists' attention since the advent of quantum mechanics. Atomic wave functions

$$\Psi_n(t) = \mathcal{U}_n \exp(-iE_n t) \tag{1}$$

under the influence of monochromatic electromagnetic radiation with wave vector $\vec{K}$ and polarization $\vec{e}$

$$\vec{\mathcal{E}}(\vec{z},t) = \vec{\mathcal{E}}_{\vec{K},\vec{e}}\, e^{i\vec{K}\vec{z}-i\omega t} + \text{C.C.} \tag{2}$$

undergo great changes when the frequency of radiation field $\omega$ is close to the atomic transition $E_{nm} = E_n - E_m$. In dipole approximation the interaction energy of the field with the atom can be written in the following form

$$V = -\vec{d}\,\vec{\mathcal{E}}(\vec{z},t) \tag{3}$$

For perturbed wave function we get

$$\Phi_1 = \Psi_1 + \sum_n \left\{ \frac{\vec{d}_{n1}\vec{\mathcal{E}}_{\vec{K},e}}{\hbar(E_{n1}-\omega)} e^{-i\omega t} + \frac{\vec{d}_{n1}\vec{\mathcal{E}}^{*}_{\vec{K},e}}{\hbar(E_{n1}+\omega)} e^{i\omega t} \right\} \mathcal{U}_n e^{-iE_1 t} \tag{4}$$

where $\vec{d}_{n1}$ is dipole matrix element for $1 \rightarrow n$ transition. Near resonance only one term of series (4) is important, where the condition

$$|\varepsilon| = |E_{21} - \omega| \ll \omega \tag{5}$$

is fulfilled. In the following $\varepsilon$ will be called the detuning, or offset of resonance. Assuming that other levels are far from the resonance, expression (4) yields the expansion parameter

$$\alpha = \frac{2}{\hbar}\left|\frac{\vec{d}_{21}\vec{\mathcal{E}}_{\vec{K},e}}{\varepsilon}\right| \tag{6}$$

which for nonresonant transitions, when $\varepsilon \sim \omega$ converts into the usual parameter $\alpha'$, is equal to

$$\alpha' \sim \frac{\mathcal{E}_{laser}}{\mathcal{E}_{atom.}} \ll 1 \tag{7}$$

widely used in nonlinear optics.
Now we shall consider that laser field is substantially less than the atomic one. The condition $\alpha' \ll 1$ might be very easily fulfilled. Radiation fields when $\alpha \gtrsim 1$ will be called intense fields. For weak $\alpha \ll 1$ fields the ordinary perturbation theory is applicable. Therefore, the problem of interaction of radiation field near resonance with the atoms reduces to that of interaction with two-level system. If using (4) we calculate the radiation of atom in external field (that means we calculate the scattering of intense field by the atom) we obtain the well known Weisskopf expression for resonance fluorescence. The probability of scattering of photons with wave factor $\vec{K}'$ and polarization $\vec{e}'$ into a solid angle is given by the formula

$$dW_{\vec{K}',\vec{e}'} = \frac{|\sum(\vec{d}_{2i}\vec{\mathcal{E}}_{\vec{K},e})(\vec{e}'^{*}\vec{d}_{12})|^2}{(E_{21}-\omega)^2 + \Gamma^2/4}\cdot\frac{\omega'^3 dO'}{2\pi\hbar c^3} \tag{8}$$

If state 2 is degenerated summing must be carried out over all intermediate states. In reality the two-level nondegenerate system can be used as a good model for our problems. If the radiation field has circular polarization, say $\sigma^+$, only two states in transition $S_{-1/2} \rightarrow P_{+1/2}$ with $\Delta m = 1$ are involved in interaction. For simplicity, we shall neglect the relaxation time of the upper state. It means that the observation time $\tau$ must be substantially less, than the relaxation time

$$\tau < \frac{1}{\Gamma} \tag{9}$$

If condition (9) is fulfilled, there is no necessity at all to take into account the relaxation by calculating the probabilities of processes of interest.
For this case the interaction of atom with radiation field is considerably simplified and can be analysed in detail on the basis of Schrödinger and Maxwell equations. In the opposite case when inequality (9) is not fulfilled, the spontaneous radiation processes or other relaxations will change the processes under consideration. In this case the density matrix formalism must be used for calculating the atom-field interaction processes. To simplify the problem and to separate the resonance phenomena from relaxation processes, we assume that (9) holds. Upper state relaxation is caused by diverse reasons. Besides Doppler and collision broadenings the relaxation time determined by spontaneous decay of upper state equals approximately to $10^{-8}$ sec, if the transition is not forbidden. This means that the laser pulse duration of incident radiation must not exceed $10^{-8}$ sec.

On the other hand, we shall discuss the influence of intensity of radiation on various phenomena. As we shall see later the intensity effects are determined by parameter

$$\gamma = \sqrt{\varepsilon^2 + \gamma_R^{\,2}} \tag{10}$$

where $\gamma_R$ is the well known Rabi frequency.
The values of parameter $\gamma_R$ are tabulated in the following table as a function of intensity

| $I = \int I_\omega d\omega$ | $10^3$ | $10^6$ | $10^8$ |
|---|---|---|---|
| $\gamma_R = 2\lvert\mathcal{E} d\rvert/\hbar$ | 0,5 | 14 | 140 |

$\gamma_R$ is in $cm^{-1}$, and $I$ in $\frac{\text{watt}}{cm^2}$, $d$ is taken equal to $5 \cdot 10^{-18}$ C.G.S.E.

Therefore, if effects due to intensity are taken into account and relaxation of upper state is neglected the following inequality must be fulfilled

$$\Gamma \ll |\varepsilon|, \gamma_R \tag{11}$$

By choosing the lasers parameters inequalities, (9) and (11) can be fulfilled. After the introductory remarks we have made let us consider two-level system in an intense applied field. The Schrödinger equation in driven field $\vec{\mathcal{E}}(\vec{r},t)$ is

$$i\hbar \frac{\partial \phi}{\partial t} = \left(H_0 - \vec{d}\vec{\mathcal{E}}(\vec{r},t)\right)\phi \tag{12}$$

The solution is expressed in the form

$$\phi = a_1 \Psi_1 + a_2 \Psi_2 \tag{13}$$

In resonant (rotating wave) approximation equation (12) reduces to the set of equations

$$\begin{aligned} i\hbar \frac{\partial a_1}{\partial t} &= -a_2 \left(\vec{\mathcal{E}}_k \cdot \vec{d}_{12}\right) \exp(-i\varepsilon t) \\ i\hbar \frac{\partial a_2}{\partial t} &= -a_1 \left(\vec{\mathcal{E}}_k \cdot \vec{d}_{21}\right) \exp(i\varepsilon t) \end{aligned} \tag{14}$$

Solution can be found in the form

$$a_1 = a_1^0 \exp(-i\lambda t),\; a_2 = a_2^0 \exp(-i(\lambda-\varepsilon)t) \tag{15}$$

Substituting (15) into (14) we obtain

$$\lambda_{1,2} = \frac{\varepsilon}{2} \mp \frac{|\varepsilon|}{2}\sqrt{1+\alpha^2} = \frac{\varepsilon}{2} \mp \frac{\gamma}{2} \tag{16}$$

To reconstruct the wave functions for a two-level system in external field, there are two different ways determined by initial conditions. In the first case the wave function is given at the time t = 0 when the light pulse has not reached the atom. We shall call this function a nonstationary wave function. In the second case the switch-on of the interaction is done very slowly (adiabatically), i.e. each of the atomic states of the unperturbed atom tends and converts into one definite state of an atom in an external field.

It can be shown that in the case of

$$\gamma\tau \gg 1 \tag{17}$$

quasi-energetic states are formed. In the opposite case we have instant switching-on of interaction, and nonstationary wave functions are generated. Analytical solutions can also be obtained in an intermediate case for some concrete pulse shape form.

Nonstationary wave functions can be written in the following form

$$\Phi_1' = A_1(t)\Psi_1 + A_2(t)\Psi_2 \qquad \Phi_1' = \Psi_1 \text{ at } t = 0$$

$$\Phi_2' = -A_2^*(t)\Psi_1 + A_2(t)\Psi_2 \qquad \Phi_2' = \Psi_2 \text{ at } t = 0, \tag{18}$$

where coefficients $A_1$ and $A_2$ are determined by formulae

$$A_1(t) = e^{-i\frac{\varepsilon}{2}t}\left(\cos\frac{\gamma}{2}t + \frac{i\,\mathrm{sgn}\,\varepsilon}{\sqrt{1+\alpha^2}}\sin\frac{\gamma}{2}t\right);$$

$$A_2(t) = e^{i\frac{\varepsilon}{2}t}\,\frac{2i(\vec{\mathcal{E}}_k\cdot\vec{d}_{21})}{\hbar\gamma}\sin\frac{\gamma}{2}t. \tag{19}$$

The $\Phi_{1,2}'$ functions form a complete set of orthonormalized eigenstates

$$\int \Phi_i'^{*}\Phi_k\, dv = \delta_{ik} \tag{20}$$

It results from the expression (20) that upper level population in the state with the function $\Phi_1'$ equals to

$$N_2(t) = \frac{\alpha^2}{1+\alpha^2}\sin^2\frac{\gamma}{2}t \tag{21}$$

The population oscillates with frequency $\gamma/2$ . This is the well known solution of Rabi [1].

In adiabatic limit the expressions for quasi-energetic functions are [2]:

$$\Phi_1 = C_1 e^{-i\bar{\lambda}_1 t - iE_1 t}(\mathcal{U}_1 + B\mathcal{U}_2 e^{-i\omega t}) \tag{22}$$

$$\Phi_2 = C_2 e^{i\bar{\lambda}_1 t - iE_2 t}(\mathcal{U}_2 - B^*\mathcal{U}_1 e^{i\omega t})$$

$C_1, C_2$ - are determined by normalization.
$B$ has the form

$$B = -\frac{\hbar\bar{\lambda}_1}{(\vec{\mathcal{E}}_k^* \cdot \vec{d}_{21})} = -\frac{\hbar\mathcal{E}}{2(\vec{\mathcal{E}}_k^* \cdot \vec{d}_{21})}\cdot(1-\sqrt{1+\alpha^2}) \tag{23}$$

For convenience, we shall sometimes use slightly different formula for the roots

$$\bar{\lambda}_{1,2} = \frac{\mathcal{E}}{2}(1 \mp \sqrt{1+\alpha^2}) \tag{24}$$

Notice that $\lambda_{1,2} = \bar{\lambda}_{1,2}$ for $\mathcal{E} > 0$, and $\lambda_{1,2} = \bar{\lambda}_{2,1}$ for $\mathcal{E} < 0$. It is clear, that the expression (24) gives the possibility to establish a unique correspondence between inperturbed atom levels and levels of an atom in an external resonant field. In the case of weak intensities, that is when $\alpha \to 0$, $\Phi_1 \to \Psi_1, \Phi_2 \to \Psi_2$. Therefore the quantities

$$W_1 = E_1 + \bar{\lambda}_1, \quad W_2 = E_2 - \bar{\lambda}_1 = E_1 + \omega + \bar{\lambda}_2 \tag{25}$$

playing the role of energy are called the quasi-energy (see for detail [3] ). In Fig. 1 and 2 the quasi-energy is drawn as a function of intensity. Instead of $W_{1,2}$, $\bar{\lambda}_{1,2}$ is widely used. also called quasi-energy and represented in Fig. 3.

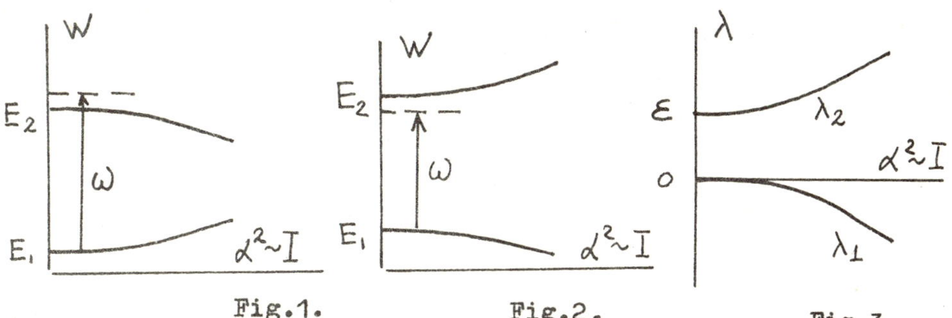


Fig.1. Quasi-energy levels for $\mathcal{E}<0$.

Fig.2. Quasi-energy levels for $\mathcal{E}>0$.

Fig.3. Quasi-energy $\lambda_{1,2}$.

Conditions $\overline{\lambda}_1 = \overline{\lambda}_2$ (quasi-energy crossing) corresponds to the case of exact resonance. Figures 1,2,3 show that under the influence of the resonant field the two-level system "goes away" from resonance.

If we quantize the exciting field, the quasienergetic functions will correspond to eigenfunctions of the total "atom + field" system.

$$\Phi_1(n_{\vec{k}}) = C_1 e^{-i[E_1+\lambda_1+n_{\vec{k}}\omega]t} [\Phi_1^\circ(n_{\vec{k}}) + B\Phi_2^\circ(n_{\vec{k}}-1)]$$
$$\Phi_2(n_k-1) = C_2 e^{-i[E_2-\lambda_1+(n_k-1)\omega]t} [\Phi_2^\circ(n_{\vec{k}}-1) - B^*\Phi_1^\circ(n_{\vec{k}})] \tag{26}$$

$$\Phi_1^\circ = \mathcal{U}_1 |0,.. n_{\vec{k}}, ..0\rangle, \quad \Phi_2^\circ = \mathcal{U}_2 |0... n_{\vec{k}}-1,..0\rangle \tag{27}$$

The variables $n_{\vec{k}'e'}$, with $\vec{k}' \neq \vec{k}$, $\vec{e}' \neq \vec{e}$ are omitted for simplicity. The energy levels of the total system "atom + field" are represented in Fig.4, as a function of atomic energy spacing $E_{21}$. There are an anticrossing energy levels at $\varepsilon = 0$ (see for detail [4]). The energy level shift in the driven field was measured in a series of experiments [5].

Using wave functions (22) for atom in an external field, we can obtain the expressions for radiated energy of "atom + + field" system. For this purpose the following calculation method can be used which gives the results analogous to quantum electrodynamics calculation. We will derive the expression for electric dipole moment of atom induced by an external field and divide it into two parts with positive and negative frequencies

$$\vec{\mathcal{D}}_{ik} = e\int \Phi_i^* \vec{z} \Phi_k dv = \vec{\mathcal{D}}_{ik}^- e^{-i\omega' t} + \vec{\mathcal{D}}_{ik}^+ e^{i\omega' t} \tag{28}$$

Substituting $\vec{\mathcal{D}}_{ik}^-$ in classical expression for probability of dipole radiation of photon with frequency $\omega'$, wave vector $\vec{k}'$ and polarization $\vec{e}'$

$$dW_{\vec{k}',\vec{e}'} = \frac{\omega'^3}{2\pi\hbar c^3} |\vec{e}'^*_{\omega'} \vec{\mathcal{D}}_{ik}^-|^2 dO' \tag{29}$$

we obtain the expression for probability of radiation of "atom + field" system. For example probability of a photon

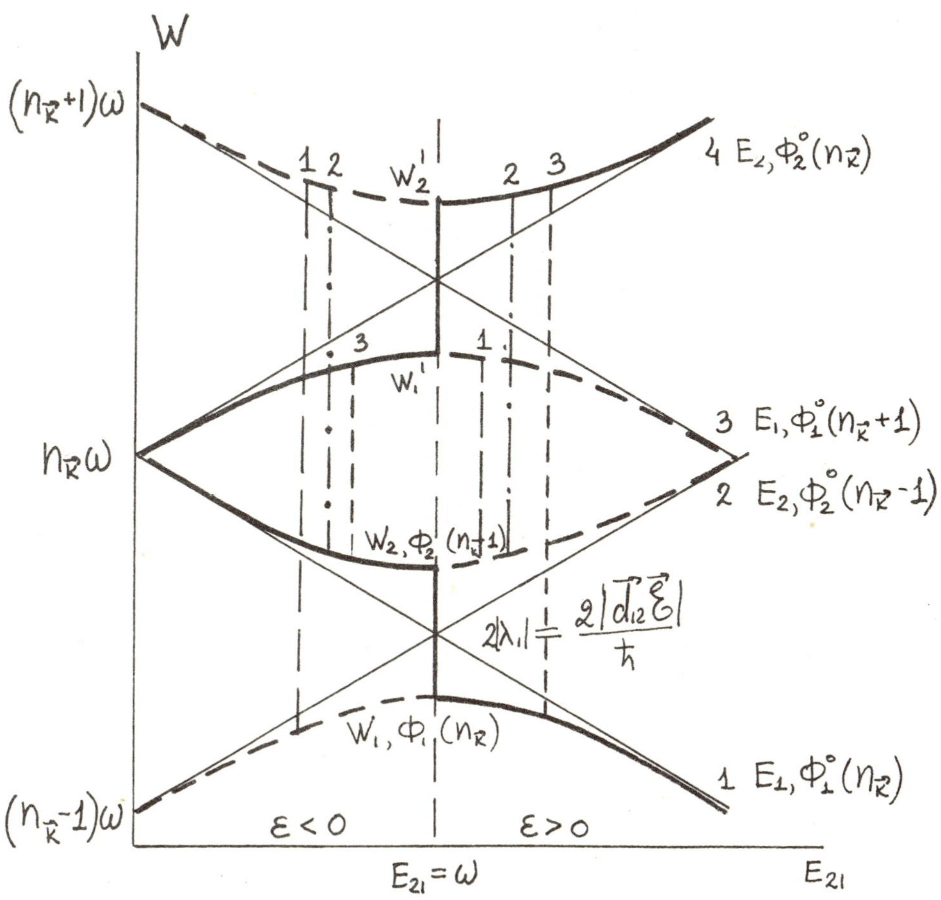


Fig.4.

Energy level of "atom + quantized field" system. The curves 1 and 2 (dotted-solid) correspond to the functions (26), solid curves 1 and 4 and dotted curves 2 and 3 correspond to the functions (22). Vertical dotted lines show the allowed transitions $\Phi_1 \rightleftarrows \Phi_2$ with an absorption or emission of a photon $\vec{k}', \omega'$. 1 - three-photon process, with emission of a photon $\omega' = \omega - \varepsilon\sqrt{1+\alpha^2}$, 2 - Rayleigh scattering $\omega' = \omega$. 3 - absorption of a photon $\omega' = \omega + \varepsilon\sqrt{1+\alpha^2}$.

radiation by dipole moment $\vec{\mathcal{D}}_{21}^{-}$ with frequency

$$\omega' = 2\omega - E_{21} + 2\lambda_1 = \omega - \varepsilon\sqrt{1+\alpha^2} \tag{30}$$

is given by formula [2]

$$dW_{\vec{k}';\vec{e}'} = \frac{\omega'^3}{8\pi\hbar c^3}\,\frac{(1-\sqrt{1+\alpha^2})^2}{1+\alpha^2}\,|\vec{e}'^*\vec{d}_{12}|^2(1+n_{\vec{k}';\vec{e}'})\,dO'. \tag{31}$$

The item

$$n_{\vec{k}';\vec{e}'} = \frac{8\pi^3 c^2}{\hbar\,\omega'^3}\, I_{\vec{k}';\vec{e}'} \tag{32}$$

determines the induced process of the so-called "three-photon radiation". The intensity of radiation is expressed in terms of spectral density $I_{\vec{k}';\vec{e}'}$ by the following formula

$$I = \int I_{\vec{k}';\vec{e}'}\, d\omega'\, dO'. \tag{32'}$$

From the quoted expression it follows, that at $\alpha \ll 1$ the probability is proportional to $\alpha^4 \sim I^2$ or to $\varepsilon^4$, that is it is absent in the second approximation of usual perturbation theory. The third order (31) describes the following process: absorbing two photons from the laser beam, the atom jumps from state 1 to state 2 and emits the "three-photon frequency" radiation (30).

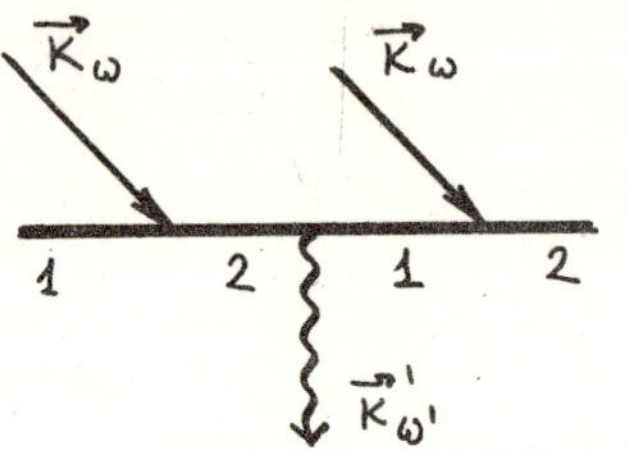


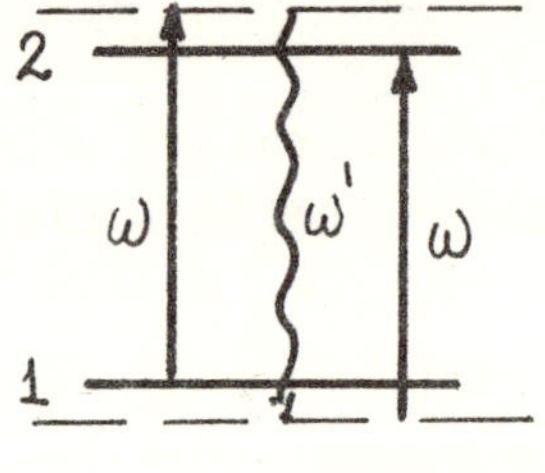


Fig.5 illustrates the process of three-photon radiation schematically.

At $\alpha \gg$ the probability of three-photon radiation is turned into the ordinary one-photon radiation probability, determined by the matrix element of transition $\vec{d}_{21}$ with 1/4 coefficient. The process is induced and at large $n_{\vec{k}';\vec{e}'}$ values may exceed several times the probability of spontaneous transition 2 $\longrightarrow$ 1. This unusual example illustrates a situation in nonlinear resonance spectroscopy rather well. Near resonance that is at $\alpha \gg 1$, the higher order perturbation processes become the same order as the first order perturbation theory processes. It is an unusual situation in quantum electrodynamics and causes a series of interesting peculiarities. In optical range the process of three-photon scattering was observed in [6] for the first time. In Fig.6 the three-photon line scattering frequency depending upon the intensity of resonant field is shown.

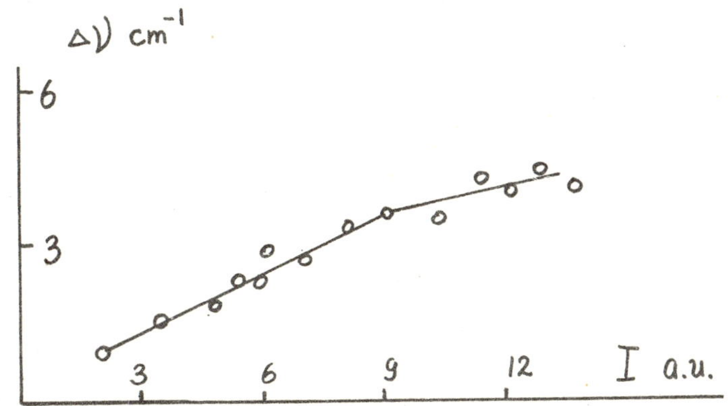


Fig.6.

The frequency shift of $\Delta\nu = \frac{\lambda_1}{\pi}$ three-photon scattering depending on the intensity of resonant radiation with $\mathcal{E}$ = 21cm$^{-1}$; $\alpha^2 \sim 0,6$; $I_\omega \sim 30^{Mw}/cm^2$.

Certainly, besides this unusual process in the two-level system the usual process also takes place (Rayleigh scattering and absorption of "atom + field" system). We shall not write down the corresponding probabilities [2]. It is only worth mentioning that Rayleigh scattering has always manifested itself as a spontaneous process. Unfortunately, the processes mentioned here are not investigated in the region where condition (9)' is fulfilled.

In the alternate limiting case, when a condition reverse to (9) is fulfilled, a great number of investigations on resonance fluorescence are available [7]. We shall not dwell on them here, as for the interpretation it is necessary to take into account all the possible processes of atomic level broadening and use the matrix density formalism, which makes our treatment essentially complicated. Let us consider the dielectric constant of atoms in the presence of resonant field. Using formula (28) we calculate the induced dipole moment in the ground $\Phi_1$ state from which the expression for the dielectric constant follows

$$\mathcal{E}_+(\omega) = 1 + \frac{4\pi |\langle 2|d_+|1\rangle|^2}{\hbar \varepsilon \sqrt{1+\alpha^2}} N_1 \tag{33}$$

where $N_1$ is the number of atoms in state $\Phi_1$ . The pulse group velocity is as follows

$$\vartheta = \frac{\partial \omega}{\partial k} = \frac{c}{\sqrt{\mathcal{E}(\omega)} + \omega \frac{\partial \sqrt{\mathcal{E}(\omega)}}{\partial \omega}} \tag{34}$$

$$\frac{1}{\vartheta} = \frac{\sqrt{\mathcal{E}(\omega)}}{c} + \frac{\mathcal{E}(\omega) - 1}{2c\sqrt{\mathcal{E}(\omega)}} \cdot \frac{\omega}{\varepsilon} \cdot \frac{1}{1+\alpha^2} \tag{35}$$

Elegant experiments of Grischkowsky for light velocity determination in resonant medium [8] were carried out in rubidium vapors using scheme levels pictured in Fig.7.

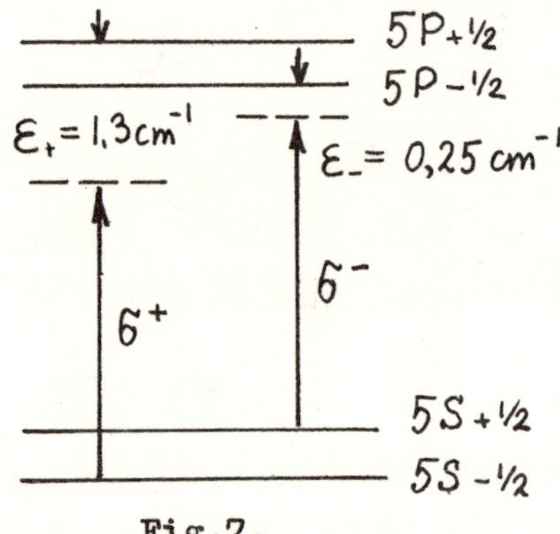


Fig.7.

In spite of the fact that in the mentioned experiments $\sqrt{\mathcal{E}(\omega)} - 1 \sim 6 \cdot 10^{-4}$, the multiplier $\omega/\varepsilon$ amounts the values $0{,}3 \cdot 10^5$ and light rate $\vartheta_-$ was $\vartheta_- \sim c/9$, and $\vartheta_+ = 0{,}75$ . The pulse $\tau$ duration in the experiments [8] was changed from $5 \cdot 10^{-9}$ to $1{,}8 \cdot 10^{-8}$, whereas the radiation decay time forms 2,8 . $10^{-7}$, and the collision damping time forms $\sim 10^{-6}$sec, that is the condition (9) was fulfilled. In the quoted experiments the condition (11) was fulfilled, where $\Gamma$ - is the level width and $\gamma_{R+} \sim 1{,}2 \cdot 10^{9}$ hertz, and $\gamma_{R-} \sim 2{,}5 \cdot 10^{8}$hertz correspondingly. Condition (18) $\varepsilon\tau \gg 1$ is valid, which means that quasi-energetic functions (22) must be used.

The experimental results are plotted in Fig.8.

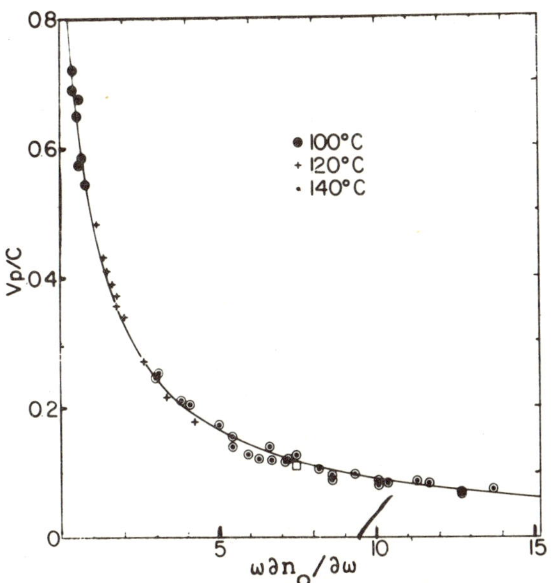


Fig.8. The dependence of pulse velocity in resonant medium upon parameter $\omega \frac{\partial n_0}{\partial \omega}$, where $n_0 = \sqrt{\varepsilon_1(\omega)}$.

It is interesting to note, that during the passage through medium, being in the state of $\Phi_2$, the group velocity will be larger than the velocity of light. This interesting situation has not been investigated in detail by anyone. Evidently, the detailed analysis will show, that this problem is similar to the light passage in the anomalous dispersion region [9].

Resonant interaction analysis mentioned above is also applicable for more complicated systems. Consider the four-level system interacting with the resonant field with frequency $\omega$ (Fig.9). Unlike the two-level model, in which under the field influence the levels are shifted away from resonance, for three and more level systems the dynamic Stark shifts caused by an intensive field can lead to the enhancement of levels and even to their passing through resonance. In terms of quasi-energies this means that quasi-energy levels can be drawn together to the distances substantially less that the initial detunings [10]. This phenomena will be named below as the self-induced resonance (SIR). The Schrodinger equation solution for the system shown in Fig. 3 will be chosen in the following form:

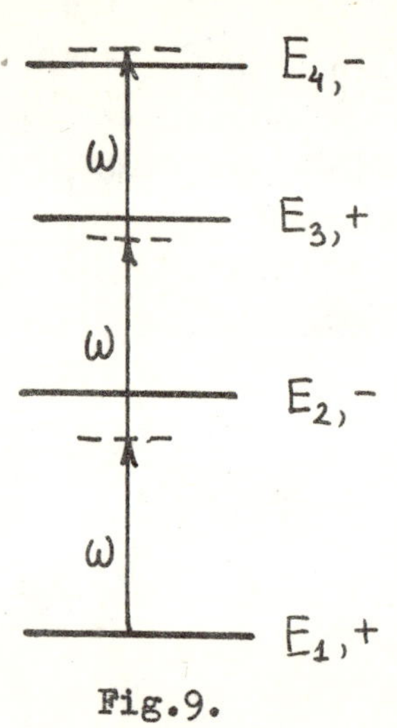

Fig.9.

$$\Phi_s = e^{-i(E_1+\lambda_s)t} \sum_{m=1}^{4} a_m^{(s)} U_m e^{-i(m-1)\omega t}, \tag{36}$$

where $U_m$ are nonperturbed wave functions which satisfy the following equation

$$H_0 U_m = E_m U_m . \tag{37}$$

The equations for the determination of amplitudes $a_m^{(s)}$ in resonant (RW) approximation are

$$\begin{aligned} -\lambda a_1 + V_{12}^{+} a_2 &= 0 \\ V_{21}^{-} a_1 - (\lambda - \varepsilon_{21}) a_2 + V_{23}^{+} a_3 &= 0 \\ V_{32}^{-} a_2 - (\lambda - \varepsilon_{31}) a_3 + V_{34}^{+} a_4 &= 0 \\ V_{43}^{-} a_3 - (\lambda - \varepsilon_{41}) a_4 &= 0, \end{aligned} \tag{38}$$

where

$$\mathcal{E}_{m1} \doteq E_{m1} - (m-1)\omega \tag{39}$$

are resonsnce detunings. The equation for quasi-energy is as follows

$$\begin{vmatrix} -\lambda & V_{12}^{+} & 0 & 0 \\ V_{21}^{-} & -(\lambda-\mathcal{E}_{21}) & V_{23}^{+} & 0 \\ 0 & V_{32}^{-} & -(\lambda-\mathcal{E}_{31}) & V_{34}^{+} \\ 0 & 0 & V_{43}^{-} & -(\lambda-\mathcal{E}_{41}) \end{vmatrix} = 0 \tag{40}$$

The roots of the forth order equation (40) together with (38) fully determines the mutually orthogonal wave functions $\Phi_S$ ( $S$ = 1,2,3,4). The $a_m^{(s)}$ coefficients are equal to

$$\begin{aligned} a_2^{(s)} &= \frac{\lambda_s}{V_{12}^{+}}\, a_1^{(s)} \\ a_3^{(s)} &= \frac{\lambda_s(\lambda_s-\mathcal{E}_{21})-|V_{21}^{-}|^2}{V_{12}^{+}V_{23}^{+}}\, a_1^{(s)} \\ a_4^{(s)} &= \frac{V_{43}^{-}}{V_{12}^{+}V_{23}^{+}} \cdot \frac{\lambda_s(\lambda_s-\mathcal{E}_{21})-|V_{21}^{-}|^2}{\lambda_s-\mathcal{E}_{41}}\, a_1^{(s)} \end{aligned} \tag{41}$$

and $a_1^{(s)}$ is determined by the normalization condition with an accuracy to arbitrary phase.

$$\sum_{m=1}^{4} |a_m^{(s)}|^2 = 1 \tag{42}$$

With that the procedure of deriving the full orthonormalized set of functions of the four-level system in driven resonant field is completed. In general case, at arbitrary acceptable parameter values (detuning and field strength), the roots $\lambda_s$ of equation (40) are complicated, and the $a_m^{(s)}$ coefficients and, correspondingly, the wave functions become cumbersome. Therefore, it is preferable to use numerical calculations. However, in some physically interesting limiting cases the roots are simplified, and it is possible to obtain simple analytical expressions for $\lambda_s$, $a_m^{(s)}$. Here we note that the roots of equation (40) are real and nondegenerated which means that for not vanishing field strength the quasi-energies $\lambda_s$ do not cross each other anywhere, and their limiting values at $\mathcal{E}_{laser} \to 0$ are determined by the expressions (44) given below. The coefficients do not have any peculiarities and during field switching off the wave functions are converted into the corresponding nonperturbed ones $\Psi_s = \mathcal{U}_s \exp(-iE_s t)$ .

Now we start analyzing eq.(40) in the limit of perturbation theory, that is when dynamic Stark shifts are substantially less than initial resonance detunings. Solving the equation at

$$|V_{ik}| \ll |\varepsilon_{je}|, \tag{43}$$

we obtain the following values for quasi-energy

$$\begin{aligned} \lambda_1 &= -\frac{|\bar{V_{21}}|^2}{\varepsilon_{21}} \\ \lambda_2 &= \varepsilon_{21} + \frac{|\bar{V_{21}}|^2}{\varepsilon_{21}} - \frac{|\bar{V_{32}}|^2}{\varepsilon_{32}} \\ \lambda_3 &= \varepsilon_{31} + \frac{|\bar{V_{32}}|^2}{\varepsilon_{32}} - \frac{|\bar{V_{43}}|^2}{\varepsilon_{43}} \\ \lambda_4 &= \varepsilon_{41} + \frac{|\bar{V_{43}}|^2}{\varepsilon_{43}}. \end{aligned} \tag{44}$$

In this approximation the $\Phi_s$ functions coincide with the wave functions of ordinary third order perturbation theory, taking into account the resonance terms only. For example, for the wave function $\Phi_1$ the coefficients $a_m^{(1)}$ have the form

$$a_1 \simeq 1,\ a_2 \simeq -\frac{\bar{V_{21}}}{\varepsilon_{21}},\ a_3 \simeq \frac{V_{21}\bar{V_{32}}}{\varepsilon_{21}\varepsilon_{31}},\ a_4 \simeq -\frac{\bar{V_{21}}\bar{V_{32}}\bar{V_{43}}}{\varepsilon_{21}\varepsilon_{31}\varepsilon_{41}}. \tag{45}$$

Equation (40) is also solved easily in the opposite limiting case when strong interaction for all transitions $1 \to 2$, $2 \to 3$, $3 \to 4$ takes place. If we put in equation (40) all $\varepsilon_{m+1,m} = 0$, then for the roots determination we obtain the biquadratic equation which yields the following values for quasi-energy.

$$\lambda_s = \mp\sqrt{\frac{|V_{21}^-|^2+|V_{32}^-|^2+|V_{43}^-|^2}{2} \mp \sqrt{\left(\frac{|V_{21}^-|^2+|V_{32}^-|^2+|V_{43}^-|^2}{2}\right)^2 - |V_{21}^- V_{43}^-|^2}} \qquad (46)$$

The roots ordering has to be done as follows. As long as the roots do not cross each other anywhere the mutual quasi-energy distribution order determined by the primary detunings is preserved at any field values. If for example for the initial detunings the following relations are fulfilled $\varepsilon_{41} < 0 < \varepsilon_{21} < \varepsilon_{31}$ , then the roots ordering for (46) has to be done in such a way, that the quasi-energies by their values could have been arranged in the same order i.e., the signs before the roots should be chosen as

$$\lambda_4(-+) < \lambda_1(--) < \lambda_2(+-) < \lambda_3(++). \qquad (47)$$

In the same way the ordering for arbitrary initially detunings are available. Simple analytical solutions are also obtainable in the cases of strong two-photon and three-photon resonance when other resonances are weak. We will consider in detail the case of strong three-photon resonance, when the following conditions are fulfilled

$$|\varepsilon_{41}|, |V_{m+1,m}| \ll |\varepsilon_{21}|, |\varepsilon_{31}|, |\varepsilon_{32}|. \qquad (48)$$

According to (48) the states 2 and 3 are taken into account by using the perturbation theory while the three-photon resonance - "exactly". The values of $\lambda_2$ and $\lambda_3$ quasi-energies have the form (44) corresponding to the perturbation theory, and for $\lambda_1$ and $\lambda_4$ we obtain

$$\lambda_{1,4} = \frac{\varepsilon_{41}}{2}\left\{1-\gamma_1+\gamma_3 \mp \sqrt{(1+\gamma_{eff.})^2 + 4|\mathcal{V}|^2/\varepsilon_{41}^2}\right\}, \qquad (49)$$

where

$$\gamma_1 = \frac{|V_{21}^-|^2}{\varepsilon_{21}\varepsilon_{41}}, \; \gamma_3 = \frac{|V_{43}^-|^2}{\varepsilon_{43}\varepsilon_{41}}, \; \gamma_{eff.} = \gamma_1 + \gamma_3 \qquad (50)$$

are the parameters characterizing the relations between Stark shifts and three-photon detuning, and

$$\mathcal{W} = \frac{V_{21}^{-} V_{32}^{-} V_{43}^{-}}{\varepsilon_{21} \varepsilon_{31}} \qquad (50')$$

is the effective matrix element.

In Fig.10 the quasi-energy as a function of the field intensity at different signs $\gamma_{eff.}$ is plotted.

From the expression (49) it is clear that the minimum distance between quasi-levels is observed at

$$1 + \gamma_{eff.} = 0, \qquad (51)$$

and is equal to

$$|\lambda_1 - \lambda_4|_{min} = 2|\mathcal{W}|_{I_{cr}} \qquad (52)$$

For given $\varepsilon_{ik}$ and dipole matrix element, equation (51) determines the critical intensity $I_{cr.}$ when the (52) holds. In this case it is obvious that condition (17) brought above may be broken down, and quasi-energetic functions are not valid anymore. The calculations show that in order to make the passage of the critical point (51) adiabatic, it is necessary to fulfil the following inequality

$$\frac{1}{\tau} \ll |\mathcal{W}|^2_{I_{cr}} / |\varepsilon_{41}| \qquad (53)$$

If this condition is fulfilled atoms, initially being in state $\Psi_s$, during the field switching on transform into the corresponding state $\Phi_s$ .

Using the obtained values of quasi-energies and the expressions for the $a_m^{(s)}$ coefficients it is possible to calculate the atomic level populations in states $\Phi_s$ ( $s = 1,2,3,4$). At weak fields the populations are described by the ordinary perturbation theory expressions of (45) type. We will write down the expression for population of upper 4-th level in state $\Phi_1$ . The latter is formed from the ground nonperturbed state $\Psi_1 = U_1 \exp(-iE_1 t)$ at adiabatic switching on of the field.

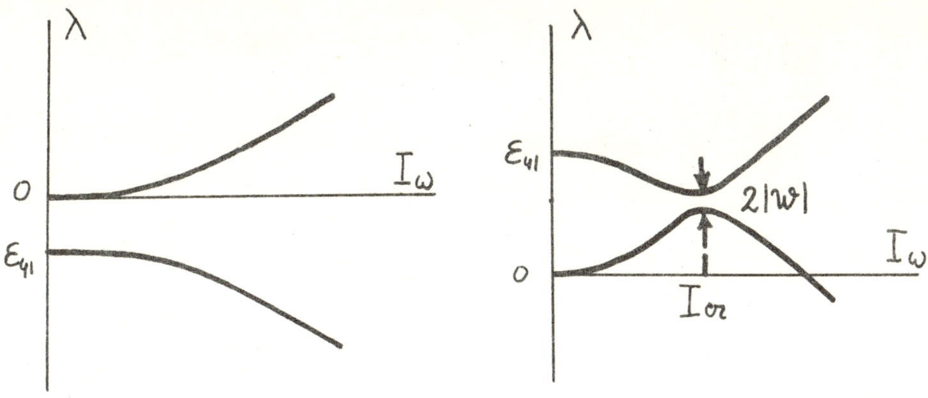


Fig.10.

a) $\gamma_{eff.} > 0$ b) $\gamma_{eff.} < 0$

The population of the 4-th level relative to the first one is expressed by:

$$\left|\frac{a_4}{a_1}\right|^2 = \frac{\varepsilon_{41}^2}{4|w|^2}\left\{1+\gamma_{eff.}-\sqrt{(1+\gamma_{eff.})^2+4|w|^2/\varepsilon_{41}^2}\right\}^2. \quad (54)$$

We will consider the two possible cases

I. $\gamma_{eff} > 0$

$$\left|\frac{a_4}{a_1}\right|^2 = \frac{|w|^2}{\varepsilon_{41}^2(1+\gamma_{eff.})^2} \ll 1 \quad (55)$$

a) $\gamma_{eff.} \ll 1$ , $|a_4|^2/|a_1|^2 \sim I_\omega^3$ (55')

b) $\gamma_{eff} \gg 1$ $|a_4|^2/|a_1|^2 \sim I_\omega$ (55'')

II. $\gamma_{eff} < 0$

a) $|\gamma_{eff.}| < 1$ $I_\omega < I_{cr}$ (perturbation theory)

$$\left|\frac{a_4}{a_1}\right| = \frac{|w|^2}{\varepsilon_{41}^2(1+\gamma_{eff.})^2} \ll 1 \quad (56)$$

b) $|1+\gamma_{eff.}| \ll 2|W/\varepsilon_{41}|$ (that is near $I_{cr}$).

$$\left|\frac{a_4}{a_1}\right| \simeq 1 \qquad (56')$$

c) $|\gamma_{eff.}| > 1$ , $I > I_{cr}$

$$\left|\frac{a_4}{a_1}\right|^2 = \frac{\varepsilon_{41}^2 (1+\gamma_{eff})^2}{|W|^2} \gg 1 \qquad (56'')$$

So then at corresponding choice of the sign $\varepsilon_{41}(\gamma_{eff.})$ the atom, being initially in the $\Psi_1$ state, during the field switching on is transformed into the quasi-energetic state $\Phi_1$ with the population inversion. The population inversion takes place at the leading front of the pulse and exists only in the presence of the field with intensity greater than $I_{cr}$ (Fig.II). When the field is switched off (at the back front) atoms transform into the nonperturbed $\Psi_1$ state, if during the pulse time an essential decay of the state $\Phi_1$ because of the transitions $\Phi_1 \to \Phi_j$ ( $j$ = 2,3,4) does not take place.

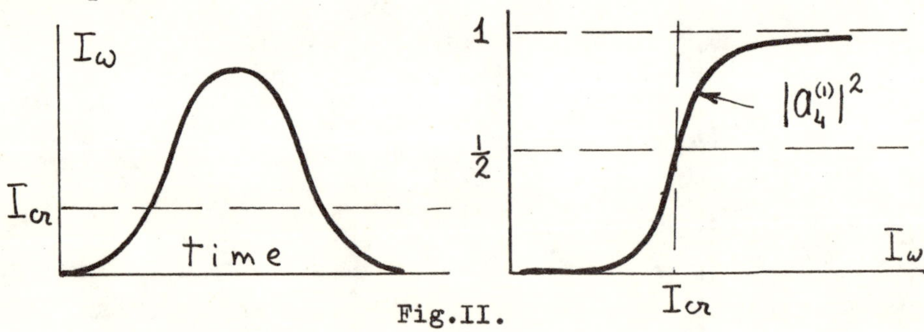


Fig.II.

The obtained effect bears some resemblence to the well known phenomenon of 180° inversion of population at adiabatic rapid passage of resonance in the two-level system which was also recently observed in optical region [II] . In one-photon A.R.P. experiments the resonance passing is realized due to the level shift by means of external field (Stark or Zeeman effects), or by means of the fast frequency modulation of the laser. In the system with the level number $m > 2$ the dynamic Stark shifts caused by intense field may provide a passage through resonance. The threshold field strength for such a self-induced "passage" is determined by minimum primary detunings which, in their turn, are restricted by the width of radiation incident on the atom.

Consider now photons scattering in the four-level system. We will observe atom scattering which "initially" existed in the ground nonperturbed state $\Psi_1$ . The scattering problem reduces to the radiation (or emission) problem of the electrons of atoms with the wave functions $\Phi_S$ ( $S$ = 1,2,3) being, in accordance with our choice, in the initial state of the system which is described by the wave function $\Phi_1|n_{\vec{k}}\dots n_{\vec{k}'}\rangle$. We assume, that photons of scattered field exist in the space so that not only the spontaneous but also the induced radiation is taken into account. To carry out perturbation theory calculations in dipole approximation it is necessary to calculate matrix elements of the type

$$M \sim \langle \dots n_{\vec{k}'} \pm 1 \dots | \int \Phi_j^* (\vec{d}\vec{\mathcal{E}}') \Phi_1 \, dv dt | \dots n_{\vec{k}'} \dots \rangle, \qquad (57)$$

where

$$\vec{\mathcal{E}}' = \sqrt{\frac{2\pi}{V}} \sum_{\vec{k}', \omega', \vec{e}'} \sqrt{\omega'} \left( \hat{C}_{\vec{k}'\omega'} \vec{e}' e^{-i\omega' t} + e.c. \right). \qquad (58)$$

$\hat{C}_{\vec{k}'}$ - is the destruction operator of photon with wave vector $\vec{K}'$ and frequency $\omega'$; $\vec{e}'$ - is the polarization vector.

Let us consider processes when the atomic system state in driven field is not changed. In this case, besides Rayleigh scattering (or radiation with the frequency exactly equal to the frequency of the driven field) a radiation at the frequency of third harmonic $\Omega = 3\omega$ occurs. It is clear, that for transitions $\Phi_1|\dots n'\dots\rangle \to \Phi_1|\dots n'+1\dots\rangle$ only a coherent spontaneous radiation occurs (because induced scattering processes with radiation and absorption compensate each other) which contributes to the refraction index of medium.

The probability of Rayleigh spontaneous scattering is given by the following formula

$$dW_\omega = \frac{\omega'^3}{2\pi\hbar c^3}\left|\left(a_1^* a_2 \vec{d}_{12} + a_2^* a_3 \vec{d}_{23} + a_3^* a_4 \vec{d}_{34}\right)\vec{e}_\omega^{\,*\prime}\right|^2 dO' \tag{59}$$

and the probability of the third harmonic spontaneous radiation by

$$dW_\Omega = \frac{\Omega^3}{2\pi\hbar c^3}\left|a_1 a_4 \left(\vec{d}_{14} \cdot \vec{e}_\Omega^{\,*\prime}\right)\right|^2 dO' \tag{60}$$

The considered processes are plotted graphically in Figures 12 and 13.

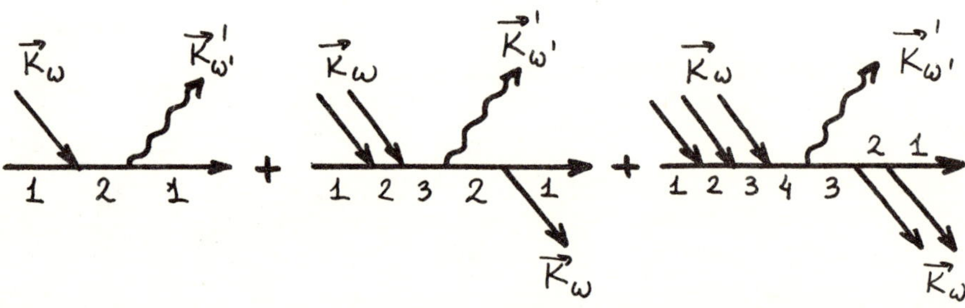


Fig.12. The process of Rayleigh scattering takes place through different channels which interferate.

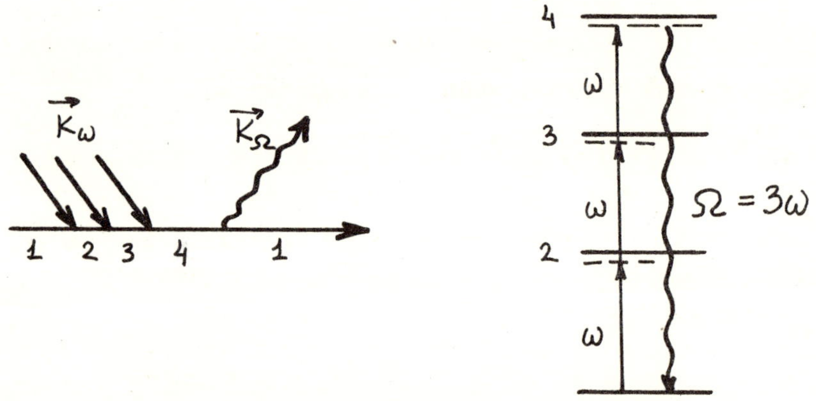


Fig.13. The process of photons scattering with the emission on third harmonic is shown.

Let us consider in detail the third harmonic emission process. In the limit of perturbation theory the expression (60) leads to the usual expression

$$dW_{\Omega}=\frac{\mathcal{E}_{\omega}^{6}(3\omega)^{3}}{2\pi\hbar c^{3}}\cdot\frac{|(\vec{e}_{\omega}\vec{d}_{21})(\vec{e}_{\omega}\vec{d}_{32})(\vec{e}_{\omega}\vec{d}_{43})(\vec{e}^{\,\prime *}_{\Omega}\vec{d}_{14})|^{2}}{\varepsilon_{21}^{2}\,\varepsilon_{31}^{2}\,\varepsilon_{41}^{2}}\,dO' \tag{61}$$

In the presence of strong three-photon resonance the value $|a_1 a_4|^2$ is expressed by

$$|a_1 a_4|^2=\frac{|\mathcal{V}|^2}{\varepsilon_{41}^{2}(1+\gamma_{eff.})^2+4|\mathcal{V}|} \tag{62}$$

Consider the possible cases:

I. $\gamma_{eff}>0$

As the following condition is always fulfilled

$$\varepsilon_{41}^{2}(1+\gamma_{eff.})^2\gg 4|\mathcal{V}|^2 \tag{63}$$

we obtain

$$|a_1 a_4|^2=\frac{|\mathcal{V}|^2}{\varepsilon_{41}^{2}(1+\gamma_{eff.})^2} \tag{64}$$

In the case of weak fields, when $\gamma_{eff}\ll 1$

$$|a_1 a_4|^2=|\mathcal{V}|^2/\varepsilon_{41}^{2}\sim I_{\omega}^{3} \tag{64'}$$

and the probability (60) transforms into the perturbation theory expression (61). By increasing the intensity $I_{\omega}$, when $\gamma_{eff}\gg 1$

$$|a_1 a_4|^2=\frac{|\mathcal{V}|^2}{\varepsilon_{41}^{2}\gamma_{eff.}^{2}}\sim I_{\omega} \tag{64''}$$

II. $\gamma_{eff}<0$

At weak fields when $|\gamma_{eff}|\ll 1$ the expression for $|a_1 a_4|^2$

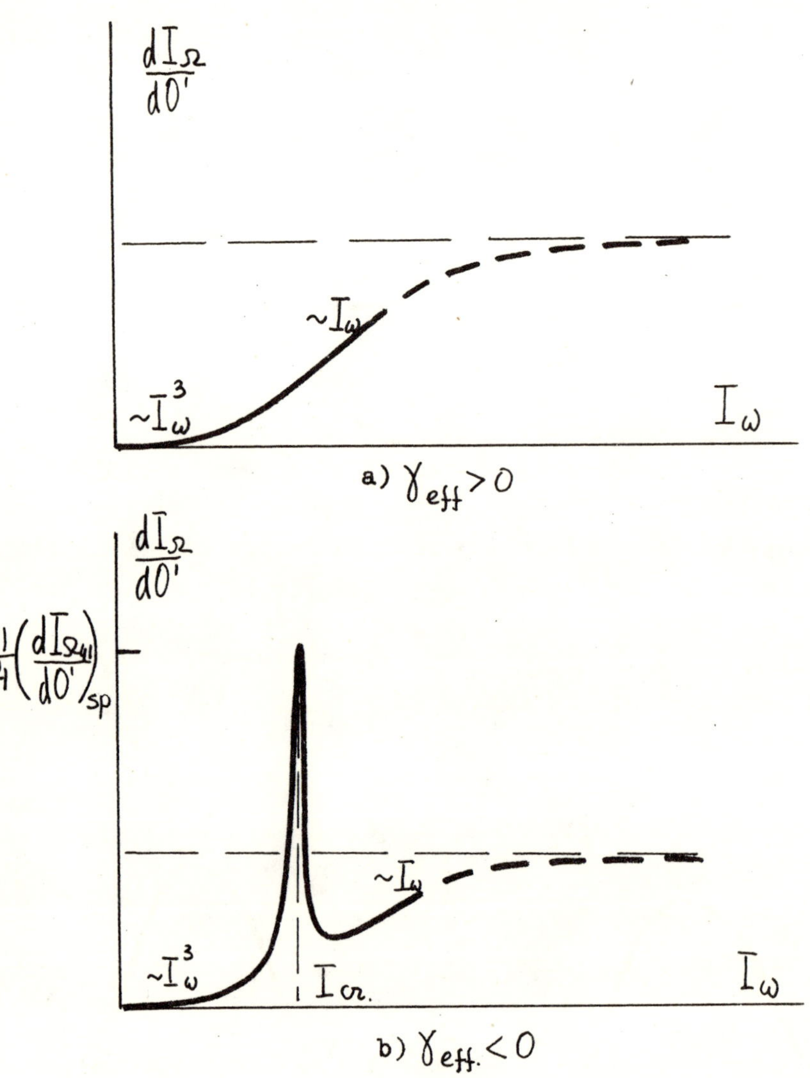


Fig.14.

The third harmonic intensity as a function of the pump intensity at different signs of $\gamma_{eff}$.

coincides with (64').

In the critical point when $1+\gamma_{eff}\simeq 0$, the expression (62) yields

$$|a_1 a_4|^2 \simeq \frac{1}{4}. \tag{65}$$

When intensity still increases we obtain (64'') for $|a_1 a_4|^2$

In Fig.14ab the dependence of the third harmonic intensity upon laser field intensity is shown. Maximum probability achieved at the critical point equals to

$$dW_{\Omega}^{max} = \frac{(3\omega)^3}{8\pi\hbar c^3} |\vec{e}_{\Omega}^{\,\prime *}\vec{d}_{14}|^2 dO'. \tag{66}$$

Hence, at self-induced resonance conditions, the probability of the third harmonic generation in the order of magnitude is comparable to the probability of spontaneous emission from the upper level to the lower. Since at pulse propagation the critical field intensity value is achieved in two points ( on the ascending slope and then on the descending slope), the third harmonic pulse with two sharp spikes will be formed (Fig. 15).

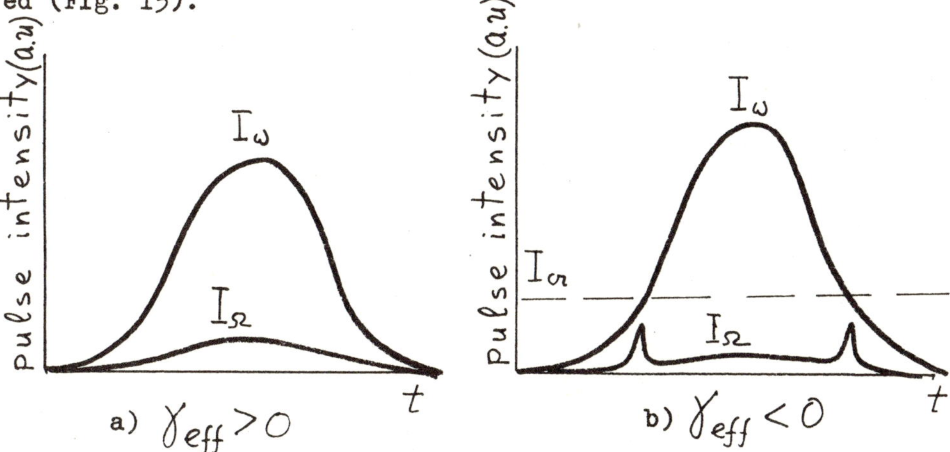


Fig.15. The expected curves for the third harmonic pulses at different signs of $\gamma_{eff}$ are presented.

Such a temporal structure of third harmonic pulse points to the fact of the self-induced adiabatic "passage" of resonance.

Let us consider briefly the cases of very strong fields when all detunings may be neglected. Using the quasi-energy values (46) from (41), (42) it is easy to get the populations $|a_1|^2$ and $|a_4|^2$, which do not depend on the field and are determined only by the matrix elements of the system. In general, however, for the arbitrary matrix element values they are complicated, that is why we consider a simple system which expresses the character of the vibrational levels of the molecules. The system parameters are:

$$d_{21} = d, \quad d_{32} = d\sqrt{2}, \quad d_{43} = d\sqrt{3}, \tag{67}$$

and $E_{21} > E_{32} > E_{43}$,

i.e. the frequency of transition decreases (molecular levels unharmonicity) with the level number increase. For the given system the probabilities of the third harmonic emission depending on the pump frequency are

I. $\omega < E_{41}/3$; $\quad \omega > E_{21}$;

$$dW_\Omega = \frac{3-\sqrt{6}}{48} \cdot |\vec{e}_\Omega^{\,*} \vec{d}_{14}|^2 \frac{\Omega^3}{2\pi\hbar c^3} dO'. \tag{68}$$

II. $E_{41}/3 < \omega < E_{21}$

$$dW_\Omega = \frac{3+\sqrt{6}}{48} \cdot |\vec{e}_\Omega^{\,*} \vec{d}_{14}|^2 \frac{\Omega^3}{2\pi\hbar c^3} dO'. \tag{69}$$

As we see, the probabilities $dW_\Omega^{I}$ and $dW_\Omega^{II}$ differ for about ten times. The same calculations were carried out for the other values of matrix elements. The following table presents the values of $|a_1 a_4|^2$ for some correlations between matrix elements of the transitions.

| $d_{21}/d$ | $d_{32}/d$ | $d_{43}/d$ | $\lvert a_1 a_4\rvert^2 \left(\frac{E_{41}}{3} < \frac{E_{31}}{2} < E_{21}\right)$ | |
|---|---|---|---|---|
| | | | $\omega < \frac{E_{41}}{3};\ \omega > E_{21}$ | $\frac{E_{41}}{3} < \omega < E_{21}$ |
| 1 | 1 | 1 | $\frac{3-\sqrt{5}}{40}$ | $\frac{3+\sqrt{5}}{40}$ |
| 1 | $\sqrt{2}$ | $\sqrt{3}$ | $\frac{3-\sqrt{6}}{48}$ | $\frac{3+\sqrt{6}}{48}$ |
| $\sqrt{3}$ | $\sqrt{2}$ | 1 | | |
| 1 | $\sqrt{2}$ | 1 | $\frac{2-\sqrt{6}}{24}$ | $\frac{2+\sqrt{6}}{24}$ |

It is seen, that when the condition $E_{41}/3 < \omega < E_{21}$ holds, the probability of the third harmonic is greater than in the two other cases. For the given system this result is general, and does not depend on the relations between matrix elements. Fig.16 gives the dependence of the third harmonic intensity on the frequency and intensity of the pumping field. The experimental data obtained in [12] can be possibly explained by the formulas presented in this paper.

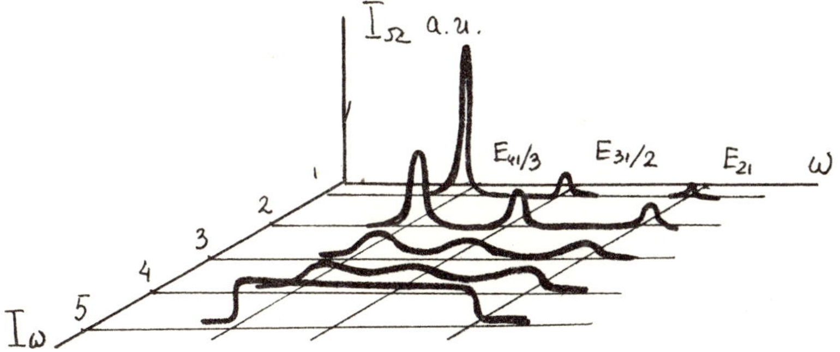


Fig. 16. The dependence of $I_\Omega$ on $I_\omega, \omega$ is demonstrated. In sections (1,2,3,4,5) the frequency dependence at different pumping intensities $I_\omega$ is shown. In sections parallel to the surface $I_\omega O I_\Omega$ the curves correspond to the graphs of Figures 14 a,b.

Now we consider processes which follow the transitions $\Phi_1 \to \Phi_j$ ( $j$ = 2,3,4). We will see later that the processes with the emission and absorption occurs at different frequencies, therefore stimulated emission at frequencies

$$\omega_j = \omega + (\lambda_1 - \lambda_j) \tag{70}$$

$$\Omega_j = 3\omega + (\lambda_1 - \lambda_j) \tag{71}$$

became possible where $j$ = 2,3,4.

The probability of photon emission with the frequencies close to $\omega$ , is determined by the following expression:

$$dW_{\omega_j} = \frac{\omega_j^3}{2\pi\hbar c^3} \left| \left( a_1^{(j)*} a_2^{(1)} \vec{d}_{21} + a_2^{(j)*} a_3^{(1)} \vec{d}_{32} + a_3^{(j)*} a_4^{(1)} \vec{d}_{43} \right) \vec{e}_{\omega_j}^{\,\prime *} \right|^2 (1 + n_{\omega_j}) dO', \tag{72}$$

and the probability of photon emission with the frequency $\Omega_j \simeq 3\omega$ is

$$dW_{\Omega_j} = \frac{\Omega_j^3}{2\pi\hbar c^3} \left| a_1^{(j)*} a_4^{(1)} \right|^2 \left| \vec{e}_{\Omega_j}^{\,\prime *} \vec{d}_{14} \right|^2 (1 + n_{\Omega_j}) dO'. \tag{73}$$

The corresponding processes in the lowest order of the perturbation theory are shown in Figures 17-18.
The frequencies of the absorbed photons in transition $\Phi_1 \to \Phi_j$ are

$$\begin{aligned} \omega_j' &= \omega - (\lambda_1 - \lambda_j) \\ \Omega_j' &= 3\omega - (\lambda_1 - \lambda_j). \end{aligned} \tag{74}$$

The probabilities of absorption processes can be easily received from the expressions (72), (73) if substitution $j \rightleftarrows 1$, $\vec{e}^{\,*} \to \vec{e}^{\,\prime}$ and $1 + n \to n'$ is performed.

As we see, the absorption spectrum does not coincide with the emission spectrum, therefore the reabsorption of generated photons in the given medium is impossible, provided the difference of frequencies $2|\lambda_1 - \lambda_j|$ in accordance with the condition (53) given before is greater than all of the widths.

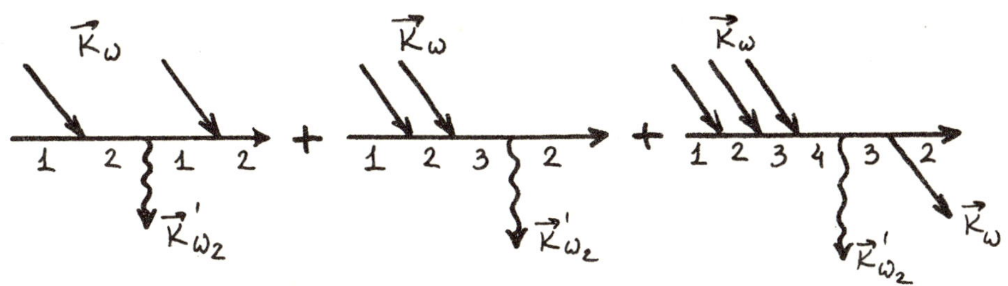


Fig.17a. $\Phi_1 \rightarrow \Phi_2$.

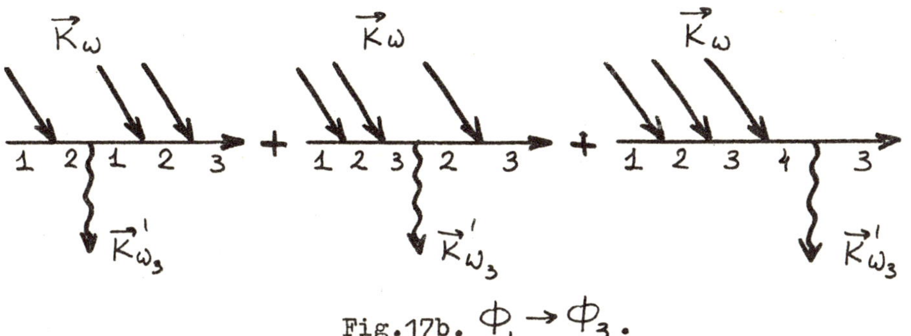


Fig.17b. $\Phi_1 \rightarrow \Phi_3$.

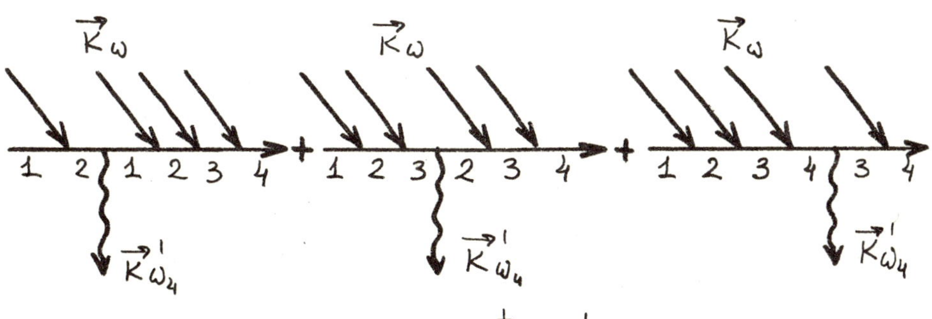


Fig.17c. $\Phi_1 \rightarrow \Phi_4$.

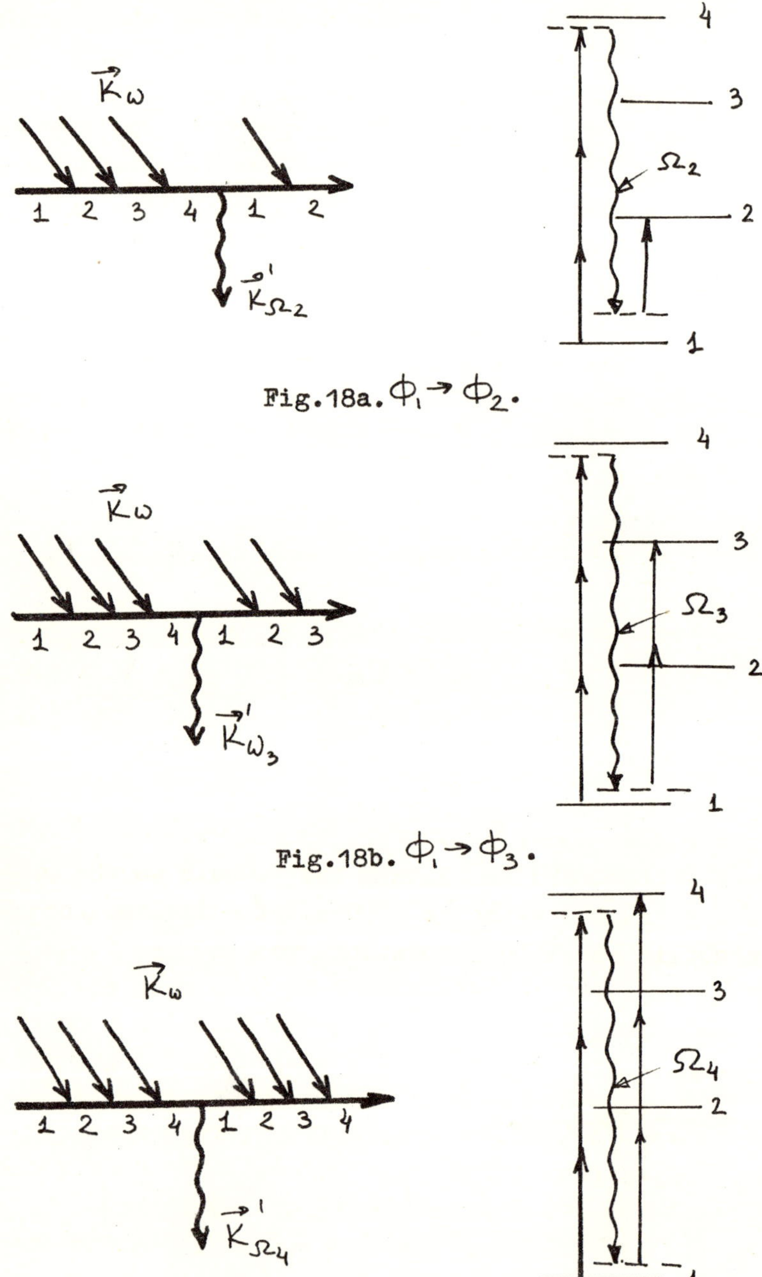


Fig.18a. $\Phi_1 \rightarrow \Phi_2$.

Fig.18b. $\Phi_1 \rightarrow \Phi_3$.

Fig.18c. $\Phi_1 \rightarrow \Phi_4$.

Consider the expression (73) for probability of photon emission with the frequency $\Omega_j \simeq 3\omega$ in detail. The relative intensity of the fine structure lines of the third harmonics $\Omega_j$ ( $j$ = 2,3,4) are determined by the relation of $|a_1^{(j)}|^2$. Since the process of emission is stimulated, only one of the components (with the largest value $|a_1|^2$ ) will be important. This leads to the transitions $\Phi_1 \to \Phi_j$ only through one channel. Thus, in the medium only one of the components will be generated. According to (73), the photon emission probability is proportional to $|a_4^{(1)}|^2$ and $|a_1^{(j)}|^2$ and its maximum value will be determined by the maximum possible populations of the 4-th level in $\Phi_1$ and the 1-st level in $\Phi_j$ . When self-induced three-photon resonance condition holds the probability of the third harmonic component emission is given by the expression (73) in which $|a_4^{(1)}|^2 \simeq 1$ and $|a_1^{(j)}|^2$'s have the form:

$$
\begin{aligned}
j = 2 \quad & |a_1^{(2)}|^2 \simeq \frac{|V_{21}^-|^2}{\mathcal{E}_{21}^2} \\
j = 3 \quad & |a_1^{(3)}|^2 \simeq \frac{|V_{21}^- V_{32}^-|^2}{\mathcal{E}_{21}^2 \mathcal{E}_{31}^2} \qquad (75) \\
j = 4 \quad & |a_1^{(4)}|^2 \simeq 1
\end{aligned}
$$

Thus, under the SIR condition the probability of the seven-photon process (Fig.18b) is maximum and equals to the probability of the usual one-photon process of emission. Accordingly, probabilities of the processes presented in Fig.18a, 18b are increasing. It is evident that in case when all the resonances are strong ($\mathcal{E}_{m+1,m} \to 0$ ) the emission probability of all components of the third harmonic are of the same magnitude and approximately equal to the probability of one-photon processes. For arbitrary values of parameters the quasi-energy values, populations and probabilities of the considered processes can be obtained by numerical methods.

We note that the results obtained may be spread on the arbitrary four-level system interaction with three different fields when each of the fields connects only two levels.

## R E F E R E N C E S


1. I.I.Rabi. Phys.Rev., 51, 632 (1937).
2. M.L.Ter-Mikaelian, A.O.Melikian, Zh.Eksp.Teor.Fiz. 58, 281, (1970).
3. V.I.Ritus. Zh.Eksp.Teor.Fiz. 51, 1544 (1966). Y.B.Zeldovich. Zh.Eksp.Teor.Fiz. 51, 1544 (1966)., Usp.Fiz.Nauk. 110, 139 (1974).
4. G.Cohen-Tannoudji, Lecture Notes École D'été de Physique Théorique de Dargass Paris 1967., A,Kastler. Physics Today 20, 34 (1967).
5. A.M.Bonch-Bruevich, V.A.Khodovoy. Usp.Fiz.Nauk. 93, 71, (1967)., P.Liao, Bjorkholm J.Phys.Rev.Lett. 34, 1, (1975).
6. N.N.Badalian, V.A.Iradian, M.E.Movsessian. Sh.Eksp.Teor. Fiz. Pis'ma 8, 518, (1968).
7. C.R.Stroud. Phys.Rev.A , 3, 1044 (1971)., H.Walther, W.Hart, W.Rasmussen, R.Schieder. Zeit.f.Physik A, 278, 205 (1976)., F.Wu, S.Ezekiel, M.Ducloy, B.Mollow. Phys. Rev.Lett. 38, 1077 (1977)., R.E.Grove, F.Y.Wu, S.Ezekiel. Phys.Rev.A, 15, 227 (1977).
8. D.Grischkowsky. Phys.Rev.A, 7, 2096 (1973)., D.Grischkowsky, E.Courtens, J.Armstrong. Laser Spectroscopy Edited by R.Brewer, A.Mouradian. Plenum Press, New York 1974.
9. L.Brillouin. Wave Propagation and Group Velocity. Academic Press, New York 1965.
10. M.L.Ter-Mikaelian. "Nelineynaya Resonansnaja Optika". Preprint IFI-74-11. Yerevan-1974. D.Grischkowsky and M.M.T.Loy. Phys.Rev.A, 12,1117 (1975). M.L.Ter-Mikaelian, M.A.Sarkissian, Preprint IFI-75-26, Ashtarak-1975., M.A.Sarkissian. Doklady Akademii Nauk Arm.SSR. LXY, N1, 23 (1977).
11. M.M.T.Loy. Phys.Rev.Lett., 32, 814, (1974).
12. H.Kildal and T.F.Deutsch. IEEE J.Quant.Elect., vol.QE - - 12, 429 (1976).
13. M.L.Ter-Mikaelian. IV Rochester Conference on Coherence and Quantum Optics. Rochester University 1977. M.L.Ter-Mikaelian, M.A.Sarkissian. Abstract. "Nonlinear Resonance Conversion of Laser Radiation". Krasnoyarsk-1977.

# INVESTIGATION OF COLLISIONS BY NONLINEAR SPECTROSCOPY METHODS

S.G. Rautian

Institute of Automation and Electrometry,
Siberian Branch, USSR Academy of Sciences,
Novosibirsk, 630090, USSR

## 1. Introduction

A lot of nonlinear spectroscopy effects in gases are presently known, which are closely connected with various collision processes. The interaction of oscillation modes in gas lasers may be given as an example. Under certain circumstances the collisions substantially intensify this interaction which results in sharp changing the oscillation mode distribution of the power up to generation of a single-mode regime. The relaxation processes have influence on the radiation pulsations of molecular absorbing cell lasers, as well. There are numerous nonlinear polarization effects which are very sensitive to deorientating (or depolarizing) collisions. In particular, the radiation polarization state of an isotropic resonator gas laser is predetermined by the relationship between relaxation rates of alignment and orientation. From the viewpoint of interest, gyrotropy, dichroism and birefringence effects induced by a laser field, may be useful. A number of magnetic field effects depending sharply on the collision processes may be noted too: Hanley's effect, radiation polarization and spectrum of a laser placed in magnetic field, etc. Note finally various manifestations of saturation effect.

Not all of the phenomena listed above and those which are not mentioned, are understood so perfectly that it will be possible to speak about their application for collision investigation. The situation with polarization effects and the so called nonlinear resonances is, perhaps, the most favourable. Nonlinear polarization effects have already been described in reviews and monographs (see, e.g., [1-4] and also the papers by M.P. Tschaika[5] and by M.I. Dijakonov and V.I. Perel[6]), and they are not discussed below. As to the collision action on nonlinear resonances, a considerable advance has been made recently, and it is this problem to which my report is devoted to.

With the pressures comparatively small the gas emission and absorption spectra may acquire sharp structure or nonlinear resonances, caused by gas interaction with a rather powerful coherent field. These resonances occur in frequency dependence of the laser radiation power, in various versions of the probe field spectroscopy, in case of magnetic scanning, in spontaneous emission of atoms immersed in the external field, etc.

The shape and amplitude of a sharp spectral structure depend on relaxation processes taking place in gas. In this respect the nonlinear resonances are analogous to "usual" spectral lines. As opposed to the latter, however, for the nonlinear resonances not only the dipole moment relaxation is significant, but also the atom (molecule) level, velocity and orientation distribution. The matter is that the coherent external field provides substantially non equilibrium atom distributions, collisions lead to the elimination of the nonequilibrium and thereby result in nonlinear resonances. Thus the analysis of nonlinear resonance contours allows to investigate cross sections and differential sections of elastic, inelastic and deorientating processes. To explain the abovesaid consider atom velocity distribution for the levels $m, n$ interacting in resonance with a monochromatic plane wave. Distract our attention of the velocity change in elastic processes. The nonequilibrium part of the distribution $\Delta N_j(\vec{v})(j=m,n)$ is then proportional to

$$\Delta N_j(\vec{v}) \backsim \frac{CW(\vec{v})}{\Gamma^2+(\omega-\omega_{mn}-\vec{k}\vec{v})^2}; \quad W(\vec{v})=(\sqrt{\pi}\,\bar{v})^{-3}e^{-\vec{v}^2/\bar{v}^2}, \quad (1.1)$$

where $\Gamma$, $\omega_{mn}$ are the natural line width and the Bohr frequency for the $m-n$ transition; $\vec{k}, \omega$ are the wave vector and the field frequency; the coefficient $C$ is proportional to the wave intensity. Distribution (1.1), as a function of the velocity $(\vec{v})$ projection on the wave vector, has the half-width $\Gamma/k$ and the ratio between it and the mean thermal velocity $\bar{v}$ may amount to the value of about $10^{-1}$ - $10^{-5}$. Formula (1.) describes thus a disk-shaped atomic beam with equilibrium distribution for the velocity projections being perpendicular to the wave vector, and collimated for the projection of $\vec{v}$ onto $\vec{k}$ (Fig. 1). The collimation degree is equal to $\Gamma/k\bar{v}$.

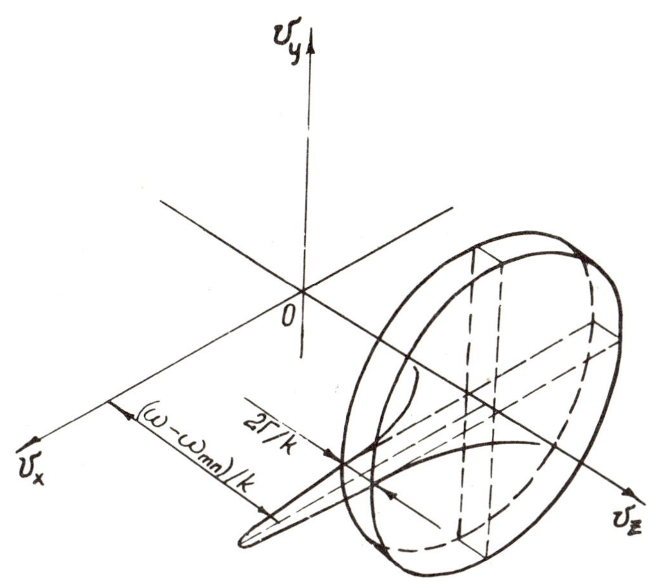


Fig. 1. A disk-shaped atomic beam equivalent to non-equilibrium component of the velocity distribution.

Let us take into account the velocity change in collisions. The smallest deviation angle caused by the diffraction effects is equal in the order of magnitude to the relation between the de Broglie wavelength $\lambda\!\!\!^{-}$ and interaction radius $\rho_W$ :

$$\vartheta_{dif} \sim \lambda\!\!\!^{-}/\rho_W = \hbar/(m\bar{v}\rho_W) = 4{,}9\cdot 10^{-8}/[\rho_W\sqrt{MT}], \qquad (1.2)$$

where $M,T$ are molecular weight and absolute temperature. If $M = 10$, $T$ = 300°K, $\rho_W = 10^{-7}$ cm, then $\vartheta_{dif} \approx 10^{-2}$. Thus at pressures small enough and for small radiation width transitions the excited atoms may have velocity collimation sufficiently good for studying the most fine scattering effects.

It should be particularly noted that the experiments with nonlinear resonances are performed with both ground and excited states. The latter fact is very important as there are well-known difficulties in applying the atomic beam method to excited states and the immediate experimental data on the interaction potentials and scattering indicatrices for excited states are rather limited.

## 2. Collision Integrals and Scattering Amplitudes

Adequate description of nonlinear resonance contours is provided by a quantum kinetic equation with a Boltzmann collision integral as an impact approximation is applicable for low pressures.

A number of works [7-13] have appeared recently, where the relationship between the collision integral and scattering amplitudes characterizing the result of binary collision of radiating and perturbing particles, is formulated both in general and rigorous manner. A reliable theoretical basis is thereby created for the nonlinear resonance investigation.

The quantum kinetic equation for the density matrix $\rho$ may be written as follows

$$i\hbar \frac{\partial}{\partial t}\rho = \left[H + V(t), \rho\right] + i\hbar R + i\hbar S. \tag{2.1}$$

Here $H$ is the Hamiltonian of the radiating particle $a$ ; $V(t)$- its interaction with the external field. The terms $R$ and $S$ describe spontaneous radiation processes and collisions. The collision integral $S$, which is of special interest for us, is presented by the operator expression

$$i\hbar S = Tr_b\left\{T(\rho \times \rho_b)\Omega^{+} - \Omega(\rho \times \rho_b)T^{+}\right\}, \tag{2.2}$$

where the trace is calculated over the perturbing particle $b$ variables ( $\rho_b$ is its density matrix). The Møller operator $\Omega$ is determined as follows

$$\Omega = \lim_{t' \to -\infty} \Omega(t, t'); \tag{2.3}$$

$$i\hbar \frac{\partial}{\partial t}\Omega(t,t') = \left[H + H_b + V(t) + V_b(t), \Omega(t,t')\right] + W_{ab}\Omega(t,t') + i\hbar\delta(t-t'), \tag{2.4}$$

where $H_b$ is the Hamiltonian of the particle $b$ ; $V_b(t)$ and $W_{ab}$ denote its interactions with the external field and the particle $a$ . Finally, the matrix $T$ is defined by the ratio

$$T = W_{ab}\,\Omega \,. \tag{2.5}$$

Expression (2.2) for the collision integral is the most general one. It comprises any processes taking place during collisions and described within the framework of the impact approximation: elastic and inelastic processes, deorientation, phase shift of an atomic oscillator, excitation, rotational relaxation, etc. Possible influence of the external field on the scattering amplitudes (terms $V(t)$ , $V_b(t)$ in eq. (2.4) ), polarization and nonequilibrium distribution of perturbing particles are taken into account, as well. The expression for $S$ is rather complicated due to a large number of phenomena, and one or another simplifying assumption is adopted when solving particular problems. An assumption of spatial homogeneity of the problem and perturbance isotropy is most generally used. For such conditions the matrix element $S_{mn}(\vec{v},\vec{r},t)$ of the collision integral is of the form

$$S_{mn}(\vec{v},\vec{r},t) = -\nu_{mn}\,\rho_{mn}(\vec{v},\vec{r},t) + \sum_{m_1 n_1} \int d\vec{v}_1\, A(mn\vec{v}\,|\,m_1 n_1 \vec{v}_1)\,\rho_{m_1 n_1}(\vec{v}_1,\vec{r},t), \tag{2.6}$$

where $m, n, \ldots$ are quantum number sets, determining internal states of the particle $a$ . The departure rate $\nu_{mn}$ and the kernel $A(mn\vec{v}\,|\,m_1 n_1 \vec{v}_1)$ can be expressed in terms of a scattering amplitude $f$ :

$$\nu_{mn} = \langle f(m\beta\vec{u})m\beta\vec{u} - f^*(n\beta\vec{u}\,|\,n\beta\vec{u})\rangle ; \tag{2.7}$$

$$A(mn\vec{v}\,|\,m_1 n_1 \vec{v}_1) = \langle\!\langle f(m\beta\vec{u}\,|\,m_1\beta_1\vec{u}_1) f^*(n\beta\vec{u}\,|\,n_1\beta_1\vec{u}_1)\rangle\!\rangle . \tag{2.8}$$

In this equation brackets denote averaging of the values enclosed over the relative velocities $\vec{u}, \vec{u}_1$ and internal states of the particles $b$ (quantum numbers $\beta, \beta_1$ ); in this case the averaging is performed with the weights

$$(2\pi\hbar / i\mu)\rho_b(\beta, \vec{v}-\vec{u}); \quad \mu = m m_b/(m+m_b); \tag{2.9}$$

$$2\delta(u^2-u_1^2+\frac{2}{\mu}\Delta E)\delta\left[\vec{v}-\vec{v}_1-\frac{\mu}{m}(\vec{u}-\vec{u}_1)\right]\rho_b(\beta_1,\vec{v}_1-\vec{u}_1) \qquad (2.10)$$

where $m$ and $m_b$ are the masses of the particles $a$ and $b$ ; $\Delta E = E_m + E_\beta - E_{m1} - E_{\beta 1}$.

The great majority of theoretical and experimental works is devoted to finding integral characteristics of the departure rate $\nu_{mn}$, and the arrival rate $\tilde{\nu}(mn/m_1n_1)$

$$\tilde{\nu}(mn|m_1n_1)=\int A(mn\vec{v}'/m_1n_1\vec{v})d\vec{v}' \qquad (2.11)$$

or the differences $\nu_{mn}-\tilde{\nu}(mn|mn)$ . Characteristics of this kind are significant when analyzing spectral line broadening, transport phenomena in gases, excitation processes, fluorescence depolarization and some other problems (see, e.g. [3,14-17] ). The values listed are defined by effective cross-sections of corresponding processes and are strongly averaged characteristics of the interaction potentials. Here we are especially interested in the most differential characteristics of the collision integral, that is in the kernels as the functions of $\vec{v}$ and $\vec{v}_1$. The properties of kernels are rather little investigated and we will dwell on them.

When analyzing potential possibilities of finding the scattering indicatrix from the data on nonlinear resonances the primary is the question to what extent the features of differential cross-sections are masked by averaging over perturbing particle states and particularly over their velocities.

Consider a rather simple case of elastic scattering in a central-symmetrical potential during collisions with the structureless particles $b$ . It is convenient to take a module of the relative velocity $u$ and that of the relative velocities difference $|\vec{u}_1-u|$ before and after the collision as the differential cross-section arguments

$$\sigma_{mn}(u\;|\vec{u}-\vec{u}_1|)\equiv f(m\vec{u}/m\vec{u}_1)f^*(n\vec{u}/n\vec{u}_1).$$

According to the momentum conservation law, $m(\vec{v}-\vec{v}_1)=$ $=\mu(\vec{u}-\vec{u}_1)$. The angular dependence of the differential cross-section we are interested in does not undergo averaging

$$A_{mn}(\vec{v}/\vec{v}_1)\equiv A(mn\vec{v}/mn\vec{v}_1)=\langle\langle\sigma(u,\frac{m}{\mu}|\vec{v}-\vec{v}_1|)\rangle\rangle . \qquad (2.12)$$

Suppose that the differential cross-section depends only on $|\vec{u}-\vec{u}_1|$ (e.g. Born approximation). Then from (2.12) it follows [18] that

$$A_{mn}(\vec{v}/\vec{v}_1)=N_b v_b 4\pi\sigma\left(\frac{m}{\mu}|\vec{v}-\vec{v}_b|\right)C(\vec{v}/\vec{v}_1), \qquad (2.13)$$

where $N_b$ is the density of the particles $b$. Thus in the conditions given the differential cross-section is singled out as an independent factor. The factor $C(\vec{v}/\vec{v}_1)$ results from the integration of expression (2.10) over $\vec{u}$, $\vec{u}_1$ and is presented by the formula

$$C(\vec{v}/\vec{v}_1)=\frac{\Delta v}{|\vec{v}-\vec{v}_1|}(\sqrt{\pi}\,\Delta v)^{-3}\exp\left\{-\left[\frac{\vec{n}(\vec{v}-\gamma\vec{v}_1)}{\Delta v}\right]^2\right\}, \qquad (2.14)$$

$$\Delta v=\frac{2\mu}{m}\bar{v}_b;\ \gamma=1-\frac{2\mu}{m};\ \vec{n}=\frac{\vec{v}-\vec{v}_1}{|\vec{v}-\vec{v}_1|};\ \bar{v}_b^2=\frac{2k_B T_b}{m_b}. \qquad (2.15)$$

It is seen from (2.14) that the value $\Delta v$ is a characteristic scale for $C(\vec{v}/\vec{v}_1)$. The ratio between this value and mean thermal velocity $\bar{v}=\sqrt{2k_B T/m}$ of the particle $a$ is equal to

$$\Delta v/\bar{v}=2\sqrt{T_b/T}\left[\sqrt{m/m_b}+\sqrt{m_b/m}\right]^{-1}, \qquad (2.16)$$

i.e. $\Delta v/\bar{v}$ depends on $\sqrt{m/m_b}$ and $\sqrt{T/T_b}$. If we distract our attention of collisions with the electrons ($\sqrt{m_b/m}\sim 10^{-2}$), then $\Delta v\sim\bar{v}$.

As it is known, the real differential cross section contains an almost isotropic component and a sharply directed selective one. A considerable contribution is made to the latter by difraction effects characterized by the width which is substantially less than $\bar{v}$:

$$\delta v_{dif}\sim \vartheta_{dif}\,\bar{v}\sim 10^{-2}\,\bar{v}\ll \Delta v\sim\bar{v}. \qquad (2.17)$$

That is why the behaviour of the selective part of the kernel is given mainly by the angular dependence of the differential cross-section, and averaging over the perturbing particle velocities does not play a significant role. The conclusion drawn relates to diagonal $(m=n)$ and nondiagonal $(m\neq n)$ kernels, as well as to more complicated cases when the differential cross-section, being a sharp function of $|\vec{u}-\vec{u}_1|$, depends also on $u$. This dependence will somehow change the factor analogue $C(\vec{v}/\vec{v}_1)$ in (2.13), but its scale will coincide with $\Delta v$ as before.

To analyze of the kernel isotropic part it is possible not to take into account the differential cross-section dependence on $|\vec{u}-\vec{u}_1|$, i.e. this part of the kernel is defined by statistical factor $C(\vec{v}|\vec{v}_1)$ and its analogues, when $\sigma_{mn}(u)$ depends on $u$. In consequence the isotropic part of the kernel is of the width approximately equal to $\Delta v$.

In the majority of nonlinear spectroscopy problems the external field takes the form of a plane wave and it causes nonequilibrium of the distribution only for the velocity $\vec{v}$ projection onto the wave vector $\vec{\kappa}$. The velocity distribution over orthogonal components $\vec{v}_\perp$ is believed to be a Maxwell one. Owing to the reason mentioned above the nonlinear resonance contours are defined by one-dimensional kernels of the form

$$A(mnv|m_1n_1v') = \int A(mn\vec{v}|m_1n_1\vec{v}')W(\vec{v}_\perp')d\vec{v}_\perp d\vec{v}_\perp' ; \qquad (2.18)$$

$$W(\vec{v}_\perp) = (\sqrt{\pi}\,\bar{v})^{-2}\exp|-\vec{v}_\perp^{\,2}/\bar{v}^2| ; \quad \bar{v}^2 = 2\kappa_B T/m. \qquad (2.19)$$

A change-over to one-dimensional kernels means, obviously, averaging over $\vec{v}_\perp$ in a "disk-shaped" atomic beam" shown in Fig. 1. Let the characteristic width $\delta u$ of the differential cross-section selective part, as a function $|\vec{u}-\vec{u}_1|$, be considerably less than $\bar{u} \equiv \sqrt{\bar{v}^2+\bar{v}_b^2}$. If, besides, the conditions $\bar{v} \ll \bar{v}_b$ or $\bar{v} \gg \bar{v}_b$ (light and heavy perturbing particles $b$) are fulfilled then from formula (2.18) it follows [18]

$$A_{mn}(v|v_1) = A_0\frac{2}{\bar{u}^2}\int_0^\infty e^{-u^2/\bar{u}^2}u\,du\int_{m|v-v_1|/\mu}^\infty \sigma_{mn}(u,\eta)d\eta ; \quad A_0 = \frac{2\sqrt{\pi}m}{\mu\bar{u}}N_b ; \qquad (2.20)$$

$$A_{mn}(v|v_1) = A_0\frac{2}{\sqrt{\pi}\,\bar{u}}\int_0^\infty e^{-u^2/\bar{u}^2}du\int_0^\infty \sigma_{mn}(u,\sqrt{\eta^2+m^2(v-v_1)^2/\mu^2})\,d\eta . \qquad (2.21)$$

Thus, a certain averaging of the differential cross-sections is performed for selective components of one-dimensional kernels. It can be seen, however, that a characteristic width of kernels defined by eqs. (2,20), (2.21) is equal in the order of its magnitude to $\delta v = \mu\delta u/m$, i.e. it is given by the differential cross-section properties.

A qualitative presentation of the isotropic part of one-dimensional kernel gives an example of a classical hard sphere scattering, i.e. $\sigma_{mn}(u,|\vec{u}-\vec{u}_1|) = \sigma = \mathrm{const}$ [18]:

$$A_{mn}(v/v_1)=\frac{A_1}{2}\left\{1-\Phi\left(\frac{|v-v_1|}{\Delta v}+\frac{v_1}{v_b}\frac{v-v_1}{|v-v_1|}\right)+\right.$$

$$\left.+\exp\left[-x\left(x-2\frac{v_1}{\bar{v}_b}\right)\right]\left[1-\Phi\left(\alpha\frac{|v-v_1|}{\Delta v}-\frac{v_1}{v_b}\frac{v-v_1}{|v-v_1|}\right)\right]\right\}; \qquad (2.22)$$

$$A_1=N_b\frac{\bar{v}_b}{\Delta v}4\pi\sigma;\ x=2\frac{\bar{v}\bar{v}_b}{\bar{u}^2}\frac{v-v_1}{\Delta v};\ \alpha=\frac{\bar{v}_b^2-\bar{v}^2}{\bar{u}^2};\ \Delta v=\frac{2\mu}{m}\bar{v}_b, \qquad (2.23)$$

where $\Phi(z)$ is a probability integral. The calculation shows that when $T=T_b$ and $0<m/m_b<5$ the width of the kernel (2.22) differs from not more than by 1.8 times. Hence the isotropic part of the kernel is described well by a "strong collision" model

$$A_{mn}(v/v_1)=\tilde{\nu}_{mn}W(v), \qquad (2.24)$$

which is widely used when solving specific problems of linear and nonlinear spectroscopy. In this case all information on scattering resides in the value of the arrival rate $\nu_{mn}$.

The basic regularities noted above and considered in more detail in [18,19] are well illustrated by the model problem about hard sphere scattering. Fig. 2 gives diagrams of a

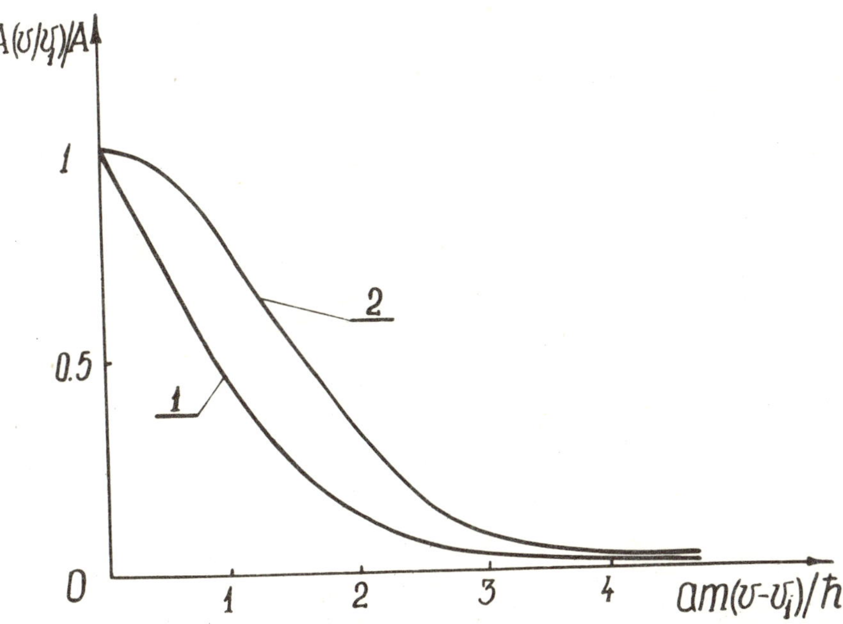


Fig. 2. Plot of one-dimensional kernel for scattering at hard sphere. 1 - $m \gg m_b$ ; 2 - $m \ll m_b$ .

selective part of an appropriate kernel for $m \gg m_b$ and $m \ll m_b$ (curves 1 and 2); their half-widths (at the half-height are equal to 0.910 $\hbar/ma$ and 1.573 $\hbar/ma$, where $a$ is a sphere radius. The isotropic part of one-dimensional kernel, as it is said above, is described by expression (2.22).

Besides hard sphere scattering, known are the results of kernel calculations for the resonance excitation exchange resulting from the dipole-dipole interaction [20,21] for resonance radiation trapping [22,24], for the Lennard-Jones potential in the eikonal approximation [18] and for inelastic atom scattering in the Born-Bethe approximation [25] (see also section 5). Due to insufficient investigation of collision integrals the empirical kernel approximations are used almost in all publications, except works [22-26] where the collision integral parameters are calculated on the basis of microscopic concepts.

We will not dwell on a wide variety of phenomena associated with the collision influence on nonlinear resonance characteristics and will consider two problems only: nonlinear pressure dependence for the resonance widths in the case of elastic selective scattering and contours of the nonlinear resonances caused by inelastic and deorientating collisions.

## 3. Nonlinear resonance contour in the case of elastic selective scattering

As it is known, in the absence of velocity changes in collisions (the relaxation constant model) the nonlinear resonances are described by one or several Lorentzians with widths proportional to the density of perturbing particles. This fundamental result borrowed from linear spectroscopy served as the only basis for interpreting experimental data on the nonlinear resonance broadening. However, as early as 1966, it was shown theoretically that selective scattering affects the resonance shapes, makes them assymetrical and conditions nonlinear dependence of their widths and shifts [27]. This circumstance manifests itself in two ways: firstly, in the velocity distribution changes, as it has been stated above, and, secondly., in the velocity change introducing a frequency modulation which with the collision rates high enough, is eliminated and transforms the decay law of a dipole moment and the form of corresponding correlation function.

Various aspects of this problem were discussed in Refs. [26, 28-33]. The phenomenon was discovered and investigated for rotational-vibrational molecule transitions [34-40]. As an example Fig. 3 shows the values of a nonlinear resonance half-width for the line $P(20)$ of the transition 00°1-10°0 of a molecule $CO_2$ according to the measurements [38]. Curve 2 represents the result of a theoretical calculation described below.

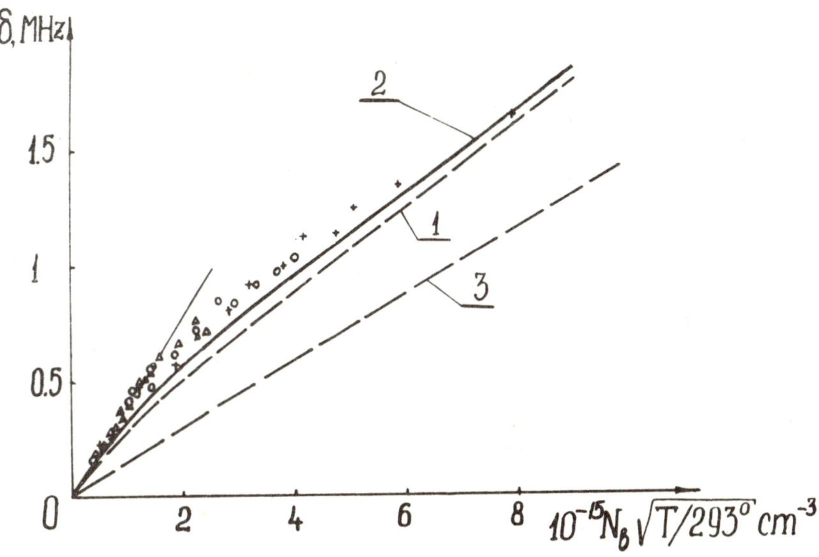


Fig. 3. The resonance half-width dependence on perturbing particle concentration. The points are experimental data [38]; 1 is a half-width of the function $B(\Omega)$; 2 is a half-width of the function $I(\Omega)$; 3 is a half-width $\Gamma$.

Distracting our attention of the level degeneracy and using the isotropic perturbance model we obtain a kinetic equation with a subtractive one-dimensional kernel

$$\left(\frac{\partial}{\partial t}+v\frac{\partial}{\partial z}+\gamma_j+\gamma_l+\nu_{jl}\right)\rho_{jl}(v,z,t)=-i\left[V(z,t),\rho(v,z,t)\right]_{jl}+$$
$$+\int A_{jl}(v-v_1)\rho_{jl}(v_1,z,t)dv_1+\delta_{jl}q_j(v);\quad j,l=m,n. \qquad (3.1)$$

Here $2\gamma_{j,l}$ denotes the rate of spntaneous decay of the level $j,l=m,n$; $V(z,t)$ is an electromagnetic field interaction Hamiltonian; the term $\delta_{jl}\, q_j(v)$ describes excitation of combinating levels.

Due to the kernel dependence on the difference $v-v_1$ the solution of system of equations (3.1) can be expressed in quadratures; here interaction with the field is taken into account by the method of successive approximations. If the field is a plane monochromatic wave, the nonlinear resonance contour (e.g. the Lamb dip) is described by the function[26,32]

$$I(\Omega)=\tau_1 B(\Omega)+\int_{-\infty}^{\infty} F(v)B(\Omega-\kappa v/2)\,dv\,; \qquad (3.2)$$

$$\Omega=\omega-\omega_{mn}\,; \quad \tau_1=\tau_{1m}+\tau_{1n}\,; \quad \tau_{1j}=(2\gamma_j+\nu_{jj})^{-1}. \qquad (3.3)$$

The factor $F(v)$ is related to stationary velocity distributions on the levels $m,n$ with monokinetic excitation, i.e. to Green functions of diagonal equations (3.1):

$$F(v)=\sum_j F_j(v)\,; \quad F_j(v)=\frac{\kappa}{2\pi}\tau_{1j}^2\int_{-\infty}^{\infty} A_{jj}(\tau)\left[1-\tau_{1j}A_{jj}(\tau)\right]^{-1} e^{-i\kappa v\tau}\,d\tau; \qquad (3.4)$$

$$A_{jj}(\tau)=\int_{-\infty}^{\infty} A_{jj}(\Delta v)\exp(i\kappa\Delta v\tau)\,d(\Delta v). \qquad (3.5)$$

The function $B(\Omega)$ is proportional to the Fourier transformation of $\Phi^2(\tau)$ where $\Phi(\tau)$ is the correlation function of the dipole moment

$$B(\Omega)=\mathrm{Re}\int_0^{\infty}\Phi^2(\tau)e^{2i\Omega\tau}\,d\tau\,; \qquad (3.6)$$

$$\Phi(\tau)=\exp\left\{-(\Gamma+i\Delta)\tau-\tilde{\nu}_{mn}\int_0^{\tau}\langle 1-e^{i\kappa\Delta v\tau'}\rangle_{\Delta v}\,d\tau'\,; \qquad (3.7)$$

$$\Gamma+i\Delta=\gamma_m+\gamma_n+\nu_{mn}-\tilde{\nu}_{mn}\,; \quad \tilde{\nu}_{mn}=\int_{-\infty}^{\infty} A_{mn}(\Delta v)\,d(\Delta v). \qquad (3.8)$$

The angular brackets in (3.7) denote averaging over the velocity jumps $\Delta v$ with the weight $A_{mn}(\Delta v)/\tilde{\nu}_{mn}$

The term $\tau_1 B(\Omega)$ in formula (3.2) is determined only by the dipole moment decay law depending also on elastic scattering which, in time terms, causes frequency modulation: the value $\kappa\Delta v\tau'$ (in the integral term of expression (3.7) ) is a phase change during the time $\tau'$ caused by the Doppler frequency alteration; averaging of the frequency jumps is carried out in accordance with the kernel $A_{mn}(\Delta v)$ depending on scattering properties in both states $m,n$ .

The integral term in (3.2) is a convolution of the atom

line contour with a given velocity and of the velocity distribution $F(v)$ . That is why characteristic features of the contour $I(\Omega)$ of nonlinear resonance depend on the relationship between widths of the functions $B(\Omega)$ and $F_j(v)$ . Let $S$ denote a half-width of one-dimensional kernel $A_{mn}(\Delta v)$ . It can be shown (see, e.g.[41]) that the half-width of the function $F_j(v)$ will be equal to

$$S_j\sqrt{1+n_j}\,; \quad n_j \equiv \tilde{\nu}_{jj}/(2\gamma_j+\nu_{jj}-\tilde{\nu}_{jj}). \tag{3.9}$$

The difference $\nu_{jj}-\tilde{\nu}_{jj}$ is obviously equal to the rate of quenching the state $j$ due to inelastic processes, and therefore the value $n_j$ characterizes an efficient number of elastic collisions occuring during the life time $1/(2\gamma_j+\nu_{jj}-\tilde{\nu}_{jj})$ of an atom on the level $j=m,n$ . For a model exponential kernel, in particular, we have

$$A_{jj}(\Delta v)=\frac{\tilde{\nu}_{jj}}{2S_j}\exp\left[-\frac{|\Delta v|}{S_j}\right];\; F_j(v)=\frac{n_j}{2S_j\sqrt{1+n_j}}\exp\left[-\frac{|v|}{S_j\sqrt{1+n_j}}\right] \tag{3.10}$$

As to $B(\Omega)$ assume for simplicity that the collisions are accompanied by such a strong phase modulation that the condition $|\mathrm{Re}\,\tilde{\nu}_{mn}| \ll \mathrm{Re}\,\nu_{mn}$ is fulfilled and thus the nondiagonal collision integral kernel $A_{mn}$ may be neglected. At the same time quenching can be small $(\nu_{jj}-\tilde{\nu}_{jj} \ll \nu_{jj})$ Such conditions are realized, e.g. in case of sharply different adiabatic perturbances of the levels $j=m,n$ , which is characteristic for atoms. From (3.6), (3.7) it follows then

$$B(\Omega)=1/2\,\mathrm{Re}\left[\gamma_m+\gamma_n+\nu_{mn}-i\,\Omega\right]^{-1}, \tag{3.11}$$

i.e. $B(\Omega)$ is of a Lorentz shape as it should be in the absence of frequency modulation.

Let the radiation relaxation constants $\gamma_m$, $\gamma_n$ be small as compared to $kS_j\sqrt{1+n_j}$. Then within the range of the smallest pressures the contour of the line $B(\Omega)$ of atoms with a given velocity is considerably narrower that the velocity distribution $F(v)$ , and the nonlinear resonance has the structure shown in Fig. 4a, where the solid contour corresponds to $B(\Omega)$ and the dotted one - to the convolution. With the growth of pressure and frequency $\nu_{mn}$ a relative contribution of a wider component grows, and the width of a sharper one increases (Fig. 4b). Note that $\nu_{mn}$ increases more rapidly than $kS_j\sqrt{1+n_j}$ with the increase in pressure ( $n_j$ does not practically depend on the pressure when $\nu_{jj}-\tilde{\nu}_{jj} >> 2\gamma_j$ ). Thus when the pressures high enough the widths of two resonance components become comparable ( $Re\,\nu_{mn} \approx \kappa S_j \sqrt{1+n_j}$ , Fig. 4c).

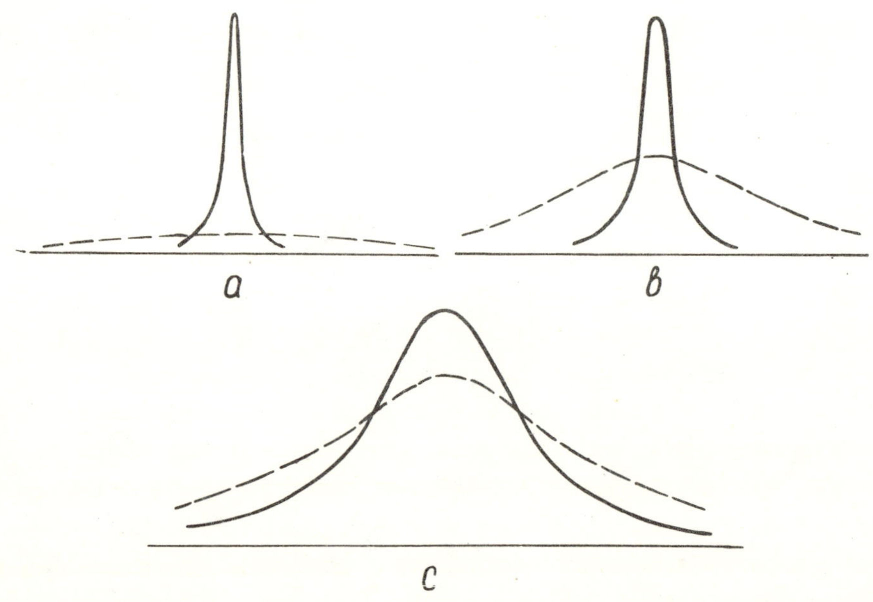


Fig. 4. Nonlinear resonance structure during elastic scattering.

Let the rates of departure $\nu_{mn}$ and arrival $\tilde{\nu}_{mn}$ be comparable now (the phase modulation and quenching are small). As it is known, this is the situation typical for vibrational-rotational transitions (see, e.g.[42]). In these conditions the general picture of resonance contour deformation is conveyed by Fig.4, but the fact that the form of the function $B(\Omega)$ and the rate of its width growth are changed should be taken into account. When pressures are small the frequency deviation $\kappa S$ exceeds the sum $\gamma_m+\gamma_n+Re\,\nu_{mn}$; it is possible to neglect the exponent $exp[i\kappa\Delta v\tau']$ in formula (3.7), and for $B(\Omega)$ expression (3.11) is true as above. With high pressures when the line width exceeds the frequency deviation $\kappa s \sqrt{Re\,\tilde{\nu}_{mn}/\Gamma}$ occuring during the correlation time $1/\Gamma$ , it is possible to neglect the whole integral term in formular (3.7) and

$$B(\Omega)=1/2\,Re\left[\Gamma-i\,(\Omega-\Delta)\right]^{-1};\quad \Gamma>>\kappa s\sqrt{Re\,\tilde{\nu}_{mn}/\Gamma} \qquad (3.12)$$

Thus the countour width for molecules with the given velocity $B(\Omega)$ varies between $\gamma_m+\gamma_n+\mathrm{Re}\,\nu_{mn}$ and $\gamma_m+\gamma_n+\mathrm{Re}\,(\nu_{mn}-\tilde{\nu}_{mn})$ as a result of frequency modulation suppression, i.e. a peculiar manifestation of the Dicke effect [43,44]. Let us stress that this transformation is carried out under the conditions meeting the equality

$$\Gamma=\gamma_m+\gamma_n+\mathrm{Re}\,(\nu_{mn}-\tilde{\nu}_{mn})\approx\kappa s\sqrt{n},\quad n\equiv\mathrm{Re}\,\nu_{mn}/\Gamma, \qquad (3.13)$$

i.e. under much the same conditions when two resonance components have very similar width (see Fig. 4c).

Thus each of the two factors considered, that is the frequency modulation suppression (Dicke effect) and the change of molecule velocity distribution, determines a nonlinear dependence of the resonance width on the perturbing particle concentration much to the same extent. As experiment [35] and calculations[27,32] show the nonlinear pressure dependence is also characteristic of the line shift due to the same reasons, and the shift, as a rule, depending on the pressure is weaker with its small values. The last mentioned circumstance is of great importance for the problem of gas laser frequency stabilization [27,35]

The results of calculations of the resonance half-width made in accordance with the considerations presented for the Lennard-Jones potential [26], are given in Fig. 3. Dotted curve 1 corresponds to the half-width of the function $B(\Omega)$, solid curve 2 corresponds to the half-width of the total resonance $I(\Omega)$. Straight line 3 conforms to the half-width of the line with very high pressures $(\Gamma\gg\kappa s\sqrt{n})$. From Fig. 3 one can see the degree of agreement between experimental and theoretical data and the relative role of various effects. It should be noted that the departure rate $\nu_{mn}$ and the kernel width $s$ depend on the same parameter, that is on the Weisskopf radius $\rho_W$:

$$\mathrm{Re}\,\nu_{mn}\sim 2\pi\rho_W^2 N_b\bar{v}\,;\qquad s\sim\hbar/m\rho_W\,. \qquad (3.14)$$

That is why only the tangent of inclination at origin of coordinates serves as a fitting parameter of a theoretical curve in Fig. 3. In the light of all said the agreement between experimental and calculation data may be considered to be satisfactory. Thus a conclusive evidence was obtained that by nonlinear resonances it is really possible to investigate such delicate effects as diffraction ones in scattering.

The question surely arises just what interaction potential or differential cross-section characteristics can be obtained from the experiments of this kind. In ideal conditions we could speak about calculation of the kernel and differential cross-section from the nonlinear resonance contour and from its width and shift dependence on the pressure. Because of the nonzero measuring errors and due to a number of other reasons, however, more realistic is to define only the Weisskopf radius and effective width of an elastic scattering differential cross-section.

Within the scope of the eikonal method,being applicable under gas-kinetic conditions,the phase jump $\varphi(\rho)$ caused by the interaction $W(r)$ in the order of magnitude is equal to

$$\varphi(\rho) \sim W(\rho)\rho/\hbar u. \tag{3.15}$$

For the Weisskopf radius defined by the condition $\varphi(\rho_W)=1$, the interaction energy accounts for a very small value:

$$W(\rho_W)=\hbar u/\rho_W \sim 10^{-16} CGS \sim 1\, cm^{-1}. \tag{3.16}$$

That is why the diffraction component of scattering determined by the impact parameters exceeding the Weisskopf radius is sensitive to the potential features in very periphere of the interaction region. In contrast, the the very pekinetic gas theory (diffusion, viscosity, etc.) give reliable information for the impact parameters where the interaction is relatively strong, and the judgement about the periphere is based on extrapolation considerations. Thus, the nonlinear resonance investigation at low pressures provides a reliable method for measuring scattering potential at long distances. To study internal parts of the interaction region by methods of nonlinear spectroscopy of resonances, our attention should be directed to inelastic and deorientating processes.

## 4. Nonlinear Resonances Induced by Collisions

A differential cross-section of inelastic atom scattering may be investigated by nonlinear spectroscopy methods in the following way. A plane monochromatic wave being resonant to a particular transition $m-n$ travels through gas and provides nonequilibrium components $\Delta N_m(\vec{v}), \Delta N_n(\vec{v})$ in the velocity distribution for atoms (molecules) at levels $m, n$ (Fig. 5).

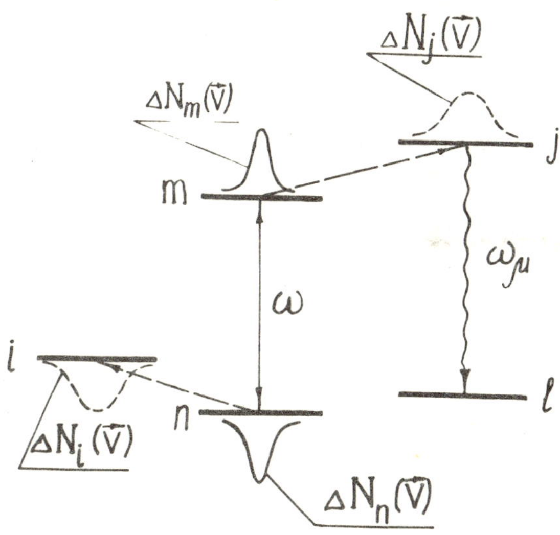


Fig. 5. Scheme of radiation and collision processes, causing collision resonance. Shown are only nonequilibrium components of the velocity distribution.

During collisions with atoms of the same sort or of buffer gas a population transfer is performed, and nonequilibrium components of the distribution $\Delta N_j(\vec{v}), \Delta N_j(\vec{v})$ are formed at the levels $j, i$, as well (the dotted arrows show such processes in Fig. 5). The components $\Delta N_j$, $\Delta N_i$ can be studied by various methods of nonlinear spectroscopy, i.e. by the probe field method, according to the spectrum of spontaneous emission from the levels $j, i$, the Lamb dip method, etc. The levels $m, j$ and $n, i$ may present magnetic sublevels of combinating states and we will speak then about deorientating collisions. In this case the static magnetic and electric fields allows to spectrally single the original resonance for the $m-n$ transition out of the resonances for the $j-l$, and etc. transitions, induced by collisions. Separation of the resonances indicated in case of magnetic sublevel population transfer can be realized using polarization spectroscopy methods.

Consider the simplest situation where the change of atom velocity during the lifetime of the atom at the level may be neglected. Under these conditions $\Delta N_j(\vec{v})$ is proportional to the number of transitions per unit time, i.e.

$$\Delta N_{jj}(\vec{v}) \backsim \int A(jj\vec{v}\,|\,mm\vec{v}_1)\Delta N_m(\vec{v}_1)\,d\vec{v}_1 . \qquad (4.1)$$

The distribution $\Delta N_j(\vec{v})$ together with the kernel of interest $A(jj\vec{v}\,|\,mm\vec{v}_1)$ can be measured by analizing a spontaneous or stimulated emission spectrum, or an absorption one for the $j-l$ transition. It is easily shown that when observation is collinear with the direction of the initial wave $(\omega)$ propogation the line contour $I_{jl}(\omega_\mu)$ for the transition $j-l$ is given by the relation

$$I_{jl}(\omega_\mu) = \frac{\Gamma}{\pi}\int_{-\infty}^{\infty} A(jjv\,|\,mmv_1)\left\{\Gamma^2+\left[\Omega_\mu-\kappa_{jl}(v-v_1)\right]^2\right\}^{-1} d(v-v_1). \qquad (4.2)$$

The function $A(jjv|mmv_1)$ represents onedimensional differential kernel, descibing the inelastic scattering $m \to j$ . In formular (4.2) the designations are introduced

$$\Gamma=\Gamma_{jl}+\kappa_{jl}\Gamma_{mn}/\kappa_{mn}; \quad \Omega_\mu=\omega_\mu-\omega_{jl}\pm\kappa_{jl}(\omega-\omega_{mn})/\kappa_{mn}, \qquad (4.3)$$

where $\Gamma_{jl}$, $\Gamma_{mn}$ are Lorentzian widths for the transitions $j-l$, $m-n$; $\omega_\mu$ - is a frequency of the recorded radiation being resonant to the transition $j-l$; $\kappa_{jl}$, $\kappa_{mn}$ - are corresponding wave numbers; the signs $\pm$ relate to counter-running and unidirectional waves. Thus, determination of the kernel amounts to usual procedure of finding out one of the factors in the convolution. If the kernel half-width exceeds considerably the value $\Gamma/\kappa_{jl}$ , the contour form $I_{jl}(\Omega_\mu)$ duplicates the kernel. Expression (4.2) is not true when collisions change their velocities during the atom lifetime at the levels $n$ , $m$ , $j$ . It is not difficult, in principle, to take this factor into account.

Among collision resonances of the type described, historically the first discovered and investigated were the resonances induced by inelastic scattering of excited neon atoms on the discharge plasma electrons [25] . The scheme of experiment is given in Fig.6. Radiation of one-frequency helium-neon laser 1 ( $\lambda$ = 6328 A ) was absorbed by neon atoms in the gas-discharge tube 2. The levels $m$ , $n$ , $j$ , $l$ were as follows:

$$n=2p_4(3p'[3/2]_2);\ m=3s_2(5s'[1/2]^o_1);\quad \lambda=6328\,\text{Å};$$
$$j=4s_1''''(4d'[5/2]_2);\ l=2p_7(3p'[3/2]_1);\quad \lambda=5657\,\text{Å}. \qquad (4.4)$$

Contour of the line 5657 Å was investigated with the Fabry-Perot interferometer 3 (the half-width of the apparatus function is 40 MHz). After the reduction to an ideal instrument and zero pressure extrapolation, the resonance half-width appears to be equal to

$$\delta=100\pm 20\,MHz=(0,63\pm 0,13)\cdot 10^9 s^{-1}. \qquad (4.5)$$

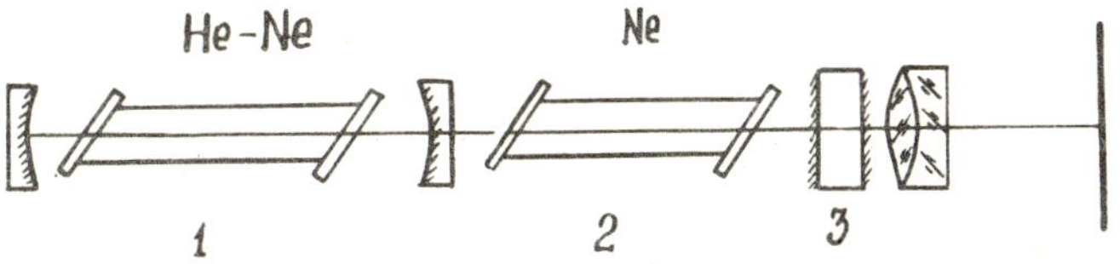


Fig. 6. Scheme of experiment on investigation of neon atom inelastic scattering on electrons.

To evaluate the kernel half-width use not a very rough estimate assuming $\delta$ being equal to the sum of Lorentzian half-width $\Gamma$ from (4.3) and the kernel half-width $\kappa_{jl}s$ (in Doppler units):

$$\delta=\Gamma_{jl}+\kappa_{jl}\Gamma_{mn}/\kappa_{mn}+\kappa_{jl}s. \qquad (4.6)$$

On account of extrapolation to zero pressure the values $\Gamma_{jk}$ in formula (4.6) have the meaning of radiation widths: $2\Gamma_{jk}=\Gamma_j+\Gamma_k$ where $\Gamma_j$, $\Gamma_k$ are the radiation widths of the levels. Substituting the known values of $\Gamma_i$ for $i=m,n,l$ [45,46] we find

$$\kappa_{jl}s+\frac{1}{2}\Gamma_l=(88\pm 20)MHz=(0,55\pm 0,13)10^9 s^{-1}. \qquad (4.7)$$

The value of $\Gamma_l$ is unknown. Radiation transition from $l$ to the ground state is, however, forbidden and $\Gamma_l$ can never be of the order of $10^{-9}$ $s^{-1}$. So an anomalously great half-width value can be interpreted as a result of the neon atom velocity change in case of their inelastic scattering on gas-discharge plasma electrons.

Compare the experimental data presented in this work with the results of calculation of the kernel $A(jjv/mmv_1)$

in the Born-Bethe approximation. For the atom and electron collision the differential cross-section of inelastic scattering $m \to j$ is given by the expression [47] (Born approximation)

$$\sigma(u_1|\vec{u}-\vec{u}_1|) = \frac{u}{u_1}\left|\frac{2e^2}{\mu|\vec{u}-\vec{u}_1|^2}\langle j|\sum_\alpha \exp[-i\mu|\vec{u}-\vec{u}_1|x_\alpha/\hbar]|m\rangle\right|^2 \tag{4.8}$$

$$u^2 = u_1^2 + 2(E_m - E_j)/\mu .$$

If the transition $m-j$ is optically forbidden the exponential factor in the Bethe approximation in formula (4.8) should be expanded to a quadratic term and the differential cross-section appears to be independent of the $|\vec{u}-\vec{u}_1|$ (isotropic scattering in center-of-mass-system). If the condition $\varepsilon = (E_j - E_m)/\kappa_B T_l \ll 1$ is also fulfilled (and it is fulfilled in case of transitions between the neon levels, $\varepsilon \sim 10^{-2}$), it is possible to put $u = u_1$ into (4.8) and for three-dimensional and one-dimensional kernels we obtain expressions (2.13) and (2.22). Taking into account that the electron velocity ($\bar{v}_l \sim 10^8$ cm/s) considerably exceeds the atomic one, from (2.22) we find

$$A(jjv/mmv_1) = A_1\left[1 - \Phi\left(\frac{|v-v_1|}{\Delta v}\right)\right], \quad \Delta v = 2\frac{m_e}{m}\bar{v}_e . \tag{4.9}$$

The half-width of function (4.9) is equal to

$$S_2 = 0{,}477\Delta v = 0{,}954\frac{m_e}{m}\bar{v}_e ; \qquad \bar{v}_l = \sqrt{2\kappa_B T_e/m_e} . \tag{4.10}$$

Thus, with isotropic scattering in center-of-mass-system the momentum $ms_2$, equal approximately to a thermal momentum of the electron $m_e\hat{v}_e$ , is transfered to an atom.

For the optically allowed transition $m-j$ the linear term of the exponent expansion in formula (4.8) makes the main contribution to the differential cross-section, and scattering is sharply selective (inverse proportionality to $|\vec{u}-\vec{u}_1|^2$ ). The calculation shows [25] that in this case a one-dimensional kernel is of the form

$$A(jjv/mmv_1) = A_2\left\{e^{-\varepsilon}\left[1-\Phi\left(\frac{\varepsilon}{4y}-y\right)\right] - 1 + \Phi\left(\frac{\varepsilon}{4y}+y\right)\right\}; \tag{4.11}$$

$$y = |v - v_1|/\Delta v .$$

The half-width of such kernel is equal to

$$S_1 = 0{,}522\,\varepsilon\,\Delta v = 1{,}043\,\varepsilon\frac{m_e}{m}\bar{v}_e = 2{,}086\,\frac{E_j - E_m}{m\bar{v}_e} . \tag{4.12}$$

Unlike isotropic scattering for the optically allowed transitions $m-j$ the kernel width is determined by the energy transferred to internal degrees of atomic freedom and appears to be less by the factor

$$\varepsilon = (E_j - E_m)/\kappa_B T_e \ll 1 .$$

The transition $m-j$ between the neon levels considered above is forbidden, and expression (4.10) should be used for estimation of $S$. For $T_l$ = $10^5$ K we have $\kappa_{jl} S_2$ = 74 MHz which is in a good agreement with experimental value (4.7).

Thus, it can be considered that an isotropic scattering during inelastic collision really takes place in the considered case.

As far as we know, until now experimental results of work[25] are the only direct data on differential cross-section of excited atom inelastic scattering. All the information available relates to the ground or metastable states, either it is of inderect character.

In the example analyzed the inelastic process is caused by collisions with electrons, and the nonequilibrium distribution is produced as a result of single-quantum absorption. Needless to say that both of these circumstances are far from being specific for the presented method as such. In Refs.[48,49], e.g. the transition of populations between the neon $3s$ and $2p$ levels, respectively, was investigated, and in this case luminescence modulated due to the laser radiation absorption at transitions having no common levels with those invistigated, served as an indicator of inelastic collisions with atoms. The measurement of the spectrum of this luminescence would allow to determine not only the rate of transition but the differential cross-sections as well.

Nonlinear resonances induced by collisions can be applied to the investigation of a differential cross-section of the atom deorientation[50]. A scheme of processes in three-level versions of the probe field method for the levels with moments $J_n = 0$, $J_m = 0$, $J_l = 1$ is most simple (Fig. 7a). Linearly polarized radiation (a quantization axis directed along the electric field) being resonant to the transition $m-n$ produces nonequilibrium components $\Delta N_{jM}(\vec{v})$ only at the levels $M=0$ of the states $j = m$. The deorientating collisions transfer the component $\Delta N_{mM}(\vec{v})$ to the levels $M = \pm 1$ of the state $m$ with some change in the velocity.

It will be recalled that between the levels $J_m=1-J_l=1$ the transition $M=0-M=0$ is forbidden. Hence for transition $m-l$ a spectrum of the probe field or spontaneous radiation of the same polarization (wavy arrows in Fig. 7a) will contain a nonlinear resonance broadened according to the scattering indicatrix with deorientation $M=0 \to M=\pm 1$, and described by the relation similar to (4.2). An analogous scheme of the

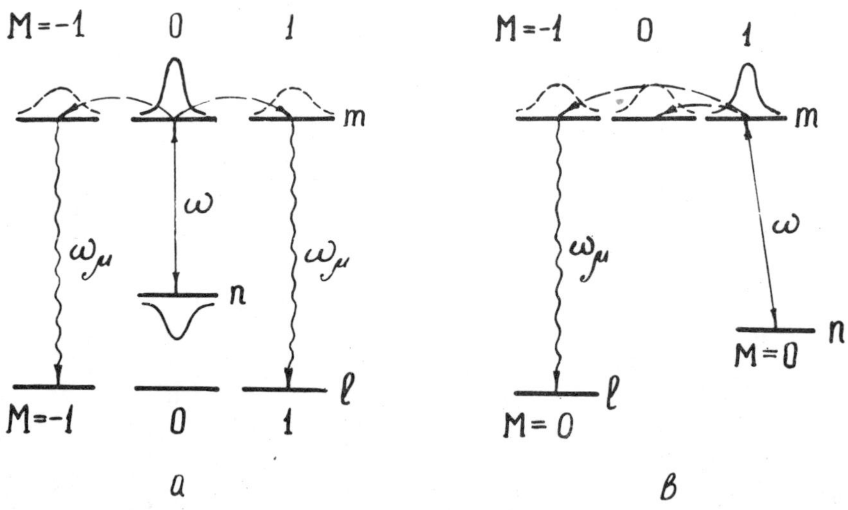


Fig. 7. Scheme of transitions for deorientating resonances.

experiment with left and right circle polarized fields interacting with three levels $J_n=0$, $J_m=1$, $J_l=0$ (a quantization axis is directed along the wave vector) is shown in Fig. 7b. Here it is obviously possible to measure differential cross-section of the deorientation $M=1 \to M=-1$ in the state $m$.

The nonlinear polarization spectroscopy methods are useful in more complex cases as well where state moments exceed 1. Let, e.g., two collinear monochromatic waves ($\omega$ and $\omega_\mu$) interact with coupled transitions $m-n$ and $m-l$ (Fig. 8). The nonlinear resonance in the probe field spectrum ($\omega_\mu$) depends, in general, on the angle between the polarization planes of the waves $\omega$ and $\omega_\mu$. We will alternately register the nonlinear part of the gain (or absorption) of the probe field for polarizations, parallel and perpendicular to polarization of the field $\omega$. Under certain conditions the difference of the given gains appears to be related only to deorientating collisions and in their absence it vanishes [50]. It is very important that the conditions mentioned, concerning the ratio of the probe field orthogonal polarization intensities, are determined only by moments of the states $n, m$ and $l$ and not by peculiarities of the collisions. A similar situation only a more complex one takes place in two-level versions of the probe field method [50].

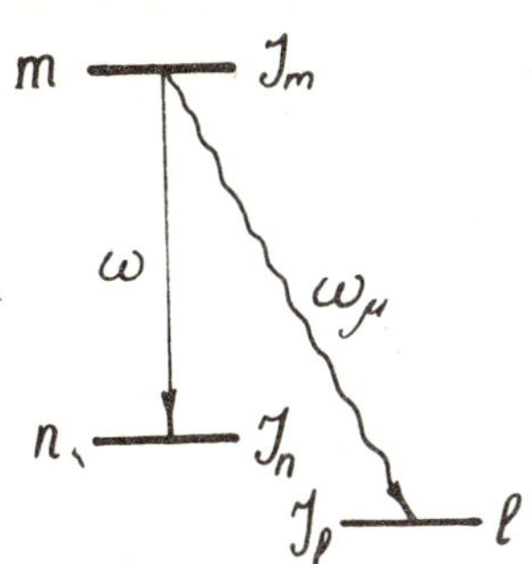


Fig. 8. Radiation processes at coupled transitions.

Collision resonances were probably rediscovered independently of [25] in the set of works [51-54] dealing with nonlinear spectroscopy of vibrational-rotational molecule transitions. In Refs. [51,52] a bichromatic field ( $\omega_\mu - \omega$ = = 30 MHz) was absorbed by gaseous fluoromethane $CH_3F$ placed in a static electric field (band $\nu_3$ , transition $(J, K) = 4.3) - (5.3)$, 1055 cm$^{-1}$). Change in the static field put the transitions between magnetic sublevels in resonance (Fig. 9, linear Stark effect). Besides "usual" three-level resonances for which the conditions of formation are denoted by solid arrows, some more resonances of less intensity were observed relating to the transference of the populations between the magnetic sublevels: in Fig. 9 the transfer is shown with half-round arrows, optical transitions-with broken ones. For the latter the resonance conditions are fulfilled with other values of the static field. A similar technique was used in Ref. [53] where collision resonances were also discovered for vibrational-ratational lines of ammonia, fluoromethane, and formaldehyde.

In experiments [51-53] they did not manage to register any additional resonance broadening which could be explained as a result of scattering during deorientation. In contrast, a nonlinear resonance induced by inelastic transitions between rotational levels of a molecule $CO_2$ and broadened substantially as a result of the velocity change was discovered in work [54]. The collision resonance width ( $\sim 10^3$ cm/s) appeared to be about 10 times of the diffraction width ( $\sim 10^2$ cm/s), but still less than the thermal velocity ( $\sim 3\cdot10^4$ cm/s).

The difference between the results of refs. [51-53] and [54] in respect to resonance broadening due to velocity change is, probably, related to the scattering potential features. In case of deorientation resonances [51-53] the polar molecule collisions were investigated, for which long range and sharply anisotropic potentials are characteristic. As for ref.[54] the rotational relaxation of a nonpolar molecule $CO_2$ in it resulted from collisions with $H_2$, He, Kr, $CO_2$ having no dipole moment, as well.

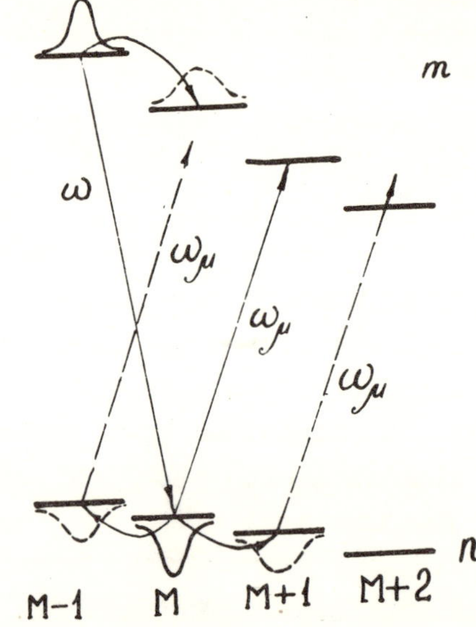


Fig. 9. Scheme of deorientating resonance formation for dipole molecules placed in static electric field.

Under these conditions the interaction potentials decrease sharply with the distance and have rather low anisotropy. For them inelastic processes and deorientation must be attended by scattering into greater angles.

Up to now the collision resonances have been studied rather little concerning both purely spectroscopic aspects and collision physics. In our opinion, the investigation presented in this paper shows promise of the given trend and it is quite reasonable to expect a considerable progress in the nearest future.

## 5. Conclusion

Besides the methods considered a lot of other methods based on measuring the effective cross-sections of one or another processes have been developed in nonlinear spectroscopy. Substantial for these methods is not the velocity distribution, but the complete nonequilibrium part of the level population and the laws of its relaxation caused by collisions.We have not discussed this range of questions, as it has been studied much better, is to some extent arranged and, besides, has much in common with traditional methods of linear spectroscopy. As to nonlinear resonances and their form dependence on differential cross-sections of various elestic and inelastic collisions,this problem is an integral part of nonlinear spectroscopy. It was built up in the last few years, and was rather little discussed. At the same time the examples considered convince us of enormous potentialities of the nonlinear spectroscopy in the field which is interesting for various branches of physics.

Of course, methods of nonlinear resonance spectroscopy alongside with other methods, are far from being universal and have well-known shortcomings. Among them is the "disk-shapeness"of an atom beam formed by plane waves. As a result, nonlinear resonance contours present information on differential cross-sections somehow averaged (see formulas 2.20) and (2.21) ). The shortcoming mentioned could be eliminated by applying laser beams crossing perpendicularly.

Considerable difficulties in investigation of excited atom and molecule scattering which were not overcome by the traditional beam techniques gave rise to a questionable point of view that theoretical results are not worse than direct experimental data, and therefore experiments present no interest at all. Still, since in practice "Nature always gives a third answer", I am sure that nontrivial results will be obtained after a proper development of measuring and laser technique, and the viewpoint mentioned will be refuted.

## References


1. M.P. Tschaika, Interferentsia virozhdennih atomnih sostojani, L., izd. LGU, 1975.
2. M.I. Dyakonov, S.A. Fridrikhov, UFN, 90, 565 (1966).
3. E.E. Nikitin, A.I. Burstein, in book:Gas lasers, Novosibirsk, "Nauka", 1977, s. 7.
4. E.B. Aleksandrov, UFN, 107, 595 (1972).

5. M.P. Tschaika, Report at VI International Conference on Atom Physics, Riga, 1978.
6. M.I. Dyakonov, V.I. Perel, Report at VI International Conference on Atom Physics, Riga, 1978.
7. R.F. Snieder, J. Chem. Phys., 32, 1061 (1960).
8. V.A. Alekseev, T.L. Andreeva, I.I. Sobelman, J. Eksp.Teor. Fiz. 62, 614 (1972); Preprint FIAN N 124 (1971).
9. A. Tip, Physica 52, 493 (1971); 53, 183 (1971).
10. E. Smith, J. Cooper, W.R. Chappell, T. Dillon, J. Quant. Spectr. Radiat. Transfer 11, 1547, 1567 (1971).
11. E. Smith, J. Cooper, W.R. Chappell, T. Dillon, J. Stat. Phys. 3, 401 (1971).
12. P.R. Berman, Phys. Rev. A5, 927 (1972); A6, 2157 (1972).
13. E.G. Pestov, S.G. Rautian, J. Eksp. Teor. Fiz. 64, 2032 (1973).
14. I.I. Sobelman, Vvedenie v theoriju atomnikh spektrov, M., Fizmatgiz, 1963.
15. H.R. Griem, Plasma Spectroscopy, MCGraw-Hill Book Company, New York, San Francisco, Toronto, London. (see transl.: G. Grim, Spectroskopia plazmi, M., "Mir", 1963).
16. J.O. Hirschfelder, Ch.F. Curtiss, R.B. Bird. Molecular Theory of Gases and Liquids , John Wiley and Sons, Inc., New York, Chapman and Hall, Lim., London. (see transl.: G. Girshfelder, Ch. Kurtis, R. Berd, Molekuljarnaja teoria gazov i zhidkostei, M., IL, 1961).
17. B.M. Smirnov, Atomnie stolknovenia i elementarnie processi v plazme. M., Atomizdat, 1968.
18. A.P. Kolchenko, S.G. Rautian, A.M. Shalagin, Preprint IAF SO AN SSSR IAF 46-72 (1972).
19. S.G. Rautian, G.I. Smirnov, A.M. Shalagin, Nelineinie rezonansi v spektrakh atomov i molekul, Novosibirsk, "Nauka", 1978.
20. A.P. Kazantsev, J. Eksp. Teor. Fiz. 51, 1751 (1966).
21. A.I. Vainstein, V.M. Galitski, in book:Voprosi teorii atomnikh stolknoveni, M., Atomizdat, 1968, p. 39.
22. I.M. Beterov, Yu.A. Matjugin, S.G. Rautian, V.P. Chebotaev, J. Eksp. Teor. Fiz. 58, 1243 (1970); Report at All-Union conference on gas quant. gener. phys. Novosibirsk, 1969.

23. M.I. Dyakonov, V.I. Perel, J. Eksp. Teor. Fiz. 58, 1090 (1970); Report at All-Union conference on gas quant. gener. phys., Novosibirsk, 1969.
24. V.I. Perel, I.V. Rogova, J. Eksp. Teor. Fiz. 61, 1814 (1971).
25. S.N. Atutov, A.G. Nikitenko, S.G. Rautian, E.G. Saprikin, Let. to J. Eksp. Teor. Fiz. 13, 232 (1971).
26. V.P. Kochanov, S.G. Rautian, A.M. Shalagin, J. Eksp. Teor. Fiz. 72, 1359 (1977); Preprint ISAN N 2 (1977).
27. S.G. Rautian, J. Eksp. Teor. Fiz. 51, 1176 (1966); Preprint FIAN N 133 (1976).
28. A.P. Kolchenko, A.K. Popov, S.G. Rautian, E.A. Cherkasov, Report at All-Union conference on gas quant. gener. phys., Novosibirsk, 1969.
29. Yu.A. Vdovin, V.M. Galitski, V.M. Ermatchenko, Report at All-Union conf. on gas quant. gener. phys., Novosibirsk, 1969.
30. T. Hänsch, P. Toschek, IEEE J. Quant. Electronics QE-5, 61 (1969).
31. P.R. Berman, W.E. Lamb, Jr., Phys. Rev. A2, 2435 (1970); A4, 319 (1970).
32. V.A. Alekseev, T.L. Andreeva, I.I. Sobelman, J. Eksp. Teor. Fiz. 64, 813 (1973).
33. J.-L. Le Gouet, P.R. Berman, Phys. Rev., A17, 52 (1978).
34. S.N. Bagaev, E.V. Baklanov, V.P. Chebotaev, Preprint IFP SO AN SSSR N 15 (1970).
35. S.N. Bagaev, E.V. Baklanov, V.P. Chebotaev, Letters to J. Eksp. Teor. Fiz. 16, 15 (1972).
36. S.N. Bagaev, E.V. Baklanov, V.P. Chebotaev, Preprint IFP SO AN SSSR N 22 (1972).
37. V.S. Letokhov, V.P. Chebotaev, Printsipi nelineinoi lazernoi spektroskopii, M., "Nauka", (1975).
38. L.S. Vasilenko, V.P. Kochanov, V.P. Chebotaev, Report at All-Union conf. on molec. spectr. of high and super-high resolution, Novosibirsk, 1974; Optics Commun. 20, 409 (1977).
39. S.N. Bagaev, E.V. Baklanov, V.P. Chebotaev, Letters to J. Eksp. Teor. Fiz. 16, 344 (1972); Preprint IFP SO AN SSSR N 25 (1972).

40. T.W. Meyer, C.K. Rhodes, H.A. Haus, Phys. Rev. A12, 1993 (1975).
41. A.P. Kolchenko, A.A. Pukhov, S.G. Rautian, A.M. Shalagin, J. Eksp. Teor. Fiz. 63, 1173 (1972).
42. V.A. Alekseev, I.I. Sobelman, J. Eksp. Teor. Fiz. 55, 1874 (1968).
43. R. Dicke, Phys. Rev. 89, 472 (1953).
44. S.G. Rautian, I.I. Sobelman, UFN 90, 209 (1966).
45. W.R. Bennett, Jr., P.I. Kindelman, Phys. Rev. 149, 38 (1966).
46. J.Z. Kloze, Phys. Rev. 141, 181 (1966); 188, 45 (1969).
47. L.D. Landau, E.M. Lifshitz, Kvantovaya mekhanika, M., "Nauka" (1974).
48. J.H. Parks, A. Javan, Phys. Rev. 139, 1351 (1965).
49. K. Fukuda, R. Nakata, M. Suemitzu, J. Phys. Soc. Japan. 38, 294 (1975).
50. S.N. Atutov, G.A. Radionov, S.G. Rautian, E.G. Saprikin, A.M. Shalagin, Letters to J. Eksp. Teor. Fiz. 3, 1335 (1977). S.G.Rautian,A.M.Shalagin, Preprint IAE,N 84
51. R.G. Brewer, R.L. Shoemaker, and S. Stenholm, Phys. Rev. 33, 63 (1974).
52. R.L. Shoemaker, S. Stenholm, R.G. Brewer, Phys. Rev. A10, 2037 (1974).
53. J.W.C. Johns, A.R.W. Mc Kellar, T. Oka, and M. Römheld, J. Chem. Phys. 62, 1488 (1975).
54. T.W. Meyer, and C.K. Rhodes, Phys. Rev. Lett. 32, 637 (1974).

# QUANTUM BEATS

E.B.Alexandrov

State Optical Institute, USSR, Leningrad

The intense studies on phenomena of quantum beats have been started in optics in early sixties. These phenomena were predicted approximately at the same time and independently by Podgoretski in the USSR and by Kastler, Dodd and Series abroad. The first experiments in this area were carried out in the USSR by the author of this review and very soon by Corney and Series in England, by the group of Kastler in France and by many others /1/. The detailed theory of quantum beats has been worked out by Perel, Konstantinov, Diakonov, Cohen-Tannoudji, Nedelec and others.

The history of the study of optical quantum beats contains two definite stages. At the first pre-laser stage the preliminary, mainly demonstrative experiments were carried out. The advantages of beat phenomena as a spectroscopic method with sub-Doppler resolution were found at the same time. However, the wide use of these new methods was limited by specific technical problems. Later, the interest to the quantum beats faded owing to the famous ideas of nonlinear laser spectroscopy based on the progress of dye lasers. But it was the progress of dye lasers that revived quantum beat spectroscopy on the second-laser-stage.

In this paper the experiments of the first stage of beat development will be very briefly reviewed. Recent experiments using lasers will be considered in more detail and a comparison of beat spectroscopy with nonlinear laser spectroscopy will be given. As it is shown later the beat spectroscopy has substantial advantages as far as the measuring of close splitting of energy levels is concerned.

The literature related to the pre-laser stage is selected arbitrarily: each statement is illustrated by the earliest papers. For a complete reference survey on quantum beats see /1/. While citing the papers on laser beats the author has aspired to completeness.

1. The interference of atomic states

The base of quantum beats is the general quantum mechanical principle of the addition of probability amplitudes of transition from the initial state to the final one while the way of transition is not specified /3/. Assume that an atom is exposed to some excitation process which transfers the atom from the state $|0\rangle$ to the state $|3\rangle$ through the intermediate states $|1\rangle$ and $|2\rangle$ with different energies $E_1$ and $E_2$. If the original excitation may transfer the atom to the state $|1\rangle$ as well as to $|2\rangle$ then an intermediate state with an indefinite energy will be realized. A part of the atom wave function $\psi$ corresponding to the excited state is described by superposition of states $|1\rangle$ and $|2\rangle$ :

$$\psi(t) = C_1|1\rangle e^{-i\omega_1 t} + C_2|2\rangle e^{-i\omega_2 t}. \qquad (1)$$

Amplitudes $C_1$ and $C_2$ depend on excitation conditions.

The subsequent transition from state $\psi$ to the final state $|3\rangle$ as a result of perturbation $\hat{V}$ is accomplished by the modulation of transition probability P(t) at the frequency $\omega_{12} = \omega_1 - \omega_2 = \frac{E_1 - E_2}{\hbar}$ :

$$P(t) \sim |\langle 3|\hat{V}|\psi\rangle|^2 = [A + B\cos(\omega_{12}t + \varphi_0)]\, e^{-\Gamma t}. \qquad (2)$$

Factor $e^{-\Gamma t}$ describes phenomenologically the radiative damping of states $|1\rangle$ and $|2\rangle$. For the sake of simplicity damping rate $\Gamma$ is assumed to be the same for both of the states. Quantities A,B and $\varphi_0$ may be expressed through amplitudes $C_i$ and matrix elements $\langle 3|\hat{V}|i\rangle$.

The energy of the final state $|3\rangle$ may exceed the energy of the states $|1\rangle$, $|2\rangle$ , or, vice versa, level 3 may be lower than 1 and 2. The case of beat in the radiation transition to the lower state may be treated classically due to the interference of optical harmonics $\omega_1$ and $\omega_2$ emitted simultaneously by a single atom.

The beat in a single transition is unobservable because a single realization can show almost nothing as to the probability distribution. To demonstrate the beat, the experiment should reveal the statistics of many identical processes of the type (2). Their identity is secured by fixation of the energy splitting $\omega_{12}$ of the excitation moment (which is assumed to be 0 in (2) and of the initial phase $\varphi_0$. It is obvious that a continuous excitation provides the uniform distribution of the phase of the elementary beats which leads to the disappearance of beats with a single exception of the beat with zero frequency $\omega_{12}$ =0. This is a well known effect of the level crossing. Apart from this case to observe the collective effect of quantum beat it is necessary to ensure the definite phase synchronism of elementary processes (2). It may be provided in several ways which lead to the different phenomena of beats.

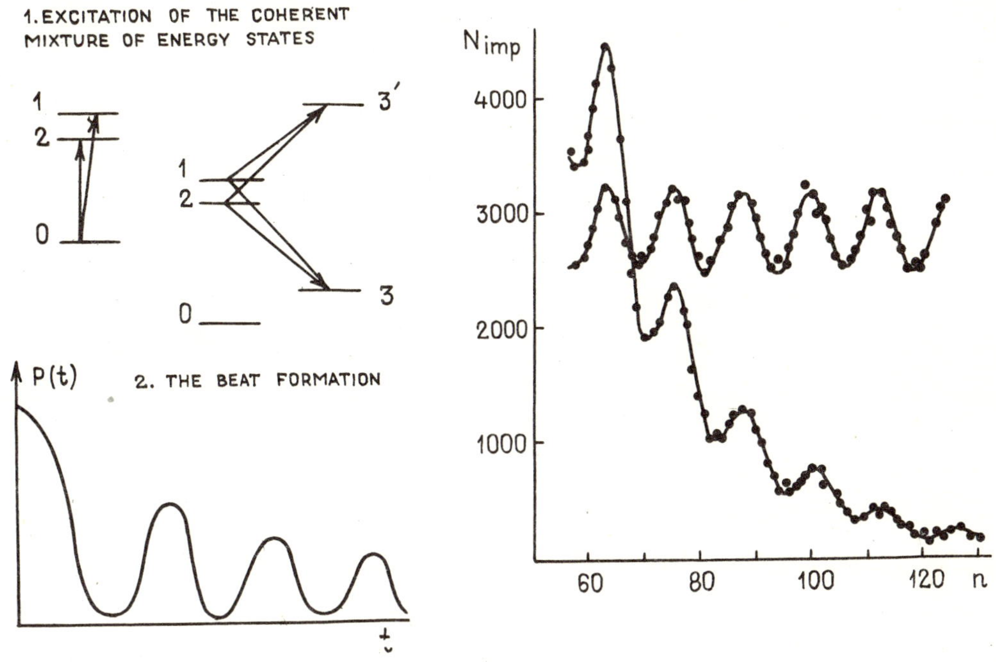


Fig.1. Scheme of the beat formation.

Fig. 2.

To conclude the introduction, I would like to emphasize that in this review the interference of energy states connected with each other by no interaction will be considered. Upon excitation the states under study evolve without any transitions between them. The beats reveal the inherent properties of unperturbed quantum system with indefinite energy. This differentiates the free beats from the relative phenomena of coherency of atomic states arising, e.g. in the double resonance experiments /4/.

2. Types of the beats phenomena

a. The beats under pulsed excitation

It is the most illustrative discovery concerning beats which arise when the ensemble of atoms (molecules) is excited during the interval which is short in comparison with the period of a beat. Then the process (2) may be observed directly provided the initial phase $\varphi_0$ is fixed. When the interference of the states with different angular momenta and/or their projections takes place the definitness of the initial phase is ensured by anisotropic excitation as well as by anisotropic registration/5/. The identity of atoms ensures the identity of splitting $\omega_{12}$, because under condition $\omega_{12} \gg \omega_1, \omega_2$ frequency $\omega_{12}$ is practically not subjected to any Doppler spread which makes the beats very attractive as a tool of the spectroscopy of close energy structures.

The pulsed beat in a spontaneous emission was observed for the first time in works /6,7/ on the fluorescence of the atomic vapours of Cd and Hg. The beat originated due to the interference of m = $\pm$ 1 components of Zeeman's splitting of $^3P_1$ level. These beats may be classically treated as the electric dipole precession in a magnetic field. For exciting optical pulse formation in /6,7/ the electrooptical shutters were used. Poor light intensity made it necessary to store very many signals of beat to obtain convincing results.

It has been proved later that beat is excited easier by electron impact /8/. In paper /9/ electronically excited beats were proposed for an accurate measurement of g-factors of atomic states. Fig.2 shows the beat of magnetic sublevels of the $3^1D_2$ state of He. An exciting electron beam was directed normally to the magnetic field. The time distribution of spontaneously emitted photons was obtained by the multichannel counting of delayed coincidences of photoelectrons.

The quantum beats in the pre-laser stage were most widely used in beam-foil spectroscopy of atoms and ions. The beam-foil method may be briefly described as follows. The fast ion beams are converted to the beam of excited atoms when passing through a carbon foil. The thickness of the foil being of the order of $10^3$ Å the time excitation is extremely short. The fast movement of radiating atoms provides natural scanning of the beats with the resolving time of about $10^{-10}$ - $10^{-11}$ s. This makes it possible to investigate the interference of the sublevels separated by the energy gap up to 1 $cm^{-1}$.

By means of beam-foil spectroscopy beats of hyperfine and fine components of atomic states have been observed. In the case of atomic hydrogen the interference of different electronic configurations was investigated as well 3/10/. The references on beam-foil quantum beats may be found in /11/.

b. Resonance of beats

In pulsed beats the interference of energy separated states reveals itself as an optical transient phenomenon. There are also several stationary manifestations of the interference of energy states. The most important one is the resonance of beats arising when the system is periodically excited at a frequency close to that of a free beat. It reveals itself in the appearance of the stationary modulation of the intensity of observed transition. For instance, the intensity of spontaneous emission proves to be modulated at any frequency $\Omega$ regardless of the inertia of spontaneous emission. Out of the resonance $\Omega \neq \omega_{12}$ the modulation of fluorescence vanishes as soon as condition $\Omega \gg \Gamma$ is fulfilled. The resonance of beats in the fluorescence of atomic vapour of Cd was investigated for the first time in /12,13/. As a matter of fact, the resonant modulation of the atomic vapour absorption was observed earlier in work /14/ dealing with the optical pumping of Rb. This work was undertaken on the basis of the classical conception of spin precession.

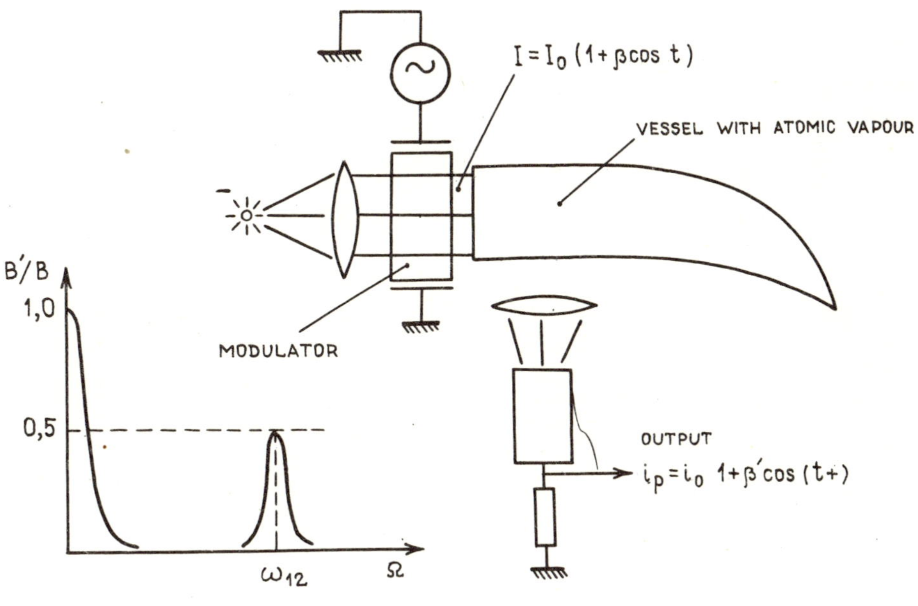


Fig.3.

Fig.3 shows the sketch of experimental set-up for the observation of beats resonance. The resonance of beat as a method of spectroscopy successfully competes with the double resonance /15/. While the radio-frequency field perturbs the atomic states causing a shift and broadening of levels and the appearance of additional multiquantum resonance, the modulated excitation does not alter the atomic states (as long as laser intensities are not consumed). Furthermore, the investigation of a large spacing of wide levels requires the r.f. field of great power which may be difficult to provide or it may be incompatible with the system under study.

At pre-laser stage the beat resonance was episodically applied in spectroscopy, see, for example, /16,17/ and it was borrowed later by nuclear physics where term "stroboscopic observation of nuclear larmor procession" for it was used /18/.

c. Other types of resonances of beats

The synchronization of elementary beats may be also achieved by modulation of initial phase $\varphi_0$ of the process (2) through the modulation, say, of the plane of polarization of the exciting light /19,20/. The resonance of this type has new features: the modulation intensity arises at the integer frequency as compared to the frequency $\Omega$ of phase oscillation the amplitude of the harmonic number $k$ going through the maximum under the condition $k\Omega = \omega_{12}$.

The parametric resonance /21/ may be also attributed to the resonances of beats. This resonance arises when the energy interval $\omega_{12}$ separating the interfering states is modulated at frequencies $\Omega = \omega_{12}/n$ (n is integer) and manifests itself in the appearance of modulation of the intensity of radiation (absorption) at the frequencies $k\Omega$, $k$ being integer factor. The phenomenon of parametric resonance is still more complicated than phase resonance of beats.

The modulation of the energy gap $\omega_{12}$ is practically accomplished by means of electric and magnetic fields. It is important to emphasize that these fields do not induce any transition between interfering states (in contrast with the double resonance) affecting only the relative phase of states. Therefore, the width of the parametric resonance does not include the field parameters.

The last type of beat resonances is the relaxational resonance caused by the modulation of the state width /22/. The phase and relaxational resonances have not found any application in the spectroscopy until now. The parametric resonances (with other related resonances of coherency) are widely used nowadays in measuring magnetic fields /23,24/.

3. Laser beat spectroscopy

The appearance of the widely tuned dye lasers opened new possibilities for observing quantum beats. The pulsed induced beats are most widely used. The basic element of a typical laser beat spectrometer is the dye laser pumped by the pulsed nitrogen laser. The use of two and more dye lasers with a common pumping source makes it possible to obtain a sequence of synchronized pulses at different frequencies and to get a stepwise excitation of high lying atomic states including those which cannot be excited by a single-photon process.

a. "Magnetic and electrical" beats

The advantages of the new technique were demonstrated first in /25/ where the beats at the frequency 3 MHz were observed due to the interference of the magnetic components of the state $6^3P_1$ ytterbium. The beat was registered in the spontaneous emission at 5556 Å directly on the screen of an oscilloscope. A single pulse produced about $10^5$ photons at the cathode of photomultiplier tube in comparison with $10^3$ photons in an earlier paper /6/ and several photons per pulse in /7/ where the total time of beat summation was 15 hours.

"Magnetic" beats under laser excitation were observed in /26/ with atoms of Ba and Ca and in /27/ with atoms of Na.

The first application of beats under pulse excitation for the investigation of molecules was demonstrated in /28/ where Zeeman beats in fluorescence from the rotational sublevels of the states $B^3{}_{o}^{+}$ of $I_2^{127}$ were observed.

The beats caused by the Stark splitting of $6^1P_1$ state of Ba were investigated in /29/.

b. "Fine and hyperfine" beats

The excitation with short light pulses in combination with a fast transient analyser makes it possible to observe beats at the frequencies of the order of $10^8$ Hz. This opens the possibilities to investigate fine and hyperfine structures of atomic levels. In /30/ the hyperfine structure (hfs) of $7^2P_{3/2}$ state of Cs has been studied. The dye laser was successively turned on the hfs components of resonance line 4555 Å ($62\ S_{1/2}$ $-7^2P_{3/2}$ ) and the fluorescence signal was detected by the strobe oscilloscope. The oscillograms were Fourier-analysed to find the beat frequencies (see the energy diagram in Fig.4).

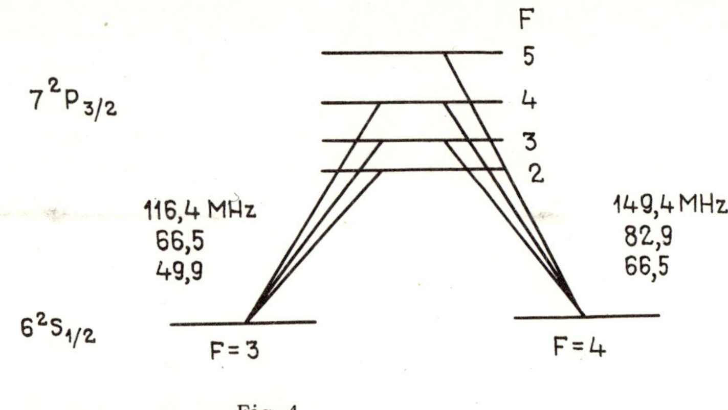


Fig.4.

The same method, though with the two-step excitation was used in /31/ for investigation of hfs of $n^2D_{3/2}$ states of Na with n varying between 8 and 10.

The interference of fine-structure components of $n^2D$ levels of Na for n = 8 - 16 was investigated by several authors /32,33/. The investigated levels were populated through the intermediate state $3P_{3/2}$ by two synchronized laser pulses. The fluorescence signal was detected on the transitions back to the 3 P state. Figure 5 shows beat patterns averaged through many excitation pulses. Figure 6 shows the spectra of beat patterns for n=11, 16 and gives the possibility to evaluate the accuracy of fine splitting determination.

In /34/ the fine component beats of the nD state of Na (n=10-12) were used to determine the probability of high excited atoms in a small static electric field.

A very interesting modification of the beat method was developed in /35/ for determining hfs of $3P_{1/2}$ state of Na. A coherent superposition of hfs components was induced by pulse excitation at 5896 Å ($3S_{1/2}$ - $3P_{1/2}$). After the first excitation pulse with the time shift t the second light pulse was generated which induced the transition $3P_{1/2}$ - $20S_{1/2}$. With a proper choice of the light polarization the probability of second transition depended harmonically on the duration t. The transition to the state $20^2S_{1/2}$ was detected by ionization of atoms in a rather weak electric field.

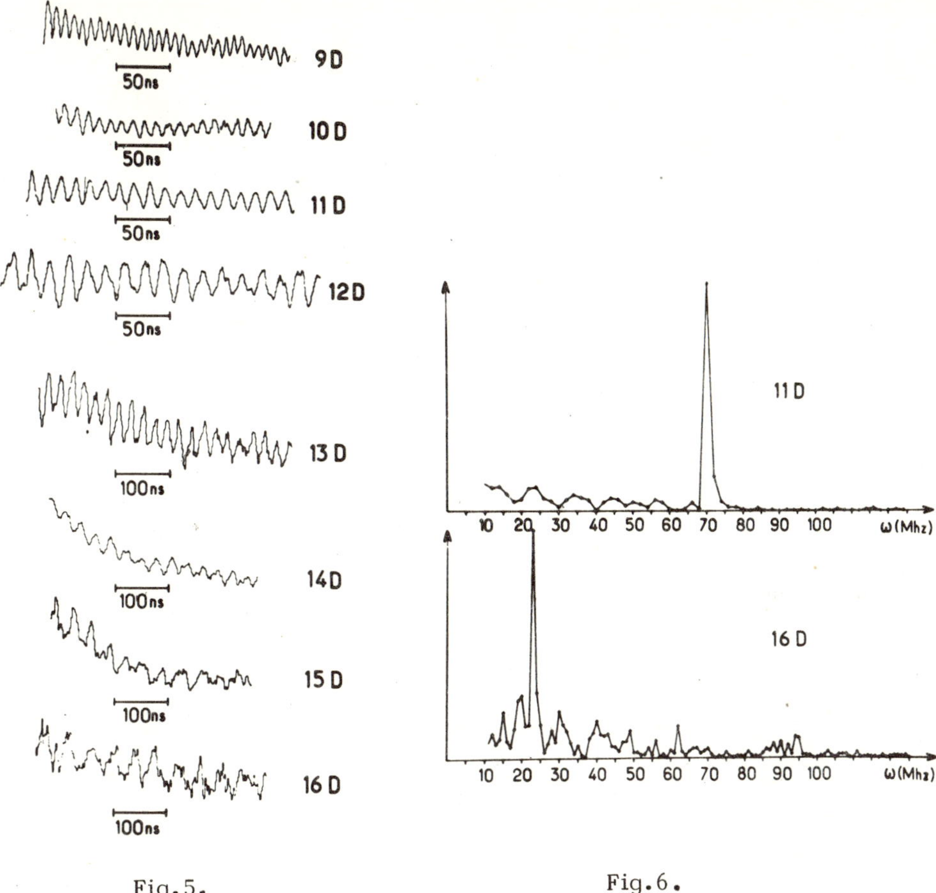


Fig.5.

Fig.6.

The ion output dependence on the interval t revealed the beat at the frequency 193 MHz. Thus, this work showed that the restriction of the time resolution of photodetecting device had been removed. The final time resolution is limited only by exciting pulse duration.

The quantum beats were observed not only in spontaneous and induced transitions, but also in superradiance, i.e., in cooperative sponatneous emission. The superradiance emission differs from the ordinary spontaneous one by the directness and high intensity which are particularly important for the studying of the infrared transitions because of the reduction of the photodetector efficiency and attenuation of the spontaneous emission. The beat in the superradiance of $NH_2D$ at 10.6 was observed in /36/ while investigating the photon echoes and optical nutation. In /37/ by means of beats in superradiance at 2.9 (7P - 7S) of Cs the hfs frequencies of $7P_{1/2,\ 3/2}$ levels were determined. It should be noted that beats in the superradiance may not be considered as a result of free evolution of quantum states which now is subjected to considerable light field perturbation.

The use of lasers widened the possibilities of the application of the resonance of beats, as narrow beams of laser radiation may be rather simply modulated by electro-optical shutters. Besides, lasers may continuously generate in a mode lock way, which is equivalent to the modulation at the high mode, difference frequency. This kind of modulation was used for the first time for observation of the magnetic beat resonance in Ne /38/. Nonlinear modification of beat resonance termed "mode crossing" was developed in /39,40/. These experiments have much in common with earlier experimental work /14/ where the spin precession of Rb by modulated light was induced.

New possibilities were open when the technique of fast atomic beam was combined with the laser excitation. The crossing of an atomic beam with a well collimated laser beam leads to a pulse excitation during the time interval as small as $10^{-10}$ s. The laser excitation has the advantage of high selectivity. The review of beat spectroscopy of this type is presented in /41/ where very elegant experiments with beats at 4554 Å of $Ba^+$ are described. The recent work /42/ along this line is devoted to the redetermination of the hfs splitting of the $3^2P_{3/2}$ state of Na. Fig.7 reproduces the picture of beats at 5890 Å revealing by means of Fourier analysis four frequencies from 33 to 100 MHz. In this work the unprecedented accuracy of hfs constant determination has been achieved.

Fig.7.

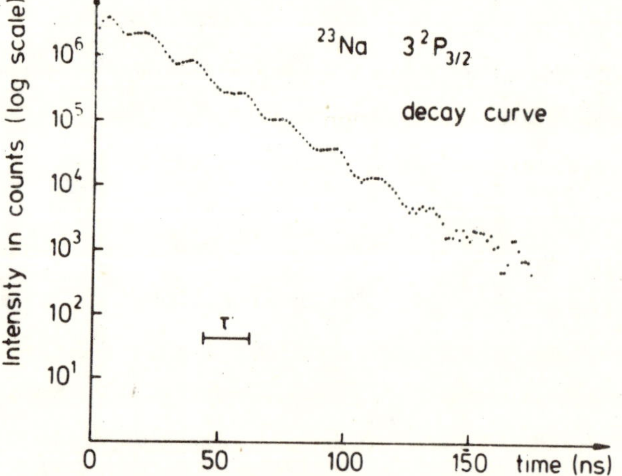


Time resolved decay of $^{23}Na$ I $3\ ^2p_{3/2}$ after laser excitation of a 150 keV Na beam.

## 4. Comparison between beat spectroscopy and nonlinear laser spectroscopy

The methods of nonlinear laser spectroscopy allow to measure the frequencies of optical transitions without limitations of Doppler broadening of spectral lines. Knowing the width and structure of a transition with additional information about the states of the transition one may reconstruct the structure and width of a single state. The methods of laser spectroscopy are universal and under favourable conditions allow to solve the problems related to the quantum beat spectroscopy as well as more general ones - absolute measurements of transition frequencies of independent quantum systems. But in the region of overlapping, i.e., when measuring with subdoppleral resolution the close structure and width of a quantum state, the quantum beat spectroscopy has several distinct advantages.

1. The beat spectroscopy provides direct information about parameters of the state. On the contrary, to interpret the single state parameters on the basis of nonlinear spectroscopy data it is necessary, as a rule, to get the additional information about the second state of the transition.

2. Being basically nonlinear, the laser spectroscopy desturbes the system under investigation. The induced changes of the system parameters may be significant. In contrast with these complications the beat phenomena are linear at their foundation. Laser modification of beats may be complicated by nonlinearity but, at any rate, the beats under pulse excitation are free from an outer perturbation.

Nonlinearity of laser spectroscopy may cause errors when seeking the weak transition near the strong one. It is demonstrated in /36/ where the hfs of excited state of the molecule of $NH_2D$ is determined by means of beat spectroscopy. The discovered splitting was not detected in the earlier works by the saturated absorption method. The saturation parameter of a transition is proportional to the fourth power of the transition strength. In the case of $NH_2D$ the ratio of transition strength is about 5 and the ratio of saturation parameters is about 500. Therefore, at the laser intensity sufficient for the weak line saturation the radiation broadening of a neighbouring strong line may greatly exceed the line separation. On the contrary, the beat signal is proportional to the product of the line strength, therefore, a strong transition helps to discover the weak one.

3. The spectral resolution of laser spectroscopy is limited by the width of the laser line which typically exceeds 1 MHz. The resolution of beat spectroscopy is practically unlimited. This advantage turns to be essential when analysing the states with a small width. In this respect it is worth comparing /32/ and /33/ where the beats of line splitting of the levels nD of Na were measured and /43,45/, where the fine splitting of other nD states was determined by two-photon Doppler free spectroscopy. Beats provided a tenfold accuracy.

4. The laser beat spectroscopy does not require from lasers such stability and width as the laser spectroscopy does. For pulse beat spectroscopy the only condition is the overlapping of the laser line with the Doppler broadenes absorption line.

Thus, there are definite reasons to conclude that the measurements of the state width and close structure by beat spectroscopy are preferable.

## REFERENCES


1. See review papers G.W.Series, Physica 33, 138, 1967; E.B.Alexandrov, Usp.Fiz.Nauk 107, 595, 1972 /Sov.Phys.Uspekhi 15, 436, 1973/; S.Haroche in High Resolution Laser Spectroscopy ed.K.Shimoda, Springer Verlag 1976, p.253
2. V.S.Letochov, V.P.Chebotajev "Nonlinear Laser Spectroscopy"Springer Verlag Berlin Heidelberg New York 1977
3. R.P.Feynman, R.B.Leighton, M.Sands "The Feynman lectures on physics" v.8, 1965
4. G.W.Series "Physics of the one- and two-electron atoms" North Holland 1969, p.268
5. M.I.D'yakonov, Opt.spectr. 19, 662, 1965
6. E.B.Alexandrov, Opt.spectr. 17, 957, 1964
7. J.N.Dodd, D.M.Warington, R.D.Kaul, Proc.Phys.Soc.84, 176, 1964
8. T.Hadeishi, W.Neirenberg, Phys.Rev.Lett.,14,891,1965
9. S.A.Bagaev, V.B.Smirnov, M.P.Chaika, Opt.spectr.41,166,1976
10. A.Gaupp, H.J.Andra, J.Macek, Phys.Rev.Lett.,32,268,1974
11. "Beam-foil spectroscopy" v.1,2, New York,London,1976 ed.I.A.Sellin, D.J.Pegg
12. E.B.Alexandrov Opt.spectr. 14, 436, 1963

13. A.Corney, G.W.Series Proc.Phys.Soc.,83, 213, 1964

14. W.E.Bell, A.L.Bloom, Phys.Rev.Lett.,6, 280, 1962

15. G.W.Series Rept.Progr.Phys.,22,280, 1959

16. T.Hadeishi Phys.Rev.Lett.,21,957,1968

17. R.L.Barger Phys.Rev.,154, 94, 1967

18. J.Christiansen, H.-E.Mahnke, E.Rechnagel, D.Kiegel, G.Schatz, G.Weyer, W.Witthuhn Phys.Rev.,C1, 613, 1970

19. E.B.Alexandrov, Opt.spectr.,19, 452, 1865

20. G.Chapman, Proc.Phys.Soc.,92, 1070, 1967

21. E.B.Alexandrov, O.V.Konstantinov, V.I.Perel, V.A.Khodovoi, Zh. Exper.Teor.Fiz., 45, 503, 1963 /Sov.Phys.JETP 18, 346, 1964/

22. L.N.Novikov, L.G.Malychev, ZhETF Pis.Red.,20, 177, 1974

23. A.I.Okunevitch Zh.Eksp.Teor.Fiz. 66, 1578, 1974

24. Dupont-Roc J., Haroche S., Cohen-Tannoudji Phys.Lett.,28A ,638,1969

25. R.E.Slocum, B.I.Marton IEEE Trans.Magn.,9, 221, 1973

26. W.Gornik, D.Kaiser, W.Lange, J.Luther, H.H.Schulz Opt.Communs, 6, 327, 1972

27. P.Schenck, H.S.Pilloff Bull.Amer.Phys.Soc.,20, 678, 1975

28. P.Schenck, R.C.Hibborn, H.Metcalf Phys.Rev.Lett.,31, 189, 1973

29. R.Wallenstein, J.A.Paisner, A.L.Schawlow Phys.Rev.Lett.,32,1333, 1974

30. A.Hese, A.Renn, H.S.Schweda Opt.Communs.,20,385,1977

31. S.Haroche, J.A.Paisner, A.L.Schawlow Phys.Rev.Lett.,30,948,1973

32. J.S.Deech, R.Luypaert, G.W.Series J.Phys.,B8,1408,1975

33. S.Haroche, M.Gross, M.P.Silverman Phys.Rev.Lett.,33,948,1973

34. C.Fabre, M.Gross, S.Haroche Opt.Communs.,13,393,1975

35. C.Fabre, S.Haroche Opt.Communs.,15,254,1975

36. T.W.Ducas, M.G.Littman, M.I. Zimmerman Phys.Rev.Lett.,35,1752, 1975

37. R.L.Shoemaker, F.A.Hopf Phys.Rev.Lett.,33,1527,1974

38. Q.H.F.Veehen, H.M.J.Hikspoor, H.M.Gibbs Phys.Rev.Lett.,38,764, 1977

39. E.I.Ivanov, M.P.Chaika Opt.spectr.,29,124,1970

40. H.R.Schlossberg, A.Javan Phys.Rev.,150,267,1966

41. H.R.Schlossberg, A.Javan Phys.Rev.Lett.,17,1242,1966

42. H.J.Andra "Atomic Physics" - 4" N.Y.,L., 1974 Plenum Press p.635

43. T.H.Krist, P.Kuske, A.Gaupp, W.Wittman, H.J.Andra Phys.Lett., 61A ,94, 1974

44. F.Biraben, B.Cagnac, G.Grinberg Phys.Lett., 48A, 469, 1974

45. M.M.Salour Opt.Communs., 18, 377, 1976

## SELECTIVE MULTISTEP LASER ACTION UPON ATOMS AND MOLECULES

N.V. Karlov

P.N. Lebedev Physical Institute, USSR Academy of Sciences, Moscow

### 1. Introduction

Recently the resonant interaction of the intense laser radiation with the matter has become more and more important. The well known progress in the research, development and commercial production of continuously tunable lasers created a possibility to carry out selective photoprocesses and use them practically. The most convincing results in the realization of selective processes have been gained for laser isotope separation generally reviewed in many good review papers. It is sufficient to mention here the papers published by such a well known journal as "Uspekhi Fizicheskikh Nauk"/1-4/.

For the time being two methods of the selective laser action upon the matter are most developed. These are the selective photoionization of atoms and selective IR-dissociation of molecules. The general idea of realization of these processes is that of dividing the process as a whole into two stages. The first stage is essentially resonant, and thus under some conditionns selective, excitation of a rather big amount of atoms and molecules. The second stage, in contrast to the first one, should lead to the irreversible change in the physical properties of the previously selectively excited particles. It is evident that the first stage, the stage of the resonant action, should be radiative. The second stage generally speaking, could be nonresonant and even nonradiative. But more common are radiative processes of photoionization mainly because of their higher universality and easier realization.

The radiative processes of selective photoionization and photodissociation are necessarily multistep ones. To realize these processes conserving the selectivity it is necessary to create the conditions for multiple cascade absorbtion of the laser radiation quanta by an atom or a molecule in a complicated system of many energy levels of the real objects which are acted upon. Therefore here arise the questions related to the spectroscopy of atomic and molecular excited states. As a rule, experimental techniques of classical spectroscopy do not answer the questions how and by which energy level sequence it is better to achieve the energy gain by an atom or a molecule sufficient for the irreversible change of the state or structure of a microparticle under question. So, the realization of the selective action is impossible without a corresponding spectroscopical study. The results of the study of this kind are often interesting by themselves.

The aim of this paper is to describe the results of the investigation of two-step selective photoionization of atoms and selective multistep IR-dissociation of polyatomic symmetric molecules carried out recently in the oscillation laboratory of the P.N. Lebedev Physical Institute of the Academy of Sciences of the USSR under the general leadership of Academician A.M. Prokhorov.

## 2. Two-step selective photoionization of atoms

The process of the selective ionization of atoms is based on the resonant transition saturation by laser radiation intense enough. Spontaneous emission and collisions are forcing atoms to return to the initial state. The photoionization of excited atoms could compete successfully with the process of the atom return to the initial state if the ionizing intensity was high enough. If the ionizing radiation quantum energy (being less than the ionization potential of an unexcited atom) is sufficient to ionize the excited atoms the photoionization can be selective. Thus, the necessary condition for the selective two-step photoionization is

$$V_{ion} > h\nu_{ion} > V_{ion} - h\nu_{exc} \qquad (1)$$

where $V_{ion}$ is the ionization potential of the atom, the quantum energy of the resonant excitation (the first step), $h\nu_{exc}$ - the quantum energy of the ionizing radiation (the second step).

Rough estimates for laser radiation intensities necessary

for effective ionization are

$$I_{exc} > h\nu_{exc}/\tau\sigma_{exc} \; ; \; I_{ion} > h\nu_{ion}/\tau\sigma_{ion} \, , \qquad (2)$$

where $\tau$ is the excited level lifetime, $\sigma_{exc}$ and $\sigma_{ion}$ are the cross-sections of the said process.

Taking into account the resonant excitation energy transfer and resonant charge exchange due to collisions in the binary atomic mixture we get the separation coefficient describing the ionization selectivity in the form

$$\alpha = \frac{\frac{1}{N_A}\left(1 + \frac{I_{ion}\sigma_{ion}}{h\nu_{ion}}\tau\right)}{\langle\sigma_+ v\rangle\tau_+\left(1 + \frac{I_{ion}\sigma_{ion}}{h\nu_{ion}}\tau\right) + \langle\sigma_* v\rangle\tau\left(1 + \langle\sigma_+ v\rangle\tau_+ N_B\right)} \, . \qquad (3)$$

Here $N_A$ is the density of the atoms to be selected, $N_B$ of those not to be selected, $\sigma_+$ - the cross-section of the charge exchange, $\sigma_*$ - the cross-section of the energy transfer, $v$ - the relative velocity of the colliding atoms, $\tau_+$ - the ion extractation time.

In the case of the binary collisions the selectivity is decreasing as $1/N_A$. Being far from the ionization saturation the selectivity decrease is caused by the charge exchange and by excitation transfer. Under strong ionization saturation it is possible to avoid the resonant energy transfer influence. Then

$$\alpha = 1/\langle\sigma_+ v\rangle\tau_+ N_A \, . \qquad (4)$$

The above brief discussion implies the need to know the values of $\sigma_*$, $\sigma_+$, $\sigma_{ion}$. Experimental technique developed for laser isotope separation in metall vapours allows us to solve the problems shown above as well as a big series of other spectroscopic problems. The investigations have been done mainly with the rare earth elements /5-9/.

We have been studying the rare earth elements mainly because they constitute a good series of similar in properties but still different atoms which are heavy enough. Physical and chemical properties of rare earths are similar but their isotope content and natural abundance are different. Optical spectra, isotopic shifts ionization potentials of the rare earths are well known. As a whole, the rare earth elements make possible to carry out a representative series of laser laser isotope separation experiments in atomic vapour by selective two-step photoionization. The experiments have been done for neodimium, samarium, europium, gadolinium, dysprosium, erbium and ytterbium.

A vacuum stainless steel chamber provided with optical windows and means for metall vapour production and control over its density has been used. The measuring head of a dynamic quadrupole mass spectrometer was inserted into the chamber.

For the first (selective) step we used a CW bye laser, 200 mW power, 5500-6100 Å tuning range, monochromaticity - $10^{-3}$ $cm^{-1}$. For the ionization we used a $N_2$-laser (3371 Å), an eximer XeCl-laser (3080 Å) or a mercury lamp. The transitions between the electron configurations 6 $s^2$ andn 6s6p have been used for the selective excitation of all investigated elements. Fig. 1-3 show the examples of the mass-spectroscopic display of the separation results. These figures present extremely different situations. Europium has two isotopes of almost equal natural abundance (Fig. 1). Fig. 2 shows sequential separation of each of 5 gadolinium isotopes alongwith the first-step laser tuning. Fig. 3 shows the laser enrichment of the erbium by $^{162}$Er isotope with natural abundance of 0.1%.

The estimate of the excited state photoionization cross-section $\sigma_{ion}$ for all the cases considered is of the order of $10^{-17} cm^2$. The direct measurement in the case of ytterbium photoionization from the $^3P_1$ excited state gives the value of $1.3 \cdot 10^{-17} cm^2 \pm 10\%$ (the ionization potential being exceeded by 0.05 ev). This experiment has been done under the condition of the ionizing intensity saturation. Every excited atom in the illuminated volume of the atomic vapour has been ionized. Such a technique allows to measure the excited state ionization cross-section without knowledge of the absolute number of the excited atoms. A similar technique has been used to ionize rubidium $6^2P_{3/2}$ and $6^2P_{1/2}$ states with a ruby laser. The values of $1.7 \cdot 10^{-17} cm^2$ and $1.5 \cdot 10^{-17} cm^2$ correspondingly have been obtained /10/. Of alkaline metall group lithium has been investigated as well /11/. At the ionizing quantum

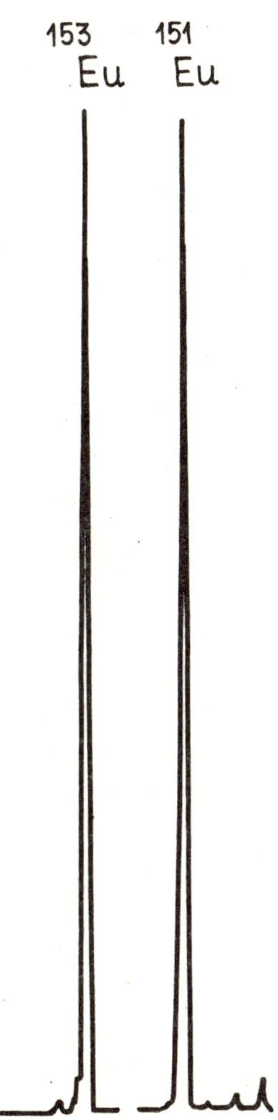


Fig. 1. Laser isotope separation of europium.

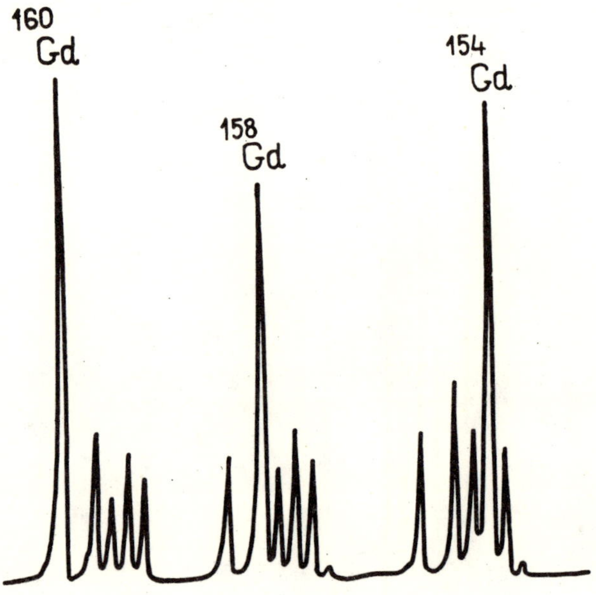


Fig. 2.Laser isotope separation of gadolinium.

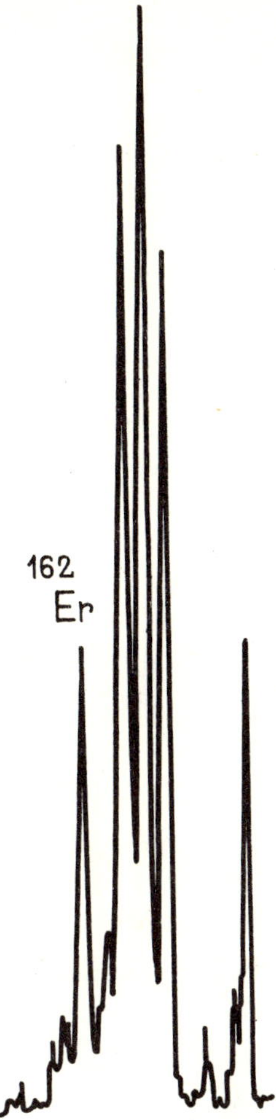


Fig. 3. Laser enrichment of erbium by the $^{162}Er$ isotope.

energy (3371 Å) excess over the ionization potential by 1250 $cm^{-1}$ the excited state ($^2P_{1/2,3/2}$) photoionization cross-section estimate is $10^{-17}$ $cm^2$. The values of the same order of magnitude have been obtained for the rare gases outside the autoionization resonances /12/. Such a small value of the $\sigma_{ion}$ at nonresonant photoionization makes important the search for the resonant autoionization transitions, which could be quite narrow and have the cross-sections higher by one - two orders of magnitude, as it has been shown experimentally /12/ for argon and krypton. The search for the narrow autoionization resonances demands to use the tunable lasers of high monochromaticity for the ionization step. In the two-step process very short wavelength lasers should be used, which is possible, in principle, with the eximer lasers but might be difficult to realize. So it is necessary to pass from the two-step photoionization to the multistep one. In the latter case three or more lasers are used for cascade resonant excitation up to the energy level approaching the continuum. This technique allows to look for, to study and to use narrow autoionization resonances using reliable and available dye lasers. The multistep ionization carried out by means of 3 - 4 monochromaticity tunable lasers is very general and applicable for the study of any element of the periodic table.

As it has been mentioned, the selectivity of multistep photoionization of atoms is deteriorating because of the resonant excitation energy transfer taking place in the course of interatomic collisions. The energy transfer processes have been very well studied in the physics of atomic collisions /13/. However, as a rule, the concrete data on the cross-section values and especially for heavy atoms are absent. The processes of the kind become significant at relatively high density of colliding atoms. At high density the traditional spectroscopic investigation becomes difficult because of radiation replenishment. The selective laser excitation allows the direct observation of the transfer processes and measurement of their cross-sections /14, 15/.

For heavy atoms it is convenient to study the excitation transfer between the isotopes in the atomic vapour. In this case the condition for the resonant transfer is fulfilled very well because the energy deficit is extremely small, and atoms are similar enough but distinguishable spectroscopically. In the course of the experiment the change of the atomic vapour luminescence has been observed while the atomic vapour density was increasing, and only one of the isotopes has been selectively excited. Luminescence line broadening, the line shift and appearence of the luminescence of a radiatively nonexcited isotopes have been observed. Essentially, this is a modified resonant fluorescence technique.

For europium at the intercombination transition $^8S_{7/2} - {}^6P_{7/2}$ ( $\lambda$ = 5765 Å) in the density range $5 \cdot 10^{12} cm^{-3}$ - $5 \cdot 10^{14}$ $cm^{-3}$ the resulting cross-section for $^{153}Eu$ - $^{151}Eu$ transfer is equal to $1.3 \cdot 10^{-13} cm^2$. Fig. 4 gives the examples of records showing the appearance of the radiatively nonexcited isotope $^{151}Eu$ luminescence line at the increase of atomic vapour density. The study of the line broadening, line shift, line-shape change could provide valuable data for the investigation of the collisional processes between identical atoms. The selectivity and the precision of the technique could be suffisiently improved if the indication scheme would include, besides the luminescence analysis, an ionizing step and selective mass-filters measuring ionic current. The use of pulsed lasers with a controlled delay could allow to study the temporal behaviour of energy migration processes.

The above-discussed technique could be used only to measure the excitation energy transfer between the isotopes. Fig. 5 shows the lithium luminescence spectrum change caused by the excitation energy transfer between the components of the $^2P_{3/2}$ - $^2P_{1/2}$ doublet by the collisions with the buffer gas atoms (helium). The laser excites selectively the $^2P_{1/2}$ component. The cross-section of this process turned out to be $(2.3 \pm 0.3) \cdot 10^{-15}$ $cm^2$.

The selective multistep photoionization of atoms is a very powerful technique of spectroscopic study. The laser isotope separation technique shows that combining the selectivity of mass-spectrometer measuring the photoionization current of ions with required mass with the selectivity of laser action we can enhance significantly the possibilities of laser high resolution spectroscopy.

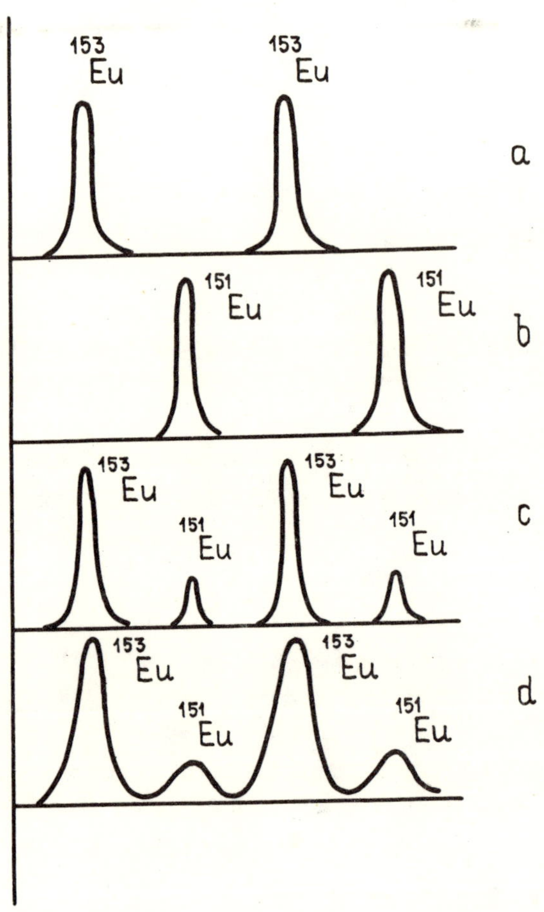


**Fig. 4.** Collisional excitation energy transfer between $^{153}$Eu - $^{151}$Eu - atoms.

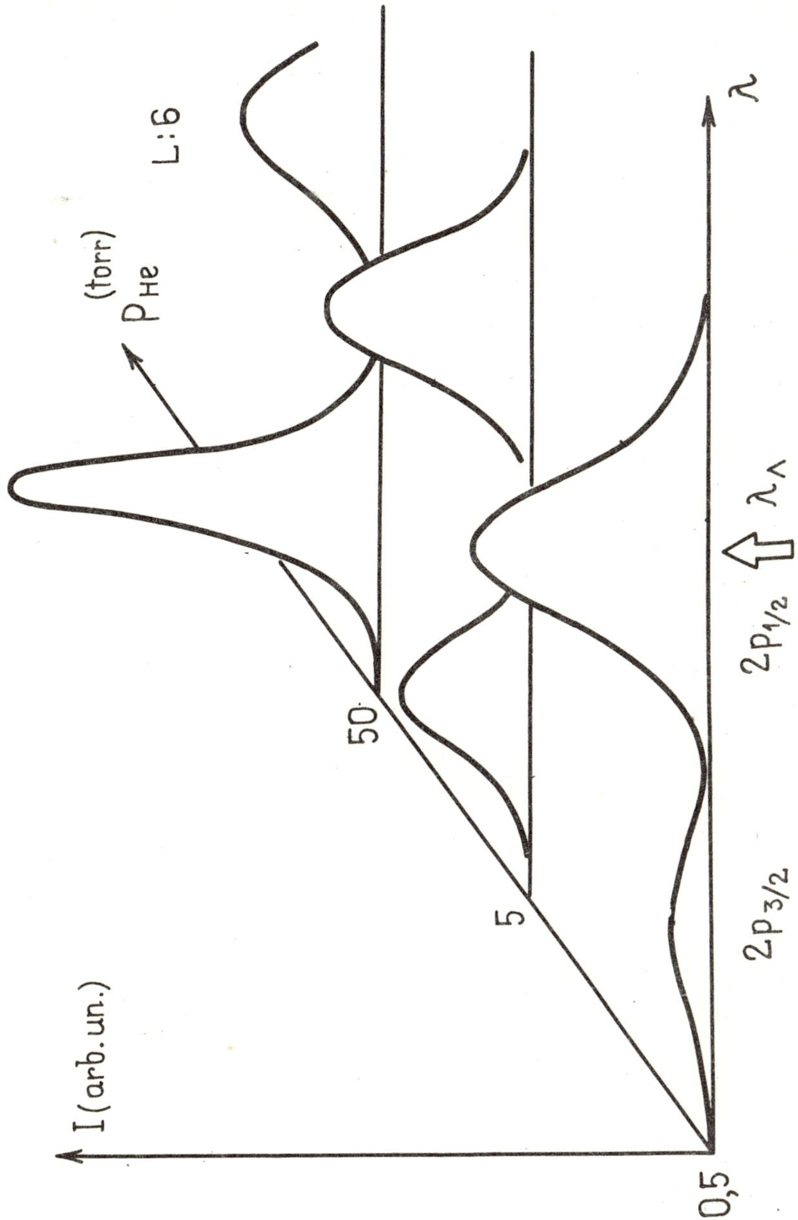


Fig. 5. The excitation energy transfer between the components of the lithium doublet $P_{1/2}$ - $P_{3/2}$ while colliding with helium.

Experimentally the HFS of separate isotopes of dysprosium and erbium taken in natural abundance has been studied /16/. The experimental setup is quite the same as the one used for laser isotope separation. The ions of an element under study obtained by the selective photoionization have been lead by an ionic lense system to the measuring head of dynamic mass-spectrometer. The mass-spectrometer was tuned to a definite mass number and served as a tunable selective mass-filter. Scanning the frequency of the exciting laser we recorded thus only the ions corresponding to the excitation spectrum of the prechosen isotope. It is necessary to have the mass-spectrometer resolution equal to 1 a.u.m., which is easy to obtain.

Fig. 6 shows the example of the dysprosium odd isotopes HFS spectrum.

The spectral resolution of the technique is determined by the tunable laser oscillation line width and (or) by the Doppler line width of the atomic beam (vapour). The sensitivity of the technique is determined by the sensitivity of the mass-spectrometer.

The technique is of universal character, and it can be applied to any isotope of any element of the periodic table. Its application should be very effective in the cases of low abundant and unstable isotopes which are difficult to be obtained in microscopic quantities. Thus, for the time being the realization and high selectivity of the multistep photoionization of atoms and its high effectivity for laser isotope separation and high resolution spectroscopy are firmly established. Due to its universality the selective photoionization is applicable for the action upon all the isotopes of all the elements of the D.I. Mendeleev Periodic Table including rare and unstable atoms as well as exotic atoms made artificially.

### 3. Selective IR-dissociation

For the time being the collisionless dissociation of polyatomic molecules by the infrared laser radiation is one of the most interesting effects of the resonant interaction of the intense laser radiation with matter. The most interesting feature of this phenomenon is the ability of polyatomic symmetric molecules, as for example $SF_6$, $SiF_4$ $OsO_4$ and $BCl_3$, to gain the high energy often significally exceeding the dissociation limit in the process of collisionless absorption of a big number of radiation quanta of low energy each.

The polyatomic molecule absorbs many IR quanta collisionlessly, i.e. purely rediativaly, but this is not a multiphoton process in the common sense of the word. The process of the absorption of many quanta by a polyatomic molecule is a process of multistep, multiple quantum absorption. So it should be labelled as a multistep or multiphoton absorption.

The main experimental features of a multiphoton process are rather well known. Recently they were discussed in detail at the International Conference on Multiphoton Processes held at the Rochester University (USA) /17-21/.

We divide the process as a whole into three stages, this division being to some extent arbitrary. The first stage is a frequency-selective passage through the sequence of some, but few vibrational levels of the molecule. The second stage is the effective energy gain by a previously excited molecule in the zonal structure of the quasicontinuous spectrum of the upper vibrational states, characteristic of the highly excited polyatomic molecules. The third stage is the dissociation of highly excited (overexcited) molecules.

For the first stage it is characteristic to devastate collisionlessly the population of many rotational states at the vibrational excitation of low-lying vibrational levels, which means the collisionless excitation of a polyatomic molecule from its ground vibrational state independently of its rotational state by a laser IR-field of moderate intensity ($\sim 10^5$w/cm$^2$) /28/. This effect, being very important for the realization of effective dissociation, is a result of a complicated structure of low-lying vibrational states of the molecules to be discussed here. This structure allows to fulfil the conditions for two- and three-photon resonances. Let us consider the $SF_6$ molecule with a fixed rotational quantum number $J$ being in the ground vibrational state. The transition 0 - 1 of the $V_3$-mode of this molecule has three rotational branches (P,Q,R). The second and the third vibrational states of the $V_3$-mode under the influence of the anharmonicity removing the degeneracy are split to three and four vibration-

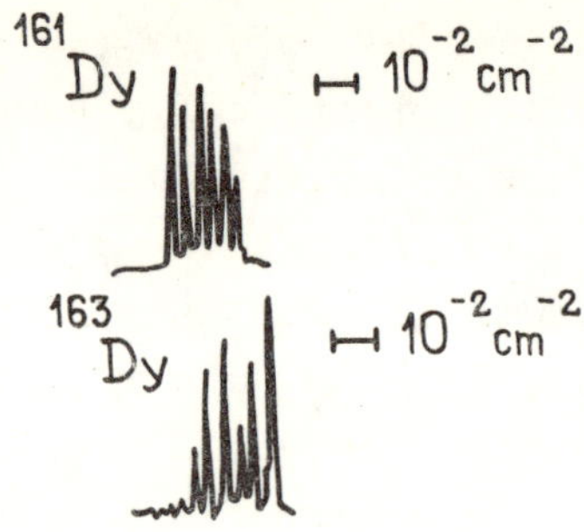


Fig. 6. HFS spectra of some dysprosium isotopes.

Table 1

| $v$ | $\varepsilon$ - correction to the harmonic approximation energy | Level symmetry | |
|---|---|---|---|
| | | $T_d$ | $O_h$ |
| 0 | $0$ | $A_1$ | $A_{1g}$ |
| 1 | $\alpha$ | $F_2$ | $F_{1u}$ |
| 2 | $4\alpha+4\gamma$<br>$4\alpha-2\gamma$<br>$2\alpha+\beta$ | $A_1$<br>$E$<br>$F_2$ | $A_{1g}$<br>$E_g$<br>$F_{2g}$ |
| 3 | $3\alpha+3\beta$<br>$5\alpha+2\beta-2\gamma$<br>$7\alpha+\beta+\gamma\pm\sqrt{(\beta+\gamma-2\alpha)^2+24\gamma^2}$ | $A_1$<br>$F_1$<br>$2\times F_2$ | $A_{2u}$<br>$F_{2u}$<br>$2\times F_{1u}$ |
| 4 | $12\alpha+2\beta+2\gamma\pm2\sqrt{(\beta+\gamma-2\alpha)^2+24\gamma^2}$<br>$12\alpha+2\beta-\gamma\pm\sqrt{(2\beta-4\alpha-\gamma)^2+24\gamma^2}$<br>$10\alpha+3\beta-6\gamma$<br>$8\alpha+4\beta+3\gamma\pm\sqrt{(2\alpha-\beta+3\gamma)^2+24\gamma^2}$ | $2\times A_1$<br>$2\times E$<br>$F_1$<br>$2\times F_2$ | $2\times A_{1g}$<br>$2\times E_g$<br>$F_{1g}$<br>$2\times F_{2g}$ |

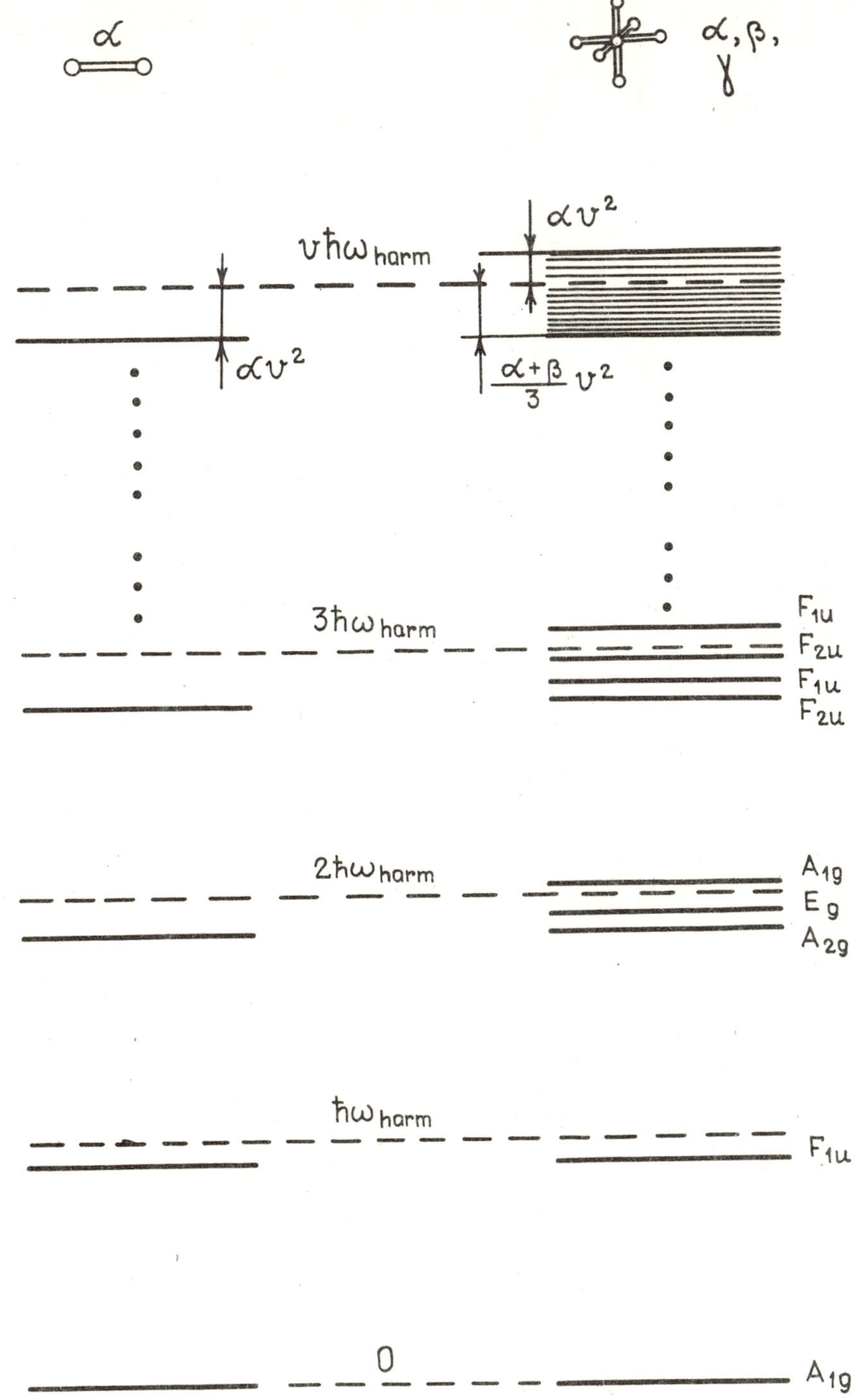


Fig. 7. The energy levels of the $SF_6$ molecule $v_3$ - mode.

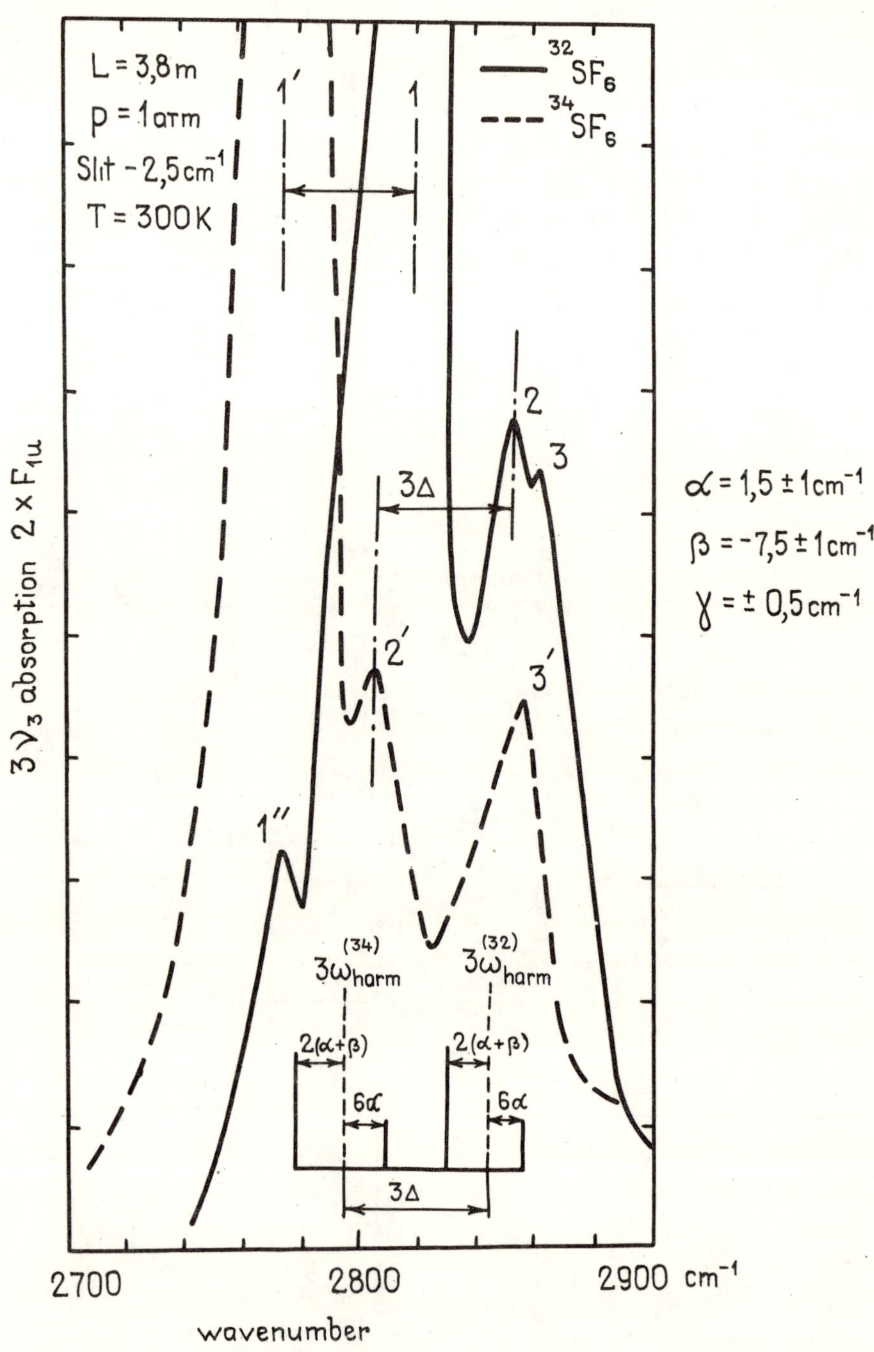


Fig. 8. Absorption spectrum of $3\nu_3$ vibration of the molecules $^{32}SF_6$ and $^{34}SF_6$.

al levels correspondingly. Each of these states could be a terminal state for the two- or three-photon resonant transitions. Each two-photon resonance has five rotational branches ($\Delta J = 0, \pm 1, \pm 2$), the three-photon - seven branches ($\Delta J = 0, \pm 1, \pm 2, \pm 3$). As a result, for the ground state $SF_6$ molecule with a fixed $J$ we have 3 single-photon, 15 two-photon and 28 three-photon resonances. The fine structure of these resonances is determined by the value of the Coriolis splitting of the rotational-vibrational states.

So the polyatomicity of the molecule is very important. Fig.7 shows the difference in the excited states spectra of diatomics and symmetrical polyatomics, as well as the necessity to take into consideration three anharmonicity constants $\alpha$, $\beta$ and $\gamma$ for the latter case. In Table 1 the corresponding energy levels are listed. The most important characteristics are the anharmonicity constant values. They have been measured while observing the transition from the ground state into the third vibrational state for the molecules $^{34}SF_6$ and $^{32}SF_6$, i.e. observing the $3V_3$ absorption band. Fig. 8 shows the absorption spectrum. This spectrum gives the values $\alpha = 1.5 \pm 1$ cm$^{-1}$, $\beta = -7.5 \pm 1$ cm$^{-1}$, $\gamma = 0.5$ cm$^{-1}$. The opposite sign and different magnitude of $\alpha$- and $\beta$-constants are of principal importance. The observation of the bluecomponent (as related to the harmonic energy position $3V_3$) of the doublet $F_{1U}$ is also essential.

At the said values of $\alpha$, $\beta$ and $\gamma$ the partial amount of molecules able to undergo two- and three-photon transitions $v = 0 \rightarrow v = 2$ and $v = 0 \rightarrow v = 3$ was theoretically shown to be rather big. Fig. 9 shows the results of the calculations explaining the experimental results /18/ of the collisionless devastation of the vibrational ground state independently of rotational quantum number $J$. This phenomenon has been investigated more carefully for cooled (140$^0$K) and warm (300$^0$K) $SF_6$ gases using optoacoustic detector and line-tuned pulsed $CO_2$-laser. Fig. 10 shows the results for the warm and cooled gases. A significant difference is seen. The existence of the spectral structure in the cooled gas case corresponds to our theoretical model. The disappearance of the structure in the warm gas is due to the hot bands. Fig. 10 shows also the nonlinear dependence of the structural peaks on the energy fluence.

The investigation of spectral and energy dependencies of the kind shows clearly the leading role of two- (three-) photon resonances for the excitation of low-lying vibrational levels.

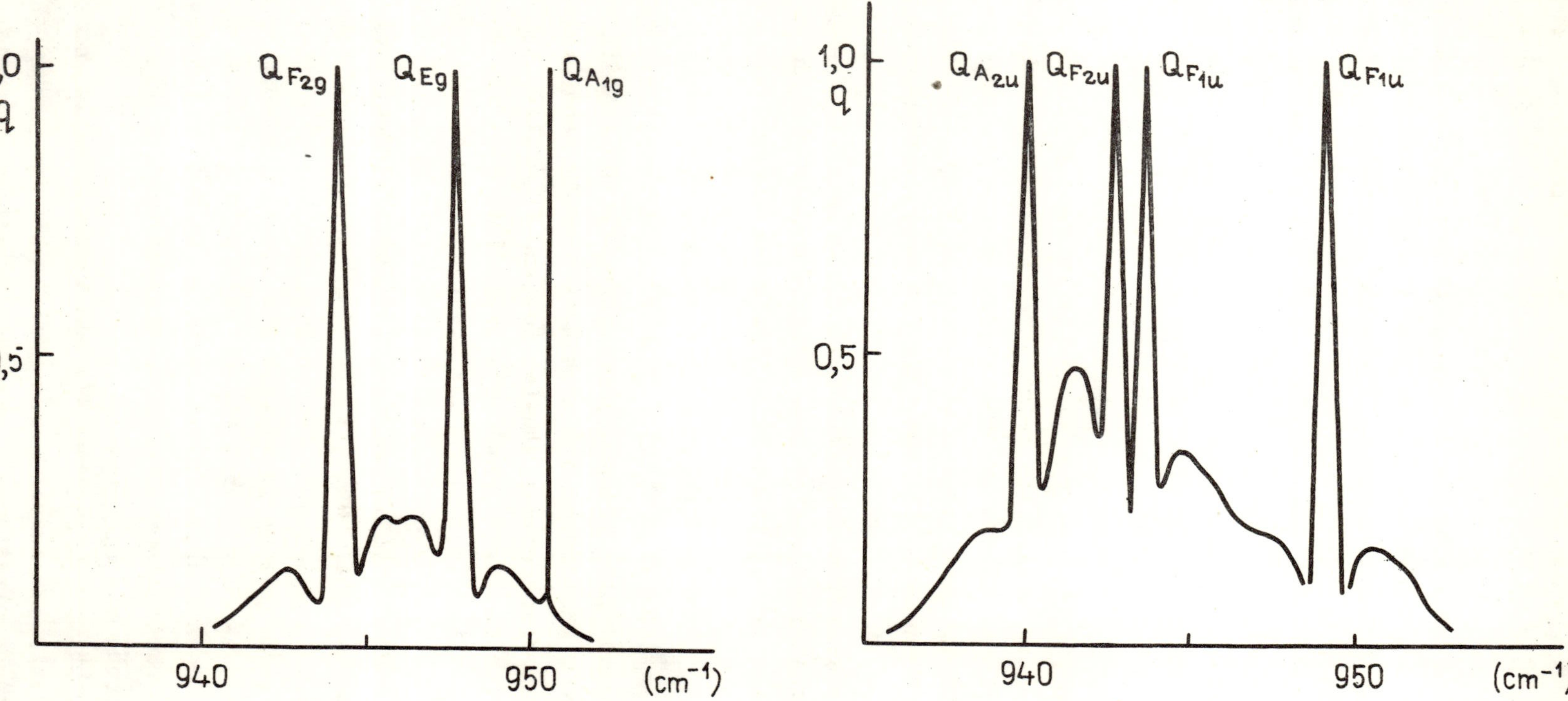


Fig. 9. The partial number of molecules able to undergo two- and three photon resonances $v = 0 \rightarrow v = 2$; $v = 0 \rightarrow v = 3$.

From the point of view of isotopical selectivity the search for the narrow resonance $Q_{A_{Ig}}$ (Fig. 9) seems to be very promising.

The second stage of the process discussed now is the energy build-up in the zonal structure of the high vibrational states quasi-continuum. The vibrational quasi-continuum does exist due to the high density of vibrational states including combination vibrations and harmonics for polyatomic molecule excited strongly enough /22/. The zonal structure in the molecules of the $O_h$- or $T_d$-symmetry has to be formed because of the excited levels degeneration removal by anharmonicity /25/. If the anharmonicity constants $\alpha$ and $\beta$ have the opposite sign, and $|\beta| > |\alpha|$, the zones overlap the harmonic energy position. Experimentally the zonal structure has been discovered /26/ using two-frequency technique. By this technique the radiation at frequency $\omega_1$, resonant to the linear absorption spectrum of the molecule, excited the low-lying vibrational levels. The intensity of this radiation should be relatively low. Simultaneously the molecule has to be irradiated by an IR-laser field at some frequency $\omega_2$. Being taken separately, the fields at frequencies $\omega_1$ and $\omega_2$ do not dissociate the molecules. The combined action of these fields leads to dissociation. Obviously, nonresonant $\omega_2$-field is absorbed by the zonal structure of the quasi-continuum.

Fig. 11 shows the dissociation rate dependence on the second field frequency $\omega_2$ in the case of the $SiF_4$-molecule. Similar results have been obtained for $SF_6$. The effect of a sizable red shift of the second field frequency resulting in the significant enhancement of the dissociation rate is clearly seen. This effect has been more carefully studied for the $SF_6$-molecule using optoacoustic detector. The first field frequency has been fixed at $\omega_1 = 947.8\ cm^{-1}$, the energy fluence was a changeable parameter. The second field frequency was tuned in a wide range from $\omega_2 = 885\ cm^{-1}$ to $\omega_2 = 954.6\ cm^{-1}$. The second field energy fluence was constant ($0.5\ J/cm^2$). Fig. 12 shows the dependence of the optoacoustic signal, proportional to the absorbed energy, on the frequency $\omega_2$ in the case of the warm gas for the two-frequency excitation $S_\Sigma$ and for the excitation only by the first frequency field ($S_1$) or by the second one ($S_2$) separately. The quantity $S - S_1 - S_2$ represents the energy deposit by the second field in the presence of the first one and thus gives the spectral dependence of the energy absorbed by the system of higher vibrational levels. The red shift is clearly seen in this spectrum. The red shift value is increasing with the increase

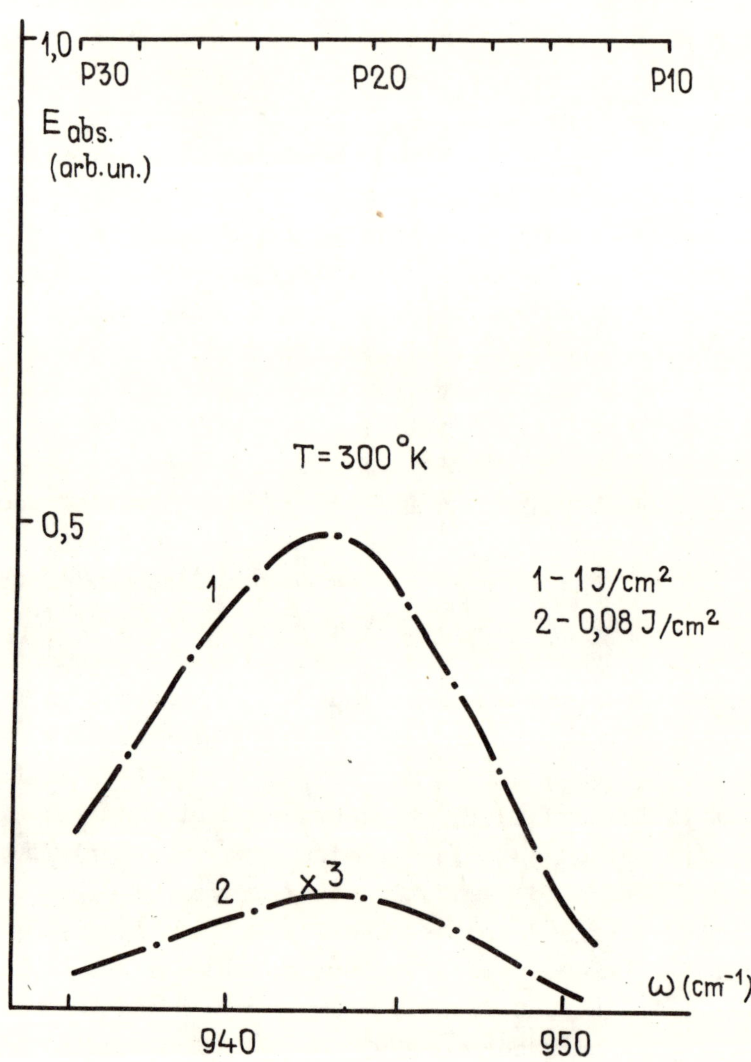


Fig. 10 a. Absorbed energy spectral dependence for warm $SF_6$ gas at moderate irradiation intensity.

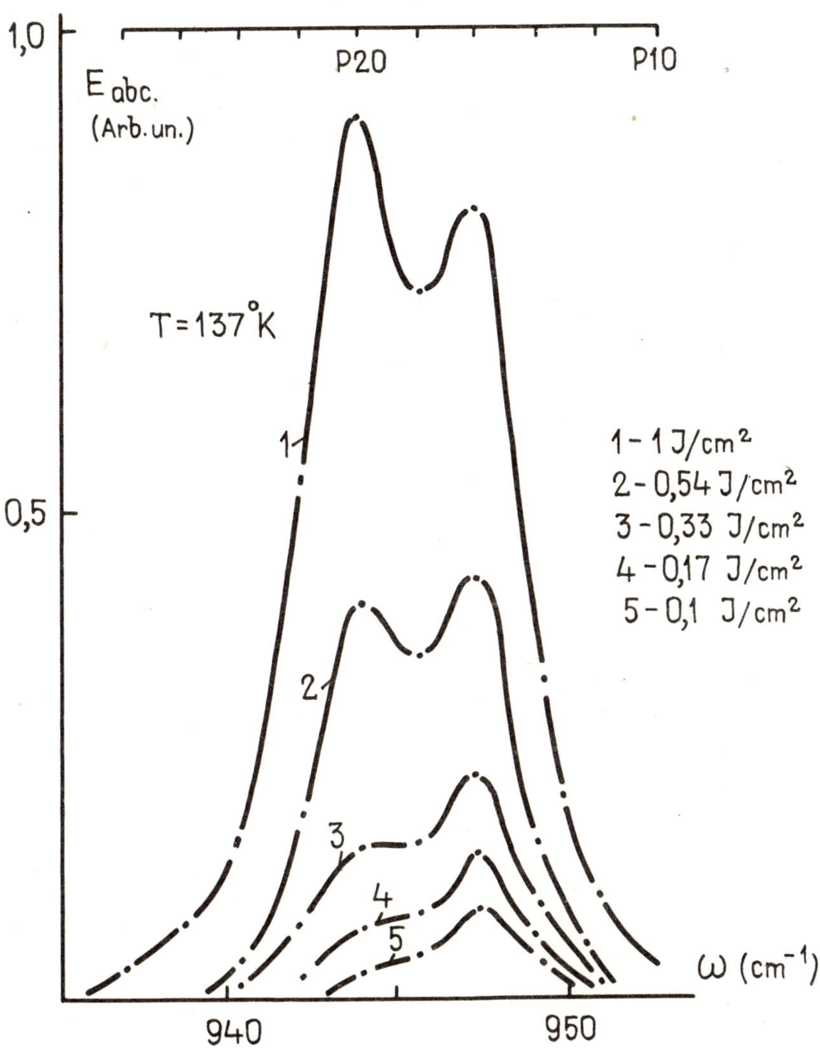


Fig. 10 b. Absorbed energy spectral dependence for cooled $SF_6$ gas at moderate irradiation intensity.

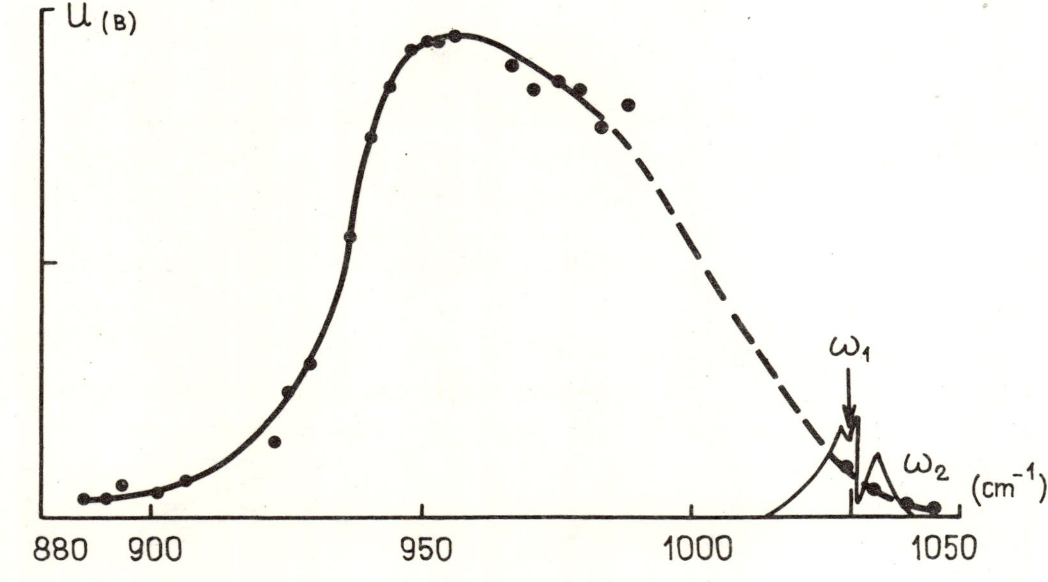


Fig. 11. The red shift effect for $SiF_4$ molecule.

of the first field intensity and at the fluence of 1 J/cm$^2$ reaches 30 ÷ 35 cm$^{-1}$. At the first field ( $\omega_1$ = 947.8 cm$^{-1}$) energy fluence of 1 J/cm$^2$ averaged absorbed energy is 3 ÷ 4 quanta per molecule. At the cooling gas down to 136°K (Fig. 13) we have essentially the same picture. But by excluding the hot band we get the structure related to the second field absorption by the lower vibrational levels. The structure disappears at larger detunings $\omega_2 - \omega_1$ , because here the second field excites molecules already excited strongly by the first field, i.e. the molecules have previously reached rather high vibrational states, where the structure is very complicated and self-averaged. The red shift is going up with the increase of the field energy fluence, because the second field absorption starts from a higher energy state.

The experimental results described above show a very good agreement with the theoretical assumptions of the zonal structure of the higher vibrational level quasi-continuum spectrum of the symmetric polyatomic molecule, shown in Fig. 14.

Fig. 15 shows the dissociation probability versus second field frequency for the warm gas. It is clearly seen that the dissociation yield curve does not repeat the absorption curve and is strongly red-shifted. At maximum available detuning of 62 cm$^{-1}$ ( $\omega_2$ = 886 cm$^{-1}$) in one laser pulse 4% of molecules to be irradiated are dissociated. For the time being it is difficult to explain these results unambigously. First, such a great red shift could be related to the resonant field breaking of an effective intramolecular bond of the highly excited molecule. Second, there is the possibility of the resonant excitation and then further dissociation of the $SF_6$-fragment according to the scheme $SF_5 \rightarrow SF_4 + F$ /20/. But the most likely reason for such a great red shift seems to be the resonant excitation of higher, situated between V = 8 and V = 25, vibrational levels of the $V_3$-mode by the second field. The red shift value of 62 cm$^{-1}$ indicates that for the further energy gain in the quasi-continuum zonal structure the starting level should be the zone corresponding to $v \geqslant 8$ (Fig. 14). For the first field energy fluence of 1 J/cm$^2$ molecules with v = 8 do exist, but they are few. They are mostly situated in the excited states v = 2 ÷ 5, thus determining the absorption spectrum.

The above-discussed results allow to conclude that in the process of collisionless dissociation the excitation energy is localized in the vibrational mode to be excited ($v_3$) at least

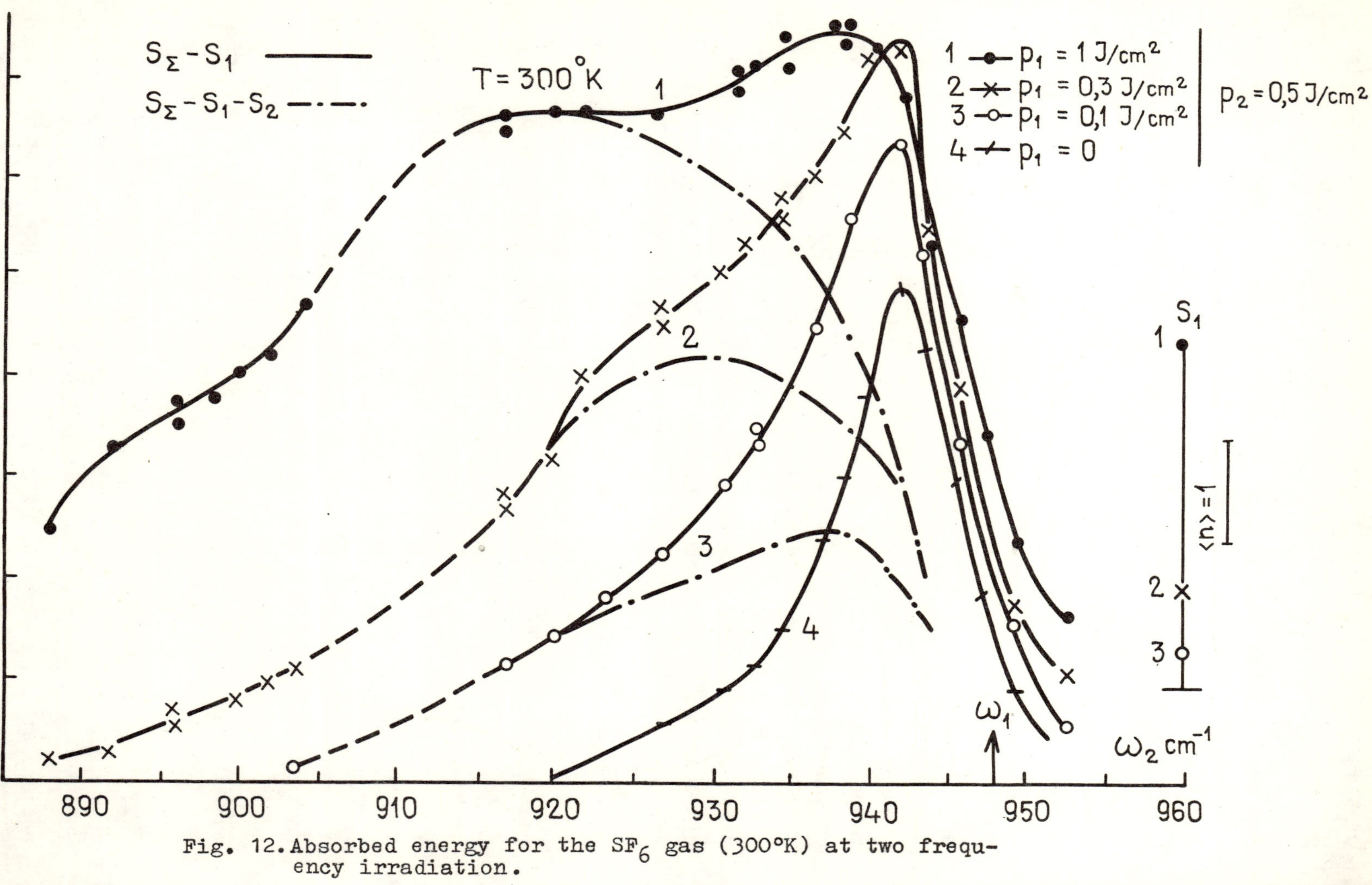


Fig. 12. Absorbed energy for the $SF_6$ gas (300°K) at two frequency irradiation.

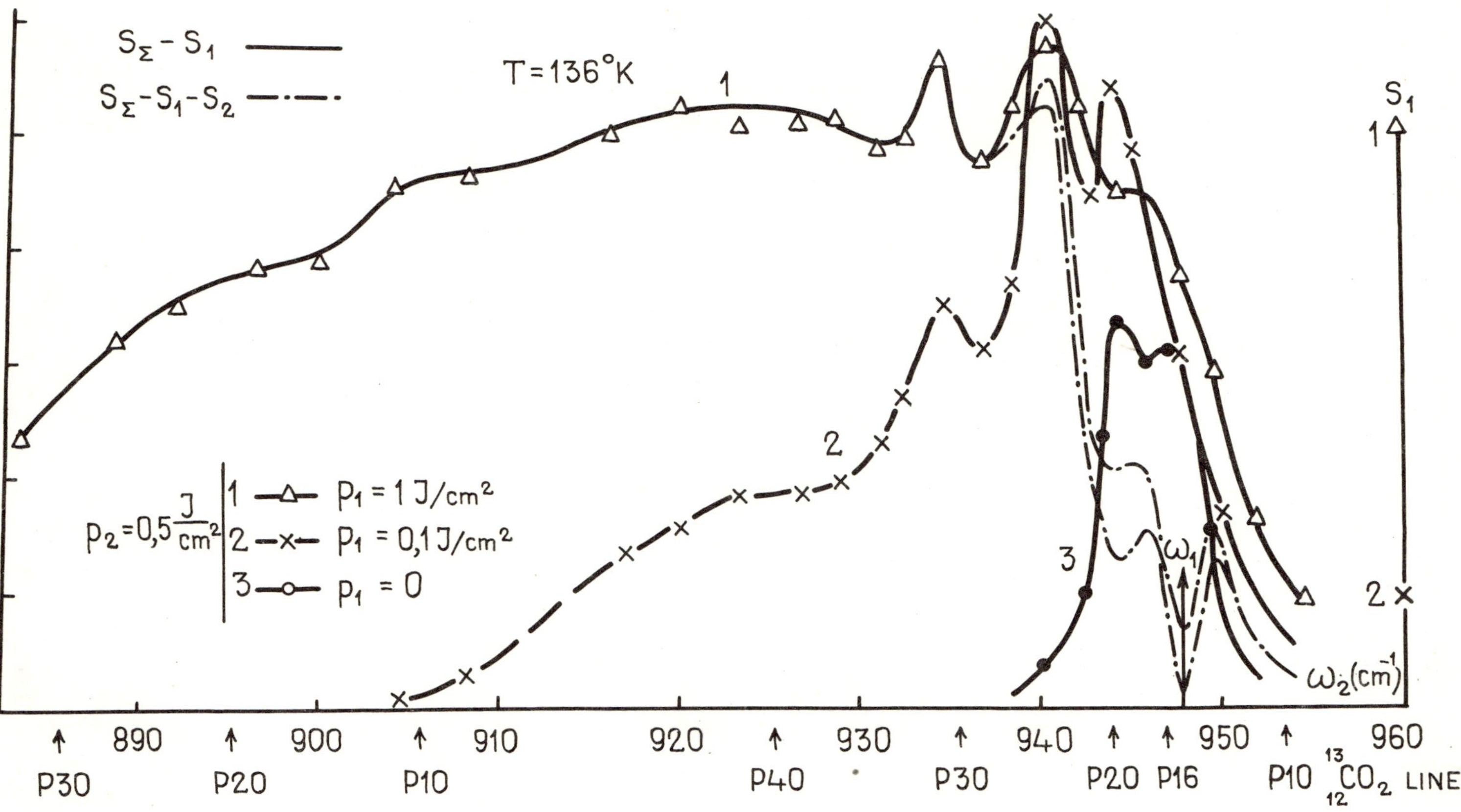


Fig. 13. Absorbed energy for the $SF_6$ gas (136°K) at two frequency irradiation.

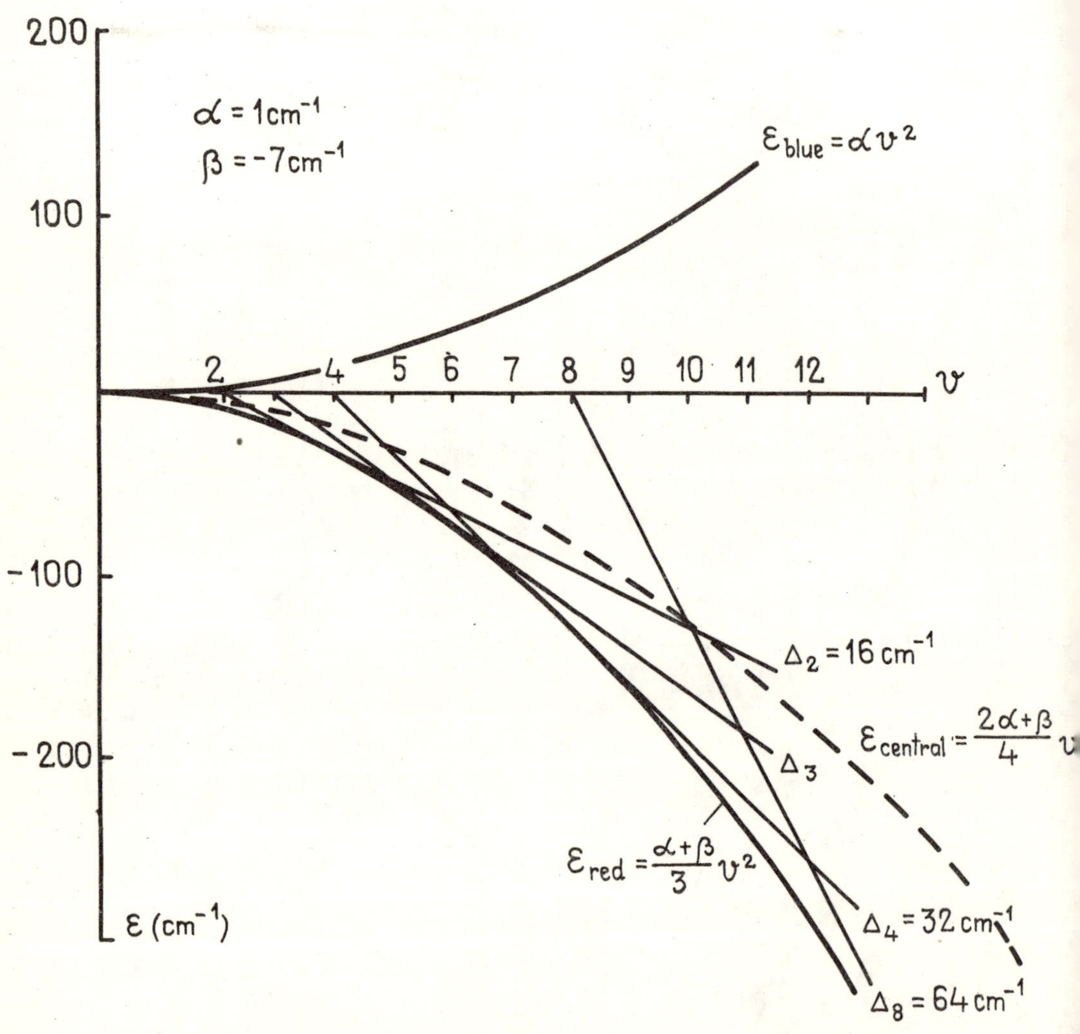


Fig. 14. The zonal structure of the $SF_6$ molecule $v_3$ mode higher vibrational states quasicontinuum.

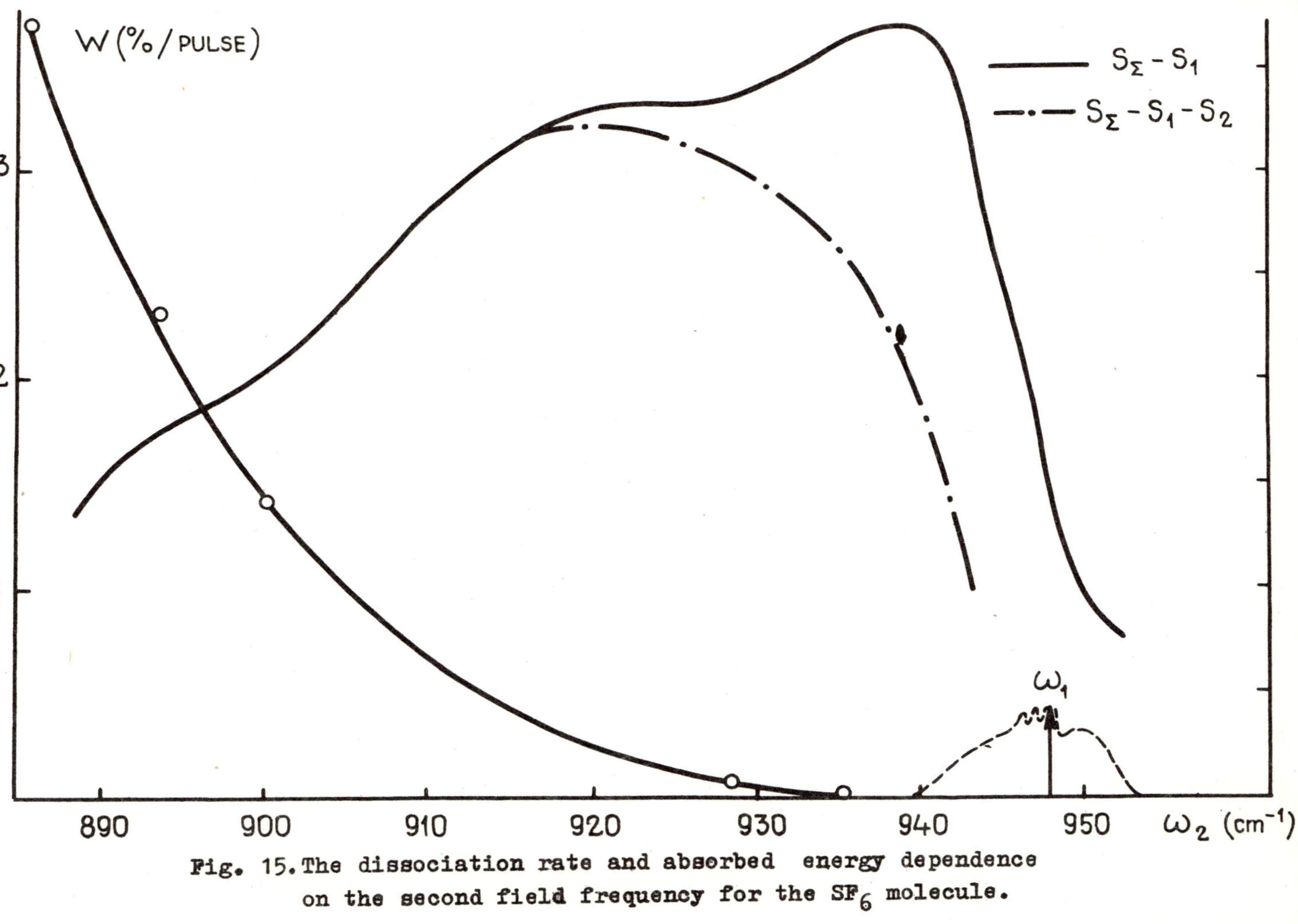


Fig. 15. The dissociation rate and absorbed energy dependence on the second field frequency for the $SF_6$ molecule.

up to $v = 8$.

## 4. Conclusion

The results presented above prove the realizationability of isotopically selective irreversible action of the laser radiation upon the matter. The experimental techniques developed in the course of the irreversible selective action investigations are very important in the spectroscopy of atoms and molecules.

The selective photoionization of atomes, being physically more clear, is the most general technique for laser isotope separation. In addition, this process turned out to be the powerful technique for spectroscopical investigations giving the means to register with high precision the hyperfine structure spectra of the odd isotopes of any element in the natural abundant isotope mixture, to study the processes of excitation energy transfer in the heavy metal vapour and so on.

The selective multistep IR-dissociation of polyatomic molecules seems to be a new phenomenon far from being trivial in its nature. The elementary act of dissociation is caused by the absorption by the molecule of great number of IR-laser radiation quanta. This purely radiative process is effective at relatively moderate intensities of radiation in spite of the strong anharmonicity of molecular vibrations. Careful investigations of this process is especially interesting for the molecular and chemical physics.

## 5. Acknowledgement

The author is grateful to A.M. Prokhorov under general leadership and with permanent collaboration with whom the investigations constituting the basis of this paper have been carried out. The author is grateful to V.M. Akulin, S.S. Alimpiev, B.B. Krynetskii, V.A. Mishin, B.G. Sartakov, O.M. Stelmakh, E.M. Khokhlov whose results constitute the theme of the review discussion given in this paper.

## REFERENCES

1. N.V.Karlov, A.M.Prokhorov. UFN 118, 583,(1976).
2. N.G.Basov, E.M.Belenov, V.A.Isakov, E.N.Markin, A.N.Oraevsky, V.I.Romanenko. UFN, 121, 427, (1977).
3. N.V.Karlov, A.M.Prokhorov. UFN, 123, 98,(1977).
4. V.S.Letokhov, UFN, 125 , 57 (1978)
5. N.V.Karlov, B.B.Krynetsky, A.A.Mishin, A.M.Prokhorov, A.D.Savel'ev, V.V.Smirnov. Pis'ma v Zh.TF, 2, 961,(1976).
6. N.V.Karlov, B.B.Krynetsky, V.A.Mishin, A.M.Prokhorov, A.D.Savel'ev, V.V.Smirnov. Kvantovaya elektronika, 3, 2486 (1976).
7. N.V.Karlov, B.B.Krynetsky, B.A.Mishin, A.M.Prokhorov. Opt. Comm. 21, 384, (1977).
8. V.N.Ischenko, N.V.Karlov, B.B.Krynetsky, V.N.Mishin, V.A.Mishin, A.M.Rashev. Pis'ma v Zh.TF, 3, 1041,(1977).
9. N.V.Karlov, B.B.Krynetsky, V.A.Mishin, A.M.Prokhorov. Appl. Optics, 17, 856,(1978).
10. R.V.Ambartsumyan, V.M.Apatin, V.S.Letokhov, A.A.Makarov, V.I.Mishin, A.A.Puretsky, N.P.Furzikov, ZhETF 70, 1660 (1976).
11. N.V.Karlov, B.B.Krynetsky, O.M.Stelmakh. Kvantovaya elektronika 4, 2275 (1977).
12. R.F.Stebbings, F.B.Dunning, R.D.Rundel. Proc. IV$^{th}$ Conf. Atom.Phys. July 22-26, 1974, Heidelberg FRG Atomic Physics 4 page 713, Plenum Press No4 L 1975.
13. B.M.Smirnov. Atomnye stolknoveniya i elementarnye protsessy v plasme. Moscow, Atomizdat 1968.
14. N.V.Karlov, B.B.Krynetsky, V.A.Mishin, A.M.Prokhorov, Pis'ma v ZhETF, 25, 535 (1977).
15. N.V.Karlov, B.B.Krynetsky, V.A.Mishin. Kvantovaya elektronika 5, 877, (1978).
16. N.V.Karlov, B.B.Krynetsky, V.A.Mishin, A.M.Prokhorov, Pis'ma v ZhETF, 25, 318,(1977).
17. C.D.Cantrell, H.W.Galbraith, J.R.Ackerhalt. Multiphoton Processes. Proceedings of an International Conference at the University of Rochester, Rochester, No4, June 6-9, 1977, edited by J.Eberly and P.Lambropoulos, John Wiley, No4, 1978, p 331.

18. V.S.Letokhov ibid p.331.
19. I.I.Tugov ibid p.349.
20. E.R.Grant, P.A.Schulz, A.S.Sudbo, M.J.Coggiola, Y.R.Shen, Y.T.Lee ibid p.359.
21. N.V.Karlov ibid p.371.
22. V.M.Akulin, S.S.Alimpiev, N.V.Karlov, L.A.Shelepin, ZhETF 69, 836, (1975).
23. V.M.Akulin, S.S.Alimpiev, N.V.Karlov, N.A.Karpov, Yu.N.Petrov, A.M.Prokhorov, L.A.Shelepin. Pis'ma v ZhETF, 22, 100,(1975).
24. V.M.Akulin, S.S.Alimpiev, N.V.Karlov, B.G.Sertakov, L.A.Shelepin. ZhETF, 71, 454,(1976).
25. V.M.Akulin, S.S.Alimpiev, N.V.Karlov, B.G.Sertakov. ZhETF, 72, 88,(1977).
26. V.M.Akulin, S.S.Alimpiev, N.V.Karlov, A.M.Prokhorov, B.G.Sertakov, E.M.Hokhlov. Pisma v ZhETF, 25, 428,(1977).
27. V.M.Akulin, S.S.Alimpiev, N.V.Karlov, B.G.Sertakov. ZhETF, 74, 490(1978).
28. S.S.Alimpiev, V.N.Bagatashvili, N.V.Karlov, V.S.Letokhov, V.V.Lobko, A.A.Makarov, B.G.Sertakov, E.M.Hokhlov. Pis'ma v ZhETF, 25, 582,(1977).

# LASER DETECTION OF SINGLE ATOMS

V.I.Balikin, G.I.Bekov, V.S.Letokhov, V.I.Mishin

Institute of Spectroscopy of Academy of Sciences
USSR, Troitsk, Moscow region, USSR

The development of methods of laser detection of microscopic amounts of the substance has been paid considerable attention to of late. In principle one atom is the limit of detection, since it still contains complete spectral information about the substance. Therefore the development of methods of single atom detection becomes an actual problem. Among the most perspective methods applicable for the detection are: the method of laser excitation of resonance fluorescence which makes it possible to obtain the maximum number of photons scattered by one atom and the method of selective stepwise ionization of atoms by laser radiation enabling to convert practically every atom into an ion /1/.

The whole problem of single atom detection can be conventionally subdivided into three interdependent problems: 1) accumulation of the element and obtaining of free atoms; 2) transference of atoms into the detection volume 3) detection of atoms.

As far as our problem is concerned, up to now the detection of atoms has been most elaborated: a single atom interacting with a laser beam provides a selective signal (photons /2/, ions /3/), surpassing the level of noises. This signal allows to fix the flight of a single atom through the laser beam. The methods allowing to carry out the above mentioned detection procedure are reported in this review.

## 1. Fluorescence detection of single atoms

### 1.1. Statement of the problem

The peculiarity of the method of resonance fluorescence is that one and the same atom can interact many times with laser radiation scattering in all directions photons the frequency of which coinsides with that of the exciting ones.

If the resonance transition of the atom corresponds to the two level system, then during the time T, when the atom flies through the light beam, the atom will emit

$$N = \frac{T}{\tau_{21}(1/\eta + 1)} \qquad (1)$$

photons due to the spontaneous decay of the upper level ( $\tau_{21}$ is the life of the upper state, $\eta$ is the ratio of populations of the upper and low states). When the laser radiation intensity exceeds the saturation intensity, populations of the levels become equal. In this case the number of photons emitted is

$$N_{max} = \frac{T}{2\tau_{21}} \qquad (2)$$

(degeneration of the states is not taken into account).
Estimate, for example, the maximum amount of photons emitted spontaneously by a sodium atom. The average velocity of the thermal motion of Na atoms with temperature $50^{o}C$ is $5.10^{4}$ $cm.sec^{-1}$. The decay time of the excited state is $1.6 \cdot 10^{-8}$ sec. According to /2/ the sodium atom will emit $N_{max}$=250 photons on the pathway $h$ =0.4 cm. It is possible to accumulate about $N_{det}$=20 photons on the photomultiplier cathode if the solid angle of collection is one sterad. The quantum efficiency of the best photomultipliers at the wavelength $\lambda$ =589.0 nm is 10-20%. Thus 2-4 photoelectrons can be obtained from the photomultiplier cathode when the atom flies through the light beam.

In the method under consideration the sensitivity is limited by the background signal (the signal which appears as a result of laser radiation scattering on the cell details). This is due to the fact that registration of photons emitted by the atoms is carried out at laser radiation frequency. The number of background photons reaching the photomultiplier cathode during the intersection of a laser beam by the atom is

$$N_{phot} = \xi \frac{P_{laser} T}{\hbar \omega}, \qquad (3)$$

where $P_{laser}$ is the laser radiation power introduced into the cell, $\xi$ is the parameter defining the cell quality. The cell used in our experiments had the parameter $\xi$ =$10^{-14}$. During the time of the atom flight through the beam $10^{-2}$ background photons reached the photomultiplier cathode.

Thus, the signal power from the atom in the method of resonance laser fluorescence can exceed considerably the background signal power, and to detect one atom it is necessary to pick out the useful signal, which appears as accidentally in time as the atom appears in a beam.

In order to pick out the fluorescence signal we have used a two-channel system of registration operating in the conditions of coincidences. The fluorescence signal was registered by two photomultipliers. At laser radiation intensity providing the absorption saturation, the atom re-emitted a sufficient number of photons, so that during the presence of the atom in a beam at least one single-electronic pulse was formed at every photomultiplier output. The appearance of pulses at the output of the photomultipliers within this time period was considered to be a coinciding event. Such a scheme of registration first of all allowed to reduce essentially the influence of the background and, secondly, to establish the fact that the atom was present in the laser beam.

1.2. Cyclic excitation of sodium atoms

**Fig.** 1 shows a hyperfine structure of the ground and excited states of a Na atom. The possible transitions between the components of a hyperfine structure, the Doppler contour of the $D_2$-absorption line and a mode structure of the dye laser are illustrated in the same figure.

The only transition, which corresponds to the two-level scheme and provides the cyclic interaction of the atom with a laser field, is the transition $3^2S_{1/2}$ (F=2) $\rightarrow 3^2P_{3/2}$ (F'=3). This is due to the fact that according to the rules of selection the atom from (F'=3)-state can transit only into a lower state F=2. This transition can be excited by single mode laser radiation, however, because of the absorption at the wing of the line, the $3^2P_{3/2}$-state with F'=2 will be also excited. As a result after several dozens of cycles the atom will be in (F=1)-state and stop interacting with a laser field. The influence of the hyperfine structure of the ground state was eliminated in our experiment in the following way. The atom was excited from both sublevels of the ground state by the wideband laser radiation ($\Delta\nu_{Laser}$ =0.16 cm$^{-1}$) consisting of 11 modes being 416 MHz apart each other. The laser radiation intensity was chosen so that every mode could broaden the components of the excited state to such an extent as to

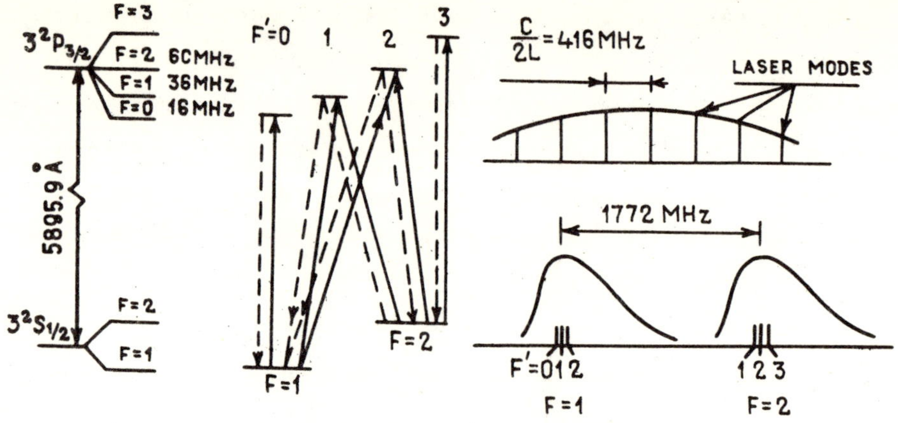


Fig. 1. Energy level diagram of the sodium atom ground and resonance states. Spontaneous and induced transitions. Laser radiation structure and absorption lines.

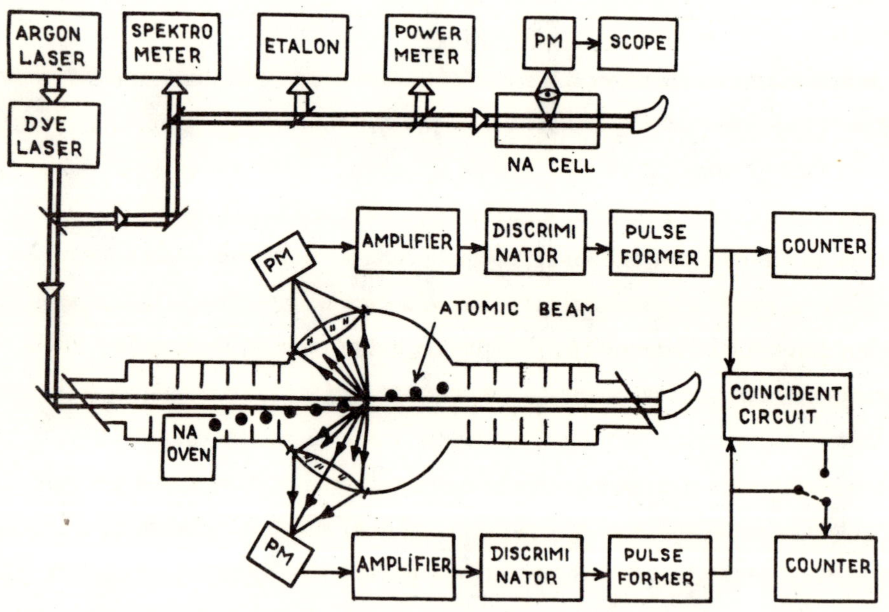


Fig. 2. The experimental set-up for a single Na atom detection by the method of resonance fluorescence.

provide their overlapping. This was achieved at the radiation intensity in every mode being 0.18 W/cm$^2$. In this case the broadening of the hyperfine structure components was 35 MHz while the total width of the upper state was 100 MHz. If now sodium atoms, which are, for example, in the ground state at F=1, interact with one of the modes of laser radiation, after several cycles they will be in the ground state at F=2 which is 1772 MHz apart from the F=1 state. The interaction with a laser field will not be interrupted in this case, since the nearest mode will be at a distance of 108 MHz from the absorption line of these atoms and will again produce their effective excitation.

1.3. Experimental set-up

The experimental set-up is shown in Fig.2. The CW dye-laser M-375 was used to excite sodium atoms. The narrowing of the laser spectrum up to 0.16 cm$^{-1}$ width was carried out by an additional quartz 2 mm etalon. The laser radiation power was 0.2 W. The laser frequency tuning to the D-line was accomplished by means of a spectrograph and an additional cell containing sodium vapours.

The beam of sodium atoms was formed when the vapours escaped from the oven through the channel of 0.8 mm in diameter and 6 mm in length. The oven temperature could be changed from the room one up to 100$^{\circ}$C. The atomic beam intersected the laser one at a distance of 130 mm from the oven at the angle 8$^{\circ}$. The laser radiation was focused into the cell. In this case the diameter of a laser beam in the region of interaction was 0.57 mm. The pathway of the atoms in a light beam was 4 mm.

The resonance fluorescence from the atoms was collected by short focus lenses on the cathodes of two photomultipliers. The signals from the multipliers arrived consequently at the amplifier, discriminator, pulse former and then at the pulse counter or at the scheme of coincidence. The number of pulses at the output of one channel of the scheme of coincidences were recorded by the counter. The time constant of the registration system was chosen to be equal to the average flight time of an atom through the laser beam and was 20 mksec. The circuit of coincidences was set in if the interval between the pulses did not exceed 20 mksec. For the maximum elimination of the background radiation the cell, which was made of aluminium alloy, was blackened. The windows were placed at Brewster's angle. Diaphragms of variable diameter were placed outside and inside the cell. As the result the ratio of the number of background photons reaching the cathode to the number of photons of laser radiation arriving at the cell has been obtained to be $10^{-14}$.

## 1.4. Results of the experiment

The dependence of the number of pulses per 10 sec in one channel (the upper curve) and in the scheme of coincidence (the lower curve) on the intensity of laser radiation are shown in Fig.3. The dependences have a typical character of the saturation curves with saturation intensity $I_{sat} = 10\ W/cm^2$.

When the laser radiation intensity is less than the saturation intensity, the signal in one channel is larger than that in the circuit of coincidences. This is explained by the fact that the atom flying through the laser beam emits the number of photons which is not sufficient for their simultaneous registration by two channels. In this case the probability of the atom registration in every channel is less than one. Since the probability of registration in the circuit of coincidence is equal to the product of probabilities of registration in every channel, it will be less than the probability of registration in one channel. When the laser radiation power is increased, the rate of photon emission by the atom grows as well. Beginning with the intensity $10\ W/cm^2$ each photomultiplier receives the signal from one and the same atom and in this case the number of coincidence pulses becomes equal with the number of pulses in one channel. The probability of registration of the signal from the atom in the scheme of coincidence becomes equal to one. Fig.4 shows the dependence of the flow of sodium atoms through the registration volume at the oven temperature. The minimum detectable signal corresponds to the oven temperature which is $45^oC$. In this case the flow of 10 atoms/psec is detected, that corresponds to $10^{-4}$ atoms in the registration volume. The limit of detection is caused by the background fluctuations contributing to the number of pulses in the scheme of coincidence.

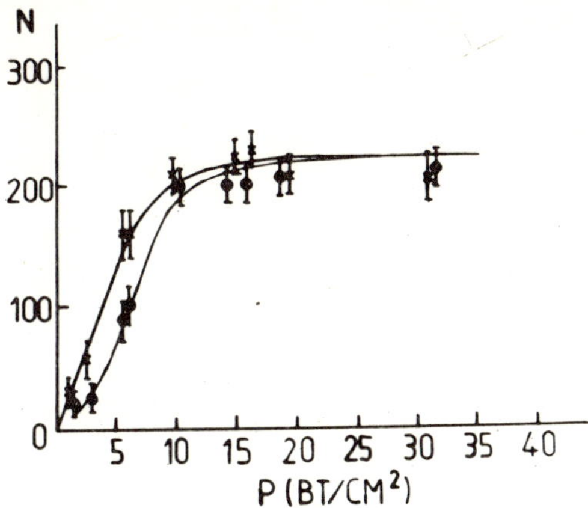


Fig.3. The dependence of the number of pulses per 10 sec in one channel (the upper curve) and in the scheme of coincidence (the low curve) on laser radiation intensity.

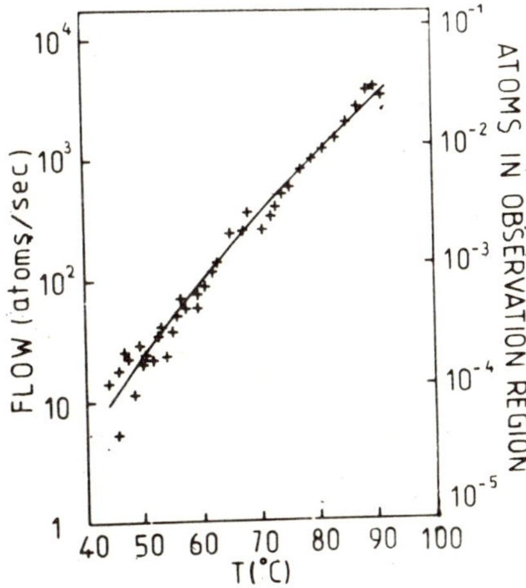


Fig.4. The dependence of sodium atom flow through the observation region on the oven temperature.

### 1.5. Ways of the method improvement

The main factors restricting the realization of the maximum sensitivity of the fluorescence method are:

1) the presence of the background radiation;
2) the insufficient quantum efficient of the cathode, and the geometry factor which prevents inevitably some photons emitted by the atom from reaching the photocathode.

One of the ways to overcome the first factor is the usage of metastable levels for cyclic multiphoton excitation of atoms. At the same time it will be possible to use this method for the cases where the atom cannot be considered as a two-level system.

To illustrate this Fig.5a shows the scheme of low levels of a Ba atom and transitions used for multiphoton cyclic excitation of the atom: $6s^2\,{}^1S_0 - 6s\,6p\,{}^1P_1^\circ$ (the wavelength 5535 Å) and $6s\,5d\,{}^1D_2 - 6p\,5d\,{}^1P_1^\circ$ (5826 Å). The ratio of rates for the ${}^1P_1 - {}^1S_0$ and ${}^1P_1 - {}^1D_2$-transitions is $\simeq$ 2.4 therefore at saturation of ${}^1P_1 - {}^1S_0$-transition by laser radiation the atom in the average after 24 cycles is in the metastable state with the lifetime 0.5 sec. The laser radiation with the wavelength 5826 Å transfers the atom from $6s5d\,{}^1D_2$-state into $6p5d\,{}^1P_1^\circ$ which decays into the ground state during $12.3 \cdot 10^{-9}$ sec. The atom detection can be performed at the wavelength 3501 Å.

Fig.5b shows the ellipsoid reflection system for the collection of photons emitted by the atom which will be used in our experiment. The atom emitting photons is in one focus of the ellipsoid, the photomultiplier registering photons is in the other focus. Such a system makes it possible to obtain a signal of 4-5 photoelectrons from every flying Na atom and discriminate it from the single-electron background.

## 2. Photoionization detection of single atoms by laser radiation

The method of stepwise selective photoionization of the atom by laser radiation can be applied for detection of atoms with high spectral resolution and high sensitivity. In this case the atom is excited into the intermediate state, then it is ionized by the radiation of a laser with another frequency. If the detection is performed in the vacuum (the most interesting case from the point of view of maximum spectral resolution), it is necessary to send pulses of laser radiation with repetition rate 50 kHz to ionize every atom intersecting the beam (the motion velocity $5 \cdot 10^4$ cm/sec and the interaction pathway 1 cm). Effective ionization of most of atoms can be performed at average power of laser radiation available in laboratory conditions ($P \approx 1$ W) if the atom is ionized through the highlying (Rydberg) states. Such a method has been proposed and investigated in works /4/. In these works the non-resonance photoionization from the intermediate state is substituted for the process of resonance excitation of an atom from the same state into a high-lying state with the subsequent ionization by the electric field pulse. The ionization efficiency in such process is close to one. Since the excitation on all the steps is resonant, not very high energy density of laser pulses which is available by means of the existing dye lasers, is required for the absorption saturation at these transitions.

The results of the investigation of selective stepwise ionization of sodium and ytterbium atoms through the Rydberg states for detection of single atoms are presented in this work.

### 2.1. Description of the set-up Technique of the experiment realization

Atoms in a beam (Fig.6) were excited into the Rydberg states by two or three tunable dye lasers. The laser beams intersected the atomic beam between the two electrodes where the electric field pulse was sent to. Ions that appeared as a result of selective photoionization were extracted through a slit in one of the electrodes and registered by the secondary emission multiplier. Dye laser pumping was carried out by a nitrogen transversal discharge laser (power - 350 kW, the pulse duration - 10 nsec, the pulse repetition rate - 12 Hz). The dye lasers consisted of the following elements: a mirror with reflection factor 4÷20%, a cell with the dye solution, a telescope with 25 fold magnification, a diffraction grating 1200 grooves/mm in the first order. The lasing spectrum width of the dye lasers was $\Delta\nu_{las} = 1$ cm$^{-1}$, the pulse duration $\tau_L = 7$ nsec.

Laser beams were directed into a vacuum chamber. The residual pressure in the chamber did not exceed $10^{-6}$ torr. The concentration of atoms in the excitation volume varied when the oven temperature was changed and could be $10^8$ atoms/cm$^3$. Concordant spark gap with a cable line started by nitrogen laser radiation was used as a generator of electric field pulses. Such a system made it possible to form a single voltage pulse at a concordant load. This pulse of a rectangular shape with the duration 10 nsec and amplitude up to 20 kV was synchronized with the nitrogen laser pulse with the accuracy up to 5 nsec. The length of a transmitting cable line was chosen so that the voltage pulse appeared at the electrodes in 20 nsec after laser pulses.

The ions which appeared as a result of ionization of the Rydberg atoms by the electric field, obtained an impulse in this field in the direction of the field intensity vector (Fig.6). The value of the velocity corresponding to this impulse was larger than the velocity of atoms in a beam by two orders. Therefore the ion motion practically occurred in the direction perpendicular to the direction of the atomic beam motion. Reaching the electrode with zero potential the ions were extracted through the slit in this electrode by the cathode electric field of the secondary-emission multiplier /SEM/, which had the potential - 4 kV. The maximum energy obtained by the ion in the region between the electrodes did not exceed 0.3 kev and in the field of SEM it was 4 kev. Therefore ion energy variation at the change of electric field intensity did not practically influence upon the output characteristics of SEM.

The excitation volume was a cylinder with the diameter 0.3÷1 mm and the length 3÷5 mm. The size of the slit, through which the ions were extracted, was 3x10 $mm^2$. Thus, the geometry of the atomic and laser beams and of the slit allowed to extract practically all the ions from the interelectrode region. In our experiment the efficiency of ion registration by the secondary emission multiplier was close to one. The signal from SEM was amplified and registered either by the oscilloscope or by the system of count of ions consisting of a discriminator, a standard pulse former and a frequency meter. The schemes of energy levels of sodium (a) and ytterbium atoms (b) and the transitions used for excitation are shown in Fig.7a,b. Sodium atoms were excited from the ground $3^2S_{1/2}$ state into the resonant $3^2P_{3/2}$ one by the laser with wavelength 5890 Å. The wavelength of the second laser was tuned within the range 4245÷4160 Å, that provided the excitation of the Rydberg states $s$ and $d$ of Na atom with the principal quantum number $n$ = 12÷16 from the $3^2P_{3/2}$ - state. The three-step scheme of excitation was used for ytterbium. The first-step laser with the wavelength $\lambda_1$ = 5556.5 Å transferred ytterbium

a)

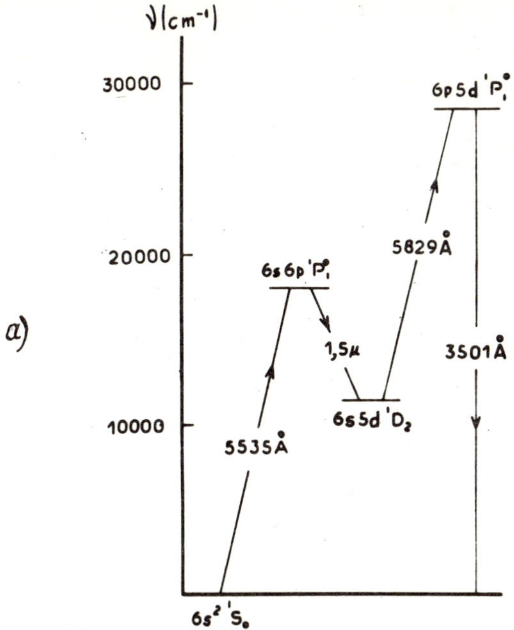


b)

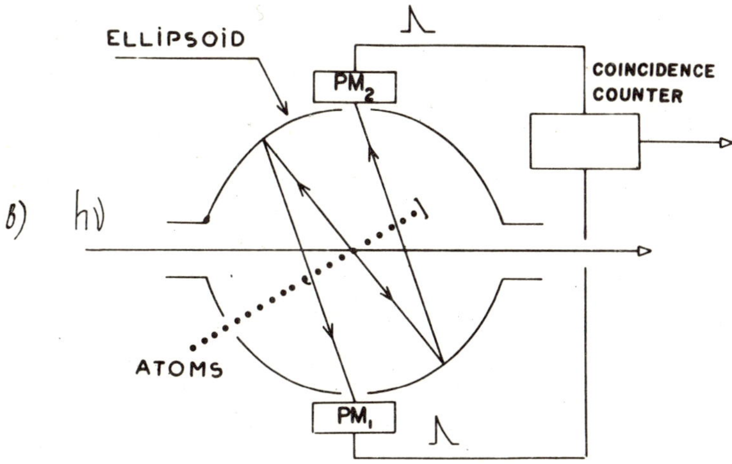


Fig.5. a) Diagram of Ba atom low energy levels and transitions used for multiple cyclic excitation.

b) The ellipsoid reflector used in the fluorescence method of single atom detection.

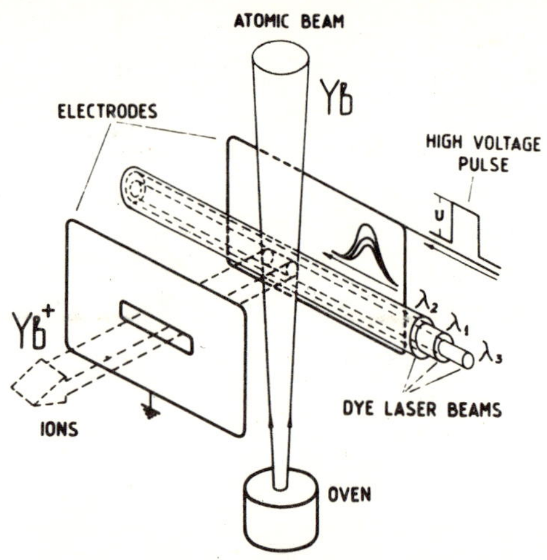


Fig.6. The experimental set-up for laser photoionization detection of single atoms through Rydberg states

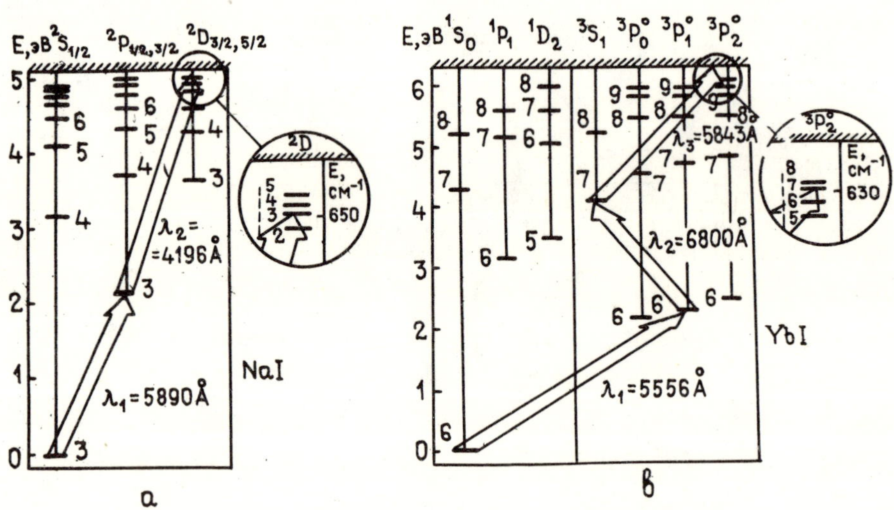


Fig.7. Sodium (a) and ytterbium (b) atom energy level diagrams and transitions used.

atoms from the ground $6\,^1S_0$ state into a metastable $6^3P_1^\circ$ state. The second laser with $\lambda_2$ = 6799.6 Å was used for the excitation of a strong transition $6^3P_1^\circ - 7^3S_1$. Tuning the wavelength of the third laser radiation in the range 5950÷5770 Å, it was possible to excite the ytterbium atoms into $P$ - states with n = 14÷20.

### 2.2. Excitation and ionization of the maximum number of atoms

The dependence of ion yield on the intensity of an electric field was investigated for high-lying $P$-states of an ytterbium atom (n = 16÷20). The strength of the critical electric field measured for ytterbium as well as for other elements is proportional to $(n^*)^{-4}$, where n* is the principal effective quantum number.

A very important feature of the method used is the choice of the principal quantum number $n$ of the Rydberg state. For effective excitation of the atoms into a high-lying state, it is necessaty to provide the saturation of the chosen transitions. The cross-section of the atom excitation on the last step falls proportionally to $(n^*)^{-3}$. On the other hand, the critical electric field increases proportionally to $(n^*)^{-4}$, if n decreases. The case will be probably optimum, when the fields with strength available in laboratory scale are applied. The $13\,^2D_{3/2}$ state for sodium ( $\mathcal{E}_{crit}$ = 14 kV/cm ) and the $17\,^3P_2$ state for ytterbium ( $\mathcal{E}_{crit}$ = 11.5 kV/cm) satisfy this condition.

Fig.8 shows the dependence of the yield of ytterbium ions on the pulse energy density of the lasers of the first, second and third steps and on the electric field strength (similar dependences were obtained for sodium as well).Laser pulse energy densities necessary for the saturation of the chosen transitions were determined from these dependences. They are $E_1 = 8\cdot10^{-5}$ J/cm$^2$, $E_2 = 10^{-6}$ J/cm$^2$, $E_3 = 3\cdot10^{-3}$ J/cm$^2$ respectively.The energy density of the transition saturation can be estimated by the formula $E_{sat} = \frac{h\nu}{2\sigma}$, $\sigma = \frac{g_a}{g_b}\cdot\frac{\lambda^2}{2\pi}\cdot\frac{A_{ab}}{\Delta\omega}$, where $\sigma$ is the cross-section of radiation transition a → b; $h\nu$ and $\lambda$ are the quantum energy and the transition wavelength, $g_a$ and $g_b$ are degeneration factors of the states, $A_{ab}$ is the Einstein coefficient and $\Delta\omega$ is the width of the transition spectral line. The Einstein coefficients of the transitions $6^3P_1 - 6^1S_1$ and $7^3S_1 - 6^3P_1$ are well known: $A_{6p-6S} = 1.2\cdot10^6$ sec$^{-1}$, $A_{7s-6p} = 10^8$ sec$^{-1}$. The

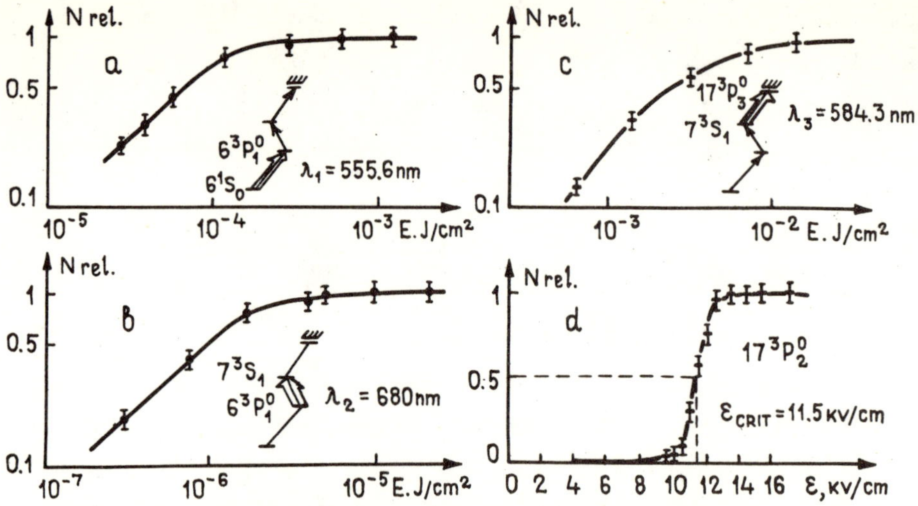


Fig. 8. The dependence of the number of ytterbium ions on the laser energy density of the first (a), second (b), third (c) steps and the electric field strength (d).

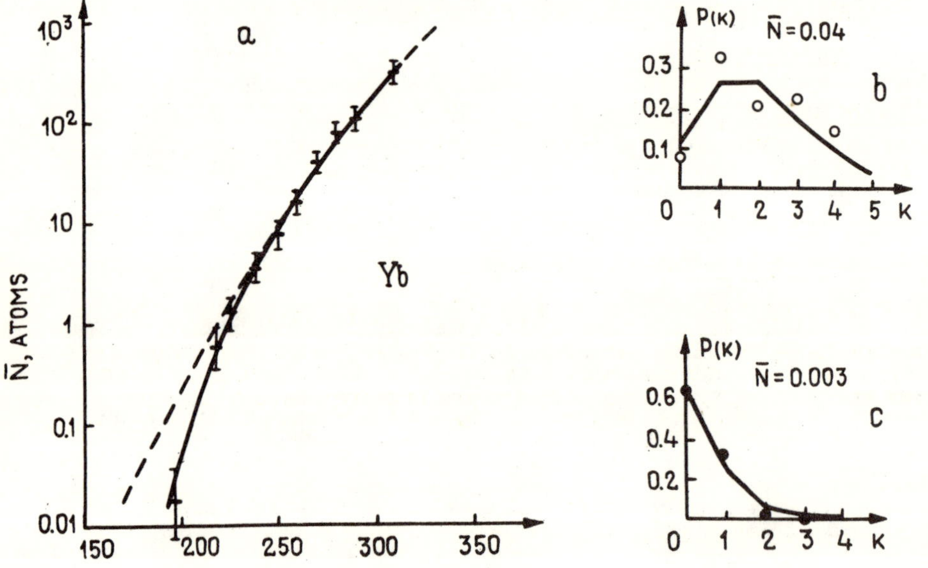


Fig. 9 a) The dependence of the number of atoms in the observation region on oven temperature;

b) Distribution of the number of counts during the observation time 10 sec; the average number of atoms in the observation region N = 0.04;

c) the same as in (b) during 20 sec; the average number of atoms in the observation region N = 0.003.

absorption line width in the atomic beam used in the experiment is $\Delta\nu_{abs}$ = 0.01 $cm^{-1}$. The corresponding transition cross-sections are $\sigma_{6s-6p}$ = $1.5\cdot10^{-12}$ $cm^2$, $\sigma_{6p-7s}$ = $6\cdot10^{11}$ $cm^2$. Laser pulse energy densities, required for the saturation, are determined by the expression $E = E_{sat}\cdot\frac{\Delta\nu_{Laser}}{\Delta\nu_{absor}}$. For laser radiation $\Delta\nu_{Laser}$ = 1 $cm^{-1}$ saturation energies are $E_1^{th}$ = $2\cdot10^{-5}$ $J/cm^2$, $E$ = $5\cdot10^{-7}$ $J/cm^2$ respectively.

Laser pulse energy exceeded the saturation energy and corresponded to the work conditions on the plateau (Fig.8) in all the steps. In this case the atoms populated the states evenly in accordance with their degeneration factor. The 5/12 of all the atoms which were in the excitation volume were in 17 $^3P_2^o$ state. Fig.8d shows that at the field strength $\mathcal{E} \gtrsim$ 14 kV/cm all the excited atoms are ionized. Thus, the probablity of the atom excitation in chosen conditions into the Rydberg states and hence the probability of the atom ionization is 1/2.

### 2.3. Dependence of atom concentration in a beam on temperature

#### Calibration of the system

Fig.9a shows the dependence of the number of atoms in the registration volume on temperature (dotted curve) which is calculated for the beam of ytterbium atoms in the usual way. The dependence of ytterbium ion yield on the oven temperature obtained at laser pulse energy densities ensuring saturation is shown in the same figure. The experimental and calculated curves are parallel if the oven temperature is more than 250°C. The experimental curve is matched with the theoretical one so that they coinside at these temperature values. The difference between the experimental curve and the calculated one at lower temperatures is probably caused by the fact that the atomic vapour in the oven in these conditions is not saturated. At the temperatures lower than 230°C a strong signal unstability is displayed which is expressed in the appearance of pulses of practically equal amplitude with the pulse repetition rate sharply decreasing if the temperature is reduced. It is worth noting that the background ion pulses (the pulses which were not connected with the selected ion) were practically absent. Therefore a strong ion signal instability is caused by the fluctuations of the number of atoms in the excitation volumewhich at low beam density can be of the same order as the average number of atoms in this volume

### 2.4. Fluctuations of the number of atoms in the volume of registration

It is possible to make the density of the atomic beam so low that the probability of simultaneous reaching the excitation volume by two atoms will be much less than the probability of reaching it by one atom. In this case the system of registration will react mostly upon the appearance of only one atom. The probability of registration of $k$ ions during T time by the system is determined by the Poisson distribution

$$P(k) = \frac{\bar{k}^k}{k!} e^{-\bar{k}},$$

where $\bar{k}$ is the average number of ions, registered during the observation time.

Fig.9(b,c) shows values of the registration probability of $k$ ions during the observation time obtained for ytterbium atoms at different values of the average number of $N$ atoms in the volume of excitation. The Poisson distributions (continuous lines) calculated for the same average values of $\bar{k}$ are presented for comparison. As Fig.9 shows, the experimental distributions are approximating to the Poisson ones.

## 3. Discussion, conclusions, applications

### 3.1. Potentialities of the methods and their comparison

The sensitivity of the method of laser resonance fluorescence can lead to the detection of one sodium atom. To achieve this it is necessary to increase the efficiency of collection of photons scattered by the atom or to excite the atoms by two- or one-frequency radiation so that the influence of a hyperfine structure is eliminated.

It is easy to estimate the excitation selectivity of the atom with the absorption line placed in the vicinity of absorption lines of other atoms. It will be $(\Delta\nu/\Delta\nu_L)^2$, where $\Delta\nu$ is the distance between the lines, $\Delta\nu_L$ is the width of the laser or the absorption line. It is easy to obtain the excitation selectivity more than $10^6$ with $\Delta\nu_L = 10^{-3}$ cm$^{-1}$ and the detuning $\Delta\nu = 1$ cm$^{-1}$.

In spite of the relative simplicity of the method it should be noted that its application is rather limited. This is explained by the fact that in order to get the two-level system it is necessary to excite the resonance state lying to the ground state closer than to other states. Here the wavelength of the corresponding transition should be in the range 420-800 nm, where the CW dye lasers are operating with the power sufficient for transition saturation ( $P_L$ = 0.2 W). Alkaline and alkaline earths present the examples of the elements that can be detected by the method of resonance fluorescence. As a rule the atoms of heavy elements have such a complex energy spectrum that the transitions from the excited state to the other low-lying states are possible. As a result of these transitions the cyclic recurrence interaction of the atom with the laser field is lost.

Among all methods used for detection of single atoms the most perspective one is the method of selective stepwise atom ionization by laser radiation. This method is universal, i.e. it is applicable for most elements of the periodic table. It is possible to ionize practically every atom by means of this method. If the laser pulse energy exceeds the saturation energy, the ionization yield is close to one /4/. Using the modern methods of registration of charged particles the appearing ion can be registered with efficiency also approaching to one.

The atom can be excited into a high-lying state in two or three steps by the radiation of pulsed dye lasers synchronized with each other. The choice of the excitation scheme depends on a concrete atom. For instance, the two-step excitation scheme is convenient for the atoms of alkaline metals. For the heavy elements with ionization potential more than 6 ev the three step excitation scheme is preferable. At excitation of the atom up to the ionization level the selectivity is achieved at every step. The resulting selectivity of the atom ionization may be of the order $10^{16}$ - $10^{20}$ if the excitation selectivity at one step is $10^{5}$ - $10^{10}$

Excitation and ionization can be carried out both in the atom beam in the vacuum and in a buffer gas. The first experiment on single atom detection by the photoionization method in a buffer gas has been accomplished in work /5/. Cesium atoms in a gas under the pressure of several hundred torr were excited into $7^2P_{3/2}$ -state by the pulsed dye laser radiation with the flash-lamp pumping and were ionized by the same radiation from the excited state. The electron that appeared was detected by the discharge in a gas. The ionization cell represented the ionization proportional counter. The system sensitivity was sufficient to make stable registration of one cesium atom that reached the laser beam. The essential drawback of this method of atom detection is the usage of a buffer gas leading to a collisional broadening of the absorption line and hence to the disappearance of the isotopic shift and hyperfine splitting. The second essential drawback is that the selectivity of atom detection of one and the same element is not high. Nevertheless, the method described in /5/ can be successfully applied for atom detection in a foreign gas with concentration of $10^{18}$ atoms/cm$^{3}$.

High spectral resolution can be achieved when atoms are detected in a beam. As it is shown in this review, the application of the Rydberg atom ionization method allows to decrease the energy of the ionization radiation by several orders in comparison with ionization directly into continuum. This provides effective detection of atoms moving with thermal velocity along the direct trajectories. The application of the copper vapour laser (the pulse repetition rate to 50 kHz) for the pumping of dye lasers will make it possible to create conditions for intercepting every atom flying into a laser beam and carry out their ionization. The third step has the largest energy capacity. If a laser with the radiation line width is about the absorption line width to be used at this step, the saturation energy density will be $3 \cdot 10^{-5}$ J/cm$^{2}$ and the average dye laser power with the pulse repetition rate 20 kHz will not exceed 1 W/cm$^{2}$.

### 3.2. Application of methods

Without any doubt laser methods of single atom detection will be widely used in various fields of physics and engineering: determination of admixtures in ultra-pure materials, fine measurements of the relative isotope abundance, measurements of the atom motion velocity, etc. The application of laser optical methods in nuclear physics for detecting atom appearance with high sensitivity and measuring the average radius of a nucleus is especially perspective.

Classical spectroscopy of super-high resolution has been used for a long time to measure the deformation of nuclei. The data thus obtained are considered to be the most reliable. The measurements are based on the fact that the position of the energy level of the atom optical electron depends on nucleus charge distribution. The accuracy of the measurement of atomic level shifts caused by the nucleus deformation increases considerably with laser application. Such measurements can be carried out for a small number of atoms ($10^3 - 10^{10}$ atoms) up to single atoms.

The atom energy level shifts $\delta$ E (in the first approximation) is proportional to the relative root-mean-square change of a radius of the charged part of the nucleus:

$$\delta E \sim \delta \langle r^2 \rangle / R^2,$$

where $R$ is the nucleus radius. The addition or elimination of one neutron in spherical nuclei leads to the relative root-mean-square change of the radius of the charge distribution equal to

$$\delta \langle r^2 \rangle / R^2 = \frac{2}{5A} \cdot$$

If, for instance, consider the ytterbium atom necleus to be spherical $\delta \langle r^2 \rangle / R$ = 0.2%, when the nucleus is changed by one neutron. Such a nucleus deformation leads to the isotopical shift $\Delta E_{isot}$ = 0.02 $cm^{-1}$ on the line 555.6 mm. The existing pulsed lasers provide radiation with the spectrum width of the order $10^{-3}$ $cm^{-1}$ and less; the shifts $\lesssim 10^{-4}$ $cm^{-1}$ can be measured in this case. Thus, the nucleus deformation can be determined with accuracy $10^{-3}$%, i.e. the magnitude $\frac{\delta \langle r^2 \rangle}{R^2} \lesssim 10^{-5}$ can be measured.

Such high sensitivity and accuracy of laser methods of single atom detection make it possible to conduct experiments on systematic measurement of root-mean-square sizes of various nuclei of isomers and isotopes, for instance, of isotopes with neutron excess or deficiency appearing at the advancement to nucleus instability borders and trace the nucleus form change in these cases.

Experiments on the searchings for atoms with super dense nuclei are also quite possible. Line shifts for such systems are expected to be tens of $cm^{-1}$, the number of atoms being $10^3$. Laser methods of single atom detection can be probably applied for solving the problem whether spontaneously decaying nuclear shape isomers exist, for instance, in $224^m$Am.

## References

1. V.S.Letokhov, in "Frontiers in Laser Spectroscopy", ed. by R.Balian, S.Horoche, S.Liberman (North-Holland Publ.Co.,1977), vol.2, p.771; Usp.Fiz.Nauk, 118, 199, 1976; in "Tunable Lasers and Applications", ed. by Mooradian, T.Jaeger, P.Stokseth (Springer-Verlag, 1976), vol.3, p.122.
2. V.I.Balikin, V.S.Letokhov, V.I.Mishin, V.A.Semchishen, Pis'ma Zh. ETF, 26, 492, 1977.
3. G.I.Bekov, V.S.Letokhov, V.I.Mishin, Pis'ma Zh. ETF, 27, 52, 1978.
4. L.N.Ivanov, V.S.Letokhov, Kvantovaya elektronika, 2, 585, 1975; R.V.Ambartzumian, G.I.Bekov, V.S.Letokhov, V.I.Mishin, Pis'ma Zh. ETF, 21, 595, 1975;
   G.I.Bekov, V.S.Letokhov, V.I.Mishin, Pis'ma Zh. ETF, 73, 152, 1977.
5. G.S.Hurst, M.H.Nayfeh, J.P.Young, Appl. Phys.Lett., 30, 229, 1977; Phys.Rev.A, 15, 2283, 1977.

# COHERENT PHENOMENA IN SUPERHIGH RESOLUTION SPECTROSCOPY

V.P.Chebotayev

Institute of Thermophysics
Siberian Branch of the USSR Academy of Sciences
630090 Novosibirsk 90, USSR

Coherent phenomena play an important role in atomic and molecular spectroscopy. Hanle's method, the method of quantum beats, a radiation echo, nutations, and others are widely used in studies of relaxation processes and superhigh resolution spectroscopy. The application of lasers has opened up qualitatively new possibilities of these traditional methods and made them general-purpose. Recent achievements in this field are surveyed in /1, 2/. Over the last years the new methods have been developed in the optical superhigh resolution spectroscopy, which are underlain by coherent processes. The present report discusses the new methods of nonlinear laser superhigh resolution spectroscopy based on the application of coherent phenomena and includes the following parts: Introduction, Coherent phenomena in saturated absorption spectroscopy, Coherent phenomena at collisions, Two-photon resonances, and Macroscopic transfer of medium polarization in a gas.

At the nonlinear resonant interaction of fields with a gas there appear narrow lines in absorption spectra.These are due to population saturation effects and coherent phenomena. Under the action of fields the change in velocity distribution of atoms occurs in a gas with an inhomogeneous broadening. In this case a sharp structure preceding the non-equilibrium velocity distribution of particles at working levels appears in the line of absorption of a probe signal. These are the phenomena associated with population effects that underlie the well-known method of saturated absorption spectroscopy. The processes due to populations are therefore not sensitive to phases of probability amplitudes of atomic states. The processes that depend on the phases of probability amplitudes of a wave function are usually associated with coherent phenomena.

A phase of an average dipole moment is connected with the phases of amplitude. The dipole moment arises in the interaction of fields with atoms. The interaction of a probe wave with an atom depends not only on the dipole moment that arises directly under the action of the probe wave in consideration of level populations but also on the dipole moment caused by the other fields. The latter results in qualitative changes in the absorption line of the probe signal. We shall relate these changes to coherent processes.

Coherent phenomena appear when absorption (emission) occurs in the interaction with the dipole which results from the interaction of an atom with other fields. Here various situations may arise. The induction of the dipole moment in the system may be continuous if two- and three-level atoms interact with the fields simultaneously. In the other cases the probe wave interacts with the dipole that remains in the atom after its interaction with pumping fields, which underlies the method of separated fields.

An important peculiarity of coherent processes in the optical band should be noted. In this band the field dimensions where various phenomena are observed are much larger than a wavelength. This results in the fact that the coherent phenomena which take place in a single atom may be observed if they are reflected in the properties of the medium as a whole. This was demonstrated very well when developing the method of separated optical fields where the elimination of the influence of random motion of particles is very important.

## Coherent Phenomena in Saturated Absorption Spectroscopy

First let us consider the resonance interaction of pumping and probe waves with a two-level gas. The case where probe and pumping waves with the same frequency propagate in the opposite directions plays an important role in saturated absorption spectroscopy. In the line center the two waves interact with the same atoms. Their velocity projections onto the axis of propagation of the wave are equal $V_z = 0$. Due to saturation of level population difference for absorption of the probe wave, as indicated in /3/, there appears a resonance dip in the line center. The first experiments on observation of a resonance dip of a weak oppositely travelling wave in an external cell were carried out in /4/ and /5/. In 1970-71 several research groups carried out important investigations in Ne /6/, $PF_5$ /7/, and in $CH_4$ /8/ that indicated the efficiency of the method of high resolution spectroscopy. If the probe field frequency is scanned at the fixed frequency of the probe field pumping, the dip shape is independent of the probe wave direction. For the oppositely travelling waves the resonance is placed at the frequency $\omega' = 2\omega_0 - \omega$, $\omega_0$ is the transition frequency, $\omega$ is the strong field frequency. For unidirectional waves the probe wave resonance is placed at the pumping wave frequency.

Now let us see how the coherent effects are taken into account by using the results of the works /9, 10/. Note that in the interaction of two waves with an atom the dipole moment arises not only at the frequency of two fields but also at the combination frequency. Consideration of this interaction will result in the following changes in a line shape of the probe wave. In a very strong field the absorption of the probe oppositely travelling wave tends to a constant quantity that depends on level relaxation constants. (In the population model the absorption tends to zero). (Fig. 1). For the probe wave of the same direction the absorption tends to zero in the frequency range $|\Delta| < \frac{pE}{\hbar}$ , where $p$ is an off-diagonal matrix element, E is the strong field amplitude (Fig. 2). The manifestation of coherent effects depends on the ratio

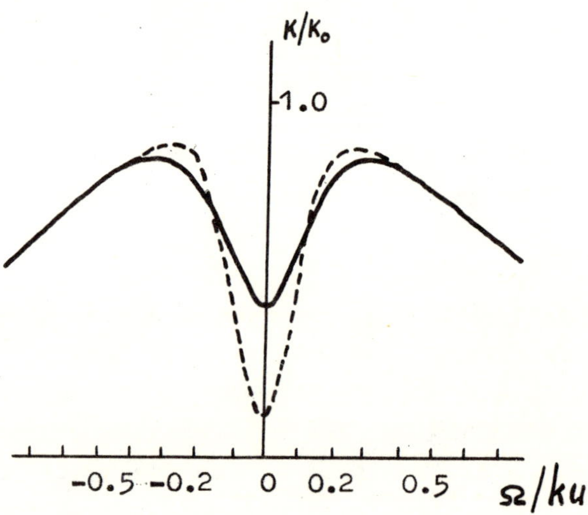


Fig. 1. Shape of the absorption line for a weak probe wave in the presence of a strong counter-running wave for the saturation parameter G = 100. The calculation was carried out with the coherence effects (the solid curve, $\gamma = \Gamma$) and rate-equations approximation (the dotted curve, $\gamma/\Gamma \to 0$). $\Gamma$/ku = 0.02.

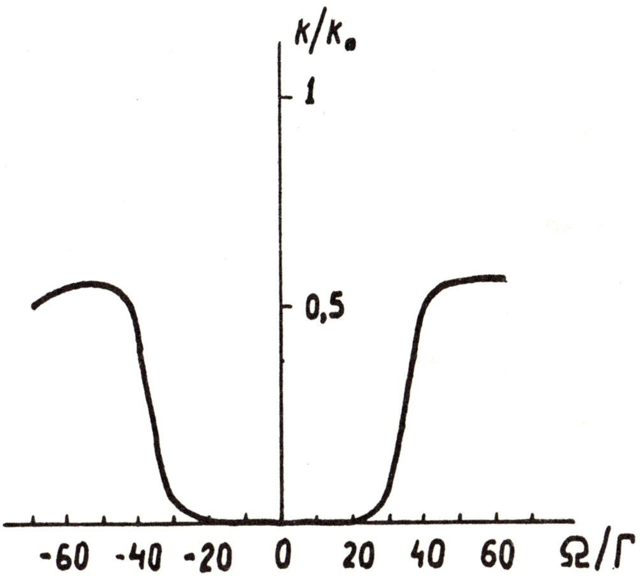


Fig. 2. Shape of the absorption line of a probe wave in the presence of a strong parallel wave for the saturation parameter $G = 10^3$. Curve corresponds to $\gamma_1/\gamma_2 = 1$.

of level relaxation constants. It is best noticed at equal relaxation constants of working levels. At low intensities of the probe oppositely travelling wave absorption, consideration of the coherent effects introduces no qualitative changes. The structure arising in the absorption line consists of three resonances whose widths correspond to those of individual levels and to the transition width. Note that the physical processes responsible for the coherent phenomena in strong fields are associated with a dynamic Stark effect.

## Coherent Phenomena at Collisions

In the presence of collisions a nonlinear resonance is broadened and, as shown in /9, 10/, the above-mentioned coherent phenomena are less manifested in the background of population effects. The randomization of the atomic dipole moment phase and the change in the atomic velocity are known to take place at collisions. This results in usual resonance broadening. A great number of works discusses the influence of these phenomena. In a low pressure gas there has been found a qualitatively new for optics phenomenon, a nonlinear dependence of resonance broadening on density /11/. This phenomenon is associated with coherence conservation in scattering at small angles. Note that the phase randomization of the atomic dipole moment is accompanied by the change in the atomic velocity. In the optical band, as a rule, scattering amplitudes on upper and lower levels are different. Thus,collisions are always accompanied by the phase randomization. Fig. 3 shows the nonlinear dependence of a resonance width in methane taken from /11/. However in scattering at small angles the scattering amplitudes on upper and lower levels may be identical and there is no phase randomization. This situation appears to be frequent for vibrational-rotational transitions because the energy of interaction of colliding particles is little dependent on the vibrational level excitation. With no

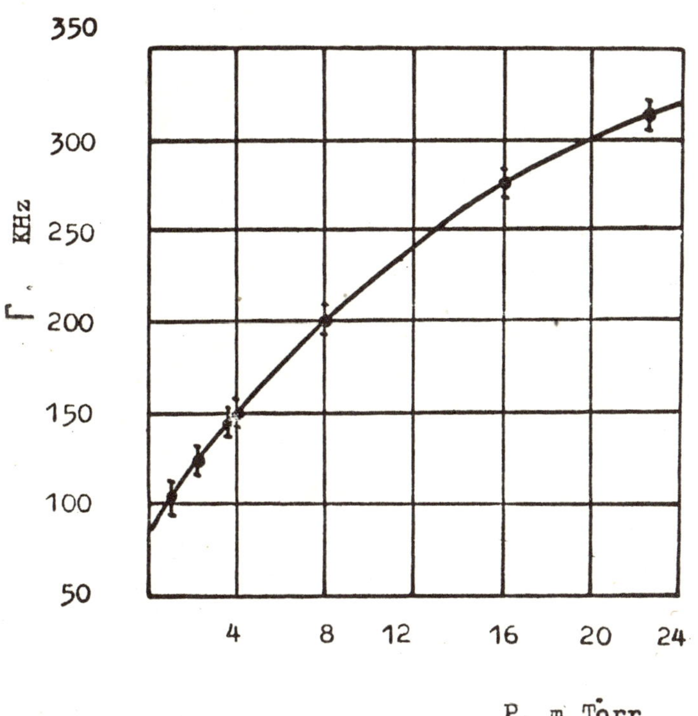


Fig. 3. Dependence of Lamb dip width in $CH_4$ at $\lambda = 3.39\ \mu m$ on pressure.

phase randomization the collisions are known to result in narrowing a Doppler line contour. This phenomenon is used to obtain narrow lines at the transitions of a hyperfine structure of atoms of alkaline metals whose wavelengths lie within the microwave range. If there is no phase randomization in collisions, the qualitative picture of the behaviour of a nonlinear resonance strongly differs from the conventional behaviour of a line contour.

Depending on the conditions the collisions with no phase randomization may either broaden the resonance or have no influence on its width. It can be easily seen from the qualitative picture given in /11/. If the Doppler line shift in scattering at the angle $\theta$ equal to $ku\theta$ ( $k$ is a wavenumber, $u$ is an average thermal atomic velocity) is more than a homogeneous linewidth $\Gamma$, after collision a particle does not interact with the field. In this case a Lamb dip of $\Gamma$ wide has a broad underlying with the width $ku\theta$. The first direct observation of this phenomenon was performed in /12/.

Fig. 4 shows the dip shape in $CH_4$ at a pressure of 1 mTorr. When adding helium there appears a wide underlying with the width of about 5 MHz caused by scattering at small angles.In spite of phase randomization absence, these collisions will broaden the resonance since they decrease the time of the particle-field interaction. So even in scattering at a small angle and with narrow resonance widths, the collisions behave as "strong". At the resonance width $\Gamma > ku\theta$ the collision with no phase randomization does not broaden the resonance. In this region the broadening is determined by some other processes. The different collisional resonance broadening may be therefore observed depending on the width. The change in gas density (the width $\Gamma$ changes) results in the nonlinear dependence of the resonance width on pressure. It is very important that at low density when the resonance broadening is nonlinearly dependent on density, the resonance width is

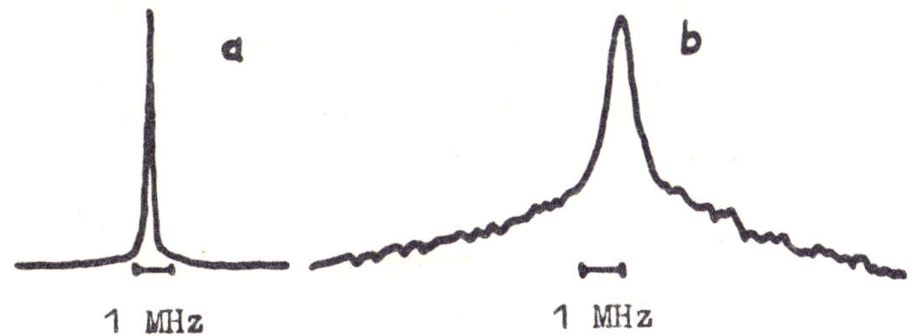


Fig. 4. Resonance shape in $CH_4$.

a) $P_{CH_4}$ = 1 mTorr, $P_{He}$ = 0;

b) $P_{CH_4}$ = 1 mTorr, $P_{He}$ = 20 mTorr.

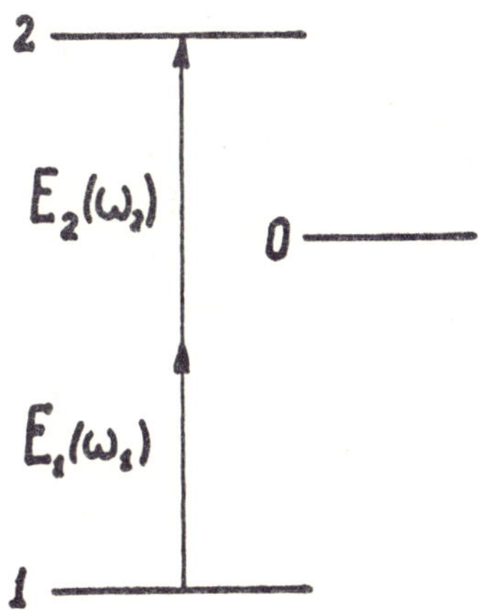


Fig. 5. Scheme of two-photon absorption.

directly related to the total scattering cross-section. This permits measurements of the resonance by spectroscopic methods. It should be noted that the similar phenomena cannot be observed at phase-randomizing collisions. A strict theoretical consideration of the above-described phenomenon was made in /13, 14/ on the basis of quantum theoretical equations /15/. Recently the nonlinear dependence at low gas pressures was experimentally observed in many works.

Note that the other kind of the nonlinear dependence may occur in collisions with phase randomization due to atoms diffusion over velocities. This nonlinear dependence that is not associated with coherent phenomena was described in details in /16/.

## Two-Photon Resonances

Now we shall consider the phenomena that are due to only coherent processes. These includes two-photon resonances in separated optical fields.

Two-photon Doppler-free resonances result from absorption (emission) of two photons propagating in opposite directions /17/. Doppler-free two-photon spectroscopy is widely used and has become one of the methods of nonlinear laser spectroscopy (see /18/). In a quasi-classical approximation two-photon absorption can be represented as a process that occurs with participation of an intermediate level (see Fig. 5). At the initial moment there is population of a lower level. An atom interacts with two oppositely travelling waves $E_1$ and $E_2$ with frequencies $\omega_1$ and $\omega_2$. In accordance with the equations for a density matrix, the following processes may take place:

$$1)\quad n_1 \xrightarrow{E_1} d(\omega_1) \xrightarrow{E_2} d(\omega_1+\omega_2) \xrightarrow{E_1} d(\omega_2) \xrightarrow{E_2} n_2$$

$$2)\quad n_1 \xrightarrow{E_1} d(\omega_1) \xrightarrow{E_1} n_o \xrightarrow{E_2} d(\omega_2) \xrightarrow{E_2} n_2 \, .$$

We think the first process is due to coherent phenomena, the second one - due to the effects of level population. Far from resonance the second process is inessential. The transition of a particle to the level 2 is due to coherent process. Under the conditions of resonance the role of the processes depends on the ratio of level relaxation constants (see /19/). In the system of the centre of inertia of a moving atom the field $E_1$ produces polarization at the frequency $\omega_1 + kv$. Owing to nonlinearity an off-diagonal matrix element appears at a double frequency of Doppler-free shift (an atom perceives the frequency of the field $E_2$ with Doppler shift $\omega_2 - kv$). When the field frequencies are identical, the field $E_2$ also interacts with an atom at the $1 \longrightarrow 0$ transition, the field $E_1$ at the $0 \longrightarrow 2$ transition. Since the processes are simultaneous, owing to interference the resulting process of absorption is multiplied by four and not doubled.

As is well known, two-photon absorption in the line centre occurs as though atoms are immovable or move in the direction perpendicular to the wave propagation. The elimination of Doppler broadening opens up possibilities for observing a number of coherent phenomena which usually cannot be observed in a gas due to the influence of inhomogeneous broadening caused by the random motion of particles. Two-photon resonances in separated fields have been considered in /20/. The observation of this phenomenon in pulsed fields /21/ and in spatially separated fields /22/ was recently reported at the Vavilov Conference on Nonlinear Optics (Novosibirsk, June 1977). B.Cagnac reported about the observation of the effect of nutation in two-photon absorption /23/.

## Macroscopic Transfer of Medium Polarization in a Gas

The method of separated optical fields, free of the Doppler effect, was used for obtaining narrow resonances for two-level atoms /24/ and two-photon absorption /20, 25/.

Recently, several research groups reported independent observations of absorption resonances in spatially separated and pulsed coherent fields /21,22,26,27/. Some theoretical problems were treated in detail /28/. The method of separated fields is usually associated with the idea of Ramsey /29/ to obtain a narrow microwave resonance. At the same time there is a number of other well-known phenomena such as superradiance, quantum echo, which use time-separated pulsed fields /1/. However, the basic ideas underlying the method of separated fields and quantum echo are different. So it was not a mere chance that these methods were considered separately, though coherent effects underlay both methods. In the course of development of the method of separated optical fields we discovered a new phenomenon, coherent radiation in separated fields (CRSF) /30/, coherent Raman scattering in separated fields of several frequencies /31/. These phenomena have many features that are common with the phenomenon of radiation echo, superradiance, and resonance coherent scattering. The scheme for observing the phenomena is given in Fig. 6. Two standing waves resonantly interact with a rarefied gas (the free path length of a particle is much longer than the distance between beams). Then continuous coherent radiation appears in a gas at the distances 2L, 3L, and so on from the first beam; the radiation intensity has a sharp peak at the centre of the line whose width is inversely proportional to the flight time.

As will be shown below, the phenomenon described is based on the same processes which are responsible for obtaining a resonance in three separated beams of standing waves for two-level atoms /24/. The appearance of an absorption resonance in the third beam is associated with macroscopic polarization transfer at the distance 2L from the first beam. In accordance with Maxwell's equations, polarization gives rise to radiation. The formation of macroscopic polarization at the distance can be recorded by observing both the absorption resonance and coherent radiation. Thus, the

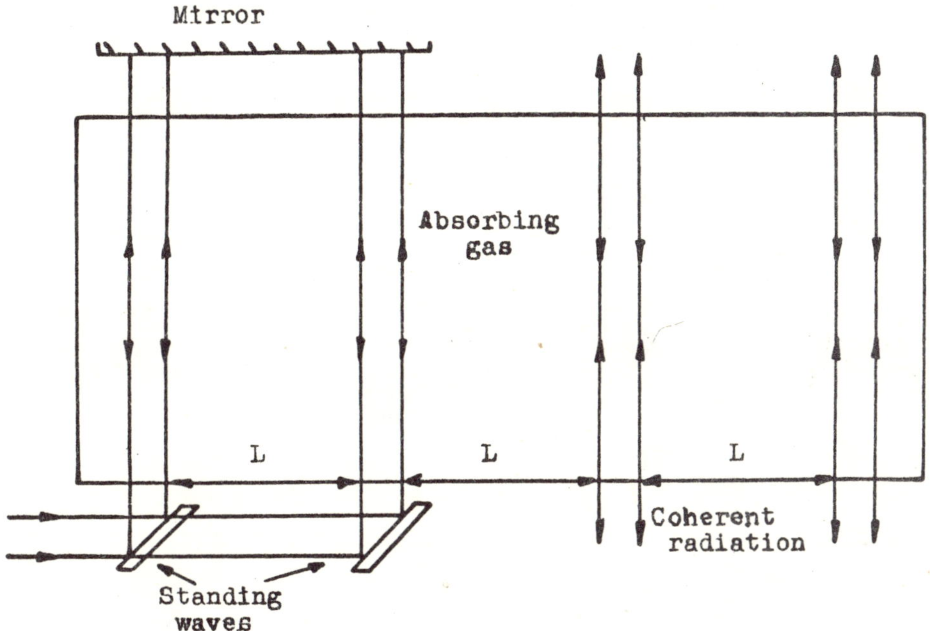


Fig. 6. Scheme for observing coherent radiation in spatially separated fields.

absorption resonance and coherent radiation in a separated field have the same nature. The coherent radiation has new interesting properties some of which are closely related to those of superradiance, coherent resonant scattering and quantum echo. The attractive feature of the phenomenon described in this paper is that it naturally unites such fundamental phenomena in nonlinear optics as the Ramsey resonance, on the one hand, and superradiance and quantum echo, on the other.

We shall be interested in the atomic polarization P(x, z, t) which is produced by moving atoms after interaction with two beams at points with z and x coordinates. The polarization at the time t is produced by the atoms which interacted with the first and second beams at the preceding times $t_1 = t - x/U$ and $t_2 = t - (x - L)/U$, respectively.

Let us consider a gas of atoms which resonantly interact with the field of two separated standing waves

$$E(x, z, t) = 2E(x, z) \cos\omega t,$$

$$E(x, z) = E_1 g(x) \cos(kz + \varphi_1) + E_2 g(x - L) \cos(kz + \varphi_2),$$

where

$$g(x) = \begin{cases} 1 \text{ at } 0 < x < a \\ 0 \text{ at } x < 0,\ x > a \end{cases}.$$

The equations for the atomic probability amplitudes are of the form

$$\dot{b} = \frac{i}{\hbar} d_{21} E(x, z) e^{-i\Omega t}$$

$$\dot{a} = \frac{i}{\hbar} d_{12} E(x, z) e^{i\Omega t} b,$$

where $a$ and b are the probability amplitudes for the

lower 1 and upper 2 levels, respectively, $d_{21}$ is the dipole moment of transition, and $\Omega = \omega - \omega_{21}$ is the frequency detuning from the transition frequency $\omega_{21}$.

We are interested in the polarization at the distance x from the first beam. It requires to find a dipole moment introduced by an atom into the point x,z at the time t, and to average it over the atomic velocity (averaging at the fixed z corresponds to that over the point $z_1$ of the atom entering the first beam). For the sake of simplicity we shall consider the beams to be narrow, i.e., $kV_z\tau \ll 1$, $\Omega\tau \ll 1$, $k = 2\pi/\lambda$, where $\tau = a/U$ is the time of flight through a beam, U is the atomic velocity along the x axis. The interaction with the first beam is considered in the first order of perturbation theory, with the second one - in the third order. The initial condition at the moment when atom enters the first beam $t_1 = t - x/U$ is $b = 0$, $a = 1$. The probability amplitudes after the atomic flight through the first beam are of the form

$$b_1 = G_1 \tau \cos(kz_1 + \varphi_1) e^{-i\Omega t}, \qquad a_1 = 1 \tag{1}$$

and those after the second beam are

$$b_2 = b_1 + G_2 \tau\, e^{-i\Omega t} \cdot \cos(kz_2 + \varphi_2) - |G_2|^2 \frac{\tau^2}{2} \cos^2(kz_2 + \varphi_2) b_1 ,$$

$$a_2 = 1 - G_2^* \tau\, e^{i\Omega t_2} \cos(kz_2 + \varphi_2) b_1 - |G_2|^2 \frac{\tau^2}{2} \cos^2(kz_2 + \varphi_2),$$

where $G_i = id_{21}E_i/\hbar$, $i = 1, 2$; $z_1$ is the coordinate of the atom entering the first beam, $z_2 = z_1 + V_zT$ - the second one $t_2 = t_1 + T$, $T = L/U$ is the time of flight of the atom between the first and the second beams (Fig. 7). After the nonlinear interaction with the second field an atomic dipole moment $d_2(t) = d_{12}b_2\, a^*\exp(-i\,\omega_{21}t) + c.c.$ has terms quadratic in $G_2$:

$$d_2(t,G_2^2) = -\Big\{ d_1 \frac{|G_2|^2}{4}\tau^2\big[2 + \exp(2ikz_2 + 2i\varphi_2)$$

$$+ \exp(-2ikz_2 - 2i\varphi_2)\big]$$

$$+ d_{12}b_1^* \frac{G_2^2}{4}\tau^2\big[2 + \exp(2ikz_2 + 2i\varphi_2)$$

$$+ \exp(-2ikz_2 - 2i\varphi_2)\big]$$

$$\exp(-2i\Omega T)\Big\}\exp(-i\,\omega_{21}t) + c.c. \qquad (2)$$

$$d_1(t) = d_1\exp(-i\,\omega_{21}t) + c.c.$$

$$= d_{12}b_1\exp(-i\,\omega_{21}t) + c.c.$$

where $d_1(t)$ is the atomic dipole moment after the interaction with the first beam. The first term corresponds to the interaction with the standing-wave field, the second one - with the travelling-wave field. A phase jump of the dipole moment is $\pm 2kz_2$ in the first term and $-2\Omega T - 2k(z_1 \pm z_2)$ in the second one (for discussion we assume $_1 = \ _2 = 0$). After passing the second beam the atom comes to the point z which is at the distance x from the first beam. Then its dipole moment $d(x, z, V_z, t)$ with the required accuracy is equal to

$$d(x, z, V_z, t) = \Big\{G_1\tau\cos(kz - \frac{V_z}{U}kx + \varphi_1)\, e^{i\Omega x/U} +$$

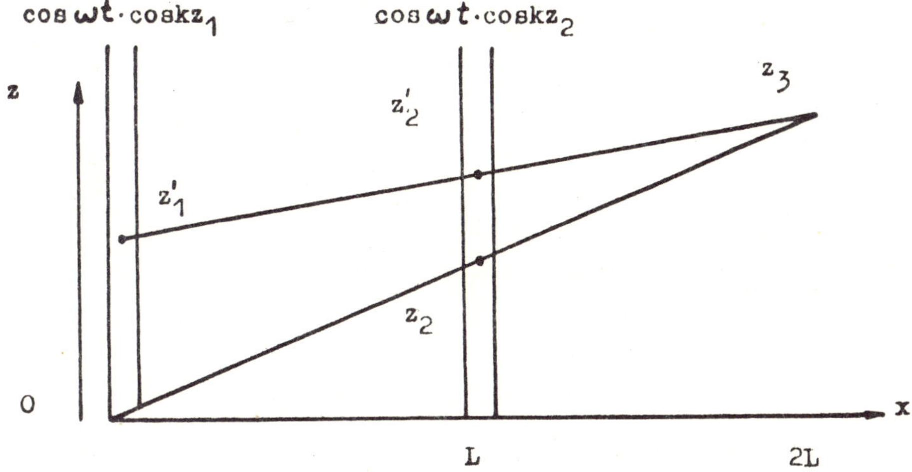


Fig. 7. Explanation for the appearance of polarization at the distance.

$$+ G_2 \tau e^{i\Omega(x-L)/U} \cos\left[kz - \frac{V_z}{U} k(x-L) + \varphi_2\right]$$

$$- G_1 |G_2|^2 \frac{\tau^3}{4} \left[ e^{i\Omega x/U} + e^{i\Omega(x-2L)/U} \right]$$

$$\left[\cos\left(kz - \frac{V_z}{U} k(x - 2L) + 2\varphi_2 - \varphi_1\right)\right.$$

$$+ \cos\left(3kz - \frac{V_z}{U} k(3x - 2L) + 2\varphi_2 + \varphi_1\right)$$

$$\left.\left. + 2\cos\left(kz - \frac{V_z}{U} kx + \varphi_1\right)\right]\right\} e^{-i\omega t} d_{12}. \qquad (3)$$

Polarization at the point z is found by averaging $d(x, z, V_z, t)$ over $V_z$ with the distribution $f(V_z)$, it is necessary for the wide distribution to keep only the third term in (3) (averaging over velocities gives the total number of particles N)

$$P(x, z, t) = -d_{12}G_1 |G_2|^2 \frac{\tau^3}{4}$$

$$\left[e^{i\Omega x/U} + e^{i\Omega(x-2L)/U}\right]$$

$$\int_{\infty}^{\infty} dV_z f(V_z) \cos\left[kz - k\frac{V_z}{U}(x - 2L) + 2\varphi_2 - \varphi_1\right] e^{-i\omega t}. \qquad (4)$$

The atomic polarization appears at the frequency $\omega$, while the dipole moment of individual atoms in the drift region naturally depends on the transition frequency $\omega_{21}$.

We should point out an important property of the system described (Fig. 6). The polarization appears only near x = 2L in a very narrow region $\Delta x \sim U/k\, \Delta V_z$ and $\Delta x \ll L$, where $\Delta V_z$ is a characteristic width of the distribution function $f(V_z)$. Thus, we have "focusing" of the first spatial harmonic at the distance x = 2L. The power absorbed by the third probe beam per unit length is determined by the expression ($x = 2L$)

$$W = 2\mathrm{Re}\left\{ -\frac{i\omega}{\ell} \int_0^{\ell} P(z,t)\, E_3(t)dz \right\}. \tag{5}$$

Substituting (4) into (5) we obtain expressions for the absorption resonance which are similar to those in /24/. Now in accordance with Maxwell's equations one can determine the radiation resulting from the polarization $P(z,t)$. The Maxwell equation for the field in the medium is of the form

$$\frac{1}{c^2}\frac{\partial^2 E}{\partial t^2} - \Delta E = -\frac{4\pi}{c^2}\frac{\partial^2 P}{\partial t^2} \tag{6}$$

The polarization $P(z,t)$ can be expressed in the form

$$P(z,t) = \left[P_+(z)e^{ikz} + P_-(z)e^{-ikz}\right] e^{-i\omega t} + \text{c.c.}, \tag{7}$$

where $P_+(z)$ and $P_-(z)$ correspond to the polarization amplitudes of the traveling waves.

$$P_\pm(z) = \begin{cases} P_\pm, & 0 < z < \ell \\ 0, & z < 0,\ z > \ell . \end{cases}$$

The solution of (7) can be found in the form

$$E(z,t) = (E_+e^{ikz} + E_-e^{-ikz})\, e^{-i\omega t} + \text{c.c.}$$

The solution at $z > \ell$ and $z < 0$ gives the amplitude of coherent radiation in a gas

$$E_\pm = 2\pi i k \ell P_\pm \tag{8}$$

Substituting (4) into (8) we obtain the amplitude of the field $E_\pm$ in the form

$$E_{\pm} = -2\pi i k \ell d_{12} G_1 |G_2|^2 \frac{\tau^3}{8}$$

$$(e^{2i\Omega T} + 1)\, e^{\pm i(2\varphi_2 - \varphi_1)}\, N . \qquad (9)$$

The field amplitude contains two terms. The first term depends on the frequency detuning $\Omega$ . This term has the same nature as the Ramsey resonance. By averaging over the velocity U, the first term gives the intensity of resonance at the line centre which is similar to the absorption resonance in its shape. The frequency dependence arises from the interaction of atoms with the standing - wave fields. The phase jump in the dipole moment of an atom after interaction with the second wave is equal to $\pm 2kz_2$ and associates with the two-quantum process of absorption and emission of photons from oppositely travelling waves. The term independent of frequency comes from atomic interaction with the travelling waves. The theory of coherent radiation in separated fields of arbitrary intensity was developed by Baklanov, Dubetsky and Semibalamut. In the higher orders perturbation theory, the coherent radiation appears at the distances 3L, 4L, and so on. Unlike the former case, it arises at the centre of the line.

We shall present a simple qualitative explanation of the appearance of polarization produced by spatially separated beams, which is associated only with coherent effects. Let us consider atoms flying out of the point $z_1$. After interaction with the field $E_1$ they have the dipole moment $d_1(t) = d_1 \exp(-i\omega_{21} t + ikz_1')$. Since the atoms have different velocity projections $V_z$, in certain points $z'$, $z''$ and others along the $z_3$ axis all the atoms have the same phase. Averaging of the dipole moment along the z axis is equivalent to that over velocities $V_z$. This averaging gives zero polarization because there is no spatial harmonic in the dipole moments of particles. It is necessary for the spatial harmonic of medium polarization that the phase of the atomic dipole moment flying out of the point $z_1'$ and into the point $z_2'$ should have a phase increase equal to $k(2z_2' - 2z_1') = 2kV_zT$ or $-2kz_2$.

As can be easily seen from (2), after interaction with the second standing-wave field, a field nonlinear addition to the dipole moment undergoes the phase jump of $2k(z_2'-z_1')$ and $-2kz_2'$. Thus, in the point $z_3$ the phases of the dipole moments will be $\pm kz_3$. Thus, owing to the phase jump in the second field a spatial polarization harmonic arises at the distance 2L from the first beam.

We shall briefly describe the principal properties of coherent radiation under investigation. They are as follows:

1) Coherent and continuous radiation takes place at the field frequency rather than at the transition frequency

2) The radiation intensity is proportional to the square of absorption (i.e., to the square of the number of excited atoms).

3) The radiation intensity is a sharp function of the frequency detuning relative to the centre of transition with a width which is inversely proportional to the time of flight between the beams (see Fig. 8).

4) Under the action of the field with an arbitrary line-width, the system operates as a narrow-band filter with a relative width of $10^{-11}$ to $10^{-12}$.

5) If both beams represent standing waves, then coherent radiation in separated fields (CRSF) occurs in two opposite directions. At the distances 3L, 4L, etc. radiation occurs only at the line centre. It is due to the interaction of standing waves*.

---

* If the first beam is a travelling wave, then at the distance 2L the radiation occurs in two directions. The intensity of the wave propagating in the same direction is independent of frequency. The oppositely travelling wave appears only in the line centre. The radiation which occurs in different directions corresponds to two different processes.

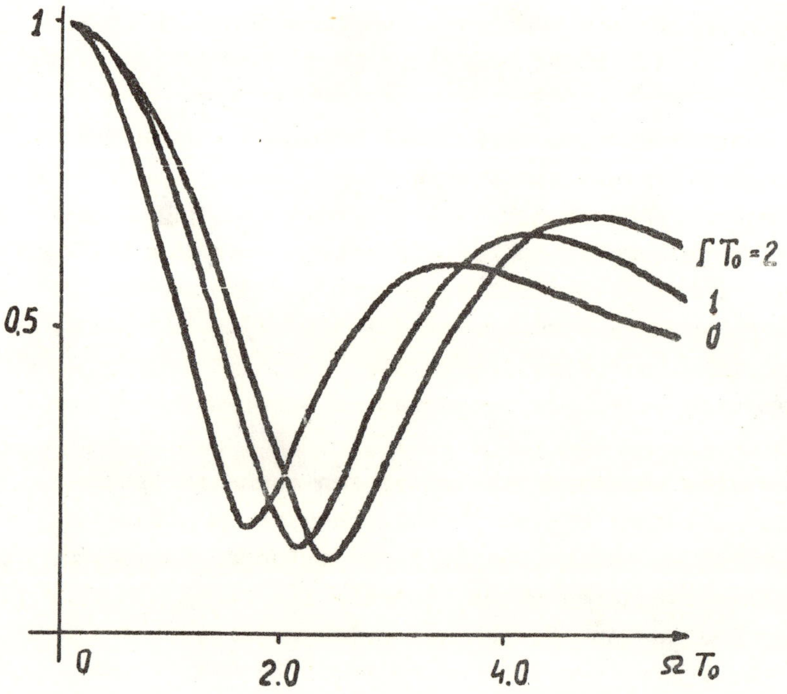


Fig. 8. The dependence of the field amplitude (in relative units) of coherent radiation at the distance 2L at various values of $\Gamma T$ ( $\Gamma$ is a line halfwidth) on the frequency detuning $\Omega$ at the optimum field. Data from /34/.

6) The atomic polarization is transferred for the time of atomic flight with a velocity of $10^4$ to $10^6$ cm/s. So the polarization and coherent radiation are delayed with respect to the inducing radiation. This is the coherent delay system first realized in the optical region. As is known, in the SHF range a delay effect is performed on ultrasonic waves and characterized by the ratio of sound and light speeds.

The described properties make CRSF a new interesting object of investigations. However, some properties have formal resemblance with some well-known phenomena. Those who deal with resonance fluorescence can consider CRSF as spatially shifted resonant scattering separated from a pumping field and thus open up new possibilities for investigations. The quadratic dependence on atomic density at coherent preparation of an ensemble of atoms enables one to consider it as the first realization of continuous superradiance. The action of the spatially separated beams is equivalent to that of two pulsed coherent standing-wave fields which are separated in time at the distance equal to the time of atomic flight between the beams. So CRSF in pulsed standing-wave fields resemble the effect of quantum echo, though its physical nature is different. Finally it should be noted that the first observations of the phenomenon described were simultaneously carried out in /32/ and /33/ for spatially separated and pulsed fields, respectively.

Fig. 9 shows the dependence of intensity of coherent radiation in methane on frequency. Experiments were carried out at a pressure of about $10^{-4}$ Torr, the distance between optical beams was 3 cm; the beam diameter was 1 cm;the radiation intensity was $10^{-15}$W.

Fig.10. depicts the dependence of radiation intensity with pulsed excitation in $SF_6$. It is seen that the intensity oscillates when tuning the frequency of a $CO_2$ laser. The first experiments indicated a high efficiency of the

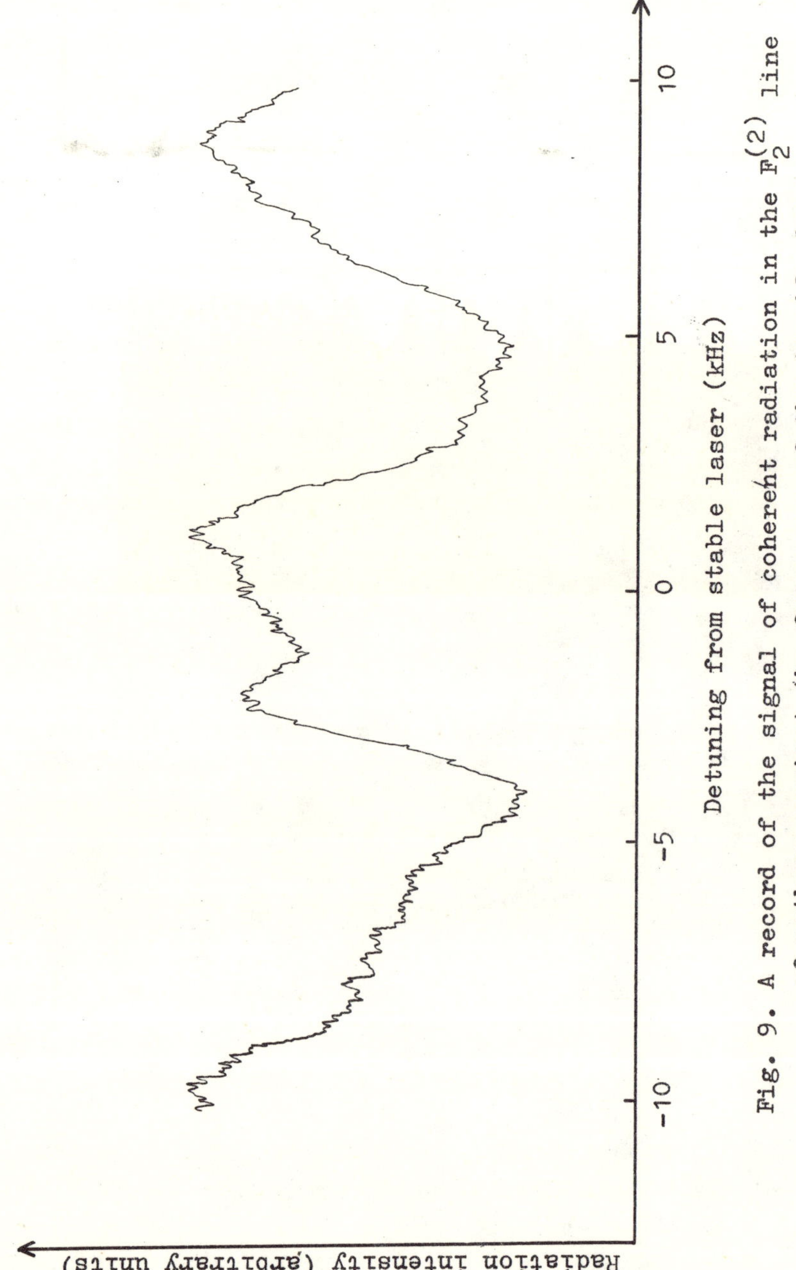


Fig. 9. A record of the signal of coherent radiation in the $F_2^{(2)}$ line of methane against the frequency of the tunable laser.

Fig. 10. The intensity change of the coherent radiation with the laser frequency detuning. The time between pulses is 0.4 μs.

method for investigating relaxation processes and obtaining supernarrow resonances.

Acknowledgement. The author is indebted to Drs. Ye. V. Baklanov and N. G. Nikulin for helpful discussions.

References

1. H,Walther. In: Topics in Applied Physics, vol. 2, ed. by H.Walther (Springer-Verlag, Berlin, Heidelberg, New York, 1976).
2. S.Haroche. In: Topics in Applied Physics, vol. 13, ed.by K.Shimoda (Springer-Verlag, Berlin, Heidelberg, New York, 1976).
3. V.S.Letokhov, V.P.Chebotayev. Pisma ZhETF 9, 364(1969).
4. N.G.Basov, I.N.Kompanetz, O.N.Kompanetz, V.S.Letokhov, V.V.Nikitin. Pisma ZhETF 9, 568 (1969).
5. Yu.A.Matyugin, B.I.Troshin, V.P.Chebotayev. Optika i Spektroskopiya 31, 111 (1971); Digest Gas Laser Physics Symp., Novosibirsk, USSR (June 1969).
6. P.W.Smith, T.Hänsch. Phys.Lett. 26, 740 (1971).
7. C.Borde. C.R.Acad.Sci.Paris 271, 371 (1970).
8. E.E.Uzgiris, J.L.Hall, R.L.Barger. Phys.Rev.Lett. 26, 289 (1971).
9. Ye.V.Baklanov, V.P.Chebotayev. ZhETF 60, 551 (1971).
10. Ye.V.Baklanov, V.P.Chebotayev. ZhETF 61, 922 (1971).
11. S.N.Bagayev, Ye.V.Baklanov, V.P.Chebotayev. ZhETF Pisma Red. 16, 15 (1972).
12. S.N.Bagayev, A.S.Dychkov, V.P.Chebotayev. Abstracts 6th Intern.Conf. on Atomic Physics, Riga, USSR (August 1978) p. 11.
13. V.A.Alekseyev, T.L.Andreyeva, I.I.Sobel'man. ZhETF 64, 813 (1973).
14. Ye.V.Baklanov. Optika i Spektroskopiya 38, 24 (1975).
15. V.A.Alekseyev, T.L.Andreyeva, I.I.Sobel'man. ZhETF 62, 614 (1972).
16. A.P.Kol'chenko, A.A.Pukhov, S.G.Rautian, A.M.Shalagin. ZhETF 63, 1173 (1972).
17. L.S.Vasilenko, V.P.Chebotayev. A.V.Shishayev. Pisma ZhETF 12, 161 (1970) /JETP Lett. 12, 113 (1970)/.

18. N.Bloembergen, M.D.Levenson. In: Topics in Applied Physics, vol. 13, ed. by K.Shimoda (Springer-Verlag, Berlin, Heidelberg, New York, 1976).

19. V.P.Chebotayev. In: Topics in Applied Physics, vol. 13, ed. by K.Shimoda (Springer-Verlag, Berlin, Heidelberg, New York, 1976).

20. Ye.V.Baklanov, B.Ya.Dubetsky, V.P.Chebotayev. Appl.Phys. 11, 201 (1976).

21. M.M.Salour, C.Cohen-Tannoudji. Phys.Rev.Lett. 38, 757 (1977).

22. V.P.Chebotayev, A.V.Shishayev, L.S.Vasilenko, B.Ya.Yurshin. Appl.Phys. 15, 43 (1978).

23. B.Cagnac. Report at 5th Vavilov Conference on Nonlinear Optics. Novosibirsk, USSR (June 1977).

24. Ye.V.Baklanov, B.Ya.Dubetsky, V.P.Chebotayev. Appl.Phys. 9, 171 (1976).

25. Ye.V.Baklanov, V.P.Chebotayev. Appl.Phys. 12, 97 (1977).

26. J.C.Bergquist, S.A.Lee, J.L.Hall. Phys.Rev.Lett. 38, 159 (1977).

27. R.Teets, J.Eckstein, T.W.Hänsch. Phys.Rev.Lett. 38, 760 (1977).

28. B.Ya.Dubetsky. Kvantovaya Elektronika 3, 1258 (1976); B.Ya.Dubetsky, V.M.Semibalamut. Kvantovaya Elektronika 5, 176 (1978).

29. N.F.Ramsey. Molecular Beams (Oxford University Press, New York, London, 1956).

30. V.P.Chebotayev. Appl.Phys. 15, 219 (1978).

31. V.P.Chebotayev. Vestnik MGU 19, 159 (1978).

32. S.N.Bagayev, V.P.Chebotayev, A.S.Dychkov. Appl.Phys. 15, 209 (1978).

33. V.P.Chebotayev, N.M.Dyuba, M.N.Skvortsov, L.S. Vasilenko. Appl.Phys. 15, 319 (1978).

34. Ye.V.Baklanov et al. Report at 5th Vavilov Conference on Nonlinear Optics. Novosibirsk, USSR (June 1977).

# LASER SPECTROSCOPY IN MOLECULAR BEAMS

W. Demtröder

Fachbereich Physik, Univ. Kaiserslautern
675 Kaiserslautern, Germany

## Introduction

For a long time the development of spectroscopic techniques and the progress achieved in the physics of molecular beams had proceeded side by side without too much mutual interaction. Although in the microwave and radiofrequency region a combination of both techniques had proved to be very successful in providing detailed information about molecules in their groundstates (Rabi-method) (1), only a few attempts had been made in the optical region with the goal to achieve narrow spectral lines with reduced Doppler-width (2). A real synthesis of optical spectroscopy and molecular beam techniques had to wait for the development of intense monochromatic tunable lasers.

For some years laser spectroscopy in molecular beams has now developed rapidly and the results obtained up to now in many laboratories have proved this method to be very productive. The combination of advantages from both fields with the amplification of experimental capabilities will certainly open new aspects to studies of molecular structure and interatomic potentials.

In this review I will briefly discuss the following aspects of laser spectroscopy in molecular beams.

1. High resolution spectroscopy with reduced Doppler-width.
2. Investigations of atoms and molecules under collision-free conditions.
3. Spectroscopic analysis of supersonic beams.
4. Production and spectroscopy of loosely bound van der Waals-molecules.
5. Spectroscopy of collision processes in crossed beams.

These different points shall be illustrated by some examples, which represent only a small selection from a great variety of interesting experiments performed in several laboratories.

## 1. High Resolution Spectroscopy with Reduced Doppler-Width

When the monochromatic laser beam is crossed perpendicularly with a collimated molecular beam the Doppler-width of the molecular absorption lines is reduced by a factor $\sin \vartheta$ where $\vartheta$ is the divergence angle of the molecular beam, which depends on the width b of the collimating aperture and its distance d from the orifice (Fig.1). With b = 10 cm and d = 10 cm the collimation ratio b/d is already 1 : 100 and the Doppler-width is decreased from a typical figure of 1000 MHz to about 10 MHz, which is for many transitions already below the natural line width. Because of the low molecular density attainable in a highly collimated beam and the short absorption pathlength the relative absorption of the incident laser power may be in some cases as low as $10^{-12}$. Such small attenuation cannot be measured directly, but can be conveniently detected by the laser induced fluorescence emitted from the absorbing molecules. Monitoring the fluorescence intensity as a function of the laser frequency $\omega_L$, while the laser is tuned across the absorption spectrum yields an "excitation spectrum" $I_A(\omega_L)$ which is a true image of the absorption spectrum, as long as the detector has constant sensitivity over the wavelength range of the fluorescence spectrum.

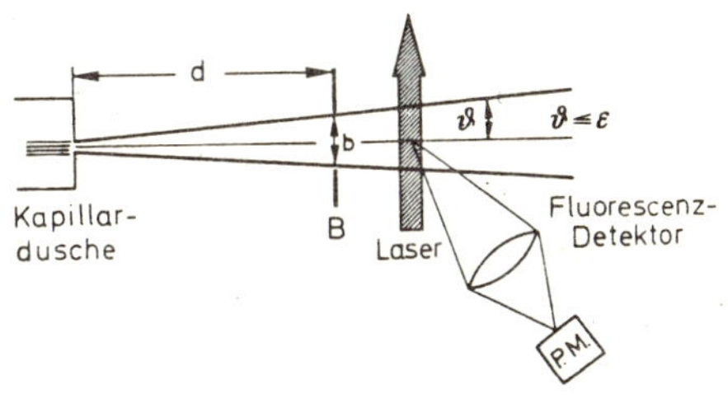


Fig. 1

The advantages of collimated atomic beams for high resolution spectroscopy with reduced Doppler-width have been used to resolve hyperfine-structures in complex atomic spectra (3,4). The high sensitivity of laser induced fluorescence allows to detect even low concentrations of short lived radioactive isotopes and to measure their hyperfine-spectrum (5).

For molecules the optical spectra are generally more densily structured because of the many rotational-vibrational levels in both electronic states, which can combine in an optical transition. Fig.2 shows as example an excitation spectrum of $Cs_2$-molecules excited in a beam with $\vartheta$ = 0.03 by a single mode argon laser, tunable around $\lambda$ = 476,5 nm (6). Since the Doppler-broadened absorption lines ($\Delta\nu_D \approx$ 700 MHz) overlap completely, they cannot be resolved using Doppler-limited spectroscopy.

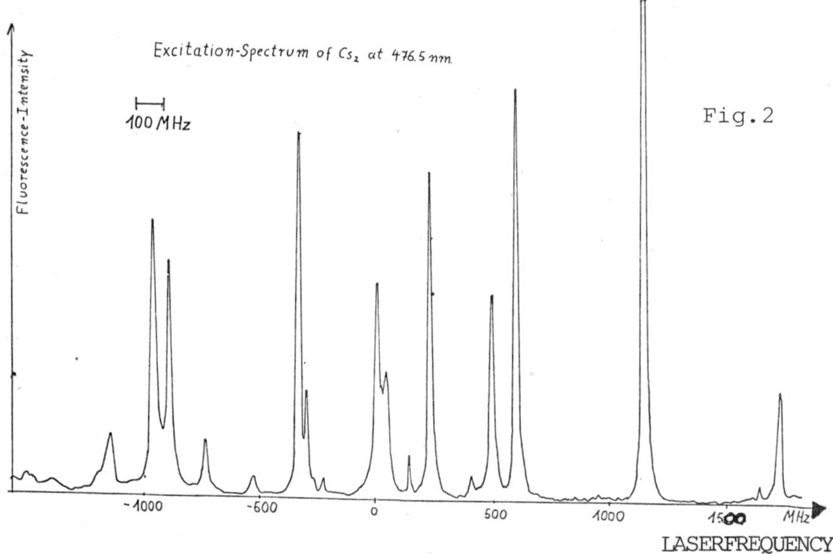


Fig.2

For polyatomic molecules the electronic spectra with their numerous rotational-vibrational lines which generally have additional fine-or hyperfine-structure and suffer pertubations from coupling with other transitions, are even more complex and Doppler-free techniques are essential to resolve the lines (7).

Fig.3 illustrates this by a small section of the $NO_2$-excitation spectrum in a frequency interval of 4 GHz which corresponds to about 4 times the Doppler-width!

The spectrum of Fig.3 was excited with a tunable argon laser around 488 nm, crossed with a molecular beam, with a collimation ratio of 1:80. The resultant line width is about 12 MHz.

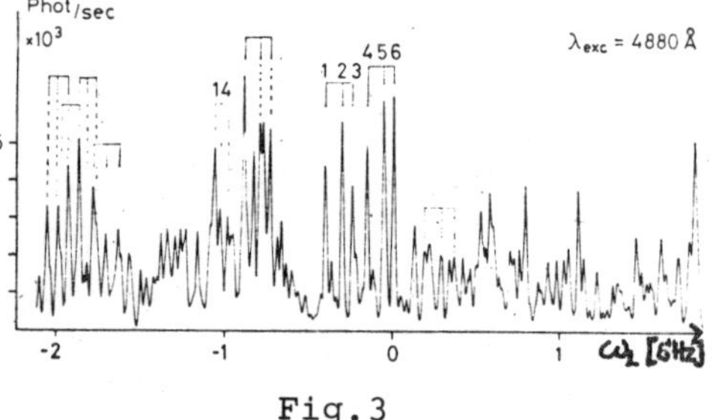


Fig.3

An interesting modification of this technique of Doppler-width reduction which is applicable to fast atomic-or ion-beams has been proposed by Kaufmann (8) and proved by Otten et al. (9) who measured the hyperfine structure of short lived radioactive isotopes produced on line with a mass separator. When the ions are accelerated to several KeV, their velocity spread, resulting from their thermal velocity distribution, reduces considerably and Doppler-reduced laser spectroscopy can be performed with the laser beam <u>parallel</u> to the ion beam. This can be immediately seen as follows: Assume two ions with thermal velocity components $v_{th} = o$ resp. $v_{th} = v_o$ in the beam direction. After being accelerated by a voltage V, their final kinetic energies are

$E_1 = (m/2)\cdot v_1^2 = e\cdot V$ and $E_2 = (m/2)\cdot v_2^2 = e\cdot V + (m/2)\cdot v_{th}^2$

Subtracting both equations yields for the final velocity difference

$$(v_1 - v_2) = v_o^2 / \sqrt{8eV/m} = \left(\frac{E_{th}}{4\cdot eV}\right)^{1/2} \cdot v_o$$

The initial thermal velocity spread $v_o$ is therefore reduced by a factor, depending on the ratio of thermal energy to acceleration energies. Assume $E_{th} = 0.1$ eV, $eV = 10$ keV, this factor becomes about $1,5\cdot 10^{-3}$, which means a reduction of the Doppler-width parallel to the beam by a factor of about 1000, provided the acceleration voltage is sufficiently stable.

The technique has the further advantage, that the ions can be tuned into resonance with a fixed frequency laser by "Doppler-tuning", because their velocity parallel to the laser beam can be changed with the acceleration-voltage (10). Through charge-exchange the ions can be neutralized without a great change of their velocity. The technique can be therefore applied also to neutral atoms or molecules, or to negative ions.

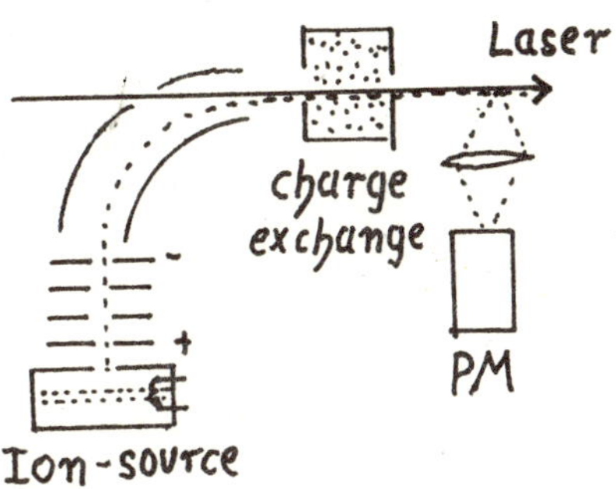


Fig.4

The attainable resolution in molecular spectroscopy can be even enhanced by using a "Rabi-technique", well known from radio frequency spectroscopy in molecular beams (Fig.5, upper part). In the laser spectroscopic version of this technique the magnets A and B are replaced by two laser beams, which cross the molecular beam perpendicularly (Fig.5, lower part). The pump-beam saturates a molecular transition $E_i \rightarrow E_k$ and depletes the population of $E_i$. This is monitored by the fluorescence excited by the probe beam, which is proportional to the population $N_i$ $(E_i)$. When a radio-frequency field between both laser beams is tuned to a transition between sublevels of $E_i$ (e.g. hyperfine-components) the population of the initially depleted

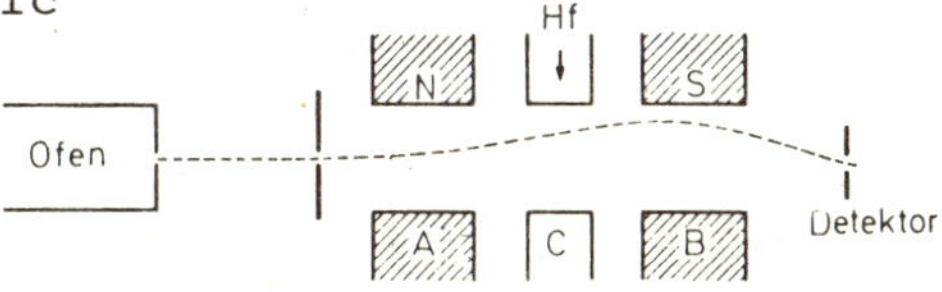


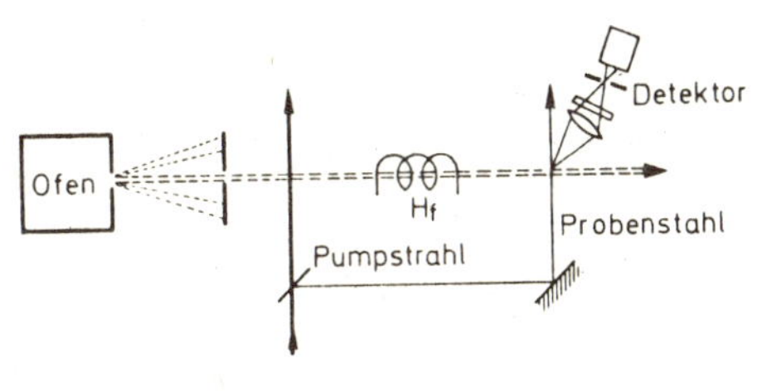


Fig.5

levels increases, which is monitored by the probe-fluorescence. This technique has been used by S. Penselin and his group in Bonn to measure quadrupole moments of atoms in ground states and excited metastable states (11,12). As an excellent example of the capability of this technique the very small hyperfine splitting of a vibrational-rotational level (v"=0,J"=28) of the $Na_2$- molecule has been measured by Rossner et al.(13). The splitting is about 100 KHz and therefore even small compared to the natural linewidth of the optical transitions, which is 20 MHz!

For complex spectra the assignment of the many lines in an excitation spectrum, although resolved, is very tedious and time consuming. The identification can be greatly facilitated by using an optical-optical double resonance technique (Fig.6) with a pump laser 1 and a probe laser 2 (14). The pump laser is stabilized on the transition 1 ⟶ 2 and its intensity is chopped at a frequency $f_1$. Because of saturation the population density $N_1$ is also modulated with the frequency $f_1$. When the probe laser is tuned across the molecular absorption spectrum, it will produce an excitation spectrum. However, only those lines are intensity modulated with $f_1$, which start from level 1, or from level 2, where the modulation phase is opposite for the two cases. This means, that the excitation spectrum of the probe laser detected with a lockin amplifier tuned to $f_1$ includes only lines with the common level 1. The frequency distance between these lines corresponds to the energy separation of the different upper levels. This double-resonance method allows to perform unambiguous spectroscopy of excited states and can be regarded as an inversion of the laser induced fluorescence technique, where from a selectively pumped upper level the different fluorescence lines are observed. Fig.7 shows for illustration such a double resonance spectrum obtained in our laboratory with two single mode argon lasers crossed with an $NO_2$-beam. The lower part shows an unmodulated excitation spectrum of the probe laser, the upper part the modulated double-resonance spectrum which indicates that two lines of the lower spectrum share a common lower level (15).

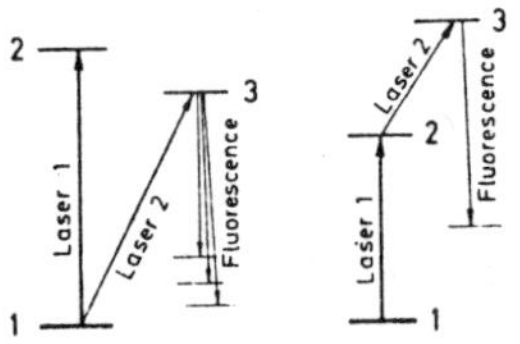

Fig.6

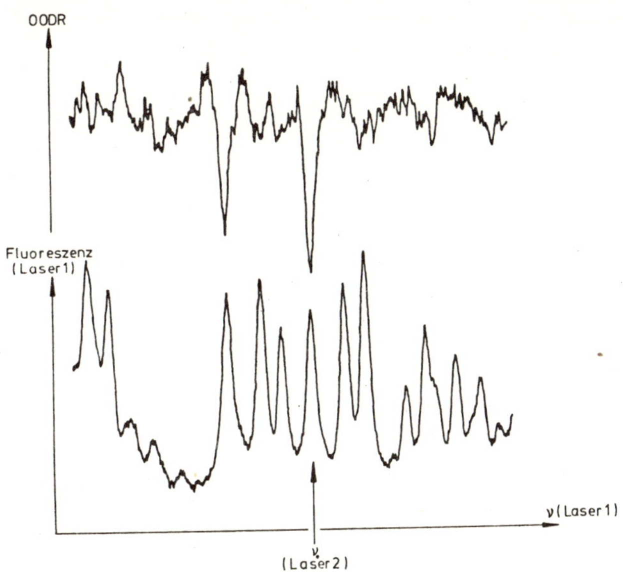


Fig.7

## 2. Investigation of Atoms and Molecules under Collision-Free Conditions

Another aspect of laser spectroscopy in molecular beams at a low background pressure in the vacuum chamber is the fact that the molecules interacting with the laser field can be investigated under collision free conditions. This means that the mean free path $\Lambda = v \cdot \tau$ which a molecule with velocity $v$ travels during the mean time $\tau$ between successive collisions is large compared to the interaction region of the molecules with the radiation field. The collision time $\tau$ is then also large compared to the spontaneous lifetime and to the interaction time of the molecule with the laser field. Pressure broadening of spectral lines or collision induced radiationless transitions can be then neglected. This point is essential for long spontaneous lifetimes or for studies of internal transitions in excited states of large molecules. Some examples shall illustrate this:

a) When exciting $NO_2$ molecules in a molecular beam with a laser-pulse (duration 10 $\mu$s) from a single mode laser the decay of the fluorescence exhibits for many selectively excited upper levels a nonexponential decay (16). This may be explained by assuming that the laser excites non-stationary Born-Oppenheimer-states which can decay rapidly into other states of the same total energy but with a different electronic state. In this case the excitation energy of the free molecule can fluctuate between several degenerate levels of different electronic states before it is emitted by the fluorescence. The question of radiationless transitions and the nature of the excited state has been investigated by Zewail et al. (17) with the technique of coherent optical spectroscopy in a molecular beam. The method uses three optical pulses ($\pi/2$, $\pi$ and $\pi/2$) and detects the photon-echo on the spontaneous emission at right angles to the exciting beam. The echo appears as a temporally burnt hole in the emission and allows to probe the optical dephasing of molecules undergoing radiationless transitions. For diatomic molecules, such as $I_2$ the dephasing time $T_2$ was found to be 0.5 of the spontaneous lifetime $T_1$ (depopulation time), indicating that no intramolecular dephasing destroys the optical coherence of the prepared state. For large molecules this is not necessarily true. A combination of life-time-measurements, level-crossing experiments, (18, 19) and optical-radio frequency double-resonance experiments (20) indicate, that the dephasing time $T_2$ for many excited $NO_2$-level is shorter than the life-time $T_1$ by more than one order of magnitude.

b) In the absence of collisions the line width of molecular transitions induced by a laser beam which crosses the molecular beam, may be limited by the transit time $T_3$ of the molecules through the laser beam. If $T_3$ is smaller than the lifetime $T_1$, it determines the interaction time of the molecules with the radiation field. This limitation can be overcome by transferring the old "Ramsay-method" (21) where molecules in a molecular beam interact coherently with two widely separated fields, to the optical region (22). This allows a spectral resolution which is limited by the travel time between the radiation zones rather than by the transit time through each zone, and ultra-narrow resonances can be achieved in principle. In praxis one has to assure that the phase relations established by the coherent excitation of the atoms in the first zone are preserved during the travel time between the zones. Because of the thermal velocity spread the travel time is different for different molecules and the phases of the prepared atoms are washed out at the second zone This problem can be solved either by using Doppler-free two-photon excitation (22,23) in both zones where all molecules from the whole Doppler-profile contribute to the two-photon transition, or by allowing the molecules to pass through three equidistant laser beams (23,24). The experimental perfection of this exciting technique might lead to a new class of phenomena accessible to such investigations with ultrahigh resolution. This technique has been demonstrated by Chebotayev and his group in Novosibirsk, by Hall et al. in Boulder and by Salour in France. One possible application lies in the possibility to distinguish between velocity changing collisions and phase perturbing collisions, when observing molecules which have suffered one collision between the second and third radiation zone. The resultant phase shift can be detected by shifting the phase of the third zone. Velocity changing collisions cause a frequency shift, which is large compared to the ultra-narrow line width, even for collisions with large impact diameter (25).

## 3. Spectroscopic Analysis of Supersonic Beams

For all scattering experiments in crossed atomic or molecular beams the knowledge of the velocity distribution and the internal state distribution is essential to derive elastic or inelastic cross-sections from measured rates of scattered particles. Several laser spectroscopic methods can be applied to determine the characteristic parameters of molecular beams and their changes with oven pressure and temperature. Of particular interest is the transition region between a thermal effusive beam and a supersonic beam where different internal cooling of rotational, vibrational and translational temperatures starts.

The velocity distribution can be measured by the Doppler shift and Doppler broadening of absorption lines. The laser beam is split into to partial beams (Fig.8a). Beam 1 crosses the molecular beam perpendicularly, and beam 2 under a small angle $\alpha$ .

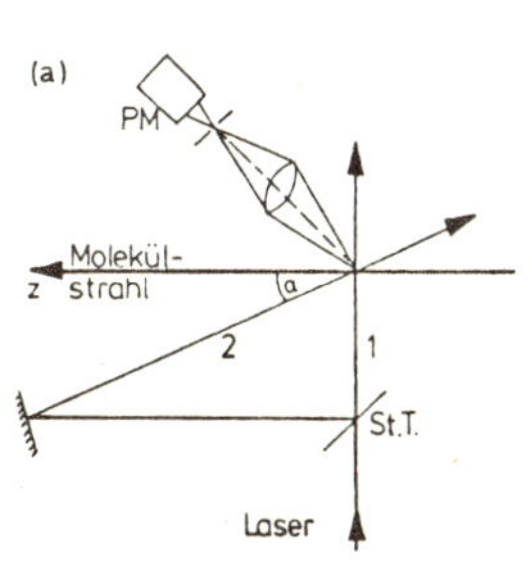


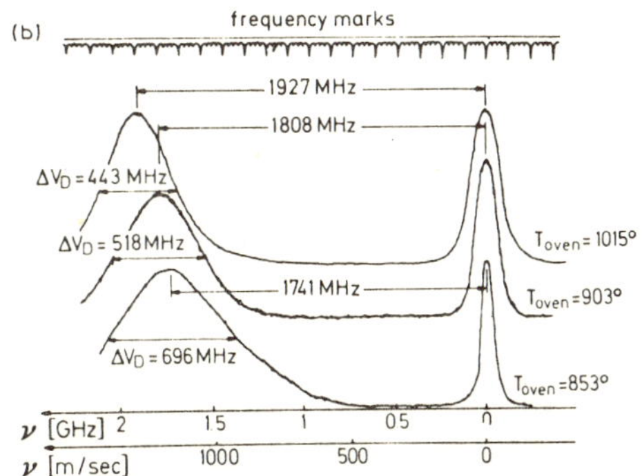


Fig.8

A molecule moving with velocity $\bar{v}$ can absorb the monochromatic laser beam with frequency $\omega_L$ and wave vector $\vec{k}$ if the Doppler-shifted frequency $\omega_L - \vec{k} \cdot \vec{v} = \omega_0$ in the moving coordinate frame of the molecule coincides with the molecular absorption frequency $\omega_0$. When the laser frequency $\omega_L$ is tuned, the beam 1 is absorbed at $\omega_L = \omega_0$ because $\vec{v} \perp \vec{k}$, and the absorption profile has a reduced Doppler-width. Beam 2 is absorbed at $\omega_L = \omega_0 + \vec{k} \cdot \vec{v}$ and shows a Doppler-broadened profile corresponding to the velocity distribution of the molecules in the absorbing state $E_i$.
Fig.8 shows such a velocity-distribution for different oven pressures, depending on the oven temperature T. One can see the increase of the most probable velocity and the narrowing of the velocity distribution with increasing oven temperature T, indicating the transition region from a thermal effusive beam to a supersonic beam (26).

The internal state distribution $N(v_i, J_i)$ can be derived from the intensities of excitation lines which are proportional to the population density of the absorbing state $(v_i, J_i)$. One method compares the relative intensities of two transitions $(v_i'', J_i'') \rightarrow (V_n', J_n')$ and $(v_k'', J_k'') \rightarrow (V_m', J_m')$ for molecules in the beam and for molecules in a cell at thermal equilibrium when the population densities follow a Boltzmann distribution. Another method relies on the saturation of molecular transitions, because the intensity of a completely saturated transition does not depend on the transition probability but solely on the number of absorbing molecules passing the laser beam per unit time.

Bergmann et al. (27) have used a method to determine molecular beam parameters and to study molecular formation in supersonic beams, which is based on a combination of optical pumping

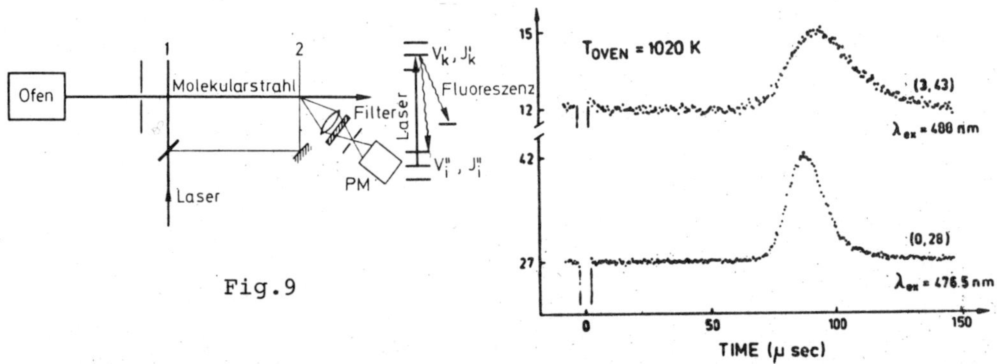


Fig.9

and time of flight measurements. The molecular beam is crossed perpendicularly by the pump laser (Fig.9a), which is tuned to a molecular transition ($v_i''$, $J_i''$ $\longrightarrow$ $v_n'$, $J_n'$) and which depletes the lower level ($v_i''$, $J_i''$) due to saturation. When the pump beam is interrupted by a chopper, for a short time interval $\Delta t$ (a few $\mu$s) a pulk of molecules in level ($v_i''$, $J_i''$) can pass the intersection zone without being pumped. Because of their velocity spread these molecules arrive at the crossing point with the probe beam at different times and are detected by the increase of fluorescence induced by the probe beam (Fig.9b). This method therefore allows to measure velocity distributions of molecules in definite quantum states.

Such measurements and many other investigations performed in several laboratories give the following result: In supersonic beams internal cooling takes place which means that the internal energy (thermal translational, rotational and vibrational energy of the molecules) is partly converted into one-dimensional kinetic energy of the beam molecules parallel to the beam axis. The mean velocity increases while the width of the velocity distribution decreases. Although the different degrees of freedom are not at thermal equilibrium, it has become common use to attribute translational temperatures as well as rotational or vibrational temperatures to the molecules in a beam. These different temperatures may differ widely depending on the beam parameters and the molecular species.

## 4. Production and Spectroscopy of Loosely Bound Van der Waals Molecules in Supersonic Beams

Most effective cooling of molecules can be achieved with seeded supersonic beams (28) where a small percentage of wanted molecules is mixed with a noble gas at high pressures in the oven (up to many atmospheres). By adiabatic expansion through a small nozzle into the vacuum the internal energy is nearly completely converted by inelastic collisions with the carrier gas atoms into directed one-dimensional kinetic energy. Translational temperatures below 1 K and rotational temperatures of a few degrees Kelvin have been achieved, while the vibrational temperature is in the range 20-200$^{\circ}$K, depending on the molecule and the beam parameters (29).

Such conditions are very advantageous for spectroscopic investigations, because of two reasons:

a) The extremely low temperature allows the formation of loosely bound molecules which have only a very shallow potential mini-

mum in their electronic groundstate and therefore a small dissociation energy D. These molecules would dissociate immediately at temperatures T with $kT > D$.
The spectroscopy of such van der Waals molecules opens the possibility to study long range potentials, up to now only accessible by scattering experiments, with the much higher spectroscopic accuracy (30;31).

b) Assuming thermal equilibrium at least within one degree of freedom, the population density N(v,J) in a level with vibrational quantum number v and rotational number J at a vibrational temperatur $T_v$ and a rotational temperature $T_R$ can be described by

$$N(v,J) = (2\cdot J+1)\cdot \exp(-E_{rot}/k\cdot T_R)\cdot \exp(-E_{vib}/kT_v)$$

Because of the low vibrational and rotational temperatures the population density is almost completely accumulated in the lowest vibrational level and in very few rotational levels with small qunatum numbers J. This simplifies the absorption spectrum considerably because the manifold of absorbing states is greatly reduced.

Wharton, Levy and Smalley (28) have applied this technique to facilitate the assignment of complex molecular spectra. In a supersonic He-beam (oven pressure 91 at) seeded with iodine the spectrum of the iodine molecules $I_2$ becomes less complicated and the only vibronic bands appearing with noticeable intensity are those starting from v"=o or v"=1 (32). The simplification is particularly pronounced for the complex $NO_2$-spectrum. The seeded beam technique allowed to oberserve and assign about 140 vibronic bands which could not be resolved in cell absorption measurements (32).

Besides the spectral lines of the $NO_2$-molecules the excitation spectrum shared features due to the absorption of the He-$NO_2$-van der Waals-molecule, which was formed in the supersonic beam despite its low binding energy. Further investigations of alcali-noble gas excimer-molecules, which may be attractive candidates for tunable lasers have been performed on Na-Ar and Na-He-excimers (30.31).

The spectroscopic analysis of excitation-and fluorescence spectra of such excimer-molecules allows very accurate determination of the attractive and the repulsive part of the interaction potential.

This shall be illustrated with the example of the NaK-molecule (34,35) which has a bound singlet-state and a repulsive triplet state $^3\Sigma^+$ with a shallow minimum with a well depth of about 200 $cm^{-1}$.

The fluorescence originates from an upper $^3\Pi_u$-state which is perturbed by the $^1\Pi$ u-state and can be therefore populated by optical pumping with an argon laser from the stable $^1\Sigma_o^+$-groundstate. The fluorescence to the repulsive part of the lower $^3\Sigma^+$-state is continiuous, with an intensity modulation determined by the vibrational wave function in the emitting state (Franck Condon principle). The discrete lines in the spectrum (see Fig.11) start at the dissociation limit. The energy level scheme is shown in Fig.10

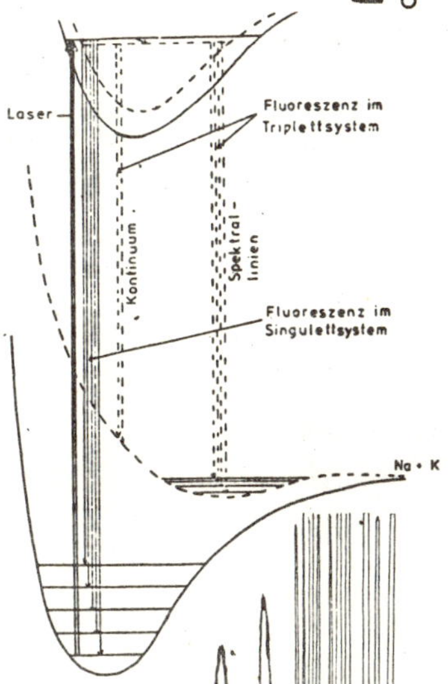

Fig.10

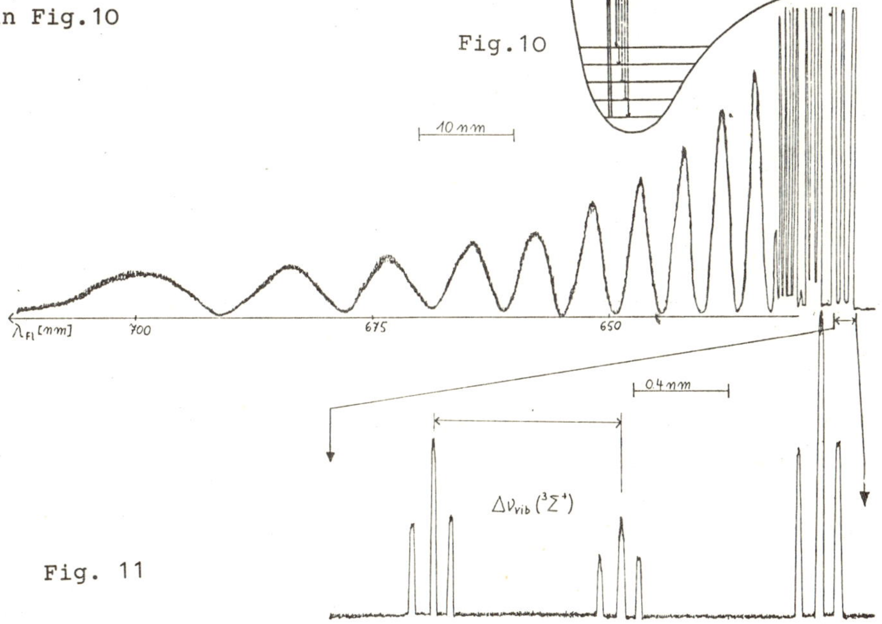

Fig. 11

The measurement of the discrete lines allows an accurate determination of the shallow minimum using the RKR-potential-method. From the modulation of the continuum the repulsive part of the potential can be obtained. Note that the wavelength $\lambda_c$ where the discrete lines start, gives an accurate measure for the dissociation energy.

## 5. Spectroscopy Investigation of Collision Processes in Crossed Beams

In conventional scattering experiments using crossed atomic or molecular beams, the number of molecules scattered per second into the solid angle $d\Omega$ around the scattering angle $\vartheta$ is measured as a function of $\vartheta$ and in dependence on the relative velocity. Definite velocity intervals are selected by mechanical velocity-selectors. Inelastic collisions with vibrational or rotational energy transfer can be generally detected only indirectly through the corresponding loss of translational energy. Laser spectroscopy opens the way to spectroscopic measurements of differential elastic and inelastic cross sections between atoms and molecules in definite quantum states. Using a pump laser, which saturates a molecular transition $(v_i", J_i") \rightarrow (v_m', J_m')$ and depletes the level $(v_i", J_i")$, and a probe laser which detects scattered molecules in the same level $(v_i", J_i")$ or in another level $(v_k", J_k")$ which have been scattered by an angle $\vartheta$, differential cross sections for elastic or inelastic collisions

$$A + M(v_i", J_i") \longrightarrow M(v_i", J_i", \vartheta)$$
$$\longrightarrow M(v_k", J_k", \vartheta)$$

can be separately measured. (See Fig. 12). Such experiments have been performed for $Na_2$-molecules and He or Ne as collision partners by Dr. Bergmann and his group in Kaiserslautern (36) and I refer to the poster session at this conference, where Dr. Bergmann has presented his results. The probe laser represents a <u>state selective</u> detector which can be tuned to the desired transition $(v",J") \longrightarrow (v',J')$ and which uses the induced fluorescence as a very sensitive monitor for the population of scattered molecules in the level $(v",J")$.

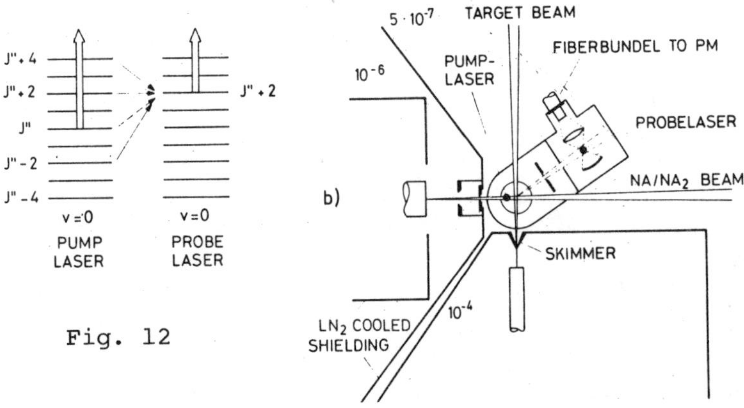


Fig. 12

As I have shown above, such spectroscopic techniques allow to measure the velocity distribution, N(u) and the population distribution $N(v_i", J_i")$. Using optical pumping, molecules in a specified state and with a specified velocity component can be marked and used for scattering experiments, where every parameter needed for the evaluation of differential inelastic cross sections can be determined.

The different aspects of laser spectroscopy in molecular beams which I could briefly discuss in such a short review talk, represent only a selection from a large number of possible applications in this rapidly expanding field. Many groups which I could not mention here, are engaged in this research and new results certainly will be published in the near future.

References

1. N.F. Ramsey: Molecular Beams (Clarendon Press Oxford 1956)
2. R.W. Stanley; J.Opt.Soc.Am.56,350 (1966)
3. W. Lange, J. Luther and A.Steudel, Advances in At.and Mol. Physics.Eds.D.R. Bates and B. Bederson, Academic Press,Vol.10, 173 (1974)
4. P. Jaquinot:"Atomic Beam Spectroscopy" in High Resolution Laser Spectroscopy,Topics in Appl.Phys.Vol.13,Edit.K.Shimoda (Springer,Berlin,Heidelberg,New York 1976)
5. G. Huber,C.Thibault,R.Klapisch,H.T. Duons,J.L.Vialle,J.Pinard, P.Juncar,P.Jaquinot;Phys.Rev.Lett.34,1209 (1975)
6. G.Höning,M.Czajkowski,W.Demtröder;J.Chem.Phys.to be published
7. R.Schmiedl,I.R.Bonilla,F.Paech and W.Demtröder;J.Mol.Spectrosc.68, 236 (1977)
8. S.L.Kaufman, Opt.Commun.17, 309 (1976)
9. E.W.Otten,Atomic Physics 5,Plenum Press (1977) p.239
10. M.Dufay and M.L.Gaillard;in Laser Spectroscopy III Eds.J.L. Hall and J.L.Carlsten,(Springer,Berlin,Heidelberg,New York 1977) p.231
11. M.Dubke,W.Jitshin and G.Meisel;Phys.Lett.65A, 109 (1978)
12. W.Ertmer, B.Hofer; Z.Physik A276, 9 (1976)
13. S.D.Rossner,R.A.Holt,T.D.Gaily;Phys.Rev.Lett.35,185 (1975)
14. M.E.Kaminsky,R.T.Hawkins,F.V.Kowalski and L.A.Schawlow;Phys. Rev.Lett. 36,671 (1976)
15. J.Foth,H.J.Vedder,W.Demtröder;Chem.Phys.Lett.to be published
16. F.Paech,R.Schmiedl and W.Demtröder;J.Chem.Phys.63,4369(1975)

# Spectroscopic and ionization properties of atomic Rydberg states

S. FENEUILLE and P. JACQUINOT*

Laboratoire Aimé Cotton, C.N.R.S. II, Bâtiment 505, 91405 Orsay, France

Our knowledge of the different properties of high lying Rydberg atomic states has made considerable progress since the advent of tunable lasers and of sensitive methods of detection. About one year ago an International Colloquium was held in Aussois (France) essentially on this subject and a good image of the status of the field at that time can be gained from the Proceedings of that Colloquium [1]. Since then new progress has been made, partly reflected in the contributed papers presented at this Conference.

Some of these new attainments, particularly on collisional properties, are beyond the scope of this lecture. Nevertheless the title is too ambitious since it deals with two important but different properties of Rydberg states, i. e. spectroscopic, and ionization properties. The first ones include positions, fine structures, transition probabilities and radiative lifetimes as well as "structure" of the levels in magnetic or electric fields. On this last point the separation of "spectroscopic" properties from ionization properties is less clear since in a sufficiently high electric field Stark effect leads to field ionization. Moreover if a "Stark" level is photoexcited and ionizes it is difficult to say whether this is a photoionization or a field ionization.

Among the different review papers given at the Aussois Colloquium two, respectively given by each of us, were very close to the subject of this lecture, the first one for the case of rubidium [2], the second one for the case of alkaline earths [3]. It is unnecessary to repeat here all that was already written there about one year ago. Only the most significant results or progress in the understanding of the phenomena will be discussed here.

* The lecture, originally **scheduled** to be given by P. Jacquinot, has not been presented at the Conference. This text has been written by S. Feneuille and P. Jacquinot.

As to the spectroscopic properties it seems to us that not many new phenomena have been observed since Aussois, so that this part of the subject will be almost entirely omitted in this paper. Let us only briefly mention the most important results that have been obtained on alkaline earths, at moderate resolution. The first one [4] is a study of Ba (and Yb ) by two photon spectroscopy, and the second one [5] a study of Ca , Sr and Ba by multistep excitation with three lasers. In both cases long series were observed, quantum defects and influence of perturbing levels were measured and an extensive analysis of the data obtained was performed by use of the multichannel quantum defect theory (M. Q. D. T.).

Another set of beautiful experiments must also be only mentioned here. They deal with the structure of Rydberg states in magnetic fields intense enough for the interaction with the field to be as large as or larger than the separation between terms with different $\ell$ or n . Spectra of Ba , Sr [6] and Li [7] have been observed in a field of 47 KG in the vicinity -below and above- of the zero field ionization limit, either by conventional absorption spectroscopy or by two-photon laser spectroscopy. They clearly show the $\ell$-mixing and n-mixing regions as well as the quasi Landau structure on both sides of the ionization limit. A detailed study of the diamagnetic structure of the sodium levels around n=28 in a field of 60 KG was also carried on with an atomic beam by laser spectroscopy [8] and successfully compared with a calculation based on a numerical diagonalization including spherical basic states from n=25 to n=31 .

This paper will thus be essentially devoted to the ionization of Rydberg states and more precisely to their ionization by an electric field. In order to present a self contained paper on this question we must admit some duplication with [2]. The problem of field ionization is not separable from the study of Stark effect. In particular, it is well known that, in the presence of a D. C. electric field, any atom in any state would lose its electron if one would be able to wait for a sufficiently long time. Properly speaking, this system has no bound states. However, the corresponding ionization probability exhibits a threshold behaviour [9][10]. Below the threshold, the ionization rate is so small (i. e. smaller than the reciprocal of the radiative lifetime) that on can observe quasi bound or pseudo stable Stark structures in atomic spectra perturbed by weak electric fields. This threshold behaviour is related to the existence of a saddle-point in the potential energy surface of an electron interacting simultaneously with a Coulomb field an a D.C. field. This is illustrated on Fig. 1. For a given field $F$ , the energy of the saddle-point is given in atomic units by :

$$E_c \simeq E_0 - 2\,F^{1/2}$$

where $E_0$ is the potential ionization of the unperturbed atom. In other words, the atom ionizes significantly, for a given field $F$ , only if its energy is higher than $E_c$ and, for a given energy $E$ , only if the D.C. field is larger than a critical value $F_c$ , given by :

$$F_c = (E_0 - E)^2/4 \quad .$$

It is clear that reasonable values of $F_c$ can be obtained only for atoms sufficiently excited, that is to say for atoms in highly excited Rydberg states. For such states, $F_c$ can be written :

$$F_c = (2n^*)^{-4}$$

where $n^*$ is the effective principal quantum number of the considered state. In fact, this ionization phenomenon has been extensively used during the last three years for detecting atoms in Rydberg states, but it appeared gradually that, after all, this phenomenon was not so simple as previously described and therefore, several research groups began to study it for itself.

A typical set-up for such studies is illustrated on Fig. 2. An atomic beam is transversally illuminated by an exciting pulsed laser beam (or several laser beams to allow multistep excitation). The interaction region is located between two plates on which a D.C. or a delayed pulsed voltage can be applied . The ions produced are accelerated on the cathode of an electron multiplier and they are counted by a digital boxcar integrator. With such a device, four type of experiments, schematized on Fig. 3, have been usually carried out.

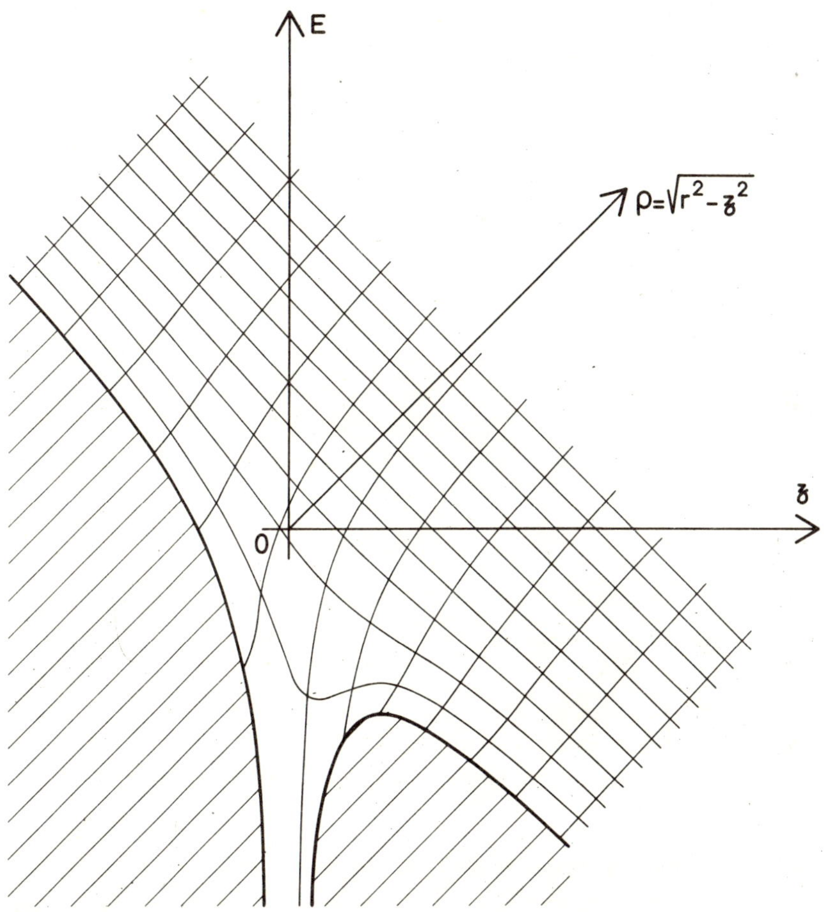


Fig. 1 - Potential energy surface of an electron interacting simultaneously with a Coulomb field and a D.C. field (only one half of the surface is represented).

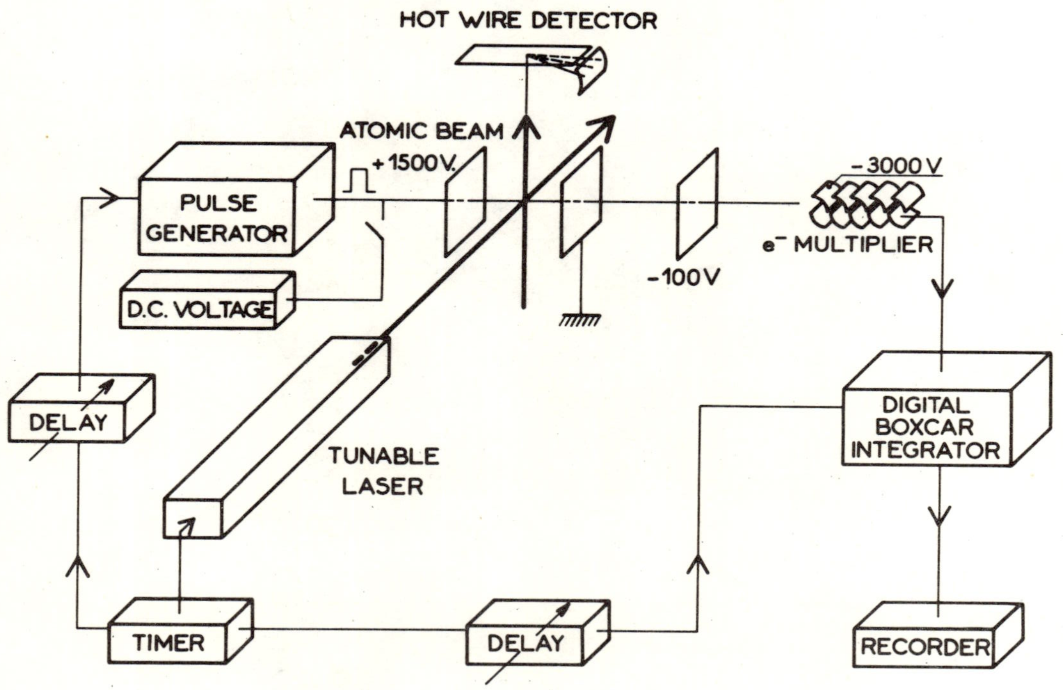


Fig. 2 - Typical set-up for studying Stark structures and field ionization properties of highly excited Rydberg states.

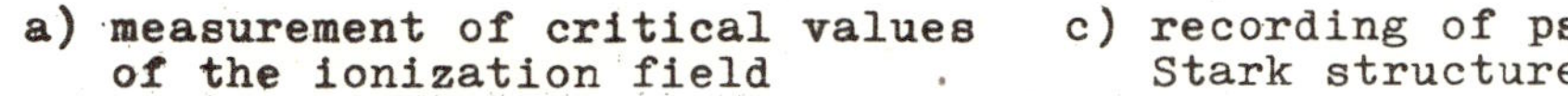

a) measurement of critical values of the ionization field

c) recording of pseudo stable Stark structures

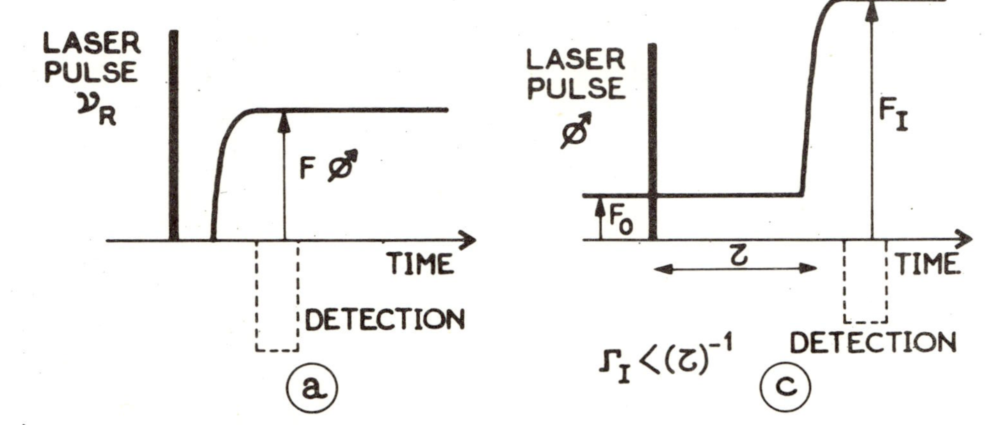


b) measurement of critical values of the ionization field

d) photoionization in the presence of a D.C. field

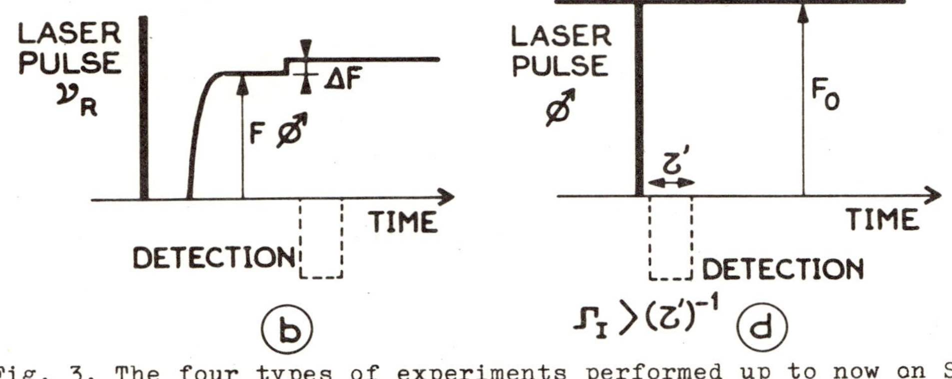


Fig. 3. The four types of experiments performed up to now on Stark structures and field ionization properties of highly excited Rydberg states.

In the first type (Fig. 3a) [11][12][13] the laser is tuned on an atomic resonance which allows one to populate selectively a given atomic level. In some cases, atomic states can be selectively excited by using adequate light polarizations. The D.C. electric field is applied only a certain time after the laser pulse. Of course, this delay must be shorter than the natural lifetime of the excited atomic state. The detection gate is open a short time after applying the field to take into account the transit time of the ions. The number of these ions is counted as a function of the strength of the D.C. field. A typical result obtained on the 40p level of rubidium [12] is given on Fig. 4. The ionization threshold appears very well defined. Moreover, in the case of rubidium, the experimental threshold are in quite good agreement with the classical saddle-point law since the quantity $16\ F_c\ n^{*4}$ remains very close to one for all the studied atomic states ( ns , np , nd ; $30 < n < 50$ ). It must be noticed however that, in many other cases, the quantity $16\ F_c\ n^{*4}$ remains constant but differs significantly from one [3]. This scaling effect is not yet quantitatively understood. Even more serious is the fact that in some cases, ionization thresholds can become quite broad and can exhibit multiple plateaus. This is illustrated on Fig. 5 on the particular case of $17\ p_{3/2}$ and $17\ p_{1/2}$ levels of sodium [13]. In such cases, it seems rather difficult to define precisely the various thresholds, in particular because of noise problems. This difficulty led recently Vialle [14] to improve the experiment so as to have a better signal to noise ratio. This is illustrated on Fig. 3b. With such a device, it is clear that ions are counted if and only if $F + \Delta F$ corresponds to a threshold value. If $F + \Delta F$ is smaller than $F_c$ no ions are produced because the field strength does not reach the critical value. On the contrary, if $F$ is larger than $F_c$ , all the excited atoms are ionized before opening the detection gate and all the produced ions are destroyed before being counted. In other words, ionization thresholds appear as peaks in the number of ions plotted as a function of $F$ . Typical results obtained by Vialle [14] are given on Fig. 6. The various thresholds are now completely resolved while, with the standard technique, only a single broadened thresholds is

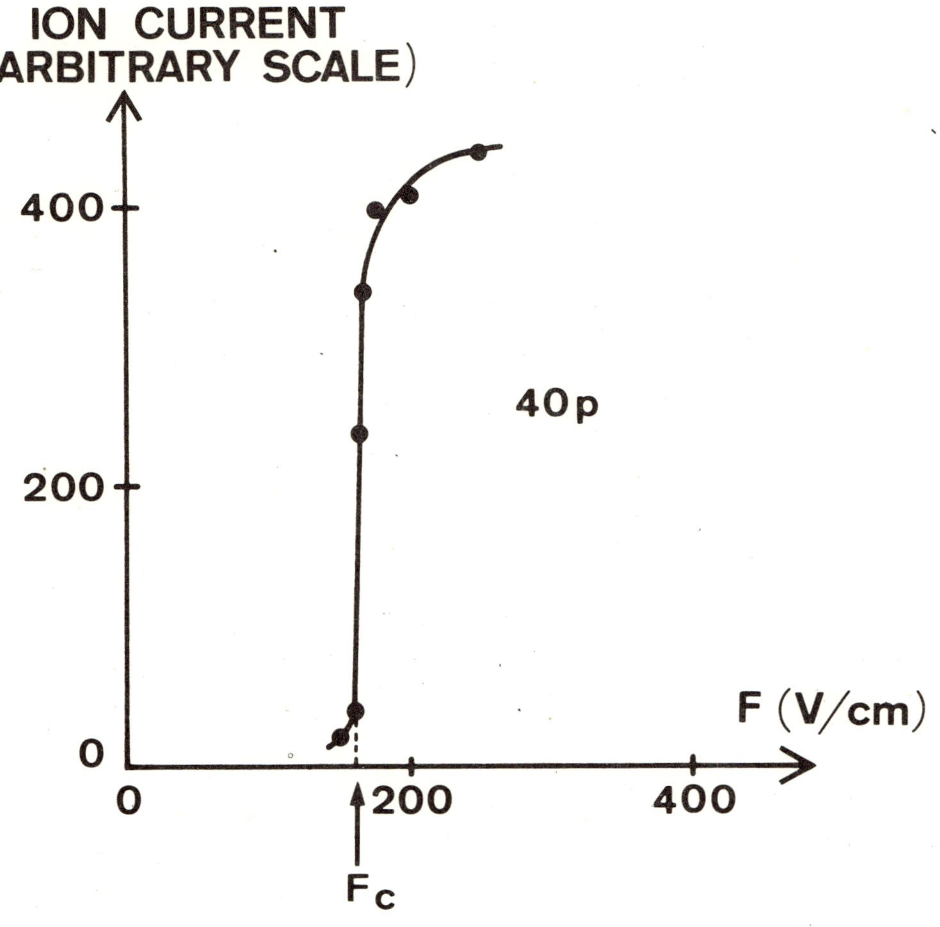


Fig. 4 - Rubidium 40p ionization threshold.

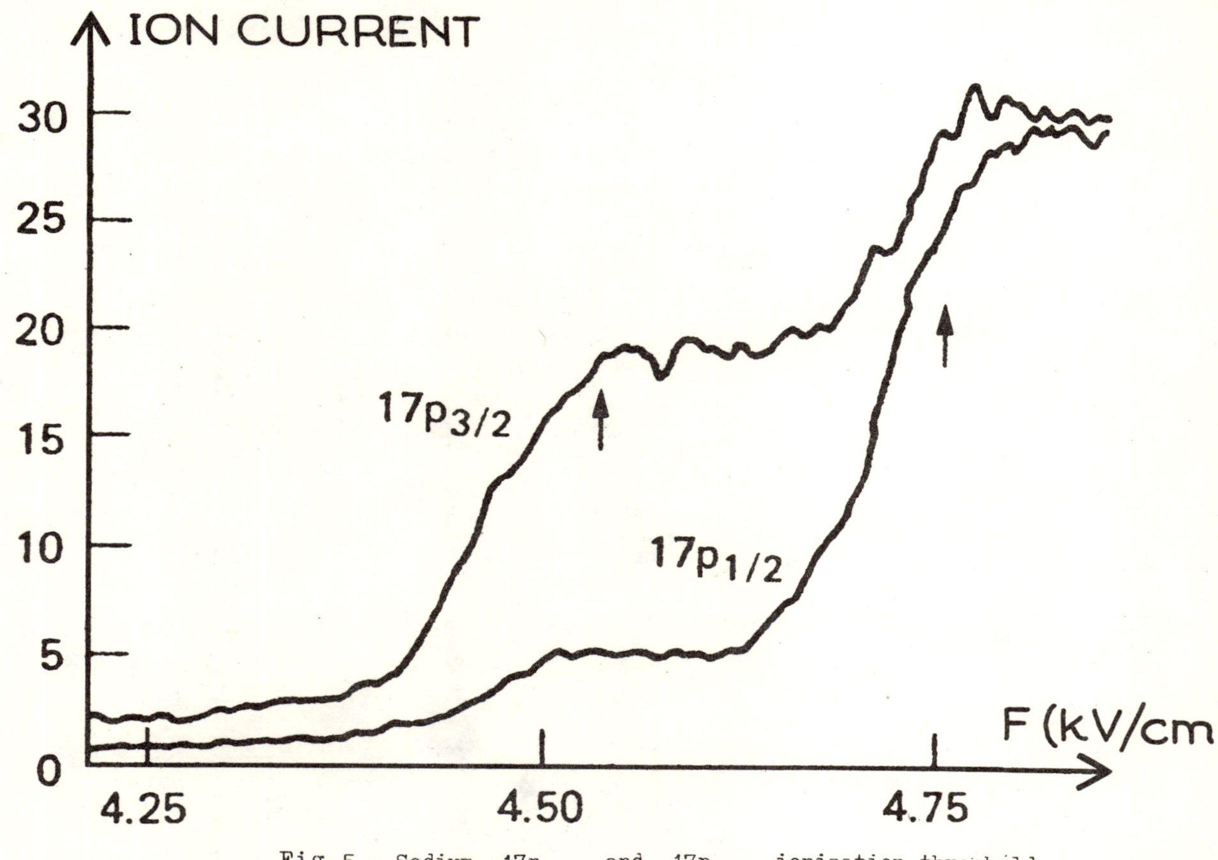


Fig.5 - Sodium $17p_{1/2}$ and $17p_{3/2}$ ionization threshold.

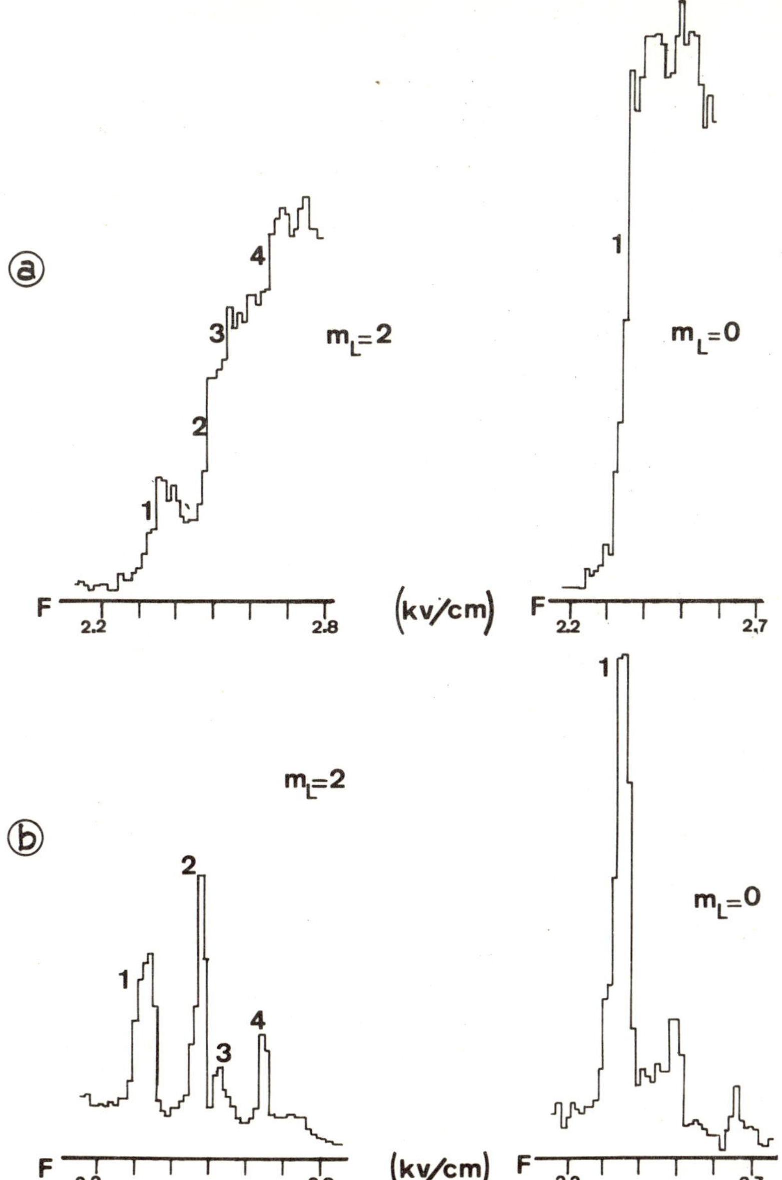


Fig. 6 - Sodium 20d ionization thresholds obtained from experiments, respectively of the types a and b of Fig. 3.

observed. A first explanation of multiple thresholds can be given by considering that they correspond to the various magnetic states of an atomic level. However, by using adequate light polarizations, it is possible to excite states with a well defined $M_L$ value, and nevertheless, even in this case, multiple thresholds can appear. A new explanation, taking into account partially diabatic effects (as in the Landau-Zenner formula) during the application of the field was recently proposed by Gallagher [13] but recent results obtained by Vialle seem to be in disagreement with this explanation. Therefore, the question can be considered as still open.

The two last types of experiments are [13][15][16][17][18] rather different from the previous ones since now the electric field value is fixed and the number of ions is counted as a function of the laser frequency which is continuously tuned. In the third one (Fig. 3c), a constant field $F_0$ is applied during the light interaction and a much more intense field is applied a certain time only after the light pulse. The detection gate is open only after applying this intense electric field which is much larger than the critical field values corresponding to the energy domain actually investigated. Thus, if resonances are observed in the number of counted ions they correspond to pseudo stable Stark structures which ionize with a rate less then $\tau^{-1}$. Of course, the results obtained depend a priori on $\tau$. However, because of threshold behaviour of field ionization probability, the dependance on $\tau$ is very weak. Typical values of $\tau$ actually used are of the order of a few mickoseconds. First, it must be noticed that, if one takes $F_0$ equal to zero, the present technique allows one to study unperturbed atomic structures. It has been extensively used in various spectroscopic experiments on Rydberg states in alkalis, in alkaline earths, in lanthanides, and in uranium. Stark structures ($F_0 \neq 0$) have also been investigated for some highly excited alkalis and alkaline earths. The most extensive study has been recently performed on lithium [17]. The most striking feature is the disappearance of any pseudo stable Stark structure as soon as the excitation energy becomes larger than $E_c$. Here again, this result is in perfect agreement with the classical saddle-point limit.

However, the latter type of experiment provides no information about the perturbed atomic structure above the saddle-point limit. In a naîve approach, one could expect a more or less flat continuum, but this was not so clear after all and the best way to solve the problem was still to perform an experiment. This one is schematized on Fig. 3d. The field $F_0$ is still applied during the light interaction but now the detection gate is open during a time $\tau'$ just after the light pulse. This means that, if resonances are detected, they correspond to states ionizing with a rate greater than $(\tau')^{-1}$. Here again, the dependance of the results on $\tau'$ is very weak and typical values of $\tau'$ actually used are of the order of a few microseconds. The first photoionization experiment of this type was carried out on rubidium at high resolution by using a single-mode pulsed dye laser [15][18]. Typical results are reported on Fig. 7. The most striking feature is of course the appearance of very narrow photoionization resonances, at least if a $\sigma$ light excitation (excited states with $M_L = \pm 1$) is used. These resonances are sufficiently narrow to be isotopically selective and in fact, their actual widths are still unknown since the observed value 50 MHz is close to the residual Doppler broadening corresponding to the relatively low collimation of the atomic beam effectively used. A more extensive study at lower resolution was recently performed on lithium [17]. The results are quite similar and their main interest is to show that these resonances appear on a very large spectral range but disappear as soon as the excitation energy becomes smaller than $E_c$. Finally, the complete Stark structure of a highly excited atom can be schematized as illustrated on Fig. 8.

Now, the problem is to understand qualitatively and, if possible, quantitatively, the latter structure. Qualitatively, there is in fact no difficulty : it is well known in classical mechanics that, if the potential energy exhibits a saddle-point, there exist closed orbits, i. e. stable states with a total energy higher than the saddle-point energy. This is illustrated on Fig. 9 in the particular case of a ball inside a tilted bowl. Furthermore, the existence of a non zero

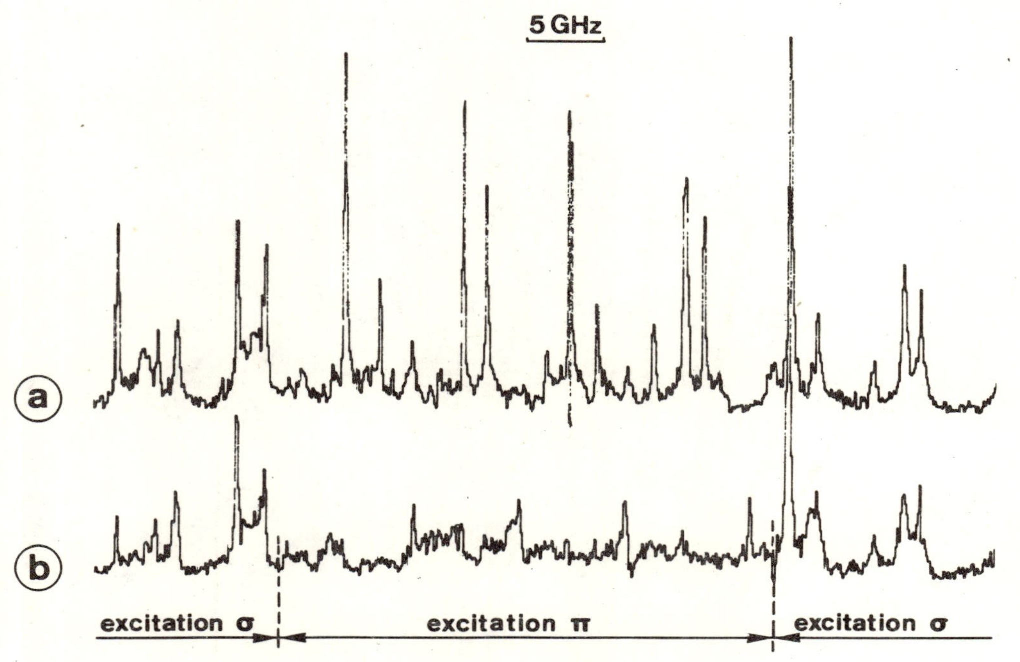


Fig. 7 - Photoionization spectrum of rubidium in the presence of an electric field (a : $\sigma$ excitation, b : $\pi$ excitation in the middle of the recording).

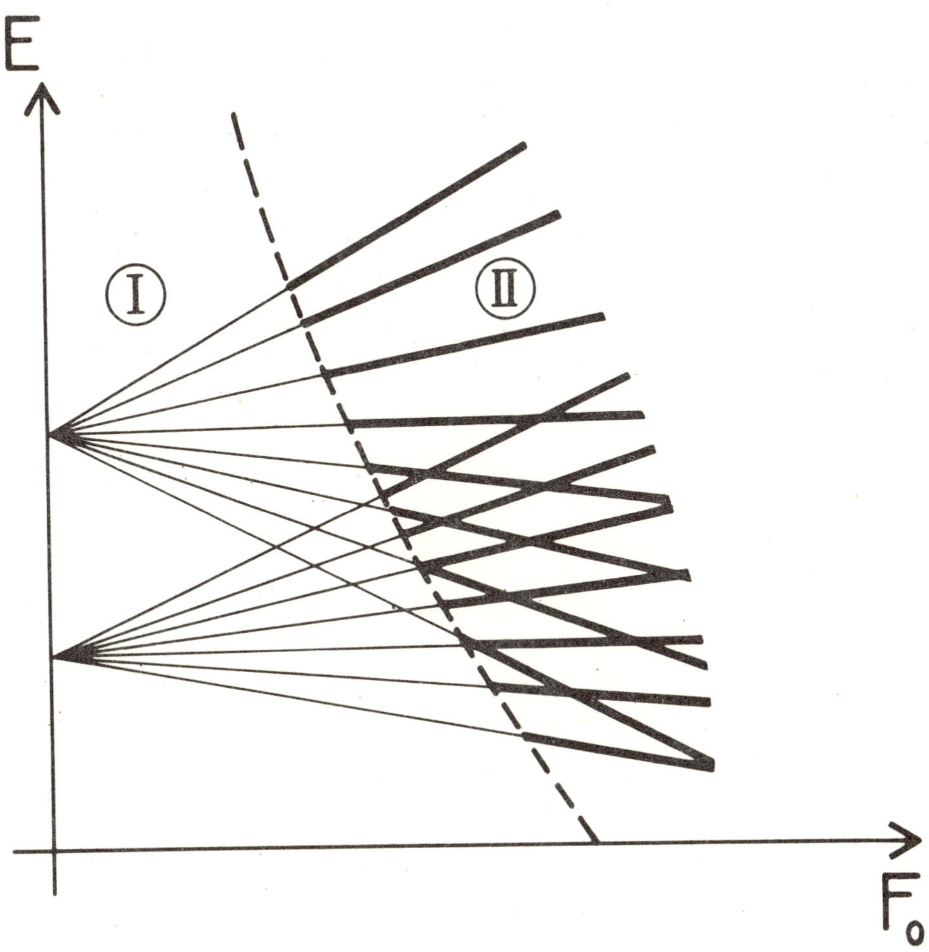


Fig. 8 - Stark structures and field ionization properties of an alkali atom. The spectrum is separated into two regions by the saddle-point limit : $E = E_0 - 2\sqrt{F_0}$ (dotted line). In region I, only pseudo stable Stark structures appear. In region II, narrow photoionization resonances are superimposed to flat continua.

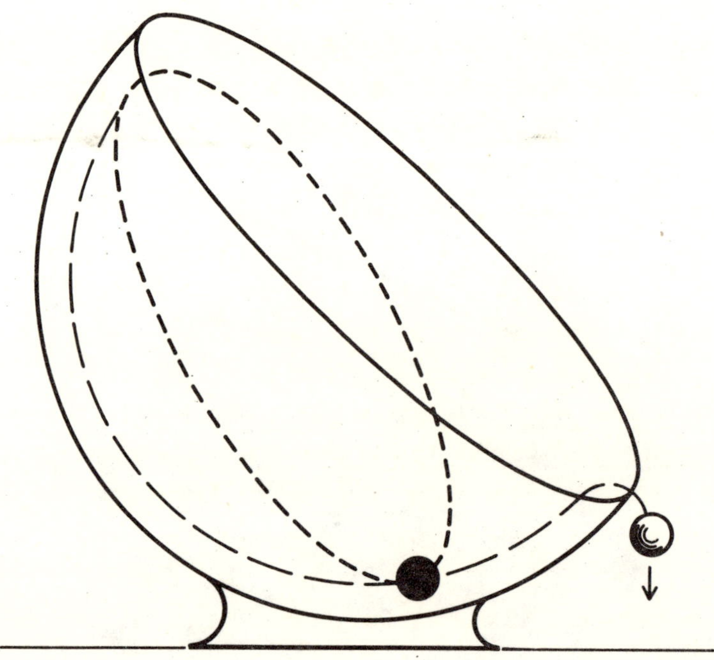

Fig. 9 - Two balls in a bowl : the black ball remains in the bowl (stable state) while the white ball escapes the bowl (ionized state).

ionization probability can be understood in terms of pure quantum effects analogue to tunneling, the stable states becoming pseudo stable ones.

Quantitatively, two main problems occur. The first one is to interpret the energy of the Stark levels (quasi bound or unstable). First of all, the case of hydrogen must be distinguished since, the corresponding Stark Hamiltonian is separable in parabolic coordinates [10] and thus the problem can be exactly solved. The only difficulties are of computational nature. For non hydrogenic atoms, and of course for hydrogen either as a test or for saving computer time, various techniques can be utilized. The most usual one is standard perturbation theory. This leads in hydrogen to well known analytic formulas [10] which provide a good agreement with experimental data as long as the electric field is sufficiently weak $(F \ll F_c)$. However, starting from bound states to describe unbound states leads to a so-called asymptotic expansion [9] which diverges beyond a certain order which depends both on the considered state and on the field strength. This difficulty with perturbation theory was recently reinvestigated by Koch [19] in hydrogen. Another technique consists to diagonalize the Stark Hamiltonian on a large basis of unperturbed atomic states. A good agreement between theoretical and experimental data was obtained for various alkalis [16][20] but here again only for weak fields. When the field increases, a larger and larger atomic basis is needed and finally, the same convergence difficulties as in standard perturbation theory are encountered. A completely different method has been proposed by various authors [21] but up to now, it has been applied only to hydrogenic states with rather low principal quantum numbers. The principle is to determine an approximate bound wave-function by minimizing for each energy the quantity :

$$\sigma(E) = \langle H \rangle^2 - \langle H^2 \rangle$$

where H is the Stark Hamiltonian. One can prove that the Stark levels (quasi bound or unstable) are given by the minima of $\sigma(E)$

considered as function of $E$. This variational approach is rather heavy but is very promising for non hydrogenic atoms.

Even more difficult is the theoretical determination of the ionization width of the Stark levels. To our knowledge, this problem has been considered up to now in hydrogen only. In this rather particular case, thanks to separability in parabolic coordinates, the system can be exactly solved. However, because of computational difficulties, exact solutions were obtained a few years ago only [22], but, in any case, most of the important features were already discovered in rather old studies using WKB approximation [23]. The most striking result of all these theoretical calculations is the presence, far above the saddle-point limit of quasi stable Stark levels. This can be understood if one remarks that, in parabolic coordinates, a saddle-point still appears but depends not only on $F$ but also on $|M_L|$ and on the separation constant $Z_2$ related to the $\eta$ coordinate. For example, for $|M_L| = 1$, the critical value of the ionization field is now given by [10]:

$$F_c^H = (E_0-E)^2/4Z_2 \qquad \text{instead of} \qquad F_c = (E_0-E)^2/4 \ .$$

Since $Z_2$ depends on $F$ in a complicated way, it is no longer possible to express $F_c^H$ as a simple function of the energy. However, one knows qualitatively that, for highly excited states, $Z_2$ can take very small values (of the ordre of $1/n$). Therefore, for some Stark states, $F_c^H$ can be much larger than $F_c$. This is illustrated on Fig. 10. Of course, because of tunneling effects, a non negligible ionization probability can appear for $F$ values smaller than $F_c^H$ but this is not related with the saddle-point limit defined in spherical coordinates. Thus, though, for hydrogen, available data are purely theoretical, while, for alkalis, they are purely experimental, one can say that the respective field ionization properties of hydrogen and of alkali atoms are quite different. This difference was recently explained by Littman et al. [17] in terms of symmetry breaking.
Even in hydrogen, above the spherical saddle-point limit, pseudo-stable

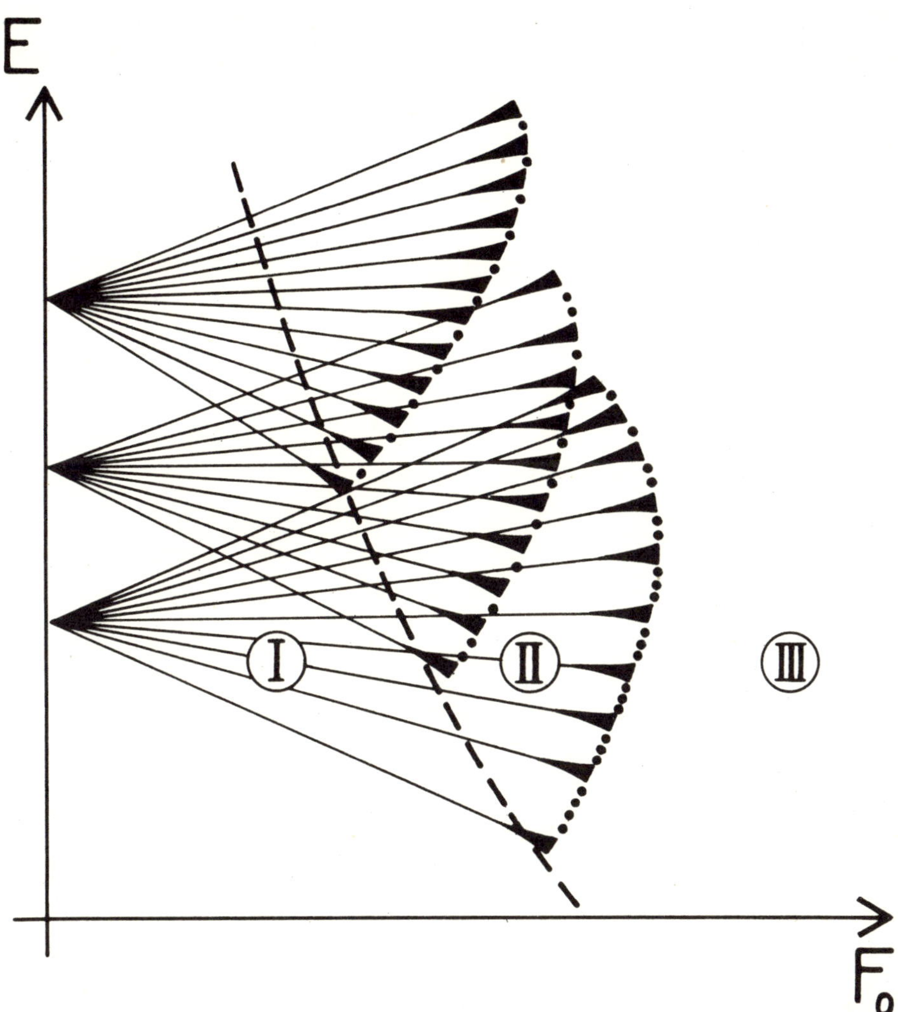


Fig.10 - Stark structures and field ionization properties of the hydrogen atom. The spectrum is separated into three regions by the saddle-point limits obtained respectively in spherical coordinates (-----) and in parabolic coordinates (-.-.-.-.-). In region I, only pseudo stable Stark structures appear ; in region II, pseudo stable Stark structures are embedded in ionization continua ; in region III, only ionization continua remain.

Stark levels are embedded in ionization continua coming from upper levels but, because of separability in parabolic coordinates which introduces supplementary good quantum numbers and therefore, selection rules, there is no interaction between the pseudo stable Stark levels and these ionization continua. In contradistinction, as soon as separability disappears, and this is the case for any non hydrogenic atom, such an interaction appears and Stark levels ionize through a mechanism similar to autoionization in many electron atomic spectra or to predissociation in molecules. Such an explanation sounds very reasonable and furthermore, it allows one to understand some characteristics of the observed photoionization spectra. So, the widths of the unstable Stark levels depend essentially on the interaction between ionization continua and pseudo stable Stark levels. This interaction itself depends only on the non hydrogenic character of the considered state. Now, it is well known that $M_L = 0$ states have the strongest non hydrogenic behaviour since they involve S states which have the largest quantum defect. Therefore, one understands very well why, in Fig. 7, resonances obtained are much broader with $\pi$ excitation than with $\sigma$ excitation [15][18]. A definite proof of the validity of the mechanism proposed by Littman et al. would certainly be the observation of a Fano profile [24]. This observation requires the use of rather high optical resolution and in principle, it should be possible in the experiment on rubidium using a single mode laser. In fact, some structures observed in this experiment with a $\pi$ excitation present some similarities with a Fano profile, but the signal to noise ratio is too poor to lead to a definite conclusion. Further experiments are certainly needed for all the alkalis especially at high resolution.

Quantitative theoretical studies would be also highly desirable in non hydrogenic atoms. In fact, many papers appeared recently [25] and various approaches, more or less sophisticated, have been proposed but none of them has been applied to physical cases, except hydrogen. Among these theoretical approach the so called "complex scaling method" is one of the most promising but, after all, all the well known techniques used in autoionization studies, such as Fano formalism, could certainly be utilized by starting from the exact Stark solutions of the hydrogen atom.

Other problems remain open both theoretically and experimentally. They concern, in particular, the Stark structure and the field ionization properties of non alkali atoms for which doubly excited valence states sometimes perturb Rydberg series. During the last three years, it thus appeared gradually that the Stark effect needed to be revisited. Recently, we learnt a lot about it but nevertheless, it is still far from being fully understood.

## R e f e r e n c e s

[1] Colloques Internationaux du C. N. R. S., n° 273, "États atomiques et moléculaires couplés à un continuum. Atomes et molécules hautement excités", C. N. R. S., Paris 1977.

[2] P. Jacquinot, S. Liberman and J. Pinard, in [1], p. 215.

[3] P. Camus, R.-J. Champeau, S. Feneuille, S. Liberman, C. Morillon and J. Pinard, in [1], p. 67.

[4] P. Camus, A. Debarre and C. Morillon, J. Phys. B 11, L1, 1978.
M. Aymar, P. Camus, M. Dieulin and C. Morillon, To be published in Phys. Rev., Oct. 1978.
M. Aymar and O. Robaux, To be published in J. Phys. B.

[5] J. A. Armstrong, J. J. Wynne and P. Esherick, Private communication. To be published in J. O. S. A. .

[6] K. T. Lu, F. S. Tomkins and W. R. S. Garton, Proc. Roy. Soc. London, A 362, 421, 1978.
R. J. Fonck, D. H. Tracy, D. C. Wright and F. S. Tomkins, Phys. Rev. Lett. 40, 1366, 1978.

[7] K. T. Lu, F. S. Tomkins, H. M. Crosswhite and H. Crosswhite, To be published in Phys. Rev. Letters.

[8] M. L. Zimmerman, J. C. Castro and D. Kleppner, Phys. Rev. Letters 40, 1083, 1978.

[9] See for example : L. D. Landau and E. M. Lifshitz, Quantum Mechanics, Pergamon Press, Oxford 1958.

[10] H. A. Bethe and E. E. Salpeter, Quantum Mechanics of One- and Two-Electron Atoms, Springer-Verlag, Berlin 1957.

[11] T. W. Ducas, M. G. Littman, R. R. Freeman and D. Kleppner, Phys. Rev. Letters 35, 366, 1975.

R. F. Stebbings, C. J. Latimer, W. P. West, F. B. Dunning and T. B. Cook, Phys. Rev. A 12, 1453, 1975.

A. F. J. Van Raan, G. Baum and W. Raith, J. Phys. B 9, L173, 1976.

[12] Duong Huong Tuan, S. Liberman and J. Pinard, Optics Commun. 18, 533, 1976.

[13] T. F. Gallagher, L. M. Humphrey, W. E. Cooke, R. M. Hill and S. A. Edelstein, Phys. Rev. A 16, 1098, 1977.

[14] J. L. Vialle, submitted to J. Phys. B ;
J. L. Vialle, to be published in Journal de Phys. Lettres, nov. 78.

[15] S. Liberman and J. Pinard, to be published in Phys. Rev.

[16] M. G. Littman, M. L. Zimmerman, T. W. Ducas, R. R. Freeman and D. Kleppner, Phys. Rev. Letters 36, 788, 1976.

[17] M. G. Littman, M. M. Kash and D. Kleppner, Phys. Rev. Letters 41, 103, 1978.

[18] S. Feneuille, S. Liberman, J. Pinard and P. Jacquinot, C. R. Acad. Sci. - Paris B284, 291, 1977.

[19] P. M. Koch, Phys. Rev. Letters 41, 99, 1978.

[20] E. Luc-Koenig, S. Liberman and J. Pinard, Phys. Rev., to be published.

[21] See for example : E. Brändas and S. Froelich, Phys. Rev. A 16, 2207, 1977.

[22] M. H. Alexander, Phys. Rev. 178, 34, 1969 ;
J. O. Hirschfelder and L. A. Curtiss, J. Chem. Phys. 55, 1395, 1971 ;
R. J. Damburg and V. V. Kolosov, J. Phys. B 9, 3149, 1976 ;
Phys. Lett. 61A, 233, 1977 ; Opt. Commun. 24, 211, 1978.

[23] C. Lanczos, Z. Physik 62, 518, 1930 ; 65, 431, 1930 ; 68, 204, 1931 ;
D. S. Bailey, J. R. Hiskes and A. C. Rivière, Nucl. Fusion 5, 41, 1965.

[24] U. Fano, Phys. Rev. 124, 1866, 1961.

[25] See for example : I. W. Herbst and B. Simon, Phys. Rev. Letters 41, 67, 1978.

# OBSERVATION OF PARITY-NONCONSERVATION IN ATOMIC TRANSITIONS

L.M.Barkov, M.S.Zolotorev

Institute of Nuclear Physics, Novosibirsk 90, USSR

The experimental search for the parity-nonconserving optical rotation in atomic bismuth vapour has been performed in Oxford, Seattle and Novosibirsk /1-3/. We present here new results of the Novosibirsk Group.

Fig. 1 presents the experimental scheme for optical rotation measurement in bismuth vapour at 6477 Å MI transition line $6p^3\,{}^4S_{3/2} - 6p^3\,{}^2D_{5/2}$. Spectra-Physics model 375 dye laser was modified to produce a single frequency light beam with power up to 15 mW. The modification permits also to modulate the wavelength with the frequency 1 kHz. The modulated light passes in turn through the prism polarizer, the bismuth vapour cell and prism analyzer, after which two beams with orthogonal polarization and different intensity are detected with the help of two photomultipliers. The signal $V_1$, from the PM1, which detect the intensity of the bright beam, was used to keep with the help of feed back scheme the bismuth line exactly at the centre of the scanned frequency interval. In this case the signal from the second photomultiplier

$$V_2 = V_1\theta^2 = V_1(\theta_0+\psi)^2 \simeq V_1\theta_0^2(1+2\psi/\theta_0)$$

contains the first harmonic of scanning frequency only through $\psi$. Here $\theta_0 = 10^{-3} \div 10^{-2}$ rad is the angle between analyzer and polarizer axes, and $\psi$ is an angle of the optical rotation by bismuth vapour due to the parity-nonconserving interaction of electrons with nucleus. The expected value of is $\sim 10^{-7}$ rad, so $\psi \ll \theta_0$. After the substraction of $V_1$ from $V_2$ (signals $V_1$ and $V_2$ had been equalized in amplitude), the substracted signal was phase detected on the first harmonic of the scanning frequency, while the second harmonic was used to uphold the equality of the subtracted signals. The first harmonic of $\psi$ was found from the difference of measurement at $+\theta_0$ and $-\theta_0$.

The prism polarizer and analyzer were used as entrance and exit windows of the bismuth vapour cell. In this case the absence of additional substance between the polarizer and analyzer results in reduction of spurious rotation in the system. To make the windows safe from the bismuth vapour action both ends of the cell were connected with the buffer helium filled vessel. It allowed also to stabilize the vapour density inside the cell. The spatial distribution of bismuth vapour density along the light beam path was found from Faraday rotation measurements by switching on in turn one of the six sections of the solenoid placed inside the Permalloy magnetic shielding. The bismuth vapour cell had a double magnetic screen and average magnetic field inside the cell was reduced to $2\cdot10^{-5}$ Gs. The construction of the bismuth oven permits to work with the temperatures up to 1200°C. At these temperature the partial pressures of the molecular and atomic bismuth are approximately equal and vibrational and rotational molecular spectra masks the hyperfine structure of the atomic line under investigation.

Fig. 2a presents the absorption spectrum of the bismuth vapour. The histogram bin corresponds to 400 MHz frequency interval between longitudinal modes of laser resonator. The measurements of Faraday rotation was performed with the help of Faraday cell. The results of the measurements are presented in Fig. 2b and its comparison with the results of the theoretical calculations/4/ allowed to identify unambiguously the hyperfine structure of 6477 A line of atomic bismuth. The results of Faraday rotation are consistent also with those of Oxford group (P.Sandars, private communication).

The measurements of optical rotation were performed in three runs on five dipole transitions of hyperfine structure of 6477 Å atomic bismuth line (lines 1, 3, 7, 12 and 18 in Fig. 2a), on three quadrupole transitions (lines 2, 10 and 17 in Fig. 2a) and on four molecular lines (A, B, C, D in Fig. 2a). In these measurements the scanning amplitude was equal to one-two Doppler widths, the pressure of bismuth vapour corresponded to 20 ÷ 30 Torr, the effective length of bismuth vapour changed from 30 to 40 cm and the angle $\theta_0$ from $2\cdot10^{-3}$ to $10^{-2}$ rad.

The average angle of optical rotation measured on five

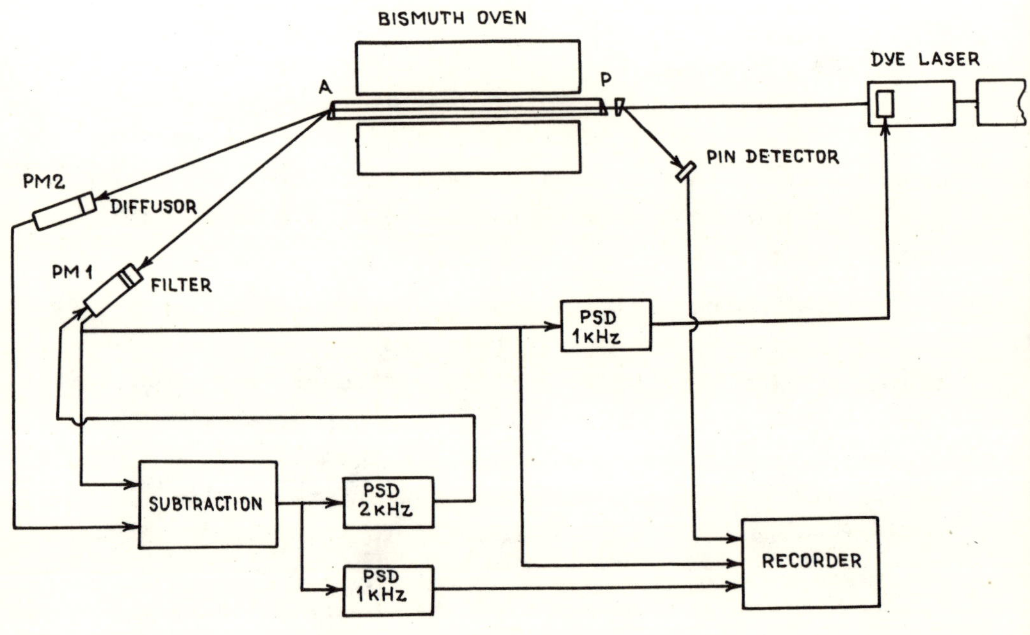


Fig. 1. The scheme of the experiment.
P - prism polarizer, A - prism analyzer.
PM1, PM2 - photomultipliers.

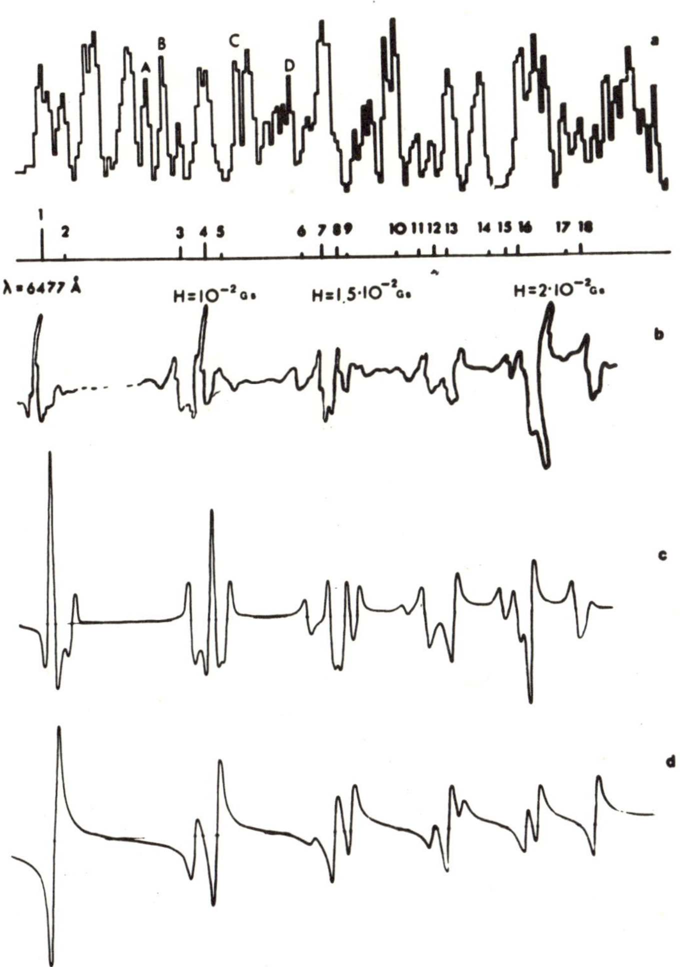


Fig. 2. a - Observed absorption spectrum;
b - observed Faraday rotation;
c - the theoretical curve for Faraday rotation;
d - predicted on the basic of Weinberg-Salam model optical activity of the bismuth vapour according to the work /5/.

dipole transitions was found to be $\overline{\psi}_{exp} = (-3.1 \pm 0.5) \cdot 10^{-8}$ rad. The measurements of optical rotation with magnetic field permitted to perform the absolute calibration of optical path in atomic bismuth. The comparison of the results of the measurements with the theoretical predictions obtained in work /5/ on the basis of Weinberg-Salam model gave $\psi_{exp}/\psi_{W-S} = 1.1 \pm 0.3$. The measurements performed on seven control lines gave the optical rotation $(1.0 \pm 0.5) \cdot 10^{-8}$ rad compatible with the prediction $0.2 \cdot 10^{-8}$ rad for the lines lying between dipole transition lines of atomic bismuth.

The results of this experiment are in contradiction with the results of Oxford /1/ and Seattle /2/ groups who observed zero effect on 6477 Å and 8757 Å bismuth lines.

We are indebted to late Prof. G.I.Budker for his support of this work and to Dr. I.B.Khriplovich for his contribution to this work and collaboration.

## References

1. P.E.G.Baird et al. Phys. Rev. Lett. 39, 798 (1977).
2. L.L.Lewis et al. Phys. Rev. Lett. 39, 795 (1977), E.N.Fortson. Talk at "Neutrino 78" conference.
3. L.M.Barkov, M.S.Zolotorev. Pis'ma ZhETF, 26, 379 (1978).
4. V.N.Novikov, O.P.Sushkov, I.B.Khriplovich. Optika i Spektroskopiya, 43, 621 (1977); 45, 413 (1978).
5. V.N.Novikov, O.P.Sushkov, I.B.Khriplovich. ZhETF, 71, 1665 (1976) Sov. Phys. JETP 44, 872 (1976) .

PARITY NON-CONSERVATION IN ATOMIC BISMUTH

P.E.G. Baird

Clarendon Laboratory, Parks Road, Oxford, England.

## INTRODUCTION

Over the last year or so a number of developments have taken place in the search for parity non-conservation (PNC) in neutral weak current interactions both in atoms and in high energy physics experiments. Firstly, the recent announcement by the S.L.A.C. group[(1)] that PNC has been detected in electron-proton and electron-deuteron scattering at energies around 20 GeV seems to support more or less unambiguously the Weinberg-Salam model[(2)] Secondly, new results have been obtained in the three atomic physics experiments working with bismuth vapour. In particular the Novosibirsk group[(3)] claim to have seen PNC at the level predicted by the Weinberg-Salam model (together with the atomic calculation of Novikov, Sushkov and Khriplovich[(4)]). The position is at present intriguing since experiments started several years ago at Oxford (648 nm transition in Bi) and Seattle (876 nm transition in Bi) and which were the first to reach the necessary sensitivity[(5)] have still failed to produce a result inconsistent with zero PNC. Thus a question mark still hangs over the atomic physics results, and the discrepancy between the Oxford and Novosibirsk results for the same atomic line must be resolved. It seems likely, however, that the situation will be clarified in the next few months as further refinements are made to the experiments with a corresponding reduction in the possible systematic error. Furthermore, experiments started more recently should also yield results[(6)].

Of course, experiments on atomic hydrogen ($^1H$ or $^2H$) will provide the ultimate test of neutral weak current interactions in atoms[(7)] - any doubts about the precision of the atomic calculation then disappear - but such experiments are difficult and it is probably realistic to expect results only in a year or so[(8)]. While the main effort is going into fine structure transitions using an atomic beam of metastable hydrogen, a different and alluring possibility has been suggested by Drukarev and Moskalev[(9)]. This is a two-photon transition ($1s_{\frac{1}{2}}$ - $2p_{\frac{1}{2}}$) which though difficult should be feasible: intra-cavity frequency doubling in a ring laser system could well provide several milliwatts of tunable radiation around 2430 A. The experiments can then be carried out on a cell of ground state hydrogen atoms, produced in a separated discharge region, for example. This would be substantially different from experiments now in progress on hydrogen and should be free from some of the background problems associated with them.

In the following I shall describe our progress in Oxford on the 648 nm and 876 nm transitions of bismuth; in many aspects the various optical rotation experiments at Seattle, Novosibirsk and Oxford are similar. I will give first a brief resume of the aim and method, followed by some details of the Faraday effect; this is used in the calibration of the sensitivity of the apparatus to PNC rotation and in any case is of general spectroscopic interest. Finally I will summarize our progress to date and compare results with the latest theoretical values.

## OPTICAL EXPERIMENTS

Inclusion of a parity violating neutral weak current interaction between the electrons and nucleons in an atom can be considered in the following way. The parity violating part of the atomic Hamiltonian can be divided into two parts[10]: a nuclear spin dependent term and a nuclear spin independent term. Estimates of the size of the optical effects these produce give values of R (see below) of $10^{-8}$ to $10^{-10}$ for the former. A discussion of possible atomic beam experiments to detect such an effect has been given by Loving and Sandars[11].

For the second part, however, Bouchiat and Bouchiat[12] have noted that there is an additional enhancement factor, $Q_W$ (which is a model dependent quantity of the order of the atomic mass number A), making this part scale approximately as $Z^3$; this is why heavy atoms such as bismuth, caesium, thallium and lead have attracted so much attention.

The optical effects due to the parity violating interaction come about because admixtures of opposite parity states influence atomic transition probabilities. Thus, for example, a pure magnetic dipole (M1) transition has a degree of electric dipole (E1) character mixed in by the weak interaction. Interference between the E1 and M1 transitions causes optical rotation and circular dichroism. The degree to which this occurs is given most conveniently in terms of the quantity,

$$R = \frac{\langle E1 \rangle^{ind}}{\langle M1 \rangle}$$

which is directly proportional to the difference in the refractive indices for right and left circularly polarised light.

For the optical dichroism experiments forbidden M1 transitions are used to enhance the effect. Thus the experiment on caesium in Paris(13) examines the 6s - 7s transition while the experiment on thallium being carried out in Berkeley(14) uses the 6p - 7p transition. Electric fields are used in these experiments to increase the transition rate in a controlled way(13).

For the optical rotation experiments it is usually advantageous to study an allowed M1 transition. The rotation follows a sharp dispersive shape through the M1 resonance with peak rotations of $\pm$ R/2 with one absorption length of vapour. For atomic bismuth there are several allowed M1 transitions between the levels of the ground configuration, $6p^3$. The lines at 648 nm and 461 nm are overlaid by molecular bismuth lines causing loss of signal; the 876 nm transition is virtually free from such background.

The basic scheme employed in Oxford to detect such optical rotation has been described before(15). Briefly, a beam from a tunable dye laser (with its wavelength set close to the M1 resonance under investigation) passes through an oven about 1 metre long containing the atomic vapour. Rotation of the plane of polarisation is measured using a Faraday modulator(16) (i.e. a water cell with a surrounding solenoid); this and the oven are between crossed polarisers. Angles are measured by picking out and amplifying the fundamental frequency. The sensitivity can be determined by applying a known angle to the system (e.g. by uncrossing the polarisers by a known amount). However, to determine R experimentally the atomic number density has to be measured. In most experiments this can be done by looking at the Faraday effect in a known magnetic field; the Faraday rotation is dependent on the atomic number density in the same way as the PNC rotation so that

$$\frac{\phi_{PNC}}{\phi_{Faraday}} = \text{Constant}/B$$

where the constant is evaluated from the theory of the Faraday effect(17) and includes a lineshape factor. More details of this are given in the next section. Finally, exploitation of the wavelength dependence through the resonance is used to pick out the PNC effect from other optical rotations (i.e. from those due to optical elements between the polarisers and any residual Faraday effect once the applied fields have been turned off).

## THE FARADAY EFFECT IN BISMUTH

First let us make a few statements about the transitions we are dealing with. These are predominantly M1 in character although there is certainly a detectable E2 component, which does not produce any PNC rotation but does produce a Faraday rotation; it must be taken into account in the calibration procedure. Further, since the spin of $^{209}Bi$ is 9/2 there are many hyperfine structure (hfs) components some of which are only partially resolved because of Doppler broadening at the operating temperature of the oven (ca 1500 K). In 648 nm the hfs spans about 60 GHz while that in 876 nm spans only 25 GHz.

An additional complication is the overlying molecular absorption; this makes it extremely difficult to identify the atomic lines in the absorption spectrum. Fortunately, however, only the atomic transitions are found to give rise to a sizeable Faraday rotation, as expected theoretically.

The Faraday rotation, then, arises from a magnetic field induced difference in the refractive indices for the right- and left-handed circularly polarised light; the field is applied axially, that is along the direction of propagation of the light. In the weak field limit these changes in the refractive indices can be traced back to three distinct mechanisms:

(1) Changes in the transition frequency with magnetic field;

$$\omega_{ij} = \omega_o - B/\hbar \{ \langle j|\mu_B|j\rangle - \langle i|\mu_B|i\rangle \}$$

(2) Mixing of hyperfine structure states with the same $M_F$ but different F;

$$|j'\rangle = |j\rangle - \sum_{k\neq j} \frac{\langle k|\mu_B|j\rangle}{\hbar\omega} |k\rangle B$$

(3) The different thermal equilibrium populations in the sublevels of each ground state hyperfine level.

Rotation due to (1) may be shown to be symmetric about line centre while the contributions arising from mechanisms (2) and (3) are antisymmetric. Mechanism (3) when evaluated for the temperature and fields typically used turns out to be about $10^{-4}$ of the size of (1) and (2), which are themselves comparable. When analysing experimental data, we thus consider only these two. Figures (1) and (2) show computer fits to experimental data

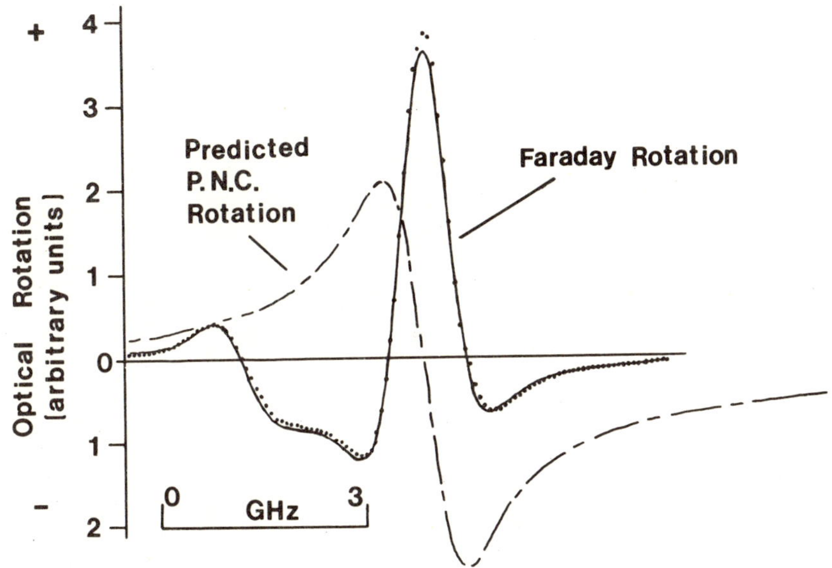


Figure 1. Faraday rotation and predicted P.N.C. rotation for the leading hyperfine component, F = 7 - F' = 6, for the 648 nm M1 transition of bismuth. The dots show the observed Faraday effect while the solid curve is the best fit obtained in the computer analysis of the data.

for both the lines 648 nm and 876 nm of bismuth. In obtaining the best fit to the data only the Lorentzian and Gaussian half-widths of the components of the profile and the ratio, $\chi = \langle E2\rangle / \langle M1\rangle$ have been allowed to vary; the hfs splittings have been fixed at the values determined by other authors[18, 19].

Note that the optical rotation is particularly sensitive to $\chi$ since there are interference terms in the Faraday effect which depend linearly on $\chi$, in contrast to the relative

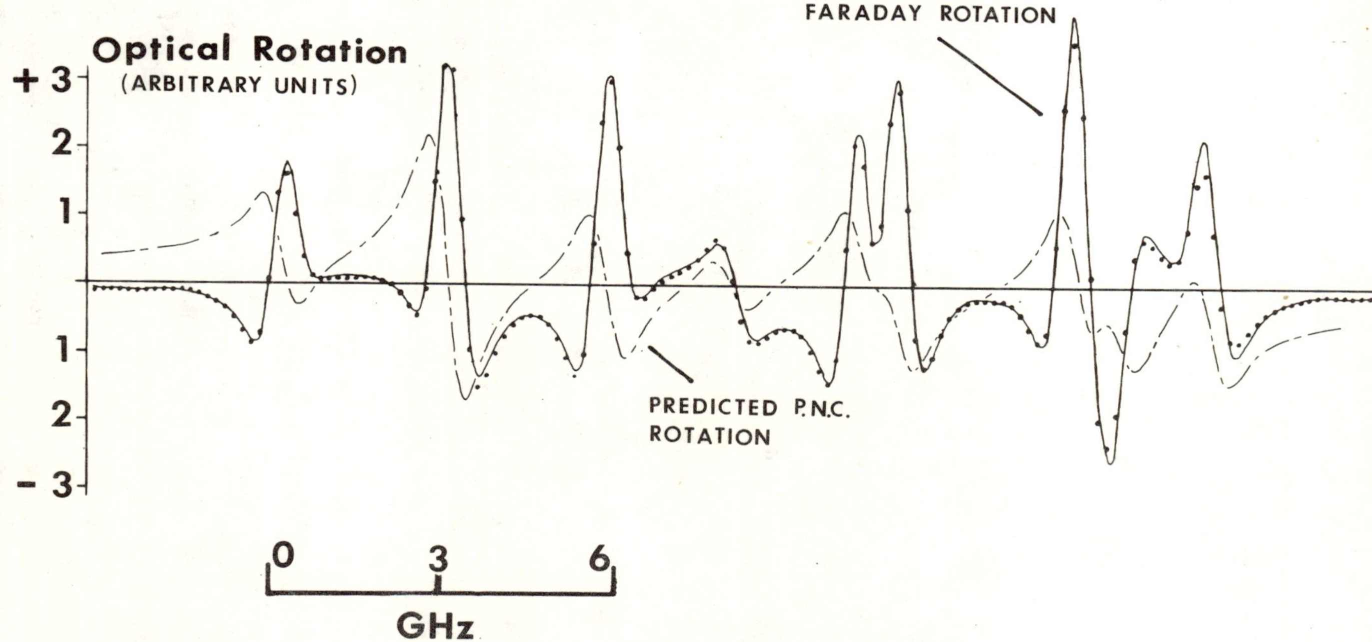


Figure 2. The Faraday effect and predicted PNC rotation for all hfs components of the 876 nm transition of bismuth. The dots show the observed Faraday rotation while the solid curve displays the best fit obtained in the computer analysis.

intensities of the components arising from different multipole transitions. Thus we have determined from $\chi$ the ratio of the Einstein A coefficients for the 648 nm and 876 nm transitions; they are respectively 22 $\pm$ 4% and 1 $\pm$ 0.7%[17]. These figures agree well with theoretical predictions[20].

THE OXFORD EXPERIMENTS

The experiment operating at 648 nm has now been running for about three years. In that time many changes and improvements have been made; it was found to be relatively straightforward to construct an apparatus sensitive to the original prediction of $R = -3 \times 10^{-7}$ but for higher sensitivity systematic effects due to microscopic changes in the laser beam as the wavelength is scanned pose a serious restriction. Unfortunately the overlapping molecular spectrum limits the usable bismuth vapour pressure so that we are forced to reduce the systematic effects to increase the sensitivity. The inherent signal to noise ratio due to the electronic system and the laser intensity fluctuations is very good (a standard deviation of $1 \times 10^{-8}$ radians can be obtained in 60 s!) and the variation with time of the apparent or 'residual' angle (seen in the absence of bismuth vapour) has until now been the major limitation. Thus over the last year we have devoted a lot of time to pin-pointing the causes of the residual angle. We now believe we understand the mechanisms involved, and this should enable us to reduce the scatter of our results in the future.

We have already described[15] the double oven arrangement which allows a rapid comparison of angles with and without bismuth; more recently, a PDP-8 computer has been incorporated into the experiment. Signal analysis is now carried out entirely digitally, and the results should give a cross-check on the previous data-handling technique. One advantage of this development is that the system becomes very flexible. The computer not only generates the modulation frequencies but is programmed to pick out the relevant components in the signal; changes can be made simply by modifying the control programme.

Our latest results for 648 nm show some evidence in support of an effect at the $-5 \times 10^{-8}$ level. To make a more definite statement would require a detailed discussion of the systematic effects in order to quote a realistic value for the error on the result.

At 876 nm, the absence of a molecular background permits the use of higher vapour pressures. In this experiment a Coherent Radiation 599 dye laser, pumped by a krypton laser is used to provide tunable continuous wave output at the infra-red wavelength. The complete hfs pattern in this line can be covered in one continuous scan and in this case the wavelength dependent rotation through the entire structure is being investigated. At present we are limited in this experiment by stray magnetic fields, but a special mu-metal shield is currently being constructed. Thus at the moment we can only say that we see no evidence for PNC in the 876 nm line at about $10^{-6}$ level; new results should be available soon.

Developments in the atomic theory[21] used to predict the value of the optical rotation from the Weinberg-Salam model suggest that R is probably significantly smaller than originally anticipated due to atomic shielding. A recent calculation performed at Oxford has yielded the following values:

$$R = -11 \times 10^{-8} \qquad \text{for } 648 \text{ nm}$$
$$R = -8 \times 10^{-8} \qquad \text{for } 876 \text{ nm}$$

with $\sin^2\theta_W = 0.25$; we should compare this with the semi-empirical value of

$$R = -18 \times 10^{-8} \qquad \text{for } 648 \text{ nm}$$

quoted by Novikov et al.[4].

The present situation is thus that there is some disagreement between the atomic physics experiments, but it has been demonstrated that it is possible to reach the sensitivity required to give a critical test of weak neutral current theories. The experiments are developing rapidly, and the position will certainly be clarified within a year or so; however, there remains the uncertainty in the heavy elements due to the atomic calculations. The hydrogen experiments should remove this difficulty.

REFERENCES


1. C.Y. Prescott et al. Phys. Lett. 77B, 347, 1978
2. S. Weinberg Phys. Rev. Lett. 19, 1264, 1967
   A. Salam Proc. 8th Nobel Symposium ed. N. Svartholm, Almkvist and Wiksell, Stockholm (1968)
3. L.M. Barkov and M.S. Zolotorev Zh. Exkp. Teor. Fiz. Pis'ma 26, 379, 1978
4. V.N. Novikov, O.P. Sushkov and I.B. Khriplovich ZhETF 71, 1665, 1976
5. P.E.G. Baird et al., E.N. Fortson et al. Nature 264, 528, 1976
6. G. Feinberg Nature 271, 509, 1978
7. R.R. Lewis and W.L. Williams Phys. Lett. 59B, 70, 1975
8. R.W. Dunford, R.R. Lewis and W.L. Williams Submitted to Phys. Rev. A 1978
9. E.G. Drukarev and A.N. Moskalev ZhETF 73, 2060, 1977
10. C. Bouchiat J. Phys. G 3, 183, 1977
11. C.E. Loving and P.G.H. Sandars J. Phys. B 10, 2755, 1977
12. M.A. Bouchiat and C.C. Bouchiat Phys. Lett. 48B, 111, 1974
13. M.A. Bouchiat and L. Pottier Phys. Lett. 62B, 327, 1976
14. S. Chu, E.D. Commins and R. Conti Phys. Lett. 60A, 96, 1977
15. P.E.G. Baird et al. Phys. Rev. Lett. 39, 798, 1977
16. The Novosibirsk experiment does not employ a Faraday modulator but simply uses the polarisers themselves to modulate the angle of rotation.
17. G.J. Roberts et al Submitted for publication to J. Phys. B
18. S. Mrozowski Phys. Rev. 62, 526, 1942
    Phys. Rev. 69, 169, 1946
19. D.A. Landmann and A. Lurio Phys. Rev. A1, 1330, 1970
20. R.H. Garstang J. Res. Nat. Bureau of Standards 68A, 1, 61, 1946
21. M.J. Harris, C.E. Loving and P.G.H. Sandars to be published in J. Phys. B.

SUBJECT INDEX


Adiabatic representation of the three-body problem, 183-186
Alignment of excited atoms, 423-433
Angular dependence of elastic scattering, 256,257
Anticrossing spectroscopy, 435-459
Asymptotic theory of atomic collisions, 267,268
Atomic fine and hyperfine structures
  Li, 442-443
  H,$He^+$, 443-447
  $Mg^+$, 447-449
  He, 449-451
  Stark effect measurements, 451-455
  quadratic Zeeman effect, 455-456
Atomic hydrogen, experiments, 623,654
Atomic polarization moments, 410-421
Atomic Rydberg states, 626-645
Atoms and molecules under collision-free conditions, 617-619
Atoms in external resonant field, 462-491
Auger electron, 289
  transitions, 289-305
Auger width, 128
Autoionizing rate coefficient, 161

Bethe-Salpeter equation, 112,114
Bimolecular exchange reaction, 28
Boltzman transport equation, 361-364
Breit interaction, 123,125
Brillouin's theorem, 81

Classical approximation for ionization by electron impact, 149,150
Coherent optical spectroscopy in a molecular beam, 618
Coherent phenomena in saturated absorption
  spectroscopy, 589-590
  at collisions,585,590-594
Collision processes in crossed beams, 624-625
Collisions, inelastic
  atom-atom, 15-24
  atom-molecule, 24-30
  electron-atom, 53,54
Collisions of mesic atoms, 187-189, 201
Collisional broadening, 36
Collisional integral, 496
Configuration
  interaction, 45,50; 78-90, 93
    calculation, 78-90, 93
Contact interaction, 94
Correlation (electron), 77
  in the $3d^n$ shell, 83-87
Cross sections, excitation for $C^{5+}$, 157
Cross sections, differential for elastic and inelastic collisions, 624
Cross sections, ionization for Na,Ne,Ag, 141,148,150
Cross section oscillations, 255,256
Crossing signal, 436-451
Cyclic excitation by laser radiation, 567-569
Cyclic multiphoton excitation, 572

Damping constant, 36
Density matrix, 496
Deorientating processes, 494,516

Dielectronic recombination, 161
Dipole approximation, 260-262
Dirac-Hartree-Fock
approximation, 95
Direct excitation, 162
Doppler effect, 21,22
Doppler tuning, 615
Double resonance spectrum,
616,617

Eikonal method, 508
Electric field influence on the
resonance, 259-260
Electric quadrupole
interaction, 46
Electron density, 103,104,105
Energy level shifts caused by
nucleus deformation forces,
583
Energy transfer
in atomic collisions, 15-24
in atom-molecule collisions,
24-30
rate constants, 17
Excimer molecules, 622
Excitation of the Rydberg
states, 574
Excitation spectrum, 613,614,616
Extended Hartree-Fock method, 85

Feynman diagrams, 114-117,121
Feshbach resonances in
photodetachment cross
sections, 234-237
Fluorescence detection of single
atoms, 565-572
Fock V.A., 67-76
Franck-Condon factor, 245

Gell-Mann and Low theory, 119-124
Genealogical coefficients, 97,98
Generalized helical
moments, 414-417
Green function
one-electron 113-116,118,119
two-electron, 114-116,119

Hanle effect, 423,435
Hartree-Fock
approximation, 77,93
Hartree-Fock-Pauli
approximation, 93-95
Hyperfine structure, 34,41,569
Hyperfine structure levels of
mesic atoms, 189-191

Interaction between molecular
hydrogen and helium
atoms, 329-332
Intercombination transitions, 58
Interference of atomic
states, 207-210; 522-523
Ionization of atom by electron
impact, 134-150
Ionization saturation, 537
Ionization threshold, 627,631-636

Lamb dip width, 590-592
Lamb shift, 123; 218-221
measuring, 218-221
Laser
bound-free, 355-364
CO, 346-354
$CO_2$, 342-346
rare-gas-halide, 358-361
mercury monohalide, 364,365
spectroscopy, 614,612-625
Laser detection of single
atoms, 565-583
Level crossing, 423,435-437
Lifetimes
of atoms, 33-61
of resonance levels of atoms,38
of rare-earth elements, 53-61

Macroscopic transfer of medium
polarization in a gas,
585,595-609
Magnetic dipole interaction, 46
Mesic atoms, 183
hyperfine structure levels,
189,191
Mesic molecule
energy levels, 191,192
nonresonant formation, 192-195
resonant formation, 195-198
Metastability exchange, 317-329
Method of separated fields, 595-609

Molecules
destruction by photodissociation, 372-374
direct radiative association, 369-372
dissociation of formaldehyde, 386-389
dissociation processes of polyatomic molecules, 383-409
direct radiative association, 369-372
formation by inverse predissociation, 375-376
isotopic effects, 389-392
multiphoton photodissociation, 393
multiphoton dissociation of $SF_6$, 397-403
probability for photo-fragmentation, 396-397
spontaneous radiative dissociation, 374-375
Multi-configuration Hartree-Fock method, 78-90

Negative hydrogen ion
energy and width, 225-228
experimental observation, 228-237
Negative ions
$He^-$, 237,238
$Br^-$, 240
$Kr^-$, $Xe^-$, 241,242
$N_2^-$, 242
transition probabilities,241-245
Nonexponential decay, 617
Nonlinear resonances,494-6,502-8
Nonlinear resonances induced by collisions, 508-516
Nuclear fusion reactions
catalysis by $\mu^-$-mesons, 198,199

Optical orientation of atoms, 308
Orbit-orbit interaction, 94
Oscillator strengths, 48, 49, 51; 87,88,108
for Cr I, 87
as a function of Z, 88,89;105

Pair correlation, 78,81,84
Parity nonconservation, 648-652; 653-661
in atomic bismuth, 648-652
in atomic hydrogen, 653,654
in atomic transitions, 648-652
bismuth vapour absorption spectrum, 649,651
Faraday rotation, measurement, 649,651
Faraday effect, 655-659
optical rotation, measurement, 648-652; 654,657,660
Weinberg-Salam model, 651,652; 653,660
Parity nonconservation in multicharged ions, 128,129
Penning interaction, 309-317
Photodetachment
$H^-$, 234,249,251
$H^-$ in electric field, 235-237
Photodissociation, 28-30
Photoionization, 79
Population inversion, 480-491
Pre-dissociation, 28,29

Quadrupole moments, measurement, 616
Quantum beats, 521-532
beats under pulsed excitation, 527-530
laser spectroscopy, 527-530
resonance, 525-526
Quasi-energies, 466-491

Rabi-technique, 615
Radiation diffusion, 424
Radiative width, 125
Radio-frequency spectroscopy in molecular beams, 615
Rare-gas excimers, 355-358
Relativistic energy corrections, 121-123, 126,127
Relativistic perturbation theory, 111-124
Relaxation matrix, 411
Resonances
in atomic collisions, 17
Feshbach, in photodetachment cross sections, 234-237

Retardation, 114,115,117

Saddle point, 628,632,637,642
Satellite structure, 159-174
Scattering of electrons by hydrogen
elastic, 229
excitation, 230
shape resonance, 232-234
Selective IR-dissociation
535,536; 546-562
Selective photoionization of atom,
535-546
Semiclassical approximation
for excitation by electron
impact, 151-156
for ionization by electron
impact, 136-139, 146
Sensitized fluorescence,
15,21,22,26
Simultaneous excitation of
magnetic resonance, 314
Single-photon decay, 301
Size of the ion as a function of Z,
103,104
Spectroscopy
Doppler-reduced laser, 614
Doppler-limited, 613
high resolution with reduced
Doppler width, 613-617
in molecular beams, 612-625
radio-frequency, 615
in supersonic beams, 619-621
anticrossing, 635-659
Spin-contact interaction, 94,95
Spin-flip process, 202
Spin-orbit interaction, 58,94,95
Spin-spin interaction, 94,95
Stark structure,
636-637, 639,643,645
Supersonic beams
spectroscopic analysis,619-621
loosely bound Van der Waals
molecules, 621-624

Tensorial operators, 97-99
Three-body problem
adiabatic representation,
183-186
Threshold behaviour of e-H
scattering, 249

Transition
correlated, 292,301
hypersatellite, 292,301
Transition of population, 513
Transition probability, 100-102,161
measuring, 33-61
relativistic corrections,
102,118,129
negative ions, 241-245
Two-center problem, 185
Two-level approximation, 186
Two-photon
Doppler-free resonances,
585, 594-595

Vacuum polarization
in $\mu$ -mesic molecules, 200

Wavelength calculation, 107

Шестая международная конференция по атомной физике. Труды
Председатель конференции А. М. Прохоров, редактор Р. Я. Дамбург

Рига, издательство «Зинатне» Академии наук Латвийской ССР

На английском языке
I SBN 0-306-40217-3